THE PRE-CALCULUS PROBLEM SOLVER®

REGISTERED TRADEMARK

A Complete Solution Guide to Any Textbook

Staff of Research and Education Association
Dr. M. Fogiel, Director

special chapter reviews by
Dennis Smolarski, Ph.D.
Associate Professor of Mathematics
Santa Clara University, Santa Clara, California

Research and Education Association
61 Ethel Road West
Piscataway, New Jersey 08854

THE PRE-CALCULUS PROBLEM SOLVER®

Printed in the United States of America

Library of Congress Catalog Card Number 98-66573

International Standard Book Number 0-87891-556-7

WHAT THIS BOOK IS FOR

Students have generally found pre-calculus a difficult subject to understand and learn. Despite the publication of hundreds of textbooks in this field, each one intended to provide an improvement over previous textbooks, students continue to remain perplexed as a result of the numerous conditions that must often be remembered and correlated in solving a problem. Various possible interpretations of terms used in pre-calculus have also contributed to much of the difficulties experienced by students.

In a study of the problem, REA found the following basic reasons underlying students' difficulties with pre-calculus taught in schools:

(a) No systematic rules of analysis have been developed which students may follow in a step-by-step manner to solve the usual problems encountered. This results from the fact that the numerous different conditions and principles which may be involved in a problem, lead to many possible different methods of solution. To prescribe a set of rules to be followed for each of the possible variations, would involve an enormous number of rules and steps to be searched through by students, and this task would perhaps be more burdensome than solving the problem directly with some accompanying trial and error to find the correct solution route.

(b) Textbooks currently available will usually explain a given principle in a few pages written by a professional who has an insight in the subject matter that is not shared by students. The explanations are often written in an abstract manner which leaves the students confused as to the application of the principle. The explanations given are not sufficiently detailed and extensive to make the student aware of the wide range of applications and different aspects of the principle being studied. The numerous possible variations of principles and their applications are usually not discussed, and it is left for the students to discover these for themselves while doing exercises. Accordingly, the average student is expected to rediscover that which has been long known and practiced, but not published or explained extensively.

(c) The examples usually following the explanation of a topic are too few in number and too simple to enable the student to obtain a thorough grasp of the principles involved. The explanations do not provide sufficient basis to enable a student to solve problems that may be subsequently assigned for homework or given on examinations.

The examples are presented in abbreviated form which leaves out much material between steps, and requires that students derive the omitted material themselves. As a result, students find the examples difficult to understand--contrary to the purpose of the examples.

Examples are, furthermore, often worded in a confusing manner. They do not state the problem and then present the solution. Instead, they pass through a general discussion, never revealing what is to be solved for.

Examples, also, do not always include diagrams/graphs, wherever appropriate, and students do not obtain the training to draw diagrams or graphs to simplify and organize their thinking.

(d) Students can learn the subject only by doing the exercises themselves and reviewing them in class, to obtain experience in applying the principles with their different ramifications.

In doing the exercises by themselves, students find that they are required to devote considerably more time to pre-calculus than to other subjects of comparable credits, because they are uncertain with regard to the selection and application of the theorems and principles involved. It is also often necessary for students to discover those "tricks" not revealed in their texts (or review books), that make it possible to solve problems easily. Students must usually resort to methods of trial-and-error to discover these "tricks", and as a result they find that they may sometimes spend several hours to solve a single problem.

(e) When reviewing the exercises in classrooms, instructors usually request students to take turns in writing solutions on the boards and explaining them to the class. Students often find it difficult to explain in a manner that holds the interest of the class, and enables the remaining students to follow the material written on the boards. The remaining students seated in the class are, furthermore, too occupied with copying the material from the boards, to listen to the oral explanations and concentrate on the methods of solution.

This book is intended to aid students in pre-calculus to overcome the difficulties described, by supplying detailed illustrations of the solution methods which are usually not apparent to students. The solution methods are illustrated by problems selected from those that are most often assigned for class work and given on examinations. The problems are arranged in order of complexity to enable students to learn and understand a particular topic by reviewing the problems in sequence. The problems are illustrated with detailed step-by-step explanations, to save the students the large amount of time that is often needed to fill in the gaps that are usually found between steps of illustrations in textbooks or review/outline books.

The staff of REA considers pre-calculus a subject that is best learned by allowing students to view the methods of analysis and solution techniques themselves. This approach to learning the subject matter is similar to that practiced in various scientific laboratories, particularly in the medical fields.

In using this book, students may review and study the illustrated problems at their own pace; they are not limited to the time allowed for explaining problems on the board in class.

When students want to look up a particular type of problem and solution, they can readily locate it in the book by referring to the index which has been extensively prepared. It is also possible to locate a particular type of problem by glancing at just the material within the boxed portions. To facilitate rapid scanning of the problems, each problem has a heavy border around it. Furthermore, each problem is identified with a number immediately above the problem at the right-hand margin.

To obtain maximum benefit from the book, students should familiarize themselves with the section, "How To Use This Book," located in the front pages.

To meet the objectives of this book, staff members of REA have selected problems usually encountered in assignments and examinations, and have solved each problem meticulously to illustrate the steps which are difficult for students to comprehend. Special gratitude is expressed to them for thier efforts in this area, as well as to the numerous contributors who devoted brief periods of time to this work.

Gratitude is also expressed to the many persons involved in the difficult task of typing the manuscript with its endless changes, and to the REA art staff who prepared the numerous detailed illustrations together with the layout and physical features of the book.

The difficult task of coordinating the efforts of all persons was carried out by Carl Fuchs. His conscientious work deserves much appreciation. He also trained and supervised art and production personnel in the preparation of the book for printing.

Finally, special thanks are due to Helen Kaufmann for her unique talents to render those difficult border-line decisions and constructive suggestions related to the design and organization of the book.

Max Fogiel, Ph.D.
Program Director

HOW TO USE THIS BOOK

This book can be an invaluable aid to students in pre-calculus as a supplement to their textbooks. The book is subdivided into 55 chapters, each dealing with a separate topic. The subject matter is developed beginning with algebraic, trigonometric, exponential, logarithmic functions and their graphs and extending through linear and quadratic equations, analytic geometry, elementary statistics, differentiation and integration, determinants, matrices, and systems of equations. An extensive number of applications have been included, since these appear to be more troublesome to students.

TO LEARN AND UNDERSTAND A TOPIC THOROUGHLY

1. Refer to your class text and read the section pertaining to the topic. You should become acquainted with the principles discussed there. These principles, however, may not be clear to you at that time.

2. Then locate the topic you are looking for by referring to the "Table of Contents" in the front of this book, "The Pre-Calculus Problem Solver."

3. Turn to the page where the topic begins and review the problems under each topic, in the order given. For each topic, the problems are arranged in order of complexity, from the simplest to the more difficult. Some problems may appear similar to others, but each problem has been selected to illustrate a different point or solution method.

To learn and understand a topic thoroughly and retain its contents, it will be generally necessary for students to review the problems several times. Repeated review is essential in order to gain experience in recognizing the principles that should be applied, and select the best solution technique.

TO FIND A PARTICULAR PROBLEM

To locate one or more problems related to a particular subject matter, refer to the index. In using the index, be certain to note that the numbers given there refer to problem numbers, not page numbers. This arrangement of the index is intended to facilitate finding a problem more rapidly, since two or more problems may appear on a page.

If a particular type of problem cannot be found readily, it is recommended that the student refer to the "Table of Contents" in the front pages, and then turn to the chapter which is applicable to the problem being sought. By scanning or glancing at the material that is boxed, it will generally be possible to find problems related to the one being sought, without consuming considerable time. After the problems have been located, the solutions can be reviewed and studied in detail. For this purpose of locating problems rapidly, students should acquaint themselves with the organization of the book as found in the "Table of Contents".

In preparing for an exam, it is useful to find the topics to be covered in the exam from the "Table of Contents," and then review the problems under those topics several times. This should equip the student with what might be needed for the exam.

CONTENTS

> *Every chapter begins with "Basic Attacks and Strategies for Solving Problems in this Chapter."*

Chapter No.   **Page No.**

SECTION 1: ALGEBRA

SECTION 2: PLANE TRIGONOMETRY

SECTION 3: ANALYTIC GEOMETRY

SECTION 4: INTRODUCTION TO CALCULUS

SECTION 5

SECTION 1 — COLLEGE ALGEBRA

CHAPTER 1

THE NUMBER SYSTEM

> **Basic Attacks and Strategies for Solving Problems in this Chapter. See pages 1 to 9 for step-by-step solutions to problems.**

Real numbers include all numbers commonly used and can be subdivided into five major categories. *Positive integers* (otherwise called *natural* or *whole* or *counting numbers*) begin with 1 and increase by 1 at each step (i.e., 1, 2, 3, ...). *Negative integers* are simply the negatives of the positive integers. *Zero* is an integer between the positive numbers and the negative numbers and is usually considered separate from both groups. *Rational numbers* are numbers formed from a *ratio* of two integers (i.e., a *fraction* such as $1/3$). (Note that integers may also be equal to certain rational numbers, such as $3 = 3/1 = 9/3$). *Irrational numbers* are those which cannot be written as fractions (such as $\sqrt{2}$).

Imaginary numbers are multiples of $i = \sqrt{-1}$ such as i, $2i$, or $-3\pi i$.

Complex numbers are sums (or differences) of real and imaginary numbers, such as $1 + 3i$ or $\sqrt{2} - \pi i$.

Those positive integers which are divisible only by themselves and 1 (or their negatives) are called *prime*. That is, x is prime if the only integer values of r for which x/r is an integer are $r = 1$ and $r = x$. These two values of r always yield integer quotients, since $x/x = 1$, an integer, and $x/1 = x$, also an integer.

If numbers are identified with a location on a graph, the distance between any number and the number zero (called the *origin*) is said to be the *absolute value* of that number. The absolute value of a real number is merely the number (if it is positive or zero) or the number without the minus sign (if it is negative). The operation of finding the absolute value of a number a is indicated by vertical lines surrounding the number, i.e., $|a|$ indicates the absolute value of a. It is a mistake to assume that

$$|a + b| \text{ and } |a| + |b| \quad = \quad WRONG$$

are equal since, in general, they are not equal. For example,

$$|3 + (-2)| = |3 - 2| = |1| = 1 \quad \checkmark \quad RIGHT$$

while

$$|3| + |-2| = 3 + 2 = 5. \quad \checkmark \quad CORRECT$$

Large numbers are frequently expressed in *scientific notation,* that is, they are written as two numbers multiplied together: the *abscissa* (or characteristic), a number usually between 1 and 10 (or between 0.1 and 1.0), and the *magnitude,* which is 10 raised to an appropriate power. Thus,

$$123 = 1.23 \times 10^2 \text{ since } 1.23 \times 10^2 = 1.23 \times 100 = 123.$$

Rational numbers are also expressed as a combination of an integer and a decimal number (i.e., a number between 0 and 1 that begins with a decimal point). Thus, $5/4 = 1.25$. In some cases, the decimal number never terminates, e.g., $1/3 = 0.33333\ldots$. In cases like these, the number is sometimes written with a bar over the group of numbers which is repeated indefinitely, such as $1/3 = 0.\overline{3}$.

To convert a rational number written as a fraction into a (possibly repeating) decimal, long division is performed until there is no remainder or until a repeating pattern appears in the quotient. It must be checked (by inspecting the remainder) that this repeating pattern will continue to repeat. For example,

$$1/4 = .25 \text{ and } 1/7 = .142857142857\ldots = .\overline{142857}, \text{ but } 112/1000 = .112,$$

and we cannot stop after .11 and conclude that $112/1000 = .111\ldots = .\overline{1}$. This would give the wrong answer.

To convert a rational number written as a repeating decimal into a fraction, one multiplies the repeating decimal (call it x) by a power of 10, in which the power is equal to the number of repeating digits. The original number x is subtracted from the new number (which should produce a simple equation with no decimal fractions in it). Solving for x produces the final result. For example, let $x = .\overline{142857}$. Multiply by 10^6 (since the repeating decimal has 6 digits in it), and then subtract x from $10^6 x$, giving

$$10^6 x - x = 142857.\overline{142857} - .\overline{142857} = 999{,}999 x = 142{,}857.$$

We can solve for x giving $142{,}857/999{,}999$ or $1/7$ (after arithmetic simplification).

One type of rational number, in which the denominator in the fractional form is 100, is called a *percent* (derived from the Latin phrase *per centum,* meaning "divided [through] by a hundred"). Thus, 5 percent (written 5%) means $5/100$, that is, $1/20$ or 0.05.

Step-by-Step Solutions to Problems in this Chapter, "The Number System"

● **PROBLEM** 1-1

(a) Is zero a natural number?

Is zero an integer?

Is zero a positive number or a negative number?

Is zero an odd number or an even number?

(b) Give three examples of each of the following:

(1) Integers

(2) Rational Numbers

(3) Irrational Numbers

(4) Natural Numbers

(5) Prime Numbers

(6) Complex Numbers

(c) What is a real number?

Solution: (a) Zero is not a natural number. The natural numbers begin with 1 and continue up through positive infinity, i.e. 1,2,3,4,5,6,7,.... Zero is an integer. Zero divides the positive integers from the negative integers on a number line. An example of a set of integers is ...,-5,-4,

1

-3,-2,-1,0,1,2,3,4,5.... Zero is an even number as 2,4,6, and 8. Zero is neither a positive number nor a negative number.

(b) (1) An integer is a whole number (not a fraction or a decimal) which can be either positive or negative. Examples of integers are 1,5, and -9.

(2) Rational Numbers are those numbers which can be represented as $\frac{p}{q}$ where p and q are integers, and $q \neq 0$. Examples of rational numbers are 8/5, 8/12, and -4/3.

(3) Irrational Numbers are those numbers which are not rational and cannot be represented as $\frac{p}{q}$. Examples of irrational numbers are $-\sqrt{2}$, π, and $\sqrt[5]{-3}$.

(4) Examples of natural numbers are 2,7, and 9.

(5) A prime number is a number which is divisible by only ± 1, and plus or minus itself. Examples of prime numbers are 2,3, and 13.

(6) A Complex Number is a number of the form a+bi, where a and b are real numbers and $i = \sqrt{-1}$. The real part of a complex number is a, and the imaginary part of a complex number is bi. Examples of complex numbers are $6+\sqrt{3}i$, 4.5-2i, and -9+16i.

(c) The Real Numbers consist of all positive and negative rational and irrational numbers, and zero.

● **PROBLEM 1-2**

Classify each of the following numbers into as many different sets as possible. Example: real, integer, rational....

(1) 0

(2) 9

(3) $\sqrt{6}$

(4) $\frac{1}{2}$

(5) $\frac{2}{3}$

(6) 1.5

Solution: (1) Zero is a real number and an integer.

(2) 9 is real, rational, natural number, and an integer.

2

(3) $\sqrt{6}$ is an irrational, real number.

(4) $\frac{1}{2}$ is a rational, real number.

(5) $\frac{2}{3}$ is a rational, real number.

(6) 1.5 is a rational, real number, a decimal.

● PROBLEM 1-3

Find the absolute value for each of the following

    (1)   zero

    (2)   4

    (3) $-\pi$

    (4)   a, where a is a real number.

Solution: The absolute value of a number is represented by 2 vertical lines around the number, and is equal to the given number, regardless of sign.

    (1) $|0| = 0$

    (2) $|4| = 4$

    (3) $|-\pi| = \pi$

    (4) for $a > 0$ $|a| = a$
             for $a = 0$ $|a| = 0$
             for $a < 0$ $|a| = -a$,

i.e.
$$|a| = \begin{cases} a & \text{if } a > 0 \\ 0 & \text{if } a = 0 \\ -a & \text{if } a < 0 \end{cases}$$

● PROBLEM 1-4

Use scientific notation to express each number.
  (a) 4,375   (b) 186,000   (c) 0.00012   (d) 4,005

Solution: A number expressed in scientific notation is written as a product of a number between 1 and 10 and a power of 10. The number between 1 and 10 is obtained by moving the decimal point of the number (actual or implied) the required number of digits. The power of 10, for a

3

number greater than 1, is positive and is one less than the number of digits before the decimal point in the original number. The power of 10, for a number less than 1, is negative and is one more than the number of zeros immediately following the decimal point in the original number. Hence,

(a) $4,375 = 4.375 \times 10^3$ (b) $186,000 = 1.86 \times 10^5$

(c) $0.00012 = 1.2 \times 10^{-4}$ (d) $4,005 = 4.005 \times 10^3$

● **PROBLEM 1-5**

Express $\dfrac{6,400,000}{400}$ in scientific notation.

Solution: In order to solve this problem, we express the numerator and denominator as the product of a number between 1 and 10 and a power of 10. This is known as scientific notation. Thus

$$6,400,000 = 6.4 \times 1,000,000 = 6.4 \times 10^6$$
$$400 = 4 \times 100 = 4 \times 10^2$$

Thus,

$$\frac{6,400,000}{400} = \frac{6.4 \times 10^6}{4.0 \times 10^2}$$

Since $\dfrac{ab}{cd} = \dfrac{a}{c} \cdot \dfrac{b}{d} : = \dfrac{6.4}{4.0} \times \dfrac{10^6}{10^2}$

Since $\dfrac{a^x}{a^y} = a^{x-y} : = 1.6 \times 10^4$

● **PROBLEM 1-6**

Write $\dfrac{2}{7}$ as a repeating decimal.

Solution: To write a fraction as a repeating decimal divide the numerator by the denominator, until a pattern of repeated digits appears.

$$2 \div 7 = .285714285714...$$

Identify the entire portion of the decimal which is repeated. The repeating decimal can then be written in the shortened form:

$$\frac{2}{7} = .\overline{285714}$$

● **PROBLEM 1-7**

Find the common fraction form of the repeating decimal 0.4242....

<u>Solution:</u> Let x represent the repeating decimal.

$$x = 0.4242...$$

$$100x = 42.42...\quad \text{by multiplying by 100}$$

$$\underline{\quad x = 0.42...\quad}$$

$$99x = 42 \qquad \text{(1) by subtracting x from 100x}$$

Divide both sides of equation (1) by 99.

$$\frac{99x}{99} = \frac{42}{99}$$

$$x = \frac{42}{99} = \frac{14}{33}$$

The repeating decimal of this example had 2 digits that repeated. The first step in the solution was to multiply both sides of the original equation by the 2nd power of 10 or 10^2 or 100. If there were 3 digits that repeated, the first step in the solution would be to multiply both sides of the original equation by the 3rd power of 10 or 10^3 or 1000.

● **PROBLEM 1-8**

Find $0.25\overline{25}$ as a quotient of integers.

<u>Solution:</u> Let $x = 0.25\overline{25}$. (1) Multiply both sides of this equation by 100:

$$100x = 100(0.25\overline{25})$$

Multiplying by 100 is equivalent to moving the decimal two places to the right, and since the digits 25 are repeated we have:

$$100x = 25.25\overline{25} \qquad (2)$$

Now subtract equation (1) from equation (2):

$$100x = 25.25\overline{25}$$
$$\underline{-\quad x = \quad 0.25\overline{25}}$$
$$99x = 25.0000$$

or $99x = 25$ (3)

(Note that this operation eliminates the decimal) Dividing both sides of equation (3) by 99:

5

$$\frac{99x}{99} = \frac{25}{99}$$

$$x = \frac{25}{99}$$

Therefore,

$$0.25\overline{25} = x = \frac{25}{99}$$

Also, note that the given repeating decimal, $0.25\overline{25}$, was multiplied by 100 or 10^2, where the power of 10 (which is 2) is the same as the number of repeating digits (namely, 2) for this problem. In general, for problems of this type, if the repeating decimal has n repeating digits, then the repeating decimal should be multiplied by 10^n.

● **PROBLEM 1-9**

Write the repeating decimal $14.\overline{23}$ as a quotient of two integers, $\frac{p}{q}$.

<u>Solution</u>: Let $x = 14.\overline{23}$. (1) Multiply both sides of this equation by 100:

$$100x = 100(14.\overline{23})$$

$$100x = 1423.\overline{23} \tag{2}$$

Subtract equation (1) from equation (2):

$$100x = 1423.\overline{23}$$
$$-\underline{\quad x = \quad 14.\overline{23}}$$
$$99x = 1409.00$$

or $\qquad 99x = 1409$ $\qquad\qquad\qquad\qquad\qquad$ (3)

(Note that this operation eliminates the decimal.) Dividing both sides of equation (3) by 99:

$$\frac{99x}{99} = \frac{1409}{99}$$

$$x = \frac{1409}{99}.$$

Therefore,

$$14.\overline{23} = x = \frac{1409}{99}.$$

Also, note that the given repeating decimal, $14.\overline{23}$, was multiplied by 100 or 10^2, where the power of 10 (which is 2) is the same as the number of repeating digits (namely, 2) for this problem. In general, for problems of this type, if the repeating decimal has n repeating digits, then the repeating decimal should be multiplied by 10^n.

(a)   Compute the value of

     (1)   90% of 400

     (2)   180% of 400

     (3)   50% of 500

     (4)   200% of 4

(b)   What percent of

     (1)   100 is 99.5?

     (2)   200 is 4?

Solution: (a)   The symbol % means per hundred, therefore $5\% = \frac{5}{100}$.

     (1)   90% of 400 $= \frac{90}{100} \times \frac{400}{1} = 90 \times 4 = 360$

     (2)   180% of 400 $= \frac{180}{100} \times \frac{400}{1} = 180 \times 4 = 720$

     (3)   50% of 500 $= \frac{50}{100} \times \frac{500}{1} = 50 \times 5 = 250$

     (4)   200% of 4 $= \frac{200}{100} \times \frac{4}{1} = 2 \times 4 = 8$

  (b)   (1) $99.5 = x \times 100$

         $99.5 = 100x$

        $.995 = x$; but this is the value of

                 x per hundred. Therefore

                 x = 99.5%

    (2) $4 = x \times 200$

        $4 = 200x$

      $.02 = x.$ Again this must be changed to percent, so x = 2%.

Express

   (1) 1.65   as a percentage.

   (2) 0.7   as a fraction.

   (3) $-\dfrac{10}{20}$ as a decimal.

   (4) $\dfrac{4}{2}$ as an integer.

<u>Solution</u>: (1) $1.65 \times 100 = 165\%$

              (2) $0.7 = \dfrac{7}{10}$

              (3) $\dfrac{-10}{20} = -0.5$

              (4) $\dfrac{4}{2} = 2$

Show that the sum of any positive number and its reciprocal cannot be less than 2.

<u>Solution:</u> Express the given example as a mathematical statement, recalling the reciprocal of $x = 1/x$. Write "the sum of any positive number and its reciprocal cannot be less than 2" as

$$a + \frac{1}{a} \nleq 2,$$

where a represents the positive number. Thus, the sum of any positive number and its reciprocal can be greater than or equal to 2. We are to show that

$$a + \frac{1}{a} \geq 2, \quad a > 0.$$

If this relation is true, the following inequalities will also hold:

$a^2 + 1 \geq 2a$, Multiply both members of the inequality
                  by a,

$a^2 - 2a + 1 \geq 0$, Transpose $2a$

$(a - 1)^2 \geq 0$,   Factor

Now this simple relation is easily shown to be true, for, whether $a - 1$ is positive, negative, or zero, its square must be non-negative. This is therefore a suitable starting point, and our synthesis, constituting the actual proof, is as follows. Since

$$(a - 1)^2 \geq 0,$$

for the reason just stated, expansion of the left member gives us the equivalent relations

$$a^2 - 2a + 1 \geqq 0 , \qquad \text{Expanding}$$

$$a^2 + 1 \geqq 2a , \qquad \text{Transposing } 2a$$

$$a + \frac{1}{a} \geqq 2 , \qquad \text{Divide both sides of the inequality by } a, \text{ with } a > 0 .$$

We see, incidentally, that the equality holds only if $a = 1$; if $0 < a < 1$, or if $a > 1$, the inequality holds.

CHAPTER 2

DEFINITIONS AND NOTATIONS OF SETS AND SET OPERATIONS

> **Basic Attacks and Strategies for Solving Problems in this Chapter. See pages 10 to 15 for step-by-step solutions to problems.**

A *set* is a collection of items. When the items of a set are indicated, the collection is usually enclosed by curly braces, e.g., $\{1, 2\}$ indicates the set of two integers, 1 and 2. The order of elements in a set is immaterial and no more than one copy of any element can exist in a set, thus $\{1, 1, 2\}$ and $\{2, 1\}$ indicate the same set as in the previous sentence. A set A is a *subset* of another set B if every item in A is also in B. This is written $A \subset B$. Two sets are equal if each is a subset of the other. Thus, any set is always a subset of itself. A is a *proper* subset of B if A is a subset of B and there is at least one element in B not in A. A set A is *infinite* if there is a proper subset C of A such that there is a one-to-one correspondence between elements of A and elements of C. If a set is not infinite, it is *finite*. The set of all subsets of a set, including the empty set, is called the *power* set.

A set is *closed (under some operation)* if, when the operation is performed on elements of the set, the result is also in the set. Thus, the integers are a closed set under addition (since the sum of any two integers is another integer), but the integers are not closed under division (since, for example, $^1/_3$ is not an integer).

Two special sets are the *null* (or *empty*) set $\emptyset$, which is the set containing *no* elements, and the *universal* set, **U**, a set containing all elements of the classification being considered. Note that the null set is a subset of every set and every set is a subset of the universal set.

There are three fundamental binary operations on sets: *union*, indicated by $\cup$, *intersection*, indicted by $\cap$, and set *difference*, indicated by the usual minus sign. The union of two sets A and B, $A \cup B$, is the new set consisting of all elements which were either in A or B or in both. The intersection of A and B, $A \cap B$, is the new set consisting of all elements which were originally both in A as well as being in B. The difference of A and B, $A - B$, is the set of those elements of A that are not in B.

There is one unary operation on a set, *complement*, indicated by a single quote mark or an over-bar. The complement of a set A, written A' or $\overline{A}$, is the set of all elements which are in the appropriate universe but not in A.

There are several laws applicable to sets.

IDENTITY LAWS

$A \cup \emptyset = A$

$A \cap \emptyset = \emptyset$

$A \cup U = U$

$A \cap U = A$

IDEMPOTENT LAWS

$A \cup A = A$

$A \cap A = A$

COMPLEMENT LAWS

$A \cup A' = U$

$A \cap A' = \emptyset$

$(A')' = A$

$\emptyset' = U; \quad U' = \emptyset$

COMMUTATIVE LAWS

$A \cup B = B \cup A$

$A \cap B = B \cap A$

ASSOCIATIVE LAWS

$(A \cup B) \cup C = A \cup (B \cup C)$

$(A \cap B) \cap C = A \cap (B \cap C)$

DISTRIBUTIVE LAWS

$A \cup (B \cap C) = (A \cup B) \cap (A \cup C)$

$A \cap (B \cup C) = (A \cap B) \cup (A \cap C)$

DE MORGAN'S LAWS

$(A \cup B)' = A' \cap B'$

$(A \cap B)' = A' \cup B'$

Sets of one type can be used to create sets of a more complicated type by means of the binary operation of *cross product*, indicated by ×. The cross product of two sets produces a set of *ordered pairs* in which the first element of any pair is taken from the first set of the product and the second element from the second. More formally,

$$A \times B = \{(a, b) \mid a \in A \text{ and } b \in B\}.$$

This is sometimes called the *Cartesian Product* of A and B.

Sets can be subdivided into a group of smaller sets. If the subdivision is such that (1) the union of all the subsets taken together form the original set (exhaustive, i.e., every element of the original set is in some subset in this group), and (2) no two subsets have a common element (disjoint), then this subdivision is called a *partition*.

Step-by-Step Solutions to Problems in this Chapter, "Definitions and Notation of Sets and Set Operations"

● PROBLEM 2-1

Is the set of all natural numbers from 1 to 10 a closed system under addition?

Solution: For {1,2,3,... 10} to be closed with respect to addition, the sum of any two numbers in this set must also be a member of this set. The set of all natural numbers from 1 to 10, inclusive, is therefore not a closed system under addition for it would not be correct to say that given any two numbers in the set there is a number in the set called their sum. For instance, 4 and 7 are in the set but their sum, 11, is not.

● PROBLEM 2-2

State the laws of set operations.

Solution: If S is an algebra of sets and if A, B, C,...,$\emptyset$, U, ... are elements of S, then the following hold for $\cup$, $\cap$, and '.

IDENTITY LAWS

1a. $A \cup \emptyset = A$ 1b. $A \cap \emptyset = \emptyset$
2a. $A \cup U = U$ 2b. $A \cap U = A$

IDEMPOTENT LAWS

3a. $A \cup A = A$ 3b. $A \cap A = A$

COMPLEMENT LAWS

4a. $A \cup A' = U$ 4b. $A \cap A' = \emptyset$
5a. $(A')' = A$ 5b. $\emptyset' = U;\ U' = \emptyset$

COMMUTATIVE LAWS

6a. $A \cup B = B \cup A$ 6b. $A \cap B = B \cap A$

ASSOCIATIVE LAWS

7a. $(A \cup B) \cup C = A \cup (B \cup C)$
7b. $(A \cap B) \cap C = A \cap (B \cap C)$

DISTRIBUTIVE LAWS

8a. $A \cup (B \cap C) = (A \cup B) \cap (A \cup C)$
8b. $A \cap (B \cup C) = (A \cap B) \cup (A \cap C)$

DE MORGAN'S LAWS

9a. $(A \cup B)' = A' \cap B'$ 9b. $(A \cap B)' = A' \cup B'$

● **PROBLEM 2-3**

Give the definitions of a finite and an infinite set, and two examples of finite sets and infinite sets.

<u>Solution</u>: By definition, a set S is infinite, provided that it has a proper subset P such that there exists a one-to-one correspondence between S and P. A set is finite if it is not infinite.

The empty set and the singleton set are two examples of the finite sets. This is because the empty set does not have any proper subset, and the singleton set (a set which consists of one element alone) has only one proper subset. This is the empty set, which cannot be put into one-to-one correspondence with the singleton set itself.

The set of all real numbers and the set of all positive odd integers are two examples of the infinite sets.

● **PROBLEM 2-4**

If $A = \{2,3,5,7\}$ and $B = \{1,-2,3,4,-5,\sqrt{6}\}$, find
(a) $A \cup B$ and (b) $A \cap B$.

<u>Solution</u>: (a) $A \cup B$ is the set of all elements in A or in B or in both A and B, with no element included twice in the union set.

$A \cup B = \{1,2,-2,3,4,5,-5,\sqrt{6},7\}$

(b) $A \cap B$ is the set of all elements in both A and B.

$A \cap B = \{3\}$

Sometimes two sets have no elements in common. Let $S = \{3,4,7\}$ and $T = \{2,-4,6\}$. What is the intersection of S and T? In this

11

case S ∩ T has no elements. Hence S ∩ T = ∅, the empty set. In that case, the sets are said to be disjoint.

The set of all elements entering a discussion is called the universal set, U, When the universal set is not given, we assume it to be the set of real numbers. The set of all elements in the universal set that are not elements of A is called the complement of A, written Ā .

● PROBLEM 2-5

Find A – B and A – (A∩B) for

A = {1,2,3,4} and

B = {2,4,6,8,10}

Solution: The relative complement of subsets A and B of the universal set U is defined as the set

A – B = {x, |x∈ A and x∉ B}. Note that it is not

assumed that B⊆A. Hence,

A – B = {1, 3}.

To find A – (A∩B), first find A∩B. The set A∩B is a set of elements that are common to both A and B, so A∩B = {2, 4}.

Therefore, A – (A∩B) = {1, 3}.

● PROBLEM 2-6

U = {1,2,3,4,5,6,7,8,9,10}, P = {2,4,6,8,10}, Q = {1,2,3,4,5}.
Find (a) P̄ and (b) Q̄ .

Solution: P̄ and Q̄ are the complements of P and Q respectively. That is, P̄ is the set of all elements in the universal set, U, that are not elements of P, and Q̄ is the set of elements in U that are not in Q. Therefore,
(a) P̄ = {1,3,5,7,9}; (b) Q̄ = {6,7,8,9,10}

● PROBLEM 2-7

Given the universal set U = {2,4,6, ..., 12}

(a) Find the set S = {x ∈ U|x²-5x+6 = 0}

(b) Find the set S if U is changed to be
U = {0,1,2,3, ..., 10}.

Solution: (a) Solving the equation $x^2 - 5x + 6 = 0$,

$$(x-2)(x-3) = 0$$

$$x = 2$$

$$x = 3$$

Thus, the solution set is {2, 3}.

Since the universal set is U = {2,4,6 ... 12}, S = {2}. Note that 3 ∉ S, although 3 is one of the two solutions of the describing equation, since S cannot have any element that is not in U.

(b) The solution set of the given equation is {2, 3}. Since both elements in the solution set are in U, S = {2, 3}.

● PROBLEM 2-8

Show that the complement of the complement of a set is the set itself.

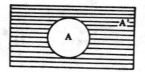

Solution: The complement of set A is given by A'. Therefore, the complement of the complement of a set is given by (A')'. This set, (A')', must be shown to be the set A; that is, that (A')' = A. In the figure the complement of the set A, or A', is the set of all points not in set A; that is, all points in the rectangle that are not in the circle. This is the shaded area in the figure. Therefore, this shaded area is A'. The complement of this set, or (A')', is the set of all points of the rectangle that are not in the shaded area; that is, all points in the circle, which is the set A. Therefore, the set (A')' is the same as set A; that is,

$$(A')' = A.$$

● PROBLEM 2-9

Let M = {1, 2} and N = {p, q}. Find (a) M × N, (b) N × M, and (c) M × M.

Solution: (a) M×N is the set of all ordered pairs in which the first component is a member of M and the second component is a member of N. Thus,

M × N = { (1,p), (1, q), (2, p), (2, q)}.

Note that the number of elements in M is 2,

     the number of elements in N is 2,

     and the number of elements in M × N = 2 × 2 = 4.

(b) N ×M is the set of all ordered pairs in which the first component is a member of N and the second component is a member of M. Thus,

    N × M = { (p, 1), (q, 1), (p, 2), (q, 2)}.

Once again note that the number of elements in N × M is 2 × 2 = 4.

(c) M × M is the set of all ordered pairs in which both components are members of M. Thus,

    M × M = {(1, 1),(1, 2), (2, 1), 2, 2)}.

Here too, the number of elements in M × M is 2 × 2 = 4.

● **PROBLEM 2-10**

List all the subsets of C = {1,2}.

Solution: {1}, {2}, {1,2}, $\emptyset$, where $\emptyset$ is the empty set. Each set listed in the solution contains at least one element of the set C. The set {2,1} is identical to {1,2} and therefore is not listed. $\emptyset$ is included in the solution because $\emptyset$ is a subset of every set.

● **PROBLEM 2-11**

Find the power set of the "tripleton" set S = {a,b,c}, and the number of elements of P(S).

Solution: The subsets of S are: the empty set ϕ, {a}, {b}, {c}, {a,b}, {a,c}, {b,c}, and the set S itself.

By definition, the power set of the set S is the set of all subsets of S.

Therefore, the power set is

    P(S) = {ϕ,{a},{b},{c},{a,b},{a,c},{b,c},{a,b,c} }.

The number of elements of set P(S) is found according to the following:

If A is a set with n(A) = m, then $n(P(A)) = 2^m$. For this problem, n(S) = 3, m = 3, and $n(P(S)) = 2^m = 2^3 = 8$. Hence, P(S) has 8 elements.

Find four proper subsets of P = {n: n ε I, -5 < n ≤ 5}.

<u>Solution:</u> P = {-4, -3, -2, -1, 0, 1, 2, 3, 4, 5}. All
these elements are integers that are either less than or
equal to 5 or greater than -5. A set A is a proper subset
of P if every element of A is an element of B and in addition
there is an element of B which is not in A.
     (a) B = {-4, -2, 0, 2, 4} is a subset because each
element of B is an integer greater than -5 but less than or
equal to 5. B is a proper subset because 3 is an element
of P but not an element of B. We can write 3 ε P but
3 ∉ B.
     (b) C = {3} is a subset of P, since 3 ε P. However,
5 ε P but 5 ∉ C. Hence, C ⊂ P.
     (c) D = {-4, -3, -2, -1, 1, 2, 3, 4, 5} is a proper
subset of P, since each element of D is an element of P,
but 0 ε P and 0 ∉ D.
     (d) φ ⊂ P, since φ has no elements. Note that φ is
the empty set. φ is a proper subset of every set except
itself.

Given the set S = {1,2,3,4,5,6}, find a partition of S.

<u>Solution</u>: A partition of a set S is a subdivision of the set
into subsets that are disjoint and exhausive, i.e., every
element of S must belong to one and only one of the subsets.
Each subset in the partition is also called a cell.

    Therefore, $(S_1, S_2, \ldots, S_n)$ is a partition of S if

      (a) $S_i \cap S_j = \phi$ (where φ is the empty set) for all $i \neq j$
(the cells are disjoint), and

      (b) $S_1 \cup S_2 \cup S_3 \cup \ldots \cup S_n = S$ (the cells are exhausive).

    Hence, one of the partitions of S is

     { {1,2,3},{4},{5,6} } .

    The partition { {1},{2},{3},{4},{5},{6} } is a partition
into unit sets.

CHAPTER 3

FUNDAMENTAL ALGEBRAIC LAWS AND OPERATIONS WITH NUMBERS

> **Basic Attacks and Strategies for Solving Problems in this Chapter. See pages 16 to 23 for step-by-step solutions to problems.**

When a numerator n is divided by a denominator d, it produces an integer quotient q and a remainder r (either of which may be zero). In other words,

$$n/d = q$$

with a remainder r, or

$$n/d = q + r/d.$$

For example, 4/3 gives 1 with a remainder of 1, or $4/3 = 1 + 1/3$.

Algebraic expressions are evaluated based on the laws of operator precedence and the laws of parenthesization. The *laws of operator precedence* tell which operations are evaluated before which other operations. The standard order is exponentiation, then multiplication and division, then addition and subtraction. When multiple operators of the same precedence occur together, they are evaluated left to right (except for exponentiation which is evaluated right to left). The standard order of precedence may be superseded by using sets of parentheses. If parentheses are found in any expression, the operations in the innermost set are performed first and then that set of parentheses is removed and the evaluation continues. Thus,

$$(3 + 4)5 = 35$$

(since the parentheses force $3 + 4$ to be evaluated first), while

$$3 + 4 \cdot 5 = 23$$

(since without parentheses multiplication is performed before addition).

Many algebraic computations involve *factoring*, a procedure based on the *distributive law of multiplication over addition*, i.e.,

$$a \cdot (b + c) = a \cdot b + a \cdot c.$$

Factoring turns a sum (or a difference) into a product (or a quotient). The hope is that the rewritten expression (mathematically equivalent to the original expression) may be more useful in obtaining some sort of answer. For example, $2a + 3a$ can be factored into $(2 + 3)a$, and then the first factor can be summed to obtain $5a$.

Evaluation of an algebraic expression (containing *variables*) is accomplished by a simple substitution of assigned values of variables in the expression. For example, $3a + 4b$, given that $a = 2$ and $b = 3$, produces

$$3 \cdot 2 + 4 \cdot 3 = 6 + 12 = 18.$$

When algebraic expressions involve the sum (or difference) of fractions (with different denominators), a *common denominator* must first be found. Since fractions are usually written in reduced form (i.e., any common factors in the numerator or denominator have been eliminated), finding a common denominator involves multiplying the numerator and denominator of each fraction by an appropriate quantity so that the denominators of both fractions are equal (i.e., common). For example,

$$1/2 + 1/3 = (1 \cdot 3) / (2 \cdot 3) + (1 \cdot 2) / (3 \cdot 2)$$

$$= 3/6 + 2/6 = (3 + 2) / 6 = 5/6.$$

Note that the first fraction is multiplied by 1 written as 3/3 while the second fraction is multiplied by 1 written as 2/2. However, both operations result in fractions with a denominator of 6.

If the denominator of a fraction is itself a fraction, the rule

$$\frac{1}{a/b} = \frac{b}{a}$$

is applied. Thus,

$$2/(3/4) = 2 \cdot 1/(3/4) = 2 \cdot (4/3) = 8/3.$$

A fraction in which the numerator and/or the denominator is also a fraction is sometimes called a *complex fraction*.

The *real number line* is a horizontal line that can be used to visualize the location of certain numbers relative to others. Zero is usually prominently indicated (called the *origin*) and positive numbers are placed to the right of the zero while negative numbers go on the left. Usually, some sort of grid ("tick") marks are used to indicate an appropriate quantity (such as every integer, every fifth integer, multiples of 0.5, etc.).

● **PROBLEM 3-1**

Find the quotient q and remainder r upon dividing 575 by 21.

Solution: If we divide 575 by 21, we obtain

$$
\begin{array}{r}
27 \\
21{\overline{\smash{\big)}\,575}} \\
\underline{42} \\
155 \\
\underline{147} \\
8
\end{array}
$$

The quotient is 27 and the remainder is 8.

Check: To check the quotient and the remainder obtained, multiply the quotient by the divisor and then add the remainder to this product. The sum should be equal to the dividend.

$$(27)(21) + 8 = 567 + 8 = 575.$$

Hence, the sum is equal to the dividend, 575. Therefore, the quotient and the remainder obtained are correct.

● **PROBLEM 3-2**

Evaluate $2 - \{5 + (2 - 3) + [2 - (3 - 4)]\}$

Solution: When working with a group of nested parentheses, we evaluate the innermost parenthesis first.

Thus, $2 - \{5 + (2 - 3) + [2 - (3 - 4)]\}$

$$= 2 - \{5 + (2 - 3) + [2 - (-1)]\}$$

$$= 2 - \{5 + (-1) + [2 + 1]\}$$

$$= 2 - \{5 + (-1) + 3\}$$

$$= 2 - \{4 + 3\}$$

$$= 2 - 7$$

$$= -5.$$

Simplify $4[-2(3 + 9) \div 3] + 5$.

<u>Solution:</u>   To simplify means to find the simplest expression. We perform the operations within the innermost grouping symbols first. That is $3 + 9 = 12$.

Thus, $4[-2(3 + 9) \div 3] + 5 = 4[-2(12) \div 3] + 5$

Next we simplify within the brackets:

$$= 4[-24 \div 3] + 5$$

$$= 4 \cdot (-8) + 5$$

We now perform the multiplication, since multiplication is done before addition:

$$= -32 + 5$$

$$= -27$$

Hence, $4[-2(3 + 9) \div 3] + 5 = -27$.

Calculate the value of

$$5 - \left[6 \times \left[1.5 \times 10\% + (-0.25 + 4\,\tfrac{3}{10}) \div \tfrac{4}{8} \right] - \tfrac{3}{9} \times 6 \div |-6.7 + 3| \right.$$

$$\left. + 4 \div \left(1 + \frac{1}{2+\frac{1}{4}} \right) \right]$$

to the fifth decimal place.

<u>Solution:</u> The rule for solving this type of problem is to work from the innermost parenthesis out. The order of operations are multiplication and division, then subtraction and addition.

The first step in this problem is to change everything to decimals.

$$5 - [6 \times [1.5 \times 0.1 + (-0.25 + 4.3) \div 0.5] - 0.33 \times 6 \div |-6.7 + 3| + 4 \div (1.44444)]$$

The second step is to reduce the parentheses.

$$5 - [6 \times [1.5 \times 0.1 + (4.05) \div 0.5] - 2 \div 3.7 + 4 \div (1.44444)]$$

$$= 5 - [6 \times [0.15 + 8.1] - 0.540541 + 2.769231]$$

$$= 5 - [6 \times (8.25) - 0.540541 + 2.769231]$$

$$= 5 - [49.5 + 2.228690]$$

$$= 5 - 51.72869$$

$$= -46.72869$$

● **PROBLEM 3-5**

(a) Add, 3a + 5a

(b) Factor, 5ac + 2bc.

Solution: (a) To add 3a + 5a, factor out the common factor a. Then,

$$3a + 5a = (3 + 5)a = 8a.$$

(b) To factor 5ac + 2bc, factor out the common factor c. Then,

$$5ac + 2bc = (5a + 2b)c.$$

● **PROBLEM 3-6**

Evaluate $p = \dfrac{(a - b)(ab + c)}{(cb - 2a)}$

when a = +2, b = $-\frac{1}{2}$, and c = $-$ 3.

Solution: Inserting the given values of a, b, and c

$$p = \frac{[+2 - (-\frac{1}{2})][(+2)(-\frac{1}{2}) + (-3)]}{[(-3)(-\frac{1}{2}) - 2(+2)]}$$

$$= \frac{[+2 + \frac{1}{2}][-1 - 3]}{[+1\frac{1}{2} - 4]}$$

$$= \frac{(2\frac{1}{2})(-4)}{-(2\frac{1}{2})}$$

The 2½ in the numerator cancels the 2½ in the denominator.

$$p = \frac{-4}{-1}$$

Multiplying numerator and denominator by - 1

$$p = \frac{+\ 4}{+\ 1}$$

$$p = +\ 4.$$

● **PROBLEM 3-7**

Calculate the value of each of the following expressions:

(1) $||2 - 5| + 6 - 14|$

(2) $|-5| \cdot |4| + \dfrac{|-12|}{4}$

(3) $1.6\% + 18\% + 12(26 - (1-3) \div |-2|)$

(4) $\dfrac{1}{6} \times 1.25 - (12.5 + 4\frac{1}{2}) \div 50\%$

<u>Solution</u>: Before solving this problem, one must remember the order of operations: parenthesis, multiplication and division, addition and subtraction.

(1) $||-3| + 6 - 14| = |3 + 6 - 14| = |9 - 14| = |-5| = 5$

(2) $(5 \times 4) + \dfrac{12}{4} = 20 + 3 = 23$

(3) $0.016 + 0.18 + 12(26 - (-2) \div 2)$
$0.016 + 0.18 + 12(26 - (-1))$
$0.196 + 12(27) = 0.196 + 324 = 324.196$

(4) $\dfrac{1.25}{6} - (12.5 + \dfrac{9}{2}) \div 0.5 = \dfrac{1.25}{6} - \left(\dfrac{34}{2}\right)\left(\dfrac{2}{1}\right)$

$\dfrac{1.25}{6} - 34 = \dfrac{1.25}{6} - \dfrac{204}{6} = -33.792$

● **PROBLEM 3-8**

Simplify the following expression: $1 - \dfrac{1}{2 - \frac{1}{3}}$.

<u>Solution:</u> In order to combine the denominator, $2 - \frac{1}{3}$, we must convert 2 into thirds. $2 = 2 \cdot 1 = 2 \cdot \dfrac{3}{3} = \dfrac{6}{3}$. Thus

$$1 - \dfrac{1}{2 - \frac{1}{3}} = 1 - \dfrac{1}{\frac{6}{3} - \frac{1}{3}} = 1 - \dfrac{1}{\frac{5}{3}}$$

Since division by a fraction is equivalent to multiplication by that fraction's reciprocal

$$1 - \frac{1}{\frac{5}{3}} = 1 - (1)\left(\frac{3}{5}\right) = 1 - \frac{3}{5} = \frac{5}{5} - \frac{3}{5} = \frac{2}{5}$$

Therefore,
$$1 - \frac{1}{2 - \frac{1}{3}} = \frac{2}{5}.$$

Simplify $\dfrac{\frac{2}{3} + \frac{1}{2}}{\frac{3}{4} - \frac{1}{3}}$.

Solution: A first method is to just add the terms in the numerator and denominator. Since 6 is the least common denominator of the numerator, $\left(\frac{2}{3} + \frac{1}{2}\right)$, we convert $\frac{2}{3}$ and $\frac{1}{2}$

into sixths:

$$\frac{2}{3} = \frac{2}{3} \cdot 1 = \frac{2}{3} \cdot \frac{2}{2} = \frac{4}{6} \text{ and } \frac{1}{2} = \frac{1}{2} \cdot 1 = \frac{1}{2} \cdot \frac{3}{3} = \frac{3}{6}$$

Therefore $\quad \frac{2}{3} + \frac{1}{2} = \frac{4}{6} + \frac{3}{6} = \frac{7}{6}$

Since 12 is the least common denominator of the denominator, $\left(\frac{3}{4} - \frac{1}{3}\right)$, we convert $\frac{3}{4}$ and $\frac{1}{3}$ into twelfths:

$$\frac{3}{4} = \frac{3}{4} \cdot 1 = \frac{3}{4} \cdot \frac{3}{3} = \frac{9}{12} \text{ and } \frac{1}{3} = \frac{1}{3} \cdot 1 = \frac{1}{3} \cdot \frac{4}{4} = \frac{4}{12}$$

Therefore $\quad \frac{3}{4} - \frac{1}{3} = \frac{9}{12} - \frac{4}{12} = \frac{5}{12}$

Thus, $\dfrac{\frac{2}{3} + \frac{1}{2}}{\frac{3}{4} - \frac{1}{3}} = \dfrac{\frac{7}{6}}{\frac{5}{12}}$

Division by a fraction is equivalent to multiplication by the reciprocal hence $\dfrac{\frac{7}{6}}{\frac{5}{12}} = \frac{7}{6} \cdot \frac{12}{5}$

Cancelling 6 from the numerator and denominator:

$$= \frac{7}{1} \cdot \frac{2}{5} = \frac{14}{5}$$

A second method is to multiply both numerator and denominator by the least common denominator of the entire fraction. Since we have already seen that L.C.D. of the numerator is 6 and the L.C.D. of the denominator is 12, and

12 is divisible by 6, we use 12 as the L.C.D. of the entire fraction. Thus

$$\frac{\frac{2}{3} + \frac{1}{2}}{\frac{3}{4} - \frac{1}{3}} = \frac{12\left(\frac{2}{3} + \frac{1}{2}\right)}{12\left(\frac{3}{4} - \frac{1}{3}\right)}$$

Distribute:

$$= \frac{12\left(\frac{2}{3}\right) + 12\left(\frac{1}{2}\right)}{12\left(\frac{3}{4}\right) - 12\left(\frac{1}{3}\right)}$$

$$= \frac{4 \cdot 2 + 6}{3 \cdot 3 - 4} = \frac{8 + 6}{9 - 4} = \frac{14}{5}$$

● **PROBLEM 3-10**

If $a = 4$ and $b = 7$ find the value of $\dfrac{a + \dfrac{a}{b}}{a - \dfrac{a}{b}}$.

Solution: By substitution, $\dfrac{a + \dfrac{a}{b}}{a - \dfrac{a}{b}} = \dfrac{4 + \dfrac{4}{7}}{4 - \dfrac{4}{7}}$.

In order to combine the terms we convert 4 into sevenths:

$$4 = 4 \cdot 1 = 4 \cdot \frac{7}{7} = \frac{28}{7} .$$

Thus, we have:

$$\frac{\frac{28}{7} + \frac{4}{7}}{\frac{28}{7} - \frac{4}{7}} = \frac{\frac{32}{7}}{\frac{24}{7}} .$$

Dividing by $\dfrac{24}{7}$ is equivalent to multiplying by $\dfrac{7}{24}$. Therefore,

$$\frac{4 + \frac{4}{7}}{4 - \frac{4}{7}} = \frac{32}{7} \cdot \frac{7}{24}$$

Now, the 7 in the numerator cancels with the 7 in the denominator. Thus, we obtain: $\dfrac{32}{24}$, and dividing numerator and denominator by 8, we obtain: $\dfrac{4}{3}$.

Therefore $\dfrac{a + \frac{a}{b}}{a - \frac{a}{b}} = \dfrac{4}{3}$ when $a = 4$ and $b = 7$.

● PROBLEM 3-11

Plot the value 1/2, 125%, –8.4, –3/2, and 2 on the real number line.

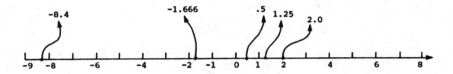

Solution: Before plotting these values, it is recommended to change all the above values to either decimals or fractions.

1/2 = 0.5

125% = 1.25

–8.4 = –8.4

–3/2 = –1.5

2 = 2.0

● PROBLEM 3-12

Classify each of the following statements as true or false. If it is false, explain why.

(1) $|{-120}| > 1$

(2) $|4{-}12| = |4| - |12|$

(3) $|4{-}9| = 9 - 4$

(4) $|12{-}3| = 12 - 3$

(5) $|{-}12a| = 12|a|$

Solution: (1)   true

(2)   false, $|4{-}12| = |4| - |12|$

$|{-}8| = 4 - 12$

$8 \neq -8$

22

In general, $|a+b| \neq |a| + |b|$

(3)   true

(4)   true

(5)   true

CHAPTER 4

FUNDAMENTAL ALGEBRAIC LAWS AND OPERATIONS WITH ALGEBRAIC EXPRESSIONS

Basic Attacks and Strategies for Solving Problems in this Chapter. See pages 24 to 39 for step-by-step solutions to problems.

Among other things, one should remember that all the algebraic rules which apply to operations with numbers also apply to operations with variables and other algebraic expressions. In particular, factoring and division by fractions (mentioned in the introduction to the previous chapter) are very common operations in algebraic expressions.

An *algebraic expression* is one in which there appear numbers and variables (possibly raised to some rational power) combined by the algebraic operations of addition, subtraction, multiplication and division. Such an expression does not contain any trigonometric functions or expressions containing a variable as an exponent. A *monomial* is an algebraic expression in which there appear only integers and variables raised to a positive power multiplied together. The *degree of a monomial* is the sum of the powers of the variables it contains. For example, the degree of $3x^2y^3$ is 5. A *polynomial* is any sum (and difference) of monomials that are integral and rational terms. A *trinomial* is a polynomial containing *three* monomials. A *multinomial* differs from a polynomial in that a multinomial may have component monomials which are non-integral or non-rational. A *non-integral term* is one in which a constant or variable is raised to a negative power. A *non-rational term* is one in which a constant or variable is raised to a non-integral power.

When attempting to simplify algebraic expressions (i.e., expressions with unevaluated variables), one searches for the largest common expression among the terms and factors it out. For example, in $4xy^2 + 5xy^2$, the expression xy^2 is common to both terms. It can be factored out to obtain $9xy^2$. In $2xy + 3y$, only y is common so that this can be transformed to $(2x + 3) \cdot y$ if desired. One can check the accuracy of a simplification by substituting arbitrary values for each of the variables. For example, letting

$x = 1$ and $y = 2$ in $2xy + 3y = (2x + 3) \cdot y$,

we see that the left side reduces to

$$2 \cdot 1 \cdot 2 + 3 \cdot 2 = 4 + 6 = 10$$

and the right side reduces to

$$(2 \cdot 1 + 3) \cdot 2 = 5 \cdot 2 = 10,$$

suggesting (but not proving) that the simplification was performed correctly. An actual proof would require one to check every set of possible values (this is not feasible).

When asked to "solve" an equation ("for x"), one performs appropriate algebraic manipulations until x appears alone on one side of the equals sign and all other variables and constants appear on the other side. The variable x cannot be on both sides of the equation — only on one side.

When an equation involves an absolute value, one should remember that the absolute value is in reality shorthand for two possible expressions, one positive and the other negative. For example, if $|a| = 3$, a can either be 3 or –3. Therefore, when removing the absolute value signs (and replacing them with parentheses), one compensates by creating two new equations, one of which includes a negative sign before the expression formerly inside absolute value bars. For example, when solving

$$|x + 1| = 2 \quad \text{for} \quad x,$$

we remove the absolute value signs and replace the single equation with two equations:

$$x + 1 = 2 \tag{1}$$

and $\quad -(x + 1) = 2.$ (2)

Equation (1) gives the result of $x = 1$, while equation (2) gives $x = -3$. The original equation $|x + 1| = 2$ thus has two possible solutions:

$$x = 1 \quad \text{or} \quad -3.$$

Simplification of algebraic fractions is often more common than the simplification of numeric fractions. The procedure is the same in either case. One begins by factoring both the numerator and denominator. Common factors in the numerator(s) and denominator(s) are then eliminated since any term divided by itself equals one.

When adding or subtracting two polynomials, one can only combine terms in which the variables (and their powers) are identical. Thus,

$$x^2 + 3xy^2 \quad \text{and} \quad 2x + xy$$

yields all four terms since no two terms contain both the same variables along with the same powers!

When multiplying two polynomials (binomials, i.e., algebraic expressions with two [*bi-*] terms), one usually uses the distributive law twice. For example

$$(ax + b) \cdot (cx + d) \quad \text{equals} \quad ax \cdot (cx + d) + b \cdot (cx + d)$$

which equals

$$ax \cdot cx + adx + bcx + bd$$

which can be simplified to

$$acx^2 + (ad + bc)x + bd.$$

Sometimes this is referred to as the *FOIL* method which is an acronym for First Outer Inner Last (terms of the original multiplied expression). After multiplication, the first term of the answer (*ac*) is composed of the first terms of the two initial expressions, the next term (*ad*) is composed of the outer two terms (i.e., the first and last term) of the two initial expressions, etc.

Long division of two polynomials is similar to long division of real numbers. Instead of being concerned about ordering numbers according to powers of 10, we must be concerned about ordering variables by the appropriate powers of the variable. Thus, if we divide $2x^2 - 2$ by $1 + x$, both expressions must first be re-written so that they are expressed in descending powers of x and all powers less than the degree of the polynomial must be explicitly indicated, i.e.,

$$2x^2 + 0x - 2 \quad \text{and} \quad x + 1.$$

After this, we can set up the typical long division grid and perform the division. Since x will divide $2x^2$ evenly $2x$ times, we write a $2x$ above the $0x$ term and multiply $2x$ times $x + 1$, writing the result $(2x^2 + 2x)$ under $2x^2 + 0x$ and then subtracting. The next term $(+ 1)$ is brought down and the process is repeated. The process is concluded when the divisor is of a greater degree than the partial dividend. At that point, there may or may not be a remainder left.

$$
\begin{array}{r}
2x - 2 \\
x + 1 \overline{\smash{\big)}\ 2x^2 + 0x - 2} \\
\underline{2x^2 + 2x} \\
-2x - 2 \\
\underline{-2x - 2} \\
0
\end{array}
$$

Step-by-Step Solutions to Problems in this Chapter, "Fundamental Algebraic Laws and Operations with Algebraic Expressions"

● **PROBLEM** 4-1

Express each of the following as a single term.

(a) $3x^2 + 2x^2 - 4x^2$ (b) $5axy^2 - 7axy^2 - 3xy^2$

Solution: (a)   Factor x^2 in the expression.

$$3x^2 + 2x^2 - 4x^2 = (3 + 2 - 4)x^2 = 1x^2 = x^2.$$

(b)   Factor xy^2 in the expression and then factor a.

$$5axy^2 - 7axy^2 - 3xy^2 = (5a - 7a - 3)xy^2$$
$$= [(5-7)a - 3]xy^2$$
$$= (-2a - 3)xy^2.$$

● **PROBLEM** 4-2

Simplify $x = a + 2[b - (c - a + 3b)]$.

Solution: When working with several groupings, we perform the operations in the innermost parenthesis first, and work outward. Thus, we first subtract $(c - a + 3b)$ from b:

$$x = a + 2[b - (c - a + 3b)] = a + 2(b - c + a - 3b)$$

Combining terms,

$$= a + 2(-c + a - 2b)$$

distributing the 2,

$$= a - 2c + 2a - 4b$$

combining terms,

$$= 3a - 2c - 4b$$

To check that $a + 2[b - (c - a + 3b)]$ is equivalent to $3a - 2c - 4b$, replace a,b, and c by any values. Letting a = 1, b = 2, c = 3, the original form a + 2[b - (c - a + 3b)] = 1 + 2[2 - (3 - 1 + 3·2)]

$$= 1 + 2[2 - (3 - 1 + 6)]$$
$$= 1 + 2[2 - 8]$$
$$= 1 + 2(-6)$$
$$= 1 + (-12)$$
$$= -11$$

The final form, 3a - 2c - 4b = 3(1) - 2(3) - 4(2) = 3 - 6 - 8

$$= -11$$

Thus, both forms yield the same result.

● **PROBLEM 4-3**

Solve for x when $|5 - 3x| = -2$.

Solution: This problem has no solution, since the absolute value can never be negative and we need not proceed further.

● **PROBLEM 4-4**

Solve for x in $|2x - 6| = |4 - 5x|$.

Solution: There are four possibilities here. 2x - 6 and 4 - 5x can be either positive or negative. Therefore,

$$2x - 6 = 4 - 5x \qquad (1)$$
$$-(2x - 6) = 4 - 5x \qquad (2)$$

and

$$2x - 6 = -(4 - 5x) \qquad (3)$$
$$-(2x - 6) = -(4 - 5x) \qquad (4)$$

Equations (2) and (3) result in the same solution, as do equations (1) and (4). Therefore it is necessary to solve only for equations (1) and (2). This gives:

$$x = \frac{10}{7} , -\frac{2}{3} .$$

● **PROBLEM 4-5**

Solve for x when $|2x - 1| = |4x + 3|$.

Solution: Replacing the absolute sysmbols with equations that can be handled algebraically according to the conditions implied by the given equation, we have:

$$2x - 1 = 4x + 3 \quad \text{or} \quad 2x - 1 = -(4x + 3) \ .$$

Solving the first equation, we have $x = -2$; solving the second, we obtain $x = -\frac{1}{3}$, thus giving us two solutions to the original equation. (We could also write: $-(2x - 1) = -(4x + 3)$, but this is equivalent to the first of the equations above.)

● **PROBLEM 4-6**

Perform the following division: $1 \Big/ \dfrac{x + y}{x^2}$.

Solution: Division by a fraction is equivalent to multiplication by that fraction's reciprocal. Hence,

$$\frac{1}{\dfrac{x + y}{x^2}} = 1 \cdot \frac{x^2}{x + y} = \frac{x^2}{x + y}.$$

● **PROBLEM 4-7**

Perform the indicated operation:
$$\frac{3a - 9b}{x - 5} \cdot \frac{xy - 5y}{ax - 3bx}$$

Solution: According to our definition of multiplication, we need only to write the product of the numerators over the product of the denominators. The only remaining step is that of reducing the fraction to lowest terms by factoring the numerator and denominator of the answer and simplifying the result.

$$\frac{3a - 9c}{x - 5} \cdot \frac{xy - 5y}{ax - 3bx} = \frac{(3a - 9b)(xy - 5y)}{(x - 5)(ax - 3bx)}$$

Factor out 3 from $(3a - 9b)$ and y from $(xy - 5y)$. Also, factor out x from $ax - 3bx$.

$$= \frac{3(a - 3b)y(x - 5)}{(x - 5)x(a - 3b)}$$

Group the same terms in numerator and the denominator.

$$= \frac{3y}{x} \cdot \frac{a - 3b}{a - 3b} \cdot \frac{x - 5}{x - 5}$$

Cancel like terms.

$$= \frac{3y}{x} \cdot 1 \cdot 1$$
$$= \frac{3y}{x}$$

This procedure could have been abbreviated in the following manner:

$$\frac{3a - 9b}{x - 5} \cdot \frac{xy - 5y}{ax - 3bx} = \frac{3(a - 3b)}{x - 5} \cdot \frac{y(x - 5)}{x(a - 3b)} = \frac{3y}{x}$$

26

Combine into a single fraction in lowest terms.

(a) $\dfrac{6(a + 1)}{a + 8} - \dfrac{3(a - 4)}{a + 8} - \dfrac{2(a + 5)}{a + 8}$

(b) $\dfrac{7x - 3y + 6}{x + y} - \dfrac{2(x - 4y + 3)}{x + y}$

(c) $\dfrac{5x + 2}{x - 6} - \dfrac{3(x + 4)}{x - 6} - \dfrac{x - 7}{x - 6}$

Solution: Noting $\dfrac{a}{x} + \dfrac{b}{x} + \dfrac{c}{x} = \dfrac{a+b+c}{x}$ (where a,b,c are any real numbers and x any non-zero real number), we proceed to evaluate these expressions:

(a) $\dfrac{6(a + 1)}{a + 8} - \dfrac{3(a - 4)}{a + 8} - \dfrac{2(a + 5)}{a + 8} = \dfrac{6(a+1) - 3(a-4) - 2(a+5)}{a+8}$

Distributing, $= \dfrac{6a + 6 - 3a + 12 - 2a - 10}{a+8} = \dfrac{6a-3a-2a+6+12-10}{a+8}$

$$= \dfrac{a + 8}{a + 8} = 1.$$

(b) $\dfrac{7x - 3y + 6}{x + y} - \dfrac{2(x - 4y + 3)}{x + y} = \dfrac{7x - 3y + 6 - 2(x-4y+3)}{x + y}$

Distributing, $= \dfrac{7x - 3y + 6 - 2x + 8y - 6}{x + y} = \dfrac{7x-2x-3y+8y+6-6}{x + y}$

$$= \dfrac{5x + 5y}{x + y} = \dfrac{5(x + y)}{x + y} = 5.$$

(c) $\dfrac{5x + 2}{x - 6} - \dfrac{3(x + 4)}{x - 6} - \dfrac{x - 7}{x - 6} = \dfrac{5x + 2 - 3(x+4) - (x-7)}{x-6}$

Distributing, $= \dfrac{5x+2-3x-12-x+7}{x-6} = \dfrac{5x-3x-x+2-12+7}{x-6} = \dfrac{x - 3}{x - 6}.$

Combine and simplify $1 + \dfrac{1}{1 + \dfrac{1}{1 - x}}$.

Solution: First combine the terms in the denominator. Recall $1 = (1-x)/(1-x)$. Thus,

$$1 + \dfrac{1}{1 + \dfrac{1}{1 - x}} = 1 + \dfrac{1}{\dfrac{1 - x}{1 - x} + \dfrac{1}{1 - x}}$$

$$= 1 + \dfrac{1}{\dfrac{1 - x + 1}{1 - x}}$$

$$= 1 + \cfrac{1}{\cfrac{2 - x}{1 - x}}$$

Division by a fraction is equivalent to multiplication by its reciprocal, thus

$$= 1 + 1 \cdot \frac{(1 - x)}{(2 - x)}$$

$$= 1 + \frac{1 - x}{2 - x}$$

Recall $1 = \frac{2 - x}{2 - x}$, therefore,

$$1 + \cfrac{1}{1 + \cfrac{1}{1 - x}} = \frac{2 - x}{2 - x} + \frac{1 - x}{2 - x}$$

$$= \frac{2 - x + 1 - x}{2 - x}$$

$$= \frac{3 - 2x}{2 - x} \; .$$

● **PROBLEM** 4-10

Simplify the complex fraction $\cfrac{\cfrac{1}{x} - \cfrac{1}{y}}{\cfrac{1}{x^2} - \cfrac{1}{y^2}}$.

Solution: Add both fractions of the numerator together using the rule: $\frac{a}{b} + \frac{c}{d} = \frac{ad + bc}{bd}$; and obtain $\frac{y - x}{xy}$. Similarly for the denominator, obtain:

$$\frac{y^2 - x^2}{x^2 y^2}$$

Now invert the fraction in the denominator and multiply by the numerator:

$$\frac{y - x}{xy} \cdot \frac{x^2 y^2}{y^2 - x^2} = \frac{(y - x)}{xy} \cdot \frac{(xy)(xy)}{(y - x)(y + x)}$$

$$= \frac{(y - x)(xy)(xy)}{xy(y - x)(y + x)}$$

$$= \frac{xy}{y + x} \quad .$$

Simplify $\dfrac{\frac{2}{x} + \frac{3}{y}}{1 - \frac{1}{x}}$.

Solution: A first method is to just add the terms in the numerator and denominator, obtaining

$$\frac{\frac{2}{x} + \frac{3}{y}}{1 - \frac{1}{x}} = \frac{\frac{2y}{xy} + \frac{3x}{xy}}{\frac{x}{x} - \frac{1}{x}} = \frac{\frac{2y + 3x}{xy}}{\frac{x - 1}{x}}$$

Since dividing by fraction is equivalent to multiplying by its reciprocal,

$$= \frac{2y + 3x}{xy} \cdot \frac{x}{x - 1} = \frac{2y + 3x}{y(x - 1)}$$

A second method is to multiply both numerator and denominator by the least common denominator of the entire fraction, in this case xy:

$$\frac{\frac{2}{x} + \frac{3}{y}}{1 - \frac{1}{x}} = \frac{xy\left(\frac{2}{x} + \frac{3}{y}\right)}{xy\left(1 - \frac{1}{x}\right)}$$

Distributing, $\qquad = \dfrac{xy\left(\frac{2}{x}\right) + xy\left(\frac{3}{y}\right)}{xy(1) - xy\left(\frac{1}{x}\right)}$

Cancelling like terms,

$$= \frac{2y + 3x}{xy - y}$$

Using distributive law,

$$= \frac{2y + 3x}{y(x - 1)} .$$

Combine $\dfrac{1}{6x} + \dfrac{1}{3y} - \dfrac{3x + 2y}{12xy}$ into a single fraction.

Solution: Since both 6x and 3y are factors of 12xy, the least common denominator (the LCD) of the given fractions is 12xy. Thus, we wish to convert the given fractions to equal fractions having 12xy as a denominator. We can accomplish this by multiplying each member of the first fraction by 2y and each member of the second by 4x. We thereby obtain

$$\frac{1}{6x} + \frac{1}{3y} - \frac{3x + 2y}{12xy} = \frac{2y \cdot 1}{2y(6x)} + \frac{4x \cdot 1}{4x(3y)} - \frac{3x + 2y}{12xy}$$

$$= \frac{2y}{12xy} + \frac{4x}{12xy} - \frac{3x + 2y}{12xy}$$

$$= \frac{2y + 4x - (3x + 2y)}{12xy}$$

$$= \frac{2y + 4x - 3x - 2y}{12xy}$$

$$= \frac{2y + x - 2y}{12xy}$$

$$= \frac{2y - 2y + x}{12xy}$$

$$= \frac{x}{12xy} \, .$$

Cancelling x from numerator and denominator,

$$= \frac{1}{12y}$$

Thus, $\quad \frac{1}{6x} + \frac{1}{3y} - \frac{3x + 2y}{12xy} = \frac{1}{12y} \quad .$

● **PROBLEM 4-13**

Simplify $\quad \dfrac{\dfrac{1}{x - 1} - \dfrac{1}{x - 2}}{\dfrac{1}{x - 2} - \dfrac{1}{x - 3}} \, .$

Solution: Simplify the expression in the numerator by using the addition rule:

$$\frac{a}{b} + \frac{c}{d} = \frac{ad + bc}{bd}$$

Notice bd is the Least Common Denominator, LCD.

We obtain $\dfrac{x - 2 - (x - 1)}{(x - 1)(x - 2)} = \dfrac{-1}{(x - 1)(x - 2)}$ in the numerator.

Repeat this procedure for the expression in the denominator:

$$\frac{x - 3 - (x - 2)}{(x - 2)(x - 3)} = \frac{-1}{(x - 2)(x - 3)}$$

We now have

$$\frac{\dfrac{-1}{(x-1)(x-2)}}{\dfrac{-1}{(x-2)(x-3)}} \, ,$$

which is simplified by inverting the fraction in the denominator and multiplying it by the numerator and cancelling like terms

$$\frac{-1}{(x-1)(x-2)} \cdot \frac{(x-2)(x-3)}{-1} = \frac{x-3}{x-1} \, .$$

● PROBLEM 4-14

A) If $x = \dfrac{c - ab}{a - b}$, find the value of the expression $a(x + b)$.

B) Also, if $x = \dfrac{c - ab}{a - b}$, find the value of the expression $bx + c$.

<u>Solution:</u> A) Substituting $x = \dfrac{c - ab}{a - b}$ for x in the expression $a(x + b)$,

$$a(x + b) = a\left(\frac{c - ab}{a - b} + b\right) \qquad (1)$$

Obtaining a common denominator of $a - b$ for the two terms in parenthesis; equation (1) becomes:

$$a(x + b) = a\left[\frac{c - ab}{a - b} + \frac{(a-b)b}{a - b}\right]$$

Distributing the numerator of the second term in brackets:

$$a(x + b) = a\left[\frac{c-ab}{a-b} + \frac{ab-b^2}{a-b}\right] = a\left[\frac{c-ab+ab-b^2}{a-b}\right]$$

$$= a\left[\frac{c-b^2}{a-b}\right]$$

$$a(x + b) = \frac{a(c-b^2)}{a-b}$$

B) Substituting $x = \dfrac{c-ab}{a-b}$ for x in the expression $bx + c$,

$$bx + c = b\left(\frac{c-ab}{a-b}\right) + c$$

$$= \frac{b(c-ab)}{a-b} + c \qquad (2)$$

Obtaining a common denominator of $a - b$ for the two terms on the right side of equation (2):

$$bx + c = \frac{b(c-ab)}{a-b} + \frac{(a-b)c}{a-b}$$

31

Distributing the numerator of each term on the right side:

$$bx + c = \frac{bc - ab^2}{a - b} + \frac{ac - bc}{a - b}$$

$$= \frac{bc - ab^2 + ac - bc}{a - b} = \frac{-ab^2 + ac}{a - b}$$

$$= \frac{ac - ab^2}{a - b}$$

Factoring out the common factor of a from the numerator of the right side:

$$bx + c = \frac{a\left(c - b^2\right)}{a - b}$$

● PROBLEM 4-15

Find the value of the polynomial $3x^2y - 2xy^2 + 5xy$ when $x = 1$ and $y = -2$.

Solution: Replace x by 1 and y by -2 in the given polynomial to obtain,

$$3x^2y - 2xy^2 + 5xy = \left[3(1)^2 - (-2)\right] - \left[2(1)(-2)^2\right]$$

$$+ \left[5(1)(-2)\right]$$

$$= \left[(3)(-2)\right] - \left[(2)(4)\right]$$

$$+ \left[(5)(-2)\right]$$

$$= -6 - 8 - 10$$

$$= -24.$$

Thus, when $x = 1$ and $y = -2$, the polynomial

$$3x^2y - 2xy^2 + 5xy = -24.$$

● PROBLEM 4-16

Add $\left(3xy^2 + 2xy + 5x^2y\right) + \left(2xy^2 - 4xy + 2x^2y\right)$.

Solution: Use the vertical form, align all like terms, and apply the distributive property.

$$3xy^2 + 2xy + 5x^2y$$
$$2xy^2 - 4xy + 2x^2y$$
$$\overline{}$$
$$(3 + 2)xy^2 + (2 - 4)xy + (5 + 2)x^2y$$

Thus, the sum is $5xy^2 - 2xy + 7x^2y$.

From the sum of $6x^2 + 4xy - 8y^2 - 11$ and $3x^2 - 4y^2 + 8 + 5xy$ subtract $xy - 10 - 5x^2 + 7y^2$.

Solution: First find the sum of $6x^2 + 4xy - 8y^2 - 11$ and $3x^2 - 4y^2 + 8 + 5xy$. Adding these two polynomials together:

$$\left(6x^2 + 4xy - 8y^2 - 11\right) + \left(3x^2 - 4y^2 + 8 + 5xy\right)$$

$$= 6x^2 + 4xy - 8y^2 - 11 + 3x^2 - 4y^2 + 8 + 5xy$$

Grouping like terms together,

$$\left(6x^2 + 4xy - 8y^2 - 11\right) + \left(3x^2 - 4y^2 + 8 + 5xy\right)$$

$$= \left(6x^2 + 3x^2\right) + \left(4xy + 5xy\right) + \left(-8y^2 - 4y^2\right) + (-11 + 8)$$

$$= 9x^2 + 9xy + \left(-12y^2\right) + (-3) = 9x^2 + 9xy - 12y^2 - 3 \qquad (1)$$

Now subtract $xy - 10 - 5x^2 + 7y^2$ from the resultant sum, which is the right side of equation (1), or $9x^2 + 9xy - 12y^2 - 3$. Then,

$$\left(9x^2 + 9xy - 12y^2 - 3\right) - \left(xy - 10 - 5x^2 + 7y^2\right) =$$

$$= 9x^2 + 9xy - 12y^2 - 3 - xy + 10 + 6x^2 - 7y^2 \qquad (2)$$

Grouping like terms together, equation (2) becomes:
$$\left(9x^2 + 9xy - 12y^2 - 3\right) - \left(xy - 10 - 5x^2 + 7y^2\right)$$

$$= \left(9x^2 + 5x^2\right) + (9xy - xy) + \left(-12y^2 - 7y^2\right) + (-3 + 10)$$

$$= 14x^2 + 8xy + \left(-19y^2\right) + 7$$

$$= 14x^2 + 8xy - 19y^2 + 7,$$

which is the final answer.

Subtract $3x^4y^3 + 5x^2y - 4xy + 5x - 3$ from the polynomial $5x^4y^3 - 3x^2y + 7$.

<u>Solution:</u> $\left(5x^4y^3 - 3x^2y + 7\right) - \left(3x^4y^3 + 5x^2y - 4xy + 5x - 3\right)$

$= \left(5x^4y^3 - 3x^2y + 7\right) + \left(-3x^4y^3 - 5x^2y + 4xy - 5x + 3\right)$

$= 5x^4y^3 + \left(-3x^4y^3\right) - 3x^2y + \left(-5x^2y\right) + 4xy - 5x + 7 + 3$

$= 2x^4y^3 - 8x^2y + 4xy - 5x + 10$

The column form may also be used for subtraction. Here we align the like terms and subtract the coefficients.

$$5x^4y^3 - 3x^2y \qquad\qquad + 7$$
$$- \left[3x^4y^3 + 5x^2y - 4xy + 5x - 3\right]$$
$$\overline{2x^4y^3 - 8x^2y + 4xy - 5x + 10}$$

● **PROBLEM 4-19**

Multiply $(4x - 5)(6x - 7)$.

<u>Solution:</u> We can apply the FOIL method. The letters indicate the order in which the terms are to be multiplied.

F = first terms
O = outer terms
I = inner terms
L = last terms

Thus,

$(4x - 5)(6x - 7) = (4x)(6x) + (4x)(-7) + (-5)(6x) + (-5)(-7)$
$= 24x^2 - 28x - 30x + 35 = 24x^2 - 58x + 35.$

Another way to multiply algebraic expressions is to apply the distributive law of multiplication with respect to addition. If a,b, and c are real numbers, then a(b+c) = ab + ac. In this case let a = (4x-5) and b + c = 6x - 7.

$(4x - 5)(6x - 7) = (4x - 5)(6x) + (4x - 5)(-7)$

Then, apply the law again.

$(4x - 5)(6x - 7) = (4x)(6x) - (5)(6x) + (4x)(-7) + (-5)(-7)$
$(4x - 5)(6x - 7) = 24x^2 - 30x - 28x + 35$

Add like terms.

$(4x - 5)(6x - 7) = 24x^2 - 58x + 35.$

● **PROBLEM 4-20**

Show that $(3x - 2)(x + 5) + 15 = 3x^2 + 13x + 5$ is an identity.

<u>Solution:</u> An equation in x is an identity if it holds for all real values of x. Thus, the given equation is an identity since for each x ε R,

$$(3x - 2)(x + 5) + 15 = 3x^2 + 13x - 10 + 15$$

$$= 3x^2 + 13x + 5$$

● **PROBLEM** 4-21

Simplify $(5ax + by)(2ax - 3by)$.

<u>Solution:</u> The following formula can be used to simplify the given expression:

$$\left(N_1 + N_2\right)N_3 = N_1N_3 + N_2N_3$$

where N_1, N_2 and N_3 are any three numbers. Note that the distributive property is used in this formula. N_1 is replaced by 5ax and by replaces N_2. Also, N_3 is replaced by (2ax - 3by). Therefore:

$(5ax + by)(2ax - 3by) = 5ax(2ax - 3by) + by(2ax - 3by)$

Use the distributive property to simplify the right side of the above equation:

$(5ax + by)(2ax - 3by) = 5ax(2ax) + 5ax(-3by) + by(2ax) + by(-3by)$

$$= 10a^2x^2 - 15abxy + 2abxy - 3b^2y^2$$

$$= 10a^2x^2 - 13abxy - 3b^2y^2$$

It is often convenient to arrange the two factors vertically as we do in ordinary arithmetic. Hence, the problem is an ordinary multiplication problem.

$$
\begin{array}{r}
5ax + by \\
\times \quad 2ax - 3by \\
\hline
-15abxy - 3b^2y^2 \\
+ \quad 10a^2x^2 + 2abxy \\
\hline
10a^2x^2 - 13abxy - 3b^2y^2
\end{array}
$$

Note that this answer (i.e., product) is the same as the answer obtained above.

● **PROBLEM** 4-22

Expand $(x + 3y - 5t)^2$.

<u>Solution:</u> It is sometimes convenient to group two or more terms and treat them as a single term. x + 3y will be considered as one term. Hence,

$$(x + 3y - 5t)^2 = [(x + 3y) - 5t]^2$$

$$= [(x + 3y) - 5t][(x + 3y) - 5t]$$

Apply the FOIL method:

$$(x + 3y - 5t)^2 = (x + 3y)^2 - 5t(x + 3y) - 5t(x + 3y) + (-5t)^2$$

$$= (x + 3y)^2 + 2(x + 3y)(-5t) + (-5t)^2$$

$$= (x + 3y)(x + 3y) + 2(x + 3y)(-5t) + (-5t)^2$$

$$= x^2 + 3xy + 3xy + 9y^2 + (2x + 6y)(-5t) + 25t^2$$
$$= x^2 + 6xy + 9y^2 - 10tx - 30ty + 25t^2 \ .$$

● **PROBLEM 4-23**

Find the product
$$\left(2x^2 - 3xy + y^2\right)(2x - y)$$

Solution: Multiplication of polynomials can be carried out very much the same way we multiply numbers. One polynomial is written under the other, and then multiplied term by term. Like terms in the product are arranged in columns and added.

$$
\begin{array}{r}
2x^2 - 3xy + y^2 \\
2x - y \\
\hline
4x^3 - 6x^2y + 2xy^2 \\
- 2x^2y + 3xy^2 - y^3 \\
\hline
4x^3 - 8x^2y + 5xy^2 - y^3
\end{array}
$$

We are applying the Distributive Law in the following way:

$$\left(2x^2 - 3xy + y^2\right)\left(2x - y\right) = \left(2x^2 - 3xy + y^2\right)\left(2x\right) = 4x^3 - 6x^2y + 2xy^2$$
$$+ \left(2x^2 - 3xy + y^2\right)\left(-y\right) = \quad\quad -2x^2y + 3xy^2 - y^3$$
$$\left(2x^2 - 3xy + y^2\right)\left(2x-y\right) = 4x^3 - 8x^2y + 5xy^2 - y^3$$

● **PROBLEM 4-24**

Divide $(37 + 8x^3 - 4x)$ by $(2x + 3)$.

Solution: Arrange both polynomials in descending powers of the variable. The first polynomial becomes: $8x^3 - 4x + 37$. The second polynomial stays the same: $2x + 3$. The problem is: $2x + 3\sqrt{8x^3 - 4x + 37}$. In the dividend, $8x^3 - 4x + 37$, all powers of x must be included. The only missing power of x is x^2. To include this power of x, a coefficient of 0 is used; that is, $0x^2$. This term, $0x^2$, can be added to the dividend without changing the dividend because $0x^2 = 0$ (anything multiplied by 0 is 0).

Now to accomplish the division we proceed as follows: divide the first term of the divisor into the first term of the dividend. Multiply the quotient from this division by each term of the divisor and subtract the products of each term from the dividend. We then obtain a new dividend. Use this dividend, and again divide by the first term of the divisor, and repeat all steps again until we obtain a re-

36

mainder which is of degree lower than that of the divisor or zero. Following this procedure we obtain :

$$
\begin{array}{r}
4x^2 - 6x + 7 \\
2x + 3\overline{\smash{\big)}\ 8x^3 + 0x^2 - 4x + 37} \\
\underline{8x^3 + 12x^2} \\
-12x^2 - 4x + 37 \\
\underline{-12x^2 - 18x} \\
14x + 37 \\
\underline{14x + 21} \\
16
\end{array}
$$

The degree of a polynomial is the highest power of the variable in the polynomial.

The degree of the divisor is 1. The number 16 can be written as $16x^0$ where $x^0 = 1$. Therefore, the number 16 has degree 0. When the degree of the divisor is greater than the degree of the dividend, we stop dividing.

Since the degree of the divisor in this problem is 1 and the degree of the dividend (16) is 0, the degree of the divisor is greater than the degree of the dividend. Therefore, dividing is stopped and the remainder is 16. Therefore, the quotient is $4x^2 - 6x + 7$ and the remainder is 16.

In order to verify this, multiply the quotient, $4x^2 - 6x + 7$, by the divisor, $2x + 3$, and then add 16. These two operations should total up to the dividend $8x^3 - 4x + 37$. Thus,

$$\left(4x^2 - 6x + 7\right)(2x + 3) + 16 =$$

$$8x^3 - 12x^2 + 14x + 12x^2 - 18x + 21 + 16 =$$

$$8x^3 - 4x + 37,$$

which is the desired result.

● **PROBLEM** 4-25

Divide $3x^5 - 8x^4 - 5x^3 + 26x^2 - 33x + 26$ by $x^3 - 2x^2 - 4x + 8$.

Solution: To divide a polynomial by another polynomial we set up the divisor and the dividend as shown below. Then we divide the first term of the divisor into the first term of the dividend. We multiply the quotient from this division by each term of the divisor, and subtract the products of each term from the dividend. We then obtain a new dividend. Use this dividend, and again divide by the first term of the divisor, and repeat all steps again until we obtain a remainder which is of degree lower than that of the divisor or = zero. Following this procedure we obtain:

$$
\begin{array}{r}
3x^2 - 2x + 3 \\
x^3-2x^2-4x+8 \ \overline{\smash{\big)}\ 3x^5 - 8x^4 - 5x^3 + 26x^2 - 33x + 26}
\end{array}
$$

$$
\begin{array}{r}
\underline{3x^5 - 6x^4 - 12x^3 + 24x^2} \\
- 2x^4 + 7x^3 + 2x^2 - 33x + 26 \\
\underline{- 2x^4 + 4x^3 + 8x^2 - 16x} \\
3x^3 - 6x^2 - 17x + 26 \\
\underline{3x^3 - 6x^2 - 12x + 24} \\
- 5x + 2
\end{array}
$$

Thus, the quotient is $3x^2 - 2x + 3$ and the remainder is $- 5x + 2$.

● **PROBLEM** 4-26

Find the quotient and remainder when $3x^7 - x^6 + 31x^4 + 21x + 5$ is divided by $x + 2$.

<u>Solution:</u> To divide a polynomial by another polynomial we set up the divisor and the dividend as shown below. Then we divide the first term of the divisor into the first term of the dividend. We multiply the quotient from this division by each term of the divisor, and subtract the products of each term from the dividend. We then obtain a new dividend. Use this dividend, and again divide by the first term of the divisor, and repeat all steps again until we obtain a remainder which is of degree lower than that of the divisor, or which is zero. Following this procedure we obtain:

$$
\begin{array}{r}
3x^6-7x^5+14x^4+3x^3-6x^2+12x-3 \\
x + 2 \ \overline{\smash{\big)}\ 3x^7- x^6 \quad\quad +31x^4 \quad\quad\quad\quad +21x + 5} \\
\underline{3x^7+6x^6} \\
-7x^6 \quad\quad +31x^4 \quad\quad\quad\quad +21x + 5 \\
\underline{-7x^6-14x^5} \\
14x^5+31x^4 \quad\quad\quad\quad +21x + 5 \\
\underline{14x^5+28x^4} \\
3x^4 \quad\quad\quad\quad +21x + 5 \\
\underline{3x^4+6x^3} \\
-6x^3 \quad +21x + 5 \\
\underline{-6x^3-12x^2} \\
12x^2+21x + 5 \\
\underline{12x^2+24x} \\
- 3x + 5 \\
\underline{- 3x - 6} \\
11
\end{array}
$$

Thus the quotient is $3x^6 - 7x^5 + 14x^4 + 3x^3 - 6x^2 +12x - 3$, and the remainder is 11.

Give 2 examples of each of the following expressions.

(1) Algebraic

(2) A monomial

(3) A trinomial

(4) A multinomial

(5) A non-integral term

(6) A non-rational term

Solution: (1) $4x^2 - 3xy + 9y^4$, $2a^5b^7$

(2) $7x^3y^5$, $6xyz^3$

(3) $3x^9y^3 + 6x^2y + 15x$, $3x^5 + 6xy^3 - 3z$

(4) $7x + 6z$, $\dfrac{4x^3}{y} + \dfrac{3y}{z} + \dfrac{7z}{x}$

(5) $\dfrac{4}{x}$, $\dfrac{-9}{z^2}$

(6) $\sqrt{3}\,y$, $6\sqrt[3]{z}$

(1) What is the difference between a polynomial and a multinomial?

(2) What is the degree of a monomial?

Solution: (1) A polynomial consists only of multinomials or monomials in which every term is integral and rational (i.e., $4\sqrt{y} + 3$ is <u>not</u> a polynomial).

(2) The degree of a monomial is the sum of all the exponents in the expression. (for example, $3x^2y^4$: degree = 6).

CHAPTER 5

FACTORING EXPRESSIONS

> **Basic Attacks and Strategies for Solving Problems in this Chapter. See pages 40 to 55 for step-by-step solutions to problems.**

The *Least Common Multiple*, or *LCM*, of two or more numbers is a number that is an integral multiple of each, and also is the *least* such multiple. Thus, 6, 12, 18 and 24 are all multiples of 2 and 3, but 6 is the LCM since there is no multiple of both 2 and 3 that is less than 6. One way to find the LCM is to rewrite the initial numbers as powers of prime factors. Then the LCM is the product of the highest powers of all primes present in the factorizations. The LCM of

$$20 = 2^2 \cdot 5 \quad \text{and} \quad 50 = 2 \cdot 5^2 \text{ is } \quad 2^2 \cdot 5^2 = 100.$$

The *Greatest Common Divisor*, or *GCD*, of two or more numbers is a number that is a *divisor* of each and also is the *greatest* of all such divisors. Thus, 1, 2, 3 and 6 divide 6 and 1, 3, 5 and 15 divide 15. The common divisors of 6 and 15 are 1 and 3 and the GCD is 3. If the GCD of two integers is one, then the two numbers are said to be *relatively prime*. For example, GCD (2,3) = 1, which indicates that 2 and 3 are relatively prime.

Factoring polynomials (already mentioned in the introduction to Chapter 3) is an application of the distributive law. If each term in a polynomial has a common factor, the largest such common factor should be determined first and then factored out. For example, in

$$4x^2y - 4xy^2,$$

the largest common factor is $4xy$. When this is factored out, the polynomial can be rewritten as

$$4xy \cdot (x - y).$$

Many polynomials do not have a common factor for all terms. In these cases, the terms with a common factor can be factored first, in the hope that a subsequent factoring might be possible. For example,

$$xy + 3x + 2y + 6 = x(y + 3) + 2(y + 3) = (x + 2)(y + 3).$$

Sometimes, the terms need to be rearranged first to group those with a common factor together.

Factoring quadratics, that is, polynomials of degree 2, sometimes requires special techniques. In general,

$$(x + a)(x + b) = x^2 + (a + b)x + ab.$$

Thus, when given a quadratic,

$$x^2 + px + q,$$

we must find a and b such that

$$p = a + b \quad \text{and} \quad q = ab.$$

One usually begins by determining all the factors of the constant term $q = ab$ and seeing which factors sum to p. For example, given

$$x^2 - 5x + 6,$$

we know that

$$6 = -3 \cdot -2 (= a \cdot b) \quad \text{or that} \quad 6 = -6 \cdot -1 (= a \cdot b).$$

Using the first set of values for a and b,

$$p(= a + b) = -5,$$

which is the coefficient of x in the quadratic we were given. Thus, we can rewrite the quadratic as

$$x^2 - 3x - 2x + 6$$

and factor it into $(x - 3)(x - 2)$. On the other hand, given

$$x^2 - 7x + 6,$$

we would choose the second set of values for a and b to obtain the factorization $(x - 6)(x - 1)$.

There are several special cases of polynomials which reoccur with enough frequency to be particularly noted.

A *difference of squares*, $x^2 - y^2$, can be factored as $(x + y)(x - y)$. Sometimes, one must first rewrite terms to make it obvious that they are, in fact, squares. For example,

$$4x^4 - 9y^2$$

can be rewritten as

$$(2x^2)^2 - (3y)^2$$

which then factors into

$(2x^2 - 3y)(2x^2 + 3y)$.

A *sum of squares*, $x^2 + y^2$, cannot be factored with coefficients among the real numbers. If complex arithmetic is allowed, then it factors into

$(x + iy)(x - iy)$.

A *difference of cubes*, $x^3 - y^3$, factors into

$(x - y)(x^2 + xy + y^2)$.

Similarly, a *sum of cubes*, $x^3 + y^3$, factors into

$(x + y)(x^2 - xy + y^2)$.

The quadratic factors cannot be factored into simpler, linear terms (assuming real numbers as coefficients).

Every difference of equal powers, $x^n - y^n$, can be factored. One factor is always $x - y$ and the other factor can be obtained by long division.

Often, sums or differences of powers greater than 3 can be rewritten as a sum or difference of lower powers. Then, an appropriate rule can be used. For example,

$8x^6 - y^9$

can be rewritten as

$(2x^2)^3 - (y^3)^3$

and then factored as a difference of cubes. It should also be emphasized that the number 1 equals 1^n for every n, so that $x^2 - 1$ is the difference of squares (and thus can be factored into $(x + 1)(x - 1)$).

It is possible to compute the *Least Common Multiple*, or *LCM*, of two polynomials using the same procedure as for two numbers. First, each polynomial should be factored as completely as possible. Any constant multiple should be factored into prime factors. Then (as with two numbers) the LCM is the product of all the factors raised to the highest power that appears.

Problems involving fractions in which the denominators and numerators are polynomials are solved in ways similar to problems involving numeric fractions. One should remember, however, that simplification usually takes place through factoring and the subsequent cancellation of identical terms, and that cancellation is only possible when terms are multiplied together, since cancellation is dividing a quantity by itself to obtain 1. Thus,

$$\frac{(x+1)(x+2)}{(x-2)(x+1)} = \frac{(x+2)}{(x-2)}$$

since the presence of $(x + 1)$ in both the numerator and denominator "cancel" each other out (i.e., divide to produce a factor of 1). Cancellation is NOT permitted in quantities merely added together. For example, $\frac{x^2+y}{x^2-2}$ cannot be simplified even though there is an x^2 in both the numerator and denominator. Both expressions cannot be reduced any further.

Technically, $(x + 1) / (x + 1)$ equals 1 only if x does not equal $- 1$. In some applications, when simplifying polynomial fractions, one must make note of these exception cases, e.g.,

$(x^2 - 1) / (x + 1) = x - 1$ if $x \ne - 1$.

When adding or subtracting two fractions of polynomials, as with numeric fractions, one first finds a least common denominator.

When reducing a fraction of polynomials, one first factors the numerator and denominator and then cancels out equal terms.

Step-by-Step Solutions to Problems in this Chapter, "Factoring Expressions"

● **PROBLEM** 5-1

Find the Least Common Multiple, LCM, of 12, 18, 21, 25 and 35.

Solution: We want to express each number as a product of prime factors:

$$12 = 2^2(3), \quad 18 = 2(3^2), \quad 21 = 3(7),$$

$$25 = 5^2, \quad 35 = 5(7)$$

Find the LCM by retaining the highest power of each distinct factor and multiplying them together, making sure to use each factor only once regardless of the number of times it appears. Thus,

$$LCM = (2^2)(3^2)(5^2)(7) = (4)(9)(25)(7)$$

$$= 6300$$

● **PROBLEM** 5-2

Find the greatest common divisor $\{15,28\}$.

Solution: If 15 and 28 are factored completely into their respective prime factors, $15 = 3 \cdot 5$ and $28 = 2 \cdot 2 \cdot 7$

Since 1 divides every integer, and since 15 and 28 possess no common prime factors, it follows that

$$gcd\{15,28\} = 1.$$

If the gcd of two integers is 1, then the two integers are said to be relatively prime. Since $gcd\{15,28\} = 1$, the integers 15 and 28 are relatively prime.

Find the greatest common divisor of 24 and 40. Also, find the least common multiple of 24 and 40.

Solution: To find the greatest common divisor of 24 and 40, or gcd{24,40}, we write down the set of all positive integers which divide both 24 and 40. Thus we obtain the two sets

$$\{1,2,3,4,6,8,12,24\} \text{ for } 24$$
$$\{1,2,4,5,8,10,20,40\} \text{ for } 40$$

Those integers dividing both 24 and 40 are in the intersection of these two sets. Thus,

$$\{1,2,3,4,6,8,12,24\} \cap \{1,2,4,5,8,10,20,40\} = \{1,2,4,8\}$$

The largest element in this last set is 8. Thus,

$$8 = \text{gcd}\{24,40\} \ .$$

Another method for finding the gcd{24,40} is called the factoring technique. Factor the two given numbers into their prime factors.

$$24 = 2 \cdot 2 \cdot 2 \cdot 3$$
$$40 = 2 \cdot 2 \cdot 2 \cdot 5$$

The greatest common divisor of any two numbers is the largest number which divides both of those numbers. Therefore,

$$\frac{24}{40} = \frac{2 \cdot 2 \cdot 2 \cdot 3}{2 \cdot 2 \cdot 2 \cdot 5} = \frac{(8)(3)}{(8)(5)} \ . \ \text{Hence,}$$

$$\text{gcd} \ \{24,40\} = 8 \ .$$

The following technique is the definition for finding the least common multiple of 24 and 40, or lcm {24,40}. To find the lcm{24,40}, we write down the set of all positive integer multiples of both 24 and 40. Then we obtain

$$\{24,48,72,96,120,144,168,192,216,240,264,\ldots\} \text{ for } 24$$

$$\{40,80,120,160,200,240,280,\ldots\} \qquad \text{for } 40$$

The integers which are multiples of both 24 and 40; that is, common multiples of 24 and 40, are in the intersection of these two sets. This is the set {120,240,...}. The smallest element of this set is 120. Hence, lcm{24,40} = 120. Another method for finding the lcm{24,40} is called the factoring technique. Factor the two given numbers into their prime factors.

$$24 = 2 \cdot 2 \cdot 2 \cdot 3$$
$$40 = 2 \cdot 2 \cdot 2 \cdot 5$$

Now, take the different factors of the two numbers and multiply them together. The exponent to be used for each factor is the highest number of times that the factor appears in either number (24 or 40). The produce obtained will be the lcm{24,40}. Hence:

$$\text{lcm}\{24,40\} = 2^3 \cdot 3^1 \cdot 5^1 = (8)(3)(5) = (24)(5) = 120.$$

Factor   A) $4a^2b - 2ab$

              B) $9ab^2c^3 - 6a^2c + 12ac$

              C) $ac + bc + ad + bd$

<u>Solution:</u> Find the highest common factor of each polynomial.

A) $4a^2b = 2 \cdot 2 \cdot a \cdot a \cdot b$

     $2ab = 2 \cdot a \cdot b$

The highest common factor of the two terms is therefore $2ab$. Hence,
$$4a^2b - 2ab = 2ab(2a - 1)$$

B) $9ab^2c^3 = 3 \cdot 3 \cdot a \cdot b \cdot b \cdot c \cdot c \cdot c$

     $6a^2c = 3 \cdot 2 \cdot a \cdot a \cdot c$

    $12ac = 3 \cdot 2 \cdot 2 \cdot a \cdot c$

The highest common factor of the three terms is $3ac$. Then,
$$9ab^2c^3 - 6a^2c + 12ac = 3ac\left(3b^2c^2 - 2a + 4\right)$$

C) An expression may sometimes be factored by grouping terms having a common factor and thus getting new terms containing a common factor. The type form for this case is ac+bc+ ad+bd, because the terms ac and bc have the common factor c, and ad and bd have the common factor d. Then,
$$ac + bc + ad + bd = c(a + b) + d(a + b)$$

Factoring out $(a + b)$, we obtain:
$$= (a + b)(c + d).$$

Factor the following polynomials:

(a) $15 ac + 6bc - 10ad - 4bd$

(b) $3a^2c + 3a^2d^2 + 2b^2c + 2b^2d^2$.

Solution: (a) Group terms which have a common factor. Here, they are already grouped. Then factor.

    $15 ac + 6 bc - 10ad - 4bd = 3c(5a + 2b) - 2d(5a + 2b)$

Factoring out $(5a + 2b)$

    $3c(5a + 2b) - 2d(5a + 2b) = (5a + 2b)(3c - 2d),$

     (b) Apply the same method as in (a),

$$3a^2c + 3a^2d^2 + 2b^2c + 2b^2d^2 = 3a^2(c + d^2) + 2b^2(c + d^2)$$
$$= (c + d^2)(3a^2 + 2b^2).$$

● **PROBLEM 5-6**

Factor $xy - 3y + y^2 - 3x$ completely.

<u>Solution:</u> Note that the first and last terms have a common factor of x. Also note that the second and third factors have a common factor of y. Hence, group the x and y terms together and factor out the x and y from their respective two terms. Therefore,

$$xy - 3y + y^2 - 3x = (xy - 3x) + (-3y + y^2)$$

Since $(-3y + y^2) = (y^2 - 3y),$

$$xy - 3y + y^2 - 3x = (xy - 3x) + (y^2 - 3y)$$
$$= x(y - 3) + y(y - 3)$$

Now factor out the common factor $(y - 3)$ from both terms:

$$xy - 3y + y^2 - 3x = (x + y)(y - 3).$$

● **PROBLEM 5-7**

Factor $x^2 - x - 12$ over the integers.

<u>Solution:</u> A quadratic equation whose roots are a and b may be written in the form: $(x - a)(x - b) = x^2 - (a + b)x + ab = 0$. Considering $x^2 - x - 12$, the coefficient of x is -1 and the constant is -12; thus, we want to find the values for a and b such that

$$a + b = 1 \quad \text{and} \quad a \cdot b = -12$$

One of the numbers must be negative and the other one positive, since only a negative multiplied by a positive will give us a negative quantity (-12) for a · b.

After examining the possible factors of -12, we find that 4 and -3 are the desired ones since 4 + (-3) = 1. Thus, let a = 4, b = -3, and

$$x^2 - x - 12 = x^2 - (4 - 3)x + (4)(-3)$$
$$= (x - 4)(x + 3).$$

● **PROBLEM 5-8**

Factor the expression $16a^2 - 4(b - c)^2$.

<u>Solution:</u> $16 = 4^2$, thus $16a^2 = 4^2a^2$. Since $a^xb^x = (ab)^x$, $16a^2 = 4^2a^2 = (4a)^2$. Similarly $4 = 2^2$, thus $4(b-c)^2 = 2^2(b-c)^2 = [2(b-c)]^2$. Hence $16a^2 - 4(b-c)^2 = (4a)^2 - [2(b-c)]^2$. We are now dealing with the difference of two squares. Applying the formula for the difference of two squares, $x^2 - y^2 = (x+y)(x-y)$, and re-placing x by 4a and y by 2(b-c) we obtain:

$$(4a)^2 - [2(b-c)]^2 = [4a + 2(b-c)][4a - 2(b-c)]$$

$$= (4a + 2b - 2c)(4a - 2b + 2c).$$

Therefore, $16a^2 - 4(b-c)^2 = (4a + 2b - 2c)(4a - 2b + 2c)$.

● **PROBLEM 5-9**

Simplify $(x + 2y)(x - 2y)(x^2 + 4y^2)$.

<u>Solution:</u> Here we use the factoring formula $a^2 - b^2 = (a - b)(a + b)$ to rewrite the product $(x + 2y)(x - 2y)$:

$$(x + 2y)(x - 2y) = (x)^2 - (2y)^2 \quad \text{difference of two squares.}$$

$$= x^2 - 4y^2.$$

Hence, $(x + 2y)(x - 2y)(x^2 + 4y^2) = (x^2 - 4y^2)(x^2 + 4y^2)$ (1)

Now, again use the factoring formula given above to rewrite the right side of equation (1) in which $x^2 = a$ and $4y^2 = b$. Hence,

$$(x^2 - 4y^2)(x^2 + 4y^2) = (x^2)^2 - (4y^2)^2$$

$$= x^4 - 4^2y^4 \quad \text{since } (a^x)^y = a^{x \cdot y}$$

$$= x^4 - 16y^4$$

Hence, equation (1) becomes:

$$(x + 2y)(x - 2y)(x^2 + 4y^2) = x^4 - 16y^4.$$

● **PROBLEM 5-10**

Find the factors of $125m^3n^6 - 8a^3$.

<u>Solution:</u> Note that $125 = 5 \cdot 5 \cdot 5 = 5^3$. Also since $a^{xy} = (a^x)^y$, $n^6 = n^{2 \cdot 3} = (n^2)^3$. Thus $125m^3n^6 = 5^3m^3(n^2)^3$. Since $a^xb^xc^x = (abc)^x$, $5^3m^3(n^2)^3 = (5mn^2)^3$. $8 = 2 \cdot 2 \cdot 2 = 2^3$, thus $8a^3 = 2^3a^3 = (2a)^3$. Now $125m^3n^6 - 8a^3 = (5mn^2)^3 - (2a)^3$, which is the difference of two cubes. Apply the

formula for the difference of two cubes $x^3 - y^3 = (x - y)(x^2 + xy + y^2)$, replacing x by $5mn^2$ and y by 2a. Hence,

$$125m^3n^6 - 8a^3 = (5mn^2 - 2a)\left[(5mn^2)^2 + 5mn^2(2a) + (2a)^2\right].$$
$$= (5mn^2 - 2a)(25m^2n^4 + 10amn^2 + 4a^2)$$

● **PROBLEM 5-11**

Factor $(x + y)^3 + z^3$.

Solution: The given expression is the sum of two cubes. The formula for the sum of two cubes can be used to factor the given expression. This formula is:

$$a^3 + b^3 = (a + b)(a^2 - ab + b^2).$$

Using this formula and replacing a by x + y and b by z:

$$(x + y)^3 + z^3 = \left[(x+y) + z\right]\left[(x+y)^2 - (x+y)z + z^2\right]$$

● **PROBLEM 5-12**

Factor: (a) $2x^2 + 2y^2$ (b) $a^3x + b^3x$.

Solution: (a) First we factor out a 2 from this expression. Thus $2x^2 + 2y^2 = 2(x^2 + y^2)$. $x^2 + y^2$, the sum of two like even powers, cannot be factored. Thus, it is a prime expression.   Hence $2x^2 + 2y^2 = 2(x^2 + y^2)$.
        (b) First we factor out an x from this expression. Thus, $a^3x + b^3x = (a^3 + b^3)x$. $(a^3 + b^3)$ is the sum of two cubes. Applying the formula for the sum of two cubes:
$c^3 + d^3 = (c + d)(c^2 - cd + d^2)$, replacing c by a and d by b we obtain
$$a^3 + b^3 = (a + b)(a^2 - ab + b^2).$$

Hence, $a^3x + b^3x = (a + b)(a^2 - ab + b^2)x$.

● **PROBLEM 5-13**

Factor $a^4 - b^4$.

Solution: Note that $a^4 = (a^2)^2$ and $b^4 = (b^2)^2$; thus

$a^4 - b^4 = (a^2)^2 - (b^2)^2$, the difference of two

squares. Thus we apply the formula for the difference of two squares, $x^2 - y^2 = (x + y)(x - y)$, replacing x by a^2 and y by b^2 to obtain:

$$a^4 - b^4 = (a^2)^2 - (b^2)^2 = (a^2 + b^2)(a^2 - b^2).$$

Since $a^2 - b^2$ is also the difference of two squares, we once again apply the above formula to obtain:

$$a^2 - b^2 = (a + b)(a - b).$$

Therefore, $a^4 - b^4 = (a^2 + b^2)(a + b)(a - b)$.

● **PROBLEM 5-14**

Factor   A) $a^4 + 4a^2 + 4$

B) $9a^2 - 6ab^2 + b^4$.

Solution: The first example is a trinomial which is a perfect square, in the form:

$$x^2 + 2xy + y^2 = x^2 + xy + xy + y^2 = (x+y)(x+y) = (x+y)^2 .$$

For example A), replace x by a^2 and y by 2 to obtain

$$a^2 + 4a^2 + 4 = (a^2)^2 + 2 \cdot a^2 \cdot 2 + 2^2 = (a^2 + 2)^2 ,$$

The second example is a trinomial perfect square whose form is:

$$x^2 - 2xy + y^2 = x^2 - xy - xy + y^2 = (x-y)(x-y) = (x-y)^2 .$$

For example B) replace x by $3a$ and y by b^2 to obtain:

$$9a^2 - 6ab^2 + b^4 = (3a)^2 - 2(3a)(b^2) + (b^2)^2$$
$$= (3a - b^2)(3a - b^2)$$
$$= (3a - b^2)^2 .$$

● **PROBLEM 5-15**

Factor $128x^6 - 2y^6$.

Solution: We first observe that 2 may be factored from this expression. Thus

$$128x^6 - 2y^6 = 2(64x^6 - y^6).$$

Now, since $a^{b \cdot c} = (a^b)^c$, $x^6 = x^{3 \cdot 2} = (x^3)^2$ and

$$y^6 = y^{3 \cdot 2} = \left(y^3\right)^2.$$

Therefore $2\left(64x^6 - y^6\right) = 2\left[64\left(x^3\right)^2 - \left(y^3\right)^2\right]$

$$= 2\left[8^2\left(x^3\right)^2 - \left(y^3\right)^2\right]$$

$$= \left[\left(8x^3\right)^2 - \left(y^3\right)^2\right].$$

Thus, we have the difference of two squares. Applying the formula for the difference of two squares, $a^2 - b^2 = (a+b)(a-b)$, and, replacing a by $8x^3$ and b by y^3, we obtain

$$2\left[\left(8x^3\right)^2 - \left(y^3\right)^2\right] = 2\left[\left(8x^3 + y^3\right)\left(8x^3 - y^3\right)\right]$$

$$= 2\left[\left(2^3 x^3 + y^3\right)\left(2^3 x^3 - y^3\right)\right]$$

$$= 2\left[\left(2x\right)^3 + y^3\right]\left[\left(2x\right)^3 - y^3\right].$$

Now since the expressions in brackets are the sum and difference of two cubes, respectively, we apply the formulas for the sum and difference of two cubes:

$$a^3 + b^3 = (a+b)\left(a^2 - ab + b^2\right)$$

$$a^3 - b^3 = (a-b)\left(a^2 + ab + b^2\right).$$

Replacing a by 2x and b by y we have

$$2\left[\left(2x\right)^3 + y^3\right]\left[\left(2x\right)^3 - y^3\right] = 2(2x+y)\left(4x^2-2xy+y^2\right)(2x-y)\left(4x^2+2xy+y^2\right)$$

Therefore, $128x^6 - 2y^6 = 2(2x+y)\left(4x^2-2xy+y^2\right)(2x-y)\left(4x^2+2xy+y^2\right).$

● **PROBLEM 5-16**

Factor: (a) $4a^4 - b^6$

(b) $a^2 - b^2 + 2bc - c^2$

Solution: (a) First note that $4a^4 - b^6$ may be expressed as the difference of two squares: $4a^4 - b^6 = \left(2a^2\right)^2 - \left(b^3\right)^2$. Recall the formula for the difference of two squares: $x^2 - y^2 = (x-y)(x+y)$. Replacing x by $2a^2$ and y by b^3 in this formula we obtain:

$$4a^4 - b^6 = \left(2a^2\right)^2 - \left(b^3\right)^2 = \left(2a^2 - b^3\right)\left(2a^2 + b^3\right)$$

(b) Observe that $b^2 - 2bc + c^2$ is a perfect square, since the first and last terms are perfect squares and positive, and the middle term is twice the product of the square roots of the end

47

terms. Then, $b^2 - 2bc + c^2 = (b - c)^2$.

Thus, express the given algebraic expression as the difference of two squares.

$$a^2 - b^2 + 2bc - c^2 = a^2 - (b^2 - 2bc + c^2) = a^2 - (b - c)^2$$

Then apply the formula for the difference of two squares

$$(x^2 - y^2) = (x-y)(x+y),$$

replacing x by a and y by $(b-c)$:

$$a^2 - (b-c)^2 = [a - (b-c)][a + (b-c)]$$

$$= (a - b + c)(a + b - c)$$

Thus,

$$a^2 - b^2 + 2bc - c^2 = (a - b + c)(a + b - c)$$

● **PROBLEM** 5-17

Find the LCM of: $6x^2 + 24x +24$, $4x^2 - 8x - 12$, and $3x^2 + 9x + 6$.

Solution: Factor each expression completely. Constant factors should be written as a product of prime numbers.

$$6x^2 + 24x + 24 = 6(x^2 + 4x + 4) = 6(x + 2)^2 = (3)(2)(x + 2)^2$$

$$4x^2 - 8x - 12 = 4(x^2 - 2x - 3) = 4(x + 1)(x - 3)$$

$$= (2)^2(x + 1)(x - 3)$$

$$3x^2 + 9x + 6 = 3(x^2 + 3x + 2) = 3(x + 1)(x + 2).$$

Each of the factors of these expressions appears in the product known as the LCM. Each factor is raised to the highest power to which it appears in any one of the expressions. Therefore,

$$LCM = (2)^2(3)(x + 2)^2(x + 1)(x - 3).$$

● **PROBLEM** 5-18

Find the LCM of: $(x - 1)^2$, $(1 - x)^3$, $1 - x^3$.

Solution: Factor each polynomial completely. Notice in the factoring of the second and the third polynomials that -1 may be factored from the expressions first so that the terms of highest degree in the factors will have positive coefficients.

$$(x - 1)^2 = (x - 1)^2$$

48

$$(1 - x)^3 = [-1(x - 1)]^3 = (-1)^3(x - 1)^3 = -(x - 1)^3$$
$$1 - x^3 = (-1)(x^3 - 1) = -(x - 1)(x^2 + x + 1).$$

$\left(x^3 - 1\right.$ is the difference of two cubes.$\left.\right)$

Each of the factors of these expressions appears in the product known as the LCM. Each factor is raised to the highest power to which it appears in any one of the expressions. Therefore the

$$\text{LCM} = (x - 1)^3\left(x^2 + x + 1\right).$$

● **PROBLEM** 5-19

Combine $\dfrac{3x + y}{x^2 - y^2} - \dfrac{2y}{x(x - y)} - \dfrac{1}{x + y}$ into a single fraction.

Solution: Fractions which have unlike denominators must be transformed into fractions with the same denominator before they may be combined. This identical denominator is the least common denominator (L.C.D.), the least common multiple of the denominators of the fractions to be added. In the process of transforming the fractions to a common denominator we make use of the fact that the numerator and denominator of a fraction may be multiplied by the same non-zero number without changing the value of the fraction. In our case the denominators are $x^2 - y^2 = (x + y)(x - y)$, $x(x - y)$, and $x + y$. Therefore the LCD is $x(x + y)(x - y)$, and we proceed as follows:

$$\frac{3x + y}{x^2 - y^2} - \frac{2y}{x(x-y)} - \frac{1}{x+y} = \frac{3x + y}{(x+y)(x-y)} - \frac{2y}{x(x-y)} - \frac{1}{x+y}$$

$$= \frac{x(3x + y)}{x(x+y)(x-y)} - \frac{(x + y)2y}{(x+y)(x)(x-y)}$$

$$- \frac{x(x - y)}{x(x-y)(x+y)}$$

$$= \frac{3x^2 + xy}{x(x+y)(x-y)} - \frac{2xy + 2y^2}{x(x+y)(x-y)}$$

$$- \frac{x^2 - xy}{x(x+y)(x-y)}$$

$$= \frac{3x^2 + xy - \left(2xy + 2y^2\right) - \left(x^2 - xy\right)}{x(x+y)(x-y)}$$

$$= \frac{3x^2 + xy - 2xy - 2y^2 - x^2 + xy}{x(x+y)(x-y)}$$

$$= \frac{3x^2 - x^2 + xy + xy - 2xy - 2y^2}{x(x+y)(x-y)}$$

$$= \frac{2x^2 - 2y^2}{x(x+y)(x-y)}$$

$$= \frac{2\left(x^2 - y^2\right)}{x(x+y)(x-y)}$$

$$= \frac{2(x+y)(x-y)}{x(x+y)(x-y)}$$

$$= \frac{2}{x}$$

● **PROBLEM** 5-20

Simplify:
$$\frac{\dfrac{1}{a-b} + \dfrac{1}{a+b}}{1 + \dfrac{b^2}{a^2 - b^2}}$$

Solution: This is a complex fraction, a fraction whose numerator and denominator both contain fractions. To simplify it, multiply the numerator and denominator by the least common denominator, LCD. To find the LCD of several fractions, first factor each denominator into its prime factors.

$$a - b = (a - b)$$
$$a + b = (a + b)$$
$$a^2 - b^2 = (a - b)(a + b)$$

The LCD of the fractions is the product of the highest power of the different prime factors, with each prime factor being used only once. Hence (a - b)(a + b) is our LCD. Multiplying, we obtain:

$$\frac{(a-b)(a+b)\left[\dfrac{1}{a-b} + \dfrac{1}{a+b}\right]}{(a-b)(a+b)\left[1 + \dfrac{b^2}{a^2 - b^2}\right]}$$

Distributing in the numerator and denominator, and recalling that $a^2 - b^2 = (a - b)(a + b)$ we have:

$$\frac{\dfrac{(a-b)(a+b)}{(a-b)} + \dfrac{(a-b)(a+b)}{(a+b)}}{(a-b)(a+b) + \dfrac{b^2(a-b)(a+b)}{(a-b)(a+b)}} = \frac{(a+b) + (a-b)}{(a-b)(a+b) + b^2}$$

$$= \frac{a + b + a - b}{a^2 - b^2 + b^2} = \frac{2a}{a^2} = \frac{2}{a} \quad .$$

Reduce $\dfrac{4x - 20}{50 - 2x^2}$ to lowest terms.

Solution: Factor the numerator and the denominator:

$$\frac{4x - 20}{50 - 2x^2} = \frac{4(x - 5)}{2(25 - x^2)} = \frac{4(x - 5)}{2(5 - x)(5 + x)}$$

Multiply the numerator and denominator by (- 1) to reverse the sign of the factor (5 - x) in the denominator. Then divide both the numerator and denominator by 2(x - 5).

$$\frac{(-1)[4(x - 5)]}{(-1)[2(5 - x)(5 + x)]} = \frac{-4(x - 5)}{2(x - 5)(5 + x)}$$

Dividing, we obtain:

$$-\frac{2}{x + 5} .$$

Combine into a single fraction

$$\frac{2x}{x^2 - 6x + 9} - \frac{8}{x^2 - 2x - 3} - \frac{1}{x + 1} .$$

Solution: Fractions which have unlike denominators cannot be combined directly. First they must be transformed into fractions with the same denominator. This identical denominator is called the Least Common Denominator (L.C.D.).

Before we can obtain the L.C.D., we factor each individual denominator.

$$x^2 - 6x + 9 = (x-3)(x-3) = (x-3)^2$$
$$x^2 - 2x - 3 = (x-3)(x+1)$$
$$x + 1 = (x + 1)$$

To find the L.C.D., we consider all the different factors. Take the highest value of the exponent of each factor. Thus, factoring the denominators, we obtain

$$\frac{2x}{(x-3)^2} - \frac{8}{(x-3)(x+1)} - \frac{1}{x+1}$$

and the L.C.D. is $(x-3)^2(x+1)$. We shall now rewrite the three given fractions as equivalent fractions, each having the denominator $(x-3)(x+1)$. To this end, multiply numerator and denominator of the first fraction by $x + 1$, of the second fraction by $x - 3$, and of the third fraction by $(x-3)^2$. This gives

$$\frac{2x(x+1)}{(x-3)^2(x+1)} - \frac{8(x-3)}{(x-3)^2(x+1)} - \frac{(x-3)^2}{(x+1)(x-3)^2}$$

$$= \frac{2x^2+2x-(8x-24)-(x^2-6x+9)}{(x-3)^2(x+1)}$$

$$= \frac{2x^2+2x-8x+24-x^2+6x-9}{(x-3)^2(x+1)}$$

$$= \frac{2x^2-x^2+2x-8x+6x+24-9}{(x-3)^2(x+1)}$$

$$= \frac{x^2+15}{(x-3)^2(x+1)}$$

Check. The given expression should be equal to the resulting fraction for all permissible values of x, that is, for all values of x except 3 and -1 (x = 3 and x = -1 give us zero in the denominator of the fraction, which is undefined). Replacing x arbitrarily by 2,

$$\left. \frac{2x}{x^2-6x+9} - \frac{8}{x^2-2x-3} - \frac{1}{x+1} \right]_{x=2}$$

$$= \frac{2(2)}{(2)^2-6(2)+9} - \frac{8}{(2)^2-2(2)-3} - \frac{1}{2+1}$$

$$= \frac{4}{4-12+9} - \frac{8}{4-4-3} - \frac{1}{3}$$

$$= \frac{4}{1} - \frac{8}{-3} - \frac{1}{3}$$

$$= \frac{12}{3} + \frac{8}{3} - \frac{1}{3} = \frac{19}{3}$$

and

$$\left. \frac{x^2+15}{(x-3)^2(x+1)} \right]_{x=2} = \frac{(2)^2+15}{(2-3)^2(2+1)} = \frac{4+15}{(-1)^2(3)} = \frac{19}{1\cdot3} = \frac{19}{3}$$

Hence, we have shown that the given expression holds true for x = 2 in the uncombined and combined forms.

● **PROBLEM 5-23**

Reduce $\dfrac{x^2 - 5x + 4}{x^2 - 7x + 12}$ to lowest terms.

Solution: Factor the expressions in both the numerator and denominator and cancel like terms.

$$\frac{x^2-5x+4}{x^2-7x+12} = \frac{(x-1)(x-4)}{(x-3)(x-4)}$$

$$= \frac{x-1}{x-3}$$

The numerator and the denominator were both divided by x - 4.

Divide $\dfrac{y^2 + y - 20}{y - 3}$ by $\dfrac{y^2 - 16}{y^2 + y - 12}$.

Solution: Dividing by a nonzero polynomial is the same as multiplying by its reciprocal. That is,

$$\frac{y^2 + y - 20}{y - 3} \div \frac{y^2 - 16}{y^2 + y - 12} = \frac{y^2 + y - 20}{y - 3} \cdot \frac{y^2 + y - 12}{y^2 - 16}$$

Factor each numerator and denominator, where possible. Note that $y^2 + y - 20 = (y + 5)(y - 4)$

$$y^2 + y - 12 = (y + 4)(y - 3),$$

and $y^2 - 16 = y^2 - 4^2$, the difference of two squares. Using the formula for the difference of two squares, $(a^2 - b^2) = (a - b)(a + b)$, replace a by **y** and b by 4 to obtain, $(y^2 - 16) = (y - 4)(y + 4)$.

Thus, $\dfrac{y^2 + y - 20}{y - 3} \cdot \dfrac{y^2 + y - 12}{y^2 - 16}$

$$= \frac{(y + 5)(y - 4)}{y - 3} \cdot \frac{(y + 4)(y - 3)}{(y - 4)(y + 4)} \qquad (1)$$

$$= \frac{(y + 5)(y - 4)(y + 4)(y - 3)}{(y - 3)(y - 4)(y + 4)} \qquad (2)$$

$$= \frac{(y + 5)(y - 4)(y + 4)(y - 3)}{(y - 4)(y + 4)(y - 3)} \qquad (3)$$

$$= y + 5.$$

Note that in equation (2) we are dividing by $(y - 3)(y - 4)(y + 4)$. If any of these factors equal 0, then we are dividing by zero, making our fraction invalid. Thus, in order to be certain we are proceeding correctly, we must establish the following restrictions:

$$(y - 3) \neq 0, \quad (y - 4) \neq 0, \quad (y + 4) \neq 0;$$

thus, $y \neq 3$, $y \neq 4$, $y \neq -4$.

Therefore, $\dfrac{y^2 + y - 20}{y - 3} \div \dfrac{y^2 - 16}{y^2 + y - 12} = y + 5$,

and $y \neq 3, 4, -4$.

Find $\dfrac{4a^2 + 4ab + b^2}{2a^3 + 16b^3} \div \dfrac{4a^2 - b^2}{6a + 12b}$

<u>Solution:</u> Division by a fraction is equivalent to multiplication by its reciprocal, hence:

$$\frac{4a^2 + 4ab + b^2}{2a^3 + 16b^3} \div \frac{4a^2 - b^2}{6a + 12b} = \frac{4a^2 + 4ab + b^2}{2a^3 + 16b^3} \times \frac{6a + 12b}{4a^2 - b^2}$$

Factor the numerators and denominators as completely as possible.

$4a^2 + 4ab + b^2$ is called a trinomial perfect square for it is in the form $(2a)^2 + 2(2ab) + b^2$. The formula for factoring a trinomial perfect square is given by $x^2 + 2xy + y^2 = (x+y)(x+y) = (x+y)^2$.

Replacing x by $2a$ and y by b we obtain:

$$4a^2 + 4ab + b^2 = (2a + b)(2a + b) = (2a + b)^2$$

$$2a^3 + 16b^3 = 2\left(a^3 + 8b^3\right) = 2\left(a^3 + (2b)^3\right) \text{ where } a^3 + (2b)^3$$

us the sum of two cubes. The formula for the sum of two cubes is

$$x^3 + y^3 = (x + y)\left(x^2 - xy + y^2\right)$$

Replacing x by a and y by $2b$ we obtain:

$$a^3 + (2b)^3 = (a + 2b)\left(a^2 - 2ab + (2b)^2\right)$$
$$= (a + 2b)\left(a^2 - 2ab + 4b^2\right)$$

thus,

$$2a^3 + 16b^3 = 2(a + 2b)\left(a^2 - 2ab + 4b^2\right)$$

Remove the highest common factor from $6a + 12b$ which is 6. Hence,

$$6a + 12b = 6(a + 2b)$$

$4a^2 - b^2$ is the difference of two squares, $(2a)^2 - b^2$.

Applying the formula for the difference of two squares $x^2 - y^2 = (x-y)(x+y)$:

$$4a^2 - b^2 = (2a)^2 - b^2 = (2a - b)(2a + b).$$

Now, express all the denominators and numerators in their factored form, and cancel:

$$\frac{4a^2 + 4ab + b^2}{2a^3 + 16b^3} \times \frac{6a + 12b}{4a^2 - b^2} = \frac{\cancel{(2a + b)}(2a + b)}{2\cancel{(a+2b)}(a^2 - 2ab + 4b^2)} \times \frac{\overset{3}{\cancel{6}(\cancel{a + 2b})}}{(2a - b)\cancel{(2a + b)}}$$

$$= \frac{3(2a + b)}{\left(a^2 - 2ab + 4b^2\right)(2a - b)}$$

• **PROBLEM 5-26**

Perform the indicated operation,

$$\frac{x^3 - y^3}{x^2 - 5x + 6} \cdot \frac{x^2 - 4}{x^2 - 2xy + y^2}$$

<u>Solution:</u> We factor numerators and denominators to enable us to cancel terms.

$$x^3 - y^3$$

is the difference of two cubes. Thus we factor it

applying the formula for the difference of two cubes,

$$a^3 - b^3 = (a - b)(a^2 + ab + b^2),$$

replacing a by x and b by y. Thus,

$$x^3 - y^3 = (x - y)(x^2 + xy + y^2).$$

$x^2 - 5x+6$ is factored as $(x-2)(x-3)$.

$$x^2 - 4 = x^2 - 2^2,$$

the difference of two squares. Applying the formula for the difference of two squares,

$$a^2 - b^2 = (a + b)(a - b),$$

and replacing a by x and b by 2 we obtain,

$$x^2 - 4 = (x + 2)(x - 2).$$

$$x^2 - 2xy + y^2 = (x - y)(x - y).$$

Thus,

$$\frac{x^3-y^3}{x^2 - 5x + 6} \cdot \frac{x^2 - 4}{x^2 - 2xy + y^2} = \frac{(x - y)(x^2 + xy + y^2)}{(x - 2)(x - 3)}$$

$$\cdot \frac{(x + 2)(x - 2)}{(x - y)(x - y)}$$

$$= \frac{(x^2 + xy + y^2)(x - y)(x - 2)(x + 2)}{(x - 3)(x - 2)(x - y)(x - Y)}$$

$$= \frac{(x^2 + xy + y^2)(x + 2)}{(x - 3)(x -y)}$$

CHAPTER 6

EXPONENT, RADICAL AND POWER

Basic Attacks and Strategies for Solving Problems in this Chapter. See pages 56 to 90 for step-by-step solutions to problems.

An e*xponent* is a superscript following a number or expression (sometimes called the *base*) that indicates some operation to be performed on the number or expression.

An exponent that is a *positive integer* indicates *repeated multiplication*, that is, raising the base to a *power*. The value of the exponent indicates the number of copies of the base that are multiplied together. Thus, 2^3 indicates $2 \cdot 2 \cdot 2 = 8$.

An exponent that is *negative* indicates the *reciprocal* of the expression raised to a positive exponent. Thus,

$$3^{-2} = \frac{1}{3^2} = \frac{1}{9}.$$

In general,

$$x^{-a} = \frac{1}{x^a}$$

even if a is negative. Thus,

$$\frac{1}{x^{-a}} = \frac{1}{1/(x^a)} = x^a.$$

When multiplying two expressions with exponents in which the bases are the same, the exponents are merely added. Thus,

$$a^n \cdot a^m = a^{n+m}.$$

This is derived from the definition of an exponent. With this assumption, it follows that

$$a^n \cdot a^{-n} = a^{n-n} = a^0.$$

But,

$$a^n \cdot a^{-n} = a^n \cdot 1/a^n = 1.$$

From this it follows that an exponent of *zero* indicates a value of one for the expression, i.e., $a^0 = 1$, unless $a = 0$.

Similarly, when dividing two expressions with exponents in which the bases are the same, the exponents are merely subtracted. Thus,

$$\frac{a^n}{a^m} = a^{n-m}.$$

Evaluating an exponent takes precedence over multiplication and division (and other arithmetic operators). Thus, given $-a^n$, first a^n is evaluated and then the result is negated.

When a base expression is a product of numbers or terms, the exponent can be applied to each term separately, i.e.,

$$(abc)^n = a^n b^n c^n.$$

When the base expression includes an exponent, the two exponents are multiplied, i.e.,

$$(a^n)^m = a^{nm}.$$

For example,

$$(2^2)^3 = 2^2 2^2 2^2 = 2^{2+2+2} = 2^6 = 2^{2\cdot3}.$$

If $a^b = a^c$ then it must be the case that the exponents are equal, i.e., that $b = c$. Thus, if one is given that $3^x = 9$, since $9 = 3^2$, it must be true that $x = 2$.

Very large or very small numbers are sometimes written in terms of *scientific* or *exponential* notation. This notation involves a number usually between 1 and 10 (called the *mantissa*) and a multiplier, which is 10 raised to an appropriate power. Thus, 3,200,000 can be rewritten as

$$3.2 \times 1,000,000 = 3.2 \times 10^6.$$

Using this notation can significantly simplify certain computations.

Exponents that are fractions follow the same rules as exponents that are integers. The only difficulties that may arise involve the numeric computation of new exponents (since the computation now involves fractions). For example,

$$a^{1/2} \cdot a^{1/3} = a^{1/2 + 1/3} = a^{3/6 + 2/6} = a^{5/6}.$$

Expressions involving fractional exponents are defined in terms of roots of their bases. These expressions that involve the *root* symbol are also called expressions involving *radicals*. In general,

$$a^{1/n} = \sqrt[n]{a}.$$

Using this fundamental definition, it follows that

$$a^{p/n} = \sqrt[n]{a^p} = \left(\sqrt[n]{a}\right)^p.$$

Thus,

$$a^{n/n} = a = \sqrt[n]{a^n}.$$

The rules given above involving exponents can be rewritten to involve radicals. For example,

$$(ab)^n = a^n \cdot b^n$$

can be rewritten as

$$\sqrt[R]{ab} = \sqrt[R]{a}\sqrt[R]{b}.$$

Problems involving numeric bases with fractional exponents can sometimes be simplified by reducing the base to prime factors. For example,

$$8^{1/4} = (2^3)^{1/4} = 2^{3/4}.$$

Problems involving roots can often be simplified by converting to fractional exponents, e.g.,

$$\sqrt[3]{8^2} = ((2^3)^2)^{1/3} = 2^{6/3} = 2^2 = 4.$$

Sometimes, using irrational factors can further simplify certain problems, e.g.,

$$\sqrt[3]{2\sqrt{2}} = (2 \cdot 2^{1/2})^{1/3} = (2^{3/2})^{1/3} = 2^{1/2}.$$

If we limit ourselves to positive and negative real numbers, then if n is negative, $\sqrt{n}$ has no answer. If, however, we allow *imaginary* and *complex* numbers, then we can speak in terms of $i = \sqrt{-1}$, and, e.g., say that $\sqrt{-4} = 2i$.

One should also remember the general rule exemplified by $2^2 = 4$ and $(-2)^2 = 4$. The equation $x^2 = 4$ has two solutions: x can equal 2 or -2 (although oftentimes we omit mentioning the negative answer). The same can be said about any equation with an even power of any positive (or negative) number. In general, however, if one converts the equation by taking the root of both sides and thus introduces a radical, unless a negative sign precedes the sign, a positive value is assumed. This means that $\sqrt{x^2} = |x|$.

Some problems involving fractional expressions include a denominator that is not a rational number. In these situations, it is helpful to *rationalize the denominator* by multiplying both the numerator and denominator of the fraction by a quantity that converts the denominator into a rational number. For example,

$\dfrac{3}{\sqrt{2}}$ should be multiplied by $\dfrac{\sqrt{2}}{\sqrt{2}}$ to obtain $\dfrac{3\sqrt{2}}{2}$. When the denominator is a sum or difference of roots of numbers, an appropriate multiplier can often be found by first substituting and then using a formula derived from a polynomial that is a sum or difference of powers of variables. For example, given

$$2^{1/2} - 3^{1/2},$$

one can substitute x for $2^{1/2}$ and y for $3^{1/2}$. It is obvious that

$$x^2 - y^2 = 2 - 3 = -1$$

is rational. It is easy to see that one should multiply

$$x - y = 2^{1/2} - 3^{1/2} \quad \text{by} \quad x + y = 2^{1/2} + 3^{1/2}$$

to obtain the rational quantity.

● PROBLEM 6-1

Simplify the following expressions:

(a) -3^{-2}　　(b) $(-3)^{-2}$　　(c) $\dfrac{-3}{4^{-1}}$

<u>Solution:</u>

(a) Here the exponent applies only to 3.

Since $x^{-y} = \dfrac{1}{x^y}$, $\quad -3^{-2} = -(3^{-2}) = -\dfrac{1}{3^2} = -\dfrac{1}{9}$.

(b) In this case the exponent applies to the negative base.

Thus, $\quad (-3)^{-2} = \dfrac{1}{(-3)^2} = \dfrac{1}{(-3)(-3)} = \dfrac{1}{9}$.

(c) $\dfrac{-3}{4^{-1}} = \dfrac{-3}{(\frac{1}{4})^1} = \dfrac{-3}{\frac{1^1}{4^1}} = \dfrac{-3}{\frac{1}{4}}$.

Division by a fraction is equivalent to multiplication by that fraction's reciprocal, thus

$$\dfrac{-3}{\frac{1}{4}} = -3 \cdot \dfrac{4}{1} = -12,$$

and

$$\dfrac{-3}{4^{-1}} = -12.$$

● PROBLEM 6-2

Evaluate:

(a) $8\left(-\dfrac{1}{4}\right)^0$　　(b) $6^0 + (-6)^0$　　(c) $-7(-3)^0$　　(d) 9^{-1}　(e) 7^{-2}.

<u>Solution:</u> Note $x^0 = 1$ and $x^{-a} = \dfrac{1}{x^a}$ for all non-zero real numbers x,

(a) $\qquad 8\left(-\tfrac{1}{4}\right)^0 = 8(1) = 8$

(b) $\qquad 6^0 + (-6)^0 = 1 + 1 = 2$

(c) $\qquad -7(-3)^0 = -7(1) = -7$

(d) $\qquad 9^{-1} = \dfrac{1}{9^1} = \dfrac{1}{9}$

(e) $\qquad 7^{-2} = \dfrac{1}{7^2} = \dfrac{1}{49}$

● **PROBLEM 6-3**

Simplify the expression $\left(3^{-1} + 2^{-1}\right)^{-2}$.

<u>Solution:</u> Since $x^{-y} = \dfrac{1}{x^y}$, $3^{-1} = \dfrac{1}{3^1} = \dfrac{1}{3}$ and $2^{-1} = \dfrac{1}{2^1} = \dfrac{1}{2}$. Thus,

$$\left(3^{-1} + 2^{-1}\right)^{-2} = \left(\tfrac{1}{3} + \tfrac{1}{2}\right)^{-2} .$$

Now, we combine fractions, using 6 as our least common denominator:

$$= \left[\tfrac{2}{2}(\tfrac{1}{3}) + \tfrac{3}{3}(\tfrac{1}{2})\right]^{-2}$$

$$= \left(\tfrac{2}{6} + \tfrac{3}{6}\right)^{-2}$$

$$= \left(\tfrac{5}{6}\right)^{-2}$$

$$= \dfrac{1}{\left(\tfrac{5}{6}\right)^2}$$

$$= \dfrac{1}{\tfrac{25}{36}}$$

and since division by a fraction is equivalent to multiplying the numerator by the reciprocal of the denominator, we have:

$$= 1 \times \dfrac{36}{25}$$

$$= \dfrac{36}{25} .$$

● **PROBLEM 6-4**

Perform the indicated operations:

$$\left(7 \cdot 10^5\right)^3 \cdot \left(3 \cdot 10^{-3}\right)^4 .$$

<u>Solution:</u> Since $(ab)^x = a^x b^x$,

$$(7 \cdot 10^5)^3 \cdot (3 \cdot 10^{-3})^4 = (7)^3 (10^5)^3 \cdot (3)^4 (10^{-3})^4.$$

Recall that $(a^x)^y = a^{xy}$. Thus,

$$= (7^3)(10^{5 \cdot 3}) \cdot (3^4)(10^{-3 \cdot 4})$$
$$= (7^3)(10^{15}) \cdot (3^4)(10^{-12})$$
$$= (7^3)(3^4)(10^{15})(10^{-12}).$$

Since $a^x \cdot a^y = a^{x+y}$,

$$= (7^3)(3^4)\left[10^{15+(-12)}\right]$$
$$= 7^3 3^4 10^3.$$

● **PROBLEM 6-5**

Given $4^{x^2} \cdot 2^x = 8$, solve for x.

<u>Solution</u>: $4^{x^2} \cdot 2^x = 8 = 2^3$

$$(2^2)^{x^2} \cdot 2^x = 2^3$$

$$2^{2x^2} \cdot 2^x = 2^3$$

$$\frac{2^{2x^2+x}}{2^3} = 1$$

$$2^{2x^2+x-3} = 1 = 2^0$$

$$2^{2x^2+x-3} = 2^0$$

$2x^2+x-3 = 0$. Solving for x. one has

$$2x^2+x-3 = (2x+3)(x-1) = 0$$

$$x_1 = -\frac{3}{2}, \quad x_2 = 1$$

Checking the solutions by substituting x_1 and x_2 into the original equation,

$$4^{\left(-\frac{3}{2}\right)^2} \cdot 2^{\left(-\frac{3}{2}\right)} = 4^{\frac{9}{4}} \cdot 2^{\left(-\frac{3}{2}\right)}$$

$$= (2^2)^{\left(\frac{9}{4}\right)} \cdot 2^{\left(-\frac{3}{2}\right)} = 2^{\frac{9}{2}} \cdot 2^{-\frac{3}{2}} = 2^{\frac{9}{2}-\frac{3}{2}} = 2^{\frac{6}{2}}$$

$$= 2^3 = 8$$

Similarly, $4^{1^2} \cdot 2^1 = 4 \cdot 2 = 8$

Therefore, $x = -\frac{3}{2}$ and x=1 are indeed the solutions of the original equation.

Use the theorems on exponents to perform the indicated operations:

(a) $5x^5 \cdot 2x^2$ (b) $\left(x^4\right)^6$ (c) $\dfrac{8y^8}{2y^2}$ (d) $\dfrac{x^3}{x^6}\left(\dfrac{7}{x}\right)^2$.

<u>Solution:</u> Noting the following properties of exponents:

(1) $a^b \cdot a^c = a^{b+c}$ (2) $\left(a^b\right)^c = a^{b \cdot c}$ (3) $\dfrac{a^b}{a^c} = a^{b-c}$ (4) $\left(\dfrac{a}{b}\right)^c = \dfrac{a^c}{b^c}$

we proceed to evaluate these expressions.

(a) $5x^5 \cdot 2x^2 = 5 \cdot 2 \cdot x^5\, x^2 = 10 \cdot x^5 \cdot x^2 = 10x^{5+2} = 10x^7$

(b) $\left(x^4\right)^6 \quad = x^{4 \cdot 6} \quad\quad = x^{24}$

(c) $\dfrac{8y^8}{2y^2} \quad = \dfrac{8}{2} \cdot \dfrac{y^8}{y^2} \quad = 4 \cdot y^{8-2} \quad = 4y^6$

(d) $\left(\dfrac{x^3}{x^6}\right)\left(\dfrac{7}{x}\right)^2 = \left(\dfrac{x^3}{x^6}\right)\left(\dfrac{7^2}{x^2}\right) = \dfrac{x^3 \cdot 49}{x^6 \cdot x^2} = \dfrac{49x^3}{x^{6+2}} = \dfrac{49x^3}{x^8} = \dfrac{49x^3}{x^{5+3}}$

$$= \dfrac{49x^3}{x^5 \cdot x^3} \quad = \dfrac{49}{x^5}$$

Write the expression $\left(x + y^{-1}\right)^{-1}$ without using negative exponents.

<u>Solution:</u> Since $x^{-a} = \dfrac{1}{x^a}$, $y^{-1} = \dfrac{1}{y^1} = \dfrac{1}{y}$,

$$\left(x + y^{-1}\right)^{-1} = \left(x + \dfrac{1}{y}\right)^{-1}$$

$$= \dfrac{1}{x + \dfrac{1}{y}}$$

Multiply numerator and denominator by y in order to eliminate the fraction in the denominator,

$$\dfrac{y(1)}{y\left(x + \dfrac{1}{y}\right)} = \dfrac{y}{yx + \dfrac{y}{y}} = \dfrac{y}{yx + 1}$$

Thus $\left(x + y^{-1}\right)^{-1} = \dfrac{y}{yx + 1}$

Express $\left[\dfrac{a^{-2}}{b^{-3}}\right]^{-2}$ using only positive exponents.

Solution A: By the law of exponents which states that $(x)^{-n} = \dfrac{1}{x^n}$ where n is a positive integer,

$$\left[\frac{a^{-2}}{b^{-3}}\right]^{-2} = \frac{1}{\left[\frac{a^{-2}}{b^{-3}}\right]^2}.$$

Since $\left(\dfrac{x}{y}\right)^n = \dfrac{x^n}{y^n}$, $\left[\dfrac{a^{-2}}{b^{-3}}\right]^2 = \dfrac{(a^{-2})^2}{(b^{-3})^2}$. Also, since $(x^m)^n = x^{mn}$,

$(a^{-2})^2 = a^{(-2)(2)} = a^{-4}$, $(b^{-3})^2 = b^{(-3)(2)} = b^{-6}$. Hence,

$$\left[\frac{a^{-2}}{b^{-3}}\right]^{-2} = \frac{1}{\left[\frac{a^{-2}}{b^{-3}}\right]^2}$$

$$= \frac{1}{\dfrac{a^{-4}}{b^{-6}}}$$

$$= \frac{1}{\dfrac{(a^4)^{-1}}{(b^6)^{-1}}}$$

$$= \frac{1}{\left[\dfrac{a^4}{b^6}\right]^{-1}}$$

$$\frac{1}{\left[\dfrac{1}{\dfrac{a^4}{b^6}}\right]}$$

Note that division is the same as multiplying the numerator by the reciprocal of the denominator. This principle is applied to the term in brackets.

$$\left[\frac{a^{-2}}{b^{-3}}\right]^{-2} = \frac{1}{(1)\left[\dfrac{b^6}{a^4}\right]} = \left(\dfrac{1}{\dfrac{b^6}{a^4}}\right).$$

Applying the same principle to the term in parenthesis on the right side of the equation:

$$\left(\frac{a^{-2}}{b^{-3}}\right)^{-2} = \left(\frac{1}{\frac{b^6}{a^4}}\right) = (1)\left(\frac{a^4}{b^6}\right) = \frac{a^4}{b^6}.$$

<u>Solution B:</u> Since $\left(\frac{x}{y}\right)^n = \frac{x^n}{y^n}$, $\left(\frac{a^{-2}}{b^{-3}}\right)^{-2} = \frac{(a^{-2})^{-2}}{(b^{-3})^{-2}}$. Also, since $(x^m)^n = x^{mn}$, $(a^{-2})^{-2} = a^{(-2)(-2)} = a^4$, and $(b^{-3})^{-2} = b^{(-3)(-2)} = b^6$. Hence,

$$\left(\frac{a^{-2}}{b^{-3}}\right)^{-2} = \frac{a^4}{b^6}.$$

● **PROBLEM 6-9**

Evaluate the following expression: $\dfrac{12x^7y}{3x^2y^3}$

<u>Solution:</u> Noting (1) $\frac{abc}{def} = \frac{a}{d} \cdot \frac{b}{e} \cdot \frac{c}{f}$, (2) $a^{-b} = \frac{1}{a^b}$ and (3) $\frac{a^b}{a^c} = a^{b-c}$ for all non-zero real values of a,d,e,f, we

proceed to evaluate the expression:

$$\frac{12x^7y}{3x^2y^3} = \frac{12}{3} \cdot \frac{x^7}{x^2} \cdot \frac{y}{y^3} = 4 \cdot x^{7-2} \cdot y^{1-3} = 4x^5y^{-2} = \frac{4x^5}{y^2}.$$

● **PROBLEM 6-10**

Express $\dfrac{3x^{-1} - y^{-2}}{x^{-2} + 2y^{-1}}$ without negative exponents.

<u>Solution:</u> Since $x^{-a} = \frac{1}{x^a}$ for all real $x \neq 0$,

$$x^{-1} = \frac{1}{x}, \qquad y^{-2} = \frac{1}{y^2}$$

$$x^{-2} = \frac{1}{x^2}, \text{ and } y^{-1} = \frac{1}{y}; \text{ thus,}$$

61

$$\frac{3x^{-1} - y^{-2}}{x^{-2} + 2y^{-1}} = \frac{3\left(\frac{1}{x}\right) - \frac{1}{y^2}}{\frac{1}{x^2} + 2\left(\frac{1}{y}\right)}$$

$$= \frac{\frac{3}{x} - \frac{1}{y^2}}{\frac{1}{x^2} + \frac{2}{y}}$$

Multiplying numerator and denominator by the least common multiple, x^2y^2,

$$= \frac{x^2y^2\left(\frac{3}{x} - \frac{1}{y^2}\right)}{x^2y^2\left(\frac{1}{x^2} + \frac{2}{y}\right)}$$

Distributing, $= \dfrac{x^2y^2\left(\frac{3}{x}\right) - x^2y^2\left(\frac{1}{y^2}\right)}{x^2y^2\left(\frac{1}{x^2}\right) + x^2y^2\left(\frac{2}{y}\right)}$

Cancelling like terms, $= \dfrac{3xy^2 - x^2}{y^2 + 2x^2y}$

Factoring x from numerator and y from denominator,

$$= \frac{x(3y^2 - x)}{y(y + 2x^2)}$$

Thus, $\dfrac{3x^{-1} - y^{-2}}{x^{-2} + 2y^{-1}} = \dfrac{x(3y^2 - x)}{y(y + 2x^2)}$

● **PROBLEM** 6-11

Use the properties of exponents, to perform the indicated operations in
$$(2^3x^4 5^2 y^7)^5.$$

<u>Solution:</u> Since the product of several numbers raised to the same exponent equals the product of each number raised to that exponent (i.e., $(abcd)^x = a^x b^x c^x d^x$) we obtain,

$$(2^3x^4 5^2 y^7)^5 = (2^3)^5 (x^4)^5 (5^2)^5 (y^7)^5.$$

Recall that $(x^a)^b = x^{a \cdot b}$; thus

$$(2^3x^4 5^2 y^7)^5 = (2^3)^5 (x^4)^5 (5^2)^5 (y^7)^5$$

$$= (2^{3 \cdot 5})(x^{4 \cdot 5})(5^{2 \cdot 5})(y^{7 \cdot 5})$$

$$= 2^{15}x^{20}5^{10}y^{35}.$$

Perform the indicated operations, and simplify. (Write without negative or zero exponents.) Each letter represents a positive real number.

(a) $\left(7x^{-3}y^5\right)^{-2}$ (b) $\left(5x^7y^{-8}\right)^{-3}$.

Solution: Note that: (1) $(abc)^x = a^xb^xc^x$ (for all real a,b,c), (2) $a^{-x} = \dfrac{1}{a^x}$ (for all non-zero real a) and (3) $\left(a^b\right)^c = a^{bc}$ (for all real a,b,c). These will be useful in evaluating the given expressions.

(a) $\left(7x^{-3}y^5\right)^{-2} = 7^{-2}\left(x^{-3}\right)^{-2}\left(y^5\right)^{-2} = 7^{-2}(x)^{(-3)(-2)}(y)^{(5)(-2)}$

$$= 7^{-2}(x)^6(y)^{-10} = \frac{x^6}{\left(7^2\right)\left(y^{10}\right)} = \frac{x^6}{49y^{10}}.$$

(b) $\left(5x^7y^{-8}\right)^{-3} = 5^{-3}\left(x^7\right)^{-3}\left(y^{-8}\right)^{-3} = 5^{-3}(x)^{(7)(-3)}(y)^{(-8)(-3)}$

$$= 5^{-3}(x)^{-21}(y)^{24} = \frac{y^{24}}{5^3x^{21}} = \frac{y^{24}}{125x^{21}}.$$

Evaluate the following expressions:

(a) $\dfrac{-12x^{10}y^9z^5}{3x^2y^3z^6}$ (b) $\dfrac{-16x^{16}y^6z^4}{-4x^4y^2z^7}$.

Solution: Noting (a) $\dfrac{abcd}{efgh} = \dfrac{a}{e}\cdot\dfrac{b}{f}\cdot\dfrac{c}{g}\cdot\dfrac{d}{h}$, (2) $a^{-b} = \dfrac{1}{a^b}$ and (3) $\dfrac{a^b}{a^c} = a^{b-c}$ for all non-zero real values of a,e,f,g,h, we proceed to evaluate these expressions:

(a) $\dfrac{-12x^{10}y^9z^5}{3x^2y^3z^6} = \dfrac{-12}{3}\cdot\dfrac{x^{10}}{x^2}\cdot\dfrac{y^9}{y^3}\cdot\dfrac{z^5}{z^6} = -4\cdot x^{10-2}\cdot y^{9-3}\cdot z^{5-6}$

$$= -4x^8y^6z^{-1} = \frac{-4x^8y^6}{z^1}.$$

Thus $\dfrac{-12x^{10}y^9z^5}{3x^2y^3z^6} = \dfrac{-4x^8y^6}{z}.$

(b) $\dfrac{-16x^{16}y^6z^4}{-4x^4y^2z^7} = \dfrac{-16}{-4}\cdot\dfrac{x^{16}}{x^4}\cdot\dfrac{y^6}{y^2}\cdot\dfrac{z^4}{z^7}$

$$= 4x^{16-4}\cdot y^{6-2}\cdot z^{4-7} = 4x^{12}y^4z^{-3} = \frac{4x^{12}y^4}{z^3}.$$

Perform the indicated operations and simplify:

$$\left(\frac{-5b^y}{3^2 x^5}\right)^3 \left(\frac{3x^7}{5b^y}\right)^2 .$$

Solution: $\left(\frac{-5b^y}{3^2 x^5}\right)^3 \left(\frac{3x^7}{5b^y}\right)^2 = \frac{(-5b^y)^3}{(3^2 x^5)^3} \cdot \frac{(3x^7)^2}{(5b^y)^2}$ since $\left(\frac{a}{b}\right)^x = \frac{a^x}{b^x}$

$= \frac{(-5)^3 (b^y)^3}{(3^2)^3 (x^5)^3} \cdot \frac{3^2 (x^7)^2}{5^2 (b^y)^2}$ since $(ab)^x = a^x b^x$

$= \frac{-5^3 b^{3y}}{3^6 x^{15}} \cdot \frac{3^2 x^{14}}{5^2 b^{2y}}$ since $(a^x)^y = a^{x \cdot y}$

$= \frac{\left(-5^3 b^{3y}\right)\left(3^2 x^{14}\right)}{\left(3^6 x^{15}\right)\left(5^2 b^{2y}\right)}$

$= \frac{\left(3^2 x^{14}\right)\left(-5^3 b^{3y}\right)}{\left(3^6 x^{15}\right)\left(5^2 b^{2y}\right)}$ using the commutative law of multiplication

$= \left(3^{2-6}\right)\left(x^{14-15}\right)\left[-\left(5^{3-2}\right)\left(b^{3y-2y}\right)\right]$ because

$\frac{x^a}{x^b} = x^{a-b},$

$= (3^{-4})(x^{-1})(-5^1)(b^y)$

$= \frac{-5b^y}{3^4 x}$ because $x^{-a} = \frac{1}{x^a}$

$= \frac{-5b^y}{3 \cdot 3 \cdot 3 \cdot 3 x}$

$= \frac{-5b^y}{81x}$

Rewrite the value of each of the following terms in exponential form.

(1) $\frac{0.000000074}{1200000}$

(2) $0.00008072 \times 0.000043$

Solution: (1) $\dfrac{7.4 \times 10^{-8}}{1.2 \times 10^{6}} = \dfrac{7.4}{1.2} \times 10^{-14} = 6.167 \times 10^{-14}$

(2) $\dfrac{8.072 \times 10^{-5}}{4.3 \times 10^{-5}} = \dfrac{8.072}{4.3} = 1.877 \times 10^{5}$

● **PROBLEM 6-16**

Simplify $\left[\dfrac{1600 \times 10{,}000}{2000}\right]^{1/3}$.

Solution: Observe $1600 = 16 \times 100 = 16 \times 10^{2}$

$10{,}000 = 10^{4}$

$2{,}000 = 2 \times 10^{3}$.

Thus, $\left[\dfrac{1600 \times 10{,}000}{2000}\right]^{1/3} = \left[\dfrac{\left(16 \times 10^{2}\right)\left(10^{4}\right)}{2 \times 10^{3}}\right]^{1/3}$.

Using the associative property,

$$= \left[\dfrac{16 \times \left(10^{2} \times 10^{4}\right)}{2 \times 10^{3}}\right]^{1/3}$$

Recall: $a^{x} \cdot a^{y} = a^{x+y}$, $\quad = \left[\dfrac{16 \times 10^{6}}{2 \times 10^{3}}\right]^{1/3}$

● **PROBLEM 6-17**

Determine the value of

$$\dfrac{5^{3/4}\, 5^{2/3}\, 5^{-5/2}\, 5^{5/3}}{5^{1/3}\, 5^{-5/2}\, 5^{7/4}} \ .$$

Solution: Since $a^{x} \cdot a^{y} \cdot a^{z} = a^{x+y+z}$ for any real base, then,

$$\dfrac{5^{\frac{3}{4}} \cdot 5^{\frac{2}{3}} \cdot 5^{-\frac{5}{2}} \cdot 5^{\frac{5}{3}}}{5^{\frac{1}{3}} \cdot 5^{-\frac{5}{2}} \cdot 5^{\frac{7}{4}}} = \dfrac{5^{\frac{3}{4} + \frac{2}{3} - \frac{5}{2} + \frac{5}{3}}}{5^{\frac{1}{3} - \frac{5}{2} + \frac{7}{4}}}$$

The fractional **exponents** have denominators 4, 3, and 2. Their least common denominator (L.C.D.) is the least common multiple of the denominators, 12. Converting the fractional exponents to twelfths we obtain

$$\dfrac{5^{\frac{3}{4} + \frac{2}{3} - \frac{5}{2} + \frac{5}{3}}}{5^{\frac{1}{3} - \frac{5}{2} + \frac{7}{4}}} = \dfrac{5^{\frac{9}{12} + \frac{8}{12} - \frac{30}{12} + \frac{20}{12}}}{5^{\frac{4}{12} - \frac{30}{12} + \frac{21}{12}}}$$

$$= \frac{5^{\frac{7}{12}}}{5^{-\frac{5}{12}}}$$

Since $\frac{a^x}{a^y} = a^{x-y}$ for any real base $(a \neq 0)$

$$\frac{5^{\frac{7}{12}}}{5^{-\frac{5}{12}}} = 5^{\frac{7}{12} - \left(-\frac{5}{12}\right)}$$

$$= 5^{\frac{7}{12} + \frac{5}{12}}$$

$$= 5^{\frac{12}{12}}$$

$$= 5^1$$

$$= 5.$$

Therefore, $\dfrac{5^{\frac{3}{4}} \, 5^{\frac{2}{3}} \, 5^{-\frac{5}{2}} \, 5^{\frac{5}{3}}}{5^{\frac{1}{3}} \, 5^{-\frac{5}{2}} \, 5^{\frac{7}{4}}} = 5.$

● PROBLEM 6-18

Express $\left(5^{\frac{1}{2}} + 9^{\frac{1}{8}}\right) \div \left(5^{\frac{1}{2}} - 9^{\frac{1}{8}}\right)$

as an equivalent fraction with a rational denominator.

Solution: The given expression can be rewritten as,

$$\frac{5^{\frac{1}{2}} + 9^{\frac{1}{8}}}{5^{\frac{1}{2}} - 9^{\frac{1}{8}}}, \qquad \text{and since}$$

$$9^{\frac{1}{8}} = \left(3^2\right)^{\frac{1}{8}} = 3^{\frac{1}{4}} \qquad \text{we write:}$$

$$\frac{5^{\frac{1}{2}} + 3^{\frac{1}{4}}}{5^{\frac{1}{2}} - 3^{\frac{1}{4}}}.$$

To rationalize the denominator, put

$5^{\frac{1}{2}} = x$, $3^{\frac{1}{4}} = y$; then since

$$x^4 - y^4 = \left(5^{\frac{1}{2}}\right)^4 - \left(3^{\frac{1}{4}}\right)^4 = 5^2 - 3 = 25 - 3 = 22,$$

which is rational, we can write

66

$$x^4 - y^4 = (x - y)(x^3 + x^2y + xy^2 + y^3),$$

and the factor which rationalizes $x - y$, or

$5^{\frac{1}{2}} - 3^{\frac{1}{4}}$ is $x^3 + x^2y + xy^2 + y^3$, and substituting for x and y:

$$\left(5^{\frac{1}{2}}\right)^3 + \left(5^{\frac{1}{2}}\right)^2 \cdot 3^{\frac{1}{4}} + 5^{\frac{1}{2}} \cdot \left(3^{\frac{1}{4}}\right)^2 + \left(3^{\frac{1}{4}}\right)^3$$

$$= \quad 5^{\frac{3}{2}} + 5^{\frac{2}{2}} \cdot 3^{\frac{1}{4}} + 5^{\frac{1}{2}} \cdot 3^{\frac{2}{4}} + 3^{\frac{3}{4}} \; ;$$

and the rational denominator is

$$x^4 - y^4 = 5^{\frac{4}{2}} - 3^{\frac{4}{4}} = 5^2 - 3 = 22.$$

Now, since

$$x^4 - y^4 = (x - y)(x^3 + x^2y + xy^2 + y^3), \qquad \text{then}$$

$$(x - y) = \frac{x^4 - y^4}{x^3 + x^2y + xy^2 + y^3}, \qquad \text{and substituting:}$$

$$5^{\frac{1}{2}} - 3^{\frac{1}{4}} = \frac{22}{5^{\frac{3}{2}} + 5^{\frac{2}{2}} \cdot 3^{\frac{1}{4}} + 5^{\frac{1}{2}} \cdot 3^{\frac{2}{4}} + 3^{\frac{3}{4}}}$$

Therefore, the given expression

$$= \frac{\dfrac{5^{\frac{1}{2}} + 3^{\frac{1}{4}}}{22}}{5^{\frac{3}{2}} + 5^{\frac{2}{2}} \cdot 3^{\frac{1}{4}} + 5^{\frac{1}{2}} \cdot 3^{\frac{2}{4}} + 3^{\frac{3}{4}}}$$

$$= \frac{\left(5^{\frac{1}{2}} + 3^{\frac{1}{4}}\right)\left[5^{\frac{3}{2}} + \left(5^{\frac{2}{2}} \cdot 3^{\frac{1}{4}}\right) + \left(5^{\frac{1}{2}} \cdot 3^{\frac{2}{4}}\right) + 3^{\frac{3}{4}}\right]}{22}$$

$$= 5^{\frac{4}{2}} + 5^{\frac{3}{2}} 3^{\frac{1}{4}} + 5^{\frac{3}{2}}3^{\frac{1}{4}} + 5^{\frac{2}{2}}3^{\frac{2}{4}} + 5^{\frac{2}{2}}3^{\frac{2}{4}} +$$

$$+ 5^{\frac{1}{2}}3^{\frac{3}{4}} + 5^{\frac{1}{2}}3^{\frac{3}{7}} + 3^{\frac{4}{4}}$$

$$= \frac{5^{\frac{4}{2}} + \left(2 \cdot 5^{\frac{3}{2}} \cdot 3^{\frac{1}{4}}\right) + \left(2 \cdot 5^{\frac{2}{2}} \cdot 3^{\frac{2}{4}}\right) + \left(2 \cdot 5^{\frac{1}{2}} \cdot 3^{\frac{3}{4}}\right) + 3^{\frac{4}{4}}}{22}$$

$$= \frac{5^2 + 2 \left(5^{\frac{3}{2}} \cdot 3^{\frac{1}{4}} + 5^{\frac{2}{2}} \cdot 3^{\frac{2}{4}} + 5^{\frac{1}{2}} \cdot 3^{\frac{3}{4}} \right) + 3}{22}$$

$$= \frac{28 + 2 \left(5^{\frac{3}{2}} \cdot 3^{\frac{1}{4}} + 5^{\frac{2}{2}} \cdot 3^{\frac{2}{4}} + 5^{\frac{1}{2}} \cdot 3^{\frac{3}{4}} \right)}{22}$$

$$= \frac{14 + 5^{\frac{3}{2}} \cdot 3^{\frac{1}{4}} + 5 \cdot 3^{\frac{1}{2}} + 5^{\frac{1}{2}} \cdot 3^{\frac{3}{4}}}{11}$$

● **PROBLEM** 6-19

Find the value of $\sqrt[4]{-64a^4}$.

Solution: We can rewrite $\sqrt[4]{-64^4}$ as,

$$\left[64a^4 \cdot (-1) \right]^{\frac{1}{4}} = \left[\left(\underset{-}{+} 8a^2 \right)^2 \cdot (-1) \right]^{\frac{1}{4}} ,$$

by first factoring -1 from the expression under the radical, and then using the fact that $64a^4 = \left(\underset{-}{+} 8a^2 \right)^2$. Also recall that $\sqrt[4]{x} = x^{\frac{1}{4}}$. Now, since $(ab)^x = a^x \cdot b^x$, and $(a^x)^y = a^{xy}$, we write:

$$\left[(8a^2)^2 \cdot (-1) \right]^{\frac{1}{4}} = \left[(8a^2)^2 \right]^{\frac{1}{4}} \cdot (-1)^{\frac{1}{4}}$$
$$= \left[8^2 \cdot (a^2)^2 \right]^{\frac{1}{4}} \cdot (-1)^{\frac{1}{4}}$$
$$= (8^2)^{\frac{1}{4}} \cdot \left[(a^2)^2 \right]^{\frac{1}{4}} \cdot (-1)^{\frac{1}{4}}$$
$$= 8^{\frac{1}{2}} \cdot (a^2)^{\frac{1}{2}} \cdot (-1)^{\frac{1}{2} \cdot \frac{1}{2}}$$
$$= 8^{\frac{1}{2}} \cdot (a^2)^{\frac{1}{2}} \cdot \left[(-1)^{\frac{1}{2}} \right]^{\frac{1}{2}}$$

Since $x^{\frac{1}{2}} = \sqrt{x}$, and $\left(x^{\frac{1}{2}} \right)^{\frac{1}{2}} = \sqrt{\sqrt{x}}$, we write:

$$\sqrt[4]{-64a^4} = \sqrt{\underset{-}{+} 8a^2 \sqrt{-1}} = \sqrt{4a^2 \cdot 2 \cdot \underset{-}{+} \sqrt{-1}} ,$$

and since

$$\sqrt{a \cdot b \cdot c} = \sqrt{a}\sqrt{b}\sqrt{c} , \quad \sqrt[4]{-64a^4} = \sqrt{4a^2} \cdot \sqrt{2} \cdot \sqrt{\underset{-}{+} \sqrt{-1}}$$
$$= 2a \sqrt{2} \sqrt{\underset{-}{+} \sqrt{-1}} .$$

It remains to find the value of $\sqrt{\underset{-}{+} \sqrt{-1}}$.
Assume $\sqrt{+ \sqrt{-1}} = x + y\sqrt{-1}$;
then, squaring both sides of the equation we obtain:

$$+ \sqrt{-1} = (x + y\sqrt{-1})(x + y\sqrt{-1})$$

or, by performing the multiplication,

68

$$+ \sqrt{-1} = x^2 + xy\sqrt{-1} + xy\sqrt{-1} + y\sqrt{-1} \cdot y\sqrt{-1} \ .$$

Combining like terms, and recalling that $\sqrt{-1} \cdot \sqrt{-1} = -1$, we have:

$$+ \sqrt{-1} = x^2 - y^2 + 2xy\sqrt{-1} \ .$$

Let us examine the equation,

$$+ \sqrt{-1} = x^2 - y^2 + 2xy\sqrt{-1} \ .$$

This equation is only true if $x^2 - y^2 = 0$, and $2xy = 1$, because then the equation becomes:

$$+ \sqrt{-1} = 0 + {}^{-}1\sqrt{-1}$$
$$+ \sqrt{-1} = + \sqrt{-1}$$

Therefore, we have the following system of equations:

$$x^2 - y^2 = 0 \quad \text{and} \quad 2xy = 1 \ .$$

To solve for x and y we use the method of substitution. Solving for y in the second equation, $2xy = 1$, we have:

$$y = \frac{1}{2x} \ ;$$

and substituting this value into equation one, $x^2 - y^2 = 0$, we obtain:

$$x^2 - \left(\frac{1}{2x}\right)^2 = 0$$
$$x^2 - \frac{1}{4x^2} = 0$$

Multiply both sides by $4x^2$: $\quad 4x^2\left(x^2 - \frac{1}{4x^2}\right) = 4x^2(0)$

Distribute: $\qquad\qquad\qquad 4x^4 - 1 = 0$

Add 1 to both sides: $\qquad\qquad 4x^4 = 1$

Divide both sides by 4: $\qquad\quad x^4 = \tfrac{1}{4}$

Now, taking the fourth root of both sides we obtain:

$$x = \sqrt[4]{\tfrac{1}{4}}$$
$$= (\tfrac{1}{4})^{\frac{1}{4}} \ , \text{ since } \sqrt[4]{x} = x^{\frac{1}{4}}$$
$$= \frac{1^{\frac{1}{4}}}{4^{\frac{1}{4}}} \ , \text{ since } \left(\frac{a}{b}\right)^x = \frac{a^x}{b^x}$$
$$= \frac{1}{\left(2^2\right)^{\frac{1}{4}}}$$
$$= \frac{1}{2^{\frac{1}{2}}} \ , \text{ since } \left(a^b\right)^x = a^{bx}$$
$$= \frac{1}{\sqrt{2}} \ , \text{ since } x^{\frac{1}{2}} = \sqrt{x}$$

Now, since $y = \frac{1}{2x}$, by substitution:

$$y = \frac{1}{2\left(\frac{1}{\sqrt{2}}\right)} = \frac{1}{\frac{2}{\sqrt{2}}} = \frac{\sqrt{2}}{2}$$

Observing the y-value, $\frac{\sqrt{2}}{2}$, closely we see that it is equivalent to the x-value, $\frac{1}{\sqrt{2}}$. This can be seen by multiplying $\frac{\sqrt{2}}{2}$ by $\frac{\sqrt{2}}{\sqrt{2}}$ (which

equals 1 and therefore does not alter the value of the fraction).
Thus,

$$\frac{\sqrt{2}}{2} \cdot \frac{\sqrt{2}}{\sqrt{2}} = \frac{2}{2\sqrt{2}} = \frac{1}{\sqrt{2}}$$

Therefore,

$$x = \frac{1}{\sqrt{2}} , \ y = \frac{1}{\sqrt{2}} ; \text{ or } \ x = -\frac{1}{\sqrt{2}} , \ y = -\frac{1}{\sqrt{2}} ;$$

We must include the negative values for x and y as solutions to the equations, because these give us the same results in the equations $x^2 - y^2 = 0$ and $2xy = 1$ as do the positive values for x and y, since a negative value squared is positive (equation 1), and a negative multiplied by a negative is positive (equation 2). Therefore, substituting into the equation, $\sqrt{+\sqrt{-1}} = x + y\sqrt{-1}$, we have:

$$\sqrt{+\sqrt{-1}} = {\textstyle\pm} \frac{1}{\sqrt{2}} + \left({\textstyle\pm} \frac{1}{\sqrt{2}} \right)\sqrt{-1} ,$$

and factoring $\pm \dfrac{1}{\sqrt{2}}$ from both terms on the right side:

$$\sqrt{+\sqrt{-1}} = {\textstyle\pm} \frac{1}{\sqrt{2}}(1 + \sqrt{-1}).$$

Similarly, we assume $\sqrt{-\sqrt{-1}} = x - y\sqrt{-1}$, and proceeding as in the case when $\sqrt{+\sqrt{-1}} = x + y\sqrt{-1}$, we find that the x and y values are again $\pm \dfrac{1}{\sqrt{2}}$, thus:

$$\sqrt{-\sqrt{-1}} = {\textstyle\pm} \frac{1}{\sqrt{2}}(1 - \sqrt{-1})$$

Therefore,

$$\sqrt{{\textstyle\pm}\sqrt{-1}} = {\textstyle\pm} \frac{1}{\sqrt{2}}(1 {\textstyle\pm} \sqrt{-1}) ;$$

and finally, from the fact that,

$$\sqrt[4]{-64a^4} = 2a\sqrt{2}\sqrt{{\textstyle\pm}\sqrt{-1}} , \text{ and}$$

$$\sqrt{{\textstyle\pm}\sqrt{-1}} = {\textstyle\pm} \frac{1}{\sqrt{2}}(1 {\textstyle\pm} \sqrt{-1}) , \text{ we have:}$$

$$\sqrt[4]{-64a^4} = 2a\sqrt{2} \cdot \left[{\textstyle\pm} \frac{1}{\sqrt{2}}\left(1 {\textstyle\pm} \sqrt{-1} \right) \right]$$

$$= 2a\left[{\textstyle\pm} (1 {\textstyle\pm} \sqrt{-1}) \right] , \text{ by cancelling } \frac{\sqrt{2}}{\sqrt{2}} .$$

Therefore,

$$\sqrt[4]{-64a^4} = {\textstyle\pm} 2a(1 {\textstyle\pm} \sqrt{-1}) .$$

● **PROBLEM 6-20**

Find the indicated roots.

(a) $\sqrt[5]{32}$ (b) $\pm \sqrt[4]{625}$ (c) $\sqrt[3]{-125}$ (d) $\sqrt[5]{-16}$.

70

<u>Solution:</u> The following two laws of exponents can be used to solve these problems: 1) $\left(\sqrt[n]{a}\right)^n = \left(a^{1/n}\right)^n = a^1 = a$, and 2) $\left(\sqrt[n]{a}\right)^n = \sqrt[n]{a^n}$.

(a) $\sqrt[5]{32} = \sqrt[5]{2^5} = \left(\sqrt[5]{2}\right)^5 = 2$. This result is true because $(2)^5 = 32$, that is, $2 \cdot 2 \cdot 2 \cdot 2 \cdot 2 = 32$.

(b) $\sqrt[4]{625} = \sqrt[4]{5^4} = \left(\sqrt[4]{5}\right)^4 = 5$. This result is true because $\left(5^4\right) = 625$, that is, $5 \cdot 5 \cdot 5 \cdot 5 = 625$.

$-\sqrt[4]{625} = -\left(\sqrt[4]{5^4}\right) = -\left[\left(\sqrt[4]{5}\right)^4\right] = -\left[5\right] = -5$. This result is true because $(-5)^4 = 625$, that is, $(-5) \cdot (-5) \cdot (-5) \cdot (-5) = 625$.

(c) $\sqrt[3]{-125} = \sqrt[3]{(-5)^3} = (\sqrt[3]{-5})^3 = -5$. This result is true because $(-5)^3 = -125$, that is, $(-5) \cdot (-5) \cdot (-5) = -125$.

(d) There is no solution to $\sqrt[4]{-16}$ because any number raised to the fourth power is a positive number, that is, $N^4 = (N) \cdot (N) \cdot (N) \cdot (N) = $ a positive number $\neq$ a negative number, -16.

● **PROBLEM 6-21**

Simplify: (a) $\sqrt[3]{-512}$ (b) $\sqrt[4]{\dfrac{81}{16}}$ (c) $\sqrt[3]{-16} \div \sqrt[3]{-2}$.

<u>Solution:</u> (a) By the law of radicals which states that $\sqrt[n]{ab} = \sqrt[n]{a}\,\sqrt[n]{b}$ where a and b are any two numbers, $\sqrt[3]{-512} = \sqrt[3]{8(-64)} = \sqrt[3]{8}\sqrt[3]{-64}$. Therefore, $\sqrt[3]{-512} = \sqrt[3]{8}\sqrt[3]{-64} = (2)(-4) = -8$. The last result is true because $(2)^3 = 8$ and $(-4)^3 = -64$.

(b) By another law of radicals which states that $\sqrt[n]{\dfrac{a}{b}} = \dfrac{\sqrt[n]{a}}{\sqrt[n]{b}}$ where a and b are any two numbers, $\sqrt[4]{\dfrac{81}{16}} = \dfrac{\sqrt[4]{81}}{\sqrt[4]{16}}$.

Therefore, $\sqrt[4]{\dfrac{81}{16}} = \dfrac{\sqrt[4]{81}}{\sqrt[4]{16}} = \dfrac{3}{2}$. The last result is true because $(3)^4 = 81$ and $(2)^4 = 16$.

(c) By the law of radicals stated in example (b), $\sqrt[3]{-16} \div \sqrt[3]{-2} = \dfrac{\sqrt[3]{-16}}{\sqrt[3]{-2}} = \sqrt[3]{\dfrac{-16}{-2}} = \sqrt[3]{8} = 2$. The last result is true because $(2)^3 = 8$.

Find the numerical value of each of the following.

(a) $8^{2/3}$ (b) $25^{3/2}$

Solution:

(a) Since $x^{a/b} = \left(x^{1/b}\right)^a$, $8^{2/3} = \left(8^{1/3}\right)^2 = \left(\sqrt[3]{8}\right)^2 = (2)^2 = 4$

(b) Similarly, $25^{3/2} = \left(25^{1/2}\right)^3 = 5^3 = 125$.

In solving the following problem, find the value of $x+y+\sqrt{(x-y)^2}$ for x=4 and y=8. A student wrote:

$$x + y + \sqrt{(x-y)^2} = x + y + (x - y) = 2x$$

$$= 2 \cdot 4 = 8$$

Was this student's answer correct?

Solution: No, $x + y + \sqrt{(x-y)^2} = 4 + 8 + \sqrt{(-4)^2} = 12 + 4 = 16$. Note that in general,

$$\sqrt{(x-y)^2} = |x-y| = \begin{cases} x-y & \text{if } x > y \\ 0 & \text{if } x = y \\ y-x & \text{if } x < y \end{cases}$$

Simplify $5\sqrt{12} + 3\sqrt{75}$.

Solution: Express 12 and 75 as the product of perfect squares if possible. Thus, $12 = 4 \cdot 3$ and $75 = 25 \cdot 3$; and $5\sqrt{12} + 3\sqrt{75} = 5\sqrt{4 \cdot 3} + 3\sqrt{25 \cdot 3}$.

Since $\sqrt{a \cdot b} = \sqrt{a} \cdot \sqrt{b}$:
$$= [5 \cdot \sqrt{4} \cdot \sqrt{3}] + [3\sqrt{25} \cdot \sqrt{3}]$$
$$= [(5 \cdot 2)\sqrt{3}] + [(3 \cdot 5)\sqrt{3}]$$
$$= 10\sqrt{3} + 15\sqrt{3}.$$

Using the distributive law:
$$= (10 + 15)\sqrt{3}$$
$$= 25\sqrt{3}.$$

Approximate $\sqrt{23} \times \sqrt{40}$ and $\sqrt{23} \div \sqrt{40}$.

<u>Solution:</u> A four-place table of square roots gives

$$\sqrt{23} = 4.796 \qquad \text{and} \qquad \sqrt{40} = 6.325.$$

The product $\sqrt{23} \times \sqrt{40} = (4.796)(6.325)$

$$= 30.334700$$

Thus, rounding off to the nearest one hundredth, we obtain

$$\sqrt{23}\ \sqrt{40} = 30.33, \text{ approximately.}$$

For the division: $\sqrt{23} \div \sqrt{40} = 4.796 \div 6.325$

$$= 6.325\ \overline{\smash{)}4.796}$$

Thus,

```
           .7582
  6.325) 4.7960000
         4 4275
           36850
           31625
           52250
           50600
           16500
           12650
            3850
```

Hence, $\sqrt{23} \div \sqrt{40}$ is approximately 0.7582.

If a = 3 and b = 2, find $(6a - b)^{-5/4}$.

<u>Solution:</u>   Substitute a = 3 and b = 2: $(6 \cdot 3 - 2)^{-5/4}$

Perform the indicated multiplication: $(18 - 2)^{-5/4}$

$$= 16^{-5/4}$$

Since $x^{-y} = \dfrac{1}{x^y}$: $= \dfrac{1}{16^{5/4}}$

73

$$= \frac{1}{\left(\sqrt[4]{16}\right)^5}$$

Since $2^4 = 2 \cdot 2 \cdot 2 \cdot 2 = 16$,

$\sqrt[4]{16} = 2$. Hence: $\qquad = \dfrac{1}{2^5}$

$$= \frac{1}{32}$$

Determine whether each of the following expressions are true or false. If false, explain why.

(1) $x^4 \cdot x^6 = x^{24}$

(2) $\dfrac{a^6}{a^2} = a^3$

(3) $(y^4)^2 = y^6$

(4) $\dfrac{a^4}{a^{-4}} = a^{4-4} = a^0 = 1$

(5) $a^4 + a^6 = a^{10}$

(6) $\sqrt{25} = \pm 5$

(7) $\sqrt{a+b} = \sqrt{a} + \sqrt{b}$

(8) $x^{2/5} = (\sqrt[2]{x})^5$

(9) $\dfrac{1}{a^{-1}+b^{-1}} = a + b$

(10) $(a+b)^{-1} = a^{-1} + b^{-1}$

Solution: (1) False. The rule for multiplying exponential values is $a^p \cdot a^q = a^{p+q}$. Therefore,

$$x^4 \cdot x^6 = x^{10} \neq x^{24}.$$

(2) False. The rule for dividing exponential values is $\dfrac{a^p}{a^q} = a^{p-q}$. Therefore, $\dfrac{a^6}{a^2} = a^{6-2}$

$$= a^4 \neq a^3.$$

(3) False. $(y^4)^2 = y^{4 \cdot 2} = y^8 \neq y^6$

(4) False. $\dfrac{a^4}{a^{-4}} = a^{4-(-4)} = a^8 \neq 1$

(5) False. In order to add exponential values, the two values must be raised to the same exponent. Therefore,

$$a^4 + a^5 = a^4 + a^6 \neq a^{10}$$

(6) True.

(7) False. $\sqrt{a+b} \neq \sqrt{a} + \sqrt{b}$

(8) False. $x^{2/5} = \sqrt[5]{x^2}$

(9) False.

$$\frac{1}{a^{-1}+b^{-1}} = \frac{1}{\frac{1}{a}+\frac{1}{b}} = \frac{1}{\frac{b+a}{ab}}$$

$$= \frac{ab}{b+a} \neq a + b$$

(10) False $(a+b)^{-1} = \frac{1}{a+b} \neq a^{-1} + b^{-1}$

● **PROBLEM 6-28**

Simplify the quotient $\sqrt{x}/\sqrt[4]{x}$. Write the result in exponential notation.

<u>Solution:</u> Since $\sqrt[b]{n^a} = n^{a/b}$, the numerator and the denominator can be rewritten as:

$$\sqrt{x} = x^{1/2} \quad \text{and}$$

$$\sqrt[4]{x} = x^{1/4}$$

Therefore,

$$\frac{\sqrt{x}}{\sqrt[4]{x}} = \frac{x^{1/2}}{x^{1/4}} \qquad (1)$$

According to the law of exponents which states that $\frac{n^a}{n^b} = n^{a-b}$, equation (1) becomes:

$$\frac{\sqrt{x}}{\sqrt[4]{x}} = \frac{x^{1/2}}{x^{1/4}}$$

$$= x^{\frac{1}{2}-\frac{1}{4}}$$

$$= x^{\frac{2}{4}-\frac{1}{4}}$$

$$= x^{\frac{1}{4}}$$

● **PROBLEM 6-29**

Simplify: (a) $\sqrt{8x^3y}$ (b) $\sqrt{\frac{2a}{4b^2}}$ (c) $\sqrt[4]{25x^2}$.

<u>Solution:</u>(a) $\sqrt{8x^3y}$ contains the perfect square $4x^2$. Factoring

out $4x^2$ we obtain,

$$\sqrt{8x^3y} = \sqrt{4x^2 \cdot 2xy} \quad .$$

Recall that $\sqrt{ab} = \sqrt{a} \cdot \sqrt{b}$. Thus,

$$= \sqrt{4x^2} \cdot \sqrt{2xy}$$

$$= \sqrt{4}\sqrt{x^2}\sqrt{2xy} \quad .$$

Since $\sqrt{x^2} = |x|$,

$$\sqrt{8x^3y} = 2|x|\sqrt{2xy} \quad .$$

(b) $\sqrt{\dfrac{2a}{4b^2}}$ has a fraction for the radicand, but the denominator

is a perfect square.

$$\sqrt{\frac{2a}{4b^2}} = \frac{\sqrt{2a}}{\sqrt{4b^2}} \quad , \text{ since } \sqrt{\frac{a}{b}} = \frac{\sqrt{a}}{\sqrt{b}} \; ; \; \frac{\sqrt{2a}}{\sqrt{4b^2}} = \frac{\sqrt{2a}}{2|b|} \quad .$$

(c) $\sqrt[4]{25x^2}$ has a perfect square for the radicand.

$$\sqrt[4]{25x^2} = \sqrt[4]{(5x)^2} \quad .$$

Recall that $\sqrt[4]{x} = \sqrt[2]{\sqrt[2]{x}}$; hence $\sqrt[4]{(5x)^2} = \sqrt[2]{\sqrt[2]{(5x)^2}}$. Now, since

$\sqrt[2]{(5x)^2} = |5x|$, $= \sqrt[2]{|5x|}$. Since

$$|ab| = |a||b| , = \sqrt[2]{|5||x|} = \sqrt[2]{5|x|} = \sqrt{5|x|} \quad .$$

Radicals with the same index can be multiplied by finding the product of the radicands, the index of the product being the same as the factors.

● PROBLEM 6-30

Find the product $\sqrt[4]{x^3y} \cdot \sqrt[4]{xy^2}$ and simplify.

Solution: Note that $\sqrt[x]{a} \cdot \sqrt[x]{b} = \sqrt[x]{ab}$; thus,

$$\sqrt[4]{x^3y} \cdot \sqrt[4]{xy^2} = \sqrt[4]{(x^3y)(xy^2)} \quad .$$

Recall that when multiplying, we add exponents; hence

$$\left(x^3y^1\right)\left(x^1y^2\right) = \left(x^{3+1}\,y^{1+2}\right), \quad \text{and}$$

we obtain,

$$= \sqrt[4]{x^4y^3}$$

$$= \sqrt[4]{x^4\left(4\sqrt{y^3}\right)}$$

Now, since $\sqrt[4]{x^4} = \left(x^{\frac{1}{4}}\right)^4 = x^1 = x$, $\sqrt[4]{x^3y} \cdot \sqrt[4]{xy^2} = x \sqrt[4]{y^3}$.

Perform the indicated operations in the following expression and write the final result without negative or zero exponents:

$$\left(\frac{64a^{-3}b^{4/3}}{27a^{-9}b^{-14/3}}\right)^{-2/3}$$

__Solution:__ Since $\left(\frac{a}{b}\right)^n = \frac{a^n}{b^n}$

$$\left(\frac{64a^{-3}b^{4/3}}{27z^{-9}b^{-14/3}}\right)^{-2/3} = \frac{\left(64a^{-3}b^{4/3}\right)^{-2/3}}{\left(27a^{-9}b^{-14/3}\right)^{-2/3}}$$

and $(abc)^n = a^n b^n c^n$. Thus

$$\frac{\left(64a^{-3}b^{4/3}\right)^{-2/3}}{\left(27a^{-9}b^{-14/3}\right)^{-2/3}} = \frac{(64)^{-2/3}\left(a^{-3}\right)^{-2/3}\left(b^{4/3}\right)^{-2/3}}{(27)^{-2/3}\left(a^{-9}\right)^{-2/3}\left(b^{-14/3}\right)^{-2/3}}$$

Recall $(x^y)^z = x^{yz}$ thus

$$\frac{(64)^{-2/3}\left(a^{-3}\right)^{-2/3}\left(b^{4/3}\right)^{-2/3}}{(27)^{-2/3}\left(a^{-9}\right)^{-2/3}\left(b^{-14/3}\right)^{-2/3}} = \frac{(64)^{-2/3}\left(a^{2}\right)\left(b^{-8/9}\right)}{(27)^{-2/3}\left(a^{6}\right)\left(b^{28/9}\right)}$$

Since $a^6 = a^4 \cdot a^2$, cancel a^2 from numerator and denominator,

$$= \frac{(64)^{-2/3}\left(b^{-8/9}\right)}{(27)^{-2/3}\left(a^{4}\right)\left(b^{28/9}\right)}$$

and since $\frac{a^x}{a^y} = a^{x-y}$, $\frac{b^{-8/9}}{b^{28/9}} = b^{-8/9 - 28/9} = b^{-36/9} = \frac{1}{b^{36/9}} = \frac{1}{b^4}$

thus

$$\frac{(64)^{-2/3}\left(b^{-8/9}\right)}{(27)^{-2/3}\left(a^{4}\right)\left(b^{28/9}\right)} = \frac{(64)^{-2/3}}{(27)^{-2/3}\left(a^{4}\right)\left(b^{4}\right)}$$

Since

$$x^{a/b} = \left(\sqrt[b]{x}\right)^a$$

$$(64)^{-2/3} = \left(\sqrt[3]{64}\right)^{-2} = (4)^{-2}$$
$$(27)^{-2/3} = \left(\sqrt[3]{27}\right)^{-2} = (3)^{-2}$$

Thus

$$\frac{(64)^{-2/3}}{(27)^{-2/3}a^4b^4} = \frac{(4)^{-2}}{(3)^{-2}a^4b^4}$$

Recall

$$x^{-a} = \frac{1}{x^a}$$

therefore

$$(4)^{-2} = \frac{1}{4^2} = \frac{1}{16}$$

and

$$(3)^{-2} = \frac{1}{3^2} = \frac{1}{9}$$

hence

$$\frac{(4)^{-2}}{(3)^{-2}a^4b^4} = \frac{1/16}{\frac{1}{9}a^4b^4}$$

Multiply numerator and denominator by $16 \cdot 9$,

$$= \frac{16 \cdot 9(1/16)}{16 \cdot 9\left(\frac{1}{9}\right)a^4b^4}$$

$$= \frac{9}{16\,a^4b^4}$$

thus

$$\left(\frac{64a^{-3}b^{4/3}}{27a^{-9}b^{-14/3}}\right)^{-2/3} = \frac{9}{16\,a^4b^4}$$

● **PROBLEM 6-32**

Simplify $\sqrt[3]{-81x^3} - 2x\sqrt[3]{3} + 5x\sqrt[3]{24}$.

<u>Solution:</u> Rewrite the expression so it contains similar radicals.

$\sqrt[3]{-81x^3} = \sqrt[3]{(-3)^3x^3 \cdot 3} = \sqrt[3]{(-3x)^3 \cdot 3}$ by the law $(ab)^n = a^nb^n$. Also, since $\sqrt[n]{ab} = \sqrt[n]{a}\,\sqrt[n]{b}$, $\sqrt[3]{(-3x)^3 \cdot 3} = \sqrt[3]{(-3x)^3}\,\sqrt[3]{3}$. Hence, $\sqrt[3]{-81x^3} = \sqrt[3]{(-3x)^3}\sqrt[3]{3}$. Since $(\sqrt[n]{a})^n = (a^{1/n})^n = a^{\frac{1}{n} \cdot n} = a^1 = a$, $\sqrt[3]{(-3x)^3} = -3x$. Therefore, $\sqrt[3]{-81x^3} = -3x\sqrt[3]{3}$. By the same laws, $5x\sqrt[3]{24} = 5x\sqrt[3]{2^3 \cdot 3} = 5x\sqrt[3]{(2)^3}\sqrt[3]{3} = 5x(2)\sqrt[3]{3} = 10x\sqrt[3]{3}$. Therefore, $\sqrt[3]{-81x^3} - 2x\sqrt[3]{3} + 5x\sqrt[3]{24} = -3x\sqrt[3]{3} - 2x\sqrt[3]{3} + 10x\sqrt[3]{3}$

$$= -5x\sqrt[3]{3} + 10x\sqrt[3]{3} = 5x\sqrt[3]{3}.$$

Hence, $\sqrt[3]{-81x^3} - 2x\sqrt[3]{3} + 5x\sqrt[3]{24} = 5x\sqrt[3]{3}$. Note that the radical used to simplify the given expression was $\sqrt[3]{3}$.

● **PROBLEM 6-33**

Find the square root of

$$\frac{3}{2}(x - 1) + \sqrt{2x^2 - 7x - 4}.$$

<u>Solution:</u> The given expression can be rewritten as

$$\frac{3}{2}x - \frac{3}{2} + \sqrt{2x^2 - 7x - 4}.$$

We can eliminate the fractions by factoring $\frac{1}{2}$ from all the terms. To do this we must first multiply the third term by 2 so as not to change the value of this term. Thus, we obtain:

$$\frac{1}{2}\left(3x - 3 + 2\sqrt{2x^2 - 7x - 4}\right)$$

Let us examine the expression under the radical. Notice that this can be rewritten in factored form since

$$2x^2 - 7x - 4 = (2x + 1)(x - 4).$$

Substituting this in the given expression we have:

$$\frac{1}{2}\left(3x - 3 + 2\sqrt{(2x + 1)(x - 4)}\right).$$

Our aim now is to transform the expression into one which is a perfect square. This can be accomplished as follows: Rewrite the first two terms, $3x - 3$ as: $(2x + 1) + (x - 4)$, and substitute this into the expression. Thus, we obtain,

$$\frac{1}{2}\left[(2x + 1) + (x - 4) + 2\sqrt{(2x + 1)(x - 4)}\right],$$

and we are looking for:

$$\sqrt{\frac{1}{2}\left[(2x + 1) + (x - 4) + 2\sqrt{(2x + 1)(x - 4)}\right]}$$

But,

$$(2x + 1) + (x - 4) + 2\sqrt{(2x + 1)(x - 4)}$$

$$= (2x + 1) + (x - 4) + 2\sqrt{2x + 1}\sqrt{x - 4}$$

$$= (\sqrt{2x + 1} + \sqrt{x - 4})(\sqrt{2x + 1} + \sqrt{x - 4})$$

$$= (\sqrt{2x + 1} + \sqrt{x - 4})^2;$$

Therefore, our expression becomes:

$$\sqrt{\frac{1}{2}(\sqrt{2x + 1} + \sqrt{x - 4})^2}$$

$$= \sqrt{\frac{1}{2}}(\sqrt{2x + 1} + \sqrt{x - 4})$$

$$= \frac{1}{\sqrt{2}}(\sqrt{2x + 1} + \sqrt{x - 4}).$$

● PROBLEM 6-34

Find the cube root of $9\sqrt{3} + 11\sqrt{2}$.

Solution: In this problem we wish to find:

$$\sqrt[3]{9\sqrt{3} + 11\sqrt{2}}$$

Factoring $3\sqrt{3}$ from both terms under the radical, $9\sqrt{3}$ and $11\sqrt{2}$, we obtain:

$$\sqrt[3]{3\sqrt{3}\left(3 + \frac{11}{3}\frac{\sqrt{2}}{\sqrt{3}}\right)}, \qquad \text{or}$$

$$\sqrt[3]{9\sqrt{3} + 11\sqrt{2}}$$

$$= \sqrt[3]{3\sqrt{3}\left(3 + \frac{11}{3}\sqrt{\frac{2}{3}}\right)}.$$

Now, since $\sqrt[3]{3\sqrt{3}} = \sqrt{3}$ (that is, $\sqrt{3}\cdot\sqrt{3}\cdot\sqrt{3} = 3\sqrt{3}$) we can write the expression as:

$$= \sqrt{3}\left(\sqrt[3]{3 + \frac{11}{3}\sqrt{\frac{2}{3}}}\right)$$

Observe that $3 + \frac{11}{3}\sqrt{\frac{2}{3}}$ is a perfect cube, since:

$$\left(1 + \sqrt{\frac{2}{3}}\right)\left(1 + \sqrt{\frac{2}{3}}\right)\left(1 + \sqrt{\frac{2}{3}}\right)$$

$$= \left(1 + 2\sqrt{\frac{2}{3}} + \frac{2}{3}\right)\left(1 + \sqrt{\frac{2}{3}}\right)$$

$$= 1 + 2\sqrt{\frac{2}{3}} + \frac{2}{3} + \sqrt{\frac{2}{3}} + 2\cdot\frac{2}{3} + \frac{2}{3}\sqrt{\frac{2}{3}}$$

$$= \left(1 + \frac{4}{3} + \frac{2}{3}\right) + \left(2\sqrt{\frac{2}{3}} + \sqrt{\frac{2}{3}} + \frac{2}{3}\sqrt{\frac{2}{3}}\right)$$

$$= 3 + \frac{11}{3}\sqrt{\frac{2}{3}}.$$

Thus,

$$\sqrt[3]{3 + \frac{11}{3}\sqrt{\frac{2}{3}}} = 1 + \sqrt{\frac{2}{3}} \text{ , and}$$

the required cube root $= \sqrt{3}\left[1 + \sqrt{\frac{2}{3}}\right]$

$$= \sqrt{3} + \sqrt{2}.$$

Write in fractional exponent form with no denominators.

(a) $\sqrt[3]{\dfrac{x}{y}}$ (b) $\sqrt[3]{3}$ (c) $\sqrt[4]{x^2}\,\sqrt[3]{xy^{-1}}$

<u>Solution:</u> Noting that $\sqrt[b]{a} = a^{1/b}$, $\left(\dfrac{a}{b}\right)^c = \dfrac{a^c}{b^c}$,

and $a^{-b} = \dfrac{1}{a^b}$, we proceed to evaluate these expressions.

(a) $\sqrt[3]{\dfrac{x}{y}} = \left(\dfrac{x}{y}\right)^{1/3} = \dfrac{x^{1/3}}{y^{1/3}} = x^{\frac{1}{3}}y^{-\frac{1}{3}}$

(b) $\sqrt[3]{3} = 3^{\frac{1}{3}}$

(c) $\sqrt[4]{x^2}\,\sqrt[3]{xy^{-1}} = (x^2)^{\frac{1}{4}}(xy^{-1})^{\frac{1}{3}}$

$= (x^{\frac{2}{4}})(x^{\frac{1}{3}})(y^{-\frac{1}{3}})$, since $(a^b)^c = a^{bc}$
                                and $(ab^c)^d = a^d b^{cd}$

$= (x^{\frac{1}{2}})(x^{\frac{1}{3}})(y^{-\frac{1}{3}})$

$= x^{\frac{1}{2}+\frac{1}{3}}y^{-\frac{1}{3}}$, since $(x^a)(x^b) = x^{a+b}$

$= x^{\frac{3}{6}+\frac{2}{6}}y^{-\frac{1}{3}}$

$= x^{\frac{5}{6}}y^{-\frac{1}{3}}$

Rationalize $\dfrac{\sqrt[3]{3ax}}{\sqrt[3]{4a^2}}$.

<u>Solution:</u> Multiply the numerator and the denominator by the radical $\sqrt[3]{(4a^2)^2}$ to eliminate the radical in the denominator:

$$\frac{\sqrt[3]{3ax}}{\sqrt[3]{4a^2}} = \frac{\left[\sqrt[3]{(4a^2)^2}\right]\sqrt[3]{3ax}}{\left(\sqrt[3]{(4a^2)^2}\right)\sqrt[3]{(4a^2)}} = \frac{\sqrt[3]{(4a^2)^2}\,\sqrt[3]{3ax}}{\sqrt[3]{(4a^2)^3}}$$

Note the last result is true because of the law involving radicals which states that $\sqrt[3]{a} \cdot \sqrt[3]{b} = \sqrt[3]{ab}$. Also, since $\sqrt[3]{a^3} = (\sqrt[3]{a})^3 = (a^{1/3})^3 = a^1 = a$, $\sqrt[3]{(4a^2)^3} = \left(\sqrt[3]{4a^2}\right)^3 = 4a^2$. Hence,

$$\frac{\sqrt[3]{3ax}}{\sqrt[3]{4a^2}} = \frac{\sqrt[3]{(4a^2)^2}\,\sqrt[3]{3ax}}{4a^2} = \frac{\sqrt[3]{16a^4}\,\sqrt[3]{3ax}}{4a^2}.$$

Since $\sqrt[3]{ab} = \sqrt[3]{a}\sqrt[3]{b}$, $\sqrt[3]{16a^4} = \sqrt[3]{(8a^3)(2a)} = \sqrt[3]{8a^3}\sqrt[3]{2a}$

$$= \sqrt[3]{(2a)^3}\sqrt[3]{2a}.$$

Note that the last result is true because $(ab)^x = a^x b^x$; that is, $8a^3 = 2^3 a^3 = (2a)^3$. Hence:

$$\frac{\sqrt[3]{3ax}}{\sqrt[3]{4a^2}} = \frac{\sqrt[3]{(2a)^3}\sqrt[3]{2a}\sqrt[3]{3ax}}{4a^2}$$

$$= \frac{2a\sqrt[3]{2a}\sqrt[3]{3ax}}{4a^2}$$

$$= \frac{2a\sqrt[3]{(2a)(3ax)}}{4a^2}$$

$$= \frac{2a\sqrt[3]{6a^2x}}{4a^2}.$$

Therefore, $\dfrac{\sqrt[3]{3ax}}{\sqrt[3]{4a^2}} = \dfrac{\sqrt[3]{6a^2x}}{2a}.$

● **PROBLEM 6-37**

Find the factor which will rationalize $\sqrt{3} + \sqrt[3]{5}$.

<u>Solution:</u> We can rewrite $\sqrt{3} + \sqrt[3]{5}$ as,

$$3^{\frac{1}{2}} + 5^{\frac{1}{3}}$$

Observe that both of the above irrational numbers, when raised to the sixth power, become rational

$$\left[\left(3^{\frac{1}{2}}\right)^6 = 3^3 = 27, \left(5^{\frac{1}{3}}\right)^6 = 5^2 = 25\right].$$

Let $x = 3^{\frac{1}{2}}$, $y = 5^{\frac{1}{3}}$; then x^6 and y^6 are both rational.

Since x^6 and y^6 are rational, so is $x^6 - y^6$ (and in fact, is equal to $27 - 25 = 2$). To find the factor which rationalizes $x + y$ ($\sqrt{3} + \sqrt[3]{5}$), we divide $x^6 - y^6$ by $x + y$, and find the quotient to be,

$$x^5 - x^4y + x^3y^2 - x^2y^3 + xy^4 - y^5.$$

Thus

$$x^6 - y^6 = (x + y)(x^5 - x^4y + x^3y^2 - x^2y^3 + xy^4 - y^5);$$

and substituting for x and y, the required factor is

$$x^5 - x^4y + x^3y^2 - x^2y^3 + xy^4 - y^5 =$$

$$\left(3^{\frac{1}{2}}\right)^5 - \left(3^{\frac{1}{2}}\right)^4 5^{\frac{1}{3}} + \left(3^{\frac{1}{2}}\right)^3 \left(5^{\frac{1}{3}}\right)^2 - \left(3^{\frac{1}{2}}\right)^2 \left(5^{\frac{1}{3}}\right)^3$$

$$+ 3^{\frac{1}{2}} \left(5^{\frac{1}{3}}\right)^4 - \left(5^{\frac{1}{3}}\right)^5 =$$

$$3^{\frac{5}{2}} - 3^{\frac{4}{2}} \cdot 5^{\frac{1}{3}} + 3^{\frac{3}{2}} \cdot 5^{\frac{2}{3}} - 3^{\frac{2}{2}} \cdot 5^{\frac{3}{3}}$$

$$+ 3^{\frac{1}{2}} \cdot 5^{\frac{4}{3}} - 5^{\frac{5}{3}}, \qquad\qquad \text{or}$$

$$3^{\frac{5}{2}} - 9 \cdot 5^{\frac{1}{3}} + 3^{\frac{3}{2}} \cdot 5^{\frac{2}{3}} - 15 + 3^{\frac{1}{2}} \cdot 5^{\frac{4}{3}} - 5^{\frac{5}{3}};$$

and the rational product, $\left(3^{\frac{1}{2}}\right)^6 - \left(5^{\frac{1}{3}}\right)^6$, of the above factor and $3^{\frac{1}{2}} + 5^{\frac{1}{3}}$ is

$$3^{\frac{6}{2}} - 5^{\frac{6}{3}} = 3^3 - 5^2 = 2.$$

● PROBLEM 6-38

Express with rational denominator $\dfrac{4}{\sqrt[3]{9} - \sqrt[3]{3} + 1}$.

Solution: The given expression can be written as,

$$\frac{4}{9^{\frac{1}{3}} - 3^{\frac{1}{3}} + 1} = \frac{4}{\left(3^2\right)^{\frac{1}{3}} - 3^{\frac{1}{3}} + 1} = \frac{4}{3^{\frac{2}{3}} - 3^{\frac{1}{3}} + 1}.$$

To ratonalize the denominator, multiply both

numerator and denominator by $\left(3^{\frac{1}{3}} + 1\right)$. This will not change the value of the fraction because

$$\frac{3^{\frac{1}{3}} + 1}{3^{\frac{1}{3}} + 1} = 1,$$

and multiplication by 1 does not change the value of any given expression.

Thus, multiplying, we obtain:

$$\frac{4\left(3^{\frac{1}{3}} + 1\right)}{\left(3^{\frac{1}{3}} + 1\right)\left(3^{\frac{2}{3}} - 3^{\frac{1}{3}} + 1\right)}$$

$$= \frac{4\left(3^{\frac{1}{3}} + 1\right)}{\left(3^{\frac{1}{3}}\right)\left(3^{\frac{2}{3}}\right) + 3^{\frac{2}{3}} - \left(3^{\frac{1}{3}}\right)\left(3^{\frac{1}{3}}\right) - 3^{\frac{1}{3}} + 3^{\frac{1}{3}} + 1}$$

$$= \frac{4\left(3^{\frac{1}{3}} + 1\right)}{3^{\frac{3}{3}} + 3^{\frac{2}{3}} - 3^{\frac{2}{3}} - 3^{\frac{1}{3}} + 3^{\frac{1}{3}} + 1}$$

$$\frac{4\left(3^{\frac{1}{3}} + 1\right)}{3 + 1} = \frac{4\left(3^{\frac{1}{3}} + 1\right)}{4}$$

$$= 3^{\frac{1}{3}} + 1.$$

● PROBLEM 6-39

Find the following products:

    a. $\sqrt{x}\,(\sqrt{2x} - \sqrt{x})$

    b. $(\sqrt{x} - 2\,\sqrt{y})(2\sqrt{x} + \sqrt{y})$

Solution: The following two laws concerning radicals can be used to find the indicated products:

    1) $\sqrt{a}\,\sqrt{b} = \sqrt{ab}$

where a and b are any two numbers

    2) $\sqrt{a^2} = (\sqrt{a})^2 = \left(a^{1/2}\right)^2 = a^{(1/2) \cdot 2} = a^1 = a$

or $\sqrt{a^2} = a$

a) Using the distributive property,

$$\sqrt{x} \ (\sqrt{2x} - \sqrt{x}) = \sqrt{x} \ (\sqrt{2x}) - \sqrt{x} \ (\ \sqrt{x})$$

Using the first law concerning radicals to further simplify this equation,

$$\sqrt{x} \ (\sqrt{2x} - \sqrt{x}) = \sqrt{x \cdot 2x} - \sqrt{x \cdot x}$$

$$= \sqrt{2x^2} - \sqrt{x^2}$$

$$= \sqrt{2} \sqrt{x^2} - \sqrt{x^2}$$

Using, the second law to simplify this equation,

$$\sqrt{x} \quad (\sqrt{2x} - \sqrt{x}) = \sqrt{2} \ (x) - x$$

$$= \sqrt{2}x - x$$

b) $(\sqrt{x} - 2 \sqrt{y})(2\sqrt{x} + \sqrt{y}) = \sqrt{x} \ (2\sqrt{x}) - 2\sqrt{y}(2\sqrt{x}) + \sqrt{x}(\sqrt{y})$

$- 2\sqrt{y} \ (\sqrt{y})$

Using the first law concerning radicals to simplify this equation,

$$(\sqrt{x} - 2\sqrt{y})(2\sqrt{x} + \sqrt{y}) = 2\sqrt{x \cdot x} - 4 \sqrt{x \cdot y} + \sqrt{x \cdot y} - 2 \sqrt{y \cdot y}$$

$$= 2 \sqrt{x^2} - 4 \sqrt{xy} + \sqrt{xy} - 2\sqrt{y^2}$$

$$= 2 \sqrt{x^2} - 3\sqrt{xy} - 2\sqrt{y^2}$$

Using the second law to simplify this equation,

$$(\sqrt{x} - 2\sqrt{y})(2\sqrt{x} + \sqrt{y}) = 2 \ (x) - 3\sqrt{xy} - 2 \ (y)$$

$$= 2x - 3\sqrt{xy} - 2y$$

● **PROBLEM 6-40**

When $x = \dfrac{3 + 5\sqrt{-1}}{2}$, find the value of $2x^3 + 2x^2 - 7x + 72$; and show that it will be unaltered if $\dfrac{3 - 5\sqrt{-1}}{2}$ be substituted for x.

85

<u>Solution:</u> When $x = \dfrac{3 + 5\sqrt{-1}}{2}$,

(eq. 1) $2x^3 + 2x^2 - 7x + 72 = 2\left(\dfrac{3+5\sqrt{-1}}{2}\right)^3 + 2\left(\dfrac{3+5\sqrt{-1}}{2}\right)^2 - 7\left(\dfrac{3+5\sqrt{-1}}{2}\right) + 72.$

Simplifying the right side of this equation:

$$\left(\dfrac{3+5\sqrt{-1}}{2}\right)^2 = \left(\dfrac{3+5\sqrt{-1}}{2}\right)\left(\dfrac{3+5\sqrt{-1}}{2}\right) = \dfrac{9 + 30\sqrt{-1} + 25(-1)}{4}$$

$$= \dfrac{30\sqrt{-1} - 16}{4}$$

$$\left(\dfrac{3+5\sqrt{-1}}{2}\right)^3 = \left(\dfrac{3+5\sqrt{-1}}{2}\right)\left(\dfrac{3+5\sqrt{-1}}{2}\right)^2 = \left(\dfrac{3+5\sqrt{-1}}{2}\right)\left(\dfrac{30\sqrt{-1} - 16}{4}\right)$$

$$= \dfrac{90\sqrt{-1} + 150(-1) - 48 - 80\sqrt{-1}}{8}$$

$$= \dfrac{10\sqrt{-1} - 198}{8}$$

Therefore, equation (1) becomes:

$$2x^3 + 2x^2 - 7x + 72 = 2\left(\dfrac{10\sqrt{-1} - 198}{8}\right) + 2\left(\dfrac{30\sqrt{-1} - 16}{4}\right) - 7\left(\dfrac{3+5\sqrt{-1}}{2}\right) + 72$$

$$= \dfrac{10\sqrt{-1} - 198}{4} + \dfrac{30\sqrt{-1} - 16}{2} - \dfrac{21 + 35\sqrt{-1}}{2} + 72$$

$$= \dfrac{10\sqrt{-1}}{4} - \dfrac{198}{4} + 15\sqrt{-1} - 8 - \dfrac{21}{2} - \dfrac{35}{2}\sqrt{-1} + 72$$

$$= 2\tfrac{1}{2}\sqrt{-1} - 49\tfrac{1}{2} + 15\sqrt{-1} - 8 - 10\tfrac{1}{2} - 17\tfrac{1}{2}\sqrt{-1} + 72$$

Associating, $= (2\tfrac{1}{2}\sqrt{-1} + 15\sqrt{-1} - 17\tfrac{1}{2}\sqrt{-1}) - 49\tfrac{1}{2} - 8 - 10\tfrac{1}{2} + 72$

$$= (17\tfrac{1}{2}\sqrt{-1} - 17\tfrac{1}{2}\sqrt{-1}) - 68 + 72$$

$$= 0 - 68 + 72$$

$$= 4$$

When $x = \dfrac{3 - 5\sqrt{-1}}{2}$,

(eq. 2) $2x^3 + 2x^2 - 7x + 72 = 2\left(\dfrac{3-5\sqrt{-1}}{2}\right)^3 + 2\left(\dfrac{3-5\sqrt{-1}}{2}\right)^2 - 7\left(\dfrac{3-5\sqrt{-1}}{2}\right) + 72.$

Simplifying the right side of this equation:

$$\left(\dfrac{3-5\sqrt{-1}}{2}\right)^2 = \left(\dfrac{3-5\sqrt{-1}}{2}\right)\left(\dfrac{3-5\sqrt{-1}}{2}\right) = \dfrac{9 - 30\sqrt{-1} + 25(-1)}{4}$$

$$= \dfrac{-30\sqrt{-1} - 16}{4}$$

$$\left(\dfrac{3-5\sqrt{-1}}{2}\right)^3 = \left(\dfrac{3-5\sqrt{-1}}{2}\right)\left(\dfrac{3-5\sqrt{-1}}{2}\right)^2 = \left(\dfrac{3-5\sqrt{-1}}{2}\right)\left(\dfrac{-30\sqrt{-1} - 16}{4}\right)$$

$$= \dfrac{-90\sqrt{-1} - 48 + 150(-1) + 80\sqrt{-1}}{8}$$

$$= \dfrac{-10\sqrt{-1} - 198}{8}$$

Therefore, equation (2) becomes:

$$2x^3 + 2x^2 - 7x + 72 = 2\left(\dfrac{-10\sqrt{-1} - 198}{8}\right) + 2\left(\dfrac{-30\sqrt{-1} - 16}{4}\right) - 7\left(\dfrac{3-5\sqrt{-1}}{2}\right) + 72$$

$$= \frac{-10\sqrt{-1}}{4} - \frac{198}{4} - 15\sqrt{-1} - 8 - \frac{21}{2} + \frac{35}{2}\sqrt{-1} + 72$$

$$= -2\tfrac{1}{2}\sqrt{-1} - 49\tfrac{1}{2} - 15\sqrt{-1} - 8 - 10\tfrac{1}{2} + 17\tfrac{1}{2}\sqrt{-1} + 72$$

Associating, $= \left(-2\tfrac{1}{2}\sqrt{-1} - 15\sqrt{-1} + 17\tfrac{1}{2}\sqrt{-1}\right) -49\tfrac{1}{2} -8 -10\tfrac{1}{2} + 72$

$$= \left(-17\tfrac{1}{2}\sqrt{-1} + 17\tfrac{1}{2}\sqrt{-1}\right) -68 + 72$$

$$= 0 - 68 + 72$$

$$= 4$$

Therefore, when $x = \frac{3+5\sqrt{-1}}{2}$, the value of the equation is 4. The equation is unaltered when $3-5\sqrt{-1}/2$ is substituted for x, since the value of the equation is also 4 for this substitution.

● **PROBLEM 6-41**

Find the product by inspection:

$$\left(\sqrt[3]{2}a + \sqrt[3]{4}b\right)\left(\sqrt[3]{4}a^2 - 2ab + 2\sqrt[3]{2}b^2\right)$$

Solution: The formula for the sum of two cubes can be used to find the product. This formula is:

$$x^3 + y^3 = (x + y)(x^2 - xy + y^2).$$

The product $\left(\sqrt[3]{2}a + \sqrt[3]{4}b\right)\left(\sqrt[3]{4}a^2 - 2ab + 2\sqrt[3]{2}b^2\right)$ corresponds to the right side of the formula for the sum of two cubes where x is replaced by $\sqrt[3]{2}a$ and y is replaced by $\sqrt[3]{4}b$. Hence,

$$\left(\sqrt[3]{2}a + \sqrt[3]{4}b\right)\left(\sqrt[3]{4}a^2 - 2ab + 2\sqrt[3]{2}b^2\right) = \left(\sqrt[3]{2}a\right)^3 + \left(\sqrt[3]{4}b\right)^3$$

$$= (\sqrt[3]{2})^3 a^3 + (\sqrt[3]{4})^3 b^3$$

since $(ab)^x = a^x b^x$. Also,

$$(\sqrt[n]{x})^n = \left(x^{\frac{1}{n}}\right)^n = x^{\frac{n}{n}} = x^1 = x, \text{ hence}$$

$$(\sqrt[3]{2})^3 = 2 \text{ and } (\sqrt[3]{4})^3 = 4.$$

Therefore $(\sqrt[3]{2}a + \sqrt[3]{4}b)(\sqrt[3]{4}a^2 - 2ab + 2\sqrt[3]{2}b^2) = 2a^3 + 4b^3$.

Draw the graph of each of the following equations:

(1) $y = 2^x$

(2) $y = (\frac{1}{2})^x$

<u>Solution</u>:

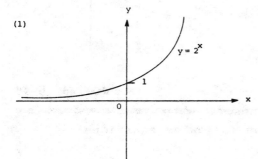

Draw the graph of the following equations:

(1) $y^2 = x^3$ for $x \geq 0$

(2) $y^3 = x^2$

(3) $y = \frac{1}{x^2}$

(4) $y = \sqrt[3]{x}$

<u>Solution</u>:

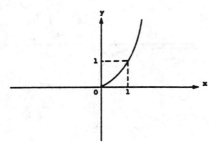

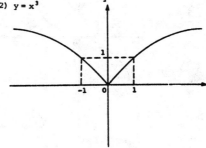

(3) $y = \frac{1}{x^2}$

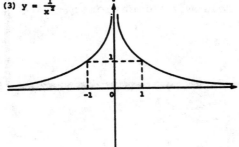

(4) $y = \sqrt[3]{x}$

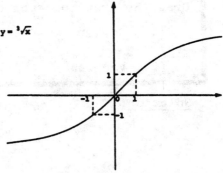

Draw the graph for each of the following equations:

(1) $y = x^3$

(2) $y = x^{-3}$

(3) $y = x^{\frac{1}{2}}$ for $x \geq 0$

(4) $y = x^{-\frac{1}{2}}$ for $x > 0$

(5) $y = 4x^4$

(6) $y = \sqrt{|x|}$

(7) $y = \sqrt{x-3}$ for $x \geq 3$

Solution:

(1)

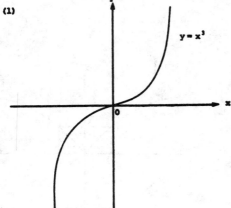

$y = x^3$

(2)

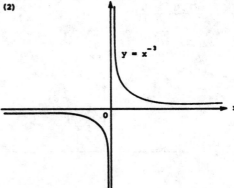

$y = x^{-3}$

(3)

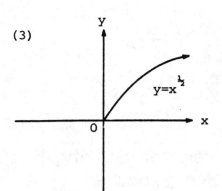

$y=x^{\frac{1}{2}}$

(4)

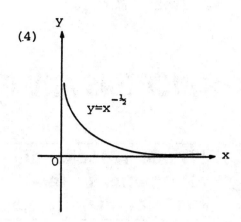

$y=x^{-\frac{1}{2}}$

(5)

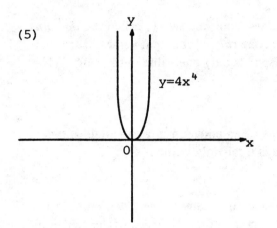

$y=4x^4$

(6)

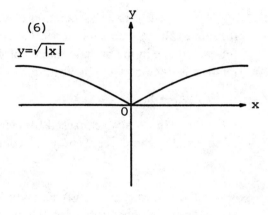

$y=\sqrt{|x|}$

(7)

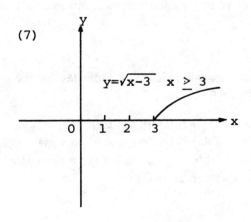

$y=\sqrt{x-3} \quad x \geq 3$

CHAPTER 7

FUNCTIONS AND GRAPHS

> **Basic Attacks and Strategies for Solving Problems in this Chapter. See pages 91 to 119 for step-by-step solutions to problems.**

A *function, f,* is an association of values from one set, X, called the *domain* (of input values), into another set, Y, called the *range* (of output values), such that for each value in the domain, there corresponds only one value in the range. This relation is often denoted by writing

$f : X \rightarrow Y.$

Functions are usually expressed by means of an equation in two variables (very often x and y where $x \in X$ and $y \in Y$) and is frequently written

$f(x) = y.$

When the values (e.g., (x, y)) satisfying such an equation are plotted on a Cartesian plane with coordinate axes, the plotted points form the *graph* of the equation. A function can be identified by testing its graph as follows: any vertical line can only cross the graph of a function at most once. For example, the graph of

$x^2 + y^2 = 1$

is a circle centered at the origin with a radius of one. The y-axis is a vertical line through this graph that crosses it twice. This indicates that the relationship between x and y cannot be called a function.

Functions are frequently written in the form where a single variable (e.g., y, called the *dependent* variable) is on one side of an equals sign and an arithmetic expression in another variable (e.g., x, called the *independent* variable) is on the other side. Often this is written as

$y = f(x)$ or $z = g(x)$

and the variable in the parentheses is the independent variable. This type of function is said to be a function of a single variable, because there is only one independent variable. It is also possible to have functions of several (independent) variables — the only restriction is that for any set of values that the independent variables take on, there is only one value for the dependent variable.

Functions of a single variable are sometimes presented as a special subclass of all binary relations in $A \times B$ where A contains the domain (possible values of the independent variable) and B contains the range (possible values of the dependent variable). A binary relation R is merely a subset of $A \times B$. If $(a, b) \in R$, the relation is sometimes written as aRb, i.e., "a is R-related to b." Not every binary relation in $A \times B$ is a function — only those sets in which no two distinct ordered pairs have the same value for the first coordinate are functions. If R is a relation in $A \times B$, then R', the *complementary relation* to R, is

$$\{(x, y) \mid (x, y) \in A \times B \quad \text{and} \quad (x, y) \notin R\}$$

and R^{-1}, the *inverse* of R, is

$$\{(y, x) \mid (x, y) \in R\}.$$

A function is *evaluated* when a specific value is substituted for the independent variable. If

$$y = f(x) = 3x^2, \quad \text{then} \quad f(2) = 3 \cdot 2^2 = 12.$$

For a specific function, certain values may not be possible in the domain or the range. For example, the range of $|x|$ does not include negative numbers, and (assuming real numbers only) the domain of $\sqrt{x-3}$ does not include any number less than 3 (since that would cause the argument of the radical to be negative).

Solving a function or *finding the roots of a function* or *finding the zeroes of a function* consists in finding those values of the independent variable(s) (e.g., x) that cause the function to evaluate to zero. In general, the function is set equal to zero, and then the equation is solved for the independent variable. For example, if

$$f(x) = 2x + 1,$$

we set $2x + 1 = 0$ and obtain

$$2x = -1 \quad \text{or} \quad x = -1/2$$

as the solution value. To solve a quadratic function

$$f(x) = ax^2 + bx + c,$$

we use the quadratic formula:

$$x = \frac{-b \pm \sqrt{b^2 - 4ac}}{2a}.$$

A *linear function* is one whose graph is a straight line. Its variables are raised only to the power one. In general, it has the form of

$$y = mx + b.$$

The value of m indicates the slope of the line and the value of b indicates where the line intersects the y-axis.

An *inverse relationship* (denoted f^{-1}) for a function f is one that takes a value, y, in the *range* of f and gives back the value, x, in the *domain* of f, such that $f(x) = y$. Sometimes the inverse relationship is single-valued and thus is a true function, but often it is many-valued. As an example, the inverse function, f^{-1} for

$$f(x) = 2x - 1, \quad \text{is} \quad f^{-1}(w) = \frac{w+1}{2}.$$

We easily see that $f(2) = 2 \cdot 2 - 1 = 3$ and that

$$f^{-1}(3) = \frac{3+1}{2} = \frac{4}{2} = 2.$$

One obtains the inverse by solving the equation $y = f(x)$ for x in terms of y.

To determine whether any point (a, b) is on a curve determined by some function or relation, one merely substitutes the values in the equation for the appropriate variables. For example, to determine whether $(1, 2)$ and $(2, 4)$ are on the curve $y = x^2$, the first coordinate of each pair is substituted for x and the second for y. The first pair yields $2 = 1$ (an impossibility), indicating that $(1, 2)$ is not on the curve, and the second pair yields $4 = 4$, indicating that $(2, 4)$ is on the curve.

When graphing relations and functions, several helpful hints should be kept in mind.

- Any straight line can be determined by only two points. Thus, to graph a linear function, one needs only to determine two points on the line, and draw a straight line through those two points.

- One should locate major points first, for example, where $x = 0$ and (if it can easily be done) where $y = 0$. Computing a small chart with key values of the independent variable (e.g., $-2, -1, 0, 1, 2$) and the computed values of the dependent (or other variable) can be helpful to locate certain points on the graph.

- One should identify any characteristic about the relation that might help in graphing, for example, the functions $y = |x|$ and $y = x^2$ (both result in non-negative values). Thus, the graph of either function should always be above the x-axis.

It is possible to combine two functions by letting the output value (dependent variable value) of one function become the input value (independent variable value) of another function. Such a combination is called a *composition* of functions. If

$$f: X \rightarrow Y \quad \text{and} \quad g: Y \rightarrow Z,$$

then $g \circ f$ is the *composition* of f and g. The domain of $g \circ f$ is X and its range is Z, i.e.,

$$g \circ f : X \to Z,$$

and $g \circ f(x)$ is defined by $g(f(x))$. For example, if

$$f(x) = 3x^2 + 1 \quad \text{and} \quad g(x) = x - 2,$$

then $\quad g \circ f(x) = (3x^2 + 1) - 2 = 3x^2 - 1$

and $\quad f \circ g(x) = 3(x - 2)^2 + 1 = 3(x^2 - 4x + 4) + 1$

$$= 3x^2 - 12x + 12 + 1 = 3x^2 - 12x + 13.$$

● **PROBLEM** 7-1

Let the domain of $M = \{(x,y): y = x\}$ be the set of real numbers. Is M a function?

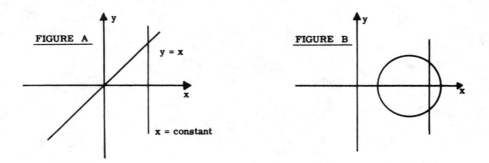

FIGURE A   y = x   x = constant

FIGURE B

<u>Solution:</u> The range is also the set of real numbers since $y = \{y|y = x\}$. The graph of $y = x$ is the graph of a line ($y = mx + b$ where $m = 1$ and $b = 0$). See fig. A. If for every value of x in the domain, there corresponds only one y value then y is said to be a function of x. Since each element in the domain of M has exactly one element for its image, M is a function. Also notice that a vertical line (x = constant) crosses the graph $y = x$ only once. Whenever this is true the graph defines a function. Consult figure B.
The vertical line (x = constant) crosses the graph of the circle twice; i.e., for each x,y is not unique, therefore the graph does not define a function.

● **PROBLEM** 7-2

If $f(x) = (x - 2)/(x + 1)$, find the function values $f(2)$, $f(\frac{1}{2})$, and $f(- 3/4)$.

<u>Solution:</u> To find f(2), we replace x by 2 in the given formula for f(x), f(x) = x - 2/x + 1; thus

$$f(2) = \frac{2 - 2}{2 + 1} = \frac{0}{3} = 0.$$

Similarly, $f(\frac{1}{2}) = \frac{\frac{1}{2} - 2}{\frac{1}{2} + 1}$.

Multiply numerator and denominator by 2,

$$= \frac{2(\frac{1}{2} - 2)}{2(\frac{1}{2} + 1)} \cdot$$

Distribute, $\quad = \frac{2(\frac{1}{2}) - 2 \cdot 2}{2(\frac{1}{2}) + 2}$

$$= \frac{1 - 4}{1 + 2}$$

$$= - \frac{3}{3} = - 1.$$

$$f(- 3/4) = \frac{- 3/4 - 2}{- 3/4 + 1} \cdot$$

Multiply numerator and denominator by 4,

$$= \frac{4(- 3/4 - 2)}{4(- 3/4 + 1)} \cdot$$

Distribute, $\quad = \frac{4(- 3/4) - 4(2)}{4(- 3/4) + 4(1)}$

$$= \frac{- 3 - 8}{- 3 + 4}$$

$$= \frac{- 11}{1}$$

$$= - 11.$$

● **PROBLEM** 7-3

If $g(x) = x^2 - 2x + 1$, find the given element in the range.

  a) g(-2)   b) g(0)   c) g(a + 1)   d) g(a - 1)

<u>Solution:</u> a) To find g(-2), substitute -2 for x in the given equation.

$$g(x) = g(-2)$$

$$= (-2)^2 - 2(-2) + 1$$

$$= 4 + 4 + 1$$

$$= 8 + 1$$

$$= 9$$

Hence, $g(-2) = 9$

      b) To find $g(0)$, substitute 0 for x in the given equation.

$$g(x) = g(0)$$

$$= (0)^2 - 2(0) + 1$$

$$= 0 - 0 + 1$$

$$= 1$$

Hence, $g(0) = 1$

      c) To find $g(a + 1)$, substitute $a + 1$ for x in given equation.

$$g(x) = g(a + 1)$$

$$= (a + 1)^2 - 2(a + 1) + 1$$

$$= (a^2 + 2a + 1) - 2a - 2 + 1$$

$$= a^2 + \cancel{2a} + 1 - \cancel{2a} - 2 + 1$$

$$= a^2 + 1 - 2 + 1$$

$$= a^2 + 0$$

$$= a^2$$

Hence, $g(a + 1) = a^2$.

      d) To find $g(a - 1)$, substitute $a - 1$ for x in given equation.

$$g(x) = g(a - 1)$$

$$= (a - 1)^2 - 2(a - 1) + 1$$

$$= (a^2 - 2a + 1) - 2a + 2 + 1$$

$$= a^2 - 2a + 1 - 2a + 2 + 1$$

$$= a^2 - 4a + 4$$

Hence, $g(a - 1) = a^2 - 4a + 4$

Find the domain D and the range R of the function $\left(x, \frac{x}{|x|}\right)$.

Solution: Note that the y-value of any coordinate pair (x,y) is $\frac{x}{|x|}$. We can replace x in the formula $\frac{x}{|x|}$ with any number except 0, since the denominator, $|x|$, can not equal 0, (i.e. $|x| \neq 0$) which is equivalent to $x \neq 0$. This is because division by 0 is undefined. Therefore, the domain D is the set of all real numbers except 0. If x is negative, i.e. $x < 0$, then $|x| = -x$ by definition. Hence, if x is negative, then $\frac{x}{|x|} = \frac{x}{-x} = -1$. If x is positive, i.e. $x > 0$, then $|x| = x$ by definition. Hence, if x is positive, then $\frac{x}{|x|} = \frac{x}{x} = 1$. (The case where $x = 0$ has already been found to be undefined). Thus, there are only two numbers -1 and 1 in the range R of the function; that is, $R = \{-1, 1\}$.

Describe the domain and range of the function
$f = (x,y) | y = \sqrt{9 - x^2}\}$ if x and y are real numbers.

Solution: In determining the domain we are interested in the values of x which yield a real value for y. Since the square root of a negative number is not a real number, the domain is restricted to those values of x which make the radicand positive or zero. Therefore x^2 cannot exceed 9 which means that x cannot exceed 3 or be less than − 3. A convenient way to express this is to write − 3 ≤ x ≤ 3, which is read "x is greater than or equal to −3 and less than or equal to 3." This is the domain of the function the range is the set of values that y can assume. To determine the range of the function we note that the largest value of y occurs when x = 0. Then $y = \sqrt{9 - 0} = 3$. Likewise, the smallest value of y occurs when x = 3 or x = − 3. Then $y = \sqrt{9 - 9} = 0$. Since this is an inclusive interval of the real axis, the range of y is $0 \leq y \leq 3$.

If $f(x) = 3x + 4$ and $D = \{x | -1 \leq x \leq 3\}$, find the range of $f(x)$.

Solution: We first prove that the value of 3x + 4 increases when x increases. If X > x, then we may multiply both sides of the inequality by a positive number to obtain an equivalent inequality. Thus, 3X > 3x. We may also add a number to both sides of the inequality to obtain an equivalent inequality. Thus

$$3X + 4 > 3x + 4.$$

Hence, if x belongs to D, the function value f(x) = 3x + 4 is least when x = -1 and greatest when x = 3. Consequently, since f(-1) = -3 + 4 = 1 and f(3) = 9 + 4 = 13, the range is all y from 1 to 13; that is,

$$R = \{y \mid 1 \leq y \leq 13\}.$$

● **PROBLEM 7-7**

Find the zeros of the function f if f(x) = 3x - 5.

Solution: The zeros of the function f(x) = 3x - 5 are those values of x for which 3x - 5 = 0:

$$3x - 5 = 0$$
$$3x = 5$$
$$x = \frac{5}{3}$$

$$f(x) = 3x - 5$$

Thus x = 5/3 is a zero of f(x) = 3x - 5, which means that the graph of f(x) crosses the x axis at the point (5/3, 0) (see figure).

● **PROBLEM 7-8**

Find the zeros of the function

$$\frac{2x + 7}{5} + \frac{3x - 5}{4} + \frac{33}{10} .$$

Solution: Let the function f(x) be equal to $\frac{2x + 7}{5} + \frac{3x - 5}{4} + \frac{33}{10}$.

95

A number, a, is a zero of a function $f(x)$ if $f(a) = 0$. A zero of $f(x)$ is a root of the equation $f(x) = 0$. Thus, the zeros of the function are the roots of the equation

$$\frac{2x + 7}{5} + \frac{3x - 5}{4} + \frac{33}{10} = 0.$$

The least common denominator, LCD, of the denominators of 5, 4, and 10 is 20. This is a fractional equation which can be solved by multiplying both members of the equation by the LCD.

$$20\left(\frac{2x + 7}{5} + \frac{3x - 5}{4} + \frac{33}{10}\right) = (20)(0)$$

$$4(2x + 7) + 5(3x - 5) + (2 \cdot 33) = 0.$$

Distributing,

$$8x + 28 + 15x - 25 + 66 = 0.$$

$$23x + 69 = 0$$

$$23x = -69$$

$$x = -3$$

Hence $x = -3$ is the zero of the given function.

● PROBLEM 7-9

Find the zeros of $f(x) = 2x^2 + 4x + 8$.

Solution: A zero of a function $f(x)$ is that value of x for which $f(x) = 0$.

Hence, to find the zeros of the given function, set $f(x) = 0$ and solve for x.

$$f(x) = 2x^2 + 4x + 8 = 0$$

$$x^2 + 2x + 4 = 0$$

$$x = \frac{-b \pm \sqrt{b^2 - 4ac}}{2a}$$

$$= \frac{-2 \pm \sqrt{4 - 4 \times 1 \times 4}}{2 \cdot 1}$$

$$= \frac{-2 \pm \sqrt{-12}}{2} = -1 \pm i\sqrt{3}$$

Therefore, the zeros of $f(x)$ are

$$-1 + i\sqrt{3} \quad \text{and} \quad -1 - i\sqrt{3}.$$

● PROBLEM 7-10

If $y = f(x) = (x^2 - 2)/(x^2 + 4)$ and $x = t + 1$, express y as a function of t.

Solution: y is given as a function of x, $y = f(x) = \dfrac{x^2 - 2}{x^2 + 4}$.

To express y as a function of t, replace x by t + 1 (since x = t + 1) in the formula for y. Thus,

$$y = f(x) = f(t + 1) = \frac{(t + 1)^2 - 2}{(t + 1)^2 + 4}$$

$$= \frac{\left(t^2 + 2t + 1\right) - 2}{\left(t^2 + 2t + 1\right) + 4}$$

$$= \frac{t^2 + 2t - 1}{t^2 + 2t + 5}$$

$$= g(t).$$

Hence, y = g(t); that is, y is now a function of t since y has been expressed in terms of t.

● **PROBLEM** 7-11

If $f(x) = x^2 - x - 3$, $g(x) = \left(x^2 - 1\right)/\left(x + 2\right)$, and $h(x) = f(x) + g(x)$, find h(2).

Solution:

h(x) = f(x) + g(x), and we are told that f(x) = $x^2 - x - 3$ and g(x) = $x^2 - 1/x + 2$; thus h(x) = $\left(x^2 - x - 3\right) + \left(x^2 - 1\right)/\left(x + 2\right)$.

To find h(**2**), we replace x by 2 in the above formula for h(x),

$$h(2) = \left[(2)^2 - 2 - 3\right] + \left(\frac{2^2 + 1}{2 + 2}\right)$$

$$= (4 - 2 - 3) + \left(\frac{4 - 1}{4}\right)$$

$$= (- 1) + \left(\frac{3}{4}\right)$$

$$= - \frac{4}{4} + \frac{3}{4}$$

$$= - \frac{1}{4} .$$

Thus, h(2) = - ¼.

$$= - \frac{1}{4} .$$

Thus, h(2) = - ¼.

Let $f(x) = 2x^2$ with domain $D_f = R$ (or, alternatively, C) and $g(x) = x - 5$ with $D_g = R$ (or C). Find (a) $f + g$ (b) $f - g$ (c) fg (d) $\frac{f}{g}$.

<u>Solution:</u> (a) $f + g$ has domain R (or C) and

$$(f + g)(x) = f(x) + g(x) = 2x^2 + x - 5$$

for each number x. For example, $(f + g)(1) = f(1) + g(1) = 2(1)^2 + 1 - 5 = 2 - 4 = -2$.

(b) $f - g$ has domain R (or C) and

$$(f - g)(x) = f(x) - g(x) = 2x^2 - (x - 5) = 2x^2 - x + 5$$

for each number x. For example, $(f - g)(1) = f(1) - g(1) = 2(1)^2 - 1 + 5 = 2 + 4 = 6$.

(c) fg has domain R (or C) and

$$(fg)(x) = f(x) \cdot g(x) = 2x^2 \cdot (x - 5) = 2x^3 - 10x^2$$

for each number x. In particular, $(fg)(1) = 2(1)^3 - 10(1)^2 = 2 - 10 = -8$.

(d) $\frac{f}{g}$ has domain R (or C) excluding the number x = 5 (when x = 5, g(x) = 0 and division by zero is undefined) and

$$\left(\frac{f}{g}\right)(x) = \frac{f(x)}{g(x)} = \frac{2x^2}{x-5}$$

for each number $x \neq 5$. In particular, $\left(\frac{f}{g}\right)(1) = \frac{2(1)^2}{1 - 5}$

$= \frac{2}{-4} = -\frac{1}{2}$.

If $D = \{ x \mid x$ is an integer and $-2 \leq x \leq 1 \}$, find the function $\{ (x, f(x)) \mid f(x) = x^3 - 3$ and x belongs to D $\}$.

<u>Solution:</u> $D = \{ -2, -1, 0, 1 \}$. Substituting these values of x in the equation $f(x) = x^3 - 3$, we find the corresponding f(x) values. Thus,

$$f(-2) = (-2)^3 - 3 = -8 - 3 = -11$$

$$f(-1) = (-1)^3 - 3 = -1 - 3 = -4$$

$$f(0) = 0^3 - 3 = 0 - 3 = -3$$

and
$$f(1) = 1^3 - 3 = 1 - 3 = -2.$$

Hence, $f = \left\{ \left(x, f(x)\right) \mid f(x) = x^3 - 3 \text{ and } x \text{ belongs to D} \right\}$

$$= \left\{ (-2, -11), (-1, -4), (0, -3), (1, -2) \right\}$$

● PROBLEM 7-14

Define and give an example of

   (1) a single-valued relation

   (2) a many-valued relation.

Solution: (1) A single-valued function is a function in which for every value of x, there is one and only one value of y. ex. $y = x^2 - 3x + 4$

(2) A many-valued function is a function in which there is more than one y for each x. ex. $y = \sqrt{x^2} \implies y = \pm x$

● PROBLEM 7-15

Answer the following questions:

   (1) What is a linear function?

   (2) Give an example of a linear function and an example of a non-linear function.

   (3) Give an example of a function of 3 variables.

Solution: (1) A straight line is formed by the graph of $y = ax + b$, where a and b are constants. Hence, $f(x) = ax + b$ is called a linear function.

(2) Linear function: $y = 6x + 3/2$

    Non-linear function: $y = 2x^2 + 3$

(3) $y = 3x^2 + 2y + z$

Let f be the linear function that is defined by the equation $f(x) = 3x + 2$. Find the equation that defines the inverse function f^{-1}.

Solution: To find the inverse function f^{-1}, the given equation must be solved for x in terms of y. Let $x = f^{-1}(y)$.

Solving the given equation for x:

$$y = 3x + 2, \text{ where } y = f(x).$$

Subtract 2 from both sides of this equation:

$$y - 2 = 3x + \cancel{2} - \cancel{2}$$

$$y - 2 = 3x.$$

Divide both sides of this equation by 3:

$$\frac{y - 2}{3} = \frac{3x}{3}$$

$$\frac{y - 2}{3} = x$$

or

$$x = \frac{y}{3} - \frac{2}{3}$$

Hence, the inverse function f^{-1} is given by::

$$x = f^{-1}(y) = \frac{y}{3} - \frac{2}{3}$$

$$\text{or } x = f^{-1}(y) = \frac{1}{3}y - \frac{2}{3}.$$

Of course, the letter that we use to denote a number in the domain of the inverse function is of no importance whatsoever, so this last equation can be rewritten $f^{-1}(u) = \frac{1}{3}u - \frac{2}{3}$, or $f^{-1}(s) = \frac{1}{3}s - \frac{2}{3}$, and it will still define the same function f^{-1}.

Show that the inverse of the function $y = x^2 + 4x - 5$ is not a function.

Solution: Given the function f such that no two of its ordered pairs have the same second element, the inverse function f^{-1} is the set of ordered pairs obtained from f by interchanging in each ordered pair the first and second elements. Thus, the inverse of the function

$y = x^2 + 4x - 5$ is $x = y^2 + 4y - 5$.

The given function has more than one first component corresponding to a given second component. For example, if $y = 0$, then $x = -5$ or 1. If the elements $(-5,0)$ and $(1,0)$ are reversed, we have $(0,-5)$ and $(0,1)$ as elements of the inverse. Since the first component 0 has more than one second component, the inverse is not a function (a function can have only one y value corresponding to each x value).

● PROBLEM 7-18

Construct the graph of the function defined by $y = 3x - 9$.

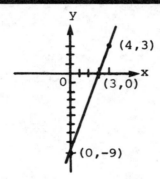

Solution: An equation of the form $y = mx + b$ is a linear equation; that is, the equation of a line.

A line can be determined by two points. Let us choose the intercepts. The x-intercept lies on the x-axis and the y-intercept is on the y-axis.

We find the intercepts by assigning 0 to x and solving for y and by assigning 0 to y and solving for x. It is helpful to have a third point. We find a third point by assigning 4 to x and solving for y. Thus we get the following table of corresponding numbers:

x	y = 3x - 9	y
0	y = 3(0) - 9 = 0 - 9 =	-9
4	y = 3(4) - 9 = 12 - 9 =	3

Solving for x to get the x-intercept:

$y = 3x - 9$

$y + 9 = 3x$

$x = \dfrac{y + 9}{3}$

When $y = 0$, $x = \dfrac{9}{3} = 3$. The three points are $(0,-9)$, $(4,3)$, and $(3,0)$. Draw a line through them (see sketch).

101

Are the following points on the graph of the equation 3x-2y=0?
(a) point (2,3)? (b) point (3,2)? (c) point (4.6)?

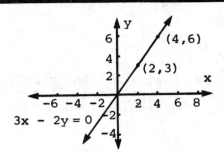

Solution: The point (a,b) lies on the graph of the equation
3x - 2y = 0 if replacement of x and y by a and b, respectively, in the given equation results in an equation which is true.

(a) Replacing (x,y) by (2,3):

$$3x - 2y = 0$$
$$3(2) - 2(3) = 0$$
$$6 - 6 = 0$$
$$0 = 0 , \text{ which is true.}$$

Therefore (2,3) is a point on the graph.

(b) Replacing (x,y) by (3,2):

$$3x - 2y = 0$$
$$3(3) - 2(2) = 0$$
$$9 - 4 = 0$$
$$5 = 0 , \text{ which is not true.}$$

Therefore (3,2) is not a point on the graph.

(c) Replacing (x,y) by (4,6):

$$3x - 2y = 0$$
$$3(4) - 2(6) = 0$$
$$12 - 12 = 0$$
$$0 = 0 , \text{ which is true.}$$

Therefore (4,6) is a point on the graph.

This problem may also be solved geometrically as follows:
draw the graph of the line 3x - 2y = 0 on the coordinate
axes. This can be done by solving for y:

$$3x - 2y = 0$$
$$-2y = -3x$$
$$y = \frac{-3}{-2}x = \frac{3}{2}x ,$$

and plotting the points shown in the following table:

x	$y = \frac{3}{2} x$
0	0
1	$3/2 = 1\frac{1}{2}$
2	3
-2	-3

(See accompanying figure.)

Observe that we obtain the same result as in our algebraic solution. The points (2,3) and (4,6) lie on the line $3x - 2y = 0$, whereas (3,2) does not.

● **PROBLEM** 7-20

Sketch the graph of the following functions:

(a) $f(x) = \begin{cases} x & x \le 4 \\ \frac{1}{2}x - 2 & x > 4 \end{cases}$

(b) $f(x) = |4x + 9|$

(c) $f(x) = \frac{|x|}{x+1}$ $x \ne -1$

Solution:

(a)

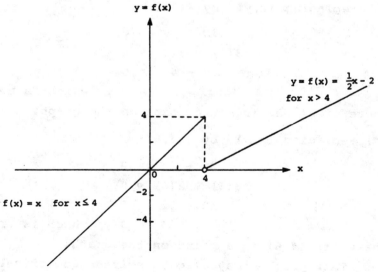

(b)

$$f(x) = |4x+9| = \begin{cases} 4x + 9 & x > -\frac{9}{4} \\ 0 & x = -\frac{9}{4} \\ -(4x+9) & x < -\frac{9}{4} \end{cases}$$

103

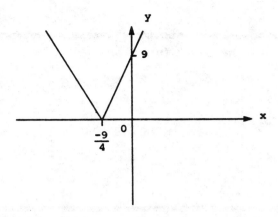

In this case we are graphing the equation of a straight line. However, because the equation is found under the absolute value signs our line never crosses the x-axis. When our values for y would have been negative the absolute value is taken - so we get a positive y.

(c)

$$f(x) = \frac{|x|}{x+1} = \begin{cases} \dfrac{x}{x+1} & x > 0 \\[2mm] 0 & x = 0 \\[2mm] \dfrac{-(x)}{x+1} & x < 0 , \quad x \neq -1 \end{cases}$$

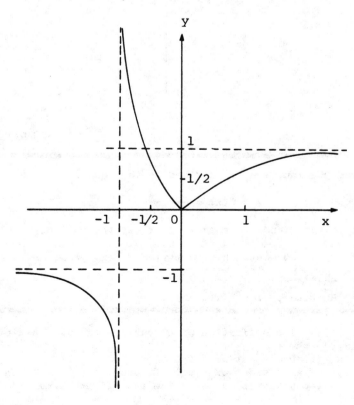

104

Draw a graph of the step function

$$y = \begin{cases} 2 & -2 \le x < -1 \\ 1 & -1 \le x < 0 \\ 0 & 0 \le x < 1 \\ 1 & 1 \le x < 2 \\ 2 & 2 \le x \le 3 \end{cases}$$

Solution:

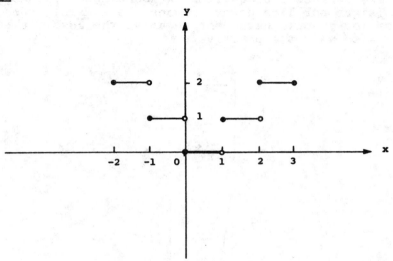

Given the function f defined by the equation

$$y = f(x) = \frac{3x + 4}{5} , \qquad (1)$$

where the domain (and the range) of f is the set R of all real numbers.

(a) Find the equation $x = g(y) = f^{-1}(x)$ that defines f^{-1} .

(b) Show that $f^{-1}\big(f(x)\big) = x$.

(c) Show that $f\big(f^{-1}(x)\big) = y$.

Solution: (a) The definition of a function f is a set of ordered pairs (x,y) where
1) x is an element of a set X
2) y is an element of a set Y, and
3) no two pairs in f have the same first element.

105

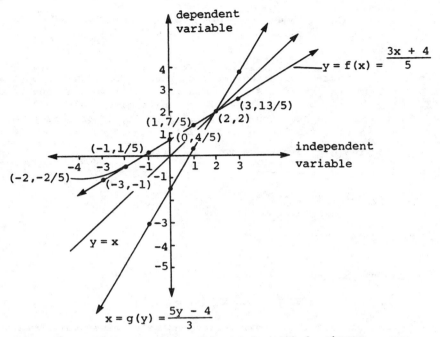

By definition, f is the infinite set of ordered pairs

$$\{(x, \frac{3x + 4}{5}) \mid x \in R\},$$

which includes (0,4/5), (2,2),(7,5),(12,8), (-3,-1), etc. Furthermore, no two ordered pairs have the same first element . That is, for each element of X a unique value of Y is assigned. For example, if x = 0, we obtain only one y value, 4/5.

We construct the following table to calculate the x and corresponding y values. Note that x is the independent variable and y is the dependent variable.

x	$\frac{3x+4}{5}$ =	y	(x,y)
-3	$\frac{3(-3)+4}{5} = \frac{-5}{5}$	-1	(-3,-1)
-2	$\frac{3(-2)+4}{5} = \frac{-2}{5}$	$\frac{-2}{5}$	$\left(-2, \frac{-2}{5}\right)$
-1	$\frac{3(-1)+4}{5} = \frac{1}{5}$	$\frac{1}{5}$	$\left(-1, \frac{1}{5}\right)$
0	$\frac{3(0)+4}{5} = \frac{4}{5}$	$\frac{4}{5}$	$\left(0, \frac{4}{5}\right)$
1	$\frac{3(1)+4}{5} = \frac{7}{5}$	$\frac{7}{5}$	$\left(1, \frac{7}{5}\right)$
2	$\frac{3(2)+4}{5} = \frac{10}{5}$	2	(2,2)
3	$\frac{3(3)+4}{5} = \frac{13}{5}$	$\frac{13}{5}$	$\left(3, \frac{13}{5}\right)$

See the accompanying figure, which shows the graph of the function f (which is also the graph of the equation $y = \frac{3x+4}{5}$). We can say that f   carries (or maps) any real number   x   into the number $\frac{3x+4}{5}$:

$$f: x \to \frac{3x + 4}{5} .$$

Now to find the inverse function, we must find a function which takes each element of the original set   Y   and relates it to a unique value of   X.   There cannot be two values of   X   for a given value of   Y   in order for the inverse function to exist.   That is, if this is true: $\left(x_1, y\right)$ and $\left(x_2, y\right)$,   then there is no f^{-1}.

To find $x = g(y)$,   we solve for   x   in terms of   y.

Given: $\qquad\qquad\qquad y = \frac{3x + 4}{5}$

Multiply both sides by   5 ,

$$5y = 3x + 4$$

Subtract   4   from both sides,

$$5y - 4 = 3x$$

Divide by   3   and solve for   x,

$$x = \frac{5y - 4}{3} = f^{-1}(x) = g(y) .$$

Choose   y   values and find their corresponding x-values, as shown in the following table.   Note that   y   is the independent variable and   x   is the dependent variable.

y	$g(y)=\frac{5y-4}{3} =$	x	(y,x)
-3	$\frac{5(-3)-4}{3}$	$\frac{-19}{3} = -6\frac{1}{3}$	$\left(-3, -6\frac{1}{3}\right)$
-2	$\frac{5(-2)-4}{3}$	$\frac{-14}{3} = -4\frac{2}{3}$	$\left(-2, -4\frac{2}{3}\right)$
-1	$\frac{5(-1)-4}{3}$	$\frac{-9}{3} = -3$	$(-1, -3)$
0	$\frac{5(0)-4}{3}$	$\frac{-4}{3} = -1\frac{1}{3}$	$\left(0, -1\frac{1}{3}\right)$
1	$\frac{5(1)-4}{3}$	$\frac{1}{3}$	$\left(1, \frac{1}{3}\right)$
2	$\frac{5(2)-4}{3}$	2	$(2, 2)$
3	$\frac{5(3)-4}{3}$	$\frac{11}{3} = 3\frac{2}{3}$	$\left(3, 3\frac{2}{3}\right)$

See graph.   Since there is only one value of   y   for each value of   x, this equation defines the inverse function f^{-1}.   The graph of f^{-1} is the image of the graph of   f   in the mirror   y = x.

(b)   Given $f(x) = \frac{3x + 4}{5} = y.$

Then perform the operation of f^{-1} on   y = f(x)   where $f^{-1}(x) = \frac{5y - 4}{3}$.   That is, substitute for   y: $\frac{3x + 4}{5}$.

$$f^{-1}\big(f(x)\big) = f^{-1}\left(\frac{3x + 4}{5}\right) = \frac{5\left(\frac{3x + 4}{5}\right) - 4}{3} = \frac{\frac{15x + 20}{5} - 4}{3}$$

$$= \frac{\frac{15x + 20 - 20}{5}}{3} = \frac{3x}{3} = x , \qquad \text{or}$$

$$f^{-1}\!\left(f(x)\right) = \frac{5f(x) - 4}{3} = \frac{5\!\left(\frac{3x + 4}{5}\right) - 4}{3} = x .$$

(c) We now perform the operation of f on $f^{-1}(x)$. Substitute for $f^{-1}(x)$: $\frac{5y - 4}{3} = x$. Note $f(x) = \frac{3x + 4}{5}$

$$f\!\left(f^{-1}(x)\right) = f\!\left(\frac{5y - 4}{3}\right) = \frac{3\!\left(\frac{5y - 4}{3}\right) + 4}{5}$$

$$= \frac{5y - 4 + 4}{5} = \frac{5y}{5} = y = f(x)$$

Comment. Since $f = \left\{\left(x, \frac{3x + 4}{5}\right)\right\}$, this function f may be thought of as a sequence of directions listing the operations that must be performed on x to get $\frac{3x + 4}{5}$. These operations are, in order: take any number x, multiply it by 3, add 4, and then divide by 5. The inverse function

$$f^{-1} = \left\{\left(y, \frac{5y - 4}{3}\right)\right\}$$

tells us to multiply by 5, subtract 4, and then divide by 3. This "undoes," in reverse order, the operations performed by f. The function f^{-1} could be called "the undoing function" because it undoes what the function f has done.

● **PROBLEM 7-23**

Graph the function $y = x^3 - 9x$.

Solution: Choosing values of x in the interval $-4 \leq x \leq 4$, we have for $y = x^3 - 9x$,

x	- 4	-3	- 2	-1	0	1	2	3	4
y	-28	0	10	8	0	-8	-10	0	28

Notice that for each ordered pair (x,y) listed in the table there exists a pair $(-x,-y)$ which also satisfies the equation, indicating symmetry with respect to the origin. To

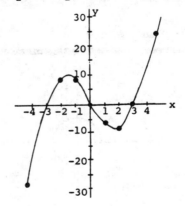

prove that this is true for all points on the curve, we substitute (-x,-y) for (x,y) in the given equation and show that the equation is unchanged. Thus

$$-y = (-x)^3 - 9(-x) = -x^3 + 9x$$

or, multiplying each member by -1,

$$y = x^3 - 9x$$

which is the original equation.

The curve is illustrated in the figure. The domain and range of the function have no restrictions in the set of real numbers. The x-intercepts are found from

$$y = 0 = x^3 - 9x$$

$$0 = x(x^2 - 9)$$

$$0 = x(x - 3)(x + 3)$$

$$x = 0 \quad | \quad x - 3 = 0 \quad | \quad x + 3 = 0$$

$$x = 3 \quad \qquad x = -3.$$

The curve has three x-intercepts at x = -3, x = 0, x = 3. This agrees with the fact that a cubic equation has three roots. The curve has a single y-intercept at y = 0 since for x = 0, $y = 0^3 - 9(0) = 0$.

● PROBLEM 7-24

Sketch the following binary relations in A × A, where A = {All real numbers}.

    (1) R_1: $x^2 + y^2 = 4$

    (2) R_2: $x = 4y^2$

    (3) R_3: $\dfrac{x^2}{4} - y^2 = 1$

Are R_1, R_2, and R_3 functions?

Solution:

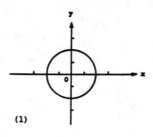

(1)

109

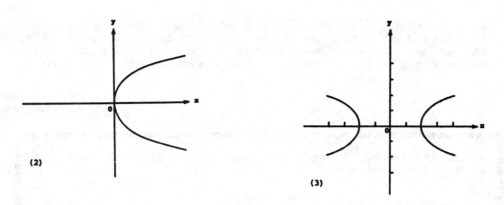

(2)

(3)

A binary relation R, from set A to set B is a subset of
A × B, and a function from A to B is a binary relation such
that with each element of A there is associated in some way
exactly one element of another set B. Thus, to determine
whether the given relations are also functions is to determine whether the set of ordered pairs of a given relation has
any two ordered pairs with the same first coordinate. This
is done by drawing some vertical lines on the graph of a relation. If no vertical lines have more than one intersection
with the graph (which implies that there are no two ordered
pairs in the given relation having the same first coordinate),
then the given relation is a function. The procedure
described above is called the vertical-line test.

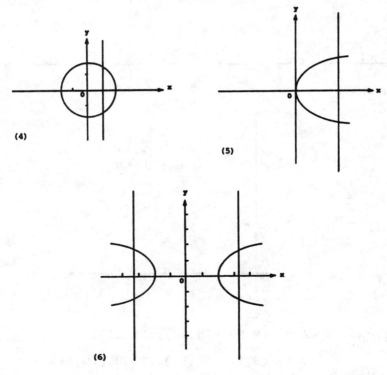

(4)

(5)

(6)

By the vertical-line test, it's seen that none of the given
relations are functions from X to Y, where X and Y are the
set of x-coordinates and y-coordinates respectively. Note

110

that relations (2) and (3) are functions from Y to X since no horizontal lines can have more than one intersection with the graphs of relations (2) and (3) whereas relation (3) is neither a function from X to Y nor a function from Y to X.

Given A = {9,10,11} and R is a relation in A × A, defined as R = {(a,b) | a = b}

(1) find R, and the complementary relation to R.

(2) draw the graphs of A × A, R, and R'.

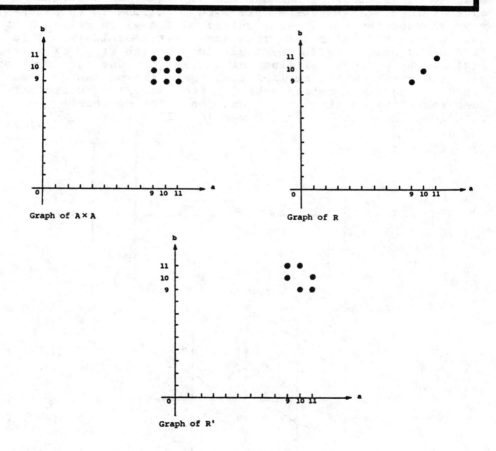

Graph of A × A

Graph of R

Graph of R'

Solution: A × A = {(9,9),(10,10),(11,11)

(9,10),(9,11),(10,9)(10,11)

(11,9)(11,10)}

From the definition of R, one obtains

111

$$R = \{(a,b) \mid a=b\} = \{(9,9),(10,10),(11,11)\} \text{ for}$$

$$R \subset A \times A.$$

By definition, if R is a relation in a cartesian-product set $S \times S$, then $R' = \{(x,y) \mid (x,y) \in S \times S \text{ and } (x,y) \notin R\}$ is the complementary relation to R.

In other words, R' is the set containing those ordered pairs (x,y) of $S \times S$ not in R. Therefore,

$$R' = \{(9,10),(9,11),(10,9),(10,11),(11,9),(11,10)\}$$

(2) The graphs of $A \times A$, R, and R' are shown in the figure.

● **PROBLEM 7-26**

Given the relation $R = \{(9,8),(10,9)(11,10)\}$ in the set $S \times S$, where $S = \{8,9,10,11\}$.

(1) Find the inverse of R, and the complementary relation to R.

(2) Find the domains and the ranges of R and R^{-1}.

(3) Sketch R, R^{-1}, and R'.

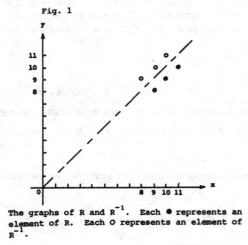

Fig. 1

The graphs of R and R^{-1}. Each ● represents an element of R. Each O represents an element of R^{-1}.

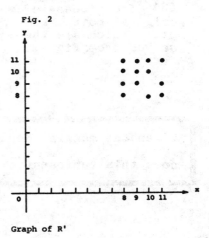

Fig. 2

Graph of R'

<u>Solution</u>: (1) Let R be a relation from set A to set B. The inverse of R, denoted by R^{-1}, is the relation from B to A consisting of those ordered pairs which, when reversed, belong to R.

$$R^{-1} = \{(y,x) \mid (x,y) \in R\}.$$

If R is a relation in a cartesian-product set $A \times A$, then the complementary relation to R, denoted by R', is the set containing those ordered pairs (x,y) of $A \times A$ not in R.

112

$$R' = \{(x,y) \mid (x,y) \notin R \quad \text{and} \quad (x,y) \in A \times A\}.$$

Note that R^{-1}, the inverse of R, is defined for a relation from any set A to any other set B. However, the complementary relation to R is defined for relations in a cartesian-product only!

The given R is $R = \{(9,8),(10,9),(11,10)\} = \{(x,y) \mid x = y+1\}$

If a relation R is represented by a certain defining condition then the corresponding condition defining R^{-1} is obtained by replacing x for y and y for x. Hence,

$$R^{-1} = \{(x,y) \mid y = x+1\} = \{(8,9),(9,10),(10,11)\}$$

$$R' = \{(8,8),(8,9),(8,10),(8,11),(9,9),(9,10),(9,11),$$

$$(10,10),(10,11),(10,8),(11,8),(11,9),(11,11)\}$$

$$= \{(x,y) \mid (x,y) \in S \times S \quad \text{and} \quad (x,y) \notin R\}$$

(2) The domain of a relation R is the set of all first elements of the ordered pairs which belong to R, and the range of R is the set of second elements. The domain of the given R is {9,10,11} and the range of R is {8,9,10}. For R^{-1}, the domain is { 8,9,10 } and the range is {9,10,11}. It is seen that as a result of the process of interchanging the components of the ordered pairs, the domain of R is the range of R^{-1} and the range of R is the domain of R^{-1}.

(3) The process of obtaining R^{-1} yields mirror images of the original points and creates correspondingly a set of ordered pairs which are the inverse relation. This is clearly observed from Fig. 1.

● PROBLEM 7-27

Given the mapping depicted in the following figure,

does this represent a function? Explain.

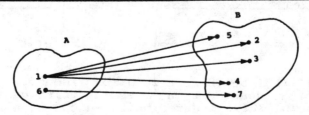

Solution: No. This mapping does not represent a function, because a function is a set of ordered pairs, no two of which have the same first coordinate, i.e. a function f associates with each element x of the domain exactly one element y of the range. Hence, a function cannot have more than one image for each element of the domain. In the given mapping from A to B, the element 1 in A has four images in B.

Which of the following graphs represent

    (a)   relations between x and y ?

    (b)   a function f: x → y ?

    (c)   a function f⁻¹: y → x ?

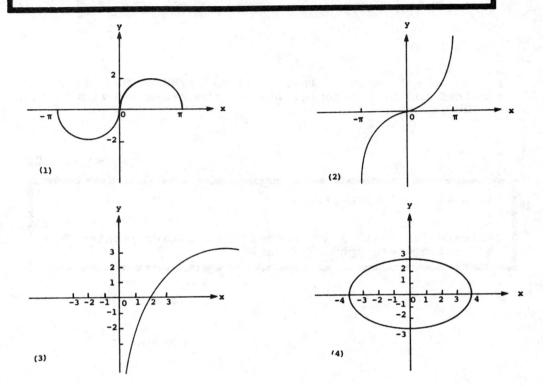

(1)   (2)

(3)   (4)

Solution: (a) If A and B are sets, then a binary relation, R, from A to B is a subset of A×B. If (x,y) ∈ R, it is said that x is R-related to y and denoted by x Ry. Here, A and B are the set of x-coordinates and y-coordinates respectively of all the points in the coordinate plane. Since each of the given graphs represents a certain set of points in the coordinate plane, each graph represents a subset of A×B. Therefore, all the given graphs represent relations between x and y.

(b) By definition, if to each element of a set A there is assigned a unique element of a set B; the collection of such assignments is called a function from A into B. Hence, graphs (1), (2) and (3) are functions f: x → y.

(c) Graph (2) and graph (3) represent functions f⁻¹: y → x. Note that graph (1) does not represent a function f⁻¹: y → x because for certain y the associated element x is not unique. Similarly graph (4) is not a function at all since for certain x (or y) the associated element y

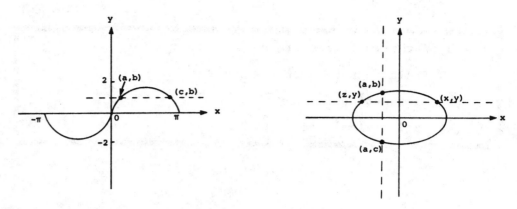

(or x) is not unique. This is illustrated by drawing a
vertical and/or horizontal line on the graphs as shown
in the figure.

● **PROBLEM** 7-29

Given the following graphs

Indicate (a) which graph represents a binary relation?
(b) which are graphs of functions?

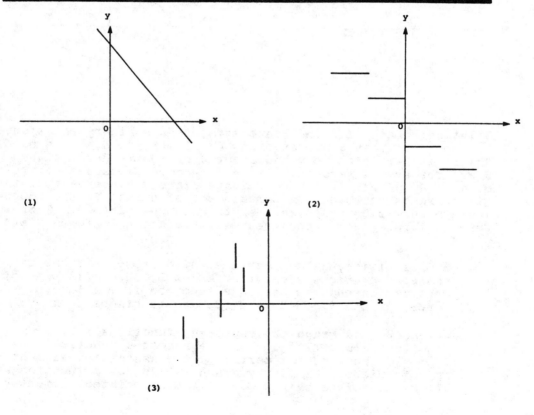

(1)

(2)

(3)

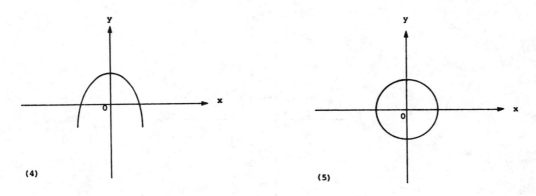

(4)   (5)

Solution: (a) By definition, a binary relation in A is a set of ordered pairs whose first and second coordinates are both members of set A. Here, A is the set of real numbers. Therefore, all the given graphs are graphs of binary relations.

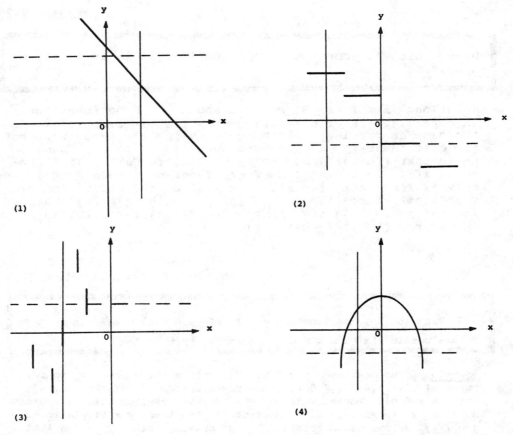

(1)   (2)

(3)   (4)

(b)   A function f from set A into set B is defined as a set of ordered pairs {(a,b) | a ∈ A and b ∈ B} no two of which have the same first coordinate. Hence, by the vertical-line test (see Figure), it's found that graphs (1), (2) and (4) are graphs of functions from X to Y where X

116

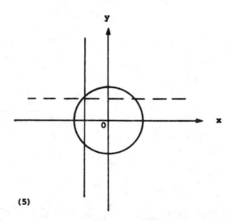

(5)

and Y are sets of x-coordinates and y-coordinates respectively. Graphs (1) and (3) are graphs of functions
from Y to X by the horizontal-line test (see Figure).

● PROBLEM 7-30

Prove that the composition of functions is an associative
operation.

Solution: Let f: A → B, g: B → C and h: C → D be functions.
Then proving that functional composition is associative is
the same as proving that $(h \circ g) \circ f = h \circ (g \circ f)$. Note that both
$(h \circ g) \circ f$ and $h \circ (g \circ f)$ are functions from A to D. Thus, to
prove that $((h \circ g) \circ f)(x) = (h \circ (g \circ f))(x)$ for all x in A (because if f: X → Y and g: X → Y are functions, then f = g if and
only if $f(x) = g(x)$ for all x in X.) By the definition of
functional composition, $((h \circ g) \circ f)(x) = (h \circ g)(f(x)) = h(g(f(x)))$
and $(h \circ (g \circ f))(x) = h(g \circ f(x)) = h(g(f(x)))$ for all x in A.
Therefore, $(h \circ g) \circ f = h \circ (g \circ f)$.

● PROBLEM 7-31

Find the inverse function of f if $y = f(x) = 2\sqrt{9 - x^2}$ and
f has domain $\{x | -3 \leq x \leq 0\}$ and range $\{y | 0 \leq y \leq 6\}$.

Solution: The equation $y = f(x)$ determines the number y in the
range of f, $\{y | 0 \leq y \leq 6\}$, from a given number x of the domain,
$\{x | -3 \leq x \leq 0\}$. Now we want to know if this equation also determines
x when y is given. If a function f is given by a simple formula
$y = f(x)$, we can often obtain f^{-1} by solving this for x so that
$x = f^{-1}(y)$. Therefore,

$$y = 2\sqrt{9 - x^2} = 2(9 - x^2)^{\frac{1}{2}}$$

Squaring both sides,

$$y^2 = 2^2\left[(9 - x^2)^{\frac{1}{2}}\right]^2 = 4(9 - x^2)$$

117

Distributing,
$$y^2 = 36 - 4x^2$$
$$y^2 - 36 = -4x^2$$
$$x^2 = \frac{y^2 - 36}{-4} = \frac{36 - y^2}{4}$$
$$x = \frac{\pm\sqrt{36 - y^2}}{2}$$

In order to obtain a function of y, we must have for each value of
y one and only one value of x. Thus, we choose only one sign.
Now the domain of the inverse function is the range of f, and the
range of the inverse function is the domain of f. Therefore, the
domain of $f^{-1}(y)$ is to be $0 \le y \le 6$ and the range is $-3 \le x \le 0$. Now
since x can assume negative values then we must consider the negative values of $\pm\sqrt{36 - y^2}$. Hence, the inverse function is

$$x = -\frac{\sqrt{36 - y^2}}{2}$$

Plot each function by selecting values in the domain and finding the
corresponding values in the range.

For $y = 2\sqrt{9 - x^2}$

Domain $\{x \mid -3 \le x \le 0\}$

x	$2\sqrt{9 - x^2}$	y
-3	$2\sqrt{9 - (-3)^2}$	0
-2	$2\sqrt{9 - (-2)^2}$	$2\sqrt{5}$
-1	$2\sqrt{9 - (-1)^2}$	$4\sqrt{2}$
0	$2\sqrt{9 - (0)^2}$	6

For $x = -\sqrt{36 - y^2}\big/2$

Domain $\{y \mid 0 \le y \le 6\}$

y	$-\sqrt{36 - y^2}/2$	x
0	$-\sqrt{36 - 0^2}/2$	-3
1	$-\sqrt{36 - 1^2}/2$	$-\sqrt{35}/2$
2	$-\sqrt{36 - 2^2}/2$	$-2\sqrt{2}$
3	$-\sqrt{36 - 3^2}/2$	$-3\sqrt{3}/2$
4	$-\sqrt{36 - 4^2}/2$	$-\sqrt{5}$
5	$-\sqrt{36 - 5^2}/2$	$-\sqrt{11}/2$
6	$-\sqrt{36 - 6^2}/2$	0

See graph

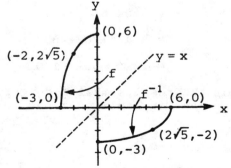

The graphs of f and f^{-1} show that if (a,b) is a point of the
graph of f, then (b,a) is a point of the graph of f^{-1}. Furthermore, this is true by the definition of an inverse function which
states that, given the function f such that no two of its ordered
pairs have the same second element, the inverse function f^{-1} is
the set of ordered pairs obtained from f by interchanging in each

118

ordered pair the first and second elements. The graphs of f and f⁻¹ are symmetric with respect to the line y = x; that is, each is the image of the other in the mirror y = x.

● PROBLEM 7-32

Let f: R → R and g: R → R be two functions, given by f(x) = 2x + 5 anf g(x) = 4x², respectively for all x in R, where R is the set of real numbers. Find expressions for the compositions (f ∘ g)(x) and (g ∘ f)(x).

Solution: Consider functions f: A → B and g: B → C - that is, where the co-domain of f is the domain of g. Then the function g∘f is defined as g∘f: A → C, where (g ∘ f)(x) = g(f(x)) for all x in A, and it is called the coposition of f and g.

In another notation,

g∘f = {(x,z) A × C | for all y ∈ B such that (x,y) ∈ f and (y,z) ∈ g}.

Therefore,

$$(f \circ g)(x) = f(g(x))$$

$$= f(4x^2)$$

$$= 2[4x^2] + 5$$

$$= 8x^2 + 5 \qquad \text{and}$$

$$(g \circ f)(x) = g(f(x))$$

$$= g(2x+5)$$

$$= 4[2x+5]^2$$

$$= 4(4x^2+20x+25)$$

$$= 16x^2 + 80x + 100$$

It's seen from the results that f∘g ≠ g∘f. This is true in general, i.e. functional compositions are not commutative operations.

CHAPTER 8

RATIOS, PROPORTIONS AND VARIATIONS

> **Basic Attacks and Strategies for Solving Problems in this Chapter. See pages 120 to 129 for step-by-step solutions to problems.**

A *ratio* is the expression of the relationship between two quantities. This relationship is usually expressed either as a fraction or by using a colon to separate the two quantities. For example, the ratio between one and two can be expressed as $^1/_2$ or as 1:2 (often read as "1 *to* 2").

The fractional notation leads to numerous rules about ratios. For example, if x:y is a:b, this is equivalent to the equation $\frac{x}{y} = \frac{a}{b}$, from which the equation $\frac{x}{a} = \frac{y}{b}$ can be derived. The "colon" notation can also be used to express two successive ratios, for example, x:y:z indicates x:y and y:z.

Problems involving ratios can often be solved by solving a simple fractional equation. For example, if it is known that the slope of the land is one foot in height for every 12 feet of horizontal distance, i.e., a ratio of 1:12, then to find the change in height for 40 feet (horizontal distance), one merely solves

$$\frac{1}{12} = \frac{x}{40} \text{ for } x.$$

The equivalence of two ratios is called a *proportion*, e.g., $\frac{a}{b} = \frac{c}{d}$. Given this proportion, several other *Laws of Proportion* also hold:

1. $ad = bc.$

2. $\dfrac{a}{c} = \dfrac{b}{d}.$

3. $\dfrac{b}{a} = \dfrac{d}{c}.$

4. $\dfrac{a+b}{b} = \dfrac{c+d}{d}.$

5. $\dfrac{a-b}{b} = \dfrac{c-d}{d}$.

6. $\dfrac{a+b}{a-b} = \dfrac{c+d}{c-d}$.

Each different quantity in a proportion is sometimes called a *proportional*, e.g., *a* is the first proportional in *a:b* = *b:c* and *c* is the *third* (since *b* appears twice).

A *constant of proportionality* is an invariant number that expresses the proportion of other varying quantities. Sometimes this is expressed by saying that "*y* varies directly as *x*." For example, in most types of soil, the change in temperature, say *T*, is proportional to the depth below surface level, say *d*, and this can be expressed by the equation

$T = \mathbf{k}d$

where **k** is some constant of proportionality.

One quantity, *y*, is said to be *inversely proportional* to another, *x* (or it is said that "*y* varies inversely as *x*"), if the relation can be expressed as $y = \mathbf{k}/x$ where **k** is a constant of proportionality.

In problems involving a constant of proportionality, it is sometimes necessary to first calculate the constant from given data, and then, using the constant, to compute other desired information.

Step-by-Step Solutions to Problems in this Chapter "Ratios, Proportions and Variations"

● PROBLEM 8-1

On a map, $\frac{3}{16}$ inch represents 10 miles. What would be the length of a line on the map which represents 96 miles?

Solution: The lengths of line segments on the map are proportional to the actual distances on the earth. If L represents the length of the line segment on the map corresponding to a distance of 96 miles, then

$$\frac{\frac{3}{16} \text{ inches}}{\text{L inches}} = \frac{10 \text{ miles}}{96 \text{ miles}}$$

$$\frac{3}{16}(96) = 10L \qquad \text{by cross multiplying}$$

$$L = \frac{(3)(96)}{(16)(10)} = \frac{3(6)}{10}$$

$$L = \frac{18}{10}$$

$$L = 1\frac{4}{5} \text{ inches.}$$

● PROBLEM 8-2

Find the ratios of x : y : z from the equations

$$7x = 4y + 8z, \qquad 3z = 12x + 11y.$$

Solution: By transposition we have

$$7x - 4y - 8z = 0$$

$$12x + 11y - 3z = 0.$$

To obtain the ratio of x : y we convert the given system into an equation in terms of just x and y. z may be eliminated as follows: Multiply each term of the first equation by 3, and each term of the second equation by 8, and then subtract the second equation from the first. We thus obtain:

$$21x - 12y - 24z = 0$$

$$\underline{- (96x + 88y - 24z = 0)}$$
$$-75x - 100y = 0$$

Dividing each term of the last equation by 25 we obtain,

$$- 3x - 4y = 0 \quad \text{or,}$$

$$- 3x = 4y.$$

Dividing both sides of this equation by 4, and by - 3, we have the proportion:

$$\frac{x}{4} = \frac{y}{-3} \, .$$

We are now interested in obtaining the ratio of y : z. To do this we convert the given system of e-quations into an equation in terms of just y and z, by eliminating x as follows: Multiply each term of the first equation by 12, and each term of the second e-quation by 7, and then subtract the second equation from the first. We thus obtain:

$$84x - 48y - 96z = 0$$

$$\underline{- (84x + 77y - 21z = 0)}$$
$$- 125y - 75z = 0.$$

Dividing each term of the last equation by 25 we obtain

$$- 5y - 3z = 0, \quad \text{or,}$$

$$- 3z = 5y$$

Dividing both sides of this equation by 5, and by - 3, we have the proportion:

$$\frac{z}{5} = \frac{y}{-3} \, .$$

From this result and our previous result we obtain:

$$\frac{x}{4} = \frac{y}{-3} = \frac{z}{5} \quad \text{as the desired ratios.}$$

If (2ma + 6mb + 3nc + 9nd)(2ma − 6mb − 3nc + 9nd)

   = (2ma − 6mb + 3nc − 9nd)(2ma + 6mb − 3nc − 9nd),

prove that a, b, c, d are proportionals.

Solution: Dividing both sides of the given equation by (2ma − 6mb − 3nc + 9nd), and then by (2ma − 6mb + 3nc − 9nd) we have

$$\frac{2ma + 6mb + 3nc + 9nd}{2ma - 6mb + 3nc - 9nd} = \frac{2ma + 6mb - 3nc - 9nd}{2ma - 6mb - 3nc + 9nd} .$$

Since the above two ratios are of the form $\frac{a}{b} = \frac{c}{d}$, we can use the Law of Proportions which states that

$$\frac{a + b}{a - b} = \frac{c + d}{c - d} .$$

Doing this we obtain,

$$\frac{2ma + 6mb + 3nc + 9nd + (2ma - 6mb + 3nc - 9nd)}{2ma + 6mb + 3nc + 9nd - (2ma - 6mb + 3nc - 9nd)} =$$

$$\frac{2ma + 6mb - 3nc - 9nd + (2ma - 6mb - 3nc + 9nd)}{2ma + 6mb - 3nc - 9nd - (2ma - 6mb - 3nc + 9nd)}$$

or, $\frac{4ma + 6nc}{12mb + 18nd} = \frac{4ma - 6nc}{12mb - 18nd}$; and factoring gives us,

$$\frac{2(2ma + 3nc)}{2(6mb + 9nd)} = \frac{2(2ma - 3nc)}{2(6mb - 9nd)} \qquad \text{or,}$$

$$\frac{2ma + 3nc}{6mb + 9nd} = \frac{2ma - 3nc}{6mb - 9nd} .$$

Now, since $\frac{a}{b} = \frac{c}{d}$ can be alternately written as $\frac{a}{c} = \frac{b}{d}$, we write:

$$\frac{2ma + 3nc}{2ma - 3nc} = \frac{6mb + 9nd}{6mb - 9nd} .$$

Now, rewriting this last proportion as,

$$\frac{2ma + 3nc + (2ma - 3nc)}{2ma + 3nc - (2ma - 3nc)} = \frac{6mb + 9nd + (6mb - 9nd)}{6mb + 9nd - (6mb - 9nd)}$$

we obtain: $\frac{4ma}{6nc} = \frac{12mb}{18nd} .$

Again, using the fact that $\frac{a}{b} = \frac{c}{d}$ can be rewritten

122

as $\frac{a}{c} = \frac{b}{d}$, we write:

$$\frac{4ma}{12mb} = \frac{6nc}{18nd} \qquad \text{or,}$$

$$\frac{a}{3b} = \frac{c}{3d} \ .$$

Thus, $\frac{a}{b} = \frac{c}{d}$ or, $a : b = c : d$.

● **PROBLEM** 8-4

(1) State the laws of proportion.

(2) Determine the value of x for

$(x-4):(x-8) = (2x-4):x$.

(3) Find the fourth proportional to a,b,98.

(4) Find the third proportional to $a:b = b:c$.

(5) Find the mean proportional in $a:b = b:c$ between a and c.

<u>Solution:</u> (1) If $\frac{a}{b} = \frac{c}{d}$, then the laws of proportion are:

(a) $ad = bc$ (b) $\frac{a}{c} = \frac{b}{d}$ (c) $\frac{b}{a} = \frac{d}{c}$

(d) $\frac{a+b}{b} = \frac{c+d}{d}$ (e) $\frac{a-b}{b} = \frac{c-d}{d}$ (f) $\frac{a+b}{a-b} = \frac{c+d}{c-d}$

(2) $\frac{x-4}{x-8} = \frac{2x-4}{x}$

$x^2 - 4x = 2x^2 + 32 - 20x$

$-x^2 + 16x - 32 = 0$

$x = \frac{-b \pm \sqrt{b^2-4ac}}{2a}$ where a=-1, b=16, c=-32

$x = \frac{-16 \pm \sqrt{256-4(-1)(-32)}}{-2}$

$= 8 \pm \frac{\sqrt{256-128}}{-2}$

$= 8 \pm \frac{\sqrt{128}}{-2} = 8 \pm (4\sqrt{2}) = 8 \pm 4\sqrt{2}$

123

(3) $\dfrac{a}{b} = \dfrac{98}{x}$

   $ax = 98b$

   $x = \dfrac{98b}{a}$

(4) In a:b = b:c, the third proportional is c.

(5) The mean proportional in a:b = b:c between a and c is b.

● **PROBLEM** 8-5

According to Hooke's Law, the length of a spring, S, varies directly as the force, F, applied on the spring. In a spring to which Hooke's Law applies, a force of 18.6 lb stretches the spring by 1.27 in. Find k, the proportionality constant.

<u>Solution:</u> The direct variation of the length of the spring, S, and the force applied on it, F, is expressed symbolically as

   F = kS, where k is the constant of proportionality.

We are given that F = 18.6 lb and S = 1.27 in. Thus, it is necessary merely to substitute the given values in the equation F = kS, obtaining

$$18.6 = k(1.27),$$

from which k = 18.6/1.27 = 14.65 lb/in.

● **PROBLEM** 8-6

The resistance R of a given size of wire at constant temperature varies directly as the length ℓ. It is found that the resistance of 100 feet of number 14 copper is 0.253 ohm. Construct a table of values for the given lengths of number 14 copper wire assuming the temperature is constant.

ℓ	25	75	125	175	225
R					

<u>Solution:</u> Direct variation implies a variable y is equal to a constant c times x; that is, y = cx. In this particular example, the resistance R varies directly as the length ℓ. Since several values of R are to be found, we use the general equation for this variation. Thus,

$$R = k\ell \qquad\qquad (1)$$

Since R = 0.253 when ℓ = 100

$$0.253 = 100k \qquad (2)$$

Solving for k

$$k = 0.00253 \qquad (3)$$

The specific equation is therefore

$$R = 0.00253\ell \qquad (4)$$

The values of R corresponding to the given values of ℓ may then be found directly from Equation 4, as follows.

ℓ	25	75	125	175	225
R	0.063	0.190	0.316	0.443	0.569

Notice that if values of ℓ were to be determined from given values R, it would be convenient to solve Equation 4 for ℓ in terms of R. Thus

$$\ell = \frac{R}{0.00253} \qquad (5)$$

or

$$\ell = 395.26R$$

● PROBLEM 8-7

If y varies inversely as the cube of x, and y = 7 when x = 2, express y as a function of x.

Solution: The relationship "y varies inversely with respect to x" is expressed as,

$$y = \frac{k}{x}.$$

The inverse variation is now with respect to the cube of x, x^3 and we have,

$$y = \frac{k}{x^3}$$

Since y = 7 and x = 2 must satisfy this relation, we replace x and y by these values,

$$7 = \frac{k}{2^3} = \frac{k}{8},$$

and we find $k = 7 \cdot 8 = 56$. Substitution of this value of k in the general relation gives,

$$y = \frac{56}{x^3},$$

which expresses y as a function of x. We may now, in addition find the value of y corresponding to any value of x. If we had the added requirement to find the value of y when x = 1.2, x = 1.2 would be substituted in the function so that for x = 1.2, we have,

$$y = \frac{56}{(1.2)^3} = \frac{56}{1.728} = 32.41$$

125

Other expressions in use are "is proportional to" for "varies directly," and "is inversely proportional to" for "varies inversely."

The weight W of an object above the earth varies inversely as the square of the distance d from the center of the earth. If a man weighs 180 pounds on the surface of the earth, what would his weight be at an altitude of 1000 miles? Assume the radius of the earth to be 4000 miles.

<u>Solution:</u> W varies inversely with d^2; therefore $W = \dfrac{k}{d^2}$

where k is the proportionality constant. Similarly,

$W_1 = \dfrac{k}{d_1^2}$, $W_2 = \dfrac{k}{d_2^2}$ and, solving these two equations for k,

$W_1 d_1^2 = k$ and $W_2 d_2^2 = k$. Hence,

$$k = W_1 d_1^2 = W_2 d_2^2$$

$$\text{or} \quad \frac{W_1 d_1^2}{W_2} = \frac{\cancel{W_2} d_2^2}{\cancel{W_2}}$$

$$\frac{W_1 d_1^2}{W_2} = d_2^2$$

$$\frac{W_1 \cancel{d_1^2}}{W_2 \cancel{d_1^2}} = \frac{d_2^2}{d_1^2}$$

$$\frac{W_1}{W_2} = \frac{d_2^2}{d_1^2} \tag{1}$$

Letting d_1 = radius of the earth, 4000, then $d_2 = 4000 + 1000 = 5000$.

Substituting the given values in Equation (1):

$$\frac{180}{W_2} = \frac{5000^2}{4000^2} = \frac{(5 \times 1000)^2}{(4 \times 1000)^2} = \frac{5^2 \times \cancel{1000}^2}{4^2 \times \cancel{1000}^2}$$

$$= \frac{5^2}{4^2}$$

$$= \frac{25}{16}$$

$$\frac{180}{W_2} = \frac{25}{16}$$

$$W_2\left(\frac{180}{W_2}\right) = W_2\left(\frac{25}{16}\right)$$

$$180 = \frac{25}{16}W_2$$

$$\frac{16}{25}\overset{36}{(\cancel{180})} = \cancel{\frac{16}{25}}\left(\cancel{\frac{25}{16}}W_2\right)$$
$$\underset{5}{}$$

$$\frac{576}{5} = W_2$$

$$115\tfrac{1}{5} \text{ pounds} = W_2$$

$$\text{or } 115.2 \text{ pounds} = W_2$$

● **PROBLEM** 8-9

The pressure of wind on a sail varies jointly as the area of the sail and the square of the wind's velocity. When the wind is 15 miles per hour, the pressure on a square foot is one pound. What is the velocity of the wind when the pressure on a square yard is 25 pounds?

<u>Solution:</u> Let p = pressure of the wind, in pounds
v = the velocity of the wind, in miles per hour
a = the area of the sail, in square feet.

Pressure, p, varies jointly as the area of the sail, a, and the square of the wind's velocity, v^2. Therefore p varies directly as the product av^2 times a proportionality constant, k. k must be determined before we can proceed to find v as desired. Use the given information a = 1 and p = 1 when v = 15 to determine the proportionality constant, k.

$$p = kav^2.$$

$$1 = k(1)(225).$$

$$k = \frac{1}{225}, \text{ value of the proportionality constant.}$$

Now we can find v using $k = \frac{1}{225}$ when p = 25 and a = 9 (1 yard = 3 feet, 1 square yard = 9 square feet).

$$p = \frac{1}{225}av^2.$$

$$25 = \frac{1}{225}(9)v^2.$$

$$v^2 = \frac{(25)(225)}{9}$$

$$v = \sqrt{\frac{(25)(225)}{9}} = \frac{(5)(15)}{3}$$

$v = 25$, number of miles per hour.

A certain beam L ft. long has a rectangular cross section b in. in horizontal width and d in. in vertical depth. It is found that, when the beam is supported at the ends, the deflection D at the center varies directly as the fourth power of L, inversely as b, and inversely as the cube of d. If the length is decreased by 10 per cent but the width kept the same, by how much should the depth be changed in order that the same deflection D be obtained?

Solution: A quantity m varies directly as another quantity n if m equals the product of a constant and n; that is, m = cn where c is the constant. Also, a quantity p varies inversely as another quantity q if p equals the product of a constant and the reciprocal of q; that is, p = c(1/q) = c/q where c is the constant. From the statement of the problem, we see that the combined variation is given by

$$D = \frac{kL^4}{bd^3} \text{ , where } k \text{ is a constant.}$$

Since corresponding values of the four variables are not known, we cannot determine the value of the constant of proportionality k. But if one set of variables is designated with the subscript 1, and the new set with the subscript 2, we have

$$D_1 = \frac{kL_1^4}{b_1 d_1^3} \text{ , } \quad D_2 = \frac{kL_2^4}{b_2 d_2^3} \text{ .}$$

Since it is desired to obtain a relationship between d_1 and d_2, carry out the following procedure in order to isolate d_1 and d_2 on different sides of the equation.

Divide D_2 by D_1:

$$\frac{D_2}{D_1} = \frac{kL_2^4/b_2 d_2^3}{kL_1^4/b_1 d_1^3} \tag{1}$$

Since division by a fraction is equivalent to multiplication by that fraction's reciprocal, equation (1) becomes:

$$\frac{D_2}{D_1} = \frac{kL_2^4}{b_2 d_2^3} \frac{b_1 d_1^3}{kL_1^4} = \frac{L_2^4 b_1 d_1^3}{b_2 d_2^3 L_1^4} = \frac{L_2^4 b_1 d_1^3}{L_1^4 b_2 d_2^3}$$

or

$$\frac{D_2}{D_1} = \frac{L_2^4 b_1 d_1^3}{L_1^4 b_2 d_2^3}$$

128

In this problem we have $D_2 = D_1$, $L_2 = 0.9L_1$, and $b_2 = b_1$. Therefore we get

$$\frac{D_1}{D_1} = \frac{\left(0.9L_1\right)^4 b_1 d_1^3}{L_1^4 b_1 d_2^3}$$

$$1 = \frac{(0.9)^4 \cancel{L_1^4}\, \cancel{b_1}\, d_1^3}{\cancel{L_1^4}\, \cancel{b_1}\, d_2^3}$$

$$1 = (0.9)^4\, \frac{d_1^3}{d_2^3}$$

Multiplying both sides by d_2^3:

$$d_2^3(1) = \cancel{d_2^3}(0.9)^4\, \frac{d_1^3}{\cancel{d_2^3}}$$

$$d_2^3 = (0.9)^4\, d_1^3$$

Take the cube root of each side:

$$\sqrt[3]{d_2^3} = \sqrt[3]{(0.9)^4 d_1^3}$$

$$d_2 = \sqrt[3]{(0.9)^4}\, \sqrt[3]{d_1^3}$$

$$d_2 = \sqrt[3]{(0.9)^4}\, d_1$$

$$d_2 = (0.9)^{4/3}\, d_1$$

$$d_2 = \left[(0.9)^{1/3}\right]^4 d_1$$

$$d_2 = \left[\sqrt[3]{0.9}\right]^4 d_1$$

$$d_2 \approx (.966)^4\, d_1$$

$$d_2 \approx .871\, d_1$$

Hence, the new depth, d_2 is only $.871\, d_1$. Subtracting $.871\, d_1$ from $1\, d_1$:

$$
\begin{array}{r}
1.000\ d_1 \\
-\ 0.871\ d_1 \\
\hline
0.129\ d_1
\end{array}
$$

Therefore, the depth has been decreased by an amount of $.129$ or

$$\frac{129}{1000} = \frac{129}{10} \times \frac{1}{100}$$

$$= \frac{129}{10}\ \%\ \text{since "hundredths means per cent"}$$

$$= 12.9\%.$$

Hence, the depth has been decreased approximately 13%.

CHAPTER 9

EQUATIONS AND GRAPHS

Basic Attacks and Strategies for Solving Problems in this Chapter. See pages 130 to 134 for step-by-step solutions to problems.

Solving an equation in a single variable (e.g., x) is the process of obtaining the value of the variable (called the *solution*) that makes the equality hold true. For example, given

$$3x = x + 7,$$

one obtains the solution by moving the expressions with the variable to one side of the equality (obtaining $2x = 7$) and eventually isolating the variable (obtaining $x = {}^7/_2$). The solution can be verified by substituting the value into the original equation and seeing if equality results.

Some equations hold true for any value of the variable, e.g.,

$$2x = x + x.$$

Such equations are called *identities*. If an equation is true for only certain values of the variable, it is sometimes called a *conditional equation*. An equation can be shown to be conditional if one can exhibit a value of the variable that makes the equality true and another value of the variable that makes the equality false.

An *equation of degree n* is an equation in which the sum of the powers of the variables in at least one term is n and in all the other terms is no greater than n. One may also speak of the *degree of an equation in a variable* or the *degree of an equation in several variables*. When specifying one variable, one merely considers all other variables as constants when computing the degree. When specifying several variables, one considers the degree of the products of those variables (if they exist), and considers all other variables as constants. A *linear equation* is an equation of degree 1. A *quadratic equation* is an equation of degree 2 and a *cubic equation* is one of degree 3.

The same suggestions given in the previous chapter for graphic functions also hold for graphing equations. It is often helpful to determine the *intercepts*, i.e., where the graph crosses the x and y axes. Sometimes it is helpful to determine if any *symmetry* exists, that is, whether the graph can be reflected about either axis or some other line. For example, if

$$y = f(x) = x^2,$$

then $f(-x) = (-x)^2 = x^2 = f(x).$

Thus, one needs only to graph the equation for positive x, and then to reflect this about the y-axis (where $x = 0$).

● **PROBLEM 9-1**

Solve, justifying each step. $3x - 8 = 7x + 8$.

Solution: $3x - 8 = 7x + 8$

Adding 8 to both members, $3x - 8 + 8 = 7x + 8 + 8$

Additive inverse property, $3x + 0 = 7x + 16$

Additive identity property, $3x = 7x + 16$

Adding $(-7x)$ to both members, $3x - 7x = 7x + 16 - 7x$

Commuting, $-4x = 7x - 7x + 16$

Additive inverse property, $-4x = 0 + 16$

Additive identity property, $-4x = 16$

Dividing both sides by -4, $x = \dfrac{16}{-4}$

                                 $x = -4$

Check: Replacing x by -4 in the original equation:

$$3x \quad - 8 \ = \ 7x \ + \ 8$$
$$3(-4) - 8 \ = 7(-4) + \ 8$$
$$- 12 - 8 \ = \ - 28 + \ 8$$
$$-20 \ = \ - 20$$

● **PROBLEM 9-2**

Determine which of the following is a conditional equation and which is an identity.

(1) $3x - (x-4) = 2(x+2)$

(2) $y + 9 = 12$

130

Solution: (1) $3x - (x-4) = 2(x+2)$

$3x - x + 4 = 2x + 4$

$2x + 4 = 2x + 4$ this is an identity;

it is true for all values of x.

(2) $y + 9 = 12$ is a conditional equation. It is

only true for $y = 3$.

● **PROBLEM 9-3**

Show that $x^2 + 2x + 5 = 20$ is a conditional equation.

Solution: A conditional equation is an equation for which there exists at least one value which may be substituted for the variable that makes the equation false, but is true for other values. It is sufficient to exhibit one replacement for x that makes the equation true and one that makes it false.

Let x = 3: $(3)^2 + 2(3) + 5 \overset{?}{=} 20$

$9 + 6 + 5 \overset{?}{=} 20$

$20 = 20$

Let x = -4: $(-4)^2 + (-4) + 5 \overset{?}{=} 20$

$16 - 8 + 5 \overset{?}{=} 20$

$13 \neq 20$

When x = -4, this value of x makes the equation false. For x = 3, the equation is **true**. Therefore, $x^2 + 2x + 5 = 20$ is a conditional equation.

Notice that we have not solved the equation in this example . An equation is solved when its solution set is completely known.

● **PROBLEM 9-4**

Give an example of each of the following:

(1) A rational integral equation of degree 5 in unknown a

(2) linear equation

(3) quadratic equation

(4) cubic equation.

131

Solution: (1) $a^5bc^2 + 3ac^2 = 2a^3b + 3c^2$

(2) $2y = 3x + 9$

(3) $x^2 - 4xz + 3z^2 = 0$

(4) $y^3 + 3y^2 - 2y + 9 = 6$

● PROBLEM 9-5

Determine the degree of the following equations in each of the indicated unknowns.

$x^2 + xy + z^4x^4 + 14 = 0$

x; y; z; x and y; x and z; v and z.

Solution: x : 4

y : 1

z : 4

x and y : = 2

x and z : = 8

y and z : = 0

● PROBLEM 9-6

Draw the graphs of the following equations:

(1) $y = 4x + 9$

(2) $y = x^2 + 2x + 9$

(3) $y = x^3$

Solution:

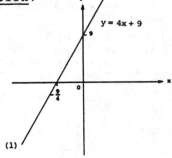

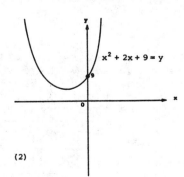

132

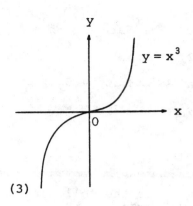

(3)

Investigate each of the following equations, determine the intercepts and symmetry, and draw the graph.

(1) $x^2 + y^2 = 9$

(2) $2x^2y - x^2 - y = 0$

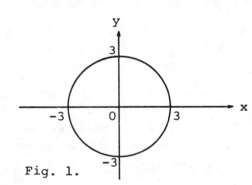

Fig. 1.

Fig. 2.

Solution: (1) $x^2 + y^2 = 9$

intercepts: x=0 y=±3   (y-intercepts)

y=0 x=±3   (x-intercepts)

symmetry: The locus is symmetric with respect to the coordinate axes and the origin.

The locus exists for all values of x where x ≥ -3 or x ≤ 3, and all values of y where y ≥ -3 or y ≤ 3. The locus consists of one closed piece. Various values of x were chosen and the values of y calculated as in the table shown below. The graph was then drawn, as shown in Fig. 1.

x	-3	-2	-1	0	1	2	3
y	0	$\pm\sqrt{5}$	$\pm\sqrt{8}$	±3	$\pm\sqrt{8}$	$\pm\sqrt{5}$	0

133

(2) $2x^2y - x^2 - y = 0$

intercepts: $x=0$ $y=0$ (y-intercept)

                $y=0$ $x=0$ (x-intercept)

symmetry: The locus is symmetric with respect to the y-axis.

Extent: $y = \dfrac{x}{2x^2-1}$, $x = \pm\sqrt{\dfrac{y}{2y-1}}$

The locus exists for all $x \neq \pm\sqrt{1/2}$ (this comes from the inequality $2x^2 - 1 \neq 0$), and for all y such that $\dfrac{y}{2y-1} \geq 0$. Find the value of y for which the numerator is equal to zero, that is $y=0$, and the value of y for which the denominator is zero, $y=\frac{1}{2}$. Therefore, the locus exists for $y \leq 0$ and for $y > \frac{1}{2}$.

Asymptotes: $x = \pm\sqrt{1/2}$; $y = 1/2$.

After choosing various values for x, and solving for y we form a table as follows

x	-2	-1	0	1	2
y	4/7	1	0	1	4/7

and the graph as seen in Fig. 2.

CHAPTER 10

LINEAR FUNCTIONS AND EQUATIONS

Basic Attacks and Strategies for Solving Problems in this Chapter. See pages 135 to 158 for step-by-step solutions to problems.

As has been mentioned in previous chapter introductions, *linear functions* (or *equations*) are functions in which all variables have a degree of one and there are no products of variables. Because of this, solving linear equations is usually easier than solving equations of a higher degree. The solution procedure is the same: one isolates the variable on one side of the equal sign and combines all constants on the other side.

Linear equations of more than one variable have multiple solutions (values for each variable present). For example, $x + y = 2$ has solutions $(x, y) = (0, 2)$ or $(1, 1)$ or $(2, 0)$, etc.

It is possible for a linear equation to have no solution. For example,

$$2x + 1 = 2x + 3$$

has no solution, since, by subtracting $2x$ from both sides, one obtains the impossibility that $1 = 3$. Similarly,

$$\frac{x}{x-1} = \frac{1}{x-1}$$

also has no solution, since the computed "solution," $x = 1$, when substituted into the original equation, produces $1/0 = 1/0$, and division by zero is never permitted.

One deals with equations of fractions with linear equations as denominators using the same rules for combining numeric fractions as seen in Chapter 3. One begins by factoring the terms (isolating constant factors present), then by finding a least common denominator, and then combining terms. When the equation has been reduced to two fractions on either side of the equals sign, one can eliminate the denominators on both sides (multiplying both sides by the least common denominator), and then solve the resulting equation according to standard means.

Whenever the process of finding a solution to an equation involves squaring both sides of the equation or multiplying by a term that includes a variable, the

135 – A

solution or solutions should always be double-checked by substituting into the original equation. For example, $\sqrt{x-2} = -2$ might seem to have the solution $x = 6$ if one squares both sides of the equation and obtains $x - 2 = 4$ or $x = 6$. However, since the radical sign normally indicates a *positive* root only, the original equation indicates an impossible situation: a positive value can never equal -2.

Linear functions and equations are graphed according to the general suggestions given for Chapter 7. It should be noted that functions that have a constant y value (e.g., $y = 3$) result in a graph that is parallel to the x-axis.

Word problems are solved by translating the information in the statement of the problems into algebraic equations. In these equations, the varying and unknown quantities are represented by variables. Every sentence that contains information relating some quantity to some other quantity should be translated into an appropriate equation. A significant number of word problems translate into linear equations. Unfortunately, there does not exist any one set of rules that can be applied equally to all such problems. It is also often possible that the same problem can be solved in more than one way, both of which are equally correct.

For example, look at the following problem.

> The sum of two numbers is 23. One number is twice the other number. What are the two numbers?

We are asked to find two numbers (the final sentence). Let us call one of these numbers n. Using the information in the second sentence of the problem statement, we decide to call the other number $2n$. From the first sentence, we derive the equation:

$$n + 2n = 23.$$

Adding the terms on the left side and solving, we obtain $n = {}^{23}/_3$. Thus, our two numbers are

$$n = {}^{23}/_3 \text{ and } 2n = 2 \cdot {}^{23}/_3 = {}^{46}/_3.$$

Note that if the problem asked for *integers* rather than *numbers*, our answer would not be a correct solution and, in fact, the problem would have no solution.

It would also be possible to begin by calling the two numbers m and ${}^m/_2$ and proceed in a similar fashion. The final numeric answers would be the same, but the equations leading to that answer would be different.

Solve each equation (find the solution set), and check each solution.
(a) $4(6x + 5) - 3(x - 5) = 0$.   (b) $8 + 3x = -4(x - 2)$.

<u>Solution:</u>   (a) $4(6x + 5) - 3(x - 5) = 0$

distributing, $24x + 20 - 3x + 15 = 0$

combining like terms,
$$21x + 35 = 0$$

adding (–35) to both sides,
$$21x = -35$$

dividing both sides by 21,
$$x = \frac{-35}{21} = \frac{-5}{3}$$

Therefore the solution set to this equation is $\left\{ \frac{-5}{3} \right\}$.

Check: Replace x by $\frac{-5}{3}$ in the equation,

$$4(6x + 5) - 3(x - 5) = 0$$
$$4\left[6\left(\frac{-5}{3}\right) + 5 \right] - 3\left[\frac{-5}{3} - 5 \right] = 0$$
$$4\left(\frac{-30}{3} + 5\right) - 3\left(\frac{-5}{3} - \frac{15}{3}\right) = 0$$
$$4(-10 + 5) - 3\left(\frac{-20}{3}\right) = 0$$
$$4(-5) + 20 = 0$$
$$-20 + 20 = 0$$
$$0 = 0$$

(b) $8 + 3x = -4(x - 2)$

distributing, $8 + 3x = -4x + 8$

adding 4x to both sides,
$$8 + 7x = 8$$

135

adding (-8) to both sides,
$$7x = 0$$

dividing both sides by 7, $x = 0$

Therefore the solution set to this equation is {0}, (not to be confused with the null set { }).

Check: Replace x by 0 in the equation,
$$8 + 3x = -4(x - 2)$$
$$8 + 3(0) = -4(0 - 2)$$
$$8 + 0 = -4(-2)$$
$$8 = 8$$

● **PROBLEM 10-2**

Solve the equation $2(x + 3) = (3x + 5) - (x - 5)$.

Solution: We transform the given equation to an equivalent equation where we can easily recognize the solution set.

$$2(x + 3) = 3x + 5 - (x - 5)$$

Distribute, $2x + 6 = 3x + 5 - x + 5$

Combine terms, $2x + 6 = 2x + 10$

Subtract 2x from both
sides, $6 = 10$

Since 6 = 10 is not a true statement, there is no real number which will make the original equation true. The equation is inconsistent and the solution set is ϕ, the empty set.

● **PROBLEM 10-3**

Solve the equation $\frac{3}{4} x + \frac{7}{8} + 1 = 0$.

Solution: There are several ways to proceed. First we observe that $\frac{3}{4} x + \frac{7}{8} + 1 = 0$ is equivalent to

$\frac{3}{4} x + \frac{7}{8} + \frac{8}{8} = 0$, where we have converted

1 into $\frac{8}{8}$. Now, combining fractions we obtain:

$$\frac{3}{4} x + \frac{15}{8} = 0$$

Subtract $\frac{15}{8}$ from both sides:

136

$$\frac{3}{4} x = \frac{-15}{8}$$

Multiplying both sides by $\frac{4}{3}$:

$$\left(\frac{4}{3}\right) \frac{3}{4} x = \left(\frac{4}{3}\right)\left(\frac{-15}{8}\right)$$

Cancelling like terms in numerator and denominator:

$$x = \frac{-5}{2}$$

A second method is to multiply both sides of the equation by the least common denominator, 8:

$$8 \left(\frac{3}{4} x + \frac{7}{8} + 1\right) = 8(0)$$

Distributing: $8 \left(\frac{3}{4}\right) x + 8 \left(\frac{7}{8}\right) + 8 \cdot 1 = 0$

$$(2 \cdot 3)x + 7 + 8 = 0$$

$$6x + 15 = 0$$

Subtract 15 from both sides: $6x = -15$

Divide both sides by 6: $x = \frac{-15}{6}$

Cancelling 3 from numerator and denominator: $x = \frac{-5}{2}$

● **PROBLEM 10-4**

Solve the equation $2\left(\frac{2}{3} y + 5\right) + 2\left(y + 5\right) = 130$.

<u>Solution:</u> The procedure for solving this equation is as follows:

$\frac{4}{3}y + 10 + 2y + 10 = 130,$ Distributive property

$\frac{4}{3}y + 2y + 20 = 130,$ Combining like terms

$\frac{4}{3}y + 2y = 110,$ Subtracting 20 from both sides

$\frac{4}{3}y + \frac{6}{3}y = 110,$ Converting 2y into a fraction with denominator 3

$\frac{10}{3}y = 110,$ Combining like terms

$y = 110 \cdot \frac{3}{10} = 33,$ Dividing by $\frac{10}{3}$

Check: Replace y by 33 in our original equation,

137

$$2 \left[\frac{2}{3}(33) + 5 \right] + 2(33 + 5) = 130$$

$$2(22 + 5) + 2(38) = 130$$

$$2(27) + 76 \overset{\cdot}{=} 130$$

$$54 + 76 = 130$$

$$130 = 130$$

Therefore the solution to the given equation is y=33.

● **PROBLEM 10-5**

Find the set indicated by
$$\left\{ x \mid \tfrac{1}{2}x - \frac{2}{3} = \frac{3}{4}x + \frac{1}{12} \right\}$$

Solution: The set indicated by $\left\{ x \mid \tfrac{1}{2}x - \frac{2}{3} = \frac{3}{4}x + \frac{1}{12} \right\}$
is the set of all x such that x makes the statement
$$\tfrac{1}{2}x - \frac{2}{3} = \frac{3}{4}x + \frac{1}{12}$$
true. Hence to obtain the required set, we must solve
the equation
$$\tfrac{1}{2}x - \frac{2}{3} = \frac{3}{4}x + \frac{1}{12} \ .$$

Since 2, 3, 4, and 12, the denominators of the
fractions are all factors of 12, we may multiply both
sides of the equation by 12 to eliminate the fractions.
Therefore, 12 is called the least common multiple (LCM).
Thus,

$$12 \left[\tfrac{1}{2}x - \frac{2}{3} \right] = 12 \left[\frac{3}{4}x + \frac{1}{12} \right]$$

Distribute, $12(\tfrac{1}{2}x) - 12\left(\frac{2}{3}\right) = 12\left(\frac{3}{4}x\right) + 12\left(\frac{1}{12}\right)$

$$6x - 8 = 9x + 1$$

Add (-9x) to both sides, $6x - 8 + (-9x) = 9x + 1 + (-9x)$

commute, $\qquad 6x + (-9x) - 8 = 9x + (-9x) + 1$

$$-3x - 8 = 1$$

Add 8 to both sides, $-3x - 8 + 8 = 1 + 8$

$$-3x = 9$$

Divide both sides by -3 to obtain,

$$x = -3.$$

Thus, our solution is $x = -3$, and the set indicated by

$\left\{ x \mid \frac{1}{2}x - \frac{2}{3} = \frac{3}{4}x + \frac{1}{12} \right\}$ is $\{x \mid x = -3\}$. Now, we check this

solution.

Check: Substitute (-3) for x in our original equation,

$$\frac{1}{2}x - \frac{2}{3} = \frac{3}{4}x + \frac{1}{12}$$

$$\left(\frac{1}{2}\right)(-3) - \frac{2}{3} = \frac{3}{4}(-3) + \frac{1}{12}$$

$$\frac{-3}{2} - \frac{2}{3} = \frac{-9}{4} + \frac{1}{12}$$

Convert each fraction into a fraction whose denominator is 12. Here we are using the fact that 12 is the least common denominator (we could also multiply both members by the LCM 12 as before). Thus,

$$\frac{6}{6}\left(\frac{-3}{2}\right) - \frac{4}{4}\left(\frac{2}{3}\right) = \frac{3}{3}\left(\frac{-9}{4}\right) + \frac{1}{12}$$

$$\frac{-18}{12} - \frac{8}{12} = \frac{-27}{12} + \frac{1}{12}$$

$$\frac{-26}{12} = \frac{-26}{12}$$

Since substitution for x by (-3) results in this equivalent equation, which is always true, (-3) is indeed a root of the equation.

● **PROBLEM 10-6**

Solve $A = \frac{h}{2}(b + B)$ for h.

Solution: Since the given equation is to be solved for h, obtain h on one side of the equation. Multiply both sides of the equation $A = \frac{h}{2}(b + B)$ by 2. Then, we have:

$$2(A) = 2\left[\frac{h}{2}(b + B)\right].$$

Therefore: $$2(A) = \frac{\cancel{2}h}{\cancel{2}}(b + B)$$

$$2A = h(b + B). \tag{1}$$

Since it is desired to obtain h on one side of the equation, divide both sides of equation (1) by $(b + B)$.

$$\frac{2A}{(b + B)} = \frac{h(b + \cancel{B})}{(\cancel{b} + \cancel{B})} .$$

Therefore:
$$\frac{2A}{b + B} = h.$$

Thus, the given equation, $A = \frac{h}{2}(b + B)$, is solved for h.

This is the form of the formula used to determine values of h for a set of trapezoids, if the area and lengths of the bases are known.

● PROBLEM 10-7

Solve the equation $a(x + b) = bx + c$ for x if $a \neq b$.

Solution:

$ax + ab = bx + c$	Distributive property
$ax + ab + (-bx) = bx + c + (-bx)$	adding $(-bx)$ to both sides
$ax + (-bx) + ab = bx + (-bx) + c$	commutative law of addition
$ax + (-bx) + ab = 0 + c$	additive inverse property
$ax + (-bx) + ab = c$	additive identity property
$ax + (-bx) + ab + (-ab) = c + (-ab)$	adding $(-ab)$ to both sides
$ax + (-bx) + 0 = c + (-ab)$	additive inverse property
$ax - bx = c - ab$	additive identity property
$(a - b)x = c - ab$	factoring out x
$x = \frac{c - ab}{a - b}$ if $a \neq b$	Dividing by $(a - b)$

If $a = b$ the denominator of this fraction is zero, and thus the fraction has no meaning.

● PROBLEM 10-8

Find a solution of the equation
$$3x + 4y + 5z = 13 \qquad (1)$$

Solution: The above equation is linear in x, y, and z. Any ordered triple (x, y, z) which satisfies it is a solution. If we chose x = 2, and y = 3, by substitution

$$6 + 12 + 5z = 13$$
$$z = -1$$

The one solution of Equation 1 is x = 2, y = 3, and z = -1.

Obviously, the number of solutions is unlimited, since any choice of values for two of the variables will determine the value of the third variable.

Find the solutions of the equation $\dfrac{4x-7}{x-2} = 3 + \dfrac{1}{x-2}$.

Solution: Assume that there is a number x such that

$$\frac{4x-7}{x-2} = 3 + \frac{1}{x-2}$$

In order to eliminate the fractions multiply both sides of the equation by $x-2$ to obtain

$$(x-2)\ \frac{4x-7}{x-2} = \left(3 + \frac{1}{x-2}\right)(x-2)$$

Thus

$$4x-7 = 3(x-2) + \frac{x-2}{x-2}$$

$$4x-7 = 3(x-2) + 1$$

$$4x-7 = 3x-6+1$$

$$4x-7 = 3x-5$$

Add $(-3x)$ to both sides, $\qquad 4x-7+(-3x) = -5$

$$x-7 = -5$$

Add 7 to both sides, $\qquad x = -5+7$

and hence $x = 2$.

We have shown that if x is a solution of the equation

$$\frac{4x-7}{x-2} = 3 + \frac{1}{x-2},$$

then $x = 2$. But if we substitute $x = 2$ in the right-hand member of the equation we obtain

$$3 + \frac{1}{0}$$

and we know that we cannot divide by zero. Hence 2 is not a solution.

Before we analyze the process which led to the conclusions that 2 was a possible solution to our equation, let us see exactly why our equation has no solution. To do this, we note that

$$3 + \frac{1}{x-2} = 3 \cdot \frac{x-2}{x-2} + \frac{1}{x-2} = \frac{3(x-2)+1}{x-2} = \frac{3x-6+1}{x-2} = \frac{3x-5}{x-2}$$

and hence that the original equation is equivalent to

$$\frac{4x-7}{x-2} = \frac{3x-5}{x-2} \tag{1}$$

Now we know that two fractions, $\dfrac{a}{b}$ and $\dfrac{c}{d}$ are equal if and only if $ad = bc$. Thus (1) holds, providing that $x \neq 2$, if and only if

$$(x-2)(4x-7) = (x-2)(3x-5) \tag{2}$$

holds. But, since $x \neq 2$, $x-2 \neq 0$, and we can divide both sides of (2) by $x-2$ and have

$$4x - 7 = 3x - 5$$

which gives $x = 2$, a contradiction. In other words, the only possible solution is a number which we knew in advance could not be a solution, and hence there are no solutions to our given equation.

Solve the equation

$$\frac{5}{x - 1} + \frac{1}{4 - 3x} = \frac{3}{6x - 8} .$$

Solution: By factoring out a common factor of -2 from the denominator of the term on the right side of the given equation, the given equation becomes:

$$\frac{5}{x - 1} + \frac{1}{4 - 3x} = \frac{3}{-2(-3x + 4)} = \frac{3}{-2(4 - 3x)} = \frac{3}{2(4 - 3x)}$$

Hence,

$$\frac{5}{x - 1} + \frac{1}{4 - 3x} = -\frac{3}{2(4 - 3x)}$$

Adding $\frac{3}{2(4 - 3x)}$ to both sides of this equation:

$$\frac{5}{x - 1} + \frac{1}{4 - 3x} + \frac{3}{2(4 - 3x)} = 0. \qquad (1)$$

Now, in order to combine the fractions, the least common denominator (l.c.d.) must be found. The l.c.d. is found in the following way: list all the different factors of the denominators of the fractions. The exponent to be used for each factor in the l.c.d. is the greatest value of the exponent for each factor in any denominator. Therefore, the l.c.d. of the given fractions is:

$$2^1(x - 1)^1(4 - 3x)^1 = 2(x - 1)(4 - 3x)$$

Hence, equation (1) becomes:

$$\frac{(2)(4-3x)(5)}{(2)(4-3x)(x-1)} + \frac{(2)(x-1)(1)}{(2)(x-1)(4-3x)} + \frac{(x-1)(3)}{(x-1)(2)(4-3x)} = 0 \quad (2)$$

Simplifying equation (2):

$$\frac{10(4 - 3x) + 2(x - 1) + 3(x - 1)}{2(x - 1)(4 - 3x)} = 0$$

$$\frac{40 - 30x + 2x - 2 + 3x - 3}{2(x - 1)(4 - 3x)} = 0$$

$$\frac{-25x + 35}{2(x - 1)(4 - 3x)} = 0$$

142

Multiplying both sides of this equation by $2(x - 1)(4 - 3x)$:

$$2(x - 1)(4 - 3x)\frac{-25x + 35}{2(x - 1)(4 - 3x)} = 2(x - 1)(4 - 3x)(0)$$

$$-25x + 35 = 0$$

Adding 25x to both sides of this equation:

$$-25x + 35 + 25x = 0 + 25x$$

$$35 = 25x$$

Dividing both sides of this equation by 25:

$$\frac{35}{25} = \frac{25x}{25}$$

$$\frac{7}{5} = x$$

Therefore, the solution set to the equation $\frac{5}{x-1} + \frac{1}{4-3x} = \frac{3}{6x-8}$ is: $\left\{ \frac{7}{5} \right\}$.

● PROBLEM 10-11

Solve the equation

$$\frac{2x}{3 + x} + \frac{3 + x}{3} = 2 + \frac{x^2}{3(x - 3)} .$$

<u>Solution:</u> In order to eliminate the fractions in this equation we multiply both members of the equation by the Least Common Denominator (the LCD). Our denominators are $(3 + x), 3$, and $3(x - 3)$. Thus the LCD is $3(3 + x)(x - 3)$. Therefore

$$[3(3 + x)(x - 3)]\left[\frac{2x}{3 + x} + \frac{3 + x}{3}\right] = [3(3 + x)(x - 3)]\left[2 + \frac{x^2}{3(x-3)}\right] .$$

Distribute:

$$[3(3 + x)(x - 3)]\left[\frac{2x}{3 + x}\right] + [3(3 + x)(x - 3)]\left[\frac{3 + x}{3}\right]$$

$$= [3(3 + x)(x - 3)](2) + [3(3 + x)(x - 3)]\left[\frac{x^2}{3(x - 3)}\right] .$$

Cancelling like terms in numerator and denominator,

$$3(x - 3)(2x) + (3 + x)(x - 3)(3 + x) = 3 \cdot 2(3 + x)(x - 3)$$
$$+ (3 + x)x^2$$

Factoring both sides of the equation,

$$6x(x - 3) + (9 + 6x + x^2)(x - 3) = 6(3x + x^2 - 9 - 3x) + (3x^2 + x^3)$$
$$6x(x - 3) + (9 + 6x + x^2)(x - 3) = 6(x^2 - 9) + (3x^2 + x^3)$$

Distributing the two terms on the left side and the one term on the right side of this equation,

143

$$\left(6x^2 - 18x\right) + \left(9x + 6x^2 + x^3\right) - \left(27 + 18x + 3x^2\right) = \left(6x^2 - 54\right) + \left(3x^2 + x^3\right)$$

Grouping terms and simplifying,

$$6x^2 - 18x + 9x + 6x^2 + x^3 - 27 - 18x - 3x^2 = 6x^2 - 54 + 3x^2 + x^3$$

$$x^3 + 9x^2 - 27x - 27 = x^3 + 9x^2 - 54$$

Subtract x^3 from both sides of this equation:

$$x^3 + 9x^2 - 27x - 27 - x^3 = x^3 + 9x^2 - 54 - x^3$$

$$9x^2 - 27x - 27 = 9x^2 - 54$$

Subtract $9x^2$ from both sides of this equation:

$$9x^2 - 27x - 27 - 9x^2 = 9x^2 - 54 - 9x^2$$

$$-27x - 27 = -54$$

Add 27 to both sides of this equation:

$$-27x - 27 + 27 = -54 + 27$$

$$27x = -27$$

Divide both sides of this equation by -27:

$$\frac{-27x}{-27} = \frac{-27}{-27}$$

Therefore, $x = 1$.

To verify that $x = 1$ is the solution to our problem, we perform the following check.

Check: Replace x by 1 in our original equation,

$$\frac{2x}{3 + x} + \frac{3 + x}{3} = 2 + \frac{x^2}{3(x - 3)}$$

$$\frac{2(1)}{3 + 1} + \frac{3 + 1}{3} = 2 + \frac{(1)^2}{3(1 - 3)}$$

$$\frac{2}{4} + \frac{4}{3} = 2 + \frac{1}{3(-2)}$$

$$\frac{2}{4} + \frac{4}{3} = 2 + \frac{1}{-6}$$

$$\frac{2}{4} + \frac{4}{3} = 2 - \frac{1}{6}$$

Multiplying both members by LCD, 12:

$$12\left(\frac{2}{4} + \frac{4}{3}\right) = 12\left(2 - \frac{1}{6}\right)$$

$$12\left(\frac{2}{4}\right) + 12\left(\frac{4}{3}\right) = 12(2) - 12\left(\frac{1}{6}\right)$$

$$6 + 16 = 24 - 2$$

$$22 = 22$$

Hence, $x = 1$ is our solution, and our solution set is $\{1\}$.

Solve the equation

$$\sqrt{x} = 7 + \sqrt{x - 7}$$

<u>Solution:</u> Squaring both sides of the given equation,

$$x = 49 + 14 \sqrt{x - 7} + x - 7$$

Simplifying

$$- 42 = 14 \sqrt{x - 7}$$

$$- 3 = \sqrt{x - 7} \qquad\qquad (1)$$

Squaring both sides of equation (1),

$$9 = x - 7$$

$$x = 16$$

Checking the root by substitution in the given equation:

$$\sqrt{16} \neq 7 + \sqrt{16 - 7}$$

$$4 \neq 7 + 3$$

Clearly x = 16 does not satisfy the given equation, and therefore the equation has no roots. The fact that the given equation has no roots could have been anticipated from equation (1), $- 3 = \sqrt{x - 7}$, since the positive root is indicated in the original equation.

Solve $\sqrt{4x + 5} + 2\sqrt{x - 3} = 17$.

<u>Solution:</u> Transpose:

$$\sqrt{4x + 5} - 17 = -2\sqrt{x - 3}.$$

Square: $4x + 5 - 34\sqrt{4x + 5} + 289 = 4(x - 3)$

$$4x + 5 - 34\sqrt{4x + 5} + 289 = 4x - 12.$$

Transpose: $-34\sqrt{4x + 5} = 4x - 12 - 4x - 5 - 289.$

Simplify: $-34\sqrt{4x + 5} = -306$

$$\sqrt{4x + 5} = 9.$$

Square: $4x + 5 = 81.$

Solve for x: x = 19.

Check: $\sqrt{4(19) + 5} + 2\sqrt{(19) - 3} \overset{?}{=} 17$

$\sqrt{81} + \qquad 2\sqrt{16} \overset{?}{=} 17$

$9 + \qquad 2(4) \overset{?}{=} 17$

$17 = 17.$

Sol.: x = 19.

● **PROBLEM 10-14**

Solve the equation

$$\frac{\sqrt{x + 1} + \sqrt{x - 1}}{\sqrt{x + 1} - \sqrt{x - 1}} = \frac{4x - 1}{2}.$$

<u>Solution:</u> We can use the following law to rewrite the given proportion: If
$\frac{a}{b} = \frac{c}{d}$, then $\frac{a + b}{a - b} = \frac{c + d}{c - d}$. Applying this law we have:

$$\frac{\sqrt{x + 1} + \sqrt{x - 1} + (\sqrt{x + 1} - \sqrt{x - 1})}{\sqrt{x + 1} + \sqrt{x - 1} - (\sqrt{x + 1} - \sqrt{x - 1})} = \frac{4x - 1 + (2)}{4x - 1 - (2)} \quad \text{or,}$$

$$\frac{2\sqrt{x + 1}}{2\sqrt{x - 1}} = \frac{4x + 1}{4x - 3}. \text{ Eliminating } \frac{2}{2} \text{ we have:}$$

$$\frac{\sqrt{x + 1}}{\sqrt{x - 1}} = \frac{4x + 1}{4x - 3}.$$

Squaring both sides of the equation gives us,

$$\left(\frac{\sqrt{x + 1}}{\sqrt{x - 1}}\right)^2 = \left(\frac{4x + 1}{4x - 3}\right)^2 \quad \text{or} \quad \frac{(\sqrt{x + 1})^2}{(\sqrt{x - 1})^2} = \frac{(4x + 1)^2}{(4x - 3)^2}.$$

Finding the above squares we obtain:

$$\frac{x + 1}{x - 1} = \frac{16x^2 + 8x + 1}{16x^2 - 24x + 9}.$$

We can again rewrite this new proportion as:

$$\frac{x + 1 + (x - 1)}{x + 1 - (x - 1)} =$$

$$\frac{16x^2 + 8x + 1 + (16x^2 - 24x + 9)}{16x^2 + 8x + 1 - (16x^2 - 24x + 9)} \qquad \text{or,}$$

$$\frac{2x}{2} = \frac{32x^2 - 16x + 10}{32x - 8} \; ;$$

therefore, $x = \dfrac{32x^2 - 16x + 10}{32x - 8} = \dfrac{2\left(16x^2 - 8x + 5\right)}{2(16x - 4)} \; ;$

thus, $\quad x = \dfrac{16x^2 - 8x + 5}{16x - 4} \; ;$

and multiplying both sides of this equation by (16x - 4) we have,

$\quad\quad x(16x - 4) = 16x^2 - 8x + 5 \quad\quad$ or,

$\quad 16x^2 - 4x = 16x^2 - 8x + 5.$

Now, combining similar terms we obtain:

$\quad 16x^2 - 16x^2 - 4x + 8x = 5 \quad\quad$ or

$$4x = 5.$$

Therefore, $x = \dfrac{5}{4}$.

● **PROBLEM** 10-15

Graph the constant function 2y = 4.

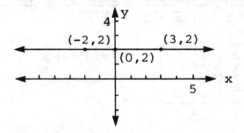

Solution: First rewrite 2y = 4 in y-form. If 2y = 4, then y = 2. Hence, g = $\{(x,y): y = 2\}$.

x	-2	0	3
y	2	2	2

For all values of x, y is equal to 2. The graph of g is a straight line with slope 0 and y-intercept (0,2).

● **PROBLEM** 10-16

Graph the function 3x - 5.

Solution: Let y = 3x - 5; then assign values to x and compute the corresponding values of y, the results being conveniently arranged in a table.

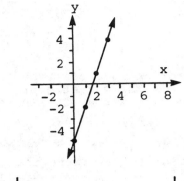

x	y = 3x - 5	y

0	y = 3(0) - 5
	  = 0 - 5
	  = -5
		-5

1	y = 3(1) - 5
	  = 3 - 5
	  = -2
		-2

2	y = 3(2) - 5
	  = 6 - 5
	  = 1
		1

3	y = 3(3) - 5
	  = 9 - 5
	  = 4
		4

The various points (x,y) are then plotted and joined by a smooth curve, which turns out to be a straight line. See the accompanying figure.

● **PROBLEM** 10-17

Graph the function defined by 3x - 4y = 12.

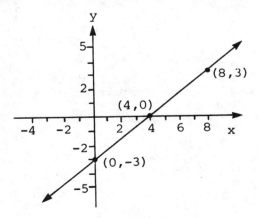

Solution: Solve for y: 3x - 4y = 12

$$-4y = 12 - 3x$$

148

$$y = -3 + \frac{3}{4}x$$

$$y = \frac{3}{4}x - 3.$$

The graph of this function is a straight line since it is
of the form y = mx + b. The y-intercept is the point
(0, -3) since for x = 0, y = b = -3. The x-intercept is

the point (4, 0) since for y = 0,

x = (y + 3) · $\frac{4}{3}$ = (0 + 3) · $\frac{4}{3}$ = 4. These two points, (0,-3)

and (4,0) are sufficient to determine the graph (see the
figure). A third point, (8,3), satisfying the equation of
the function is plotted as a partial check of the inter-

cepts. Note that the slope of the line is m = $\frac{3}{4}$. This

means that y increases 3 units as x increases 4 units anywhere along the line.

● **PROBLEM** 10-18

a) Find the zeros of the function f if f(x) = 3x - 5.
b) Sketch the graph of the equation y = 3x - 5.

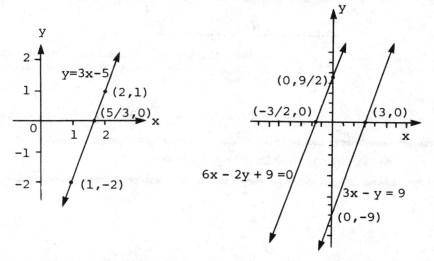

are parallel. Thus we must show that the two slopes are
equal. In standard form the equation of a line is y = mx + b,
where m is the slope.
     Putting 3x - y = 9 in standard form,

$$-y = 9 - 3x$$

$$y = -9 + 3x$$

$$y = 3x - 9.$$

Thus the slope of the first line is 3. Putting
6x - 2y + 9 = 0 in standard form,

$$-2y + 9 = -6x$$

$$-2y = -6x - 9$$

$$y = 3x + \frac{9}{2}.$$

Thus the slope of this line is also 3. The slopes are equal. Hence, the lines are parallel.

To graph these equations pick values of x and substitute them into the equation to determine the corresponding values of y. Thus we obtain the following tables of values. Notice we need only <u>two</u> points to plot a line (2 points determine a line).

$$6x - 2y + 9 = 0 \qquad\qquad 3x - y = 9$$

$$y = 3x + \frac{9}{2} \qquad\qquad y = 3x - 9$$

x	0	$-\frac{3}{2}$
y	$\frac{9}{2}$	0

x	0	3
y	-9	0

(See accompanying figure)

● **PROBLEM 10-19**

Sketch the graph of the function $H = \{(x, H(x)) \mid H(x) = 6\}$.

<u>Solution:</u> $H(x) = 6$ can be expressed in the form $H(x) = mx + b$, for in this particular example $m = 0$; hence $H(x) = 6$ can be written as $H(x) = 0 \cdot x + 6$. From this, regardless of the choice of a value for x,

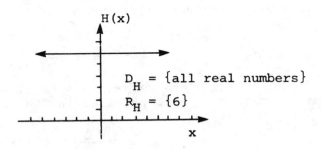

the corresponding value for $H(x)$ will be 6. When there is no domain set given, it is taken to be the largest subset of the real numbers for which the corresponding $H(x)$ value is also real. Hence the domain of H is {all real numbers} and the range is {6}. The graph of H is a horizontal line, i.e., has slope = $m = 0$ and H-intercept = $b = 6$. The graph is sketched in the figure.

Two cars traveled the same distance. One car traveled at
50 mph and the other car traveled at 60 mph. It took the
slower car 50 minutes longer to make the trip. How long
did it take the faster car to make the trip?

Solution: Step. 1. Read problem again. Step 2. If we let
t represent the number of hours the faster car travels,
we can construct the following table from the given
statements.

Note: The 50 minutes must be converted to $\frac{5}{6}$ hr.

This is done by the following:

$$\frac{50 \text{ minutes}}{60 \text{ minutes}} \times 1 \text{ hour} = \frac{50}{60} \text{ hours} = \frac{5}{6} \text{ hour}.$$

Step 3

	Distance	Rate	Time
Faster car	D	60 mph	t
Slower car	D	50 mph	t + 5/6

Note: Since the two cars travelled the same distance,
the distance for both cars is D, as indicated in the above
table.

Step 4

Formula D = rt,

where D is distance, r is rate, and t is time.

$$D = 60t$$
$$D = 50 \left(t + \frac{5}{6} \right)$$

Since the distances are the same, we can set the
two expressions for D equal as in Step 5.

Step 5

$$60t = 50 \left(t + \frac{5}{6} \right) \qquad (1)$$

Multiply each term within the parentheses by 50
to eliminate the parentheses.

$$60t = 50t + \frac{250}{6} \qquad (2)$$

Subtract 50t from both sides of equation (2).

$$60t - 50t = 50t + \frac{250}{6} - 50t$$

Therefore: $60t - 50t = \frac{250}{6}$.

Therefore: $10t = \frac{250}{6}$ (3),

Multiply both sides of equation (3) by $\frac{1}{10}$.

$$\frac{1}{\cancel{10}} (\cancel{10t}) = \frac{1}{\cancel{10}_1} \left(\frac{\cancel{250}^{25}}{6}\right)$$

$$t = \frac{25}{6} = 4\frac{1}{6} \text{ hours} = 4 \text{ hours } 10 \text{ minutes.}$$

Thus, it took the faster car 4 hours 10 minutes to make the trip.

● **PROBLEM** 10-21

Two cars are traveling 40 and 50 miles per hour, respectively. If the second car starts out 5 miles behind the first car, how long will it take the second car to overtake the first car?

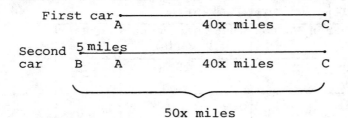

	RATE (in mph)	TIME (hours)	DISTANCE (miles)
FIRST CAR	40	x	40x
SECOND CAR	50	x	50x

Solution: Let x = number of hours it takes the second car to overtake the first car. See table.

$$\text{Distance} = \text{rate} \times \text{time}$$

Then   50x = distance second car travels in x hours,

and   40x = distance first car travels in x hours.

Since the second car must travel an additional 5 miles (from B to A in diagram).

$$40x + 5 = 50x$$

Simplify,

$$-10x = -5$$

152

Divide by -10, $x = \frac{1}{2}$, number of hours it takes the second
car to overtake the first car.

Check: $40\left(\frac{1}{2}\right) + 5 = 50\left(\frac{1}{2}\right),$

$25 = 25.$

● **PROBLEM 10-22**

A grocer mixes two grades of coffee which sell for 70 cents and 80 cents per pound, respectively. How much of each must he take to make a mixture of 50 pounds which he can sell for 76 cents per pound?

Solution: Let x = the number of pounds of 70-cent coffee. Since the mixture is to contain 50 pounds and there are x pounds of 70 cent coffee, then 50 - x = number of pounds of 80 cent coffee. [Thus the total number of pounds in the mixture is x pounds (of 70-cent coffee) + (50 - x) pounds (of 80-cent coffee) = x - x + 50 = 50 lbs, our desired amount]. Using our formula:

	Number of Pounds X	Price per Pound (in cents)	= Total Price
70¢/lb. coffee	x	70	70x
80¢/lb. coffee	50-x	80	80(50-x)
76¢/lb.	50	76	76(50)

The total price of the 70¢ coffee, 70x and total price of the 80¢ coffee, 80(50 - x) equals total price of the 76¢ coffee, (76)(50).

Therefore $70x + 80(50 - x) = (50)(76).$

Using the distributive law, we obtain:

$70x + 4000 - 80x = 3800.$

Subtract 4000 from both sides,

$70x - 80x = 3800 - 4000.$

Collect terms, $-10x = -200.$

Divide by -10, $x = 20$, number of pounds of 70-cent coffee.

Then $50 - x = 30$, number of pounds of 80-cent coffee.

153

Check: $(70)(20) + (80)(30) = (50)(76)$,

$$1400 + 2400 = 3800,$$

$$3800 = 3800.$$

How many quarts of pure alcohol must be added to 40 quarts of a mixture that is 35% alcohol to make a mixture that will be 48% alcohol?

Solution: Let x = number of quarts of pure alcohol to be added. The relationship used to set up the equation is

Amount of alcohol in new mixture = Amount of alcohol in old mixture + Amount of alcohol added

$$.48(40 + x) = (.35)(40) + x$$
$$19.2 + .48x = 14 + x$$
$$x = 10 \text{ quarts of alcohol}$$

Check: Amount of alcohol in new mixture = 14 quarts + 10 quarts = 24 quarts. New mixture contains a total of 40 + 10 = 50 quarts.

$$\frac{24}{50} = 48\% \ .$$

If a container contains a mixture of 5 gallons of white paint and 11 gallons of brown paint, how much white paint must be added to the container so that the new mixture will be two-thirds white paint?

Solution: Let x = number of gallons of white paint to be added then x + 5 = number of gallons of white paint in the final mixture. x + 16 = total number of gallons of paint in the final mixture.

The new mixture will be two-thirds white paint if the ratio of the final number of gallons of white paint to the final total number of gallons of paint is equal to $\frac{2}{3}$. The proportion is

$$\frac{x + 5}{x + 16} = \frac{2}{3}$$

$$3x + 15 = 2x + 32 \text{ by cross-multiplying}$$

$$3x - 2x = 32 - 15 \text{ by isolating x terms}$$

$$x = 17 \text{ gallons by combining like terms.}$$

Check: The final mixture would contain x + 5 = 22 gallons of white paint, and the total number of gallons of paint would be x + 16 = 33 gallons.

$$\frac{\text{final number of gallons of white paint}}{\text{final total number of gallons of paint}} = \frac{22 \text{ gal}}{33 \text{ gal}} = \frac{2}{3}.$$

● **PROBLEM 10-25**

The sum of two numbers is 23. One of the numbers is 7 more than the other number. What are the numbers?

<u>Solution:</u> Let x = one of the numbers, and x + 7 = the other number. Since we are given that the sum of the two numbers is 23,

$$x + (x + 7) = 23.$$

By the associative law of addition:

$$x + (x + 7) = 23$$

is the same as $(x + x) + 7 = 23,$

or $2x + 7 = 23.$

Subtract 7 from both sides:

$$2x = 23-7.$$

Collect terms, $2x = 16.$

Divide by 2, $x = 8$, one of the numbers.

Then solving for our other number x + 7, we substitute 8 for x.

Hence, $x + 7 = 8 + 7 = 15$, the other number.

Therefore, the two numbers are 8 and 15. We can verify this result by observing that the sum of the two numbers is indeed 23, and 15 is 7 more than 8, 8 + 7 = 15.

● **PROBLEM 10-26**

A toy savings bank contains $17.30 consisting of nickels, dimes, and quarters. The number of dimes exceeds twice the number of nickels by 3 and the number of quarters is 4 less than 5 times the number of nickels. How many of each coin are in the bank?

<u>Solution:</u> Let x = the number of nickels
Then 2x + 3 = the number of dimes
And 5x - 4 = the number of quarters
The relationship used in setting up the equation is:

Value of nickels + Value of dimes + Value of quarters = 1730.

$5x + 10(2x + 3) + 25(5x - 4) = 1730; x = 12$

There are 12 nickels, 27 dimes and 56 quarters in the bank.

Check: The nickels are worth $.60, the dimes are worth $2.70, and the quarters are worth $14.00, making a total of $17.30.

● **PROBLEM 10-27**

Can the sum of three consecutive odd integers be (a) 25? (b) 45?

<u>Solution:</u> Notice that all consecutive odd integers differ by 2:

$$1, \quad 1 + 2 = 3, \quad 3 + 2 = 5, \quad 5 + 2 = 7, \quad 7 + 2 = 9, \ldots$$

Thus, if we let x = the first consecutive odd integer
          x + 2 = the 2nd consecutive odd integer
and
  (x+2) + 2 = x + 4 = the 3rd consecutive odd integer,

(a) We take the sum of these three numbers and determine if it can be 25:

$$x + (x+2) + (x+4) = 25$$
$$3x + 6 = 25$$
$$3x = 19$$
$$x = \frac{19}{3}.$$

Since 19/3 is not an integer, there are no such odd integers.

(b) If 25 is replaced by 45, the equation takes the form

$$x + (x+2) + (x+4) = 45$$
$$3x + 6 = 45$$
$$3x = 39$$
$$x = \frac{39}{3} = 13$$
$$x + 2 = 13 + 2 = 15$$
$$x + 4 = 13 + 4 = 17$$

Thus, the three consecutive odd integers are 13, 15, 17.

In general, if the sum of three consecutive odd integers is to be the number N, then N must be an integral multiple of 3.

● **PROBLEM 10-28**

John is now 18 years old and his brother, Charles, is 14 years old. How many years ago was John twice as old as Charles?

<u>Solution:</u> Let x = the number of years ago John was twice as old as Charles.

Then 18 - x = John's age x years ago.
And 14 - x = Charles' age x years ago.
The relationship used in setting up the equation is:
    x years ago, John was twice as old as Charles

$$18 - x = 2(14 - x); \quad x = 10$$

Check: 10 years ago, John was 8 and Charles was 4. At this time, John was twice as old as Charles.

A tank can be filled in 9 hours by one pipe, in 12 hours by a second pipe, and can be drained when full, by a third pipe, in 15 hours. How long would it take to fill the tank if it is empty, and if all pipes are in operation?

Solution: Let x = the number of hours the pipes are in operation

Then $\frac{x}{9}$ = part of tank filled by first pipe

and $\frac{x}{12}$ = part of tank filled by second pipe

and $\frac{x}{15}$ = part of tank emptied by third pipe

The relationship used in setting up the equation is:

Part of tank filled by first pipe + Part of tank filled by second pipe - Part of tank emptied by third pipe = 1 Full tank.

$$\frac{x}{9} + \frac{x}{12} - \frac{x}{15} = 1$$

$$20x + 15x - 12x = 180$$

$$x = 7\frac{19}{23} \text{ hours}$$

Check: $\dfrac{7\frac{19}{23}}{9} + \dfrac{7\frac{19}{23}}{12} - \dfrac{7\frac{19}{23}}{15} = 1$

$$\left(\frac{180}{23} \cdot \frac{1}{9}\right) + \left(\frac{180}{23} \cdot \frac{1}{12}\right) - \left(\frac{180}{23} \cdot \frac{1}{15}\right) = 1$$

$$\frac{20}{23} + \frac{15}{23} - \frac{12}{23} = 1$$

A man wishes to invest a part of $4200 in stocks earning 4% dividends and the remainder in bonds paying 2½%. How much must he invest in stocks to receive an average return of 3% on the whole amount of money?

Solution: Let x = amount invested at 4%

Then 4200 - x = amount invested at 2½%

The relationship used to set up the equation is:

Income from 4% investment + Income from 2½% investment = 3% of $4200, or $126.

$$.04x + .025(4200 - x) = 126$$
$$40x + 25(4200 - x) = 126,000$$
$$40x + 105,000 - 25x = 126,000$$
$$15x = 21,000$$
$$x = 1,400$$
$$4200 - x = 2,800$$

Check: 4% of $1,400 = $56
2½% of $2,800 = $70
Total income = $126 which is 3% of $4,200

CHAPTER 11

SYSTEMS OF LINEAR EQUATIONS

> **Basic Attacks and Strategies for Solving Problems in this Chapter. See pages 159 to 180 for step-by-step solutions to problems.**

A system of linear equations or a simultaneous system of *linear equations* is any set of linear equations each of which is assumed to be satisfied by the same set (or sets) of values for the variables. For example,

$$x + y = 0$$

$$x - y = 4$$

both hold for the values $x = 2$, $y = -2$ which are considered to be the solution of this linear system.

A system of linear equations normally consists of as many equations as different variables in the various equations. Thus, if there are two variables, there should normally be two equations. If there are more equations than variables, the system is said to be *over-determined* and possibly inconsistent (i.e., no solution exists for all the equations, although one may exist for fewer equations). If there are fewer equations than variables, the system is said to be *under-determined* and one unique solution set (i.e., one unique value for each variable) cannot be determined. Even if there are the correct number of equations, it is possible that a unique solution is not possible, or that multiple solutions are possible.

In the two-dimensional case, that is, when dealing with two variables, one frequently speaks about a *2 by 2 linear system* (written 2 × 2). In such a system, each linear equation corresponds to a line in the Cartesian coordinate plane. (In the three-dimensional case, a linear equation corresponds to a *plane* in three-dimensional space, and so on in higher dimensions.) The solution of a system of linear equations corresponds to those points at which the graphs of two equations (in two variables) intersect when drawn in the plane. Thus, one can solve a 2 × 2 system graphically by carefully drawing the graphs of each linear equation and identifying the point of intersection.

There are numerous ways to solve 2 × 2 linear systems, and some of these methods can be applied to systems of a higher order.

In the *method of substitution*, one solves one equation for one variable in terms of the other, and then substitutes this value into the other equation. For example, given

$$2x + y = 0$$

$$x - 2y = 4$$

one can solve the second equation for x, obtaining $x = 2y + 4$, and then substitute this into the first equation, obtaining

$$2 \cdot (2y + 4) + y = 4y + 8 + y = 5y + 8 = 0,$$

which gives $y = -\,^8/_5$. One can substitute this back into the intermediate equation to find the corresponding value of x, i.e.,

$$x = 2 \cdot (-\,^8/_5) + 4 = -\,^{16}/_5 + 4 = \,^4/_5.$$

In the *method of elimination by adding* (or *Gaussian elimination*), one multiplies each equation by an appropriate quantity so that the new equations have corresponding terms that sum to zero. One can then add the two equations and solve the resulting equation for the remaining variable. For example, given the previous example

$$2x + y = 0$$

$$x - 2y = 4$$

one can multiply the first equation by 2 and the second one by 1 (i.e., leave it unchanged). One then obtains

$$4x + 2y = 0$$

$$x - 2y = 4$$

which can be added together to obtain $5x = 4$. This gives a solution value of $x = \,^4/_5$ as above, and the value of y can be obtained by substituting the value of x into either equation and solving again.

One can always verify a solution by substituting the answer back into the original equations to check for equality.

If we think of 2×2 linear systems as representing two lines in a plane, then clearly it is possible to have two lines that are parallel, and thus do not have any point of intersection. If we try to solve the equivalent linear system we will obtain some arithmetic impossibility, such as two different constants equaling each other.

It is also possible to have two different representations of the same line (i.e., one equation is a constant multiple of the other). In this case, there exist an infinite number of solutions. If we try to solve the equivalent linear system, we will obtain an algebraic identity, such as $x = x$ or $1 = 1$.

To solve a 3 × 3 linear system, one normally uses the method of elimination in some form. Often this results in a 2 × 2 system that can then be solved in one of the standard ways. Most often, one uses an appropriate multiple of one equation to remove one of the variables from the other two equations by adding and subtracting. After obtaining the values of two variables, one can substitute in any equation to obtain the value of the third. Since a 3 × 3 system can be considered to represent 3 planes in 3 dimensions, there is the possibility that the three planes will not all intersect in a common point or that two (or all three) planes will be parallel. In these cases, one cannot obtain any solution to the system.

Word problems can also result in systems of linear equations. Often the major difficulty occurs in trying to translate the information given in the statement of the problem into appropriate equations that can then be solved. The techniques for solving word problems suggested in the previous chapter can also be applied to problems that are solved by a system of linear equations.

Step-by-Step Solutions to Problems in this Chapter, "Systems of Linear Equations"

● PROBLEM 11-1

Solve

$$\begin{cases} y = x + 4 \\ -x = y \end{cases} \quad \text{graphically.}$$

<u>Solution</u>: First graph the two equations as shown in the figure.

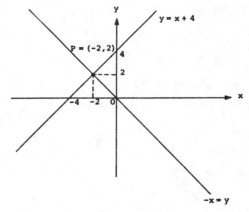

The solution of this system of equations is the intersection of the two lines. It's found graphically that the point of intersection is P(-2,2). Therefore, the solution of the given system of equations is x = -2, y = 2.

● PROBLEM 11-2

Solve the simultaneous equations 2x + 4y = 11, -5x + 3y = 5 by the method of substitution and by the method of elimination by addition.

159

<u>Solution:</u> The method of substitution involves solving for one variable in terms of the other and then substituting the obtained value into the second equation. Thus, we solve the first equation for x and substitute in the second:

$$2x + 4y = 11$$
$$2x = 11 - 4y$$

$$x = \frac{11 - 4y}{2}$$

Replacing x by $\left(\frac{11 - 4y}{2}\right)$ in the second equation,

$$-5\left(\frac{11 - 4y}{2}\right) + 3y = 5$$

$$\frac{-55 + 20y}{2} + 3y = 5$$

$$\frac{-55}{2} + 10y + 3y = 5$$

Multiply both sides by 2,

$$-55 + 20y + 6y = 10$$
$$26y = 65$$

$$y = \frac{65}{26} = \frac{5}{2} .$$

Substituting this value for y into the first equation:

$$2x + 4\left(\frac{5}{2}\right) = 11$$

$$2x + 10 = 11$$
$$2x = 1$$
$$x = \frac{1}{2} .$$

We obtain the same result by the method of elimination by addition.

$$2x + 4y = 11 \qquad\qquad (1)$$
$$-5x + 3y = 5 \qquad\qquad (2)$$

Multiplying equation (1) by 5 and equation (2) by 2 and adding the result we obtain:

$$10x + 20y = 55$$
$$- \underline{10x + 6y = 10}$$
$$26y = 65$$
$$y = \frac{65}{26} = \frac{5}{2}$$

Once again, replacing y by $\frac{5}{2}$ in equation (1):

$$2x + 4\left(\frac{5}{2}\right) = 11$$

$$2x + 10 = 11$$
$$2x = 1$$
$$x = \frac{1}{2}$$

Thus $\left\{\left(\frac{1}{2} , \frac{5}{2}\right)\right\}$ is the solution to the given system of equations.

● **PROBLEM** 11-3

Solve the equations $2x + 3y = 6$ and $4x + 6y = 7$ simultaneously.

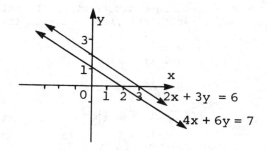

<u>Solution:</u> We have 2 equations in 2 unknowns,

$$2x + 3y = 6 \qquad (1)$$

and

$$4x + 6y = 7 \qquad (2)$$

There are several methods to solve this problem. We have chosen to multiply each equation by a different number so that when the two equations are added, one of the variables drops out. Thus

multiplying equation (1) by 2: $\quad 4x + 6y = 12 \qquad (3)$

multiplying equation (2) by -1: $\quad \underline{-4x - 6y = -7} \qquad (4)$

adding equations (3) and (4): $\qquad 0 = 5$

We obtain a peculiar result!

Actually, what we have shown in this case is that if there were a simultaneous solution to the given equations, then 0 would equal 5. But the conclusion is impossible; therefore there can be no simultaneous solution to these two equations, hence no point satisfying both.

The straight lines which are the graphs of these equations must be parallel if they never intersect, but not identical, which can be seen from the graph of these equations (see the accompanying diagram).

● **PROBLEM 11-4**

> Solve for x and y .
> $$x + 2y = 8 \qquad (1)$$
> $$3x + 4y = 20 \qquad (2)$$

<u>Solution:</u> Solve equation (1) for x in terms of y:

$$x = 8 - 2y \qquad (3)$$

Substitute $(8 - 2y)$ for x in (2):

$$3(8 - 2y) + 4y = 20 \qquad (4)$$

Solve (4) for y as follows:

Distribute: $\quad 24 - 6y + 4y = 20$

Combine like terms and then subtract 24 from both sides:

$$24 - 2y = 20$$

$$24 - 24 - 2y = 20 - 24$$

$$-2y = -4 \qquad \text{Divide both sides by } -2: \quad y = 2$$

Substitute 2 for y in equation (1):

$$x + 2(2) = 8$$
$$x = 4$$

Thus, our solution is x = 4, y = 2.

Check: Substitute x = 4, y = 2 in equations (1) and (2):

$$4 + 2(2) = 8$$
$$8 = 8$$
$$3(4) + 4(2) = 20$$
$$20 = 20$$

● **PROBLEM 11-5**

Solve the equations $2x + 3y = 6$ and $y = -(2x/3) + 2$ simultaneously.

Solution: We have 2 equations in 2 unknowns,

$$2x + 3y = 6 \qquad\qquad (1)$$
$$y = -(2x/3) + 2 \qquad\qquad (2)$$

There are several methods of solution for this problem. Since equation (2) already gives us an expression for y, we use the method of substitution. Substituting $-(2x/3) + 2$ for y in the first equation:

$$2x + 3\left(-\frac{2x}{3} + 2\right) = 6$$

Distributing,
$$2x - 2x + 6 = 6$$
$$6 = 6$$

Apparently we have gotten nowhere! The result 6 = 6 is true, but indicates no solution. Actually, our work shows that no matter what real number x is, if y is determined by the second equation, then the first equation will always be satisfied.

The reason for this peculiarity may be seen if we take a closer look at the equation $y = -(2x/3) + 2$. It is equivalent to $3y = -2x + 6$, or $2x + 3y = 6$.

In other words, the two equations are equivalent. Any pair of values of x and y which satisfies one satisfies the other.

It is hardly necessary to verify that in this case the graphs of the given equations are identical lines, and that there are an infinite number of simultaneous solutions of these equations.

● **PROBLEM 11-6**

Solve algebraically:
$$\begin{cases} 4x + 2y = -1 & \qquad (1) \\ 5x - 3y = 7 & \qquad (2) \end{cases}$$

<u>Solution:</u> We arbitrarily choose to eliminate x   first.

Multiply (1) by 5: $20x + 10y = -5$ (3)

Multiply (2) by 4: $20x - 12y = 28$ (4)

Subtract, (3) - (4): $22y = -33$ (5)

Divide (5) by 22: $y = -\frac{33}{22} = -\frac{3}{2}.$

To find x, substitute $y = -\frac{3}{2}$ in either of the original equations. If we use Eq. (1), we obtain $4x + 2(-3/2) = -1$, $4x - 3 = -1$, $4x = 2$, $x = \frac{1}{2}.$

The solution $\left(\frac{1}{2}, -\frac{3}{2}\right)$ should be checked in both equations of the given system.

Replacing $\left(\frac{1}{2}, -\frac{3}{2}\right)$ in Eq. (1):

$$4x + 2y = -1$$
$$4\left(\frac{1}{2}\right) + 2\left(-\frac{3}{2}\right) = -1$$
$$\frac{4}{2} - 3 = -1$$
$$2 - 3 = -1$$
$$-1 = -1$$

Replacing $\left(\frac{1}{2}, -\frac{3}{2}\right)$ in Eq. (2):

$$5x - 3y = 7$$
$$5\left(\frac{1}{2}\right) - 3\left(-\frac{3}{2}\right) = 7$$
$$\frac{5}{2} + \frac{9}{2} = 7$$
$$\frac{14}{2} = 7$$
$$7 = 7$$

(Instead of eliminating x from the two given equations, we could have eliminated y by multiplying Eq. (1) by 3, multiplying Eq. (2) by 2, and then adding the two derived equations.)

● **PROBLEM 11-7**

Determine the nature of the system of linear equations

$$2x + y = 6 \qquad (1)$$
$$4x + 2y = 8 \qquad (2)$$

<u>Solution:</u> These linear equations may be written in the standard form $y = mx + b$:

$$y = -2x + 6 \qquad (3-1)$$
$$\text{and } y = -2x + 4 \qquad (4-2)$$

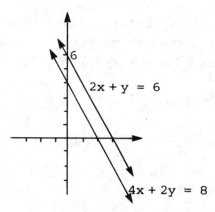

Observe that the slope of each line is m = -2, but the
y-intercepts are different, that is, b = 6 for equation
(3-1) and b = 4 for equation (4-2). The lines are therefore parallel and distinct. The graph below also indicates
that the lines are parallel. The system is therefore inconsistent, and there is no solution.

● **PROBLEM 11-8**

Determine the nature of the system of linear equations

$$x + 2y = 8$$

$$x - 2y = 2.$$

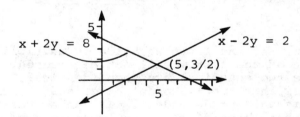

<u>Solution:</u> Add the two equations, eliminating the y-terms,
to obtain a single equation in terms of x. Values of x
satisfying this equation will yield solutions of the system.

$$x + 2y = 8 \qquad (1)$$

$$+ x - 2y = 2 \qquad (2)$$

$$\overline{2x \qquad = 10}$$

$$x \qquad = 5$$

Substituting x = 5 into Equation (1) yields
y = (8 - x)/2 = (8 - 5)/2 = $\frac{3}{2}$ or into Equation (2) yields

y = (2 - x)/(-2) = (2 - 5)/(-2) = $\frac{3}{2}$. Thus we have x = 5,

y = $\frac{3}{2}$ as the only solution of the system. Alternately, the figure indicates that the lines intersect in the point $\left(5, \frac{3}{2}\right)$. The system is therefore consistent and independent. Substitution of x = 5 and y = $\frac{3}{2}$ in both equations yields

$$5 + 2\left(\frac{3}{2}\right) = 8, \text{ or } 8 = 8$$

$$5 - 2\left(\frac{3}{2}\right) = 2, \text{ or } 2 = 2$$

so that x = 5, y = 3/2, is a solution, and the only solution of the system.

● **PROBLEM 11-9**

Determine the nature of the system of linear equations

$$x + 3y = 4 \tag{1}$$
$$2x + 6y = 8. \tag{2}$$

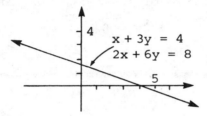

Solution: If the first equation is multiplied by 2, the solution of the system will not be altered. Note, however, that the two equations are then identical. The graph too, indicates that the lines coincide, and therefore the system is consistent and dependent. It can be verified by substitution that three of the solutions are x = 1, y = 1; x = 7, y = -1; and x = -5, y = 3.

● **PROBLEM 11-10**

Solve for x and y .

$$3x + 2y = 23 \tag{1}$$
$$x + y = 9 \tag{2}$$

Solution: Multiply equation (2) by -3:

$$-3x - 3y = -27 \tag{3}$$

Add equations (1) and (3):

$$3x + 2y = 23$$
$$\underline{-3x - 3y = -27}$$
$$-y = -4$$

$$y = 4$$

Substitute 4 for y in equation (1):

$$3x + 2(4) = 23$$

$$3x + 8 \quad = 23$$

Subtract 8 from both sides: $3x = 15$

Divide each side by 3: $\quad x = 5$

Hence our solution is, $x = 5$ and $y = 4$.

Check: Substitute 5 for x and 4 for y in equation (1):

$$3(5) + 2(4) = 23$$

$$23 = 23$$

Substitute 5 for x and 4 for y in equation (2):

$$5 + 4 = 9$$

$$9 = 9 \; .$$

● **PROBLEM 11-11**

Find the point of intersection of the graphs of the equations:
$$\begin{cases} x + y = 3, \\ 3x - 2y = 14. \end{cases}$$

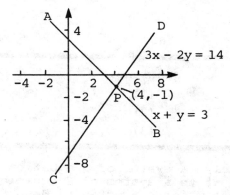

Solution: To solve these linear equations, solve for y in terms of x. The equations will be in the form $y = mx + b$, where m is the slope and b is the intercept on the y-axis.

$$x + y = 3$$
$$y = 3 - x \quad \text{subtract } x \text{ from both sides}$$

$$3x - 2y = 14 \quad \text{subtract } 3x \text{ from both sides}$$

$$-2y = 14 - 3x \quad \text{divide by } -2.$$

$$y = -7 + \frac{3}{2} x$$

The graphs of the linear functions, $y = 3 - x$ and $y = -7 + \frac{3}{2} x$, can be determined by plotting only two points. For example, for $y = 3 - x$, let $x = 0$, then $y = 3$. Let $x = 1$, then $y = 2$. The two points on this first line are $(0,3)$ and $(1,2)$. For $y = -7 + 3/2 \ x$, let $x = 0$, then

$y = -7$. Let $x = 1$, then $y = -5\frac{1}{2}$. The two points on this second line are $(0, -7)$ and $(1, -5\frac{1}{2})$.

To find the point of intersection P of

$$x + y = 3$$

and

$$3x - 2y = 14,$$

solve them algebraically. Multiply the first equation by 2. Add these two equations to eliminate the variable y.

$$
\begin{array}{rcl}
2x + 2y & = & 6 \\
3x - 2y & = & 14 \\
\hline
5x & = & 20
\end{array}
$$

Solve for x to obtain $x = 4$. Substitute this into $y = 3 - x$ to get $y = 3 - 4 = -1$. P is $(4,1)$. AB is the graph of the first equation, and CD is the graph of the second equation. The point of intersection P of the two graphs is the only point on both lines. The coordinates of P satisfy both equations and represent the desired solution of the problem. From the graph, P seems to be the point $(4,-1)$. These coordinates satisfy both equations, and hence are the exact coordinates of the point of intersection of the two lines.

To show that $(4,-1)$ satisfies both equations, substitute this point into both equations.

$$
\begin{array}{rcl}
x + y & = & 3 \\
4 + (-1) & = & 3 \\
4 - 1 & = & 3 \\
3 & = & 3
\end{array}
\qquad\qquad
\begin{array}{rcl}
3x - 2y & = & 14 \\
3(4) - 2(-1) & = & 14 \\
12 + 2 & = & 14 \\
14 & = & 14
\end{array}
$$

● PROBLEM 11-12

Solve the system of equations,

$$
\begin{array}{ll}
2x - y - 4z = 3 & (1) \\
-x + 3y + z = 10 & (2) \\
3x + 2y - 2z = 2 & (3)
\end{array}
$$

<u>Solution:</u> To solve a system of 3 equations in 3 unknowns, we first reduce it to a system of 2 equations in 2 unknowns, a process which can often be done many ways. Although various other algebraic manipulations may be used to arrive at the same result, we will employ the following method: Multiplying equation 1 by (-1) we obtain,

$$2x + y + 4z = 3 \qquad (4)$$

Adding equations (4), (2), and (3) we obtain,

$$-2x + y + 4z = -3$$

$$-x + 3y + z = 10$$

$$\underline{3x + 2y - 2z = 2}$$

$$6y + 3z = -15 \qquad (5)$$

Multiplying equation (2) by 3 we obtain,

$$-3x + 9y + 3z = -30 \qquad (6)$$

Adding equations (6) and (3) we obtain,

$$-3x + 9y + 3z = -30$$

$$\underline{3x + 2y - 2z = -2}$$

$$11y + z = -32 \qquad (7)$$

Multiplying equation (7) by (-3) we obtain,

$$-33y - 3z = 96 \qquad (8)$$

Adding equations (8) and (5) we obtain,

$$-33y - 3z = 96$$

$$\underline{6y + 3z = -15}$$

$$-27y = 81$$

$$y = -3$$

Solving for z, we replace, y by (-3) in equation (5):

$$6y + 3z = -15$$

$$6(-3) + 3z = -15$$

$$-18 + 3z = -15$$

$$3z = 3$$

$$z = 1$$

Solving for x, we replace y by (-3) and z by 1 in equation (1):

$$2x - y - 4z = 3$$

$$2x - (-3) - 4(1) = 3$$

$$2x + 3 - 4 = 3$$

$$2x - 1 = 3$$

$$2x = 4$$

$$x = 2$$

Thus the solution to this system is x = 2, y = -3, and z = 1.

Check: Replace x,y, and z by 2, -3, and 1 in each equation.

$$2x - y - 4z = 3 \qquad (1)$$

$$2(2) - (-3) - 4(1) = 3$$

$$4 + 3 - 4 = 3$$

$$3 = 3$$

$$-x + 3y + z = -10 \qquad (2)$$

$$-(2) + 3(-3) + 1 = -10$$

$$-2 - 9 + 1 = -10$$

$$-10 = -10$$

$$3x + 2y - 2z = -2 \qquad (3)$$

$$3(2) + 2(-3) - 2(1) = -2$$

$$6 - 6 - 2 = -2$$

$$-2 = -2$$

● **PROBLEM 11-13**

Find the solution set for the system:

$$3x + 4y - z = -2$$
$$2x - 3y + z = 4$$
$$x - 6y + 2z = 5$$

Solution: Adding the first and second equations, we obtain another equation without a term involving z:

$$\begin{array}{r} 3x + 4y - z = -2 \\ \underline{2x - 3y + z = 4} \\ 5x + y \quad\;\; = 2 \end{array}$$

Similarly, after multiplying through by -2 in the second equation, we can use this new equation and the third one to obtain another equation without a term involving z:

$$\begin{array}{r} -4x + 6y - 2z = -8 \\ \underline{x - 6y + 2z = 5} \\ -3x \qquad\qquad = -3 \end{array}$$

Our problem has been somewhat simplified in that not only have we obtained an equation without a term involving z, but we have obtained one without a y term.

The solution set of $-3x = -3$ is $\{1\}$. Upon substituting this into the equation $5x + y = 2$, we find that $y = -3$. Finally, upon substituting these values for x and y in either of the three equations of the system, we can obtain a value for z. If we use the first equation, $3x + 4y - z = -2$, we find that $z = -7$.
Hence the solution set for this system is $\{(1, -3, -7)\}$.

Solve for x, y and z:

$$5x + y - z = 9 ,\qquad(1)$$

$$3x + y + 2z = 17 ,\qquad(2)$$

$$x + 2y + 3z = 20.\qquad(3)$$

Solution: Subtract (2) from (1):

$$5x + y - z = 9$$
$$\underline{- (3x + y + 2z = 17)}$$
$$2x - 3z = -8 \qquad(4)$$

Multiply (2) by 2: $6x + 2y + 4z = 34.$ (5)

Subtract (3) from (5):

$$6x + 2y + 4z = 34$$
$$\underline{- (x + 2y + 3z = 20)}$$
$$5x + z = 14 \qquad(6)$$

Subtract 5x from both sides: $z = 14 - 5x$ (7)

Substitute (14 - 5x) for z in equation (4):

$$2x - 3(14 - 5x) = -8 .$$

Distribute: $2x - 42 + 15x = -8$

$$17x - 42 = -8$$

Add 42 to both sides: $17x = 34$

$$x = 2$$

Substitute 2 for x in equation (7)

$$z = 14 - 5(2) = 14 - 10 = 4 .$$

Therefore, $x = 2$, and $z = 4$.

Substitute in (1): $5(2) + y - 4 = 9$

$$10 + y - 4 = 9$$
$$6 + y = 9$$

Subtract 6 from both sides: $y = 3$

Thus, $x = 2$, $y = 3$, $z = 4$.

Check: $5(2) + 3 - 4 = 9,$

$$9 = 9.$$

$$3(2) + 3 + 2(4) = 17,$$

$$17 = 17.$$

$$2 + 2(3) + 3(4) = 20,$$

$$20 = 20.$$

● **PROBLEM 11-15**

Solve the system:
$$2a - 3b + c = 2 \qquad (1)$$
$$3a + 2b - c = 4 \qquad (2)$$
$$2a - 3b + c = 5 \qquad (3)$$

<u>Solution:</u> Observe that equations (1) and (3) are inconsistent.

$$2a - 3b + c = 2 \neq 5 = 2a - 3b + c$$

This is a contradiction; that is, $2a - 3b + c$ cannot equal both 2 and 5 at the same time. This implies that there are no values of a,b, and c which will solve this set of simultaneous equations, hence this system has no solution.

● **PROBLEM 11-16**

Two airfields A and B are 400 miles apart, and B is due east of A. A plane flew from A to B in 2 hours and then returned to A in 2½ hours. If the wind blew with a constant velocity from the west during the entire trip, find the speed of the plane in still air and the speed of the wind.

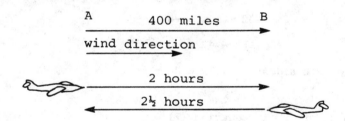

<u>Solution:</u> The essential point in solving this problem is that the wind helps the plane in flying from A to B and hinders it in flying from B to A. We therefore have the basis for two equations that involve the speed of the plane, the speed of the wind, and the time for each trip. We let

$$x = \text{speed of plane in still air, in miles per hour}$$
$$y = \text{speed of wind, in miles per hour}$$

Then, since the wind blew constantly from the west,

$$x + y = \text{speed of plane from A to B (wind helping)}$$
$$x - y = \text{speed of plane from B to A (wind hindering)}$$

The distance traveled each way was 400 miles, and so we have the following equations based on the formula distance/rate = time:

$$\frac{400}{x+y} = 2 = \text{time required for eastward trip} \qquad (8)$$

$$\frac{400}{x-y} = 2\tfrac{1}{2} = \text{time required for westward trip} \qquad (9)$$

We solve these equations simultaneously for x and y

$$400 = 2x + 2y \quad \text{multiplying (8) by } x + y \qquad (10)$$
$$800 = 5x - 5y \quad \text{multiplying (9) by } 2(x - y) \qquad (11)$$
$$2{,}000 = 10x + 10y \quad \text{multiplying (10) by 5} \qquad (12)$$
$$\underline{1{,}600 = 10x - 10y} \quad \text{multiplying (11) by 2} \qquad (13)$$
$$3{,}600 = 20x \qquad \text{adding equations (12) and (13)}$$

$$x = 180 \qquad \text{solving for } x$$
$$400 = 360 + 2y \quad \text{replacing } x \text{ by } 180 \text{ in (10)}$$
$$2y = 40$$
$$y = 20$$

Therefore the solution set of equations (8) and (9) is $\{(180, 20)\}$, and it follows that the speed of the plane in still air is 180 miles per hour and the speed of the wind is 20 miles per hour.

Check:
$$\frac{400}{180 + 20} = \frac{400}{200} = 2 \qquad \text{from (8)}$$

$$\frac{400}{180 - 20} = \frac{400}{160} = \frac{5}{2} = 2\tfrac{1}{2} \qquad \text{from (9)}$$

● **PROBLEM 11-17**

A chemist has an 18% solution and a 45% solution of a disinfectant. How many ounces of each should be used to make 12 ounces of a 36% solution?

Solution: Let x = Number of ounces from the 18% solution
And y = Number of ounces from the 45% solution

(1) $x + y = 12$
(2) $.18x + .45y = .36(12) = 4.32$

Note that .18 of the first solution is pure disinfectant and that .45 of the second solution is pure disinfectant. When the proper quantities are drawn from each mixture the result is 12 gallons of mixture which is .36 pure disinfectant, i.e., the resulting mixture contains 4.32 ounces of pure disinfectant.

When the equations are solved, it is found that

$$x = 4 \quad \text{and} \quad y = 8.$$

● **PROBLEM 11-18**

What quantities of silver 60 per cent and 82 per cent pure must be mixed together to give 12 ounces of silver 70 per cent pure?

Solution: Let x = number of ounces of 60 per cent silver, and

172

y = number of ounces of 82 per cent silver.

We use the following table to describe the given information:

	Number of ounces	% Pure Silver	Number of Ounces of Pure Silver
Silver (60%)	x	60	.60x
Silver (82%)	y	82	.82y
Silver (70%)	12	70	.70(12)

From the information obtained in the table we have the following equations:

$$.60x + .82y = .70(12) \qquad (1)$$

$$x + y = 12 \qquad (2)$$

Multiplying each term of equation (1) by 100, we obtain:

$$60x + 82y = 70(12) \qquad (3)$$

Equation (2) multiplied by 60 gives:

$$60x + 60y = (12)(60). \qquad (4)$$

Then equation (3)-(4) gives:

$$\begin{aligned} 60x + 82y &= 840 \\ -60x - 60y &= -720 \\ \hline 22y &= 120 \end{aligned} \quad ;$$

dividing both sides by 22, $\quad y = \dfrac{120}{22} = \dfrac{60}{11} = 5\dfrac{5}{11}$.

Substituting $5\dfrac{5}{11}$ for y in (2) gives

$$x + 5\dfrac{5}{11} = 12, \text{ or } \quad x + \dfrac{60}{11} = \dfrac{132}{11}$$

Therefore, $\qquad x = \dfrac{72}{11} = 6\dfrac{6}{11}$.

Thus, we must mix $6\dfrac{6}{11}$ ounces of 60 per cent pure silver and $5\dfrac{5}{11}$ ounces of 82 per cent pure silver to obtain 12 ounces of silver 70 per cent pure.

Check: Substituting $6\dfrac{6}{11}$ for x and $5\dfrac{5}{11}$ for y in (3) gives

$$\left(6\dfrac{6}{11}\right)(60) + \left(5\dfrac{5}{11}\right)(82) = (70)(12)$$

Convert $6\dfrac{6}{11}$ and $5\dfrac{5}{11}$ to fractions, $\dfrac{72}{11}(60) + \dfrac{60}{11}(82) = 840.$

Multiply, $\dfrac{4320}{11} + \dfrac{4920}{11} = 840$

Add fractions, $\dfrac{9240}{11} = 840$

$$840 = 840 \quad .$$

Substituting in (2) gives

$$6\,\dfrac{6}{11} + 5\,\dfrac{5}{11} = 12$$

$$12 = 12.$$

● **PROBLEM 11-19**

A tobacco dealer mixed 12 pounds of one grade of tobacco with 10 pounds of another grade to obtain a blend worth $54. He then made a second blend worth $61 by mixing 8 pounds of the first grade with 15 pounds of the second grade. Find the price per pound of each grade.

Solution: In this problem we have two basic relations that we can use to form two equations. We let

x = price per pound of the first grade, in dollars

y = price per pound of the second grade, in dollars

The relationship used to set up the equations is:

Value of the first grade + value of the second grade = Value of the mixture; that is,

(Number of pounds of the first grade)(Price per pound of the first grade) + (Number of pounds of the second grade)(Price per pound of the second grade) = The total cost of the mixture. Then

$12x + 10y = 54$ (1) using the numbers of pounds as coefficients and the values of the blends as constant terms

$8x + 15y = 61$ (2)

We eliminate y by subtraction:

$36x + 30y = 162$	multiplying (1) by 3
$16x + 30y = 122$	multiplying (2) by 2
$20x \qquad = 40$	equating the differences of the members
$x = 2$	solving for x
$16 + 15y = 61$	replacing x by 2 in (2)

174

$$15y = 45$$

$$y = 3$$

Therefore the solution set of equations (1) and (2) is
{(2, 3)}, and it follows that the prices of the two
grades are $2 and $3 per pound.

Check: To check the solution of a verbal problem, we re-
read the problem, substituting the values found and verify
if they make the statements true.

$$12(2) + 10(3) = 24 + 30 = 54$$

$$8(2) + 15(3) = 16 + 45 = 61$$

● **PROBLEM 11-20**

The sum of two numbers is 24; one number is 3 more than twice the
other. Find the numbers.

Solution: Let x = one of the numbers
         Let y = the other number
Since the sum of the two numbers is 24,

$$x + y = 2x \qquad\qquad (1)$$

and since one of the numbers is 3 more than twice the other,

$$x = 2y + 3 \qquad\qquad (2)$$

Thus we have 2 equations in 2 unknowns and we solve for x and y:
Since x = 2y + 3, we may replace x by (2y + 3) in equation (1),

$$(2y + 3) + y = 24$$

$$3y + 3 = 24$$

$$3y = 21$$

$$y = 7$$

To solve for x we replace y by 7 in equation (2)

$$x = 2(7) + 3$$

$$x = 14 + 3$$

$$x = 17$$

Thus the two numbers are 17 and 7.

Check: The sum of the two numbers is 24:

$$x + y = 17 + 7 = 24$$

One of the numbers is 3 more than twice the other:

$$17 = 2(7) + 3$$
$$17 = 14 + 3$$
$$17 = 17$$

Find two numbers such that twice the first added to the second equals 19, and three times the first is 21 more than the second.

Solution: Let x = the first number and y = the second number. The equations are

$$2x + y = 19 \text{ (twice the first added to the second equals 19)}$$

$$3x = y + 21 \text{ (three times the first is 21 more than the second)}$$

To solve this system

$$2x + y = 19$$

$$3x = y + 21$$

obtain all the variables on one side of the equations.

$$2x + y = 19 \qquad\qquad (1)$$

$$3x - y = 21 \qquad\qquad (2)$$

Add (2) to (1)

$$2x + y = 19 \qquad\qquad (1)$$

$$\underline{3x - y = 21} \qquad\qquad (2)$$

$$5x \quad\quad = 40 \qquad\qquad (3)$$

Divide by 5 to obtain x

$$x = 8$$

Substitute x = 8 into (1) or (2).

(1) $\quad 2x + y = 19$

$$2(8) + y = 19$$

$$16 + y = 19$$

$$y = 3$$

The solution of this system is

$$x = 8, \text{ the first number}$$

$$y = 3, \text{ the second number}$$

To check the solution, show that the two numbers satisfy the conditions of the problem.

Twice the first number is 2(8) = 16. Add this result

to the second is 16 + 3 = 19. Thus 19 = 19. Then three times the first number is 3(8) =24 which is 21 more than 3. That is 24 = 21 + 3; 24 = 24.

● PROBLEM 11-22

Find two real numbers whose sum is 10 such that the sum of the larger and the square of the smaller is 40.

Solution: Let x = the smaller number
Let y = the larger number

The sum of the numbers is 10, therefore

$$x + y = 10 \qquad (1)$$

The sum of the larger and the square of the smaller is 40, therefore

$$y + x^2 = 40 \qquad (2)$$

Solving for y in equation (1) by adding (-x) to both sides we obtain

$$y = 10 - x \qquad (3)$$

Replacing this value of y in equation (2) we obtain

$$(10 - x) + x^2 = 40$$

Adding -40 to both sides,

$$10 - x + x^2 - 40 = 0$$

$$x^2 - x - 30 = 0$$

Factoring,

$$(x - 6)(x + 5) = 0$$

Whenever the product of two numbers ab = 0, either a = 0 or b = 0. Thus, either

$$x - 6 = 0 \quad \text{or} \quad x + 5 = 0$$

and

$$x = 6 \qquad \text{or} \quad x = -5.$$

To find the corresponding y values we replace x by each of these values in equation (3):

Replacing x by 6:

$$y = 10 - 6$$
$$y = 4$$

Replacing x by -5:

$$y = 10 - (-5)$$
$$y = 10 + 5$$
$$y = 15$$

Thus the two possible solutions are (6,4) and (-5,15).
Since we assumed x to be the smaller number, and 6 is greater than 4, not smaller than it, we reject (6,4). To check if (-5,15) fits the conditions of this problem, we replace (x,y) by (-5,15) in equations (1) and (2):

$$x + y = 10 \qquad (1)$$
$$-5 + 15 = 10$$
$$10 = 10$$

$$y + x^2 = 40 \qquad (2)$$
$$15 + (-5)^2 = 40$$

177

$$15 + 25 = 40$$
$$40 = 40$$

Thus the pair of numbers whose sum is ten such that the sum of the larger and the square of the smaller is 40 is (-5,15).

● PROBLEM 11-23

The sum of the digits of a two-digit number is 9. The number is equal to 9 times the units' digit. Find the number.

<u>Solution:</u> (1) $t + u = 9$

(2) $10t + u = 9u$

If these two equations are solved simultaneously, $t = 4$ and $u = 5$ and the number is 45.

● PROBLEM 11-24

The three angles of a triangle are together equal to 180°. The smallest angle is half as large as the largest one, and the sum of the largest and smallest angles is twice the third angle. Find the three angles.

<u>Solution:</u> Let x = the smallest angle,

y = the largest angle,

z = the third angle.

From the given information we formulate the following equations:

$$x + y + z = 180 \qquad (1)$$
$$x = \frac{y}{2} \qquad (2)$$
$$x + y = 2z \qquad (3)$$

We wish to solve for x, y and z . Multiply both sides of equation (2) by 2. Thus, $2x = 2\left(\frac{y}{2}\right)$

$$2x = y$$

Now, subtract y from both sides. Thus,

$$2x - y = y - y,$$

or $\qquad 2x - y = 0$. $\qquad (4)$

Subtract $2z$ from both sides of equation (3):

$$x + y - 2z = 2z - 2z ,$$

or $\qquad x + y - 2z = 0$. $\qquad (5)$

Equation (1) multiplied by 2 gives

$$2x + 2y + 2z = 360 . \qquad (6)$$

178

Equation (5) + (6) gives

$$x + y - 2z = 0$$
$$\underline{+ (2x + 2y + 2z = 360)}$$
$$3x + 3y = 360 \qquad (7)$$

Equation (4) multiplied by 3 gives

$$6x - 3y = 0 \ . \qquad (8)$$

Equation (7) + (8) gives

$$3x + 3y = 360$$
$$\underline{+ (6x - 3y = 0)}$$
$$9x = 360 \quad ;$$

therefore, $\qquad x = 40°$

Substituting $40°$ for x in (2) gives $40 = \dfrac{y}{2}$. Multiply both sides

by 2: $80° = y$. Substituting $40°$ for x and $80°$ for y in (1) gives

$$40° + 80° + z = 180° \ .$$

Subtract 120 from both sides: $\qquad z = 60° \ .$

Thus, the angles are $40°, 80°,$ and $60° \ .$

Check: Substituting $40°$ for x, $80°$ for y, $60°$ for z gives:

$$40° + 80° + 60° = 180°$$
$$40° = \frac{80°}{2}$$
$$40° + 80° = 2(60°) \ .$$

● **PROBLEM 11-25**

A real estate dealer received $1,200 in rents on two dwellings last year, and one of the dwellings brought $10 per month more than the other. Find the monthly rental on each if the more expensive house was vacant for 2 months.

Solution: On inspecting the problem we see that there are two basic relations involved, the relation between the separate rentals, and the relation between the monthly rentals and the income per year. Since the monthly rentals differ by $10, we let

    x = monthly rental of the more expensive house in
        dollars
    y = monthly rental of the less expensive house in
        dollars

and we write $\qquad x = y + 10$

    or $\qquad x - y = 10 \qquad (1)$

Furthermore, since the first of the two houses was rented for 10 months and the other was rented for 12 months, we know that $10x + 12y$ is the total annual income. Hence

$$10x + 12y = 1,200 \qquad (2)$$

We now solve Eq. (1) and (2) simultaneously by eliminating y:

$12x - 12y = 120$	multiplying (1) by 12
$\underline{10x + 12y = 1200}$	(2) recopied
$22x \qquad = 1,320$	equating the sums of corresponding members
$x = 60$	solving for x
$60 - y = 10$	replacing x by 60 in (1)
$y = 50$	solving for y

Therefore, the solution set $\{(x, y)\} = \{(60, 50)\}$, and it follows that the monthly rentals are $60 and $50, respectively.

We check the obtained values by substituting in equations (1) and (2). Thus

$$x - y = 10 \qquad (1)$$
$$60 - 50 = 10$$
$$10 = 10$$

$$10x + 12y = 1200 \qquad (2)$$
$$10(60) + 12(50) = 1200$$
$$600 + 600 = 1200$$
$$1200 = 1200$$

● **PROBLEM 11-26**

A man invested $7,800, part at 6% and part at 4%. If the income of the 4% investment exceeded the investment at 6% by $92, how much was invested at each rate?

Solution: Let x = Amount invested at 4%

And y = Amount invested at 6%

(1) x + y = 7800

(2) .04x = .06y + 02

When the equations are solved it is found that x = $5,600, i.e., the amount invested at 4%, and y = $2,200, i.e., the amount invested at 6%.

CHAPTER 12

QUADRATIC FUNCTIONS AND GRAPHS

> **Basic Attacks and Strategies for Solving Problems in this Chapter. See pages 181 to 188 for step-by-step solutions to problems.**

As mentioned in Chapter 5, a quadratic expression is one in which the highest power of the variable is 2. The most general form of a quadratic equation is

$$ax^2 + bx + c = y.$$

Quadratic relations have graphs in the shape of *parabolas*. If the expression is in the variable x and the coefficient of x^2 is a positive number, the parabola opens upwards. If the coefficient of x^2 is negative, the parabola opens downwards. If the expression is in the variable y and the coefficient of y^2 is positive, the parabola opens to the right. If the coefficient of y^2 is negative, the parabola opens to the left.

A parabola has an *axis of symmetry*, which also goes through the vertex of the parabola. This axis divides the parabola into two sections which are mirror images of each other. For the general form

$$y = ax^2 + bx + c,$$

the axis of symmetry is the line

$$x = -{}^b/_{2a}.$$

If the form is

$$y = a(x - h)^2 + k,$$

the axis of symmetry is the line $x = h$.

The same suggestions given in Chapter 7 for graphing can also be used for graphs of quadratic relations.

There is only one parabola that will pass through any set of three points in the plane. If three points are given (i.e., three sets of values for x and y), one can substitute these values into the general quadratic equation to obtain three linear equations in a, b, and c. This system can be solved using the techniques presented in Chapter 11 to obtain the values of a, b, and c.

If the three sets of points fall on a straight line, then no parabola will pass through them and thus the quadratic equation *cannot* be solved. The same holds true if more than one of the points has the same value for the independent variable, since quadratics are functions.

● **PROBLEM** 12-1

Give an example of a quadratic function and draw its graph.

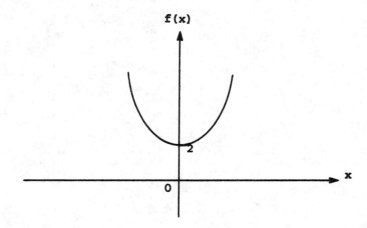

Solution: A quadratic function is a polynomial expression of degree 2.

Example:

$$f(x) = 4x^2 + 2$$

● **PROBLEM** 12-2

Draw the graph of the quadratic function

$$f(y) = x^2 + 4$$

<u>Solution</u>:

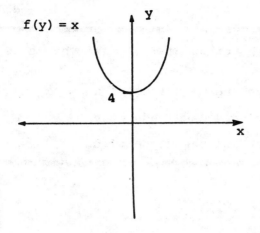

f(y) = x

4

• **PROBLEM 12-3**

Graph the function $3x^2 + 5x - 7$.

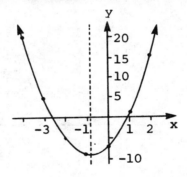

<u>Solution:</u> Let $y = 3x^2 + 5x - 7$. Substitute values for x
and then find the corresponding values of y. This is done
in the following table.

x	$y = 3x^2 + 5x - 7$
-4	21
-3	5
-2	-5
-1	-9
0	-7
1	1
2	15

These points are plotted and joined by a smooth curve in
the figure.

Determine how the graph of $f(x) = ax^2 + bx + c$ is affected by

(a)   the sign and magnitude of a

(b)   the sign and magnitude of b

(c)   the sign and magnitude of c.

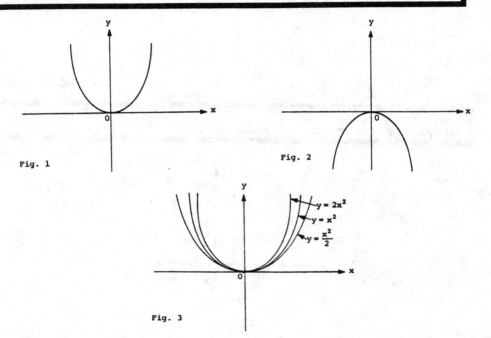

Fig. 1

Fig. 2

Fig. 3

Solution: (a)   If $a > 0$, then the graph which is a parabola opens up (see Fig. 1).

If $a < 0$, then the graph opens down (see Fig. 2).

If $a = 0$, then we do not have a quadratic, but a linear function.   As the value of a increases from $\frac{1}{2}$ to 1 to 2 etc., the graph stretches higher and becomes narrower (see Fig. 3).

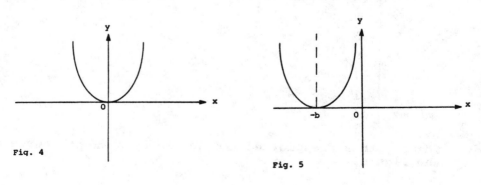

Fig. 4

Fig. 5

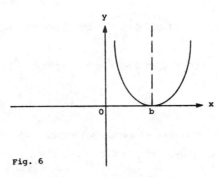

Fig. 6

(b) If b = 0, the axis of symmetry for the function is x = 0
     (see Fig. 4).
     If b > 0, then the axis of symmetry for the graph of f is
     x = $\frac{-b}{2a}$ (see fig. 5).

     If b < 0, then the axis of symmetry for the graph of f is
     x = $\frac{-b}{2a}$ (see fig. 6).

     The larger the value of |b| the further to the left or
     right of the y-axis the parabola is.

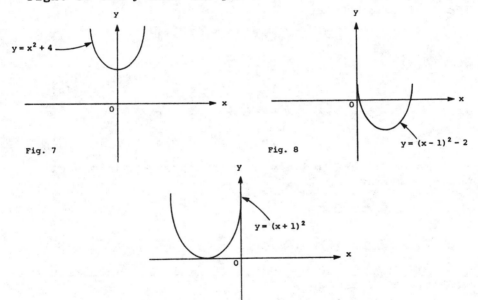

Fig. 7   Fig. 8

$y = x^2 + 4$

$y = (x - 1)^2 - 2$

$y = (x + 1)^2$

Fig. 9

(c) When written in the quadratic form

$$y = a(x-h)^2 + K$$ where a → ax^2

                               b → ahx

                               c → $ah^2 + K$

     if the value of K > 0 then the parabola is moved up K
     units (see Fig. 7).

184

If K < 0 then the parabola is moved down K units (see Fig. 8).

If K = 0 the parabola lies on y = 0 (see Fig. 9).

The graph of the quadratic function $y = ax^2 + bx + c$ is a parabola. Find the equation of a parabola passing through the points (- 1, 11), (1, 3), and (2, 5), by determining the values of a, b, and c from the given data.

Solution: Each of the three points given lies on the parabola and therefore each one must satisfy the quadratic function for a parabola,

$$ax^2 + bx + c = y.$$

For each point we substitute the coordinates of x and y into the quadratic function,

For (- 1, 11) $a(-1)^2 + b(-1) + c = 11$

$$a \quad\quad - b \quad\quad + c = 11 \quad\quad\quad (1)$$

For (1,3) $\quad a(1)^2 + b(1) + c = 3$

$$a \quad\quad + b \quad\quad + c = 3 \quad\quad\quad (2)$$

For (2,5) $\quad a(2)^2 + b(2) + c = 5$

$$4a \quad\quad + 2b \quad\quad + c = 5 \quad\quad\quad (3)$$

We have obtained a system of three equations with three unknowns.

$$a - b + c = 11 \quad\quad\quad (1)$$

$$a + b + c = 3 \quad\quad\quad (2)$$

$$4a + 2b + c = 5 \quad\quad\quad (3)$$

We can eliminate by by adding (1) and (2). We obtain a new equation in a and c.

$$a - b + c = 11 \quad\quad\quad (1)$$

$$\underline{a + b + c = 3} \quad\quad\quad (2)$$

$$2a \quad\quad + 2c = 14 \quad\quad\quad (4)$$

We have one equation in two unknowns. We need another equation in a and c before we can solve for a or c.

Eliminate b from two other equations. Let us choose (2) and (3).

$$a + b + c = 3 \quad\quad\quad (2)$$

$$4a + 2b + c = 5 \quad\quad\quad (3)$$

Multiply (2) by - 2 in order to eliminate the variable b. Then add (5) and (3)

$$- 2a - 2b - 2c = - 6 \qquad (5)$$

$$\underline{4a + 2b + c = 5} \qquad (3)$$

$$2a - c = - 1 \qquad (6)$$

Now we have two equations (4) and (6) in two unknowns, a and c.

$$2a + 2c = 14 \qquad (4)$$

$$2a - c = - 1 \qquad (6)$$

Subtract equation (6) from (4) to eliminate a.

$$2a + 2c = 14 \qquad (4)$$

$$- \underline{2a - c = -1} \qquad (6)$$

$$3c = 15 \qquad (7)$$

Divide (7) by 3

$$c = 5$$

Substitute c into either (4) or (6) to find the value of a. Choose (4)

$$2a + 2c = 14 \qquad (4)$$

$$2a + 2(5) = 14$$

$$2a + 10 = 14$$

$$2a = 4 \qquad (8)$$

Divide (8) by 2

$$a = 2$$

To find b substitute a and c into any of the three original equations (1), (2), or (3). Let us choose (2).

$$a = 2; \ c = 5$$

$$a + b + c = 3 \qquad (2)$$
$$2 + b + 5 = 3$$

$$b + 7 = 3$$

$$b = - 4$$

Therefore the solution of the original system is

$$a = 2$$

$$b = - 4$$

                    c = 5

        The equation of a parabola is
            $y = ax^2 + bx + c$
        For this particular parabola the equation is
            $y = 2x^2 - 4x + 5,$

by substituting the values of a, b, and c.

        Each of the three given points satisfy this equation.

Check

For (- 1, 11)

            $11 = 2(-1)^2 - 4(-1) + 5$

            $11 = 2 + 4 + 5$

            $11 = 11$

For (1, 3)

            $3 = 2(1)^2 - 4(1) + 5$

            $3 = 2 - 4 + 5$

            $3 = 3$

For (2, 5)

            $5 = 2(2)^2 - 4(2) + 5$

            $5 = 8 - 8 + 5$

            $5 = 5$

● PROBLEM 12-6

What is the minimum value of the expression $2x^2 - 20x + 17$?

Solution: Consider the function $y = 2x^2 - 20x + 17$. This function is defined by a second degree equation. The coefficient of its x^2 term is positive. Hence the curve is a parabola opening upward. Thus, the minimum point of this curve occurs at the vertex. The x-coordinate is equal to

$$- \frac{\text{coefficient of x term}}{2 \cdot \text{coefficient of } x^2 \text{ term}} = -\frac{b}{2a} = -\frac{(-20)}{2(2)} = \frac{20}{4} = 5.$$

For x = 5, $y = 2(5)^2 - 20(5) + 17 = -33$. Therefore the minimum value of the expression $2x^2 - 20x + 17$ for any value of x is -33. This minimum value is assumed only when x = 5.

187

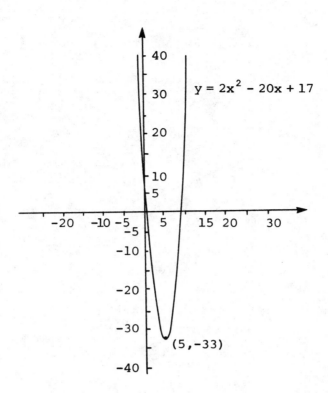

$$y = 2x^2 - 20x + 17$$

(5,-33)

CHAPTER 13

QUADRATIC EQUATIONS AND SYSTEMS OF EQUATIONS INVOLVING QUADRATICS

> **Basic Attacks and Strategies for Solving Problems in this Chapter. See pages 189 to 240 for step-by-step solutions to problems.**

Solving (or finding the roots of) algebraic expressions was mentioned in Chapters 4 and 7. When the expression is a quadratic, one can use the quadratic formula, presented in Chapter 7, or one can first factor the expression, using the techniques mentioned in Chapter 5, and then solve simpler linear equations.

Factoring a quadratic expression results in two linear factors and possibly a constant multiple, for example,

$$2x^2 + 6x + 4 = 2 \cdot (x + 2)(x + 1).$$

In order to solve an expression composed of several factors, one must remember a very basic rule of algebra: whenever a product is zero, one of the factors must be zero. In the simplest form, the rule is: if $a \cdot b = 0$ then either $a = 0$ or $b = 0$ (or both).

Thus, when solving a quadratic equation, one must solve each linear factor to obtain *two* solutions. In the example given above, one must solve $x + 2 = 0$ and and $x + 1 = 0$ to obtain the values of -2 and -1 as solutions to

$$2x^2 + 6x + 4 = 0.$$

Remembering that a quadratic can be factored into linear factors can be useful if it is necessary to create a quadratic from its roots. Since the roots of $(x - a)(x - b) = 0$ are a and b, if one is given a and b, the quadratic with those values as roots is

$$(x - a)(x - b) = x^2 - (a + b)x + ab.$$

Relations involving quadratic expressions were mentioned in Chapter 12 along with their graphs.

Sometimes fractional equations involve quadratics. In some cases, when simplifying the original fractional equation, the least common denominator is a quadratic and the resulting equation is a quadratic. Sometimes the numerators or denominators themselves are quadratics, in which case they should first be factored and (if possible) simplified before proceeding.

When given an equation involving a term with a radical, as was mentioned in Chapter 10, the process of finding a solution may involve squaring both sides. Such a process may introduce an extraneous root. In these cases, all solutions obtained should be double-checked by substituting into the original equation. If an equation involves several terms with radicals, it may be necessary to repeat the squaring process several times, each time moving one of the terms involving a radical to one side of the equals sign and leaving all other terms on the other side.

In some cases, an equation involving a term with a radical can be solved by substituting a new variable for the entire radical term and first solving a simpler equation. This may be simpler especially if the equation contains several related radicals (e.g., the square root and the cube root of the same expression).

The square of a simple linear expression, e.g., $ax + b$, produces a quadratic expression, e.g.,

$$(ax)^2 + 2abx + b^2.$$

Such a quadratic, which is merely the square of a linear expression, is often called a *perfect square*. It is often helpful to modify a quadratic expression so that it consists of an expression which is a perfect square and a constant. This process is called *completing the square*. For example, given

$$x^2 + 10x + 5,$$

the first two terms correspond to $(ax)^2 + 2abx$ in the general square with $a = 1$. Thus, $2b$ must equal 10, or b equals half of the coefficient of x, i.e., $b = 5$, half of 10. To "complete the square," a b^2 term must be added to the expression *and also subtracted* (so as not to change the value of the expression). Thus, we obtain

$$x^2 + 10x + 25 - 25 + 5.$$

The first three terms can be rewritten as a square of a linear expression and the final two terms can be simply added. We thus obtain $(x + 5)^2 - 20$.

When examining the roots of a quadratic equation by referring to the quadratic formula (given in Chapter 5),

$$x = \frac{-b \pm \sqrt{b^2 - 4ac}}{2a},$$

we can note some interesting results. If we add the two roots, the radicals sum to zero, and the sum of the roots is merely

$$\frac{(-b)+(-b)}{2a} = \frac{-2b}{2a} = \frac{-b}{a}.$$

If we multiply the two roots, we obtain the following:

$$\frac{-b+\sqrt{b^2-4ac}}{2a} \times \frac{-b-\sqrt{b^2-4ac}}{2a}$$

$$= \frac{(-b)^2-(b^2-4ac)}{2a \cdot 2a} = \frac{4ac}{4a^2} = \frac{c}{a}.$$

We can determine the sum and products of roots without ever actually calculating them, simply by referring to the coefficients a, b, and c in the general form of the quadratic.

The quantity that appears under the radical in the quadratic formula, $b^2 - 4ac$, is called the *discriminant*. By analyzing the discriminant, one may deduce certain facts about the nature of the roots of a quadratic. Let us assume that a, b, and c are integers.

- If $b^2 - 4ac > 0$ and is a perfect square, the roots are real, rational, and unequal.

- If $b^2 - 4ac > 0$ but is not a perfect square, the roots are real, irrational, and unequal.

- If $b^2 - 4ac = 0$, the roots are real, rational, and equal.

- If $b^2 - 4ac < 0$, the roots are unequal complex numbers (i.e., have an imaginary component).

It is sometimes possible to solve a system of equations in which at least one equation is quadratic by using a substitution method similar to that suggested in Chapter 11 for systems of linear equations. If the system has a linear equation, one can solve the linear equation for the independent variable of the quadratic, then substitute the value into the quadratic, and then solve the quadratic by standard means.

It should be noted that the same problems that affect linear systems also affect systems in which at least one equation is quadratic. That is, the system may have no solution (the associated graphs do not intersect), or the system may have an infinite number of solutions (two equations are actually multiples of each other). There is another possibility when quadratic equations are included in the system. Instead of only one point of intersection (as occurs with two linear equations), a quadratic and a line (or two quadratics) may intersect in two (or more) points. For example, an ellipse and a circle with the same center could easily intersect in four points.

The method of substitution may work even if the system consists of only quadratics. It may also be possible to use the elimination method.

Step-by-Step Solutions to Problems in this Chapter, "Quadratic Equations and Systems of Equations Involving Quadratics"

● PROBLEM 13-1

Solve the equation $(3x - 7)(x + 2) = 0$.

Solution: When a given product of two numbers that are equal to zero, $ab = 0$, either a must equal zero or b must equal zero (or both equal zero). So if $(3x - 7)(x + 2) = 0$, then $(3x - 7) = 0$ or $(x + 2) = 0$.

$3x - 7 = 0$ | $x + 2 = 0$

Add 7 to both sides: | Subtract 2 from both sides:

$3x = 7$ |

Divide both sides by 3: |

$x = \frac{7}{3}$ | $x = -2$

Hence $x = \frac{7}{3}$ or $x = -2$, and our solution set is $\left\{ \frac{7}{3}, -2 \right\}$.

● PROBLEM 13-2

Solve the equation $3x^2 + 5x = 0$.

Solution: Because division by zero is impossible, we must not divide by x, since x might be equal to zero. Instead of dividing by x we factor x from the left side of the equation to obtain:

$$x(3x + 5) = 0.$$

Whenever we have a situation where $ab = 0$ (the product

of two or more numbers equal to zero) either a = 0 or
b = 0. Therefore x = 0, or 3x + 5 = 0. Subtract 5 from
each side of the second equation to obtain:

$$3x = -5$$

Divide both sides by 3 to obtain x = $-\frac{5}{3}$. The

two solutions of the given equation are x = 0 and

x = $-\frac{5}{3}$.

To check the validity of the two solutions we substitute them into the given equation. Thus,

when x = 0

$$3x^2 + 5x = 0$$

$$3(0)^2 + 5(0) = 0$$

$$0 = 0$$

when x = $-\frac{5}{3}$

$$3x^2 + 5x = 0$$

$$3\left(-\frac{5}{3}\right)^2 + 5\left(-\frac{5}{3}\right) = 0$$

$$3\left(\frac{25}{9}\right) + 5\left(-\frac{5}{3}\right) = 0$$

$$\frac{25}{3} - \frac{25}{3} = 0$$

$$0 = 0$$

● **PROBLEM 13-3**

Solve the equation $x^2 + 8x + 15 = 0$.

<u>Solution:</u> Since (x + a)(x + b) = x^2 + bx + ax + ab

= x^2 + (a + b) x + ab, we may factor the given equation,
0 = x^2 + 8x + 15, replacing a + b by 8 and ab by 15.
Thus,

$$a + b = 8, \quad \text{and}$$

$$ab = 15.$$

We want the two numbers a and b whose sum is 8 and
whose product is 15. We check all pairs of numbers whose
product is 15:

(a) $1 \cdot 15 = 15$; thus a = 1, b = 15 and ab = 15.

   $1 + 15 = 16$, therefore we reject these values because $a + b \neq 8$.

(b) $3 \cdot 5 = 15$; thus a = 3, b = 5, and ab = 15.

   $3 + 5 = 8$. Therefore a + b = 8, and we accept these values.

Hence $x^2 + 8x + 15 = 0$ is equivalent to

   $$0 = x^2 + (3 + 5)x + 3 \cdot 5 = (x + 3)(x + 5)$$

Hence, x + 5 = 0   or x + 3 = 0

since the product of **these two numbers is zero, one of the** numbers must be zero. Hence, x = - 5, or x = - 3, and the solution set is X = {- 5, - 3}.

   The student should note that x = - 5 or x = - 3. We are certainly not making the statement, that x = - 5, and x = - 3. Also, the student should check that both these numbers do actually satisfy the given equations and hence are solutions.

Check: Replacing x by (- 5) in the original equation:

   $x^2 + 8x + 15 = 0$

   $(- 5)^2 + 8(- 5) + 15 = 0$

            $25 - 40 + 15 = 0$

                 $- 15 + 15 = 0$

                      $0 = 0$

Replacing x by (- 3) in the original equation:

   $x^2 + 8x + 15 = 0$

   $(- 3)^2 + 8(- 3) + 15 = 0$

            $9 - 24 + 15 = 0$

                 $- 15 + 15 = 0$

                      $0 = 0.$

● **PROBLEM 13-4**

Find the roots of $x^2 - 3x - 10 = 0$.

Solution: To find the roots of this quadratic, we factor it (put it in the form (x + a)(x + b) = 0).

Note that $(x + a)(x + b) = x^2 + (a + b)x + ab$

Thus in our quadratic, $x^2 + (-3)x + (-10)$,

$$a + b = -3 \qquad\qquad\qquad (1)$$

and $\quad ab = -10.$ $\qquad\qquad\qquad\qquad\qquad (2)$

That is, we want the two numbers a and b whose sum is (- 3), and whose product is (- 10).

To find these numbers, we can check the set of numbers whose product is (- 10):

(a) $(-10) \times (1) = -10$, therefore equation (2) is satisfied, now check these values in equation (1):
$(-10) + (1) = -9 \neq -3$ therefore we reject these values.

(b) $(-5) \times (2) = -10$, therefore equation (2) is satisfied, now checking these values in equation (1):
$(-5) + 2 = -3.$

Hence both equations are satisfied and we conclude

$$a = -5 \qquad and \qquad b = 2.$$

Thus, $\quad x^2 - 3x - 10 = x^2 + (-5 + 2)x + (-5)(2)$

$$= \big[x + (-5)\big]\big[x + 2\big]$$

$$= (x - 5)(x + 2) = 0.$$

Hence, by the fundamental property which states that if $ab = 0$, either $a = 0$ or $b = 0$, $x - 5 = 0$ or $x + 2 = 0$ and

$$x = 5 \qquad or \qquad x = -2.$$

This proves that if the equation has roots, they must be either 5 or - 2. So far we have not proved that these are roots. We can check this by substituting in the given equation. If $x = 5$, then

$$x^2 - 3x - 10 = (5)^2 - 3(5) - 10$$

$$= 25 - 15 - 10$$

$$= 25 - 25$$

$$= 0$$

Thus 5 is indeed a root of the equation.

If $x = -2$, then

$$x^2 - 3x - 10 = (-2)^2 - 3(-2) - 10$$

$$= 4 + 6 - 10$$

$$= 10 - 10$$

$$= 0$$

Thus - 2 is also a root.

Such a check not only has a logical purpose, but it also assures us that we have not made a mistake in arithmetic. We may now conclude that the solution to our equation is x = 5 or x = - 2.

Solve the following equations by factoring.

(a) $2x^2 + 3x = 0$　　　　　(c) $z^2 - 2z - 3 = 0$

(b) $y^2 - 2y - 3 = y - 3$　　(d) $2m^2 - 11m - 6 = 0$

<u>Solution</u>: (a) $2x^2 + 3x = 0$. Factoring out the common factor of x from the left side of the given equation,

$$x(2x + 3) = 0.$$

Whenever a product ab = 0, where a and b are any two numbers, either a = 0 or b = 0. Then, either

$$x = 0 \quad or \quad 2x + 3 = 0$$
$$2x = -3$$
$$x = \frac{-3}{2}$$

Hence, the solution set to the original equation $2x^2 + 3x = 0$ is: $\left\{-\frac{3}{2}, 0\right\}$.

(b) $y^2 - 2y - 3 = y - 3$. Subtract (y - 3) from both sides of the given equation:

$$y^2 - 2y - 3 - (y - 3) = y - 3 - (y - 3)$$
$$y^2 - 2y - \cancel{3} - y + \cancel{3} = \cancel{y} - \cancel{3} - \cancel{y} + \cancel{3}$$
$$y^2 - 3y = 0.$$

Factor out a common factor of y from the left side of this equation:

$$y(y - 3) = 0.$$

Thus, y = 0 or y - 3 = 0

$$y = 3$$

Therefore, the solution set to the original equation $y^2 - 2y - 3 = y - 3$ is: {0,3}.

(c) $z^2 - 2z - 3 = 0$. Factor the original equation into a product of two polynomials:

$$z^2 - 2z - 3 = (z - 3)(z + 1) = 0$$

Hence,

(z - 3)(z + 1) = 0; and z - 3 = 0 　or z + 1 = 0
　　　　　　　　　　　　　z = 3　　　　　　z = -1

193

Therefore, the solution set to the original equation $z^2 - 2z - 3 = 0$ is: $\{-1, 3\}$.

(d) $2m^2 - 11m - 6 = 0$. Factor the original equation into a product of two polynomials:

$$2m^2 - 11m - 6 = (2m + 1)(m - 6) = 0$$

Thus,
$$2m + 1 = 0 \quad \text{or} \quad m - 6 = 0$$
$$2m = -1 \qquad\qquad m = 6$$
$$m = \frac{-1}{2}$$

Therefore, the solution set to the original equation $2m^2 - 11m - 6 = 0$ is: $\left\{-\frac{1}{2}, 6\right\}$.

● **PROBLEM 13-6**

Solve $\dfrac{x}{x - 2} + \dfrac{x - 1}{2} = x + 1$.

<u>Solution:</u> First eliminate the fractions to facilitate solution. This is done by multiplying both sides of the equation by the Least Common Denominator, LCD. The LCD is obtained by multiplying the denominators of every fraction: LCD = $2(x - 2)$; and multiplying each side by this, the equation becomes:

$$2(x - 2)\left(\frac{x}{x - 2} + \frac{x - 1}{2}\right) = (x + 1)2(x - 2)$$

$$2x + (x - 1)(x - 2) = 2(x + 1)(x - 2)$$

$$2x + x^2 - 3x + 2 = 2x^2 - 2x - 4$$

$$x^2 - x - 6 = 0$$

This can be solved by factoring and setting each factor equal to zero.

$$(x - 3)(x + 2) = 0$$

$$x - 3 = 0 \qquad\qquad x + 2 = 0$$
$$x = 3 \qquad\qquad\qquad x = -2$$

Since both of these solutions are admissible values of x, they both should satisfy the original equation.

Check for x = 3:

194

$$\frac{3}{1} + \frac{2}{2} = 3 + 1$$

$$3 + 1 = 3 + 1$$

Check for $x = -2$:

$$\frac{-2}{-4} + \frac{-3}{2} = -2 + 1$$

$$\frac{1}{2} + \frac{-3}{2} = -1$$

$$-1 = -1$$

● **PROBLEM 13-7**

Solve

$$\frac{x + 1}{x^2 - 5x + 6} + \frac{x + 2}{x^2 - 7x + 12} = \frac{6}{x^2 - 6x + 8}.$$

<u>Solution</u>: Factor the denominator of each fraction to obtain

$$\frac{x + 1}{(x - 2)(x - 3)} + \frac{x + 2}{(x - 3)(x - 4)} = \frac{6}{(x - 2)(x - 4)}.$$

Obtain the Least Common Denominator, LCD, by multiplying the denominators of each fraction together and using the highest power of each factor only once, that is,

$$(x - 2)(x - 3)(x - 4)(x - 3)(x - 2)(x - 4)$$

$$\text{LCD} = (x - 2)(x - 3)(x - 4).$$

Multiply both sides of the equation by the LCD to remove all fractions and obtain:

$$(x - 2)(x - 3)(x - 4)\left[\frac{x + 1}{(x - 2)(x - 3)} + \frac{x + 2}{(x - 3)(x - 4)}\right]$$

$$= \left[\frac{6}{(x - 2)(x - 4)}\right](x - 2)(x - 3)(x - 4)$$

$$(x + 1)(x - 4) + (x + 2)(x - 2) = 6(x - 3)$$

$$(x + 1)(x - 4) + (x + 2)(x - 2) = 6x - 18$$

$$\left(x^2 - 3x - 4\right) + \left(x^2 - 4\right) = 6x - 18$$

$$2x^2 - 3x - 8 = 6x - 18$$

$$2x^2 - 9x + 10 = 0$$

$$(2x - 5)(x - 2) = 0$$

$$2x - 5 = 0 \qquad\qquad x - 2 = 0$$
$$2x = 5 \qquad\qquad\quad x = 2$$
$$x = \frac{5}{2}$$

Substituting x = 2 into the original equation shows that x = 2 is an extraneous root, since it is not an admissible value for the original equation. (It makes two of the denominators (x - 2)(x - 3) and (x - 2)(x - 4), equal to zero.)

$x = \dfrac{5}{2}$ is an admissible value of x for the

original equation and is a solution if it will satisfy the original equation.

Check: $x = \dfrac{5}{2}$

$$\frac{\dfrac{7}{2}}{\left(\dfrac{1}{2}\right)\left(-\dfrac{1}{2}\right)} + \frac{\dfrac{9}{2}}{\left(-\dfrac{1}{2}\right)\left(-\dfrac{3}{2}\right)} = \frac{6}{\left(\dfrac{1}{2}\right)\left(-\dfrac{3}{2}\right)}$$

$$\frac{14}{-1} + \frac{18}{3} = \frac{24}{-3}$$

$$-14 + 6 = -8$$

$$-8 = -8$$

● **PROBLEM** 13-8

Solve the equation $\sqrt{2x^2 - 9} = x$.

<u>Solution:</u> Squaring both sides, we have
$$2x^2 - 9 = x^2$$
$$x^2 = 9$$
$$x = 3 \quad \text{or} \quad x = -3$$

Both 3 and -3 will satisfy the equation $2x^2 - 9 = x^2$ since $2(3)^2 - 9 = 9 = (3)^2$ and $2(-3)^2 - 9 = 9 = (-3)^2$. However, -3 does

not satisfy the original equation since $\sqrt{2(-3)^2 - 9} = \sqrt{9} = 3 \neq -3$. An extraneous root was introduced by squaring. Thus the solution set is {3}.

Solve the equation $\sqrt{x^2 - 3x} = 2x - 6$.

Solution: Remove the radical by squaring both sides of the equation, and obtain:

$$\sqrt{x^2 - 3x})^2 = (2x - 6)^2 \quad \text{or}$$
$$x^2 - 3x = 4x^2 - 24x + 36$$

Writing in general form, move every term to one side of the equation.

$$3x^2 - 21x + 36 = 0$$

Dividing all terms by 3, and factoring,

$$\frac{3x^2}{3} - \frac{21x}{3} + \frac{36}{3} = \frac{0}{3}$$

$$x^2 - 7x + 12 = 0$$
$$(x - 3)(x - 4) = 0$$

The roots are: $x = 3$, $x = 4$.

Check: Substituting $x = 3$ in the original equation

$$\sqrt{9 - 9} = 6 - 6$$
$$0 = 0$$

Substituting $x = 4$ in the original equation

$$\sqrt{4} = 8 - 6$$
$$2 = 2.$$

Observe that both $x = 3$ and $x = 4$ satisfy the original equation, and there are no extraneous roots.

Find the solution of the equation
$$\sqrt{3 - 2x} = 3 - \sqrt{2x + 2}.$$

Solution: Assume that there is a number x such that $\sqrt{3 - 2x} = 3 - \sqrt{2x + 2}$. Squaring both sides, we have

$$(\sqrt{3 - 2x})^2 = (3 - \sqrt{2x + 2})^2$$

$$(\sqrt{3 - 2x})^2 = 9 - 6\sqrt{2x + 2} + (\sqrt{2x + 2})^2$$

Since $\left(\sqrt{a}\right)^2 = \left(a^{\frac{1}{2}}\right)^2 = a^{2/2} = a^1 = a$

$$(\sqrt{3 - 2x})^2 = 3 - 2x \quad \text{and} \quad (\sqrt{2x + 2})^2 = 2x + 2$$

Thus we obtain

$$3 - 2x = 9 - 6\sqrt{2x + 2} + 2x + 2$$

Adding $6\sqrt{2x + 2}$ to both sides,

$$(3 - 2x) + 6\sqrt{2x + 2} = 9 + 2x + 2$$

$$(3 - 2x) + 6\sqrt{2x + 2} = 11 + 2x$$

Adding $-(3 - 2x)$ to both sides,

$$6\sqrt{2x + 2} = 11 + 2x - (3 - 2x)$$

$$6\sqrt{2x + 2} = 11 + 2x - 3 + 2x$$

$$6\sqrt{2x + 2} = 4x + 8$$

Dividing both sides by 2,

$$3\sqrt{2x + 2} = 2x + 4$$

Squaring both sides of this new equation, we have

$$(3\sqrt{2x + 2})^2 = (2x + 4)^2$$

$$3^2(\sqrt{2x + 2})^2 = (2x + 4)(2x + 4)$$

$$9(\sqrt{2x + 2})^2 = 4x^2 + 16x + 16$$

Recall $(\sqrt{2x + 2})^2 = 2x + 2$, thus

$$9(2x + 2) = 4x^2 + 16x + 16$$

$$18x + 18 = 4x^2 + 16x + 16$$

Dividing both sides by 2,

$$9x + 9 = 2x^2 + 8x + 8$$

Adding $-(9x + 9)$ to both sides,

$$0 = 2x^2 + 8x + 8 - (9x + 9)$$

$$2x^2 + 8x + 8 - 9x - 9 = 0$$

$$2x^2 - x - 1 = 0$$

Factoring, $(2x + 1)(x - 1) = 0$

Whenever a product of two numbers $ab = 0$ either $a = 0$ or $b = 0$.
Thus, either $2x + 1 = 0$ or $x - 1 = 0$

$$x = -\tfrac{1}{2} \text{ or } x = 1$$

so that the only possible roots are $-\tfrac{1}{2}$ and 1. We must check if these
values are indeed roots. Replace x by $(-\tfrac{1}{2})$ in our original equation

$$\sqrt{3 - 2x} = 3 - \sqrt{2x + 2}$$

$$\sqrt{3 - 2(-\tfrac{1}{2})} = 3 - \sqrt{2(-\tfrac{1}{2}) + 2}$$

$$\sqrt{3 + \frac{2}{2}} = 3 - \sqrt{-\frac{2}{2} + 2}$$

$$\sqrt{3 + 1} = 3 - \sqrt{-1 + 2}$$

$$\sqrt{4} \quad = 3 - \sqrt{1}$$
$$2 = 3 - 1$$
$$2 = 2$$

Thus $(-\frac{1}{2})$ is a root.

Now replace x by 1 in our original equation

$$\sqrt{3 - 2x} = 3 - \sqrt{2x + 2}$$
$$\sqrt{3 - 2(1)} = 3 - \sqrt{2(1) + 2}$$
$$\sqrt{3 - 2} = 3 - \sqrt{2 + 2}$$
$$\sqrt{1} = 3 - \sqrt{4}$$
$$1 = 3 - 2$$
$$1 = 1$$

Therefore 1 is also a root, and the solution set is $\{-\frac{1}{2}, 1\}$.

● **PROBLEM 13-11**

Solve and check: $\sqrt{x + 10} + \sqrt[4]{x + 10} = 2.$

<u>Solution:</u> Let $y = \sqrt[4]{x + 10}$ then $y^2 = \sqrt{x + 10}$.

Substituting, the original equation may be written

$$y^2 + y - 2 = 0.$$

Factor $\qquad (y + 2)(y - 1) = 0.$

Set each factor $= 0$ to find all values of x which can make the product $= 0$

$$y + 2 = 0 \qquad \bigm| \qquad y - 1 = 0$$
$$y = -2 \qquad \bigm| \qquad y = 1$$

for $y = -2$ $\qquad\qquad$ for $y = 1$

$$y = \sqrt[4]{x + 10} = -2 \qquad\qquad y = \sqrt[4]{x + 10} = 1$$
$$x + 10 = 16 \qquad\qquad\qquad x + 10 = 1$$
$$x = 6 \qquad\qquad\qquad\qquad x = -9$$

Check: for $x = 6$: $\quad \sqrt{6 + 10} + \sqrt[4]{6 + 10} \stackrel{?}{=} 2$

$$\sqrt{16} + \sqrt[4]{16} = 2$$

$$4 + 2 \neq 2$$

This root does not check.

for $x = -9$: $\quad \sqrt{-9 + 10} + \sqrt[4]{-9 + 10} = 2$

199

$$1 + \sqrt[3]{1} = 2$$
$$1 + 1 = 2$$
$$2 = 2.$$

This root checks.

● **PROBLEM** 13-12

Solve $2\sqrt{\dfrac{x}{a}} + 3\sqrt{\dfrac{a}{x}} = \dfrac{b}{a} + \dfrac{6a}{b}$.

<u>Solution:</u> Let $\sqrt{\dfrac{x}{a}} = y;$ then $\sqrt{\dfrac{a}{x}} = \dfrac{1}{y}$;

Hence, $2y + \dfrac{3}{y} = \dfrac{b}{a} + \dfrac{6a}{b}$;

$$yab\left(2y + \dfrac{3}{y}\right) = \left(\dfrac{b}{a} + \dfrac{6a}{b}\right)yab$$

$$2y^2ab + 3ab = b^2y + 6a^2y$$

$$2aby^2 - 6a^2y - b^2y + 3ab = 0,$$

$$(2ay - b)(by - 3a) = 0; \qquad\qquad by - 3a = 0$$
$$2ay - b = 0$$
$$2ay = b \qquad\qquad\qquad\qquad by = 3a$$
$$y = \dfrac{b}{2a} , \quad \text{or} \qquad\qquad \dfrac{3a}{b} ;$$

Substitute these two values of y:
$$\sqrt{\dfrac{x}{a}} = y$$
$$\sqrt{\dfrac{x}{a}} = \dfrac{b}{2a} \qquad\qquad\qquad \sqrt{\dfrac{x}{a}} = \dfrac{3a}{b}$$

square both sides.   square both sides

$$\dfrac{x}{a} = \dfrac{b^2}{4a^2} \qquad\qquad\qquad\qquad \dfrac{x}{a} = \dfrac{9a^2}{b^2}$$

multiply both sides by a.   multiply both sides by a.

$$x = \dfrac{b^2 a}{4a^2} = \dfrac{b^2}{4a} \qquad\qquad x = \dfrac{9a^2 \cdot a}{b^2} = \dfrac{9a^3}{b^2}$$

The solution is:
$$x = \left\{ \dfrac{b^2}{4a} , \dfrac{9a^3}{b^2} \right\}$$

Solve the equation $2x^{2/5} + 5x^{1/5} - 3 = 0$.

Solution: This equation may be solved as a quadratic equation if we let $P(x) = y = x^{1/5}$. Then $x^{2/5} = y^2$ and, by substituting these expressions in the equation to be solved, we have $2y^2 + 5y - 3 = 0$. We can solve this equation for y by factoring:

$$2y^2 + 5y - 3 = 0$$
$$(2y - 1)(y + 3) = 0$$
$$(2y - 1) = 0 \quad \text{or} \quad (y + 3) = 0$$

therefore, $\qquad y = \tfrac{1}{2} \quad$ or $\qquad y = -3$

Now recall that $y = x^{1/5}$. Hence $x^{1/5} = \tfrac{1}{2}$ or $x^{1/5} = -3$. Hence, by raising both sides of each of these equations to the fifth power, we have $x = 1/32$ or $x = -243$.

Therefore the solution set is $\{1/32, -243\}$.

Complete the square in $x^2 + x - 1$.

Solution: We proceed adding the square of half the coefficient of x and, also subtracting it. That is, we write

$$x^2 + x - 1 = x^2 + x - 1 + \tfrac{1}{4} - \tfrac{1}{4}$$
$$= x^2 + x - 1 + \left(\tfrac{1}{2}\right)^2 - \left(\tfrac{1}{2}\right)^2$$

Associating, $\quad = \left[x^2 + x + \left(\tfrac{1}{2}\right)^2\right] - 1 - \left(\tfrac{1}{2}\right)^2$

$$= [x + \tfrac{1}{2}]^2 - 1 - \tfrac{1}{4}$$
$$= \left(x + \tfrac{1}{2}\right)^2 - \tfrac{4}{4} - \tfrac{1}{4}$$
$$= \left(x + \tfrac{1}{2}\right)^2 - \tfrac{5}{4}$$

Complete the square in both x and y in x^2+2x+y^2-3y.

Solution: To complete the square in x, take half the coefficient of x and square it. Add and subtract this value from the given expression. Therefore:

$$\left[\tfrac{1}{2}(2)\right]^2 = [1]^2 = 1, \text{ and } x^2+2x+y^2-3y = x^2+2x+y^2-3y+1-1.$$

Commuting, $x^2+2x+y^2-3y = x^2+2x+1+y^2-3y-1 = (x+1)^2+y^2-3y-1$. (1)

Now, take half the coefficient of y and square it. Add and subtract this value from equation (1).

$$\left[\frac{1}{2}(-3)\right]^2 = \left[-\frac{3}{2}\right]^2 = \frac{9}{4}, \text{ and}$$

$x^2 + 2x + y^2 - 3y = (x+1)^2 + y^2 - 3y - 1 + \frac{9}{4} - \frac{9}{4}$. Commuting,

$x^2 + 2x + y^2 - 3y = (x+1)^2 + y^2 - 3y + \frac{9}{4} - 1 - \frac{9}{4}$

$= (x+1)^2 + (y-\frac{3}{2})^2 - 1 - \frac{9}{4} = (x+1)^2 + (y-\frac{3}{2})^2 - \frac{4}{4} - \frac{9}{4}$. Hence, $x^2 + 2x + y^2 - 3y$

$= (x+1)^2 + (y-\frac{3}{2})^2 - \frac{13}{4}$.

● **PROBLEM 13-16**

Obtain the quadratic equation in standard form that is equivalent to $4x - 3 = 5x^2$.

Solution: The standard form of a quadratic equation is $ax^2 + bx + c = 0$. Starting with our given equation $4x - 3 = 5x^2$, we add $(-5x^2)$ to both members,

$$(4x - 3) + (-5x^2) = 5x^2 + (-5x^2)$$

$$(4x - 3) + (-5x^2) = 0$$

commuting we obtain $\quad -5x^2 + 4x - 3 = 0$

This is the required equation with a = -5, b = 4, and c = -3.

● **PROBLEM 13-17**

Find the roots of the equation $x^2 + 12x - 85 = 0$.

Solution: The roots of this equation may be found using the quadratic formula

$$x = \frac{-B \pm \sqrt{B^2 - 4AC}}{2A}$$

In this equation A = 1, B = 12, and C = -85. Hence, by the quadratic formula,

$$x = \frac{-12 + \sqrt{144 + 340}}{2} \qquad \text{or} \qquad x = \frac{-12 - \sqrt{144 + 340}}{2}$$

$$x = \frac{-12 + 22}{2} \qquad \text{or} \qquad x = \frac{-12 - 22}{2}$$

Therefore x = 5 or x = -17. This is equivalent to the statement that the solution set is {-17,5}.

Solve the equation $x^2 + 5x + 6 = 0$ by the quadratic formula.

Solution: We use the quadratic formula, which states

$$x = \frac{-b \pm \sqrt{b^2 - 4ac}}{2a} \quad , \text{ for cases}$$

where $ax^2 + bx + c = 0$. For this equation, $a = 1$, $b = 5$, $c = 6$. Therefore the solutions are

$$x = \frac{-5 \pm \sqrt{25 - 4(1)(6)}}{2 \cdot 1}$$

$$= \frac{-5 \pm \sqrt{25 - 24}}{2} = \frac{-5 \pm \sqrt{1}}{2}$$

or $\quad x_1 = \dfrac{-5 + 1}{2} = -2, \quad x_2 = \dfrac{-5 - 1}{2} = -3.$

Solve the equation $\sqrt{x + 1} + \sqrt{2x + 3} - \sqrt{8x + 1} = 0$.

Solution: Add $\sqrt{8x + 1}$ to both sides,

$$\sqrt{x + 1} + \sqrt{2x + 3} = \sqrt{8x + 1}.$$

Square both sides of the equation,

$$(\sqrt{x + 1} + \sqrt{2x + 3})(\sqrt{x + 1} + \sqrt{2x + 3}) = (\sqrt{8x + 1})^2$$

$$(\sqrt{x + 1})^2 + 2\sqrt{x + 1}\sqrt{2x + 3} + (\sqrt{2x + 3})^2 = (\sqrt{8x + 1})^2$$

Since $\sqrt{a} \cdot \sqrt{b} = \sqrt{ab}$,

$$(\sqrt{x + 1})^2 + 2\sqrt{(x + 1)(2x + 3)} + (\sqrt{2x + 3})^2 = (\sqrt{8x + 1})^2.$$

Recall: $(\sqrt{a})^2 = \sqrt{a} \cdot \sqrt{a} = \sqrt{a^2} = a.$

Thus, $\quad (\sqrt{x + 1})^2 = x + 1$

$$(\sqrt{2x + 3})^2 = 2x + 3$$

and $\quad (\sqrt{8x + 1})^2 = 8x + 1$

Substituting these values we obtain,

$$x + 1 + 2\sqrt{(x + 1)(2x + 3)} + 2x + 3 = 8x + 1.$$

Combine terms, $3x + 4 + 2\sqrt{(x + 1)(2x + 3)} = 8x + 1$

Add (-4) to both sides, $3x + 2\sqrt{(x + 1)(2x + 3)} = 8x - 3$

Add $(-3x)$ to both sides $\qquad 2\sqrt{(x + 1)(2x + 3)} = 5x - 3$

Multiply the terms within the radical,
$$2\sqrt{2x^2 + 5x + 3} = 5x - 3$$

Square both members,
$$(2\sqrt{2x^2 + 5x + 3})^2 = (5x - 3)(5x - 3).$$

Since $(ab)^2 = a^2b^2$,
$$(2)^2(\sqrt{2x^2 + 5x + 3})^2 = (5x - 3)(5x - 3)$$
$$4(\sqrt{2x^2 + 5x + 3})^2 = 25x^2 - 30x + 9.$$

Once again recall: $(\sqrt{2x^2 + 5x + 3}) = (2x^2 + 5x + 3)$.
Substituting this value, we obtain
$$4(2x^2 + 5x + 3) = 25x^2 - 30x + 9$$
Distribute,
$$8x^2 + 20x + 12 = 25x^2 - 30x + 9$$

Add $(-8x^2)$ to both sides,
$$20x + 12 = 25x^2 - 30x + 9 - 8x^2$$
$$20x + 12 = 17x^2 - 30x + 9$$

Add $(-20x)$ to both sides,
$$12 = 17x^2 - 50x + 9$$

Add (-12) to both sides,
$$17x^2 - 50x - 3 = 0$$

We can find the roots of this equation using the

quadratic formula $x = \dfrac{-b \pm \sqrt{b^2 - 4ac}}{2a}$, which applies to
the situation $ax^2 + bx + c = 0$. In our case $a = 17$, $b = -50$, and $c = -3$. Thus

$$x = \frac{50 \pm \sqrt{2500 + 204}}{34}$$

$$= \frac{50 \pm \sqrt{2704}}{34}$$

$$= \frac{50 \pm 52}{34} = \frac{50}{34} \pm \frac{52}{34}$$

$$= \frac{102}{34} \text{ and } -\frac{2}{34} ;$$

Thus, $x = 3$,
and $\quad x = -\dfrac{1}{17}$.

Check: To verify that 3 and $-1/17$ are indeed roots of
the given equation we replace x by these values in the
original equation,

$$\sqrt{x + 1} + \sqrt{2x + 3} - \sqrt{8x + 1} = 0$$

(a) Substituting 3 for x:

$$\sqrt{3 + 1} + \sqrt{2(3) + 3} - \sqrt{8(3) + 1} = 0$$
$$\sqrt{4} + \sqrt{9} - \sqrt{25} = 0$$
$$2 + 3 - 5 = 0$$
$$0 = 0$$

Thus, 3 is a root of the equation.

(b) Substitute $-\frac{1}{17}$ for x:

$$\sqrt{-\frac{1}{17} + 1} + \sqrt{2\left[-\frac{1}{17}\right] + 3} - \sqrt{8\left[-\frac{1}{17}\right] + 1} = 0$$

$$\sqrt{-\frac{1}{17} + \frac{17}{17}} + \sqrt{-\frac{2}{17} + \frac{51}{17}} - \sqrt{-\frac{8}{17} + \frac{17}{17}} = 0$$

Since $\sqrt{\frac{a}{b}} = \frac{\sqrt{a}}{\sqrt{b}}$, $\quad \sqrt{\frac{16}{17}} + \sqrt{\frac{49}{17}} - \sqrt{\frac{9}{17}} = 0$

$$\frac{4}{\sqrt{17}} + \frac{7}{\sqrt{17}} - \frac{3}{\sqrt{17}} = 0$$

$$\frac{8}{\sqrt{17}} \neq 0$$

Since substitution of x by $-1/17$ does not result in a valid equation, $-1/17$ is not a root, and our solution set is {3}.

● PROBLEM 13-20

Solve the equation $4x^2 = 8x - 7$ by means of the quadratic formula.

Solution: The quadratic formula, $x = \dfrac{-b \pm \sqrt{b^2 - 4ac}}{2a}$, applies to equations of the form $ax^2 + bx + c = 0$. If we add $(-8x + 7)$ to both sides of our given equation we obtain $4x^2 - 8x + 7 = 8x - 7 = 0$ which is an equation in the form $ax^2 + bx + c = 0$ with $a = 4$, $b = -8$, and $c = 7$. Substituting these values into the quadratic formula we obtain

$$x = \frac{-(-8) \pm \sqrt{(-8)^2 - 4(4)(7)}}{2(4)}$$

$$= \frac{8 \pm \sqrt{64 - 112}}{8}$$

$$= \frac{8 \pm \sqrt{-48}}{8}$$

Since $\sqrt{ab} = \sqrt{a} \cdot \sqrt{b}$, $\sqrt{-48} = \sqrt{-1 \cdot 48} = \sqrt{-1} \sqrt{48}$. Recall $\sqrt{-1} = i$.

Thus $\sqrt{-48} = \sqrt{-1} \sqrt{48} = i\sqrt{48}$ and

$$x = \frac{8 \pm i\sqrt{48}}{8} .$$

We can further break down this radical by noting

$$\sqrt{48} = \sqrt{16 \cdot 3} = \sqrt{16} \cdot \sqrt{3} = 4\sqrt{3} .$$

Thus,

$$x = \frac{8 \pm 4i\sqrt{3}}{8}$$

$$x = \frac{8}{8} \pm \frac{4i\sqrt{3}}{8}$$

$$x = 1 \pm \frac{i\sqrt{3}}{2}$$

Hence the solution set is $\left\{ 1 + \frac{i\sqrt{3}}{2} , 1 - \frac{i\sqrt{3}}{2} \right\}$.

We can verify that these two complex numbers are the roots of the given equation by means of the following check: We replace x by $1 + \frac{i\sqrt{3}}{2}$ in the original equation:

$$4\left(1 + \frac{i\sqrt{3}}{2}\right)^2 = 8\left(1 + \frac{i\sqrt{3}}{2}\right) - 7$$

$$4\left[1 + \frac{2i\sqrt{3}}{2} + \left(\frac{i\sqrt{3}}{2}\right)^2\right] = 8 + 8\frac{i\sqrt{3}}{2} - 7$$

$$4\left[1 + \frac{2i\sqrt{3}}{2} + i^2\left(\frac{\sqrt{3}}{2}\right)^2\right] = 1 + 4i\sqrt{3}$$

$$4\left[1 + i\sqrt{3} - \left(\frac{\sqrt{3}}{2} \frac{\sqrt{3}}{2}\right)\right] = 1 + 4i\sqrt{3}$$

$$4\left[1 + i\sqrt{3} - \frac{3}{4}\right] = 1 + 4i\sqrt{3}$$

$$4 + 4i\sqrt{3} - 3 = 1 + 4i\sqrt{3}$$

$$1 + 4i\sqrt{3} = 1 + 4i\sqrt{3}$$

Now we replace x by $1 - \frac{i\sqrt{3}}{2}$ in the original equation:

$$4\left(1 - \frac{i\sqrt{3}}{2}\right)^2 = 8\left(1 - \frac{i\sqrt{3}}{2}\right) - 7$$

$$4\left[1 - \frac{2i\sqrt{3}}{2} + \left(\frac{i\sqrt{3}}{2}\right)^2\right] = 8 - \frac{8i\sqrt{3}}{2} - 7$$

$$4\left[1 - i\sqrt{3} + i^2\left(\frac{\sqrt{3}}{2}\right)^2\right] = 1 - 4i\sqrt{3}$$

$$4\left[1 - i\sqrt{3} - 1\left(\frac{\sqrt{3}}{2}\right)^2\right] = 1 - 4i\sqrt{3}$$

$$4\left(1 - i\sqrt{3} - \frac{3}{4}\right) = 1 - 4i\sqrt{3}$$

$$4 - 4i\sqrt{3} - 3 = 1 - 4i\sqrt{3}$$
$$1 - 4i\sqrt{3} = 1 - 4i\sqrt{3}$$

Solve $3x^2 - 5x + 4 = 0$ by the quadratic formula.

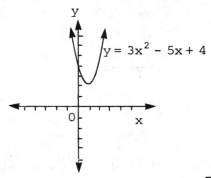

$y = 3x^2 - 5x + 4$

Solution: Recall the quadratic formula, $x = \dfrac{-b \pm \sqrt{b^2 - 4ac}}{2a}$, which

applies to equations in the form $ax^2 + bx + c = 0$. In our case
$$a = 3 \qquad b = -5 \qquad c = 4 .$$

Substituting these values into the quadratic formula we obtain,

$$x = \frac{-(-5) \pm \sqrt{(-5)^2 - 4(3)(4)}}{2 \cdot 3}$$

$$x = \frac{5 \pm \sqrt{25 - 48}}{6} = \frac{5 \pm \sqrt{-23}}{6}$$

Since $\sqrt{-23}$ is not a real number, x is not a real number, and, consequently, the equation has no real roots. The graph of $y = 3x^2 - 5x + 4$ is shown in the accompanying figure. Notice that the graph does not cross the x-axis. This is because on the x-axis $y = 0$. Hence $y = 0 = 3x^2 - 5x + 4$, which is the equation we have just shown to have no real roots.

Show that the roots of the quadratic equation $x^2 - x - 3 = 0$ are
$$x_1 = \frac{1 + \sqrt{13}}{2} \quad \text{and} \quad x_2 = \frac{1 - \sqrt{13}}{2}$$

Solution: We use the quadratic formula derived from the quadratic equation, $ax^2 + bc + c = 0$:

$$x = \frac{-b \pm \sqrt{b^2 - 4ac}}{2a}$$

For $x^2 - x - 3 = 0$, $a = 1$, $b = -1$, and $c = -3$. Replacing these values in the quadratic formula,

$$x = \frac{-(-1) \pm \sqrt{(-1)^2 - 4(1)(-3)}}{2(1)}$$

$$x = \frac{1 \pm \sqrt{13}}{2}$$

$$x_1 = \frac{1 + \sqrt{13}}{2} \qquad x_2 = \frac{1 - \sqrt{13}}{2}$$

According to the Factor Theorem: If r is a root of the equation $f(x) = 0$, i.e., if $f(r) = 0$, then $(x - r)$ is a factor of $f(x)$. x_1 and x_2 are roots of $x^2 - x - 3 = 0$. Thus,

$$\left(x - \frac{1 + \sqrt{13}}{2}\right)\left(x - \frac{1 - \sqrt{13}}{2}\right)$$

are factors, and

$$x^2 - x - 3 = \left(x - \frac{1 - \sqrt{13}}{2}\right)\left(x - \frac{1 - \sqrt{13}}{2}\right)$$

● **PROBLEM 13-23**

Find the sum and product of the roots of the equation

$$3x^2 - 2x + 1 = 0.$$

Solution: The given equation is a quadratic equation in which $a = 3$, $b = -2$, and $c = 1$. Using the quadratic formula $x = \dfrac{-b \pm \sqrt{b^2 - 4ac}}{2a}$ to find the roots of the given equation:

$$x = \frac{-(-2) \pm \sqrt{(-2)^2 - 4(3)(1)}}{2(3)} = \frac{2 \pm \sqrt{-8}}{6}$$

$$= \frac{1}{3} \pm \frac{\sqrt{4}\sqrt{-2}}{6} = \frac{1}{3} \pm \frac{2\sqrt{-2}}{6}$$

$$= \frac{1 \pm \sqrt{-2}}{3}$$

The sum of the roots is:

$$\frac{1 + \sqrt{-2}}{3} + \frac{1 - \sqrt{-2}}{3} = \frac{1 + \sqrt{-2} + 1 - \sqrt{-2}}{3} = \frac{2}{3}.$$

The product of the root is:

$$\left(\frac{1 + \sqrt{-2}}{3}\right)\left(\frac{1 - \sqrt{-2}}{3}\right) = \frac{1 + \sqrt{-2} - \sqrt{-2} - (-2)}{9}$$

$$= \frac{1}{3} .$$

● **PROBLEM 13-24**

Without solving, find the sum and product of the roots of $8x^2 = 2x + 3$.

Solution: Given a quadratic equation in standard form, $ax^2 + bx + c = 0$, the sum of the roots is given by $\frac{-b}{a}$ and the product of the roots by $\frac{c}{a}$. Adding $-(2x + 3)$ to both sides of the given equation, we obtain $8x^2 - 2x - 3 = 0$, a quadratic equation in standard form with $a = 8$, $b = -2$, and $c = -3$. Thus:

Sum of roots $= -\frac{b}{a} = \left(\frac{-2}{8}\right) = \frac{1}{4}$.

Product of roots $= \frac{c}{a} = \frac{-3}{8}$.

● **PROBLEM 13-25**

Find the sum and the product of the roots in each of the following equations: $x^2 - 3x + 2 = 0$, $2x^2 + 8x - 5 = 0$, and $\sqrt{2}x^2 + 5x - \sqrt{8} = 0$

Solution: There are two relations between the roots and coefficients of a quadratic equation. When we want to find the roots of the quadratic function, $f(x) = ax^2 + bx + c$, we set $f(x) = 0$. Then $ax^2 + bx + c = 0$. By the quadratic formula, the roots are

$$r_1 = \frac{-b + \sqrt{b^2 - 4ac}}{2a} \qquad r_2 = \frac{-b - \sqrt{b^2 - 4ac}}{2a}$$

Adding $r_1 + r_2 = \frac{-b + \sqrt{b^2 - 4ac}}{2a} + \frac{-b - \sqrt{b^2 - 4ac}}{2a}$

$$= \frac{-2b}{2a} = \frac{-b}{a}$$

Multiplying $r_1 \cdot r_2 = \left(\frac{-b + \sqrt{b^2 - 4ac}}{2a}\right)\left(\frac{-b - \sqrt{b^2 - 4ac}}{2a}\right)$

209

$$= \frac{b^2 - (b^2 - 4ac)}{4a^2} = \frac{4ac}{4a^2} = \frac{c}{a}$$

Therefore, the sum of the roots is -b/a and the product of the roots is c/a. Thus in the following tabulation -b/a is the sum of the roots, and c/a is the product of the roots.

Equation	Sum of roots	Product of roots
$x^2 - 3x + 2 = 0$ Thus a = 1, b = -3 c = 2	$-\frac{b}{a} = \frac{-(-3)}{1} = 3$	$\frac{c}{a} = \frac{2}{1} = 2$
$2x^2 + 8x - 5 = 0$ Thus a = 2, b = 8 c = -5	$-\frac{b}{a} = \frac{-8}{2} = -4$	$\frac{c}{a} = \frac{-5}{2}$
$\sqrt{2}x^2 + 5x - \sqrt{8} = 0$ Thus a = $\sqrt{2}$, b = 5 c = -58	$-\frac{b}{a} = \frac{-5}{\sqrt{2}}$	$\frac{c}{a} = \frac{-\sqrt{8}}{\sqrt{2}} = -\sqrt{4} = -2$

These two relations provide a rapid method for verifying the roots of a quadratic equation. For the first equation, $x^2 - 3x + 2 = 0$, we can solve for the roots by factoring.

$$x^2 - 3x + 2 = 0$$
$$(x - 2)(x - 1) = 0$$
$$x_1 = 2 \quad x_2 = 1$$

The sum was found by the formula to be 3.

$$x_1 + x_2 = 2 + 1 = 3$$

The product was found to be 2.

$$x_1 \cdot x_2 = 2 \cdot 1 = 2$$

Similarly for the last two equations.

$2x^2 + 8x - 5 = 0$; -b/a = -4; c/a = -5/2

$$x_1 = \frac{-8 + \sqrt{8^2 - 4(2)(-5)}}{2(2)} \qquad x_2 = \frac{-8 - \sqrt{8^2 - 4(2)(-5)}}{2(2)}$$

$$= \frac{-8 + \sqrt{104}}{4} \qquad\qquad = \frac{-8 - \sqrt{104}}{4}$$

$$x_1 + x_2 = \frac{-8 + \sqrt{104}}{4} + \frac{-8 - \sqrt{104}}{4} = \frac{-16}{4} = -4 = -b/a$$

$$x_1 \cdot x_2 = \left(\frac{-8 + \sqrt{104}}{4}\right)\left(\frac{-8 - \sqrt{104}}{4}\right) = \frac{64 - 104}{16} = \frac{-40}{16} = \frac{-5}{2} = c/a$$

$$\sqrt{2} \, x^2 + 5x - \sqrt{8} = 0$$

$$x_1 = \frac{-5 + \sqrt{25 - 4(-\sqrt{8})\sqrt{2}}}{2\sqrt{2}} \qquad x_2 = \frac{-5 - \sqrt{25 - 4(-\sqrt{8})\sqrt{2}}}{2\sqrt{2}}$$

$$= \frac{-5 + \sqrt{25 - 4(-4)}}{2\sqrt{2}} \qquad\qquad = \frac{-5 - \sqrt{25 - 4(-4)}}{2\sqrt{2}}$$

$$= \frac{-5 + \sqrt{41}}{2\sqrt{2}} \qquad\qquad = \frac{-5 - \sqrt{41}}{2\sqrt{2}}$$

$$x_1 + x_2 = \left(\frac{-5 + \sqrt{41}}{2\sqrt{2}}\right) + \left(\frac{-5 - \sqrt{41}}{2\sqrt{2}}\right) = \frac{-10}{2\sqrt{2}} = \frac{-5}{\sqrt{2}} = \frac{-b}{a}$$

$$x_1 \cdot x_2 = \left(\frac{-5 + \sqrt{41}}{2\sqrt{2}}\right)\left(\frac{-5 - \sqrt{41}}{2\sqrt{2}}\right) = \frac{25 - 41}{2 \cdot 2 \cdot 2} = \frac{-16}{8} = -2 =$$

$$= \frac{c}{a}$$

● **PROBLEM** 13-26

Find a quadratic equation whose roots are $3 + 2\sqrt{3}$ and $3 - 2\sqrt{3}$.

<u>Solution:</u> A quadratic equation is an equation of the form $ax^2 + bx + c = 0$, where a, b, and c are constants and $a \neq 0$. If both sides of this quadratic equation are divided by a, then:

$$\frac{ax^2 + bx + c}{a} = \frac{0}{a}$$

$$x^2 + \frac{b}{a}x + \frac{c}{a} = 0 \qquad\qquad (1)$$

Note that this last result is valid since $a \neq 0$. If r_1 and r_2 are the roots of a quadratic equation, then the sum of these roots, S, is,

$$S = r_1 + r_2 = \frac{-b}{a} \text{ and the product of these roots, P,}$$
is: $P = r_1 \cdot r_2 = c/a$.

Note that the coefficient of the x-term in equation (1) is $\frac{b}{a}$. In relation to the sum of the roots, S, this coefficient $= \frac{b}{a} = -\left(-\frac{b}{a}\right) = -(S) = -S$. Hence, equation (1) can be rewritten as,

$$x^2 + (-S)x + \frac{c}{a} = 0$$

or
$$x^2 - Sx + \frac{c}{a} = 0 \qquad (2)$$

Also, note that the constant term on the left side of equation (1), or $\frac{c}{a}$, is also the product, P, of the roots. Hence, equation (2) can be rewritten as:

$$x^2 - Sx + P = 0 \qquad (3)$$

The sum of the roots is:

$$S = r_1 + r_2, \text{ and here } r_1 \text{ and } r_2 \text{ are } 3 + 2\sqrt{3}, \ 3 - 2\sqrt{3}$$

Thus,

$$S = (3 + 2\sqrt{3}) + (3 - 2\sqrt{3})$$

$$= 3 + 2\sqrt{3} + 3 - 2\sqrt{3}$$

$$= 3 + 3$$

$$= 6$$

The product of the roots is:

$$P = r_1 \cdot r_2$$

$$= (3 + 2\sqrt{3})(3 - 2\sqrt{3})$$

$$= 9 + 6\sqrt{3} - 6\sqrt{3} - 4(3)$$

$$= 9 - 12$$

$$= -3$$

Then, replacing S and P by 6 and -3 respectively in equation (3):

$$x^2 - Sx + P = x^2 - 6x + (-3) = 0$$
or
$$x^2 - 6x - 3 = 0,$$

which is in the form $ax^2 + bx + c = 0$ of a quadratic equation.

● **PROBLEM 13-27**

Find the equation whose roots are the negatives of the roots of $x^2 + 7x - 2 = 0$.

<u>Solution:</u> The roots of a quadratic equation $ax^2 + bx + c = 0$ are given by the quadratic formula

For the given equation $a = 1$, $b = 7$, $c = -2$. If the roots of the given equation are r_1 and r_2, we seek an equation whose roots are $-r_1$ and $-r_2$. From the given equation, we have $r_1 + r_2 = -\frac{b}{a} = -\frac{7}{1} = -7$. Thus

$$-r_1 + (-r_2) = -(r_1 + r_2) = 7$$

and the coefficient of the first-degree term in the required equation is -7. The product of the roots of the given equation is $\frac{c}{a} = \frac{-2}{1} = -2$. Since $r_1 \cdot r_2 = (-r_1)(-r_2)$, the product of the roots of the required equation is also -2. Hence, the constant term of the required equation is -2. The required equation can be written

$$x^2 - 7x - 2 = 0.$$

$$x = \frac{-b \pm \sqrt{b^2 - 4ac}}{2a}.$$

Therefore:

$$r_1 = \frac{-b + \sqrt{b^2 - 4ac}}{2a}, \quad r_2 = \frac{-b - \sqrt{b^2 - 4ac}}{2a}$$

By adding r_1 and r_2:

$$r_1 + r_2 = \frac{-b + \sqrt{b^2 - 4ac}}{2a} + \frac{-b - \sqrt{b^2 - 4ac}}{2a} = \frac{-2b}{2a} = \frac{-b}{a}.$$

We see that the sum of the roots is:

$$r_1 + r_2 = \frac{-b}{a}.$$

Then multiply: $r_1 \cdot r_2 = \left(\frac{-b + \sqrt{b^2 - 4ac}}{2a} \right)\left(\frac{-b - \sqrt{b^2 - 4ac}}{2a} \right)$

$$= \frac{\left(-b + \sqrt{b^2 - 4ac}\right)\left(-b - \sqrt{b^2 - 4ac}\right)}{4a^2}$$

$$= \frac{b^2 + b\sqrt{b^2 - 4ac} - b\sqrt{b^2 - 4ac} - (b^2 - 4ac)}{4a^2}$$

$$= \frac{4ac}{4a^2} = \frac{c}{a}.$$

Find the value of k if one root is twice the other.
$$x^2 - kx + 18 = 0.$$

Solution: If x_1 and x_2 are the roots of the quadratic equation $ax^2 + bx + c$, then $x_1 + x_2 = -b/a$ and $x_1 \cdot x_2 = c/a$. For $x^2 - kx + 18 = 0$, a = 1, b = -k, and c = 18. Since for this quadratic, one root is twice the other, let the roots be r and 2r. Their sum is r + 2r = 3r. The sum of the roots is - b/a = k. Hence k = 3r. The product of the roots $r \cdot 2r = 2r^2$ is equal to c/a = 18.

Thus, $2r^2 = 18; \quad r^2 = 9; \quad r = \pm 3$

Therefore, $k = \pm 3 \cdot 3 = \pm 9$

Check: The roots of $x^2 - 9x + 18 = 0$ are 3, and $2 \cdot 3 = 6$.
The roots of $x^2 + 9x + 18 = 0$ are -3, and 2(-3) = -6.

If the equation $x^2 + 2(k+2)x + 9k = 0$ has equal roots, find k.

Solution: The given equation is a quadratic equation of the form $ax^2 + bx + c = 0$. In the given equation, a = 1, b = 2(k+2), and c = 9k. A quadratic equation has equal roots if the discriminant, $b^2 - 4ac$, is zero.

$$b^2 - 4ac = [2(k+2)]^2 - 4(1)(9k) = 0$$
$$4(k+2)^2 - 36k = 0$$
$$4(k+2)(k+2) - 36k = 0$$
$$4(k^2 + 4k + 4) - 36k = 0$$

Distributing, $4k^2 + 16k + 16 - 36k = 0$
$$4k^2 - 20k + 16 = 0 .$$

Divide both sides of this equation by 4:

$$\frac{4k^2 - 20k + 16}{4} = \frac{0}{4}$$

or

$$k^2 - 5k + 4 = 0 .$$

Factoring the left side of this equation into a product of two polynomials:

$$(k-4)(k-1) = 0.$$

When the product ab = 0, where a and b are any two numbers, either a = 0 or b = 0. Hence, in the case of this problem, either

$$k - 4 = 0 \quad \text{or} \quad k - 1 = 0.$$

Therefore,   k = 4   or   k = 1.

If α and β are the roots of $x^2 - px + q = 0$, find the value of (1) $\alpha^2 + \beta^2$, (2) $\alpha^3 + \beta^3$.

Solution: The roots of the given equation are α and β. Hence, $x = \alpha$ and $x = \beta$. Subtract α from both sides of the first equation:

$$x - \alpha = \alpha - \alpha = 0 ,$$

or

$$x - \alpha = 0 .$$

Subtract β from both sides of the second equation:

$$x - \beta = \beta - \beta = 0 ,$$

or

$$x - \beta = 0 .$$

Hence,

$$(x - \alpha)(x - \beta) = (0)(0) = 0,$$

or

$$(x - \alpha)(x - \beta) = 0 .$$

Also,

$$(x - \alpha)(x - \beta) = x^2 - \alpha x - \beta x + \alpha\beta = 0$$

or

$$x^2 - (\alpha + \beta)x + \alpha\beta = 0 \qquad\qquad (1)$$

Comparing the given equation with equation (1):

$$\alpha + \beta = p \quad (eq.2) , \text{ and } \quad \alpha\beta = q \quad (eq. 3)$$

Therefore, squaring both sides of equation (2):

$$(\alpha + \beta)^2 = p^2$$
$$(\alpha + \beta)(\alpha + \beta) = p^2$$
$$\alpha^2 + \alpha\beta + \alpha\beta + \beta^2 = p^2$$
$$\alpha^2 + 2\alpha\beta + \beta^2 = p^2 \qquad\qquad (4)$$

Subtract $2\alpha\beta$ from both sides of equation (4):

$$\alpha^2 + 2\alpha\beta + \beta^2 - 2\alpha\beta = p^2 - 2\alpha\beta$$
$$\alpha^2 + \beta^2 = p^2 - 2\alpha\beta$$
$$\alpha^2 + \beta^2 = p^2 - 2q ,$$

since $\alpha\beta = q$.
To obtain an expression for $\alpha^3 + \beta^3$, cube both sides of equation (2).

$$(\alpha + \beta)^3 = p^3$$
$$(\alpha + \beta)(\alpha + \beta)^2 = p^3$$
$$(\alpha + \beta)(\alpha^2 + 2\alpha\beta + \beta^2) = p^3$$

Distributing the left side of this equation:

$$(\alpha^3 + 2\alpha^2\beta + \alpha\beta^2) + (\alpha^2\beta + 2\alpha\beta^2 + \beta^3) = p^3$$

Combining terms and simplifying the left side of this equation:

$$\alpha^3 + 2\alpha^2\beta + \alpha\beta^2 + \alpha^2\beta + 2\alpha\beta^2 + \beta^3 = p^3$$
$$\alpha^3 + 3\alpha^2\beta + 3\alpha\beta^2 + \beta^3 = p^3$$
$$(\alpha^3 + \beta^3) + (3\alpha^2\beta + 3\alpha\beta^2) = p^3$$

Factor out $3\alpha\beta$ from the second term on the left side of this equation:

$$\left(\alpha^3 + \beta^3\right) + 3\alpha\beta(\alpha + \beta) = p^3$$
$$\left(\alpha^3 + \beta^3\right) + 3q(p) = p^3 ,$$

since $\alpha\beta = q$ and $\alpha + \beta = p$. Hence,

$$\left(\alpha^3 + \beta^3\right) + 3pq = p^3$$

Subtract $3pq$ from both sides of this equation.

$$\left(\alpha^3 + \beta^3\right) + 3pq - 3pq = p^3 - 3pq$$
$$\left(\alpha^3 + \beta^3\right) = p^3 - 3pq$$

Factor out p from the right side of this equation.

$$\alpha^3 + \beta^3 = p\left(p^2 - 3q\right)$$

Therefore, $\alpha^2 + \beta^2 = p^2 - 2q$, and $\alpha^3 + \beta^3 = p\left(p^2 - 3q\right)$.

● **PROBLEM 13-31**

Compute the value of the discriminant and then determine the nature of the roots of each of the following
four equations:

$$4x^2 - 12x + 9 = 0,$$

$$3x^2 - 7x - 6 = 0$$

$$5x^2 + 2x - 9 = 0$$

and $x^2 + 3x + 5 = 0.$

<u>Solution:</u> The discriminant, the term of the quadratic
formula which appears under the radical, is $b^2 - 4ac$.
It can be used to determine the nature of the roots of equations in the form $ax^2 + bx + c = 0$. Assuming a,b,c
are real numbers, then,

(1) if $b^2 - 4ac > 0$, the roots are real and unequal

(2) if $b^2 - 4ac = 0$, the roots are real and equal

(3) if $b^2 - 4ac < 0$, the roots are imaginary

Assuming a,b,c are real and rational numbers then,

(1) if $b^2 - 4ac$ is a perfect square $\neq 0$, the roots are
real, rational and unequal,

(2) if $b^2 - 4ac = 0$, the roots are real, rational,
and equal,

(3) if $b^2 - 4ac > 0$, but not a perfect square, the
roots are real, irrational and unequal,

(4) if $b^2 - 4ac < 0$, the roots are imaginary.

$$4x^2 - 12x + 9 = 0$$

Here a,b,c are rational numbers,

$$a = 4, \ b = -12 \ \text{and} \ c = 9.$$

Therefore,

$$b^2 - 4ac = (-12)^2 - 4(4)(9) = 144 - 144 = 0$$

Since the discriminant is 0, the roots are rational and equal.

(b) $3x^2 - 7x - 6 = 0$

Here a,b,c are rational numbers,

$$a = 3, \ b = -7, \ \text{and} \ c = -6.$$

Therefore,

$$b^2 - 4ac = (-7)^2 - 4(3)(-6) = 49 + 72 = 121 = 11^2$$

Since the discriminant is a perfect square, the roots are rational and unequal.

(c) $5x^2 + 2x - 9 = 0$

Here a,b,c are rational numbers,

$$a = 5, \ b = 2 \ \text{and} \ c = -9$$

Therefore,

$$b^2 - 4ac = 2^2 - 4(5)(-9) = 4 + 180 = 184$$

Since the discriminant is greater than zero, but not a perfect square, the roots are irrational and unequal.

(d) $x^2 + 3x + 5 = 0$

Here a,b,c are rational numbers,

$$a = 1, \ b = 3, \ \text{and} \ c = 5$$

Therefore,

$$b^2 - 4ac = 3^2 - 4(1)(5) = 9 - 20 = -11$$

Since the discriminant is negative the roots are imaginary.

Can the expression $16x^2 - 76x + 21$ be factored into rational factors?

Solution: To determine if this quadratic polynomial has rational factors we look at its discriminant, $b^2 - 4ac$ (this is the term that appears under the radical in the quadratic formula: $x = \dfrac{-b \pm \sqrt{b^2 - 4ac}}{2a}$, used for equations in the form of $ax^2 + bx + c = 0$).

In our example, $a = 16$, $b = -76$, $c = 21$ and our discriminant $b^2 - 4ac = (-76)^2 - 4 \cdot 16 \cdot 21 =$

$$5776 - 1344 = 4432.$$

Now, recall what the discriminant tells us about the nature of the roots:

If the discriminant (b^2-4ac) is positive or zero,
roots are real

If the discriminant (b^2-4ac) is negative
roots are complex
If the discriminant (b^2-4ac) is a perfect square
roots are rational
If the discriminant (b^2-4ac) is zero
roots are equal and rational.

Hence, roots are rational only if the discriminant is zero or a perfect square.

Looking at the column of perfect squares in a table of square roots, we note that 4,432 is not a perfect square, hence the expression $16x^2 - 76x + 21$ cannot be factored into rational factors.

Solve the system

$$xy = 24, \tag{1}$$

$$y - 2x + 2 = 0. \tag{2}$$

Solution: This system is most easily solved by the method of substitution. Solve (2) for y in terms of x:

$$y = 2x - 2.$$

Substitute $2x - 2$ for y in (1):

$$x(2x - 2) = 24,$$

$$2x^2 - 2x = 24,$$

$$2x^2 - 2x - 24 = 0,$$

$$x^2 - x - 12 = 0,$$

or factoring $(x+3)(x-4) = 0,$

Set each factor = 0 to find all values of x for which the product = 0.

$$x + 3 = 0 \quad \bigg| \quad x - 4 = 0$$
$$x = -3 \quad \bigg| \quad x = 4.$$

In Equation (1): for x = -3, (-3)y = 24 or y = -8; for x = 4, (4)y = 24 or y = 6.

In Equation (2): for x = -3, y - 2(-3) + 2 = 0 or y = -8; for x = 4, y - 2(4) + 2 = 0 or y = 6.

● PROBLEM 13-34

Solve the system

$$y = -x^2 + 7x - 5 \tag{1}$$

$$y - 2x = 2 \tag{2}$$

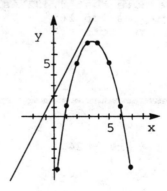

Solution: Solving Equation (2) for y yields an expression for y in terms of x. Substituting this expression in Equation (1),

$$2x + 2 = -x^2 + 7x - 5 \tag{3}$$

We have a single equation, in terms of a single variable,

to be solved. Writing Equation (3) in standard quadratic form,

$$x^2 - 5x + 7 = 0 \tag{4}$$

Since the equation is not factorable the roots are not found in this manner. Evaluating the discriminant will indicate whether Equation (4) has real roots. The discriminant, $b^2 - 4ac$, of Equation (4) equals $(-5)^2 - 4(1)(7) = 25 - 28 = -3$. Since the discriminant is negative, Equation (4) has no real roots, and therefore the system has no real solution. In terms of the graph, the figure shows that the parabola and the straight line have no point in common.

● **PROBLEM 13-35**

Solve the system

$$y = 3x^2 - 2x + 5 \tag{1}$$

$$y = 4x + 2. \tag{2}$$

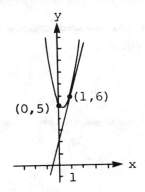

$(0,5)$ •$(1,6)$

Solution: To obtain a single equation with one unknown variable, x, substitute the value of y from Equation (2) in Equation (1),

$$4x + 2 = 3x^2 - 2x + 5. \tag{3}$$

Writing Equation (3) in standard quadratic form,

$$3x^2 - 6x + 3 = 0. \tag{4}$$

We may simplify Equation (4) by dividing both members by 3, which is a factor common to each term:

$$x^2 - 2x + 1 = 0. \tag{5}$$

To find the roots, factor and set each factor = 0. This may be done since a product = 0 implies one or all of the factors must = 0.

$$(x - 1)(x - 1) = 0 \tag{6}$$

$$x - 1 = 0 \bigm| x - 1 = 0$$

$$x = 1 \bigm| x = 1$$

Equation (5) has two equal roots, each equal to 1. For $x = 1$, from Equation (2) we have $y = 4(1) + 2 = 6$. Therefore the system has but one common solution:

$$x = 1, \; y = 6.$$

The figure indicates that our solution is probably correct. We may also check to see if our values satisfy Equation (1) as well:

Substituting in: $y = 3x^2 - 2x + 5$

$$6 \overset{?}{=} 3(1)^2 - 2(1) + 5$$

$$6 \overset{?}{=} 3 - 2 + 5$$

$$6 = 6.$$

● **PROBLEM 13-36**

Solve the system of equations

$$y = x^2 - 6x + 9 \tag{1}$$

$$y = x + 3. \tag{2}$$

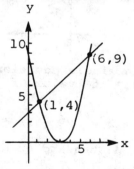

Solution: A single equation in terms of the one variable x may be obtained by the method of substitution. Substitute the value of y, x + 3, from Equation (2) for y in Equation (1).

$$x + 3 = x^2 - 6x + 9. \tag{3}$$

Writing Equation (3) in the standard form of a quadratic equation,

$$x^2 - 7x + 6 = 0. \tag{4}$$

221

Use the usual method of solving quadratic equations. Factor the equation. Then find the values of x for which each factor may = 0.

$$(x - 1)(x - 6) = 0 \qquad\qquad (5)$$

$$
\begin{array}{c|c}
x - 1 = 0 & x - 6 = 0 \\
x = 1 & x = 6
\end{array}
$$

The roots of Equation (4) are x = 1 and x = 6. Since y = x + 3, the solution of the given system is

$$x = 1, \ y = 4 \text{ and } x = 6, \ y = 9.$$

In the figure, the graph of the system, indicates that our solution is probably correct. We may prove that the solution is correct by substituting each solution in both equations of the given system, as usual. The values of y were obtained by satisfying equation (2). Now for Equation (1):

check for x = 1, y = 4	check for x = 6, y = 9
$y = x^2 - 6x + 9$	$y = x^2 - 6x + 9$
$4 \overset{?}{=} (1)^2 - 6(1) + 9$	$9 \overset{?}{=} (6)^2 - 6(6) + 9$
$4 \overset{?}{=} 1 - 6 + 9$	$9 \overset{?}{=} 36 - 36 + 9$
$4 = 4$	$9 = 9$

● **PROBLEM 13-37**

Solve the system
$$x^2 + y^2 = 10 \qquad\qquad (1)$$
$$x + 2y = 1 \qquad\qquad (2)$$

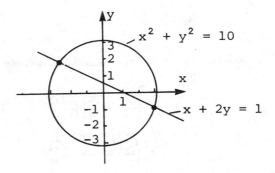

Solution: We solve the linear equation for x in terms of y by adding -2y to both sides to obtain x = 1 - 2y. We substitute the result, 1 - 2y, for x in the quadratic equation to obtain

$$(1 - 2y)^2 + y^2 = 10$$

Then we have

$$1 - 4y + 4y^2 + y^2 = 10$$

We add (-10) to both sides and combine terms:

$$5y^2 - 4y - 9 = 0$$

We factor, $(5y - 9)(y + 1) = 0$

Whenever the product of two numbers   ab = 0   either   a = 0   or   b = 0.
Thus   (5y - 9)(y + 1) = 0   implies either   5y - 9 = 0   or   y + 1 = 0
                                          5y = 9   or   y = -1

Thus, $y = \frac{9}{5}$ or y = -1 .

Substituting these values in turn in the linear equation, we find
the corresponding values for x: x + 2y = 1, for $y = \frac{9}{5}$

$$x + 2\left(\frac{9}{5}\right) = 1, \quad x = 1 - \frac{18}{5}, \quad x = \frac{-13}{5}$$

and for y = -1

$$x + 2(-1) = 1, \quad x - 2 = 1, \quad x = 3.$$

The solutions of the system are therefore

$$\left(x = -\frac{13}{5}, \ y = \frac{9}{5}\right) \quad \text{and} \quad (x = 3, \ y = -1).$$

To consider the corresponding graphs of this system, we notice
that $x^2 + y^2 = 10$ represents a circle with radius $\sqrt{10}$ and x + 2y = 1
is the line passing through the points (1,0) and (0,½).

The two points where the circle and line intersect are the solu-
tion to our problem, $\left(\frac{-13}{5}, \frac{9}{5}\right)$ and (3,-1), the points where $x^2 + y^2 = 10$
and x + 2y = 1 simultaneously.

● **PROBLEM 13-38**

Solve the system

$$x^2 + y^2 = 25, \tag{1}$$

$$x - y = 1. \tag{2}$$

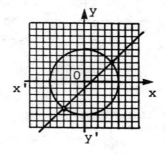

223

Solution: Solve (2) for y (the problem can be done similarly for x instead): The method of substitution is most easily employed in this example to solve the system.

$$y = x - 1. \tag{3}$$

Substitute x - 1 for y in (1):

$$x^2 + (x - 1)^2 = 25. \tag{4}$$

$$x^2 + x^2 - 2x + 1 = 25.$$

From (4) $\qquad 2x^2 - 2x - 24 = 0,$

or $\qquad\qquad x^2 - x - 12 = 0. \tag{5}$

Solve (5) by factoring: $\quad (x - 4)(x + 3) = 0$

$$x - 4 = 0 \quad x + 3 = 0$$

$$x = 4 \text{ or } -3.$$

Substituting 4 for x in (2), we obtain

$$4 - y = 1 \text{ or } y = 3.$$

Substituting -3 for x in (2), we obtain

$$-3 - y = 1 \text{ or } y = -4.$$

This gives $\quad \begin{matrix} x = 4 \\ y = 3 \end{matrix} \quad$ and $\quad \begin{matrix} x = -3 \\ y = -4 \end{matrix} \quad$ for the solutions.

Check:

for x = 4, y = 3

in Eq. (1): $\quad (4)^2 + (3)^2 = 25$

$$16 + \quad 9 = 25$$

$$25 = 25$$

in Eq. (2): $\quad (4) - (3) = 1$

$$1 = 1$$

for x = -3, y = -4

in Eq. (1): $\quad (-3)^2 + (-4)^2 = 25$

$$9 + \quad 16 = 25$$

$$25 = 25$$

in Eq. (2): $\quad (-3) - (-4) = 1$

$$-3 + \quad 4 = 1$$

$$1 = 1.$$

Graphical meaning of the two solutions. We may plot the graph for each of the equations (1) and (2). The graph of

$$x - y = 1$$

is the straight line shown in the figure, and the graph of

$$x^2 + y^2 = 25$$

is the circle there shown. To draw the graph of (1), the student may give various values to x and calculate the corresponding values for y from $y = \pm \sqrt{25 - x^2}$.

Any point on the straight line (2) has coordinates that satisfy Equation (2). Any point on the circle (1) has coordinates that satisfy Equation (1). The points (4,3) and (-3,-4) lie on both graphs and satisfy both Equations (1) and (2). That is to say, each point of intersection of the graph of (1) with the graph of (2) gives a pair of numbers that is a solution of the system.

● **PROBLEM 13-39**

Solve for x and y:
$$\begin{cases} 9x^2 - 16y^2 = 144 & (1) \\ x - 2y = 4. & (2) \end{cases}$$

Algebraic solution: We solve equation (2) for x,

$$x = 4 + 2y \tag{3}$$

and substitute this expression for x in equation (1). This gives

$$9(4 + 2y)^2 - 16y^2 = 144,$$

$$9\left(16 + 16y + 4y^2\right) - 16y^2 = 144$$

Dividing both sides by 4, $\quad 9\left(4 + 4y + y^2\right) - 4y^2 = 36.$

225

Distributing, $36 + 36y + 9y^2 - 4y^2 = 36$

Combining terms, $36 + 36y + 5y^2 = 36$

Subtracting 36 from both sides, $5y^2 + 36y = 0$

Factoring, $y(5y + 36) = 0$

Whenever the product of two numbers $ab = 0$, either $a = 0$ or $b = 0$; thus either

$$y = 0 \qquad \text{or} \qquad 5y + 36 = 0$$

$$5y = -36$$

$$y = \frac{-36}{5}$$

Thus, $y = 0$, $-\frac{36}{5}$.

Placing these values in linear equation (3):

when $y = 0$,

$$x = 4 + 2(0) = 4 + 0 = 4$$

when $y = -\frac{36}{5}$,

$$x = 4 + 2\left(-\frac{36}{5}\right) = 4 - \frac{72}{5} = \frac{20}{5} - \frac{72}{5} = -\frac{52}{5}$$

Thus the two solutions of the equations are seen to be $(4, 0)$ and $(-52/5, -36/5)$, which are then the actual coordinates of the points of intersection of the line and the hyperbola to be discussed.

Geometric solution: Construct the graph of each equation and note where the two graphs intersect. The graph of the first equation cuts the x-axis at $x = \pm 4$, and y is imaginary for any value of x between -4 and 4. The graph consists of the two curved branches in the diagram, and is a hyperbola.

The graph of the second equation is a straight line through the points $(4, 0)$ and $(0, -2)$. This line intersects the hyperbola at the points P and Q, whose coordinates are approximately $(4, 0)$ and $(-10, -7)$.

● **PROBLEM 13-40**

Obtain the solution set of

$$x^2 + 4xy - 7x = 12 \qquad (1)$$

$$3x^2 - 4xy + 4x = 15 \qquad (2)$$

<u>Solution:</u> Since the sum of the xy terms in the left members of equations (1) and (2) is zero, we proceed as follows: Adding equations (1) and (2),

$$4x^2 - 3x = 27$$

$$4x^2 - 3x - 27 = 0$$

Equations in the form $ax^2 + bx + c = 0$ can be solved using the quadratic formula, $x = \dfrac{-b \pm \sqrt{b^2-4ac}}{2a}$. In our case, a = 4, b = -3, and c = -27, thus

$$x = \frac{-\,(-3) \pm \sqrt{(-3)^2 - 4(4)(-27)}}{2(4)}$$

$$x = \frac{3 \pm \sqrt{9 + 432}}{8} = \frac{3 \pm \sqrt{441}}{8} = \frac{3 \pm 21}{8}$$

$$x = \frac{3 + 21}{8} = \frac{24}{8} = 3 \quad \text{or} \quad x = \frac{3 - 21}{8} = \frac{-18}{8} = -\frac{9}{4}$$

Thus $x = 3, -\dfrac{9}{4}$.

We find the second numbers in the solution pairs as follows: Replacing x by 3 in equation (1),

$$x^2 + 4xy - 7x = 12$$

$$3^2 + 4(3)y - 7(3) = 12$$

$$9 + 12y - 21 = 12$$

$$12y - 12 = 12$$

$$12y = 24$$

$$y = 2$$

Hence one simultaneous solution pair is (3,2). Then: Replacing x by -9/4 in equation (1),

$$x^2 + 4xy - 7x = 12$$

$$\left(\frac{-9}{4}\right)^2 + 4\left(\frac{-9}{4}\right)y - 7\left(\frac{-9}{4}\right) = 12$$

$$\frac{81}{16} - 9y + \frac{63}{4} = 12$$

Multiplying both sides by 16,

$$81 - 144y + 252 = 192$$

$$-144y + 333 = 192$$

$$-144y = -141$$

$$y = \frac{-141}{-144} = \frac{47}{48}$$

Therefore a second simultaneous solution pair is $\left(-\dfrac{9}{4}, \dfrac{47}{48}\right)$, and the simultaneous solution set is

$$\left\{(3,2), \left(-\frac{9}{4}, \frac{47}{48}\right)\right\}$$

Check: Using (3,2), we have:

From equation (1):

$$(3)^2 + 4(3)(2) - 7(3) = 9 + 24 - 21 = 33 - 21 = 12$$

From equation (2):

$$3(3)^2 - 4(3)(2) + 4(3) = 27 - 24 + 12 = 3 + 12 = 15$$

Using $\left(-\frac{9}{4}, \frac{47}{48}\right)$, we obtain:

From equation (1):

$$\left(-\frac{9}{4}\right)^2 + 4\left(\frac{-9}{4}\right)\left(\frac{47}{48}\right) - 7\left(\frac{-9}{4}\right) = \frac{81}{16} - \frac{141}{16} + \frac{63}{4} = \frac{81-141+252}{16}$$

$$= \frac{192}{16} = 12$$

From equation (2):

$$3\left(\frac{-9}{4}\right)^2 - 4\left(\frac{-9}{4}\right)\left(\frac{47}{48}\right) + 4\left(\frac{-9}{4}\right) = \frac{243}{16} + \frac{141}{16} - 9 = \frac{243+141-144}{16}$$

$$= \frac{240}{16} = 15$$

Thus, the two pairs of solutions obtained are valid.

● **PROBLEM 13-41**

Solve graphically
$$\begin{cases} x^2 + y^2 = 13, & (1) \\ y = x^2 - 1. & (2) \end{cases}$$

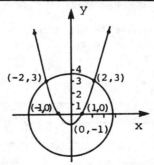

Solution: First, we must find the x and y intercepts. Set y = 0 to find the x-intercept or where the curve crosses the x-axis. Set x = 0 to find the y-intercept or where the curve crosses the y-axis. In Eq. (1), set x = 0, and find $y = \pm\sqrt{13} = \pm 3.6$. Then set y = 0, and find $x = \pm\sqrt{13}$. To get additional points, we solve for y.

$$x^2 + y^2 = 13$$
$$y^2 = 13 - x^2$$

228

$$y = \pm \sqrt{13 - x^2}$$

Then, set up a table. Choose various x values and calculate the corresponding y values. See Graph.

x	$\pm \sqrt{13 - x^2}$ =	y
-3.6	$+ \sqrt{13 - (-3.6)^2}$	≈ 0
-3	$\pm \sqrt{13 - (-3)^2}$	± 2
-2	$+ \sqrt{13 - (-2)^2}$	$+ 3$
-1	$+ \sqrt{13 - (-1)^2}$	± 3.5
0	$+ \sqrt{13 - (0)^2}$	$+ 3.6$
1	$\pm \sqrt{13 - (1)^2}$	± 3.5
2	$+ \sqrt{13 - (2)^2}$	± 3
3	$\pm \sqrt{13 - (3)^2}$	± 2
3.6	$+ \sqrt{13 - (3.6)^2}$	≈ 0

To find the domain of the relation, $+ \sqrt{13 - x^2}$, we know that the expression, $13 - x^2$, under the square root sign must be positive in order for the expression to be real, not imaginary.

$$(13 - x^2) \geq 0$$

subtract 13 from both sides, $-x^2 \geq -13$

multiply by -1 and reverse the inequality sign,

$$x^2 \leq 13$$

Take the square root of both sides.

$$|x| \leq \sqrt{13}$$

Another way to express $|b| \leq a$ is $-a \leq b \leq +a$. Thus,

$$-\sqrt{13} \leq x \leq + \sqrt{13}$$

Thus, for the relation $y = \pm \sqrt{13-x^2}$, the domain is $\{x | -\sqrt{13} \leq x \leq \sqrt{13}\}$. The curve is a circle. The general equation of a circle is $(x-h)^2 + (y-k)^2 = r^2$, where (h,k) is the center and r is the radius. In this case $(0,0)$ or the origin is the center and r^2 is 13. Therefore, the radius $= + \sqrt{13}$.

In Eq. (2), y is a quadratic function of x; hence the graph is a parabola. Set up a similar table for the quadratic function, $y = x^2 - 1$.

x	$x^2 - 1$ =	y
-3	$(-3)^2 - 1$	8
-2	$(-2)^2 - 1$	3
-1	$(-1)^2 - 1$	0
0	$(0)^2 - 1$	-1
1	$(1)^2 - 1$	0
2	$(2)^2 - 1$	3
3	$(3)^2 - 1$	8

From the graphs we read the real solutions (2,3) and (-2,3). These are points of intersection for both curves.

To find the solutions algebraically substitute equation (2) into (1).

$$x^2 + y^2 = 13 \qquad\qquad (1)$$
$$y = x^2 - 1 \qquad\qquad (2)$$

$$x^2 + (x^2 - 1)^2 = 13$$
$$x^2 + x^4 - 2x^2 + 1 = 13$$
$$x^4 - x^2 = 12$$
$$x^4 - x^2 - 12 = 0$$

Substitute z for x^2, i.e., $z = x^2$ to obtain a quadratic equation in z.

$$(x^2)^2 - (x^2) - 12 = 0$$
$$z^2 - z - 12 = 0$$
$$(z - 4)(z + 3) = 0$$
$$z - 4 = 0 \qquad z + 3 = 0$$
$$z = 4 \qquad z = -3$$

Therefore
$$x^2 = 4 \qquad x^2 = -3$$
$$x = \pm 2 \qquad x = \pm\sqrt{-3} = \pm\sqrt{3}i$$

Find the corresponding y-values by substituting into $y = x^2 - 1$.

$x = 2$	$x = -2$	$x = i\sqrt{3}$	$x = -i\sqrt{3}$
$y = (2)^2 - 1$	$y = (-2)^2 - 1$	$y = (i\sqrt{3})^2 - 1$	$y = (-i\sqrt{3})^2 - 1$
$= 3$	$= 3$	$y = i^2(3) - 1$	$= 3(i)^2 - 1$
		$= (-1)(3) - 1$	$= 3(-1) - 1$
		$= -4$	$= -4$

The algebraic solution gives $(2,3)$, $(-2,3)$, $(i\sqrt{3}, -4)$, and $(-i\sqrt{3}, -4)$. Notice that the imaginary solutions do not appear on the graph.

● **PROBLEM 13-42**

Obtain the simultaneous solution set of the equations
$$y = x^2 - 4 \qquad\qquad (1)$$
$$3x^2 + 8y^2 = 75 \qquad\qquad (2)$$
by the graphical method.

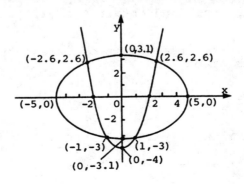

230

<u>Solution:</u> Equation (1) is in the form of the function $ax^2 + bx + c$. Its graph is a parabola. When a is positive, the curve opens upward. If it is negative, the curve opens downward. In this case $a = 1$, which is positive. Hence the graph is a parabola opening upward. We construct the parabola by means of the following table of corresponding values and show the graph in the accompanying figure.

x	$x^2 - 4 =$	y
-3	$(-3)^2 - 4 = 9 - 4 =$	5
-2	$(-2)^2 - 4 = 4 - 4 =$	0
-1	$(-1)^2 - 4 = 1 - 4 =$	-3
0	$0^2 - 4 = 0 - 4 =$	-4
1	$(1)^2 - 4 = 1 - 4 =$	-3
2	$(2)^2 - 4 = 4 - 4 =$	0
3	$(3)^2 - 4 = 9 - 4 =$	5

Equation (2) is of the type $ax^2 + by^2 = c$, with $a = 3$, $b = 8$, and $c = 75$. The graph is therefore an ellipse. To find the x-intercepts, set $y = 0$.

$$3x^2 + 8(0)^2 = 75$$

Solve for x.

$$3x^2 = 75$$
$$x^2 = 25$$
$$x = \pm 5$$

We have $(-5,0)$ and $(+5,0)$ as the x- intercepts
To find the y-intercepts, set $x = 0$

$$3(0)^2 + 8y^2 = 75$$
$$8y^2 = 75$$
$$y^2 = \frac{75}{8} = 9\frac{3}{8} = 9.375$$
$$y \approx \pm 3.1$$

We obtain $(0,3.1)$ and $(0,-3.1)$ for the y-intercepts.
We construct the graph and obtain the ellipse.

We now solve for the points of intersection, indicated on the graph.

$$y = x^2 - 4 \qquad (1)$$
$$3x^2 + 8y^2 = 75 \qquad (2)$$

Substitute the value of y in (1) into (2)

$$3x^2 + 8\left(x^2 - 4\right)^2 = 75$$

231

$$3x^2 + 8\left(x^2 - 8x^2 + 16\right) = 75$$
$$3x^2 + 8x^4 - 64x^2 + 128 = 75$$
$$8x^4 - 61x^2 + 53 = 0$$
Let $z = x^2$
$$8z^2 - 61z + 53 = 0$$

Apply the quadratic formula $z = \dfrac{-b \pm \sqrt{b^2 - 4ac}}{2a}$ with

$a = 8$, $b = -61$ and $c = 53$.

$$z = \frac{-(-61) \pm \sqrt{(-61)^2 - 4(8)53}}{2(8)} = \frac{61 \pm \sqrt{3721 - 1696}}{16}$$

$$z = \frac{61 \pm 45}{16}$$

$$z = 6.625, 1$$

$$z = x^2 = 6.625, 1$$

$$x = \pm 2.57 \approx \pm 2.6, \quad x = \pm 1$$

To solve for the y-values, substitute the x values.
$$y = x^2 - 4$$

For $x = \pm 1$
$$y = 1^2 - 4$$
$$y = -3$$

For $x = \pm 2.6$
$$y = (2.57)^2 - 4$$
$$y = 6.6 - 4 = 2.6$$

Therefore the points of intersection are
$\{(1,-3), (-1,-3), (-2.6,2.6), (+2.6,+2.6)\}$.

● **PROBLEM 13-43**

Solve $\quad xy = 3$

$\qquad\quad x^2 + y^2 = 10$

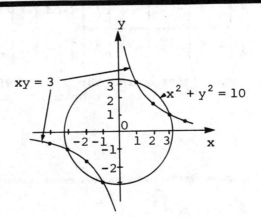

Solution: We solve the first equation for y to obtain

y = 3/x and substitute in the second equation to obtain

$$x^2 + \left(\frac{3}{x}\right)^2 = 10$$

Squaring: $x^2 + \dfrac{9}{x^2} = 10$

Multiplying both sides by x^2: $x^2\left(x^2 + \dfrac{9}{x^2}\right) = x^2\,(10)$

Distributing: $x^4 + 9 = 10x^2$

Subtracting $10x^2$ from both sides: $x^4 - 10x^2 + 9 = 0$

Factoring: $\left(x^2 - 1\right)\left(x^2 - 9\right) = 0$

Thus, $x^2 - 1 = 0$ or $x^2 - 9 = 0$

$$x^2 = 1 \qquad\qquad x^2 = 9$$

$$x = \pm\, 1 \qquad\qquad x = \pm\, 3$$

Therefore, x = 1, - 1, 3, or - 3.

Since y = 3/x, substituting these values in turn in this equation, we obtain the corresponding values for y:

x = 1	x = - 1	x = 3	x = - 3
y = 3	y = - 3	y = 1	y = - 1

To consider this system graphically, we notice that the second equation is the equation of a circle with radius $\sqrt{10}$, whereas the graph of the first equation is a hyperbola obtained from the following table. Also, the points (1, 3) (- 1, - 3), (3, 1), (- 3, - 1) belong to both the circle and the hyperbola, and the two graphs intersect at these points.

x	- 5	- 4	- 3	- 2	- 1	1	2	3	4	5
y	$-\frac{3}{5}$	$-\frac{3}{4}$	- 1	$-\frac{3}{2}$	- 3	3	$\frac{3}{2}$	1	$\frac{3}{4}$	$\frac{3}{5}$

Plotting these points, and the circle with radius $\sqrt{10}$ (approximately equal to 3.16) we have the accompanying diagram.

● **PROBLEM 13-44**

Solve for x and y: $\begin{cases} x^2 + y^2 = 25, \\ x^2 - y^2 = 7. \end{cases}$

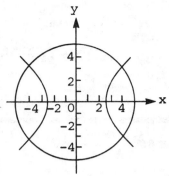

Solution: Add the following two equations to eliminate y^2.

$$(1) \quad x^2 + y^2 = 25$$
$$(2) \quad \underline{x^2 - y^2 = 7}$$
$$2x^2 = 32$$

Divide by 2
$$x^2 = 16$$

Take the square root of both sides.

$$x = \pm 4$$

Substitute x^2 into $x^2 + y^2 = 25$ to solve for y^2.

$$16 + y^2 = 25$$
$$y^2 = 25 - 16$$
$$y^2 = 9$$
$$y = \pm 3$$

$$x = 4, \quad y = -3; \qquad x = -4, \quad y = -3,$$
$$x = 4, \quad y = 3; \qquad x = -4, \quad y = 3;$$

as can be verified by substitution in the given equations.

Graphical solution. If the graph of the first equation is constructed by finding pairs of values (x,y) which satisfy the equation and plotting the corresponding points, the circle shown is obtained. In a similar manner, the hyperbola shown in the figure is obtained as the graph of the second equation. The circle and hyperbola are seen to intersect in the four points $(4,3),(-4,3),(-4,-3),(4,-3)$.

There is another way to graph these equations. The standard form of the equation of the circle whose center is at the point $c(h,k)$ and whose radius is the constant r is

$$(x - h)^2 + (y - k)^2 = r^2.$$

In this case the center is $(0,0)$ and the radius is 5. Therefore, move out 5 units from the center in all directions. The circle will then intersect the axes at $(5,0),(-5,0),(0,5)$, and $(0,-5)$. Write the hyperbola in the general form,

$$\frac{x^2}{a^2} - \frac{y^2}{b^2} = 1.$$

The hyperbola intersects one of its lines of symmetry, the x-axis, in the points $(-a,0)$ and $(a,0)$. Rewriting $x^2 - y^2 = 7$, we obtain

$$\frac{x^2}{7} - \frac{y^2}{7} = 1. \quad a^2 = 7 \text{ and } a = \pm \sqrt{7} .$$

234

Therefore the points of intersection on the x-axis are $(-\sqrt{7},0)$ and $(+\sqrt{7},0)$. This is equivalent to

$$(-2.65,0) \quad \text{and} \quad (+2.65,0).$$

● **PROBLEM 13-45**

Solve $9x^2 + 16y^2 = 288,$ (1)

$x^2 + y^2 = 25.$ (2)

Solution: Solving for y^2 first (a similar method solving for x^2 first is equally good): Multiply Equation (2) by 9 and then subtract Equation (2) from Equation (1)

$$\begin{array}{r} 9x^2 + 16y^2 = 288 \\ -9x^2 + 9y^2 = 225 \\ \hline 7y^2 = 63 \end{array}$$

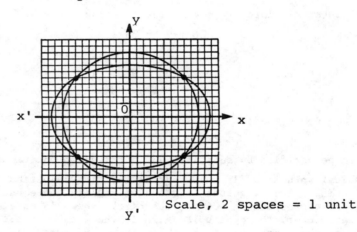

Scale, 2 spaces = 1 unit

$$y^2 = 9 \tag{3}$$

Then from Equation (2), for $y^2 = 9$, $x^2 + 9 = 25$ or

$$x^2 = 16. \tag{4}$$

From (3) and (4) $y = \pm 3$, $x = \pm 4$.

Forming all possible pairs, we have the four solutions

x = 4   x = 4   x = -4   x = -4

y = 3   y = -3   y = 3   y = -3

Check for x = 4, y = 3:

235

in Eq. (1):

$$9(4)^2 + 16(3)^2 \overset{?}{=} 288$$

$$9(16) + 16(9) \overset{?}{=} 288$$

$$144 + 144 \overset{?}{=} 288$$

$$288 = 288$$

in Eq. (2):

$$(4^2) + (3^2) \overset{?}{=} 25$$

$$16 + 9 \overset{?}{=} 25$$

$$25 = 25$$

x and y appear squared in both Eq. (1) and Eq. (2). The other pairs of values for x and y differ from the pair checked only in sign. Therefore the other pairs also satisfy Equation (1) and Equation (2).

The Equation (1) has for its locus an oval-shaped figure called an ellipse. (Fig.) The equation (2) has a circle for its locus. The four points of intersection represent graphically the four solutions.

● **PROBLEM 13-46**

Solve $x + y + z = 13$ (1),

$\quad\quad x^2 + y^2 + z^2 = 65$ (2),

$\quad\quad\quad\quad xy = 10$ (3).

<u>Solution:</u> We note that the product of two binomials is

$$(x+y)^2 = (x+y)(x+y) = x^2 + xy + xy + y^2$$
$$= x^2 + 2xy + y^2 .$$

Therefore, if we were just to add (2) and (3) we would be missing an xy term. Thus, multiplying (3) by 2,

$$2xy = 20$$

Adding this to (2),

$$x^2 + y^2 + z^2 = 65$$
$$+ \quad\quad 2xy = 20$$
$$\overline{x^2 + 2xy + y^2 + z^2 = 85}$$

$$(x+y)^2 + z^2 = 85$$

Put u for x+y; then this equation becomes
$$u^2 + z^2 = 85.$$

Also from (1),

$$u + z = 13;$$

Now to solve this system of one linear equation and one quadratic, we express one of the variables in the linear equation in terms of the other and substitute this result into the quadratic equation.

$$u^2 + z^2 = 85 \quad\quad (4)$$
$$u + z = 13 \quad\quad (5)$$

Solving (5) for z by subtracting u from both sides:

$$z = 13 - u \qquad\qquad (6)$$

Substituting this expressios for z into (4),

$$u^2 + (13 - u)^2 = 85$$
$$u^2 + 169 - 26u + u^2 = 85$$
$$2u^2 - 26u + 169 = 85.$$

Subtracting 85 from both sides,

$$2u^2 - 26u + 169 - 85 = 0$$
$$2u^2 - 26u + 84 = 0$$

Dividing by 2, $\qquad u^2 - 13u + 42 = 0$

Factor, $\qquad\qquad (u - 6)(u - 7) = 0$

Whenever the product of two factors is 0, then either one or the other must equal zero. Then, $u - 6 = 0$ or $u - 7 = 0$.

Solve for u,

$$u = 6 \quad \text{or} \quad u = 7 .$$

Substituting these values of u into equation (6) to find the corresponding z values

	$u = 6$		$u = 7$
(6)	$z = 13 - u$	(6)	$z = 13 - u$
	$z = 13 - 6$		$z = 13 - 7$
	$z = 7$		$z = 6$

whence we obtain $u = 7$ or 6; $z = 6$ or 7. Given that $z = 7$ or $z = 6$, substituting this into equation (1):

For $z = 7$		For $z = 6$
$x + y + 7 = 13$	or	$x + y + 6 = 13$
therefore $\quad x + y = 6$	or	$x + y = 7$

Thus we have

$$\left.\begin{array}{l} x + y = 7, \\ xy = 10 \end{array}\right\} \quad \text{and} \quad \left.\begin{array}{l} x + y = 6, \\ xy = 10 \end{array}\right\}$$

For $z = 6$	For $z = 7$
$\left.\begin{array}{l} x + y = 7 \\ xy = 10 \end{array}\right\}$	$\left.\begin{array}{l} x + y = 6 \\ xy = 10 \end{array}\right\}$

Solve for y or x. We choose to solve for x.

$$xy = 10$$
$$x = \frac{10}{y} \qquad\qquad (7)$$

Substituting this expression into $x + y = 7$ or $x + y = 6$:

$\left(\dfrac{10}{y}\right) + y = 7$	$\left(\dfrac{10}{y}\right) + y = 6$
$10 + y^2 = 7y$	$10 + y^2 = 6y$
$y^2 - 7y + 10 = 0$	$y^2 - 6y + 10 = 0$. Solving by
$(y - 5)(y - 2) = 0$	the quadratic formula,
$y - 5 = 0, \ y - 2 = 0$	
$y = 5 \text{ or } y = 2$	

237

For z = 6 and y = 5 substituting into (7)
$$x = \frac{10}{5} = 2.$$

For z = 6 and y = 2 substituting into (7)
$$x = \frac{10}{2} = 5.$$

$$y = \frac{-(-6) \pm \sqrt{36 - 4(1)(10)}}{2}$$

$$y = \frac{6 \pm \sqrt{-4}}{2}$$

$$= \frac{6 \mp 2\sqrt{-1}}{2}$$

$$= 3 \pm \sqrt{-1}$$

For z = 7 and y = 3 + √-1
Substituting into (7)

$$x = \frac{10}{3 + \sqrt{-1}}$$

$$x = \frac{10}{(3+\sqrt{-1})} \frac{(3-\sqrt{-1})}{(3-\sqrt{-1})}$$

$$= \frac{10}{9 - (-1)} (3 - \sqrt{-1})$$

$$= \frac{10}{10} (3 - \sqrt{-1})$$

$$= 3 - \sqrt{-1}$$

For z = 7 and y = 3 - √-1 :
Substitute into (7)

$$x = \frac{10}{3 - \sqrt{-1}} \frac{(3 + \sqrt{-1})}{(3 + \sqrt{-1})}$$

$$= \frac{10(3 + \sqrt{-1})}{9 - (-1)} = \frac{10}{10} (3 + \sqrt{-1})$$

$$= 3 + \sqrt{-1}$$

Hence, the solutions are:

x = 2	x = 5	x = 3 - √-1	x = 3 + √-1
x = 5 ;	y = 2 ;	y = 3 + √-1 ;	y = 3 - √-1
z = 6	z = 6	z = 7	z = 7

● **PROBLEM 13-47**

Eliminate x, y, z from the equations

$$y^2 + z^2 = ayz, \quad z^2 + x^2 = bzx, \quad x^2 + y^2 = cxy.$$

<u>Solution:</u> We wish to isolate a, b, and c on the right side of the three given equations. Dividing, we have:

$$\frac{y^2 + z^2}{yz} = a, \quad \frac{z^2 + x^2}{zx} = b, \quad \frac{x^2 + y^2}{xy} = c.$$

These can be rewritten as

$$\frac{y}{z} + \frac{z}{y} = a, \quad \frac{z}{x} + \frac{x}{z} = b, \quad \frac{x}{y} + \frac{y}{x} = c.$$

Multiplying together these three equations we obtain,

$$\left(\frac{y}{z} + \frac{z}{y}\right)\left(\frac{z}{x} + \frac{x}{z}\right)\left(\frac{x}{y} + \frac{y}{x}\right) = abc$$

$$\left(\frac{y}{z} \cdot \frac{z}{x} + \frac{z}{y} \cdot \frac{z}{x} + \frac{y}{z} \cdot \frac{x}{z} + \frac{z}{y} \cdot \frac{x}{z}\right)\left(\frac{x}{y} + \frac{y}{x}\right) = abc$$

$$\left[\left(\frac{y}{z} \cdot \frac{z}{x} \cdot \frac{x}{y}\right) + \left(\frac{z}{y} \cdot \frac{z}{x} \cdot \frac{x}{y}\right) + \left(\frac{y}{z} \cdot \frac{x}{z} \cdot \frac{x}{y}\right) + \left(\frac{z}{y} \cdot \frac{x}{z} \cdot \frac{x}{y}\right)\right.$$

$$+ \left(\frac{y}{z} \cdot \frac{z}{x} \cdot \frac{y}{x}\right) + \left(\frac{z}{y} \cdot \frac{z}{x} \cdot \frac{y}{x}\right) + \left(\frac{y}{z} \cdot \frac{x}{z} \cdot \frac{y}{x}\right) +$$

$$\left.\left(\frac{z}{y} \cdot \frac{x}{z} \cdot \frac{y}{x}\right)\right] = abc$$

Now, simplifying each term by multiplication, and reducing, we have:

$$1 + \frac{z^2}{y^2} + \frac{x^2}{z^2} + \frac{x^2}{y^2} + \frac{y^2}{x^2} + \frac{z^2}{x^2} + \frac{y^2}{z^2} + 1 = abc$$

or $\quad 2 + \dfrac{y^2}{z^2} + \dfrac{z^2}{y^2} + \dfrac{z^2}{x^2} + \dfrac{x^2}{z^2} + \dfrac{x^2}{y^2} + \dfrac{y^2}{x^2} = abc.$

Now, since $\frac{y}{z} + \frac{z}{y} = a$, then $\left(\frac{y}{z} + \frac{z}{y}\right)^2 = a^2$.

But we can rewrite $\left(\frac{y}{z} + \frac{z}{y}\right)^2$ as:

$$\left(\frac{y}{z} + \frac{z}{y}\right)^2 = \left(\frac{y}{z} + \frac{z}{y}\right)\left(\frac{y}{z} + \frac{z}{y}\right) =$$

$$\frac{y^2}{z^2} + \left(\frac{z}{y} \cdot \frac{y}{z}\right) + \left(\frac{y}{z} \cdot \frac{z}{y}\right) + \frac{z^2}{y^2} =$$

$$\frac{y^2}{z^2} + 1 + 1 + \frac{z^2}{y^2} = \frac{y^2}{z^2} + \frac{z^2}{y^2} + 2.$$

Similarly, since $\frac{z}{x} + \frac{x}{z} = b$, and $\frac{x}{y} + \frac{y}{x} = c$,

then $\left(\frac{z}{x} + \frac{x}{z}\right)^2 = b^2$ and $\left(\frac{x}{y} + \frac{y}{x}\right)^2 = c^2$. But writing these

we have:

$$\left(\frac{z}{x} + \frac{x}{z}\right)^2 = \left(\frac{z}{x} + \frac{x}{z}\right)\left(\frac{z}{x} + \frac{x}{z}\right) = \frac{z^2}{x^2} + 1 + 1 + \frac{x^2}{z^2}$$

$$= \frac{z^2}{x^2} + \frac{x^2}{z^2} + 2$$

and, $\left(\frac{x}{y} + \frac{y}{x}\right)^2 = \left(\frac{x}{y} + \frac{y}{x}\right)\left(\frac{x}{y} + \frac{y}{x}\right) = \frac{x^2}{y^2} + 1 + 1 + \frac{y^2}{x^2}$

$$= \frac{x^2}{y^2} + \frac{y^2}{x^2} + 2$$

Thus, $\left(\frac{y}{z} + \frac{z}{y}\right)^2 = a^2 = \frac{y^2}{z^2} + \frac{z^2}{y^2} + 2$

$$\left(\frac{z}{x} + \frac{x}{z}\right)^2 = b^2 = \frac{z^2}{x^2} + \frac{x^2}{z^2} + 2$$

$$\left(\frac{x}{y} + \frac{y}{x}\right)^2 = c^2 = \frac{x^2}{y^2} + \frac{y^2}{x^2} + 2.$$

From these three equations we obtain:

$$\frac{y^2}{z^2} + \frac{z^2}{y^2} = a^2 - 2$$

$$\frac{z^2}{x^2} + \frac{x^2}{z^2} = b^2 - 2$$

$$\frac{x^2}{y^2} + \frac{y^2}{x^2} = c^2 - 2.$$

We can now substitute into the equation:

$$2 + \frac{y^2}{z^2} + \frac{z^2}{y^2} + \frac{z^2}{x^2} + \frac{x^2}{z^2} + \frac{x^2}{y^2} + \frac{y^2}{x^2} = abc.$$

Doing this we obtain:

$$2 + \left(a^2 - 2\right) + \left(b^2 - 2\right) + \left(c^2 - 2\right) = abc;$$

therefore, $a^2 + b^2 + c^2 - 4 = abc.$

CHAPTER 14

EQUATIONS OF DEGREE GREATER THAN 2

> **Basic Attacks and Strategies for Solving Problems in this Chapter. See pages 241 to 252 for step-by-step solutions to problems.**

When dealing with equations of any degree, it is usually simpler to work with an equation in which there are no fractional coefficients. By multiplying both sides of an equation by the same constant quantity, the equality is not changed, and it is possible to change fractional coefficients into integers. For example,

$$\tfrac{2}{3}x^2 + \tfrac{1}{2}x + 2 = 1$$

can be changed (by multiplying both sides by 6) into

$$4x^2 + 3x + 12 = 6.$$

When dealing with equations of a degree greater than 2, many of the same techniques that hold for linear equations and quadratics can be applied. In general, when solving an equation, one should first factor, if possible, into only linear terms. (Some quadratic terms, such as $x^2 + 1$, cannot be factored into linear terms and must be left as quadratics.) If a cubic (for example) can be factored into $(x - a)(x - b)(x - c)$, then a, b, and c are the roots of the equation, similar to what happens for a quadratic.

It may be difficult to determine the exact roots of a higher degree equation, but it may still be possible to deduce certain facts about the roots. In general, for an equation of degree n, there are no more than n roots. If we consider the graph of an equation, a root corresponds to the point at which the graph crosses the x-axis. By calculating a table of values of the variable and values of the expression, one can determine the general locations at which the graph crosses the x-axis (where the values of the expression change from positive to negative or vice versa), and thus, the general locations of the roots.

As mentioned in Chapter 5, sometimes it is helpful to think of higher powers as powers of lower powers, e.g., $x^4 = (x^2)^2$. In this case, sometimes equations of degree 4 can be rewritten as quadratics and solved once, and then solved a second time. For example,

$$x^4 - 16 = 0$$

can be rewritten as

$$(x^2)^2 - 16 = 0$$

which can be factored into

$$(x^2 - 4)(x^2 + 4) = 0.$$

The first factor can be factored again resulting in

$$(x + 2)(x - 2)(x^2 + 4) = 0.$$

Sometimes, higher degree equations can be partially solved by guessing a solution and testing whether it satisfies the equation. If a correct solution, a, has been found, then it must be true that $x - a$ is a factor of the higher degree expression. This fact can be used to remove $x - a$ from the expression and reduce its degree to a newer expression. One can either divide the old expression by the newly found factor (as described in Chapter 4), or use the process of *synthetic division* which merely uses the coefficients. For example, if someone guessed that 2 was a correct root of $x^4 - 16 = 0$, one could reduce the polynomial via synthetic division as follows. All the coefficients of $x^4 - 16$ are listed (including the zero coefficients!) in a row and the root 2 is included in a separate place. One allows room for a second row below the first and for a sum row below these two. Synthetic division consists of computing values for the second row from left to right. The leftmost element in the second row is entered as zero.

$$
\begin{array}{rrrrr}
1 & 0 & 0 & 0 & -16 \\
0 & 2 & 4 & 8 & 16 \qquad \boxed{2} \\
\hline
1 & 2 & 4 & 8 & 0
\end{array}
$$

The elements in the first (leftmost) column are added, and the sum entered in the third (sum) row. The root is multiplied by the value just entered in the sum row and the product is placed in the second column of the second row. The numbers in the second column are then added and their sum entered in the sum row. This process is repeated for each column from left to right until the last column. If the value chosen as the root is correct, the last (rightmost) entry in the sum row should be zero. (If the remainder is not zero, its value is the remainder that would have resulted had regular long division been done.) The sum row gives the coefficients (omitting the last entry) for the reduced polynomial. (This polynomial is of one less degree than the original polynomial and is also called the *depressed* polynomial.) Thus, the reduced polynomial is

$$x^3 + 2x^2 + 4x + 8 = 0.$$

Step-by-Step Solutions to Problems in this Chapter, "Equations of Degree Greater than 2"

● **PROBLEM 14-1**

Remove fractional coefficients from the equation

$$2x^3 - \frac{3}{2} x^2 - \frac{1}{8} x + \frac{3}{16} = 0.$$

Solution: To rewrite this equation without fractional coefficients we must find a common denominator for all the terms of the equation. Observe that a common denominator is 16. Thus,

$$2x^3 - \frac{3}{2} x^2 - \frac{1}{8} x + \frac{3}{16} = 0$$

can be rewritten as:

$$\frac{2x^3}{1} - \frac{3x^2}{2} - \frac{x}{8} + \frac{3}{16} = 0 \qquad \text{or,}$$

$$\frac{32x^3 - 24x^2 - 2x + 3}{16} = 0. \text{ Multiplying both}$$

sides of the equation by 16 we obtain:

$$32x^3 - 24x^2 - 2x + 3 = 0. \text{ This is the required}$$

equation without fractional coefficients.

● **PROBLEM 14-2**

Solve the equation $x^3 - 16x = 0.$

Solution: Multiplying both sides by $\frac{1}{x}$, we have $x^2 - 16 = 0$. Factoring, we have $(x - 4)(x + 4) = 0$. Then all values of x which make this product equal to 0 satisfy either $x - 4 = 0$ or $x + 4 = 0$. Thus $x = 4$ or $x = -4$. Both 4 and -4 satisfy the original equation since

$(4)^3 - 16(4) = 0$ and $(-4)^3 - 16(-4) = 0$. However, so does the number 0, since $(0)^3 - 16(0) = 0$. From where did this root come?

There are several logical flaws in this solution. First, we cannot multiply both sides by $1/x$ if $x = 0$. But, basically, what is wrong is that $x^3 - 16x$ is not an equivalent expression to $x^2 - 16$. We may undo this error by writing $x^3 - 16x = 0$ in factored form as $x(x + 4)(x - 4) = 0$.

Then $x = 0$ or $x + 4 = 0$ or $x - 4 = 0$

Hence $x = 0$ or $x = -4$ or $x = 4$

Therefore the solutions set is $\{-4, 0, 4\}$.

● **PROBLEM 14-3**

Find all solutions of the equation $x^3 - 3x^2 - 10x = 0$.

<u>Solution:</u> Factor out the common factor of x from the terms on the left side of the given equation. Therefore,

$$x^3 - 3x^2 - 10x = x(x^2 - 3x - 10) = 0.$$

Whenever $ab = 0$ where a and b are any two numbers, either $a = 0$ or $b = 0$. Hence, either $x = 0$ or $x^2 - 3x - 10 = 0$. The expression $x^2 - 3x - 10$ factors into $(x - 5)(x + 2)$. Therefore, $(x - 5)(x + 2) = 0$. Applying the above law again:

either $x - 5 = 0$ or $x + 2 = 0$
           $x = 5$ or $x = -2.$

Hence,

$$x^3 - 3x^2 - 10x = x(x - 5)(x + 2) = 0$$

Either $x = 0$ or $x = 5$ or $x = -2.$

The solution set is $X = \{0, 5, -2\}$.

We have shown that, if there is a number x such that $x^3 - 3x^2 - 10x = 0$, then $x = 0$ or $x = 5$ or $x = -2$. Finally, to see that these three numbers are actually solutions, we substitute each of them in turn in the original equation to see whether or not it satisfies the equation $x^3 - 3x^2 - 10x = 0$.

Check: Replacing x by 0 in the original equation,

$$(0)^3 - 3(0)^2 - 10(0) = 0 - 0 - 0 = 0 \checkmark$$

Replacing x by 5 in the original equation,

$$(5)^3 - 3(5)^2 - 10(5) = 125 - 3(25) - 50$$
$$= 125 - 75 - 50$$
$$= 50 - 50 = 0 \checkmark$$

Replacing x by -2 in the original equation,

$$(-2)^3 - 3(-2)^2 - 10(-2) = -8 - 3(4) + 20$$
$$= -8 - 12 + 20$$
$$= -20 + 20 = 0 \checkmark$$

Form the equation whose roots are 2, -3, and $\frac{7}{5}$.

Solution: The roots of the equation are 2, -3, and $\frac{7}{5}$. Hence, $x = 2$, $x = -3$, and $x = \frac{7}{5}$. Subtract 2 from both sides of the first equation:

$$x - 2 = 2 - 2 = 0.$$

Add 3 to both sides of the second equation:

$$x + 3 = -3 + 3 = 0.$$

Subtract $\frac{7}{5}$ from both sides of the third equation:

$$x - \frac{7}{5} = \frac{7}{5} - \frac{7}{5} = 0.$$

Hence, $(x - 2)(x + 3)\left(x - \frac{7}{5}\right) = (0)(0)(0) = 0$ or

$$(x - 2)(x + 3)\left(x - \frac{7}{5}\right) = 0.$$

Multiply both sides of this equation by 5:

$$5(x - 2)(x + 3)\left(x - \frac{7}{5}\right) = 5(0) \text{ or}$$

$$(x - 2)(x + 3)5\left(x - \frac{7}{5}\right) = 0 \text{ or}$$

$$(x - 2)(x + 3)(5x - 7) = 0$$

$$(x^2 + x - 6)(5x - 7) = 0$$

$$5x^3 - 7x^2 + 5x^2 - 7x - 30x + 42 = 0$$

$$5x^3 - 2x^2 - 37x + 42 = 0.$$

Locate the roots of $x^3 - 3x^2 - 6x + 9 = 0$.

Solution: If we let $f(x)$ be a function, then a solution of the equation $f(x) = 0$ is called a root of the equation.

In this particular case let the function $f(x) =$

$x^3 - 3x^2 - 6x + 9$ and set it equal to zero to find its roots. When $f(x) = 0$, the graph of this equation crosses the x-axis. These x-values are the roots of the function.

To locate the roots of $x^3 - 3x^2 - 6x + 9 = 0$, we consider the function $y = x^3 - 3x^2 - 6x + 9$, assign consecutive integers from -3 to 5 to x, compute each corresponding value of y, and record the results.

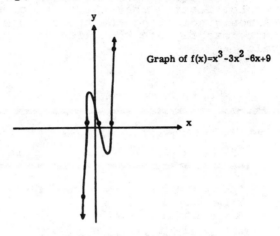

Graph of $f(x)=x^3-3x^2-6x+9$

Table of Results

x	$x^3 - 3x^2 - 6x + 9$	=	y
-3	$(-3)^3 - 3(-3)^2 - 6(-3) + 9 =$		-27
-2	$(-2)^3 - 3(-2)^2 - 6(-2) + 9 =$		1
-1	$(-1)^3 - 3(-1)^2 - 6(-1) + 9 =$		11
0	$(0)^3 - 3(0)^2 - 6(0) + 9$	$=$	9
1	$(1)^3 - 3(1)^2 - 6(1) + 9$	$=$	1
2	$(2)^3 - 3(2)^2 - 6(2) + 9$	$=$	-7
3	$(3)^3 - 3(3)^2 - 6(3) + 9$	$=$	-9
4	$(4)^3 - 3(4)^2 - 6(4) + 9$	$=$	1
5	$(5)^3 - 3(5)^2 - 6(5) + 9$	$=$	29

Since $f(-3) = -27$ and $f(-2) = 1$, there is an odd number of roots between $x = -3$ and $x = -2$. Since $f(-3) = -27$ which is negative and $f(-2) = 1$ is positive, the graph must cross the x-axis at least once. The function is continuous; thus the curve must connect the two points. To do this, the curve must cross from the negative to the positive side of the x-axis. By the definition of continuity, in order for the curve to traverse the axis it must intersect the axis. Each intersection point is called a zero or a root of the function.

Note that the curve must intersect the x-axis an odd number of times if it is to pass from the negative side to the positive, for if it traversed the axis an even number of times it would end up on the side on which it started.

Similarly, there is an odd number of roots between x = 1 and x = 2, and between x = 3 and x = 4. Further-more, since the equation is of degree 3, it has exactly three roots. Observe that the curve crosses the x-axis three times, indicating the three roots of the equation. Therefore, exactly one root lies in each of the above intervals.

● **PROBLEM 14-6**

Solve $4x^3 + 3x^2y + y^3 = 8,$

$2x^3 - 2x^2y + xy^2 = 1.$

Solution: Put $y = mx$, and substitute in both equations. Thus,

$$4x^3 + 3x^2(mx) + (mx)^3 = 8,$$

$$2x^3 - 2x^2(mx) + x(mx)^2 = 1;$$

and factoring:

$$x^3\left(4 + 3m + m^3\right) = 8 \qquad (1)$$

$$x^3\left(2 - 2m + m^2\right) = 1 \qquad (2)$$

Divide equation (1) by equation (2):

$$\frac{\cancel{x^3}\left(4 + 3m + m^3\right)}{\cancel{x^3}\left(2 - 2m + m^2\right)} = \frac{4 + 3m + m^3}{2 - 2m + m^2} = \frac{8}{1} = 8 \qquad (2a)$$

Hence, $\dfrac{4 + 3m + m^3}{2 - 2m + m^2} = 8$. $\qquad (3)$

Multiply both sides of equation (3) by $\left(2 - 2m + m^2\right)$. Then:

$$\left(2 - \cancel{2m} + m^2\right)\left(\frac{4 + 3m + m^3}{2 - \cancel{2m} + m^2}\right) = \left(2 - 2m + m^2\right)8$$

$$4 + 3m + m^3 = \left(2 - 2m + m^2\right)8 \qquad (4)$$

Distributing on the right side of equation (4):

$$4 + 3m + m^3 = 16 - 16m + 8m^2 \qquad (5)$$

Subtract $\left(16 - 16m + 8m^2\right)$ from both sides of equation (5). Then,

$$4 + 3m + m^3 - \left(16 - 16m + 8m^2\right) = 16 - 16m + 8m^2 - \left(16 - 16m + 8m^2\right)$$

$$4 + 3m + m^3 - 16 + 16m - 8m^2 = 0.$$

Combining like terms:

$$m^3 - 8m^2 + 19m - 12 = 0 \qquad (6)$$

The possible rational roots of equation (6) are all numbers $\pm p/q$ in which the values of p are the positive divisors $1,2,3,4,6,12$ of the constant term -12 and the values of q are the positive divisors of the leading coefficient 1. Hence, the only value for q is 1. Thus, the possible rational roots of equation (6) are:

and
$$\pm 1/1, \pm 2/1, \pm 3/1, \pm 4/1, \pm 6/1,$$
$$\pm 12/1, \text{ or } \pm 1, \pm 2, \pm 3, \pm 4, \pm 6, \text{ and } \pm 12.$$

Substitution of the values 1,3, and 4 for m makes equation (6) a true statement. Hence, the roots of m are 1,3, and 4. Then, $m = 1$, $m = 3$, and $m = 4$. Therefore, $m - 1 = 0$, $m - 3 = 0$ and $m - 4 = 0$. Then,

$$(m - 1)(m - 3)(m - 4) = (0)(0)(0) = 0 \qquad (7)$$

Also, $\quad (m - 1)(m - 3)(m - 4) = \Big[(m - 1)(m - 3)\Big](m - 4)$

$$= [m^2 - 4m + 3](m - 4)$$

distributing, $\quad = \left(m^3 - 4m^2 + 3m\right) - \left(4m^2 - 16m + 12\right)$

$$= m^3 - 4m^2 + 3m - 4m^2 + 16m - 12$$

$$(m - 1)(m - 3)(m - 4) = m^3 - 8m^2 + 19m - 12 \qquad (8)$$

From equations (6) and (7):

$$m^3 - 8m^2 + 19m - 12 = (m - 1)(m - 3)(m - 4) = 0$$

(i) Take $m = 1$, and substitute in either (1) or (2). From (2),

$$x^3\left(2 - 2(1) + (1)^2\right) = 1$$
$$x^3(2 - 2 + 1) = 1$$
$$x^3(0 + 1) = 1$$
$$x^3(1) = 1$$
$$x^3 = 1$$

Take the cube root of each side:

$$\sqrt[3]{x^3} = \sqrt[3]{1}$$
$$x = \sqrt[3]{1} = 1 \ .$$

Also, $y = mx = 1(1) = 1$.

(ii) Take $m = 3$, and substitute in (2):

thus $\quad 5x^3 = 1$. Then,
$$x^3 = 1/5 \ .$$

Take the cube root of each side:
$$\sqrt[3]{x^3} = \sqrt[3]{1/5}$$
$$x = \sqrt[3]{1/5}$$

and $y = mx = 3x = 3\sqrt[3]{1/5}$.

(iii) Take $m = 4$; we obtain from (2):
$$10x^3 = 1.$$

Then, $x^3 = \dfrac{1}{10}$. Take the cube root of each side:
$$\sqrt[3]{x^3} = \sqrt[3]{1/10} \ ,$$

or
$$x = \sqrt[3]{1/10} \ .$$

and
$$y = mx = 4x = 4\sqrt[3]{1/10} \ .$$

Hence the complete solution is

$$x = 1, \sqrt[3]{1/5}, \sqrt[3]{1/10}.$$

$$y = 1, 3\sqrt[3]{1/5}, 4\sqrt[3]{1/10}.$$

Solve the equation $x^4 - 5x^2 - 36 = 0$.

<u>Solution:</u> This is a fourth degree equation, but it can be solved by the same methods applied to quadratic equations.
To solve $x^4 - 5x^2 - 36 = 0$, we let $z = x^2$, substitute in the given equation, and get

$$z^2 - 5z - 36 = 0$$

This is now a quadratic equation in the variable z. We solve this equation by factoring.

$z^2 - 5z - 36 = 0$

$(z - 9)(z + 4) = 0$ Factoring

$z - 9 = 0, z + 4 = 0$ Setting both factors equal to zero

$z = 9, \quad z = -4$ Solving for z.

Hence the solution set of the equation in z is $\{-4, 9\}$. Now we replace z in $z = x^2$ by -4 and then by 9 and get

$$x^2 = -4$$

Taking the square root of each member,

$$x = \pm \sqrt{-4} = \pm \sqrt{4(-1)} = \pm \sqrt{4}\sqrt{-1}$$

$$x = \pm 2i$$

Also $x^2 = 9$

$$x = \pm 3$$

Consequently the solution set of the original equation is $\{2i, -2i\} \cup \{3, -3\} = \{2i, -2i, 3, -3\}$.

Solve $x^4 - 2x^2 - 3 = 0$ as a quadratic in x^2.

<u>Solution:</u> We can write the equation as $\left(x^2\right)^2 - 2x^2 - 3 = 0$.

Let $x^2 = z$ and we have
$$z^2 - 2z - 3 = 0 ,$$
which is a quadratic equation. We can solve a quadratic equation in the form $ax^2 + bx + c = 0$ using the quadratic formula,
$$x = \frac{-b \pm \sqrt{b^2 - 4ac}}{2a} .$$
In our case $a = 1$, $b = -2$, and $c = -3$. Thus,
$$z = \frac{-(-2) \pm \sqrt{(-2)^2 - 4(1)(-3)}}{2(1)}$$
$$= \frac{2 \pm \sqrt{4 + 12}}{2} = \frac{2 \pm 4}{2} = 1 \pm 2 .$$

Therefore, $z = 3$ or $z = -1$. Since $z = x^2$, we have $x^2 = 3$ or $x^2 = -1$. If $x^2 = 3$, $x = \pm \sqrt{3}$. If $x^2 = -1$, $x = \pm i$. Hence, the solution set of the original equation is
$$\{\sqrt{3}, - \sqrt{3}, i, -i\} .$$

● PROBLEM 14-9

Solve for x: $x^4 + 6x^3 + 2x^2 - 21x - 18 = 0$.

__Solution:__ This is a polynomial equation of the fourth degree since the power of the highest term is four. We complete the square from the x^4 and x^3 terms. This is done by taking one half the coefficient of the x^3 term and squaring it. This will then be the coefficient of the x^2 term. We obtain 9. Add and subtract $9x^2$ to maintain the same equation. Then:
$$\left(x^4 + 6x^3 + 0x^2\right) - 9x^2 + 2x^2 - 21x - 18 = 0.$$
Convert $\left(x^4 + 6x^3 + 9x^2\right)$ into a binomial square: $\left(x^2 + 3x\right)^2 -$

$9x^2 + 2x^2 - 21x - 18 = 0.$
Combine like terms: $\left(x^2 + 3x\right)^2 - 7x^2 - 21x - 18 = 0.$
Factor -7 from the second and third terms; then we obtain:
$$\left(x^2 + 3x\right)^2 - 7\left(x^2 + 3x\right) - 18 = 0.$$
By substituting $v = x^2 + 3x$, we have a quadratic equation in v:
$$v^2 - 7v - 18 = 0.$$

Factor this quadratic equation in v, in terms of a product of two binomials:
$$(v - 9)(v + 2) = 0.$$
Whenever we have a product of two numbers such that $ab = 0$, then either $a = 0$ or $b = 0$. Thus, (1) $v - 9 = 0$ or (2) $v + 2 = 0$. Substitute the expression $\left(x^2 + 3x\right)$ for v in equations (1) and (2).

Then:

(3) $x^2 + 3x - 9 = 0$ or   (4) $x^2 + 3x + 2 = 0$

Solve for x by the quadratic formula, $x = \dfrac{-b \pm \sqrt{b^2 - 4ac}}{2a}$,

from the equation $ax^2 + bx + c = 0$. For equation (3), a = 1, b = 3, c = -9. Thus,

$$x = \frac{-3 \pm \sqrt{9 - 4(1)(-9)}}{2(1)}$$

$$x = \frac{-3 \pm 3\sqrt{5}}{2}$$

For equation (4), a = 1, b = 3, c = 2. Thus,

$$x = \frac{-3 \pm \sqrt{9 - 4(1)(2)}}{2(1)}$$

$$x = \frac{-3 \pm 1}{2} = \frac{-3 + 1}{2} \text{ or } \frac{-3 - 1}{2}$$

Therefore, the four solutions are

$x = -1, -2, \dfrac{-3 + 3\sqrt{5}}{2}$, or $x = -1, -2, \dfrac{-3 + 3\sqrt{5}}{2}, \dfrac{-3 - 3\sqrt{5}}{2}$

● **PROBLEM 14-10**

Find all rational roots of the equation
$x^4 - 4x^3 + x^2 - 5x + 4 = 0$.

Solution: This is a fourth degree equation. We can solve it by synthetic division. Guess at a root by trying to find an x-value which will make the equation equal to zero. x = 4 works.

Now write the coefficients of the equation in descending powers of x. Note that if a term is missing, its coefficient is zero. In the corner box, the root 4 is placed. Bring the first coefficient down and multiply it by the root. Place the result below the next coefficient and add. Multiply the result by the root and continue as before.

1	-4	+1	-5	+4	$\lfloor 4$
	+4	0	+4	-4	
1	0	1	-1	0	

The last result is zero which indicates (x - 4) is a factor and x = 4 is a root. The other results are the coefficients of the third degree expression when (x - 4) is factored.

$$(x - 4)\left(x^3 + 0x^2 + x - 1\right) = 0$$
$$(x - 4)\left(x^3 + x - 1\right) = 0$$

To find the roots of the third degree equation, call it $g(x)$, we must set it equal to zero.

$$g(x) = x^3 + x - 1 = 0$$

Try to find where the curve of the equation crosses the x-axis which is when $y = 0$.

x	-2	-1	0	1	2
y	-11	-3	-1	1	9

It crosses the x-axis between $x = 0$ and $x = 1$. It is an irrational root.

Since the given equation is a fourth degree equation, it has 4 roots. All of the real roots, namely the rational number 4, and an irrational number between 0 and 1, have been found. Therefore, the two remaining roots are not real; that is, they are complex numbers.

● **PROBLEM 14-11**

Approximate the real roots of the equation
$$x^4 + 2x^3 - 5x^2 - 4x + 6 = 0.$$

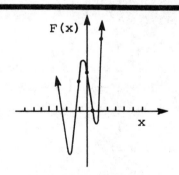

x	-4	-3	-2	-1	0	1
F(x)	70	0	-6	4	6	0

Solution: To find the real roots of the given equation, let us sketch the graph of the related polynomial function defined by
$$F(x) = x^4 + 2x^3 - 5x^2 - 4x + 6. \quad \text{(See Fig.)}$$

When $f(x) = 0$, x is a root of the equation. From the table, two real roots are $x = -3$ and $x = 1$. Reducing $F(x)$ by dividing by $(x + 3)(x - 1)$ and solving, we find that $-\sqrt{2}$ and $+\sqrt{2}$ are also roots.

Solve $x^4 + y^4 = 82$ (1),

        $x - y = 2$ (2).

Solution: Let $x = u + v$ and $y = u - v$. From equation (2)

$$x - y = 2$$
$$u + v - (u - v) = 2$$
$$u + v - u + v = 2$$
$$2v = 2$$
$$v = 1$$

Substituting in equation (1),

$$x^4 + y^4 = 82$$
$$(u + v)^4 + (u - v)^4 = 82$$

Replacing v by 1,

$$(u + 1)^4 + (u - 1)^4 = 82 \qquad\qquad (3)$$

Now, simplify the left side of equation (3),

$$(u + 1)^2 = (u+1)(u+1) = u^2 + 2u + 1$$
$$(u+1)^3 = (u+1)(u+1)^2 ,$$

or

$$(u+1)^3 = (u+1)\left(u^2+2u+1\right)$$

Distributing on the right side of this equation:

$$(u+1)^3 = \left(u^3+2u^2+u\right) + \left(u^2+2u+1\right)$$
$$= u^3 + 3u^2 + 3u + 1 .$$
$$(u+1)^4 = (u+1)(u+1)^3$$

or

$$(u+1)^4 = (u+1)\left(u^3+3u^2+3u+1\right) .$$

Distributing on the right side of this equation:

$$(u+1)^4 = \left(u^4+3u^3+3u^2+u\right) + \left(u^3+3u^2+3u+1\right)$$
$$(u+1)^4 = u^4 + 4u^3 + 6u^2 + 4u + 1 \qquad\qquad (4)$$

Also, $(u-1)^2 = (u-1)(u-1) = u^2 - 2u + 1$

$$(u-1)^3 = (u-1)(u-1)^2$$

or

$$(u-1)^3 = (u-1)\left(u^2-2u+1\right) .$$

Distributing on the right side of this equation:

$$(u-1)^3 = \left(u^3-2u^2+u\right) - \left(u^2-2u+1\right)$$
$$= u^3 - 3u^2 + 3u - 1$$
$$(u-1)^4 = (u-1)(u-1)^3$$

or

$$(u-1)^4 = (u-1)\left(u^3-3u^2+3u-1\right)$$

Distributing on the right side of this equation:

$$(u-1)^4 = \left(u^4-3u^3+3u^2-u\right) - \left(u^3-3u^2+3u-1\right)$$
$$(u-1)^4 = u^4 - 4u^3 + 6u^2 - 4u + 1 \qquad\qquad (5)$$

Using equations (4) and (5) to simplify equation (3):

$$(u+1)^4 + (u-1)^4 = \left(u^4+4u^3+6u^2+4u+1\right) + \left(u^4-4u^3+6u^2-4u+1\right)$$

$$= u^4 + 4u^3 + 6u^2 + 4u + 1 + u^4 - 4u^3 + 6u^2 - 4u + 1$$

$$= 2u^4 + 12u^2 + 2 = 82.$$

Thus, $2u^4 + 12u^2 + 2 = 82$.

Subtract 82 from both sides of this equation:

$$2u^4 + 12u^2 + 2 - 82 = 82 - 82$$

$$2u^4 + 12u^2 - 80 = 0$$

Divide both sides of this equation by 2:

$$\frac{2u^4 + 12u^2 - 80}{2} = \frac{0}{2},$$

or

$$u^4 + 6u^2 - 40 = 0$$

Factoring the left side of this equation into a product of two polynomials:

$$\left(u^2 + 10\right)\left(u^2 - 4\right) = 0.$$

Whenever a product $ab = 0$, where a and b are any two numbers, either $a = 0$ or $b = 0$. Hence, either

$$u^2 + 10 = 0 \text{ or } u^2 - 4 = 0$$

$$u^2 = -10 \text{ or } u^2 = 4$$

$$u = \pm \sqrt{-10} \text{ or } u = \pm \sqrt{4} = \pm 2$$

Using the equations relating x and y to u and v in order to solve for x and y:

when $u = \sqrt{-10}$, $x = u + v = \sqrt{-10} + 1 = 1 + \sqrt{-10}$ and

$$y = u - v = \sqrt{-10} - 1 = -1 + \sqrt{-10}.$$

when $u = -\sqrt{-10}$, $x = u + v = -\sqrt{-10} + 1 = 1 - \sqrt{-10}$ and

$$y = u - v = -\sqrt{-10} - 1 = -1 - \sqrt{-10}.$$

when $u = 2$, $x = u + v = 2 + 1 = 3$ and

$$y = u - v = 2 - 1 = 1.$$

when $u = -2$, $x = u + v = -2 + 1 = -1$ and

$$y = u - v = -2 - 1 = -3.$$

Thus, the pairs of solutions are:

$$x = 1 + \sqrt{-10}, \quad y = -1 + \sqrt{-10}$$

$$x = 1 - \sqrt{-10}, \quad y = -1 - \sqrt{-10}$$

$$x = 3, \quad y = 1$$

$$x = -1, \quad y = -3.$$

CHAPTER 15

THEORY OF EQUATIONS

> **Basic Attacks and Strategies for Solving Problems in this Chapter. See pages 253 to 260 for step-by-step solutions to problems.**

The set of polynomial forms, $F[x]$, is the set of all polynomials, $f(x)$, in one variable (or indeterminate), x, whose coefficients are elements of F. If the highest power of x in $f(x)$ is n, we say that the *degree* of $f(x)$ is n. A polynomial of degree n is said to be *monic* if the coefficient of x^n (called the *leading coefficient*) is one.

As seen in Chapter 4, it is possible to divide one polynomial by another. A precise statement of what occurs in this process is expressed in the following *Division Algorithm*. If $f(x)$ and $g(x)$ are polynomials and $g(x) \neq 0$, then there exist unique polynomials $q(x)$ (the *quotient* polynomial), and $r(x)$ (the *remainder* polynomial), such that

$$f(x) = q(x) \cdot g(x) + r(x)$$

where either $r(x) = 0$ or the degree of $r(x)$ is less than the degree of $g(x)$. As seen in Chapter 14, the method of *synthetic division* is a short-cut method of doing polynomial long division.

The remainder polynomial is a constant if the divisor polynomial $g(x)$ is a linear expression, i.e., of the form $x - r$. The value of the constant can be determined from the *Remainder Theorem*: if a polynomial $f(x)$ is divided by $x - r$, the remainder equals $f(r)$. This provides a way to evaluate $f(c)$ for an arbitrary c: merely divide $f(x)$ by $x - c$ to obtain the remainder (and this division can be done very easily by synthetic division).

As seen in Chapter 14, if r is a root of $f(x)$, the remainder should be zero (since, by definition, r is a root only if $f(r) = 0$). This fact is sometimes stated as the *Factor Theorem*: a polynomial $f(x)$ has an exact factor $x - r$ if and only if $f(r) = 0$.

Reflecting on some of the facts presented in this and other chapters, one can make certain statements about equations and their roots. If an equation has only two roots, a and b, each of multiplicity one, it is a quadratic,

$$(x - a)(x - b) = x^2 - (a + b)x + ab = 0.$$

Note that the sum of the two roots is the coefficient of x and their product is the constant term. If one root is of the form $c + \sqrt{d}$, and if the other root is of the form $c - \sqrt{d}$, the coefficient of x will be $-2c$ and the constant term will be $c^2 - d$. This is also true if c is zero or if d is negative. However, in general, even if one root is of the form $c + \sqrt{d}$, the other root can be any number (unless there are other conditions that the resulting polynomial must fulfill). If an equation has more than two roots (for example, a cubic), if all the coefficients (including the constant term) are real, then complex roots must occur in pairs. If one or more of the coefficients (or the constant) is complex, there may be a single complex root. Since, if a is a root of $f(x)$, then $x - a$ must divide $f(x)$ evenly, it must also be true that if the leading coefficient of $f(x)$ is one and if the other coefficients (along with the constant) are integers, a must be an integer dividing the constant term evenly. One can use this fact to determine possible values of roots by determining all possible (positive and negative) integer factors of the constant term. After the possible values have been determined, they can be tested by substituting into the expression to check if the possible root actually produces a value of zero.

One can determine the nature of the roots of a polynomial by using *Descartes' Rule of Signs* which counts the number of *variations* of signs in an expression. A *variation of sign* occurs whenever two successive terms have oppositive signs. The theorem asserts the following:

- The number of positive roots of $f(x) = 0$ is equal to the number of variations of signs, or is less than that number by an even integer.

- The number of negative roots of $f(x) = 0$ is equal to the number of variations of $f(-x) = 0$, or less than that number by an even integer.

- The number of complex roots is at least as large as the higher power of $f(x)$ minus the number of possible roots, positive and negative.

Define polynomial forms.

Solution: Let F be a field, $x^0 = 1 \in F$ and $x^{n+1} = x^n \cdot x$, where $n \geq 0$. The product $x^n \cdot x$ is understood to obey the same algebraic rules as if x were a member of a ring with identity. If $F[x]$ represents the set of all finite linear combinations of powers of x over F, then $F[x]$ is known as the set of all polynomial forms over F. Every member, $f(x)$, of $F[x]$ is known as a polynomial in one indeterminate, x over F. Each $f(x)$ has the form

$$f(x) = a_0 + a_1 x + a_2 x^2 + \ldots + a_{n-1}x^{n-1} + a_n x^n$$

where n is called the degree of f ($\deg.(f) = n$). a_n is known as the leading coefficient of f (if $a_n \neq 0$). A polynomial is monic if its leading coefficient is 1.

Let $f(x) = 2x^4 + x^3 + 4x^2 + 3x + 1$ and $g(x) = x^2 + 1$. Find the polynomials $q(x)$ and $r(x)$ such that $f(x) = q(x)g(x) + r(x)$.

Solution: We define the division algorithm or remainder theorem. This result is merely a precise statement of the familiar long division method. Division Algorithm: If $f(x)$ and $g(x)$ are polynomials with $g(x) \neq 0$, then there exist unique polynomials $q(x)$ and $r(x)$ such that

$$f(x) = q(x) \, g(x) + r(x) \tag{1}$$

where either $r(x) = 0$ or $\deg r < \deg g$. Suppose $\deg f \geq \deg g$. Let $f(x) = a_0 + a_1 x + \ldots + a_n x^n$, and $g(x) = b_0 + b_1 x + \ldots + b_m x^m$ where $a_n, b_m \neq 0$ and $n \geq m$. Now, form the polynomial:

$$f_1(x) = f(x) - \frac{a_n}{b_m} x^{n-m} g(x) \ . \tag{2}$$

Then, deg $f_1 <$ deg f. By induction, there exist polynomials $q_1(x)$ and $r(x)$ such that

$$f_1(x) = q_1(x) \, g(x) + r(x) \ .$$

Where either $r(x) = 0$ or deg $r <$ deg g. Substituting this into (2), and solving for $f(x)$ produces

$$f(x) = \left(q_1(x) + \frac{a_n}{b_m} x^{n-m} \right) g(x) + r(x)$$

or,

$$f(x) = q(x) \, g(x) + r(x)$$

where $q(x) = q_1(x) + \frac{a_n}{b_m} x^{n-m}$. When, for two given polynomials $f(x)$ and $g(x)$, $q(x)$ and $r(x)$ satisfy (1), we say that $f(x)$ has been divided by $g(x)$, that the quotient is $q(x)$ and that the remainder is $r(x)$. The given functions are:

$$f(x) = 2x^4 + x^3 + 4x^2 + 3x + 1 \quad \text{and} \quad g(x) = x^2 + 1 \ .$$

Now,

$$f_1(x) = f(x) - \frac{a_n}{b_m} x^{n-m} g(x) = f(x) - 2x^2 g(x)$$

or,

$$f(x) = 2x^2 g(x) + f_1(x) \ . \tag{3}$$

But,

$$\begin{aligned} f_1(x) &= (2x^4 + x^3 + 4x^2 + 3x + 1) - 2x^2(x^2 + 1) \\ &= x^3 + 2x^2 + 3x + 1 \ . \end{aligned}$$

Then,

$$f_2(x) = f_1(x) - xg(x)$$

or,

$$f_1(x) = xg(x) + f_2(x). \tag{4}$$

Also,

$$f_2(x) = x^3 + 2x^2 + 3x + 1 - x(x^2 + 1) = 2x^2 + 2x + 1 \ .$$

Then,

$$f_3(x) = 2x^2 + 2x + 1 - 2(x^2 + 1) = 2x - 1 \ .$$

Also,

$$f_2(x) = f_3(x) + 2g(x) = (2x - 1) + 2g(x) \ .$$

Substituting this into (4),

$$f_1(x) = xg(x) + (2x-1) + 2g(x).$$

Next, substituting this into (3) yields

$$f(x) = 2x^2 g(x) + xg(x) + (2x - 1) + 2g(x)$$

or,

$$f(x) = (2x^2 + x + 2) \, g(x) + (2x - 1)$$

Thus,

$$q(x) = 2x^2 + x + 2, \text{ and } r(x) = 2x - 1 \ .$$

Another sometimes useful way to compute $q(x)$ and $r(x)$ is the following: Write down $g(x) \overline{f(x)}$, including the coefficients of each x^k where k is less than or equal to the degree of the polynomial. Then, proceed as in regular rational division.

Synthetic division:

$$
\begin{array}{r}
2x^2 + x + 2 \\
x^2 + 0x + 1 \overline{\smash{\big)}\, 2x^4 + x^3 + 4x^2 + 3x + 1} \\
-(2x^4 + 0x^3 + 2x^2) \downarrow \quad \downarrow \\
\hline
0 + x^3 + 2x^2 + 3x \quad \downarrow \\
-(x^3 + 0x^2 + x) \downarrow \\
\hline
0 + 2x^2 + 2x + 1 \\
-(2x^2 + 0x + 2) \\
\hline
\end{array}
$$

Remainder: $\qquad\qquad\qquad\qquad 0 + 2x - 1$

Multiply the divisor by $2x^2$ and substract. Bring down the next term, $3x$. Now, multiply by x and subtract. When there are no more terms to bring down we are finished.

So, $q(x) = 2x^2 + x + 2$ and $r(x) = 2x - 1$. Thus, we have the same answer as found in the previous method.

$$f(x) = (2x^2 + x + 2)\, g(x) + (2x - 1).$$

● **PROBLEM** 15-3

Given $f(x) = x^8 + 2x^7 - 3x^6 + x^5 - 2x^4 - 4x^3 + x^2 - x + 6$

compute the remainder, in this case $f(3)$.

(Hint: use the remainder theorem and synthetic division.)

Solution: The remainder theorem states that if a polynomial $f(x)$ is divided by $x-r$, the remainder equals $f(r)$.

$$f(r) = \frac{f(x)}{x-r} \Rightarrow f(3) = \frac{x^8 + 2x^7 - 3x^6 + x^5 - 2x^4 - 4x^3 + x^2 - x + 6}{x - 3}$$

Now, the work necessary in dividing a polynomial $f(x)$ by $x-r$ can be shown in a 3-line process known as synthetic division.

The process for synthetic division is:

(1) arrange $f(x)$ in descending powers of x supplying zero as the coefficient for missing powers;

(2) place the synthetic divisor, r, in the first line to the left of the coefficients;

(3) carry the leading coefficient to the third line;

(4)   multiply the leading coefficient by r, the synthetic
       divisor and add this product to the second coefficient.
       (Place this sum under the second coefficient in the
       third line.)

(5)   Repeat this process of multiplication and addition until
       you reach the last coefficient. The final number in the
       third line right corner is the remainder . For our ex-
       ample r = 3, so

r = synthetic divisor

$$\underline{3|}\quad \begin{array}{ccccccccc} 1 & 2 & -3 & 1 & -2 & -4 & 1 & -1 & 6 \\ & 3 & 15 & 36 & 111 & 327 & 969 & 2910 & 8727 \\ \hline 1 & 5 & 12 & 37 & 109 & 323 & 970 & 2909 & 8733 \end{array}$$

↑
remainder

f(3) = 8733

● **PROBLEM 15-4**

Show that $x^4 + 3x^3 + 4x + 6$ is exactly divisible by x + 1.

   Hint: Use factor theorem.

Solution: The factor theorem says that a polynomial f(x) has
an exact factor x - r if and only if f(r) = 0.

$$\begin{array}{r} x^3 + 2x^2 - 2x + 6 \\ x + 1 \overline{\smash{)}\; x^4 + 3x^3 + 4x + 6} \\ \underline{x^4 + x^3} \\ 2x^3 \\ \underline{2x^3 + 2x^2} \\ -2x^2 + 4x \\ \underline{-2x^2 - 2x} \\ 6x + 6 \\ \underline{6x + 6} \\ 0 \end{array}$$

remainder = f(r) = f(-1) = 0

● **PROBLEM 15-5**

Use synthetic division to find the remainder of:

   $(3x^4 - 3x^3 + 2x^2 - x + 5) \div (x - 2)$

256

Solution:

```
Coefficients ──────→  3   -3    2   -1    5
of numerator
                          6    6   16   30
                      ─────────────────────
                      3    3    8   15  │35 → remainder
```

Hence, the remainder required is 35.

● PROBLEM 15-6

Given $f(x) = 3x^4 - 2x^3 + x^2 + 3x + 9$

evaluate by synthetic division $f(-2)$ and $f(3)$.

Solution: $f(-2) = 71$

```
                                                      coefficients
                      3    -2     1     3     9   ← of f(x)

                          -6    16   -34    62
                      ────────────────────────
                      3    -8    17   -31  │71
```

                                                      coefficients
$f(3) = 216$

```
                      3    -2     1     3     9   ← of f(x)

                           9    21    66   207
                      ────────────────────────
                      3     7    22    69  │216
```

● PROBLEM 15-7

Given that one root of $2x^3 + 4x^2 - 46x - 120 = 0$

is 5, find 2 more roots.

Solution: By synthetic division,

```
Coefficients
of equation ──────→   2    4    -46    -120     Divide
                          10     70     120     (2x³+4x²-46x-120)
                      ─────────────────────  , ──────────────────
                      2   14     24       0          (x-5)
```

therefore, $2x^2 + 14x + 24 = 0$ is the depressed equation.
Solve the depressed equation $(2x + 6)(x + 4) = 0$

$$2x + 6 = 0 \qquad x + 4 = 0$$

$$x = -3 \qquad x = -4$$

The roots of the equation are -3, -4, 5.

257

Find the equations with integral coefficients having only
the following roots:

(a)　5, 1, 12

(b)　$\frac{1}{2}$, $\frac{1}{3}$, $\frac{-1}{4}$

Solution:　(a)　(x-5)(x-1)(x-12) = 0

　　　　　　(b)　(2x-1)(3x-1)(4x+1) = 0

These equations have the form (x-h)(x-k) = 0, where h and k
are positive numbers.　These problems can be solved by setting the given root equal to zero, and then transforming the
equation to the form (x-h)(x-k) = 0.
ex.

$$x = \frac{1}{2} \Rightarrow 2x = 1 \Rightarrow (2x-1) = 0.$$

Answer the following questions about rational and real
coefficient equations:

(1)　A real coefficient equation has 2 roots, one root
is $3 - i\sqrt{6}$, what is the second root?

(2)　A rational coefficient equation has 2 roots, one
root is $-\sqrt{13}$, what is the second root?

(3)　Are the following statements correct and why?

(a)　$x^3 + 8x - 7i = 0$　has　$x = a - bi$　as one root,
therefore, the second root must be $x = a + bi$.

(b)　$x^3 + \sqrt{2}\,x^2 + 6x + \sqrt{13} = 0$ .has one root equal to
$x = a + bi$, therefore the second root is $x = a - bi$.

Solution:　(1)　The second required root is $3 + i\sqrt{6}$.　In order
for these roots to give us an equation of real coefficients,
the -i must cancel the +i, and the i^2 becomes real ($i^2 = -1$).

(2)　The second required root is $+\sqrt{13}$.　In order for these
roots to give us an equation of rational coefficients,
the $+\sqrt{13}$ must cancel the $-\sqrt{13}$, and $(\sqrt{13})^2$ becomes rational ($\sqrt{13^2} = 13$).

(3)　(a) False, the given equation contains imaginary numbers,
therefore it is not necessary that $x = a - bi$ cancel
$x = a + bi$.

(b) True, the second root must be $x = a - bi$ because the given equation contains only real coefficients, therefore the imaginary values must cancel.

● **PROBLEM 15-10**

Find the rational roots, if they exist, for the following equation:

$$x^4 - 4x^2 + 5x - 2 = 0.$$

Solution: In general, if an equation $f(x) = 0$ has integral coefficients and is in the form

$$x^n + P_1 x^{n-1} + P_2 x^{n-2} + \ldots + P_{n-1} x + P_n = 0$$

then any rational root of $f(x) = 0$ is an integer and a factor of P_n.

Therefore, the possible rational roots for the given equation are ±1 and ±2. To determine whether any of them are the root of the equation, substitute each of these possible roots into the given equation, or use synthetic division. For example, substitute -1 into the given equation.
$(-1)^4 - 4(-1)^2 + 5(-1) - 2$

$$\doteq 1 - 4 - 5 - 2 = -10 \neq 0$$

Hence, -1 is not a root of the equation. Similarly, one can show that only 1 is the root of the given equation.

● **PROBLEM 15-11**

Determine

    (a)   the equation of lowest degree with rational coefficients, one of whose roots is $\sqrt{13} - \sqrt{3}$;

    (b)   the equation of lowest degree with real or complex coefficients having the roots 4 and 4 - 9i.

Solution: (a) Let $x = \sqrt{13} - \sqrt{3}$

Squaring both sides, $x^2 = 13 - 2\sqrt{39} + 3$

$$x^2 = 16 - 2\sqrt{39}$$

$$x^2 - 16 = -2\sqrt{39}$$

Square both sides again. $x^4 + 256 - 32x^2 = 156$

259

Finally, $$x^4 - 32x^2 + 100 = 0$$

(b) $(x-4) [x - (4-9i)] = 0$

$$x^2 + 16 - 36i - 4x - 4x + 9ix = 0$$

$$x^2 - (8-9i)x + 16 - 36i = 0$$

● **PROBLEM 15-12**

Using Descartes' Rule, determine the nature of the roots of

$$3x^5 - 4x^4 + 3x^3 - 9x^2 - x + 1 = 0$$

<u>Solution</u>: Descartes' Rule of Signs says that the number of positive roots of $f(x) = 0$ is equal to the number of variations of sign, or is less than that number by an even integer. The number of negative roots of $f(x) = 0$ is equal to the number of variations of $f(-x) = 0$, or less than that number by an even integer. The number of complex roots, at least, is equal to the highest power of $f(x)$ minus the number of possible roots, positive and negative.

So for $f(x) = 3x^5 - 4x^4 + 3x^3 - 9x^2 - x + 1 = 0$

the number of variations equals 4. The number of positive roots equals 4, 2, or 0.

For $f(-x) = 3(-x)^5 - 4(-x)^4 + 3(-x)^3 - 9(-x)^2 + x + 1 = 0$

$$f(-x) = -3x^5 - 4x^4 - 3x^3 - 9x^2 + x + 1 = 0$$

The number of variations is 1, hence the number of negative roots equals 1.

The least number of complex roots equals $5 - (4 + 1) = 0$, i.e., this equation has no complex roots.

CHAPTER 16

INEQUALITIES AND GRAPHS

> **Basic Attacks and Strategies for Solving Problems in this Chapter. See pages 261 to 296 for step-by-step solutions to problems.**

Solving an inequality is similar to solving an equation, but certain safeguards must also be taken into account. The major concerns are the following:

- An inequality is not a symmetric relationship, i.e., if $a < b$, then $b \not< a$. Thus, care must be observed when performing algebraic manipulations on inequalities. Equal quantities added or subtracted from both sides of an inequality do not change the inequality, however.

- If $a < b$, then $1/a > 1/b$.

- If $a < b$, then $na < nb$ if $n > 0$, but $na > nb$ if $n < 0$.

For example, given $2 - 3x > 1$, we can subtract 2 from both sides to obtain $-3x > -1$. Dividing by -3 (i.e., multiplying by $-1/3$) reverses the inequality sign and yields $x < 1/3$.

An inequality is called *conditional* if its validity depends on the values of the variables it contains. In other words, a conditional inequality is true for certain values of the variables and false for others.

An inequality is called an *identity* if it is always true, for any value of the variables it contains.

An inequality is called *inconsistent* if it is always false, no matter what value (of allowable values) the variables take on.

Inequalities involving quadratics or expressions of higher powers require additional considerations. For example, given $x^2 > 4$, we should note that $x > 2$ is only part of the solution. It is also true that $x < -2$ provides other solution values. One can deduce this by converting the original inequality to $x^2 - 4 > 0$ which can then be factored into

$$(x - 2)(x + 2) > 0.$$

For this inequality to hold, both factors must be positive or both negative. In general, at this point, one should list every possible scenario with each factor as a separate case. For example, with the present example, one should consider the case where $x - 2 < 0$ and $x + 2 < 0$ to obtain one possible solution set, and then consider the case where $x - 2 > 0$ and $x + 2 > 0$ to obtain another possible solution set. By examining each possibility with each factor, one obtains the solutions given above. Similar techniques can be applied to situations with a greater number of factors.

A graph of an equation in two variables can be seen to subdivide the plane into several regions. In these various regions, an inequality based on the original equation holds. Such an inequality is derived by replacing the equals sign with either a less-than or a greater-than sign. Which inequality holds in which region is usually determined by choosing an arbitrary point in each region and evaluating the original expression. For example, the graph of $x^2 + y^2 = 1$ is a circle with a radius of one and centered at the origin. If one picks the origin, i.e., the point $(x, y) = (0, 0)$ and evaluates this expression, the result is

$$x^2 + y^2 = 0^2 + 0^2 = 0$$

which is less than the original constant value of 1. Thus, we can deduce that points within the circle satisfy the inequality $x^2 = y^2 < 1$.

When graphing an inequality, it is easiest to begin by graphing the related *equality*, and then determining which of the regions describes the inequality (using the techniques in the previous paragraph).

One can use graphs to determine the intersections of solution sets of various inequalities. After first determining the solution regions of individual inequalities, one can then determine the combined solution set by determining the region composed of the overlapping solution regions. This technique can be used for graphically solving a system of linear inequalities.

A *disjunction* of two statements is a compound statement created by using the connective word *or*. The solution of a disjunctive statement is the *union* of the solutions of the component parts. (Either set may be satisfied independently.)

A *conjunction* of two statements is a compound statement created by using the connective word *and*. The solution of a conjunctive statement is the *intersection* of the solutions of the component parts. (Both sets must be satisfied simultaneously.)

The shorthand notation, $a < x < b$, is, in fact, a conjunction of $a < x$ and $x < b$. It must be true that both relationships are fulfilled for this compound statement to be true.

When dealing with inequalities involving absolute values, one must remember the cautions presented in Chapter 4: an expression involving an absolute value is a shorthand for two expressions. If $|a| = 2$ then either $a = 2$ or $-a = 2$ (i.e., $a = -2$). Thus, when solving an inequality containing an absolute value sign, one should normally replace it with two expressions. For example,

$$|x - 1| < 5$$

implies that

$$x - 1 < 5 \quad \text{or} \quad -(x - 1) < 5.$$

The first implies that $x < 6$ and the second that

$$-x + 1 < 5, \quad \text{i.e.,} \quad -x < 4, \quad \text{i.e.,} \quad x > -4.$$

The sense of the original expression tells us that both of these solution sets must be satisfied together, i.e., that $-4 < x < 6$. We can test this by choosing values out of the ranges, such as $x = -5$ or $x = 7$. If the original inequality were $|x - 1| > 5$, then the solution set would be the union of $x < -4$ and $x > 6$.

Step-by-Step Solutions to Problems in this Chapter, "Inequalities and Graphs"

● **PROBLEM 16-1**

Solve the inequality $2x + 5 > 9$.

Solution:
$2x + 5 + (-5) > 9 + (-5)$.	Adding -5 to both sides.
$2x + 0 > 9 + (-5)$	Additive inverse property
$2x > 9 + (-5)$	Additive identity property
$2x > 4$	Combining terms
$\frac{1}{2}(2x) > \frac{1}{2} \cdot 4$	Multiplying both sides by $\frac{1}{2}$.
$x > 2$	

The solution set is

$$X = \{x \,|\, 2x + 5 > 9\}$$
$$= \{x \,|\, x > 2\}$$

(that is all x, such that x is greater than 2).

● **PROBLEM 16-2**

Illustrate one (a) conditional inequality, (b) identity, and (c) inconsistent inequality.

Solution: (a) A conditional inequality is an inequality whose validity depends on the values of the variables in the sentence. That is, certain values of the variables will make the sentence true, and others will make it false. $3 - y > 3 + y$ is a conditional inequality for the set of real numbers, since it is true for any replacement less than zero and false for all others.
    (b) $x + 5 > x + 2$ is an identity for the set of real numbers, since for any real valued x, the expression on the left is greater than the expression on the right.

(c)   5y < 2y + y is inconsistent for the set of non-
negative real numbers. For any x greater than 0 the sentence is always false. A sentence is inconsistent if it is
always false when its variables assume allowable values.

● **PROBLEM 16-3**

Solve the inequality $4x + 3 < 6x + 8$.

Solution: In order to solve the inequality $4x + 3$
$< 6x + 8$, we must find all values of x which make it
true. Thus, we wish to obtain x alone on one side of the
inequality.

Add -3 to both sides:

$$
\begin{array}{r}
4x + 3 < 6x + 8 \\
-\ 3 \qquad -\ 3 \\
\hline
4x < 6x + 5
\end{array}
$$

Add $-6x$ to both sides:

$$
\begin{array}{r}
4x < \quad 6x + 5 \\
-\ 6x \quad -\ 6x \\
\hline
-\ 2x < \quad 5
\end{array}
$$

In order to obtain x alone we must divide both sides
by (-2). Recall that dividing an inequality by a negative number reverses the inequality sign, hence

$$\frac{-2x}{-2} > \frac{5}{-2}$$

Cancelling $\frac{-2}{-2}$ we obtain, $x > -\frac{5}{2}$

Thus, our solution is $\left(x : x > -\frac{5}{2}\right)$ (the set of
all x such that x is greater than $-\frac{5}{2}$).

● **PROBLEM 16-4**

Solve the compound statement $4x - 5 \geq -6x + 5$.

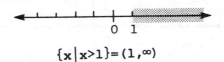

$$\{x \mid x \geq 1\} = (1, \infty)$$

Solution: To solve this compound statement we solve for x as follows:

Adding 5 to both sides of the given inequality we have:

$$4x \geq -6x + 10$$

Adding 6x to both sides: $10x \geq 10$

Multiplying both sides by $\frac{1}{10}$: $x \geq 1$

Therefore $S = \{x \mid x \geq 1\}$

The region representing this set on the number line is shown in the diagram.

You should note that the bracket in the graph includes the point 1.

● **PROBLEM 16-5**

Find the solution set of inequality $5x - 9 > 2x + 3$.

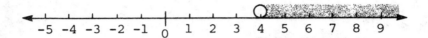

Solution: To find the solution set of the inequality $5x - 9 > 2x + 3$, we wish to obtain an equivalent inequality in which each term in one member involves x, and each term in the other member is a constant. Thus, if we add (- 2x) to both members, only one side of the inequality will have an x term:

$$5x - 9 + (- 2x) > 2x + 3 + (- 2x)$$

$$5x + (-2x) - 9 > 2x + (- 2x) + 3$$

$$3x - 9 > 3$$

Now, adding 9 to both sides of the inequality we obtain,

$$3x - 9 + 9 > 3 + 9$$

$$3x > 12$$

Dividing both sides by 3, we arrive at $x > 4$.

Hence the solution set is $\{x \mid x > 4\}$, and is pictured in the accompanying figure.

● **PROBLEM 16-6**

Solve the inequality: $x^2 > x$.

Solution: Subtracting x from both sides of the given inequality, we obtain:

263

$x^2 - x > 0$.

Factoring the left member, we have:

$x(x - 1) > 0$.

For x to satisfy this requirement it is necessary and sufficient that x and x - 1 be of the same sign. Thus, either

(i)   x > 0 and x - 1 > 0   or

(ii) x < 0 and x - 1 < 0

Case (i) is satisfied if and only if x > 1, whereas case (ii) is satisfied if and only if x < 0. Thus, the solution set is the union of the intervals $(-\infty, 0)$ and $(1, \infty)$. Expressed in another way, the solution consists of all numbers of x, which are outside of the closed interval [0, 1].

The same results may also be obtained by dividing the given inequality by x and considering the two cases of x positive and negative.

● **PROBLEM** 16-7

Given the equation: $x^2 - y^2 = 4$, describe what inequalities hold in the regions of the plane which this curve separates.

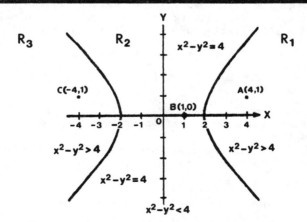

Solution: The curve divides the plane into three regions; R_1, R_2, and R_3. The region R_1 is the one containing the point A, the region R_2 is the one containing B, and R_3 the one containing C. At point A (4, 1) we have $4^2 - 1^2 > 4$, and

therefore,

$$x^2 - y^2 > 4 \quad \text{in } R_1.$$

Since $1^2 - 0^2 < 4$,

$$x^2 - y^2 < 4 \quad \text{in } R_2.$$

Similarly, $(-4)^2 - 1^2 > 4$, and

$$x^2 - y^2 > 4 \quad \text{in } R_3.$$

Solve the inequality

$$\frac{4}{x - 2} < 2.$$

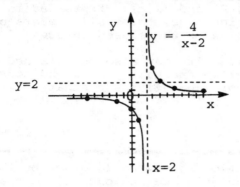

<u>Solution:</u> The inequality is meaningless for x = 2 because when x = 2 the denominator of the left member is 0, making the fraction undefined.
    If x > 2, x - 2 is positive (since x > 2 is equivalent to x - 2 > 0), and multiplication of the given inequality by x - 2 yields

$$4 < 2(x - 2)$$
$$4 < 2x - 4$$
$$8 < 2x$$
$$4 < x$$
$$x > 4.$$

Thus, the solution is the intersection of x > 2 and x > 4, x > 2 ∩ x > 4, which is $\{x \mid x > 4\}$.
    If x < 2, x - 2 is negative (since x < 2 is equivalent to x - 2 < 0), and multiplication by x - 2 yields

$$4 > 2(x - 2)$$

because multiplication by a negative number reverses an inequality. Distributing, $4 > 2x - 4$

Adding 4 to both sides,
$$8 > 2x$$

Dividing both sides by 2,
$$4 > x, \quad \text{or} \quad x < 4.$$

Thus the solution is the intersection of x < 2 and x < 4, x < 2 ∩ x < 4, which is
$$\{x \mid x < 2\}. \quad \text{Hence}$$
$$\frac{4}{x - 2} < 2$$

if x < 2 or if x > 4.
    A graphical solution of the problem (see diagram) can be obtained

by sketching the equilateral hyperbola y = 4/(x - 2) and the line
y = 2. The hyperbola may be sketched from its vertical asymptote
x = 2, its horizontal asymptote y = 0, its intercepts x = 0, y = -2,
and a few other points obtained by substitution and symmetry. It is
then possible to observe the values of x for which the hyperbola is
below the line, namely,

$$x < 2 \quad \text{and} \quad x > 4.$$

The same diagram also shows that $[4/(x - 2)] > 2$ for $2 < x < 4$.

● **PROBLEM 16-9**

Find the solution set of the disjunction

$$2 - 3x > 5 \quad \text{or} \quad 2x - 1 > 5.$$

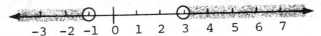

Solution: A disjunction is a compound sentence using the connective
'or'. The union of the solution sets of the two sentences comprising
the compound sentence is the solution set of the disjunction. For
this problem the solution set is

$$\{x: 2 - 3x > 5\} \cup \{x: 2x - 1 > 5\}$$

We solve each inequality independently and find the union of their
solution set:

$$2 - 3x > 5 \quad \text{or} \quad 2x - 1 > 5$$
$$-3x > 3 \quad \text{or} \quad 2x > 6$$

Solving for x, we divide both members of the inequality by a negative
number (-3). Therefore the direction of the inequality is reversed.

$$x < -1 \quad \text{or} \quad x > 3$$

The solution set of $2 - 3x > 5$ or $2x - 1 > 5$ is shown on the graph.
The unshaded circles above -1 and 3 on the number line
indicate that these values are not included in the solution set.

Compound sentences can also be formed by connecting two sentences with
the word 'and.' A compound sentence using the connective and is
called a conjunction. The solution set of a conjunction is the set of
replacements that are common to the solution sets of the sentences
making-up the conjunction (their intersection). We may write the
solution set as:

$$\{x: x > a\} \cap \{x: x < b\} = \{x: a < x < b\}.$$

● **PROBLEM 16-10**

Find the solution set of the conjunction

$$\tfrac{1}{2}x + 1 > 3 \quad \text{and} \quad x > 2x - 6$$

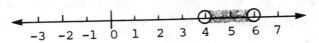

<u>Solution:</u> The solution set must be such that x satisfies both in-
equalities simultaneously. The solution set of the conjunction is

$$\{x:\ \tfrac{1}{2}x + 1 > 3\} \cap \{x:\ x > 2x - 6\}$$

$$\tfrac{1}{2}x + 1 > 3 \quad \text{and} \quad x > 2x - 6$$

$$\tfrac{1}{2}x > 2 \quad \text{and} \quad -x > -6$$

$$x > 4 \quad \text{and} \quad x < 6$$

(multiplying by negative 1 reverses the inequality).

Note: if x = 4 the sentence $\tfrac{1}{2}x + 1 > 3$ becomes false, i.e.,

$$\tfrac{1}{2}(4) + 1 = 3 > 3 \qquad \text{false}$$

and if x = 6 the sentence x > 2x - 6 becomes false

also   (6 > 2(6) - 6 = 6 false).

Therefore the solution set cannot include these two points, and
the values of x which make both sentences true simultaneously are
the inequalities x greater than but not equal to 4 and less
than but not equal to 6. That is $\{x:\ 4 < x < 6\}$. (See the number
line).

● **PROBLEM 16-11**

Graph the inequality: $y < -\tfrac{1}{2}x + 3$.

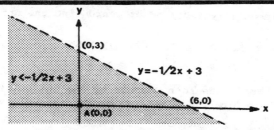

<u>Solution:</u> The graph is represented by shading the appropriate half-
plane. The plot of: $y = -\tfrac{1}{2}x + 3$, is shown by a broken line to indicate that it is not part of the graph. To determine which half-
plane is represented by the inequality, we test a point in a half-
plane. If we test point A with co-ordinates (0,0), we have:

$$0 < -\tfrac{1}{2} \cdot 0 + 3.$$
$$0 < 3$$

Therefore the shaded portion is the correct half-plane.

● **PROBLEM 16-12**

Graph the inequality: $y \geq x - 1$.

<u>Solution:</u> The graph is shown by the shaded area and includes the
line y = x - 1, which is drawn as a solid line. To verify this,
we test a point on the line and in the half-plane. Taking point

A with co-ordinates (0,0) and point B with co-ordinates (1,0), we
have:

1) $0 \geq 0 - 1$
      $0 \geq -1$

2) $0 \geq 1 - 1$
      $0 \geq 0$

The union of a half plane with the line bounding it is called a
closed half plane.

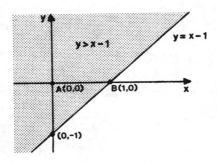

● **PROBLEM** 16-13

Graph the following inequalities and indicate the
region of their intersection:

$x \geq 1, \ y \geq 0, \ x + y \leq 4, \ 4x + 3y \leq 14.$

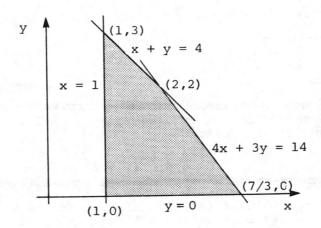

<u>Solution:</u>   The solution is the shaded region,
having the quadrilateral as its boundary. The
indicated vertices are the intersection points
of the inequalities.

Graph the following inequalities:

$$(x + 1)^2 + y^2 < 1,$$

and

$$(x + 1)^2 + y^2 > 1.$$

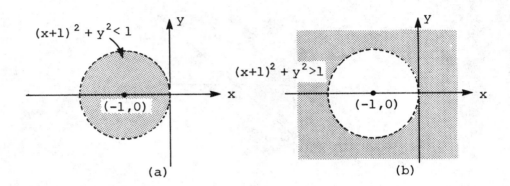

(a) (b)

$$(x + 1)^2 + y^2 = 1$$

is a circle. By testing a point inside the circle and a point outside the circle, we will determine which inequality holds.

fig. (a) is the graph of:

$$(x + 1)^2 + (y^2) < 1,$$ and

fig. (b) is the graph of:

$$(x + 1)^2 + (y^2) > 1.$$

Graph the following two inequalities and show where the two graphs coincide:

$$2 \le x < 3 \qquad \text{and} \qquad |y - 2| < \frac{1}{2}.$$

Solution: The first inequality consists of an infinite strip between the lines $x = 2$ and $x = 3$. Note that the points on the line $x = 2$ are included, but the points on $x = 3$ are not.

For the second inequality, two cases must be considered, depending on whether $(y - 2)$ is positive or negative. This may be expressed as:

$$-\frac{1}{2} < y - 2 < \frac{1}{2}.$$

Adding 2 to each term to simplify, gives:

$$\frac{3}{2} < y < \frac{5}{2}.$$

This is an infinite strip which intersects the first one in a rectangle, with vertices at the points $\left(2, \frac{3}{2}\right)$, $\left(3, \frac{3}{2}\right)$, $\left(3, \frac{5}{2}\right)$, and $\left(2, \frac{5}{2}\right)$. The result is actually a square.

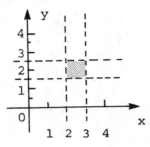

Therefore, all points inside this square satisfy both inequalities. However, in addition, the points on the boundary of the square along the line x = 2 (left boundary), except for the corners, also satisfy both inequalities.

● **PROBLEM 16-16**

Prove that if a > b > 0, then

$$\frac{1}{a} < \frac{1}{b}.$$

<u>Solution:</u> Since a and b are both positive (given), ab is positive because the product of two positive numbers is always positive. Now, since ab > 0, we may divide both sides of a > b by ab to obtain

$$\frac{a}{ab} > \frac{b}{ab}.$$

Cancelling like terms in numerator and denominator,

$$\frac{1}{b} > \frac{1}{a},$$

which is equivalent to $\frac{1}{a} < \frac{1}{b}$. To complete the proof, the student should check that the steps are reversible, as follows:

Check. $\frac{1}{b} > \frac{1}{a}$. Multiply both sides of the inequality by the least common denominator obtained by multiplying the two denominators together.

$$\frac{ab}{b} > \frac{ab}{a} = a > b.$$

● **PROBLEM 16-17**

Solve the inequality $\sqrt{x - 3} \le 2 - \sqrt{x + 1}$.

Solution:

$\sqrt{x - 3} \le 2 - \sqrt{x + 1}$	Given
$x - 3 \le 4 - 4\sqrt{x + 1} + x + 1$	Squaring
$-8 \le -4\sqrt{x + 1}$	Transposing and simplifying
$2 \ge \sqrt{x + 1}$	Dividing by -4.

Note that dividing both sides of an inequality by a negative number changes the sense of the inequality.

$4 \ge x + 1$	Squaring
$3 \ge x$ or $x \le 3$	Solving for x

Check: If $x = 3$, $\sqrt{3 - 3} = 2 - \sqrt{3 + 1}$.
        If $x > 3$, the left member, $\sqrt{x - 3}$, is positive, but the right member, $2 - \sqrt{x + 1}$, is negative. Hence, the inequality is not satisfied. Nor is the inequality satisfied if $x < 3$, for in this case the left member is not a real number. Hence, the solution set is $\{3\}$.

If the inequality were a strict inequality, the solution set would be the null set. If the left member is to be a real number, it must be positive and x must be greater than 3, but in this case the right member must be negative. However, it is impossible for a positive number to be less than a negative number.

● **PROBLEM 16-18**

What is the set $\{x < -2\} \cap \{x > 3\}$?

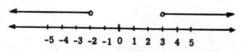

Solution: An element belongs to the intersection of two sets, if, and only if, it belongs to both of them. Thus, in order for a number to belong to our intersection, it would have to be both less than -2 and greater than 3. There is no such number, so the intersection is the empty set; that is,

$$\{x < -2\} \cap \{x > 3\} = \emptyset.$$

This can be seen from the accompanying number line representation, which illustrates that the two graphs have no points in common.

Solve $2x - 3y \geq 6$

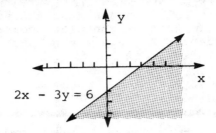

$2x - 3y = 6$

<u>Solution:</u> The statement $2x - 3y \geq 6$ means $2x - 3y$ is greater than or equal to 6. Symbolically, we have $2x - 3y > 6$ or $2x - 3y = 6$. Consider the corresponding equality and graph $2x - 3y = 6$. To find the x-intercept, set $y = 0$

$$2x - 3y = 6$$
$$2x - 3(0) = 6$$
$$2x = 6$$
$$x = 3$$

{3,0} is the x-intercept.

To find the y-intercept, set x=0

$$2x - 3y = 6$$
$$2(0) - 3y = 6$$
$$-3y = 6$$
$$y = -2$$

{0,-2} is the y-intercept.

A line is determined by two points. Therefore draw a straight line through the two intercepts {3,0} and {0,-2}. Since the inequality is mixed, a solid line is drawn through the intercepts. This line represents the part of the statement $2x - 3y = 6$.

We must now determine the region for which the inequality $2x - 3y > 6$ holds.

Choose two points to decide on which side of the line the region $x - 3y > 6$ lies. We shall try the points (0,0) and (5,1).

For (0,0)

$$2x - 3y > 6$$
$$2(0) - 3(0) > 6$$
$$0 - 0 > 6$$
$$0 > 6$$

False

For (5,1)

$$2x - 3y > 6$$
$$2(5) - 3(1) > 6$$
$$10 - 3 > 6$$
$$7 > 6$$

True

The inequality, $2x - 3y > 6$, holds true for the point $(5,1)$. We shade this region of the xy-plane. That is, the area lying below the line $2x - 3y = 6$ and containing $(5,1)$.

Therefore, the solution contains the solid line, $2x - 3y = 6$, and the part of the plane below this line for which the statement $2x - 3y > 6$ holds.

● **PROBLEM 16-20**

A livestock farmer has 500 acres to devote to grazing. He estimates that cattle require 5 acres per head and sheep require 3 acres per head. He has winter shelter facilities for 40 head of cattle and for 125 sheep. What constraints are imposed on the number of cattle and sheep he can raise?

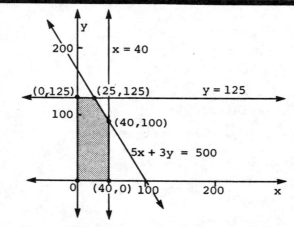

Solution: Let x represent the number of cattle raised and y the number of sheep. Since he cannot raise a negative number of either cattle or sheep, we have the constraints

$$x \geq 0 \qquad\qquad\qquad (1)$$
$$y \geq 0 \qquad\qquad\qquad (2)$$

Since 5x acres are required for the cattle and 3y acres for the sheep and there are only 500 acres available, we have

$$5x + 3y \leq 500 \qquad\qquad (3)$$

Since he can winter only 40 cattle,

$$x \leq 40 \qquad\qquad\qquad (4)$$

Since he can winter only 125 sheep,

$$y \leq 125 \qquad\qquad\qquad (5)$$

Relations (1) through (5) are the constraints.

The graph of the constraints in this example is a convex set of points. The corner points of the shaded polygon are $(0,0)$, $(40,0)$, $(40,100)$, $(25,125)$, and $(0,125)$.

Find a positive number M such that
$$|x^3 - 2x^2 + 3x - 4| \leq M$$
for all values of x in the interval [-3,2].

Solution: The expression may be written in the form:
$$|x^3 - 2x^2 + 3x - 4| \leq |x^3| + |2x^2| + |3x| + |4|,$$
and, from the rules of absolute value of products, we obtain:
$$|x^3| + |2x^2| + |3x| + |4| \leq |x|^3 + 2|x|^2 + 3|x| + 4.$$
Now, since $|x|$ can never be larger than 3, we have:
$$|x|^3 + 2|x|^2 + 3|x| + 4 \leq 27 + 2 \cdot 9 + 3 \cdot 3 + 4 = 58.$$
Therefore, the positive number M we seek is 58.

What is the largest possible value of
$$\left|\frac{x^2 + 2}{x + 3}\right|$$
if x is restricted to the interval [-4,4]?

Solution: The largest possible value for the expression results when the numerator is largest and the denominator is smallest.

The numerator is largest when the upper limit of x = 4 is substituted, as follows:
$$|x^2 + 2| \leq |x|^2 + 2 \leq 4^2 + 2 = 18.$$
We must now find a smallest value for x + 3 if x is in [-4,4]. We see that the expression is not defined for x = -3, since then the denominator would be zero, and division by zero is always excluded. Furthermore, if x is a number near -3, the denominator is near zero, the numerator has a value near 11, and the quotient is a "large" number, i.e., it approaches ∞. Hence, in this problem, there is no largest value of the given expression in [-4,4].

Solve the inequality $|5 - 2x| > 3$.

Solution: The property of absolute values states that

$|a| = +a$ or $|a| = -a$. Therefore: $|5 - 2x| = 5 - 2x$ or $-(5 - 2x)$. Thus, the given inequality becomes two new inequalities:

$$5 - 2x > 3, \quad -(5 - 2x) > 3.$$

Now, we must solve for x in both inequalities. For the first, we subtract 5 from both sides of the inequality, and then divide by -2. We must keep in mind that division or multiplication by a negative number reverses the inequality sign. Thus, for $5 - 2x > 3$ we have:

$$5 - 5 - 2x > 3 - 5$$

$$-2x > -2$$

$$\frac{-2x}{-2} > \frac{-2}{-2}$$

$$x < 1.$$

For the second inequality, we first take the negative of all the terms inside the parentheses. Thus, for $-(5 - 2x) > 3$ we have:

$$-5 + 2x > 3.$$

Now, we add 5 to both sides of the inequality, and then divide by 2. Thus, we obtain:

$$-5 + 5 + 2x > 3 + 5$$

$$2x > 8$$

$$\frac{2x}{2} > \frac{8}{2}$$

$$x > 4.$$

Therefore, the above inequality holds when $x < 1$, and when $x > 4$.

● PROBLEM 16-24

Graph $\{x: |3x - 4| \geq 2\}$.

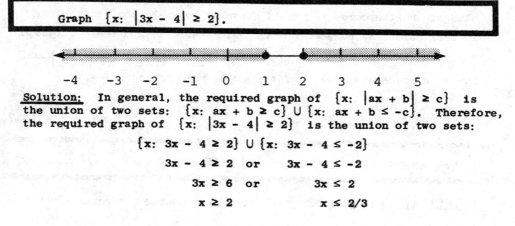

Solution: In general, the required graph of $\{x: |ax + b| \geq c\}$ is the union of two sets: $\{x: ax + b \geq c\} \cup \{x: ax + b \leq -c\}$. Therefore, the required graph of $\{x: |3x - 4| \geq 2\}$ is the union of two sets:

$$\{x: 3x - 4 \geq 2\} \cup \{x: 3x - 4 \leq -2\}$$

$$3x - 4 \geq 2 \quad \text{or} \quad 3x - 4 \leq -2$$

$$3x \geq 6 \quad \text{or} \quad 3x \leq 2$$

$$x \geq 2 \quad\quad\quad x \leq 2/3$$

275

The solution set of $\{x: |3x - 4| \geq 2\}$ is

$$\{x: x \geq 2\} \cup \{x: x \leq 2/3\}$$

The graph of the solution set is the union of two rays. Notice that the shaded circles above 2/3 and 2 on the number line indicate that 2/3 and 2 are included in the solution set.

● **PROBLEM 16-25**

Solve $|3x - 1| \leq 8$.

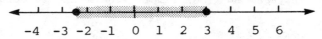

Solution: Since $|a| = a$ if $a \geq 0$ and $|a| = -a$ if $a \leq 0$. We must solve two equations

$$3x - 1 \leq 8$$

$$-(3x - 1) \leq 8 \text{ or } 3x - 1 \geq -8.$$

(Note that multiplying an inequality by a negative number, i.e., -1, reverses the inequality.)
     The solution set will be the conjunction of the solution sets of each equation; that is,

$$\{x: 3x - 1 \leq 8\} \text{ and } \{x: 3x - 1 \geq -8\}.$$

We must find

$$\{x: 3x - 1 \leq 8\} \cap \{x: 3x - 1 \geq -8\}$$

$$3x - 1 \leq 8 \text{ and } 3x - 1 \geq -8$$

$$3x \leq 9 \text{ and } \qquad 3x \geq -7$$

$$x \leq 3 \text{ and } \qquad x \geq -2\tfrac{1}{2}.$$

The solution set is $\langle x: -2\tfrac{1}{2} \leq x \leq 3 \rangle$. See the figure.

● **PROBLEM 16-26**

Find the values of x satisfying the statement $\left|\dfrac{x}{3} - 7\right| \geq 5$.

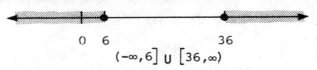

$$(-\infty, 6] \cup [36, \infty)$$

Solution: In general, $|ax + b| \geq c$ implies that $ax + b \geq c$ or $ax + b \leq -c$. Therefore, for

276

$$\left|\frac{x}{3} - 7\right| \geq 5 :$$

(1) $\frac{x}{3} - 7 \geq 5$ or   (2) $\frac{x}{3} - 7 \leq -5$

Case (1) $\frac{x}{3} - 7 \geq 5$ Case (2) $\frac{x}{3} - 7 \leq -5$

$$\frac{x}{3} \geq 12 \qquad\qquad\qquad \frac{x}{3} \leq 2$$

$$x \geq 36 \qquad\qquad\qquad x \leq 6$$

We may consider the solution set of the inequality in this example as the union of the two disjoint sets $\{x \mid x \geq 36\}$ and $\{x \mid x \leq 6\}$. This union may be represented on the number line as in the accompanying figure.

● **PROBLEM 16-27**

Solve

$$\left|\frac{x}{3} + 2\right| < 4$$

(a) $|a| < b$

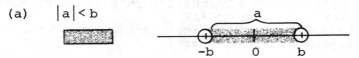

(b)

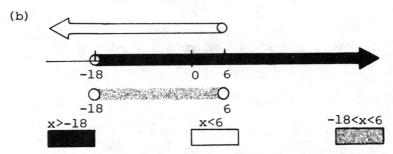

Solution: Now, if b is a nonnegative real number, then a is a real number for which $|a| < b$ if and only if $-b < a < b$. See number line (A).

We apply this rule to the given problem. Therefore, $\left|\frac{x}{3} + 2\right| < 4$ is equivalent to

$-4 < \frac{x}{3} + 2 < 4$. In other words, this inequality is satisfied if and only if both

$$\frac{x}{3} + 2 < 4 \qquad \text{and} \qquad \frac{x}{3} + 2 > -4$$

are satisfied. By adding -2 to each member of these inequalities, we get

277

$$\frac{x}{3} < 2 \qquad \text{and} \qquad \frac{x}{3} > -6$$

Hence, multiplying by 3 in each case, we note that the original inequality is satisfied by values of x that satisfy both x < 6 and x > - 18. We can observe the solution from diagram (B). Therefore, the solution set is

$$\{x| \ -18 < x < 6\} \qquad \text{or} \quad \{x|x < 6\} \cap \{x|x > -18\}.$$

● PROBLEM 16-28

Find the solution set of $|2x + 5| \leq x + 3.$

Fig. A

Key: $x \geq -\dfrac{5}{2}$ $\quad$ $x \leq -2$

Fig. B

Key: $x \geq -\dfrac{8}{5}$ $\quad$ $x < -\dfrac{5}{2}$

Solution: Case 1. If $2x + 5 \geq 0$, then $|2x + 5| = 2x + 5$ and the inequality becomes

$$2x + 5 \leq x + 3$$
$$x \leq -2$$

For Case 1 we have the simultaneous restrictions

$$2x + 5 \geq 0 \quad \text{and} \quad x \leq -2$$

or

$$x \geq -\frac{5}{2} \quad \text{and} \quad x \leq -2$$

The solution set for Case 1 is

$$X_1 = \{ x | x \geq -\frac{5}{2} \text{ and } x \leq -2\}$$
$$= \{x| -\frac{5}{2} \leq x \leq -2\}$$

This inequality holds by noting Figure A. The solution set for Case 1 is the intersection of the two inequalities on the number line. This intersection is the set

$$-2\tfrac{1}{2} \leq x \leq -2.$$

Case 2. If $2x + 5 < 0$, then $|2x + 5| = -(2x + 5)$ and the inequality becomes

$$-(2x + 5) \leq x + 3$$

Multiplying an inequality by -1 reverses the direction of the inequality.

$$2x + 5 \geq -(x + 3) = -x - 3$$
$$3x \geq -8$$
$$x \geq -\frac{8}{3}$$

For Case 2 we have

$$2x + 5 < 0 \quad \text{and} \quad x \geq -\frac{8}{3}$$

or

$$x < -\frac{5}{2} \quad \text{and} \quad x \geq -\frac{8}{3}$$

The solution set is

$$X_2 = \left\{ x \mid x < -\frac{5}{2} \quad \text{and} \quad x \geq -\frac{8}{3} \right\}$$
$$= \left\{ x \mid -\frac{8}{3} \leq x < -\frac{5}{2} \right\}$$

This inequality holds by noting Figure B. The solution set for Case 2 is the intersection of the two inequalities on the number line. This intersection is the set $-\frac{8}{3} \leq x < -\frac{5}{2}$.

Finally, the solution set, X, of the given inequality is the union of X_1 and X_2 .

$$X = X_1 \cup X_2$$
$$= \left\{ x \mid -\frac{8}{3} \leq x < -\frac{5}{2} \right\} \cup \left\{ x \mid -\frac{5}{2} \leq x \leq -2 \right\}$$
$$= \left\{ x \mid -\frac{8}{3} \leq x \leq -2 \right\}$$

● **PROBLEM 16-29**

Solve the following system graphically.
$$y - x > -3$$
$$y - 2x < 2$$
$$x + y - 3 < 0$$

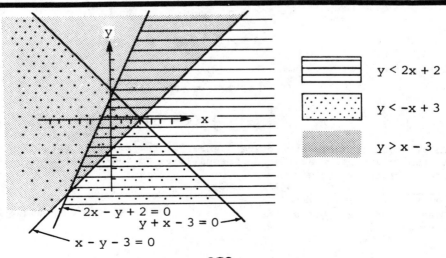

$y < 2x + 2$

$y < -x + 3$

$y > x - 3$

2x - y + 2 = 0
y + x - 3 = 0

x - y - 3 = 0

Solution: We may rewrite the system:

$$y > x - 3$$
$$y < 2x + 2$$
$$y < -x + 3$$

Graph the linear equation, $y = mx + b$, for each inequality as a straight dotted line. Thus, we graph

$$y = x - 3$$
$$y = 2x + 2$$
$$y = -x + 3$$

To determine in what region of the x - y plane the inequality holds, select points on both sides of the corresponding dotted line and substitute them into the variable statement. Shade in the side of the line whose point makes the inequality a true statement.

The graphs of the variable sentences are represented in the accompanying figure by diagonal, horizontal, and vertical shading, respectively.

The triple-shaded triangular region is the set of all points whose co-ordinate pairs satisfy all three conditions as defined by the three inequalities in the system.

● PROBLEM 16-30

Draw the graph of the given system of inequalities, and determine the coordinates of the vertices of the polygon which forms the boundary.

$y \leq 3x - 3$	(1)
$3y \leq 24 - 2x$	(2)
$2y \geq 3x - 10$	(3)
$y \geq -x + 5$	(4)

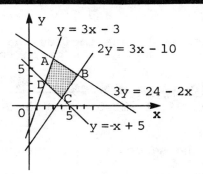

Solution: y is expressed in terms of x. For each inequality draw the corresponding equality. Choose a point on each side of each solid line to determine the area where the particular inequality holds. Shade in that region. The graph of the given system of inequalities consists of the hatched area and the four lines which form the boundary, that is, the polygon ABCD. The vertex A is found by solving the system obtained by writing Equations 1 and 2 as

$$y = 3x - 3 \tag{5}$$

$$3y = 24 - 2x \qquad\qquad (6)$$

Solving the system of equations 5 and 6, we have

$$x = 3, \qquad y = 6$$

The coordinates of the vertex A are therefore (3, 6). In a similar manner, the coordinates of B, C, and D are found to be (6, 4) (4, 1) and (2, 3), respectively.

● **PROBLEM 16-31**

Solve the inequality $(2x - 1)(x + 2) < 0$.

<u>Solution:</u> Since the two factors must be of opposite sign for their product to be negative, we have the two tentative possibilities:

$$2x - 1 < 0, \quad x + 2 > 0,$$

or $2x - 1 > 0, \quad x + 2 < 0.$

Solving the first pair of inequalities:

$$2x - 1 < 0 \qquad \text{and} \qquad x + 2 > 0$$

add 1 to both sides: | subtract 2 from both sides:

$$2x < 1 \qquad\qquad\quad |$$

divide both sides by 2: |

$$x < \tfrac{1}{2} \qquad\qquad \text{and} \qquad x > - 2$$

Thus, the first pair implies that $x < \tfrac{1}{2}$ and $x > - 2$, or $-2 < x < \tfrac{1}{2}$; the graph is as follows:

Solving the second pair of inequalities:

$$2x - 1 > 0 \qquad\qquad \text{and} \quad x + 2 < 0$$

Adding 1 to both sides: |

$$2x > 1 \qquad\qquad\quad | \quad \text{Subtracting 2 from} \\ \text{both sides:}$$

Dividing both sides by 2: |

$$x > \tfrac{1}{2} \qquad\qquad \text{and} \quad x < - 2$$

Thus, the second pair implies that $x > \tfrac{1}{2}$ and $x < - 2$; the graph is as follows:

281

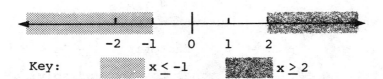

Since there is no x such that $x > \frac{1}{2}$ and $x < -2$ we reject this solution.

The complete solution is thus the solution to the first pair of inequalities, $\{x : -2 < x < \frac{1}{2}\}$.

● PROBLEM 16-32

Solve the inequality $x^2 - x - 2 \leq 0$.

Solution: Factoring the left side of the given inequality,

$$(x - 2)(x + 1) \leq 0.$$

If the product of two numbers is negative, one of the numbers is positive and the other is negative. Hence, there are two cases:

Case 1: $x - 2 \geq 0$, $x + 1 \leq 0$

Solving these two inequalities,

$$x \geq 2, \quad x \leq -1$$

Graph these new inequalities on number line (A).

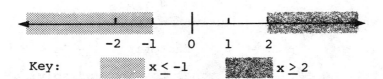

Key: ▓▓▓ $x \leq -1$ ▓▓▓ $x \geq 2$

Note that there is no value of x which satisfies both inequalities at the same time since these two inequalities do not intersect anywhere on the number line (A).

Thus $x \leq -1 \cap x \geq 2 = \emptyset$

Case 2: $x - 2 \leq 0$, $x + 1 \geq 0$.

Solving these two inequalities,

$$x \leq 2, \quad x \geq -1$$

Graph these inequalities on number line (B).

$$-2 \quad -1 \quad 0 \quad 1 \quad 2$$

Key: [shaded] $x \geq -1$ [shaded] $x \leq 2$

The interval of x which satisfies both inequalities at the same time is $-1 \leq x \leq 2$. Note that the two inequalities intersect in this interval on number line (B), that is

$$x \geq -1 \cap x \leq 2 = -1 \leq x \leq 2.$$

Hence, the solution to the inequality $x^2 - x - 2 \leq 0$ is the set:

$$\{x \mid -1 \leq x \leq 2\}$$

● **PROBLEM 16-33**

Solve the equation $\left(x^2 - 3x\right)^2 - 2\left(x^2 - 3x\right) - 8 = 0$

<u>Solution:</u> This equation is greater than degree two. However, it can be solved by the methods used for quadratic equations. An equation of the type $A[P(x)]^2 + B[P(x)] + C = 0$, where A, B, and C are real numbers, $A \neq 0$ and $P(x)$ is an expression in variable x, is said to be in quadratic form.

We let $z = x^2 - 3x$, substitute in the given equation, and get

$$z^2 - 2z - 8 = 0$$

We then complete the solution as follows:

$$(z - 4)(z + 2) = 0 \quad \text{factoring } z^2 - 2z - 8$$

Setting each factor equal to zero and solving:

$$z - 4 = 0 \qquad z + 2 = 0$$
$$z = 4 \qquad z = -2$$

Hence the solution set is $\{4, -2\}$.

Now we replace z in $z = x^2 - 3x$ with 4 and then with -2, solve each of the resulting equations, and thus obtain

$$x^2 - 3x = 4$$
$$x^2 - 3x - 4 = 0$$
$$(x - 4)(x + 1) = 0$$
$$x = 4$$

$$x = -1$$

and
$$x^2 - 3x = -2$$

$$x^2 - 3x + 2 = 0$$

$$(x - 2)(x - 1) = 0$$

$$x = 2$$

$$x = 1$$

Therefore the solution set of the given equation is
$\{4, -1\} \cup \{2, 1\} = \{4, -1, 2, 1\}$.

● **PROBLEM 16-34**

Solve the inequality
$$x - 6 > \frac{18 - 15x}{x^2 + 2x - 3}$$

Solution: We first subtract $\frac{18 - 15x}{x^2 + 2x - 3}$ from both
sides of the inequality, obtaining

$$x - 6 - \frac{18 - 15x}{x^2 + 2x - 3} > 0.$$

In order to combine terms, we convert $x - 6$ into a
fraction with $x^2 + 2x - 3$ as its denominator. Thus

$$\frac{(x^2 + 2x - 3)}{(x^2 + 2x - 3)} \cdot (x - 6) - \frac{18 - 15x}{x^2 + 2x - 3} > 0$$

Note that since $\frac{x^2 + 2x - 3}{x^2 + 2x - 3} = 1$, multiplication of

$(x - 6)$ by this fraction does not alter the value of
the inequality.

$$\frac{(x^2 + 2x - 3)(x - 6)}{x^2 + 2x - 3} - \frac{18 - 15x}{x^2 + 2x - 3} > 0$$

$$\frac{x^3 + 2x^2 - 3x - 6x^2 - 12x + 18}{x^2 + 2x - 3} - \frac{18 - 15x}{x^2 + 2x - 3} > 0$$

$$\frac{x^3 - 4x^2 - 15x + 18}{x^2 + 2x - 3} - \frac{18 - 15x}{x^2 + 2x - 3} > 0$$

$$\frac{x^3 - 4x^2 - 15x + 18 - 18 + 15x}{x^2 + 2x - 3} > 0$$

$$\frac{x^3 - 4x^2}{x^2 + 2x - 3} > 0$$

Now we factor numerator and denominator. Thus

$$\frac{x^2(x-4)}{(x-1)(x+3)} > 0.$$

We now want all values of x which make $\frac{x^2(x-4)}{(x-1)(x+3)}$

greater than zero. If $(x-1) = 0$ or $(x+3) = 0$ this fraction is undefined, thus we must place the restrictions

$$x - 1 \neq 0 \quad \text{and} \quad x + 3 \neq 0$$

or

$$x \neq 1 \quad \text{and} \quad x \neq -3.$$

Next we must eliminate all values of x which make

$\frac{x^2(x-4)}{(x-1)(x+3)}$ equal to zero (for we only want it to

be greater than zero). The numerator will be zero if $x^2 = 0$ or $x - 4 = 0$, thus $x \neq 0$ and $x \neq 4$. We now have critical values $x = -3$, $x = 0$, $x = 1$, $x = 4$.

We must test values of x in all ranges bordering on these critical values: (a) $x < -3$, (b) $-3 < x < 0$, (c) $0 < x < 1$, (d) $1 < x < 4$, (e) $x > 4$, to find the ranges in which the inequality holds:

(a) To test if the inequality holds for $x < -3$, choose any value of $x < -3$, we will use -4, and replace x by this value in the given inequality:

$$\frac{x^2(x-4)}{(x-1)(x+3)} > 0$$

$$\frac{(-4)^2(-4-4)}{(-4-1)(-4+3)} > 0$$

$$\frac{16(-8)}{(-5)(-1)} > 0$$

$$\frac{-128}{5} > 0$$

Since a negative number is not greater than zero, the range $x < -3$ is not part of the solution.

(b) To test if the inequality holds for $-3 < x < 0$, choose a value of x between 0 and -3, we will use -1, and replace x by this value in the inequality:

$$\frac{x^2(x-4)}{(x-1)(x+3)} > 0$$

$$\frac{(-1)^2(-1-4)}{(-1-1)(-1+3)} > 0$$

$$\frac{1(-5)}{(-2)(2)} > 0$$

$$\frac{-5}{-4} > 0$$

$$\frac{5}{4} > 0$$

Since 5/4 is indeed greater than zero, the range
- 3 < x < 0 is part of the solution.

(c) Testing if 0 < x < 1 is part of the solution,
we choose a value of x between 0 and 1, we will use
½, and replace x by this value in the inequality:

$$\frac{x^2(x - 4)}{(x - 1)(x + 3)} > 0$$

$$\frac{(\frac{1}{2})^2(\frac{1}{2} - 4)}{(\frac{1}{2} - 1)(\frac{1}{2} + 3)} > 0$$

$$\frac{(1/4)(- 3\frac{1}{2})}{(- \frac{1}{2})(3\frac{1}{2})} > 0$$

$$\frac{(1/4)(- 7/2)}{(- \frac{1}{2})(7/2)} > 0$$

$$\frac{- 7/8}{- 7/4} > 0$$

$$- \frac{7}{8} \cdot - \frac{4}{7} > 0$$

$$\frac{1}{2} > 0$$

Since ½ is indeed greater than zero, the range
0 < x < 1 is part of the solution.

(d) Testing if 1 < x < 4 is part of the solution,
we choose a value of x between 1 and 4. We will use 2,
and replace x by this value in the inequality:

$$\frac{x^2(x - 4)}{(x - 1)(x + 3)} > 0$$

$$\frac{(2)^2(2 - 4)}{(2 - 1)(2 + 3)} > 0$$

$$\frac{4(- 2)}{(1)(5)} > 0$$

$$\frac{- 8}{5} > 0$$

Since a negative number is not greater than zero,
the range 1 < x < 4 is not part of the solution.

(e) Testing if x > 4 is part of the solution, we
choose any value of x greater than 5, we will use 5,
and replace x by this value in the inequality:

$$\frac{x^2(x - 4)}{(x - 1)(x + 3)} > 0$$

$$\frac{(5)^2(5 - 4)}{(5 - 1)(5 + 3)} > 0$$

$$\frac{25\ (1)}{(4)(8)} > 0$$

$$\frac{25}{32} > 0$$

Since 25/32 is indeed greater than zero, the range
x > 4 is part of the solution.
　　Thus, the permissible ranges for which the inequality　　$x - 6 > \dfrac{18 - 15x}{x^2 + 2x - 3}$　　holds are

－ 3 < x < 0,　　　0 < x < 1,　　x > 4.

● **PROBLEM** 16-35

If　x　is a real quantity, prove that the expression　$\dfrac{x^2+2x-11}{2(x-3)}$

can have all numerical values except such as lie between　2　and　6.

$y \geq 6$

$y \leq 2$

Solution:　Let the given expression be represented by　y,　so that

$$\frac{x^2+2x-11}{2(x-3\ )} = y\ ;$$

then cross-multiplying and transposing, we have

$$x^2+2x-11 = 2y(x-3)$$
$$x^2+2x-11 = 2xy - 6y$$
$$x^2+2x-11-2xy+6y = 0$$
$$x^2+2x-2xy+6y-11 = 0$$
$$x^2+2x(1-y)+6y-11 = 0,$$

or

$$x^2+2(1-y)x + (6y-11) = 0 \qquad\qquad (1)$$

Equation (1) is in the form　$az^2 + bz + c = 0$,　which is a quadratic
equation. Hence, equation (1) is a quadratic equation, with　a = 1,
b = 2(1-y), and　c = 6y-11. In order that　x　may have real values,
the discriminant, b^2-4ac, must be positive; that is, in order that　x
may have real values, $[2(1-y)]^2 - 4(1)(6y-11)$　must be positive; or
$4(1-y)^2 - (24y-44)$ must be positive, i.e., $4(1-y)^2 - (24y-44) \geq 0$.
Dividing by　4　and simplifying:

$$\frac{4(1-y)^2 - (24y-44)}{4} \geq \frac{0}{4}$$

$$\frac{4(1-y)^2 - 24y+44}{4} \geq 0$$

$$(1-y)^2 - 6y + 11 \geq 0$$

$$\left(1 - 2y + y^2\right) - 6y + 11 \geq 0$$

$$y^2 - 8y + 12 \geq 0 .$$

Factoring the left side of the inequality into a product of two polynomials:

$$(y - 6)(y - 2) \geq 0 .$$

Hence, the factors of this product must both be positive or both negative, since the entire product is positive.

Case 1: Both factors positive:

$$y - 6 \geq 0 \quad \text{and} \quad y - 2 \geq 0$$
$$y \geq 6 \quad \text{and} \quad y \geq 2$$

The two inequalities, $y \geq 6$ and $y \geq 2$, mean $y \geq 6 \cap y \geq 2$, and thus yield the single inequality $y \geq 6$, since this single inequality satisfies the two inequalities.

Case 2: Both factors negative:

$$y - 6 \leq 0 \quad \text{and} \quad y - 2 \leq 0$$
$$y \leq 6 \quad \text{and} \quad y \leq 2$$

The two inequalities, $y \leq 6$ and $y \leq 2$, mean $y \leq 6 \cap y \leq 2$, and thus yield the single inequality $y \leq 2$, since this single inequality satisfies the two inequalities. Therefore, y may have real values only when $y \geq 6$ (Case 1) and $y \leq 2$ (Case 2). The real values of y are indicated on the accompanying number line. Therefore, y cannot lie between 2 and 6, but y may have any other value.

● **PROBLEM 16-36**

Find the solution set of $x^2 - 6x + 10 > 0$ by the graphical method.

<u>Solution:</u> First we graph the function $y = x^2 - 6x + 10$. Assign values to x and then calculate y-values.

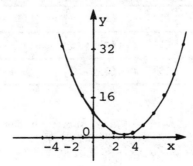

288

x	$x^2 - 6x + 10$	y
-3	$(-3)^2 - 6(-3) + 10$	37
-2	$(-2)^2 - 6(-2) + 10$	26
-1	$(-1)^2 - 6(-1) + 10$	17
0	$(0)^2 - 6(0) + 10$	10
1	$(1)^2 - 6(1) + 10$	5
2	$(2)^2 - 6(2) + 10$	2
3	$(3)^2 - 6(3) + 10$	1
4	$(4)^2 - 6(4) + 10$	2

See graph. The curve is the graph of $y = x^2 - 6x + 10$. Since the graph is entirely above the X axis, the solution set of $x^2 - 6x + 10 > 0$ is the set of all real numbers.

● **PROBLEM** 16-37

Find the solution set of $-x^2 - 4x - 5 > 0$.

<u>Solution:</u> To find the graphical solution, select values of x and find the corresponding y values. See the table. Note that $y = -x^2 - 4x - 5 = -[x^2 + 4x + 5]$.

x	$-[x^2 + 4x + 5] =$	y
-3	$-\left[(-3)^2 + 4(-3) + 5\right] = -(9 - 12 + 5) =$	-2
-2	$-\left[(-2)^2 + 4(-2) + 5\right] = -(4 - 8 + 5) =$	-1
-1	$-\left[(-1)^2 + 4(-1) + 5\right] = -(1 - 4 + 5) =$	-2
0	$-\left[(0)^2 + 4(0) + 5\right] = -(0 + 0 + 5) =$	-5
1	$-\left[(1)^2 + 4(1) + 5\right] = -(1 + 4 + 5) =$	-10
2	$-\left[(2)^2 + 4(2) + 5\right] = -(4 + 8 + 5) =$	-17
3	$-\left[(3)^2 + 4(3) + 5\right] = -(9 + 12 + 5) =$	-26

The graph of $y = -x^2 - 4x - 5$, as shown in the figure, lies entirely below the x-axis. Consequently

$$\left\{x \mid -x^2 - 4x - 5 > 0\right\} = \emptyset, \text{ the empty set.}$$

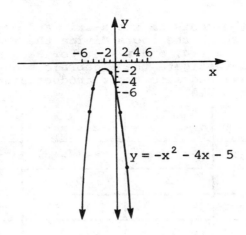

$$y = -x^2 - 4x - 5$$

Construct a graphical representation of the inequality $x^2 - 2x - 8 \leq 0$ and identify the solution set.

Solution: The graph of the relation $\{(x, y) \mid y = x^2 - 2x - 8\}$ is sketched in the figure. A table of values can be constructed.

x	$x^2 - 2x - 8$	y
- 3	$(- 3)^2 - 2(- 3) - 8$	7
- 2	$(- 2)^2 - 2(- 2) - 8$	0
- 1	$(- 1)^2 - 2(- 1) - 8$	- 5
0	$(0)^2 - 2(0) - 8$	- 8
1	$(1)^2 - 2(1) - 8$	- 9
2	$(2)^2 - 2(2) - 8$	- 8
3	$(3)^2 - 2(3) - 8$	- 5
4	$(4)^2 - 2(4) - 8$	0
5	$(5)^2 - 2(5) - 8$	7

We have to find the values of x for which $y \leq 0$ where $y = x^2 - 2x - 8$. First we consider the case $y = 0$. Factor y into $(x - 4)(x + 2)$. Set $y = 0$ and find the roots of this equation.

$x - 4 = 0 \qquad\qquad x + 2 = 0$

$x_1 = 4 \qquad\qquad x_2 = - 2$

Now, we must find where $x^2 - 2x - 8 < 0$.

We mark the roots on the x-axis and consider the regions into which the roots divide the x-axis. They are $x < -2$, $-2 < x < 4$, and $x > 4$. For each region choose an x value and investigate the algebraic signs of the factors of f(x) and also their product sign, f(x). See the following table.

Regions	$x < -2$	$-2 < x < 4$	$x > 4$
Factors of f(x)	$(x-4)(x+2)$	$(x-4)(x+2)$	$(x-4)(x+2)$
x-value	-3	0	5
Signs of factors	$(-)$ $(-)$	$(-)$ $(+)$	$(+)$ $(+)$
$\therefore$	$y > 0$	$y < 0$	$y > 0$

Thus, $y < 0$ for $-2 < x < 4$.

Furthermore, $y \leq 0$ when $-2 \leq x \leq 4$. See Graph.

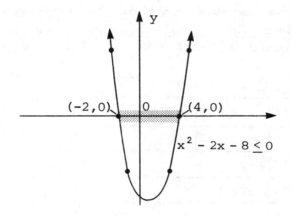

The darkened portion is the part which represents the inequality. The solution set is the interval $[-2, 4]$.

● **PROBLEM** 16-39

Use the graphical method to find the simultaneous solution set of

$$x^2 + x - 2 > 0$$

and $\dfrac{3}{4} x + \dfrac{3}{2} < 0$

<u>Solution:</u> We construct the graphs of

$$y = x^2 + x - 2 \quad \text{and} \quad y = \frac{3}{4} x + \frac{3}{2}$$

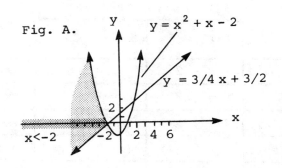

Fig. A.

$y = x^2 + x - 2$

$y = 3/4\, x + 3/2$

x<-2

$x < -2$ $-2 < x < 1$ $x > 1$

Fig. B

The shaded area represents the simultaneous solution set.

Set up tables for both equations to calculate y values.

For $y = x^2 + x - 2$

x	$x^2 + x - 2$ =	y
- 4	$(- 4)^2 + (- 4) - 2$	10
- 3	$(- 3)^2 + (- 3) - 2$	4
- 2	$(- 2)^2 + (- 2) - 2$	0
- 1	$(- 1)^2 + (- 1) - 2$	-2
0	$(0)^2 + (0) - 2$	-2
1	$(1)^2 + (1) - 2$	0
2	$(2)^2 + 2 - 2$	4
3	$(3)^2 + 3 - 2$	10

For $y = \frac{3}{4} x + \frac{3}{2}$

x	$\frac{3}{4} x + \frac{3}{2}$ =	y
-2	$\frac{3}{4}(- 2) + \frac{3}{2}$	0
0	$\frac{3}{4}(0) + \frac{3}{2}$	$\frac{3}{2}$
2	$\frac{3}{4}(2) + \frac{3}{2}$	3

See Figure A for graphs.

Now, find the region where the inequality $x^2 + x - 2 > 0$ holds. The function $f(x) = x^2 + x - 2$ can be factored into $(x + 2)(x - 1)$. Set $f(x) = 0$ and find the roots of this equation. Here $x = - 2$ and $x = 1$. Mark the roots on the x-axis and consider the regions into which the roots divide the x-axis. They are $x < - 2$, $- 2 < x < 1$, and $x > 1$. See Figure B.

For each of these regions, choose a value of x and investigate the algebraic signs of the factors of the function $f(x)$. Then look at the sign of their product, $f(x)$. The table summarizes the process.

	$f(x) = (x + 2)(x - 1)$		
Regions	$x < -2$	$-2 < x < 1$	$x > 1$
x-value	$x = -3$	$x = 0$	$x = 2$
Factors of f(x)	$(x+2)(x-1)$	$(x+2)(x-1)$	$(x+2)(x-1)$
Signs of factors	$(-)$ $(-)$	$(+)$ $(-)$	$(+)$ $(+)$
$\therefore$	$f(x) > 0$	$f(x) < 0$	$f(x) > 0$

For our problem we now know that the graph of $x^2 + x - 2$ is greater than zero, that is, above the x-axis, for $x < -2$ or $x > 1$.

Call the second function $g(x)$. Thus, $g(x) = \frac{3}{4}x + \frac{3}{2}$. We are interested in finding the values of x for which the function $g(x)$ is negative or when it is below the x-axis. Therefore,

$$\frac{3}{4} x + \frac{3}{2} < 0 \leftrightarrow \frac{3}{4} x < - \frac{3}{2} \leftrightarrow 3x < -6 \leftrightarrow x < -2$$

Hence, the solution set is $\{x \mid x < -2\}$. That is, the graph of $y = \frac{3}{4}x + \frac{3}{2}$ lies below the x-axis when $x < -2$.

Simultaneously we solve for x and we obtain $\{x \mid x < -2\}$. See Figure A.

● **PROBLEM 16-40**

Solve the inequality $(x + 2)(x - 1)(2x - 3) > 0$.

Fig. A.   Fig. B.

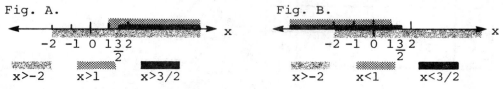

Solution: If we have a positive number which is a product of three factors, then all three factors are positive or two are negative and one is positive (since a negative multiplied by a negative is positive). The tentative possibilities are:

(1) $x+2 > 0$, and $x-1 > 0$, and $2x-3 > 0$, $\Rightarrow x > -2$, and $x > 1$, and $x > 3/2$

(2) $x+2 > 0$, $x-1 < 0$, $2x-3 < 0$, $\Rightarrow x > -2$, $x < 1$, $x < 3/2$

(3) $x+2 < 0$, $x-1 > 0$, $2x-3 < 0$, $\Rightarrow x < -2$, $x > 1$, $x < 3/2$

(4) $x+2 < 0$, $x-1 < 0$, $2x-3 > 0$, $\Rightarrow x < -2$, $x < 1$, $x > 3/2$

Then, if "and" means intersection ($\cap$), we must find,

$$(x > -2) \cap (x > 1) \cap (x > 3/2).$$

(See the number line in Figure A). Thus, the inequalities in (1) yield, as the range satisfying all three linear inequalities, $x > 3/2$.

293

For (2)
$$x > -2, \quad x < 1, \quad \text{and} \quad x < 3/2;$$

thus we must find,

$$(x > -2) \cap (x < 1) \cap (x < 3/2)$$

(See Figure B).
Thus, the inequalities in (2) yield, as the range satisfying all three linear inequalities, $-2 < x < 1$.
In Case (3), $x < -2$, $x > 1$, and $x < 3/2$, $x > 1$ is inconsistent with $x < -2$. Thus, there are no values on the number line common to all three inequalities.
For the last alternative, (4), $x < -2$, $x < 1$, and $x > 3/2$, the last inequality, $x > 3/2$, is inconsistent with $x < -2$, and $x < 1$. Thus, again the intersection of these three inequalities is the empty set.
Hence the complete solution consists of the ranges

$$-2 < x < 1 \quad \text{and} \quad x > 3/2 .$$

● **PROBLEM** 16-41

Determine the real values of x for which $\sqrt{x^3 - 3x^2 + 2x}$ is real.

Fig. A.

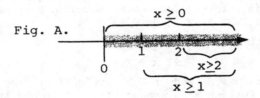

Fig. B

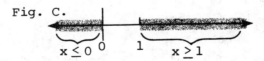

Fig. C.

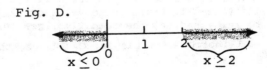

Fig. D.

Fig. E.

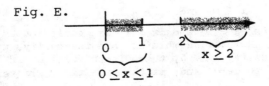

Solution: The given expression will be real for those real values of x yielding a radicand which is greater than or equal to zero. Thus, we have to solve the inequality

$$x^3 - 3x^2 + 2x \geqq 0 .$$

There is a common factor, namely x, of every term in the left side of this inequality. Hence, x is factored out from the left side:

$$x(x^2 - 3x + 2) \geqq 0$$

Factoring the expression in parenthesis into a product of two polynomials:

$$x(x - 2)(x - 1) \geqq 0 .$$

Only if all three factors are positive or two are negative and one positive will the product be positive. Thus, the following four cases result:

    Case 1: $x \geqq 0$, $x - 2 \geqq 0$, $x - 1 \geqq 0$.

    Case 2: $x \geqq 0$, $x - 2 \leqq 0$, $x - 1 \leqq 0$.

    Case 3: $x \leqq 0$, $x - 2 \leqq 0$, $x - 1 \geqq 0$.

    Case 4: $x \leqq 0$, $x - 2 \geqq 0$, $x - 1 \leqq 0$.

For Case 1: Solve the inequalities,

$$x \geqq 0, \; x \geqq 2, \; x \geqq 1.$$

These three inequalities are satisfied by the single inequality $x \geqq 2$, that is, the intersection of these three inequalities is the set $\{x \,|\, x \geqq 2\}$. This intersection can be noted in Diagram A. Hence, the solution set for Case 1 is $\{x \,|\, x \geqq 2\}$.

For Case 2: Solve the inequalities,

$$x \geqq 0, \; x \leqq 2, \; x \leqq 1.$$

The two inequalities $x \leqq 2$ and $x \leqq 1$ are satisfied by the single inequality $x \leqq 1$. Putting this inequality, $x \leqq 1$ and the remaining inequality $x \geqq 0$ together:

$$x \geqq 0 \quad \text{or} \quad 0 \leqq x, \; x \leqq 1; \text{ that is,}$$

$$0 \leqq x \leqq 1.$$

The set $\{x \,|\, 0 \leqq x \leqq 1\}$ is indicated in Diagram B. Hence, the solution set for Case 2 is $\{x \,|\, 0 \leqq x \leqq 1\}$.

For Case 3: Solving the inequalities,

$$x \leqq 0, \; x \leqq 2, \; x \geqq 1 .$$

The two inequalities $x \leqq 0$ and $x \leqq 2$ are satisfied by the single inequality $x \leqq 0$. However, combining this inequality, $x \leqq 0$, with the remaining inequality $x \geqq 1$, there is no value of x which is less than or equal to zero, and, at the same time, greater than or equal to 1. This is illustrated in Diagram C. Hence, there is no solution set for Case 3.

For Case 4: Solve the inequalities,

$$x \leqq 0, \; x \geqq 2, \; x \leqq 1.$$

The two inequalities $x \leqq 0$ and $x \leqq 1$ are satisfied by the single inequality $x \leqq 0$. However, combining this inequality, $x \leqq 0$, with the remaining inequality $x \geqq 2$, there is no value of x which is less than or equal to zero, and, at the same time, greater than or equal to 2. This is illustrated in Diagram D. Hence, there is no solution set for Case 4.

Therefore, the solution set consists of the two sets: $\{x|x \geqq 2\}$ and $\{x|0 \leqq x \leqq 1\}$; that is, the solution set is:

$$\{x|x \geqq 2\} \cup \{x|0 \leqq x \leqq 1\}$$

The solution set is pictured in Diagram E.

● **PROBLEM 16-42**

Show that $x^3 > y^3$ if $x > y$.

Solution: If we subtract y^3 from both sides of the inequality $x^3 > y^3$ we obtain:

$$x^3 - y^3 > 0$$

Recall the formula for the difference of two cubes:

$$a^3 - b^3 = (a - b)\left(a^2 + ab + b^2\right)$$

Thus, $x^3 - y^3 = (x - y)\left(x^2 + xy + y^2\right)$, by substitution.

If $x^3 - y^3 > 0$, then

$$(x - y)\left(x^2 + xy + y^2\right) > 0 \qquad\qquad (1)$$

Since we are given $x > y$, subtracting y from both sides of this inequality we obtain $x - y > 0$.

Now that we know $(x - y) > 0$ we can divide both sides of inequality (1) by $(x - y)$ without reversing the inequality:

$$\frac{(x - y)(x^2 + xy + y^2)}{(x - y)} > \frac{0}{(x - y)}$$

Cancelling $(x - y)$ in numerator and denominator we arrive at

$$x^2 + xy + y^2 > 0. \qquad\qquad (2)$$

Note that

$$x^2 + xy + y^2 = \left(x + \frac{1}{2} y\right)^2 + \frac{3}{4} y^2;$$

thus, we have written inequality (2) as the sum of two squares. The square of any number is positive, and if we add two positives we obtain a positive; hence:

$$x^2 + xy + y^2 = \left(x + \frac{1}{2} y\right)^2 + \frac{3}{4} y^2 > 0$$

Because inequality (2) is always valid, and the steps are all reversible, the given inequality, $x^3 > y^3$, has been proven.

CHAPTER 17

PROGRESSIONS AND SEQUENCES

> **Basic Attacks and Strategies for Solving Problems in this Chapter. See pages 297 to 330 for step-by-step solutions to problems.**

A sequence is a collection of items that have been given an ordering. For example, $x_1, x_2, x_3, \ldots$ is a sequence, and x_2 comes after x_1 but before x_3 in this sequence.

An *arithmetic progression* is a sequence of numbers in which the difference of any two successive terms is a constant, called the *common difference*. Thus, an arithmetic progression has the form

$$a_1 = a, \ a_2 = a + d, \ a_3 = a + 2d, \ a_4 = a + 3d, \ldots.$$

This can also be expressed as

$$a_1 = a, \ a_2 = a_1 + d, \ a_3 = a_2 + d, \ a_4 = a_3 + d, \ldots.$$

Arithmetic means are all the terms that fall between any two specified terms (called *extremes*) in an arithmetic progression.

Since an arithmetic progression is based on a constant difference, it is easy to compute a specific item in the sequence, given the item number, the initial term, and the constant difference. For example, if the first term is 2 and the constant difference is 5, then the tenth term (i.e., the ninth successor to 2) is

$$2 + 9 \cdot 5 = 2 + 45 = 47.$$

In general, if a is the first term, then the nth term is $a + (n - 1)d$, where d is the constant difference.

The sum of n consecutive terms in an arithmetic progression can be determined by the formula

$$S_n = {}^n/_2(a + l),$$

in which a is the first term and l is the last term. This can also be expressed as

$$S_n = {}^n/_2 [2a + (n - 1)d]$$

where d is the constant difference.

A *geometric progression* is a sequence of numbers in which the ratio of any two successive terms is a constant, called the *common ratio*. Thus, a geometric progression has the form

$$a_1 = a, a_2 = ar, a_3 = ar^2, a_4 = ar^3, \dots.$$

This can also be expressed as

$$a_1 = a, a_2 = a_1 \cdot r, a_3 = a_2 \cdot r, a_4 = a_3 \cdot r, \dots.$$

Geometric means are all the terms that fall between any two specified terms (called *extremes*) in a geometric progression.

The nth term of a geometric progression is $a \cdot r^{n-1}$ where r is the constant ratio.

The sum of n consecutive terms in a geometric progression can be determined by the formula

$$S_n = \frac{a(r^n - 1)}{r - 1},$$

in which a is the first term and r is the constant ratio.

The repeating decimals (mentioned in Chapter 1) can be seen as the sum of a geometric series. For example, $.\overline{3}$ can be rewritten as

$$\frac{3}{10} + \frac{3}{10^2} + \frac{3}{10^3} + \frac{3}{10^4} + \dots.$$

If the constant ratio of a geometric progression is less than one, the terms of the series decrease in magnitude. The sum of such a series is

$$S = \frac{a}{1 - r}.$$

A *harmonic progression* is a sequence of numbers whose reciprocals form an arithmetic progression.

Harmonic means are all the terms that fall between any two specified terms (called *extremes*) in a harmonic progression.

It is possible to generate sequences that cannot be classified as arithmetic, geometric, or harmonic. Nevertheless, by careful analysis, general algebraic expressions can often be obtained to describe an arbitrary term. For example, the following sequence, 1, 2, 6, 24, 120, 720, ... is a sequence based on the factorial function, $n!$, and does not fit into any other classification category.

● **PROBLEM 17-1**

If the 6th term of an arithmetic progression is 8 and the 11th term is - 2, what is the 1st term? What is the common difference?

<u>Solution:</u> An arithmetic progression is a sequence of numbers where each term excluding the first is obtained from the preceding one by adding a fixed quantity to it. This constant amount is called the common difference.

Let a = value of first term, and d = common difference

Term of
sequence: 1st 2nd 3^{thrd} 4th ... n^{th}

Value of
term: a   a+d   a+2d   a+3d ... a+(n-1)d

Use the formula for the nth term of the sequence to write equations for the given 6th and 11th terms, to determine a and d.

11th term: a + (11 - 1)d = - 2

6th term: a + (6 - 1)d = 8. Simplifying the above equations we obtain:

$$a + 10d = - 2 \qquad\qquad (1)$$
$$a + 5d = 8 \qquad\qquad (2)$$

$$5d = - 10 \text{ Subtracting (2) from (1)}$$
$$d = - 2 \text{ Substituting in (1)}$$
$$a + 10(- 2) = - 2$$
$$a = 18$$

The first term is 18 and the common difference is - 2.

Find the first term of an arithmetic progression if the fifth term is 29 and d is 3.

<u>Solution:</u> The nth term, or last term, of an arithmetic progression (A.P.) is:

$$\ell = a_1 + (n-1)d \qquad (1)$$

where a_1 = first term of the progression

d = common difference
n = number of terms
$\ell = n^{th}$ term, or last term.

Using this formula we can find the first term of an A.P. whose fifth term is 29 and d is 3. Since $\ell = a_5 = 29$, d = 3, and n = 5, substituting into equation (1) gives:

$$29 = a_1 + (5 - 1)3$$
$$29 = a_1 + 12$$
$$a_1 = 29 - 12 = 17.$$

Thus, the first term is 17.

Find the twelfth term of the arithmetic sequence

2, 5, 8, ...

<u>Solution:</u> It is given that the sequence is an arithmetic sequence. The common difference d is obtained by subtracting any term from the succeeding term;

$$d = 5 - 2 = 3$$

The twelfth term, a_{12}, can be obtained by substituting $a_1 = 2$, d = 3, and n = 12 in the expression for the nth term:

$$a_1 + (n - 1)d$$

Thus,

$$a_{12} = a_1 + (n - 1)d = 2 + (12 - 1)3 = 35$$

This can be checked by completing the sequence to the twelfth term.

2, 5, 8, 2+(4-1)3, 2+(5-1)3, 2+(6-1)3, ... , 2+(11-1)3, 35

2, 5, 8, 11, 14, 17, 20, 23, 26, 29, 32, 35

The 54th and 4th terms of an arithmetic progression are
- 61 and 64; find the 23rd term.

Solution: An arithmetic progression is a sequence of
numbers, each of which, after the first, is obtained by
adding a constant to the preceding term. This constant is
called the 'common difference'. Let a be the first term,
and d the common difference; then the sequence looks as
follows:

$$a, \ a + d, \ (a + d) + d, \ (a + d + d) + d, \ ... \qquad \text{or,}$$

$$a, \ a + d, \ a + 2d, \ a + 3d, \ ...$$

Notice that, 1st. term = $a = a + (1 - 1)d$

$$\text{2nd. term} = a + 1d = a + (2 - 1)d$$

$$\text{3rd. term} = a + 2d = a + (3 - 1)d$$

$$\text{4th. term} = a + 3d = a + (4 - 1)d$$

Thus, we obtain the formula for any term of an A.P.
(arithmetic progression). Let us call this general term
n. Then the nth. term $= a + (n - 1)d$. Therefore,

$$- 61 = \text{the 54th term} = a + 53d;$$

and $64 = \text{the } 4\text{th term} = a + 3d.$

We now solve these two equations for d by subtracting the second from the first. Thus,

$$a + 53d = - 61$$

$$\underline{a + \ \ 3d = 64}$$

$$a - a + 53d - 3d = - 61 - 64 \qquad \text{or}$$

$$50d = - 125, \text{ or } d = \frac{-5}{2}$$

Thus, substituting for d in equation two we have:

$$a + 3 \left(- \frac{5}{2}\right) = 64$$

$$a - \frac{15}{2} = 64$$

$$a = \frac{128}{2} + \frac{15}{2} = \frac{143}{2} \ .$$

Thus, the 23rd term $= a + 22d$, and substituting for a and
d we have:

$$\frac{143}{2} + 22 \left(- \frac{5}{2}\right) = \frac{143}{2} - \frac{110}{2} = \frac{33}{2} \ .$$

If the first term of an arithmetic progression is 7, and
the common difference is -2, find the fifteenth term and
the sum of the first fifteen terms.

Solution: An arithmetic progression is a sequence of
numbers each of which is obtained from the preceding one by
adding a constant quantity to it, the common difference, d.
If we designate the first term by a and the common difference by d, then the terms can be expressed as follows:

terms of
  series 1   2   3   .....   n

value of
  term   a   a + d   a + 2d $\ell = a + (n-1)d$

In this example $a = 7$, and $d = -2$. To find the fifteenth term, we have $n = 15$. The nth term is $a + (n-1)d$.
For $n = 15$, $a + (n-1)d = 7 + (15 - 1)(-2) = 7 - 28 = -21$.
To find the sum of the first fifteen terms apply the following formula:

$$S_n = \frac{n}{2}(a + \ell)$$

$$S_{15} = \frac{15}{2}[7 + (-21)] = \frac{15}{2}(-14) = -105.$$

Find the sum of the first sixteen terms of the arithmetic
series whose first term is $\frac{1}{4}$ and common difference is $\frac{1}{2}$.

Solution: The sum of the first n terms of an arithmetic
series is

$$S_n = \frac{n}{2}\left[2a_1 + (n - 1)d\right]$$

where a_1 = first term of the series

      d = common difference
      n = number of terms
      S_n = sum of first n terms

Hence, the sum of the first sixteen terms of the arithmetic
series with $a_1 = \frac{1}{4}$, $d = \frac{1}{2}$, and $n = 16$, is

$$S_{16} = \frac{16}{2}\left[2\left(\frac{1}{4}\right) + (16 - 1)\frac{1}{2}\right]$$

$$= 8\left(\frac{1}{2} + \frac{15}{2}\right)$$

$$= 8\left(\frac{16}{2}\right)$$

$$= 8(8)$$

$$= 64$$

Find the sum of the first 20 terms of the arithmetic progression -9, -3, 3, ...

<u>Solution:</u> An arithmetic progression is a sequence in which
each term after the first is formed by adding a fixed
amount, called the common difference, to the preceding term.
The common difference of -9, -3, 3, ... is 6 since $-9 + 6 = -3$,
$-3 + 6 = 3$, etc. If a is the first term, d is the common
difference, and n is the number of terms of the arithmetic
progression, then the last term (or n^{th} term) ℓ is given by

$$\ell = a + (n-1)d \qquad (1)$$

and the sum S_n of the n terms of this progression is given
by

$$S_n = \frac{n}{2}[a + \ell] \qquad (2)$$

In this example,

$$a = -9, \ d = 6, \ n = 20. \quad \text{By equation (1):}$$

$$\ell = -9 + (19)(6)$$

$$= -9 + 114$$

$$= 105.$$

By equation (2): $\quad S_{20} = \frac{20}{2}\left(-9 + 105\right)$

$$= 10(96)$$

$$= 960.$$

Thus, the sum of the first 20 terms is 960.

Find the sum of the arithmetic series

$$5 + 9 + 13 + \ldots + 401$$

<u>Solution:</u> The common difference is $d = 9 - 5 = 4$, and the
nth term, or last term, is $\ell = a + (n-1)d$, where
    a = first term of the progression
    d = common difference
    n = number of terms
    ℓ = nth term, or last term.

Hence, 401 = 5 + (n-1)4. Solving for the number of terms n, we have n = 100. The required sum is

$$S = 5 + 9 + 13 + \ldots + 393 + 397 + 401$$

Written in reverse order, this sum is

$$S = 401 + 397 + 393 + \ldots + 13 + 9 + 5$$

Adding the two expressions for S, we have

$$2S = (5 + 401) + (9 + 397) + (13 + 393) + \ldots$$
$$+ (393 + 13) + (397 + 9) + (401 + 5)$$

Each term in parentheses is equal to the sum of the first and last terms; 5 + 401 = 406. There is a parenthetic term corresponding to each term of the original series; that is, there are 100 terms. Hence,

$$2S = 100(5 + 401) = 40,600 \text{ and } S = \frac{40,600}{2} = 20,300$$

In general, the sum of the first n terms of an arithmetic series is:
$$S = \frac{n}{2}(a + \ell) = \frac{n}{2}[2a + (n-1)d]$$

For this problem,

$$S = \frac{100}{2}(5 + 401) = \frac{100}{2}[2(5) + (100-1)4] = 20,300$$

● **PROBLEM 17-9**

Find the sum of the first 100 positive integers.

Solution: The first 100 positive integers is an arithmetic progression (A.P.), because each number after the first is obtained by adding 1, called the common difference, to the preceding number. For an A.P., the sum of the first n terms is

$$S_n = \frac{n}{2}(a + \ell)$$

where a = first number of the progression
 n = number of terms
 ℓ = n^{th} term, or last term
 S_n = sum of first n terms.

Concerning the first 100 positive integers: there are 100 terms; hence n = 100. The first term is 1; hence, a = 1. The last term is 100; hence, ℓ = 100.

$$S_{100} = \frac{100(1 + 100)}{2} = 5050$$

Find the sum of the first 25 even integers.

Solution: The even integers form an arithmetic progression which is a sequence of numbers each of which is obtained from the preceding one by adding a constant quantity to it. This constant quantity is called the common difference, d. The first term of an arithmetic progression is a and the nth term is $\ell = a + (n-1)d$. In this case:

$$a = 2, \; n = 25, \; d = 2.$$

$$\ell = 2 + (25 - 1)2$$

$$= 50$$

To find the sum of the n terms of an arithmetic progression, we apply the formula

$$S_n = \frac{n}{2}(a + \ell).$$

$$S_{25} = \frac{25}{2}(2 + 50)$$

$$= 25(26)$$

$$= 650.$$

How many terms of the sequence $- 9, - 6, - 3, \ldots$ must be taken that the sum may be 66?

Solution: To solve this problem we apply the formula for the sum of the first n terms of an arithmetic progression. The formula states:

$$S_n = \frac{n}{2}[2a + (n - 1)d], \text{ where}$$

S_n = sum of the first n terms

n = number of terms
a = first term
d = common difference

We are given all of the above information except n. Therefore, by substituting for S_n, a, and d, we can solve for n. We are given that $S_n = 66$, a = $- 9$, and d = 3,

since $- 9 + 3 = - 6, \; - 6 + 3 = - 3, \ldots$

Hence, $\quad \frac{n}{2}[- 18 + (n - 1)3] = 66.$

Now, multiplying both sides of the equation by 2 we obtain: n [-18+ (n - 1)3] = 132; and simplifying the expression in brackets, we have: n (- 18 + 3n - 3) = 132, or n(3n - 21) = 132. Therefore, we have:

$3n^2 - 21n = 132$, and dividing each term by 3 we obtain:

$n^2 - 7n - 44 = 0$; factoring we have,

$(n - 11)(n + 4) = 0$;

therefore,   n = 11 or - 4.

We can reject the negative value because there cannot be a negative number of terms in the sequence, and therefore, 11 terms must be taken so that the sum of the terms is 66.

We can check this by taking 11 terms of the series. Doing this we have:

- 9, - 6, - 3, 0, 3, 6, 9, 12, 15, 18, 21;

the sum of which is 66.

● **PROBLEM 17-12**

How many terms of the sequence 26, 21, 16, ... must be taken to amount to 74?

<u>Solution:</u> To solve this problem we apply the formula for the sum of the first n terms of an arithmetic progression. The formula states:

$$S = \frac{n}{2} [2a + (n - 1)d] , \text{ where}$$

S = sum of the first n terms

n = number of terms

a = first term

d = common difference

We are given all of the above information except n. Therefore, by substituting for S, a, and d, we can solve for n. We are given that S = 74, a = 26, and d = - 5, since 26 - 5 = 21, 21 - 5 = 16, ...

Hence, $74 = \frac{n}{2} [2(26) + (n - 1)(- 5)]$, or

$\frac{n}{2} [52 + (n - 1)(- 5)] = 74$.

Simplifying, and multiplying both sides of the equation by 2, we obtain:

304

$$\frac{n}{2}(52 - 5n + 5) = 74$$

$$n(52 - 5n + 5) = 148$$

$$n(-5n + 57) = 148$$

$$-5n^2 + 57n = 148, \text{ or}$$

$$5n^2 - 57n + 148 = 0; \text{ factoring we obtain:}$$

$$(n - 4)(5n - 37) = 0;$$

therefore, $n = 4$ or $\frac{37}{5}$.

We can readily reject the value $\frac{37}{5}$, since it is not possible to have $7\frac{2}{5}$ terms. Thus, 4 terms of the sequence must be added to amount to 74.

We can check this by adding the four terms, 26, 21, 16, 11, and observing that the sum is indeed 74.

● **PROBLEM 17-13**

Insert five arithmetic means between 13 and 31.

Solution: In an arithmetic progression, the terms between any two other terms are called the arithmetic means between the two given terms. An arithmetic progression is a sequence of numbers where each is derived from the preceeding one by adding a constant quantity to it. The constant quantity is called the common difference. The first term of the A.P. is designed by a and the common difference by d. We express the terms of the series:

Term of the Series	1	2	3	4	...	n
Value of the Series	$a_1 = a$	$a_2 = a+d$	$a_3 = a_2+d$ $= (a+d)+d$ $= a+2d$	$a_4 = a_3+d$ $= a+3d$	...	$a_n =$ $a+(n-1)d$

We are concerned with seven terms here: the first term, five arithmetic means, and the last term. In order to find the arithmetic means, we need to find the common difference, d. (We know a, which is 13.) The seventh term, $31 = a + (n-1)d = 13(7-1)d = 13 + 6d$. Thus,

$$31 = 13 + 6d$$
$$18 = 6d$$
$$d = 3$$

Consequently the five arithmetic means are

$$a_2 = 13 + 3 = 16, \quad a_3 = 19, \quad a_4 = 22, \quad a_5 = 25, \quad a_6 = 28.$$

Insert 20 arithmetic means between 4 and 67.

Solution: 'Arithmetic means' are all the terms that fall between any two given terms in an arithmetic progression. The two given terms are called the extremes. Thus, in this example, including the extremes, the number of terms will be 22; so that we have to find a sequence of 22 terms in A.P., of which 4 is the first and 67 the last.

Let d be the common difference; then, since the general nth term of an A.P. = a + (n - 1)d, and 67 is the 22nd term, we have:

$$67 = a + 21d, \qquad a = \text{first term.}$$

Since the first term is 4 we obtain:

4 + 21d = 67. Solving for d we find:

$$21d = 63$$

$$d = 3.$$

Thus, the sequence is,

4, 4 + 3, (4 + 3) + 3, ...   or

4, 7, 10, 13, ..., 67

and the 20 required means are,

7, 10, 13, 16, 19, 22, 25, 28, 31, 34, 37, 40, 43, 46, 49,

52, 55, 58, 61, 64.

Determine the first four terms and 12th term of the arithmetic progression generated by F(x) = 2x + 3.

Solution: Find the terms of the progression by letting x = 1, 2, 3, ... etc.

1st term = F(1) = 2(1) + 3 = 5

2nd term = F(2) = 2(2) + 3 = 7

3rd term = F(3) = 2(3) + 3 = 9

4th term = F(4) = 2(4) + 3 = 11

12th term = F(12) = 2(12) + 3 = 27

The common difference, d, is found by subtracting one term from the one that immediately follows it.

The first term is denoted by a.

Note: For this progression a = 5 and d = 2. The coefficient of x in the linear function will always be the common difference for the arithmetic progression.

● **PROBLEM 17-16**

If an arithmetic progression is generated by the linear function F(x) = -3x + 14, what is the first term? What is the 15th term? What is the common difference?

<u>Solution:</u> 1st term = F(1) = - 3 + 14 = 11

15th term = F(15) = - 3(15) + 14 = - 31

common difference = d = - 3, the coefficient of the linear term.

The coefficient of x in a linear function will always be the common difference, d. To verify that d = - 3, find the second term and subtract the first term from it.

● **PROBLEM 17-17**

The sum of three numbers in arithmetic progression is 27, and the sum of their squares is 293; find them.

<u>Solution:</u> Let a be the middle number, d the common difference; then the three numbers are a - d, a, a + d.

Since the sum of the three numbers is 27 we have:

a - d + a + a + d = 27; or 3a = 27. Hence, a = 9, and the three numbers are 9 - d, 9, 9 + d.

Now, since the sum of the squares of the numbers is 293, we can use the following equation to solve for d:

$(9 - d)^2 + 9^2 + (9 + d)^2 = 293$.

Squaring, we obtain:

$$(81 - 18d + d^2) + (81) + (81 + 18d + d^2) = 293$$

$$2d^2 + 243 = 293$$

$$2d^2 = 50$$

$$d^2 = 25$$

$$d = \sqrt{25} = \pm 5.$$

Therefore, the three numbers are:

9 ± 5, 9, 9 ± 5 or

4, 9, 14.

● PROBLEM 17-18

If S_1, S_2, S_3, ... S_p are the sums of n terms of an arithmetic progression whose first terms are 1, 2, 3, 4, ... and whose common differences are 1, 3, 5, 7, ... respectively, find the value of

$$S_1 + S_2 + S_3 + ... S_p$$

<u>Solution:</u> We can find S_1, S_2, S_3, ..., S_p by applying the formula for the sum of the first n terms of an arithmetic progression. The formula states:

$$S = \frac{n}{2} [2a + (n - 1)d] , \text{ where}$$

S = the sum

n = the number of terms

a = the first term

d = the common difference

Thus, for S_1, a = 1 and d = 1, we have:

$$S_1 = \frac{n}{2} [2(1) + (n - 1)1] = \frac{n}{2} (2 + n - 1) =$$

$$\frac{n}{2}(n + 1) = \frac{n(n + 1)}{2} .$$

For S_2, a = 2 and d = 3; thus,

$$S_2 = \frac{n}{2} [2(2) + (n - 1)3] = \frac{n}{2} (4 + 3n - 3) =$$

$$\frac{n}{2} (3n + 1) = \frac{n(3n + 1)}{2} .$$

For S_3, a = 3 and d = 5; thus

$$S_3 = \frac{n}{2} [2(3) + (n - 1)5] = \frac{n}{2} (6 + 5n - 5) =$$

$$\frac{n}{2}(5n + 1) = \frac{n(5n + 1)}{2} .$$

Now, to find a and d for S_p we notice that a relation

308

exists between the sum and the first term, and the sum and the common difference. For S_1, the first term is 1, and the difference 1, or $2(1) - 1$. For S_2, the first term is 2, and the difference 3, or $2(2) - 1$. For S_3, the first term is 3, and the difference 5, or $2(3) - 1$. Similarly, for S_p, the first term is p, and the common difference is $(2p - 1)$. Thus,

$$S_p = \frac{n}{2} [2p + (n - 1)(2p - 1)]$$

$$= \frac{n}{2} [2p + (2pn - 2p - n + 1)] = \frac{n}{2} (2pn - n + 1).$$

Factoring n from the first two terms in the parentheses we have:

$$\frac{n}{2}[(2p - 1) n + 1].$$

Therefore, the required sum,

$S_1 + S_2 + S_3 + \ldots + S_p$ is:

$$\frac{n(n + 1)}{2} + \frac{n(3n + 1)}{2} + \frac{n(5n + 1)}{2} + \ldots + \frac{n[(2p - 1)n + 1]}{2}$$

Factoring $\frac{n}{2}$ from each term, we obtain:

$$\frac{n}{2}[(n + 1) + (3n + 1) + (5n + 1) + \ldots (\{2p - 1\}n + 1)]$$

Now, since we are adding 1, p times, we can write:

$$\frac{n}{2} [(n + 3n + 5n + \ldots + \{2p - 1\}n) + p].$$

Factoring n from the terms in the parentheses, we obtain:

$$\frac{n}{2} [n(1 + 3 + 5 + \ldots + \{2p - 1\}) + p].$$

Let us now examine the terms of the above series: $1 + 3 + 5 + \ldots + (2p - 1)$. Notice that we can apply the formula for the sum of the first n terms of an arithmetic progression, which states the following:

$$S = \frac{n}{2} [2a + (n - 1)d] = \frac{n}{2}(a + \ell).$$

In our case it is more efficient to use the form $S = \frac{n}{2}(a + \ell)$, where S = the sum

n = the number of terms

a = first term

$$\ell = \text{last term}$$

We know that $n = p$, $a = 1$, $\ell = (2p - 1)$. Thus,

$$S = \frac{p}{2}(1 + 2p - 1) \quad \text{or,} \quad S = \frac{p}{2}(2p) = p^2.$$

Therefore,

$$S_1 + S_2 + S_3 + \ldots + S_p = \frac{n}{2}[n(1 + 3 + 5 + \ldots$$

$$\ldots + \{2p - 1\}) + p] =$$

$$\frac{n}{2}\left[n(p^2) + p\right]. \quad \text{Factoring } p \text{ from both terms}$$

in the brackets we obtain: $\frac{n}{2}[p(np + 1)] =$

$$\frac{np}{2}(np + 1).$$

Note: It is of interest to observe that in the formula for the sum of n terms of an A.P.,

$$S = \frac{n}{2}[2a + (n - 1)d] = \frac{n}{2}(a + \ell),$$

we can easily derive the first formula,
$\frac{n}{2}[2a + (n - 1)d]$, from the second,

$\frac{n}{2}(a + \ell)$, as follows:

Since $n =$ the number of terms, and $\ell =$ last term, then we can use the fact that: $\ell = a + (n - 1)d$. Substituting this value for ℓ we obtain:

$$\frac{n}{2}(a + \ell) = \frac{n}{2}[a + (a + \{n - 1\}d)]$$

$$= \frac{n}{2}[2a + (n - 1)d],$$

which is precisely our first formula.

● **PROBLEM 17-19**

If the first term of a geometric progression is 9 and the common ratio is $-\frac{2}{3}$, find the first five terms.

Solution: A geometric progression (G.P.) is a sequence of numbers each of which, after the first, is obtained by multiplying the preceding number by a constant number called the common ratio, r. Thus a G.P. such as a_1, a_2,

a_3, a_4, a_5, ... or a_1, a_1r, a_2r, a_3r, a_4r, ... with $a_1 = 9$
and $r = -\frac{2}{3}$ is determined as follows:

$$a_1 = 9$$
$$a_2 = 9\left(-\frac{2}{3}\right) = -6$$
$$a_3 = (-6)\left(-\frac{2}{3}\right) = 4$$
$$a_4 = 4\left(-\frac{2}{3}\right) = -\frac{8}{3}$$
$$a_5 = \left(-\frac{8}{3}\right)\left(-\frac{2}{3}\right) = \frac{16}{9}$$

Thus, the first five terms are: $9, -6, 4, -\frac{8}{3}, \frac{16}{9}$

● **PROBLEM 17-20**

Find the next three terms of the geometric progression 1,2,4, 8,... .

Solution: First determine the common ratio of the progression by dividing any term by the term immediately preceeding it. Once the common ratio has been determined any term can be computed by multiplying the term immediately preceeding the unknown term by the common ratio. The common ratio is 2 so the 5th term is 8 X 2 = 16. The 6th term is 16 X 2 = 32. The 7th term is 32 X 2 = 64. The first seven terms of the geometric progression are:

$$1,2,4,8,16,32,64$$

● **PROBLEM 17-21**

Write the fourth, fifth, and sixth terms of the sequence with the general term

(1) $\dfrac{2n+1}{n!}$　　(2) $(-1)^{2n}\dfrac{x^{n-1}}{(2n)!}$

Solution: (1) $\dfrac{2n+1}{n!}$　where $n = 4,5,6$

$$= \frac{2(4)+1}{4!}, \quad \frac{2(5)+1}{5!}, \quad \frac{2(6)+1}{6!}$$

$$= \frac{9}{4\cdot3\cdot2\cdot1}, \quad \frac{11}{5\cdot4\cdot3\cdot2\cdot1}, \quad \frac{13}{6\cdot5\cdot4\cdot3\cdot2\cdot1}$$

$$= \frac{3}{8}, \quad \frac{11}{120}, \quad \frac{13}{720}$$

(2) $(-1)^{2n}\dfrac{x^{n-1}}{(2n)!}$　where $n = 4,5,6$

$$= (-1)^{2 \cdot 4} \frac{x^{4-1}}{(2(4))!} , \ (-1)^{2 \cdot 5} \frac{x^{5-1}}{(2(5))!} , \ (-1)^{2 \cdot 6} \frac{x^{6-1}}{(2(6))!}$$

$$= \frac{x^3}{8!} , \ \frac{x^4}{10!} , \ \frac{x^5}{12!}$$

● **PROBLEM 17-22**

The seventh term of a geometric progression is 192 and r = 2. Find the first four terms.

<u>Solution:</u> The formula for the nth term, or last term, of a geometric progression is:

$$\ell = s_1 r^{n-1}$$

where s_1 = first term of the progression

r = common ratio
n = number of terms
ℓ = nth term, or last term

Since we are given the seventh term and the common ratio of the progression we can use this information, with n = 7, to find the first term:

$$s_7 = s_1 r^{n-1}$$

$$192 = s_1 (2)^{7-1} = 2^6 s_1 = 64 s_1$$

$$s_1 = \frac{192}{64} = 3$$

Then, since a geometric progression is a sequence of numbers each of which, after the first, is obtained by multiplying the preceding number by a constant number called the common ratio,

$s_1 = 3$, $\ s_2 = 3 \cdot 2 = 6$, $\ s_3 = 6 \cdot 2 = 12$, and $s_4 = 12 \cdot 2 = 24$

● **PROBLEM 17-23**

If the 8th term of a geometric progression is 16 and the common ratio is -3, what is the 12th term?

<u>Solution:</u> A geometric progression is a sequence of numbers in which each term after the first is obtained by multiplying the preceding one by a fixed number which is called the common ratio. We express the first term of a geometric progression by a and the common ratio by r. We write the terms of the sequence in this manner:

Terms of
the progression   (1)   (2)   (3)   (4) ... (n)

312

Value of the
term   a   ar ar^2 ar^3 ... ar^{n-1}

The formula for the nth term is ar^{n-1}. We are given the 8th term which is 16 and the common ratio is -3. Then, in this case n = 8 and $ar^{n-1} = ar^{8-1} = ar^7$. Therefore, the 8th term = $16 = ar^7$

When n = 12, then the nth term is $ar^{n-1} = ar^{11}$. Therefore,
the 12th term = ar^{11}

Factor out ar^7 since we know its value.

$$ar^{11} = \left(ar^7 \right) r^4$$

Set ar^7 equal to its known value.

$$ar^{11} = 16r^4$$

Furthermore, r = -3. Then,

$$ar^{11} = 16(-3)^4 = 16 \cdot 81 = 1296$$

● **PROBLEM 17-24**

The first term of a geometric progression is 27, the nth term is 32/9, and the sum of n terms is 665/9. Find n and r.

Solution: A geometric progression (G.P.) is a sequence of numbers in which each number, after the first, is obtained by multiplying the preceding number by a constant number called the common ratio, r. The following two formulas for geometric progressions will be helpful in finding n, which is the number of terms, and r, which is the common ratio:

(1)   the nth term or last term = $\ell = ar^{n-1}$,

(2)   the sum of the first n terms = $S_n = \frac{a\left(r^n - 1\right)}{r-1}$ where a = first

term, r = common ratio, n = number of terms, ℓ = nth term, or last term, and S_n = sum of the first n terms. In this problem it is given that ℓ = 32/9, a = 27, and S_n = 665/9. Using the formula for the nth term or last term:

$$\frac{32}{9} = 27r^{n-1} \qquad\qquad (1)$$

Using the formula for the sum of the first n terms:

$$\frac{665}{9} = \frac{27\left(r^n - 1\right)}{r-1} = \frac{27r^n - 27}{r-1} \qquad (2)$$

Multiply both sides of equation (1) by r:

$$r\left(\frac{32}{9}\right) = r\left(27r^{n-1}\right)$$

313

$$\frac{32}{9} r = 27r \ r^{n-1}$$

$$\frac{32}{9} r = 27r^{1+n-1}$$

$$\frac{32}{9} r = 27r^{n}$$

Substituting $\frac{32}{9} r$ for $27r^{n}$ in equation (2),

$$\frac{665}{9} = \frac{32r/9 - 27}{r - 1}$$

$$9(r-1)\frac{665}{9} = 9(r-1)\left[\frac{32r/9 - 27}{r-1}\right]$$

$$(r-1)665 = 9[32r/9 - 27]$$

Distributing on the right side:

$$(r-1)665 = 32r - 243$$

Distributing on the left side:

$$665r - 665 = 32r - 243$$

Subtract $32r$ from both sides:

$$665r - 665 - 32r = 32r - 243 - 32r$$
$$633r - 665 = -243$$

Add 665 to both sides:

$$633r - 665 + 665 = -243 + 665$$
$$633r = 422$$

Divide both sides by 633:

$$\frac{663r}{663} = \frac{442}{633}$$

$$r = \frac{422}{633} = \frac{2(211)}{3(211)} = \frac{2}{3}$$

Hence, the common ratio = $r = \frac{2}{3}$.

Substituting $\frac{2}{3}$ for r in equation (1):

$$\frac{32}{9} = 27\left(\frac{2}{3}\right)^{n-1}$$

Multiply both sides by $\frac{1}{27}$:

$$\frac{1}{27}\left(\frac{32}{9}\right) = \frac{1}{27}\left[27\left(\frac{2}{3}\right)^{n-1}\right]$$

$$\frac{1}{27}\left(\frac{32}{9}\right) = \left(\frac{2}{3}\right)^{n-1}$$

$$\frac{32}{243} = \left(\frac{2}{3}\right)^{n-1} \qquad (3)$$

Express the fraction on the left side as a power of 2/3. Since $32 = 2^5$ and $243 = 3^5$, equation (3) becomes:

$$\frac{32}{243} = \frac{2^5}{3^5} = \left(\frac{2}{3}\right)^5 = \left(\frac{2}{3}\right)^{n-1} .$$

Hence, $5 = n - 1$.

Add 1 to both sides:

$$5 + 1 = n - 1 + 1$$
$$6 = n \ .$$

Hence, the number of terms = n = 6.

Find the sum of the first ten terms of the geometric progression: 15, 30, 60, 120, ...

Solution: A geometric progression is a sequence in which
each term after the first is formed by multiplying the
preceding term by a fixed number, called the common ratio.

If a is the first term, r is the common ratio, and n
is the number of terms, the geometric progression (G.P.)
is

$$a, \ ar, \ ar^2, \ ..., \ ar^{n-1}$$

The given G.P., 15, 30, 60, 120, ..., may be written as
15, 15(2), $15\left(2^2\right)$, $15\left(2^3\right)$... . The sum, S_n, of the first
n terms of the geometric progression is given by

$$S_n = \frac{a\left(1 - r^n\right)}{1 - r} \ , \ \text{where} \ a = \text{first term}$$
$$r = \text{common ratio}$$
$$n = \text{number of terms.}$$

Here a = 15, r = 2, and n = 10.

$$S_{10} = \frac{15\left(1 - 2^{10}\right)}{1 - 2}$$

$$= \frac{15(1 - 1024)}{-1}$$

$$= 15(1023)$$

$$= 15,345$$

Find the sum of the first four terms of the geometric

series 2 + $\left(-\frac{1}{3}\right)$ + $\frac{1}{18}$ + ...

Solution: The ratio of any number of a geometric series
to the number preceding it is constant. In this example,

the common ratio r = $\dfrac{\left(-\frac{1}{3}\right)}{2}$ = $-\frac{1}{6}$. Since the sum of the

first n terms of a geometric series is:

$$S_n = \frac{a_1\left(r^n - 1\right)}{r - 1}, \ r \neq 1$$

where a_1 = first term; r = common ratio; n = number of terms; S_n = sum of first n terms; then with $a_1 = 2$, $r = -\frac{1}{6}$, $n = 4$, the sum S_4 is

$$S_4 = \frac{2\left[\left(-\frac{1}{6}\right)^4 - 1\right]}{-\frac{1}{6} - 1}$$

$$S_4 = \frac{2\left(\frac{1}{1296} - 1\right)}{-\frac{7}{6}}$$

$$S_4 = \frac{185}{108}$$

Check: $2 + 2\left(-\frac{1}{6}\right) + 2\left(-\frac{1}{6}\right)^2 + 2\left(-\frac{1}{6}\right)^3$

$= 2 + \left(-\frac{1}{3}\right) + \left(\frac{1}{18}\right) - \left(\frac{1}{108}\right)$

$= 2 - \frac{1}{3} + \frac{1}{18} - \frac{1}{108} = \frac{185}{108}$.

● **PROBLEM** 17-27

The fourth term of a geometric progression is ½ and the sixth term is 1/8. Find the first term and the common ratio.

Solution: A geometric progression, a_1, a_2, a_3, ..., a_n, has terms with a common ratio r, so that the sequence can be expressed by

$$a_1, \ a_1 r, \ a_1 r^2, \ a_1 r^3, \ \ldots \ a_1 r^{n-1}$$

Observe that $a_2 = a_1 r$, $a_3 = a_1 r^2$, $a_4 = a_1 r^3$, ..., $a_n = a_1 r^{n-1}$. We are given that the fourth term is ½ and the sixth term is 1/8 so that

$$a_1 r^3 = \tfrac{1}{2}$$

$$a_1 r^5 = 1/8$$

Dividing the second equation by the first,

$$\frac{a_1 r^5}{a_1 r^3} = \frac{1/8}{\tfrac{1}{2}} \ .$$

Now, $a_1/a_1 = 1$, and $r^5/r^3 = r^{5-3}$; also, since division by a fraction is equivalent to multiplication by its reciprocal,

$$\frac{1/8}{\frac{1}{2}} = 1/8 \cdot 2/1$$

Therefore $r^2 = \frac{1}{4}$, and taking the square of both sides, $r = \pm \frac{1}{2}$.

Thus, there are two possible common ratios, $\frac{1}{2}$ and $-\frac{1}{2}$; and therefore there are two possible series that satisfy the given conditions:

For $r = \frac{1}{2}$, the first term, a_1, is given by:

$$a_1 = \frac{a_1 r^3}{r^3} = \frac{a_4}{r^3}.$$

Since a_4 = the fourth term = $\frac{1}{2}$, and $r = \frac{1}{2}$, the first term

$$= \frac{\frac{1}{2}}{(\frac{1}{2})^3} = 4.$$

For $r = -\frac{1}{2}$,

$$a_1 = \frac{a_4}{r^3} = \frac{\frac{1}{2}}{(-\frac{1}{2})^3} = -4.$$

● **PROBLEM** 17-28

Insert 4 geometric means between 160 and 5.

Solution: 'Geometric means' are the terms between any two given terms in a geometric progression. Thus, for this problem we have to find 6 terms in G.P. of which 160 is the first, and 5 the sixth.

We can apply the formula:

nth term = ar^{n-1}, where a = 1st term

r = common ratio

Thus, for the sixth term we have:

sixth term = $5 = 160r^{6-1}$

$$5 = 160r^5$$

Solving for r we obtain

$$r^5 = \frac{5}{160} = \frac{1}{32}$$

$$\sqrt[5]{r} = \sqrt[5]{\frac{1}{32}}$$

$$r = \frac{1}{2}$$

Now, since $r = \frac{1}{2}$ is the common ratio, we obtain each successive term of the progression by multiplication by $\frac{1}{2}$. Thus, we have:

160, 80, 40, 20, 10, 5, and the

four required means are 80, 40, 20, 10.

● **PROBLEM** 17-29

Find the first four terms of the geometric progression generated by the exponential function $f(x) = 12(3/2)^x$ if the domain of the function is the set of nonnegative integers $(0,1,2,3,...)$.

Solution: $f(0) = 12\left(\frac{3}{2}\right)^0 = 12(1) = 12$

$f(1) = 12\left(\frac{3}{2}\right)^1 = 18$

$f(2) = 12\left(\frac{3}{2}\right)^2 = 12\left(\frac{9}{4}\right) = 27$

$f(3) = 12\left(\frac{3}{2}\right)^3 = 12\left(\frac{27}{8}\right) = \frac{81}{2}$

The first four terms are 12, 18, 27, and $\frac{81}{2}$.

● **PROBLEM** 17-30

Find three numbers in geometric progression whose sum is 19, and whose product is 216.

Solution: The three numbers of the G.P. may be denoted by $\frac{a}{r}$, a, ar; then $\frac{a}{r} \times a \times ar = 216$. Carrying out the multiplication we obtain:

$$a^3 = 216$$

$$\sqrt[3]{a^3} = \sqrt[3]{216}$$

$$a = 6$$

Thus, substituting for a, we find that the numbers are $\frac{6}{r}$, 6, 6r.

Now, since the sum of the three numbers is 19, we have:

$\frac{6}{r} + 6 + 6r = 19$, and we wish to solve for r. To do this we take r as a common denominator. Thus,

$$\frac{6 + 6r + 6r^2}{r} = 19$$

$$6 + 6r + 6r^2 = 19r$$

$$6 + 6r + 6r^2 - 19r = 0$$

$$6r^2 - 13r + 6 = 0$$

Factoring, we have

$$(3r - 2)(2r - 3) = 0; \text{ hence}$$

$$r = \frac{3}{2} \text{ or } \frac{2}{3}.$$

Thus the numbers are 4, 6, 9.

● **PROBLEM 17-31**

Express $.4\overline{23}$ as a rational fraction.

Solution: We know that,

$$.4\overline{23} = .423232323\ldots$$

This is a repeating decimal in which 23 is the repeated portion of the decimal. This is indicated by the bar above the given decimal, that is, $.4\overline{23}$.

The decimal .4232323... can be rewritten as .4 + .023 + .00023 + .0000023 + ... We can easily see this by adding each term in column form. Thus, we have:

```
      .4
      .023
      .00023
   +  .0000023
      .4232323
```
; and this sum is the desired result.

Now, .4 + .023 + .00023 + .0000023 + ... can be rewritten as:

$$\frac{4}{10} + \frac{23}{1000} + \frac{23}{100000} + \frac{23}{10000000} + \ldots$$

$$= \frac{4}{10} + \frac{23}{10^3} + \frac{23}{10^5} + \frac{23}{10^7} + \ldots$$

319

Factoring $\frac{23}{10^3}$ from all terms except the first, we have:

$$.4\overline{23} = \frac{4}{10} + \frac{23}{10^3}\left[1 + \frac{1}{10^2} + \frac{1}{10^4} + \ldots\ldots\right].$$

Notice that the series $1 + \frac{1}{10^2} + \frac{1}{10^4} + \ldots$ has terms which are in a geometric progression where a = first term = 1, and r = common ratio = $\frac{1}{10^2}$.

Since r is less than 1 and the series is an infinite one, we can state that:

$$S = sum = \frac{a}{1 - r}$$

$$= \frac{1}{1 - \frac{1}{10^2}}$$

Thus, $.4\overline{23} = \frac{4}{10} + \frac{23}{10^3} \cdot \frac{1}{1 - \frac{1}{10^2}}$.

Now, we can simplify $\frac{1}{1 - \frac{1}{10^2}}$ as follows:

$$\frac{1}{1 - \frac{1}{10^2}} = \frac{1}{\left[1 - \frac{1}{100}\right]} = \frac{1}{\left[\frac{100}{100} - \frac{1}{100}\right]} = \frac{1}{\frac{99}{100}} = \frac{100}{99} .$$

Thus, substituting $\frac{100}{99}$ for $\frac{1}{1 - \frac{1}{10^2}}$, we have:

$$.4\overline{23} = \frac{4}{10} + \frac{23}{10^3} \cdot \frac{100}{99} = \frac{4}{10} + \left(\frac{23}{1000} \cdot \frac{100}{99}\right)$$

$$= \frac{4}{10} + \frac{23}{990} = \frac{396}{990} + \frac{23}{990}$$

$$= \frac{419}{990} .$$

Find the sum of the geometric series

$$30 + 10 + 3\tfrac{1}{3} + \dots + 30\left(\tfrac{1}{3}\right)^{n-1} + \dots$$

<u>Solution:</u> Rewriting the geometric series as

$$30 + 30\left(\tfrac{1}{3}\right) + 30\left(\tfrac{1}{3}\right)^2 + \dots + 30\left(\tfrac{1}{3}\right)^{n-1} + \dots,$$

it can be seen that the first term is $a_1 = 30$; the ratio is $r = \tfrac{1}{3}$. Hence, since the sum to infinity (S_∞) of any geometric progression in which the common ratio r is numerically less than 1 is given by

$$S_\infty = \frac{a_1}{1-r}, \text{ where } |r| < 1,$$

then $\quad S_\infty = \dfrac{30}{1 - \tfrac{1}{3}} = 45$

Note that the sum of the first n terms of this series differs from 45 by $\tfrac{1}{2}$ of the nth term. For example, when $n = 2$, $S_2 = 40$, $a_2 = 10$, and $\tfrac{1}{2}a_2 = 5$. Thus,

$$40 = 45 - 5$$

When $n = 3$, $S_3 = 43\tfrac{1}{3}$, $a_3 = 3\tfrac{1}{3}$, and $\tfrac{1}{2}a_3 = 1\tfrac{2}{3}$. Thus $43\tfrac{1}{3} =$
$= 45 - 1\tfrac{2}{3}$, etc.

The sum of an infinite number of terms in geometric progression is 15, and the sum of their squares is 45; find the sequence. Assume that the common ratio of the G.P. is less than 1.

<u>Solution:</u> The sum of any infinite geometric progression in which the common ratio is less than 1 is:

$$S_1 = \frac{a}{1 - r}$$

Now, squaring the terms of the sequence,

$$a, \ ar, \ ar^2, \ ar^3, \ \dots$$

we have:

$$a^2, \ (ar)^2, \ (ar^2)^2, \ (ar^3)^2, \ \dots \qquad =$$

$$a^2, \ a^2r^2, \ a^2r^4, \ a^2r^6, \ \ldots$$

This is a new infinite geometric progression with a^2 as the first term, and r^2 as the common ratio, and $r^2 < 1$. Therefore,

$$S_2 = \frac{a^2}{1 - r^2} \ .$$

We are given that the sum of the terms, or S_1, is 15, and the sum of the squares of the terms, or S_2, is 45. Thus,

$$\frac{a}{1 - r} = 15, \quad \text{and} \quad \frac{a^2}{1 - r^2} = 45$$

We must now solve for a and for r. This can be done in the following manner: Multiply both sides of the equation $\frac{a}{1 - r} = 15$ by $(1 - r)$. Thus, we obtain: $a = 15(1 - r)$.

Now, multiply both sides of the equation $\frac{a^2}{1 - r^2} = 45$ by $\left(1 - r^2\right)$. Thus, we obtain:

$$a^2 = 45\left(1 - r^2\right) \tag{1}$$

Squaring the equation $a = 15(1 - r)$ will give us a value for a^2 in terms of r. Substituting this value in equation (1) will give us an equation in r alone. Then, we can solve for r. Thus,

$$a^2 = \left(15 - 15r\right)^2, \text{ and expanding we have:}$$

$$a^2 = \left(15 - 15r\right)\left(15 - 15r\right)$$

$$= 225 - 225r - 225r + 225r^2$$

$$= 225r^2 - 450r + 225$$

Now, by substitution:

$$225r^2 - 450r + 225 = 45\left(1 - r^2\right)$$

$$225r^2 - 450r + 225 = 45 - 45r^2$$

Subtracing 45 from both sides of the equation, and adding $45r^2$ to both sides we obtain:

$$225r^2 + 45r^2 - 450r + 225 - 45 = 45 - 45 - 45r^2 + 45r^2$$

$$270r^2 - 450r + 180 = 0$$

Dividing each term by 90, we have:

$$\frac{270r^2}{90} - \frac{450r}{90} + \frac{180}{90} = \frac{0}{90}$$

$$3r^2 - 5r + 2 = 0$$

Factoring gives us:

$(3r - 2)(r - 1) = 0$, and this means that either $3r - 2 = 0$, or $r - 1 = 0$.

To solve the first equation for r we first add 2 to both sides of $3r - 2 = 0$, and then divide by 3. Thus,

$$3r - 2 + 2 = 0 + 2$$

$$3r = 2$$

$$\frac{3r}{3} = \frac{2}{3}$$

$$r = \frac{2}{3}$$

To solve the second equation we add 1 to both sides of $r - 1 = 0$. Thus,

$$r - 1 + 1 = 0 + 1$$

$$r = 1$$

Therefore, we have two values for r, $r = \frac{2}{3}$, $r = 1$. But notice that the second value, $r = 1$, can be rejected. This is so because if we substitute this into either of our original equations, $\frac{a}{1 - r} = 15$ or $\frac{a^2}{1 - r^2} = 45$, we obtain $\frac{a}{0}$, which is an undefined expression, and also because the formula for the sum, $S = \frac{a}{1 - r}$, holds only for progressions where the common ratio is less than 1.

Thus, we have $r = \frac{2}{3}$, and to solve for a we substitute this value into

$$a = 15(1 - r)$$

Thus, $a = 15\left(1 - \frac{2}{3}\right)$

$$= 15\left(\frac{3}{3} - \frac{2}{3}\right)$$

$$= 15\left(\frac{1}{3}\right) = \frac{15}{3} = 5$$

Therefore, we have a = first term of the sequence = 5, and r = common ratio = $\frac{2}{3}$. Thus the terms of the sequence are: 5, $5\left(\dfrac{2}{3}\right)$, $\left[5 \cdot \dfrac{2}{3}\right]\dfrac{2}{3}$, ... and the sequence is:

5, $\dfrac{10}{3}$, $\dfrac{20}{9}$, ...

● PROBLEM 17-34

Convert the repeating decimal .477477 ... to a fraction.

Solution: This is a geometric progression. We can compute the rational equivalent by first determining the common ratio and then using the formula for the nth partial sum of a geometric series. The common ratio is computed by dividing any term by the term immediately preceeding it. Therefore r, the common ratio, is:

$$\frac{.000477}{.477} = .001$$

Allow a_1 to be .477. Then the sum of the geometric progression, S_n, is: $a_1 + a_1 r + a_1 r^2 + \ldots + a_1 r^{n-1}$. Now we can compute S_n using the formula.

$$S_n = \frac{a_1 - a_1 r^n}{1 - r} = \frac{.477 - .477(.001)^n}{1 - r}$$

by taking the limit of S_n as $n \to \infty$. We then compute the rational expression to which this geometric progression converges.

$$\lim_{n \to \infty} S_n = \lim_{n \to \infty} \frac{.477 - .477(.001)^n}{1 - (.001)} = \frac{.477}{1 - (.001)}$$

because $(.001)^n = \left(\dfrac{1}{1000}\right)^n$ goes to zero as n goes to ∞. Then

$$\frac{.477}{1 - (.001)} = \frac{477}{999} = \frac{53}{111}$$

$\dfrac{53}{111}$ is the fractional equivalent of the repeating decimal .477477 ...

● PROBLEM 17-35

Find the 9th term of the harmonic progression $3, 2, \dfrac{3}{2}, \ldots$.

Solution: The terms of a harmonic progression that lie between two given terms are called the harmonic means between these terms. If a single harmonic mean is inserted between two numbers, it is called the harmonic mean of the numbers.

A harmonic progression (H.P.) is a sequence of numbers whose reciprocals are in arithmetic progression, (A.P.). The terms of the A.P. are $\dfrac{1}{3}, \dfrac{1}{2}, \dfrac{2}{3}, \ldots$. We find the ninth term of the A.P. and take its reciprocal to find the corresponding term in the H.P. The formula for the nth term, a_n, of an A.P. is $a_1 + (n-1)d$ where a_1 is the first term and d is the common difference, the constant quantity added to each term to form the progression. Hence, if $a_n = a_1 + (n-1)d$, $a_1 = 1/3$ and n = 9, to find d subtract the first term, 1/3, from the second term, 1/2.

$$d = \frac{1}{2} - \frac{1}{3} = \frac{3}{6} - \frac{2}{6} = \frac{1}{6}.$$

Thus,

$$a_9 = \frac{1}{3} + (9-1)\frac{1}{6} = \frac{1}{3} + 8\left(\frac{1}{6}\right) = \frac{1}{3} + \frac{4}{3} = \frac{5}{3}.$$

Therefore, the ninth term of the harmonic progression is $\frac{3}{5}$.

● **PROBLEM 17-36**

Insert 40 harmonic means between 7 and $\frac{1}{6}$.

Solution: Recall that 'means' are the terms between any two given terms of a progression. Thus, we wish to find a harmonic progression with 7 as the first term, and $\frac{1}{6}$ as the 42nd term. But a harmonic progression is a sequence of numbers whose reciprocals form an arithmetic progression (A.P.). Thus, 6 is the 42nd term of an A.P. whose first term is $\frac{1}{7}$; let d be the common difference; then, since the nth term, or last term, of an A.P. =

a + (n − 1)d, where   a = first term
n = last term
d = common difference,

we have:

$$6 = \frac{1}{7} + (42 - 1)d \qquad \text{or,}$$

$$6 = \frac{1}{7} + 41d. \quad \text{To solve for d subtract } \frac{1}{7} \text{ from}$$

both sides of the equation, and then multiply both sides by $\frac{1}{41}$. Thus,

$$6 - \frac{1}{7} = 41d$$

$$\frac{42}{7} - \frac{1}{7} = 41d$$

$$\frac{41}{7} = 41d$$

$$\frac{41}{7} \cdot \frac{1}{41} = \frac{1}{7} = d$$

Thus, the arithmetic progression is:

$$\frac{1}{7}, \frac{1}{7} + \frac{1}{7}, \quad \left(\frac{1}{7} + \frac{1}{7}\right) + \frac{1}{7}, \quad \cdots \qquad \qquad =$$

$$\frac{1}{7}, \frac{2}{7}, \frac{3}{7} \cdots \cdots, \frac{41}{7}, 6$$

Therefore, the harmonic progression is:

$$7, \frac{7}{2}, \frac{7}{3}, \cdots, \frac{7}{41}, \frac{1}{6}$$

and the 40 harmonic means between 7 and $\frac{1}{6}$ are:

$$\frac{7}{2}, \frac{7}{3}, \cdots \frac{7}{41} .$$

● **PROBLEM 17-37**

If a^2, b^2, c^2 are in arithmetic progression, show that
b + c, c + a, a + b are in harmonic progression.

<u>Solution</u>: We are given that a^2, b^2, c^2 are in arithmetic
progression. By this we mean that each new term is
obtained by adding a constant to the preceding term.

By adding (ab + ac + bc) to each term, we see
that,

a^2 + (ab + ac + bc), b^2 + (ab + ac + bc),

c^2 + (ab + ac + bc)

are also in arithmetic progression. These three terms
can be rewritten as

a^2 + ab + ac + bc, b^2 + bc + ab + ac,

c^2 + ac + bc + ab

Notice that:

a^2 + ab + ac + bc = (a + b)(a + c)

b^2 + bc + ab + ac = (b + c)(b + a)

c^2 + ac + bc + ab = (c + a)(c + b)

Therefore, the three terms can be rewritten as:

(a + b)(a + c), (b + c)(b + a), (c + a)(c + b),

which are also in arithmetic progression.

Now, dividing each term by (a + b)(b + c)(c + a),
we obtain:

$$\frac{1}{b + c}, \frac{1}{c + a}, \frac{1}{a + b}, \text{ which are in}$$

326

arithmetic progression.

Recall that a sequence of numbers whose reciprocals form an arithmetic progression, is called a harmonic progression. Thus, since $\frac{1}{b + c}$, $\frac{1}{c + a}$, $\frac{1}{a + b}$ is an arithmetic progression, $b + c$, $c + a$, $a + b$ are in harmonic progression.

● **PROBLEM 17-38**

Find the first six terms of the sequence determined by the function $g(x)$, where $x = 1, 2, 3, 4, 5, 6$.

$$g(x) = \frac{x^2}{x!}, \text{ x a positive integer}$$

<u>Solution:</u>

$$g(1) = \frac{1}{1!} = 1$$

$$g(2) = \frac{4}{2!} = \frac{4}{2} = 2$$

$$g(3) = \frac{9}{3!} = \frac{9}{6} = \frac{3}{2}$$

$$g(4) = \frac{16}{4!} = \frac{16}{24} = \frac{2}{3}$$

$$g(5) = \frac{25}{5!} = \frac{25}{120} = \frac{5}{24}$$

$$g(6) = \frac{36}{6!} = \frac{36}{720} = \frac{1}{20}$$

If we consider simplifying the function $g(x) = \frac{x^2}{x!}$ before we evaluate, we would write

$$g(x) = \frac{x^2}{(1)(2)\ldots(x - 1)x}$$

Upon dividing the numerator and denominator by x, we would obtain

$$g(x) = \frac{x}{1(2)\ldots(x - 1)} = \frac{x}{(x - 1)!} .$$

If we use this form of $g(x)$ and find $g(2)$, $g(3)$, $g(4)$, $g(5)$, $g(6)$, we obtain the same results as we did before with a little less effort. However, when we try to evaluate $g(1)$, we encounter a denominator of $0!$. Since we have already determined $g(1)$ to be 1 by using the original form of $g(x)$, $\frac{1}{0!}$ should be equal to one. The symbol $0!$ arises rather frequently and in all cases we find our results are consistent if we define it to be one. That is, $0! = 1$.

With this extension to the definition of factorial, we have:

$0! = 1$

$1! = 1$

$2! = (1)(2)$

$x! = (1)(2)...(x)$ x an integer

Applying: $g(x) = \dfrac{x}{(x-1)!}$

$g(1) = \dfrac{1}{0!} = 1$

$g(2) = \dfrac{2}{1!} = \dfrac{2}{1} = 2$

$g(3) = \dfrac{3}{2!} = \dfrac{3}{2}$

$g(4) = \dfrac{4}{3!} = \dfrac{2}{3}$

$g(5) = \dfrac{5}{4!} = \dfrac{5}{24}$

$g(6) = \dfrac{6}{5!} = \dfrac{\cancel{6}}{5 \cdot 4 \cdot \cancel{3} \cdot \cancel{2} \cdot 1} = \dfrac{1}{20}$.

● **PROBLEM 17-39**

Determine the general term of the sequence:

$$\frac{1}{2}, \quad \frac{1}{12}, \quad \frac{1}{30}, \quad \frac{1}{56}, \quad \frac{1}{90}, \quad \cdots$$

<u>Solution</u>: To determine the general term, it is necessary to find how the adjacent terms differ. In this example, it is sufficient to consider the denominator because the numerator is the same for all the terms. The difference between the first two terms is 10. For the second and third terms, the difference is 18. By continuing this process, the results are tabulated as:

10, 18, 26, 34, ...

Note that each difference is larger by 8 than for the preceding term.

Now we try to write an expression that generates the series. By inspection, each term is the product of 2 successive integers, for example:

$$\frac{1}{2} = \frac{1}{1} \cdot \frac{1}{2}, \qquad \frac{1}{12} = \frac{1}{3} \cdot \frac{1}{4},$$

$$\frac{1}{30} = \frac{1}{5} \cdot \frac{1}{6}, \qquad \frac{1}{56} = \frac{1}{7} \cdot \frac{1}{8}$$

This fact can be expressed as

$$\frac{1}{(2n - 1)(2n)} \quad ,$$

and this is the desired answer.

● **PROBLEM 17-40**

Write the general (nth) term for each of the following sequences:

   (a)   1,5,9,13....

   (b)   5,9,13,17....

   (c)   1,-1,2/3,-1/3....

   (d) $5,6,\frac{7}{2},\frac{4}{3}....$

<u>Solution</u>:   (a) $4n - 3$

                (b) $4n + 1$

                (c) $\dfrac{(-2)^{n-1}}{n!}$

                (d) $\dfrac{n+4}{(n-1)!}$

● **PROBLEM 17-41**

     $K \geq 1$

Find the value of

   (1) $\displaystyle\lim_{n\to\infty} \left(\frac{4}{n^3} + \frac{5}{n} \right)$

   (2) $\displaystyle\lim_{n\to\infty} \frac{n^2}{(n+2)^2}$

   (3) $\displaystyle\lim_{n\to\infty} 2n^k$ for $K \geq 1$

Solution:

(1) $\lim\limits_{n\to\infty} \left(\dfrac{4}{n^3} + \dfrac{5}{n}\right) = \lim\limits_{n\to\infty} \dfrac{4}{n^3} + \lim\limits_{n\to\infty} \dfrac{5}{n}$

$= 0 + 0 = 0$

(2) $\lim\limits_{n\to\infty} \dfrac{n^2}{(n+2)^2} = \lim\limits_{n\to\infty} \dfrac{n}{(n+2)} = \lim\limits_{n\to\infty} \dfrac{\frac{n}{n}}{\frac{n}{n} + \frac{2}{n}}$

$= \lim\limits_{n\to\infty} \dfrac{1}{1 + \frac{2}{n}} = \dfrac{1}{1+0} = 1$

(3) $\lim\limits_{n\to\infty} 2n^k = \lim\limits_{n\to\infty} 2n = \infty$

● **PROBLEM 17-42**

Determine the general term of the sequence:

$$\dfrac{1}{5^3}, \quad \dfrac{3}{5^5}, \quad \dfrac{5}{5^7}, \quad \dfrac{7}{5^9}, \quad \dfrac{9}{5^{11}}, \quad \ldots$$

Solution: The numerators of the terms in the series are consecutive odd numbers beginning with 1. An odd number can be represented by $2n - 1$.

In the denominators, the base is always 5, and the power is a consecutive odd integer beginning with 3.

The general term can therefore be expressed by $\dfrac{2n - 1}{5^{2n+1}}$,

and the series is generated by replacing n with n = 1, 2, 3, 4, ...

CHAPTER 18

MATHEMATICAL INDUCTION

> **Basic Attacks and Strategies for Solving Problems in this Chapter. See pages 331 to 339 for step-by-step solutions to problems.**

Mathematical induction is a method of proof frequently used to prove general formulas, such as a formula for the sum of a sequence of n numbers. This method consists of three major steps:

1. Verify that the proposed formula is true for an initial (small) value of n (e.g., $n = 1$).

2. While assuming that the proposed formula is true for a specific value of n (e.g., for $n = k$), prove that the formula is also true for the next value of n (e.g., for $n = k + 1$).

3. Conclude that (because of mathematical induction) the formula in fact does hold for all values of n.

We can see that this procedure does, in fact, prove a formula for all possible values of n by applying the second step to the first, and repeatedly applying the second step to subsequent values of n. If a formula holds for $n = 1$, the second step says it also holds for $n = 2$, and if it holds for $n = 2$, the second step says it also holds for $n = 3$, etc.

The major problem in using mathematical induction often occurs in proving the second step. One first assumes that some formula holds for $n = k$. One then adds the *same* additional term to both sides of an equation. This term makes one side obviously one term longer by the same type of term already existing on that side of the equation. However, this demands algebraic manipulation on the other side to transform it into a formula in $k + 1$ rather than merely k.

For example, suppose that

$$2 + 4 + 6 + 8 + \ldots + 2n = n(n + 1)$$

is to be proved by mathematical induction.

1. We can easily verify that the formula holds for $n = 1$, since $2 \cdot 1 = 2$

and $1 \cdot (1 + 1) = 1 \cdot 2 = 2.$

2. We assume that the formula is true for $n = k$, i.e., we assume that

    $2 + 4 + 6 + 8 + \ldots + 2k = k(k + 1).$

    We now add to both sides of the equation the quantity $2(k + 1)$ (since it is always permitted to add an equal quantity to both sides of an equation). The right side needs to be transformed into a product of a number times its successor.

    $k(k + 1) + 2(k + 1) = (k + 2)\,(k + 1) = (k + 1)\,(k + 2)$

    $= (k + 1)[(k + 1) + 1].$

    This last expression is merely the right side of the original expression with $k + 1$ in place of n. Thus, we have shown that the original formula is true for $n = k + 1$ (assuming that it is true for $n = k$).

3. Based on the fact that the original expression was true for $n = 1$ and that if it is true for $n = k$ it is also true for $n = k + 1$, we can now assert by mathematical induction that it is true for all integer values of n.

● **PROBLEM 18-1**

Prove by mathematical induction
$$1^2 + 2^2 + 3^2 + \ldots + n^2 = \frac{1}{6}n(n+1)(2n+1).$$

Solution: Mathematical induction is a method of proof. The steps are:
(1) The verification of the proposed formula or theorem for the smallest value of n. It is desirable, but not necessary, to verify it for several values of n.
(2) The proof that if the proposed formula or theorem is true for n = k, some positive integer, it is true also for n = k+1. That is, if the proposition is true for any particular value of n, it must be true for the next larger value of n.
(3) A conclusion that the proposed formula holds true for all values of n.

Proof: Step 1. Verify:

For n = 1: $1^2 = \frac{1}{6}(1)(1+1)[2(1)+1] = \frac{1}{6}(1)(2)(3) = \frac{1}{6}(6) = 1$

$1 = 1$ ✓

For n = 2: $1^2 + 2^2 = \frac{1}{6}(2)(2+1)[2(2)+1] = \frac{1}{6}(2)(3)(5) = \frac{1}{6}(6)(5)$

$1 + 4 = (1)(5)$

$5 = 5$ ✓

For n = 3: $1^2 + 2^2 + 3^2 = \frac{1}{6}(3)(3+1)[2(3)+1]$

$1 + 4 + 9 = \frac{1}{6}(3)(4)(7) = \frac{1}{6}(12)(7) = 14$

$14 = 14$ ✓

Step 2. Let k represent any particular value of n. For n = k, the formula becomes
$$1^2 + 2^2 + 3^2 + \ldots + k^2 = \frac{1}{6}k(k+1)(2k+1). \tag{A}$$

For n = k+1, the formula is
$$1^2 + 2^2 + 3^2 + \ldots + k^2 + (k+1)^2 = \frac{1}{6}(k+1)[(k+1) + 1][2(k+1) + 1]$$

$$= \frac{1}{6}(k+1)(k+2)(2k+3). \qquad \qquad (B)$$

We must show that if the formula is true for $n = k$, then it must be true for $n = k+1$. In other words, we must show that (B) follows from (A). The left side of (A) can be converted into the left side of (B) by merely adding $(k+1)^2$. All that remains to be demonstrated is that when $(k+1)^2$ is added to the right side of (A), the result is the right side of (B).

$$1^2 + 2^2 + \ldots + k^2 + (k+1)^2 = \frac{1}{6}k(k+1)(2k+1) + (k+1)^2$$

Factor out $(k+1)$:

$$1^2 + 2^2 + 3^2 + \ldots + k^2 + (k+1)^2 = (k+1)\left[\frac{1}{6}k(2k+1) + (k+1)\right]$$

$$= (k+1)\left[\frac{k(2k+1)}{6} + \frac{(k+1)6}{6}\right]$$

$$= (k+1)\frac{2k^2 + k + 6k + 6}{6}$$

$$= \frac{(k+1)(2k^2 + 7k + 6)}{6}$$

$$= \frac{1}{6}(k+1)(k+2)(2k+3),$$

since $\qquad \qquad 2k^2 + 7k + 6 = (k+2)(2k+3).$

Thus, we have shown that if we add $(k+1)^2$ to both sides of the equation for $n = k$, then we obtain the equation or formula for $n = k+1$. We have thus established that if (A) is true, then (B) must be true; that is, if the formula is true for $n = k$, then it must be true for $n = k+1$. In other words, we have proved that if the proposition is true for a certain positive integer k, then it is also true for the next greater integer $k+1$.

Step 3. The proposition is true for $n = 1,2,3$ (Step 1). Since it is true for $n = 3$, it is true for $n = 4$ (Step 2, where $k = 3$ and $k+1 = 4$). Since it is true for $n = 4$, it is true for $n = 5$, and so on, for all positive integers n.

● **PROBLEM 18-2**

Prove:

$$1\cdot 2 + 2\cdot 3 + 3\cdot 4 + \ldots + n(n+1) = \frac{n(n+1)(n+2)}{3}.$$

Solution by mathematical induction: The steps for a proof by mathematical induction are:

    I)   check validity of formula for $n = 1$

    II)   assume the formulation is true for $n = p$

    III)   prove it is true for $n = p + 1$.

I. For $n = 1$ the formula gives

$$1(1 + 1) = 1(2) = 2 = \frac{1(1+1)(1+2)}{3} = \frac{1\cdot 2\cdot 3}{3} = 2$$

which is correct and completes Step I.

II) Assume the formula is true for $n = p$:

$$1 \cdot 2 + 2 \cdot 3 + 3 \cdot 4 + \ldots + p(p+1) = \frac{p(p+1)(p+2)}{3}.$$

Prove the formula is true for $n = p + 1$, that is, prove

$$1 \cdot 2 + 2 \cdot 3 + \ldots + p(p+1) + (p+1)(p+2) = \frac{(p+1)(p+2)(p+3)}{3},$$

$(p+1)(p+2)$ is added to both members of the first equation in this step; this gives

$$[1 \cdot 2 + 2 \cdot 3 + \ldots + p(p+1)] + (p+1)(p+2) =$$

$$= \left[\frac{p(p+1)(p+2)}{3}\right] + (p+1)(p+2)$$

Factoring out $(p+1)(p+2)$,
$$= (p+1)(p+2)\left(\frac{p}{3} + 1\right)$$

$$= (p+1)(p+2)\left(\frac{p}{3} + 1\right)\left(\frac{3}{3}\right)$$

$$= \frac{(p+1)(p+2)(p+3)}{3},$$

$$= \frac{(p+1)[(p+1)+1][(p+2)+1]}{3}$$

which is of the same form as the result we assumed to be true for p terms, $p+1$ taking the place of p. Since the statement is true for $n = 1$ and $n = p + 1$ assuming it was true for $n = p$, then the statement is true for all n.

● **PROBLEM 18-3**

Prove by mathematical induction that, for all positive integral values of n,
$$1 + 2 + 3 + \ldots + n = \frac{n(n+1)}{2}.$$

<u>Solution:</u> Step 1. The formula is true for $n = 1$, since $1 = \frac{1(1+1)}{2} = 1$.

Step 2. Assume that the formula is true for $n = k$. Then, adding $(k+1)$ to both sides,

$$1 + 2 + 3 + \ldots + k + (k+1) = \frac{k(k+1)}{2} + (k+1) = \frac{(k+1)(k+2)}{2}$$

which is the value of $\frac{n(n+1)}{2}$ when $(k+1)$ is substituted for n.

Hence if the formula is true for $n = k$, we have proved it to be true for $n = k + 1$. But the formula holds for $n = 1$; hence it holds for $n = 1 + 1 = 2$. Then, since it holds for $n = 2$, it holds for $n = 2 + 1 = 3$, and so on. Thus the formula is true for all positive integral values of n.

Using mathematical induction, prove that
$$x^{2n} - y^{2n} \text{ is divisible by } x + y.$$

Solution:

(1) The theorem is true for $n = 1$, since $x^2 - y^2 = (x-y)(x+y)$ is divisible by $x + y$.

(2) Let us assume the theorem true for $n = k$, a positive integer; that is, let us assume

(A) $\qquad x^{2k} - y^{2k}$ is divisible by $x + y$.

We wish to show that, when (A) is true,

(B) $\qquad x^{2k+2} - y^{2k+2}$ is divisible by $x + y$.

Now $x^{2k+2} - y^{2k+2} = \left(x^{2k+2} - x^2 y^{2k}\right) + \left(x^2 y^{2k} - y^{2k+2}\right)$

$$= x^2\left(x^{2k} - y^{2k}\right) + y^{2k}\left(x^2 - y^2\right).$$

In the first term $\left(x^{2k} - y^{2k}\right)$ is divisible by $(x+y)$ by assumption, and in the second term $\left(x^2 - y^2\right)$ is divisible by $(x+y)$ by Step (1); hence, if the theorem is true for $n = k$, a positive integer, it is true for the next one $n = k + 1$.

(3) Since the theorem is true for $n = k = 1$, it is true for $n = k + 1 = 2$; being true for $n = k = 2$, it is true for $n = k + 1 = 3$; and so on, for every positive integral value of n.

Prove by mathematical induction that
$$1 + 7 + 13 + \ldots + (6n - 5) = n(3n - 2).$$

Solution: (1) The proposed formula is true for $n = 1$, since $1 = 1(3 - 2)$.

(2) Assume the formula to be true for $n = k$, a positive integer; that is, assume

(A) $\qquad 1 + 7 + 13 + \ldots + (6k - 5) = k(3k - 2)$.

Under this assumption we wish to show that

(B) $\qquad 1 + 7 + 13 + \ldots + (6k - 5) + (6k + 1) = (k+1)(3k+1)$.

When $(6k+1)$ is added to both members of (A), we have on the right $\qquad k(3k-2) + (6k+1) = 3k^2 + 4k + 1 = (k+1)(3k+1)$; hence, if the formula is true for $n = k$ it is true for $n = k + 1$.

(3) Since the formula is true for $n = k = 1$ (Step 1), it is true for $n = k + 1 = 2$; being true for $n = k = 2$ it is true for $n = k + 1 = 3$; and so on, for every positive integral value of n.

Prove by mathematical induction that

$$\frac{5}{1 \cdot 2 \cdot 3} + \frac{6}{2 \cdot 3 \cdot 4} + \frac{7}{3 \cdot 4 \cdot 5} + \ldots + \frac{n+4}{n(n+1)(n+2)} = \frac{n(3n+7)}{2(n+1)(n+2)} \quad .$$

Solution:

(1) The formula is true for $n = 1$, since $\dfrac{5}{1 \cdot 2 \cdot 3} = \dfrac{1(3+7)}{2 \cdot 2 \cdot 3} = \dfrac{5}{6}$.

(2) Assume the formula to be true for $n = k$, a positive integer; that is, assume

(A) $\quad \dfrac{5}{1 \cdot 2 \cdot 3} + \dfrac{6}{2 \cdot 3 \cdot 4} + \ldots + \dfrac{k+4}{k(k+1)(k+2)} = \dfrac{k(3k+7)}{2(k+1)(k+2)} \quad .$

Under this assumption we wish to show that

(B) $\quad \dfrac{5}{1 \cdot 2 \cdot 3} + \dfrac{6}{2 \cdot 3 \cdot 4} + \ldots + \dfrac{k+4}{k(k+1)(k+2)} + \dfrac{k+5}{(k+1)(k+2)(k+3)}$

$$= \frac{(k+1)(3k+10)}{2(k+2)(k+3)} \quad .$$

When $\dfrac{k+5}{(k+1)(k+2)(k+3)}$ is added to both members of (A), we have on the right

$$\frac{k(3k+7)}{2(k+1)(k+2)} + \frac{k+5}{(k+1)(k+2)(k+3)} = \frac{1}{(k+1)(k+2)} \left[\frac{k(3k+7)}{2} + \frac{k+5}{k+3} \right]$$

$$= \frac{1}{(k+1)(k+2)} \frac{k(3k+7)(k+3)+2(k+5)}{2(k+3)} = \frac{1}{(k+1)(k+2)} \frac{3k^3+16k^2+23k+10}{2(k+3)}$$

$$= \frac{1}{(k+1)(k+2)} \frac{(k+1)^2(3k+10)}{2(k+3)} = \frac{(k+1)(3k+10)}{2(k+2)(k+3)} \quad ;$$

hence, if the formula is true for $n = k$ it is true for $n = k + 1$.

(3) Since the formula is true for $n = k = 1$ (Step 1), it is true for $n = k + 1 = 2$; being true for $n = k = 2$, it is true for $n = k + 1 = 3$; and so on, for all positive integral values of n.

Let x be any real number. Show that $|\sin nx| \le n|\sin x|$ for every positive integer n.

<u>Solution:</u> Let $\{P_n\}$ be the sequence in which P_n is the statement "$|\sin nx| \leq n|\sin x|$." Recall the following theorem: If $\{P_n\}$ is a sequence of statements that possesses the properties

(i) P_1 is true and

(ii) for each index k such that P_k is true, the statement P_{k+1} is also true, then P_n is a true statement for every positive integer n. P_1 is the statement "$|\sin x| \leq |\sin x|$," which is surely true. Now we must show that the truth of the statement P_k implies that statement P_{k+1} also is true. For any k, we have

$$|\sin(k+1)x| = |\sin(kx+x)|$$
$$= |\sin kx \cos x + \cos kx \sin x|$$
$$\leq |\sin kx \cos x| + |\cos kx \sin x|$$
$$\leq |\sin kx| + |\sin x|.$$

Hence, if P_k is true $\left(\text{that is, if } |\sin kx| \leq k|\sin x|\right)$, we see that

$$|\sin(k+1)x| \leq |\sin kx| + |\sin x| \leq k|\sin x| + |\sin x|$$
$$= (k+1)|\sin x| .$$

Thus, the statement P_{k+1} ,

$$|\sin(k+1)x| \leq (k+1)|\sin x|,$$

is true whenever P_k is true. The two conditions of the above theorem are satisfied, so we conclude that statement P_n is true for any **positive integer n.**

● **PROBLEM** 18-8

Prove by mathematical induction that the number of straight lines determined by $n > 1$ points, no 3 on the same straight line, is $\frac{1}{2}n(n-1)$.

<u>Solution:</u>

(1) The theorem is true when $n = 2$, since $\frac{1}{2} \cdot 2(2-1) = 1$ and two points determine one line.

(2) Let us assume that k points, no 3 on the same straight line, determine $\frac{1}{2}k(k-1)$ lines.

When an additional point is added (not on any of the lines already determined) and is joined to each of the original k points, k new lines are determined. Thus, altogether we have $\frac{1}{2}k(k-1) + k = \frac{1}{2}k(k-1+2) = \frac{1}{2}k(k+1)$ lines and this agrees with the theorem when $n = k + 1$.

Hence, if the theorem is true for $n = k$, a positive integer greater than 1, it is true for the next one n = k + 1.

(3) Since the theorem is true for n = k = 2 (Step (1)), it is true for n = k + 1 = 3; being true for n = k = 3, it is true for n = k+1 = 4; and so on, for every possible integral value > 1 of n.

● **PROBLEM** 18-9

Prove by mathematical induction that the sum of n terms of an arithmetic progression a, a + d, a + 2d,... is $\frac{n}{2}\left[2a + (n-1)d\right]$, that is

$$a + (a+d) + (a+2d) + \ldots + \left[a+(n-1)d\right] = \frac{n}{2}\left[2a + (n-1)d\right].$$

Solution: Step 1. The formula holds for n = 1, since

$$a = \frac{1}{2}\left[2a + (1-1)d\right] = a.$$

Step 2. Assume that the formula holds for n = k. Then

$$a + (a+d) + (a+2d) + \ldots + \left[a + (k-1)d\right] = \frac{k}{2}\left[2a + (k-1)d\right].$$

Add the (k+1)th term, which is (a+kd), to both sides of the latter equation. Then

$$a + (a+d) + (a+2d) + \ldots + \left[a + (k-1)d\right] + (a+kd) = \frac{k}{2}\left[2a + (k-1)d\right] + (a+kd).$$

The right hand side of this equation $= ka + \frac{k^2 d}{2} - \frac{kd}{2} + a + kd =$

$$= \frac{k^2 d+kd+2ka+2a}{2}$$

$$= \frac{kd(k+1)+2a(k+1)}{2}$$

$$= \frac{k+1}{2}(2a+kd)$$

which is the value of $\frac{n}{2}\left[2a + (n-1)d\right]$ when n is replaced by (k+1).

Hence if the formula is true for n = k, we have proved it to be true for n = k + 1. But the formula holds for n = 1; hence it holds for n = 1 + 1 = 2. Then, since it holds for n = 2, it holds for n = 2 + 1 = 3, and so on. Thus the formula is true for all positive integral values of n.

● **PROBLEM** 18-10

Prove that the sum of the cubes of the first n natural numbers is equal to

$$\left\{\frac{n(n+1)}{2}\right\}^2.$$

<u>Solution:</u> We note by trial that the statement is true when $n = 1$, or 2, or 3 (when $n = 1$,

$$1^3 = 1 \quad \text{and} \quad \left[\frac{1(1+1)}{2}\right]^2 = \left[\frac{1(2)}{2}\right]^2 = \left(\frac{2}{2}\right)^2 = 1^2 = 1,$$

when $n = 2$,
$$1^3 + 2^3 = 1 + 8 = 9 \quad \text{and} \quad \left[\frac{2(2+1)}{2}\right]^2 = \left[\frac{2(3)}{2}\right]^2$$
$$= 3^2 = 9$$

etc.) Assume that it is true when n terms are taken; that is, suppose
$$1^3 + 2^3 + 3^3 + \ldots \text{ to } n \text{ terms} = \left\{\frac{n(n+1)}{2}\right\}^2.$$

Add the $(n+1)^{\text{th}}$ term, that is, $(n+1)^3$ to each side; then

$$1^3 + 2^3 + 3^3 + \ldots \text{ to } n+1 \text{ terms} = \left\{\frac{n(n+1)}{2}\right\}^2 + (n+1)^3$$

$$= \frac{n^2(n+1)^2}{2^2} + (n+1)(n+1)^2$$

$$= (n+1)^2\left(\frac{n^2}{4}\right) + (n+1)^2(n+1)$$

$$= (n+1)^2\left(\frac{n^2}{4} + n + 1\right)$$

$$= (n+1)^2\left(\frac{n^2}{4} + \frac{4n}{4} + \frac{4}{4}\right)$$

$$= (n+1)^2\left(\frac{n^2 + 4n + 4}{4}\right)$$

$$= \frac{(n+1)^2(n^2 + 4n + 4)}{4}$$

$$= \frac{(n+1)^2(n+2)^2}{2^2}$$

$$= \left\{\frac{(n+1)(n+2)}{2}\right\}^2 ;$$

$$= \left\{\frac{(n+1)[(n+1) + 1]}{2}\right\}^2$$

which is of the same form as the result we assumed to be true for n terms, $n + 1$ taking the place of n; in other words, if the result is true when we take a certain number of terms, whatever that number may be, it is true when we increase that number by one; but we see that it is true when 3 terms are taken; therefore it is true when 4 terms are taken; it is therefore true when 5 terms are taken; and so on. Thus the result is true universally.

Using mathematical induction, prove the binomial formula

$$(a+x)^n = a^n + na^{n-1}x + \frac{n(n-1)}{2!}a^{n-2}x^2 + \ldots + \frac{n(n-1)\ldots(n-r+2)}{(r-1)!}a^{n-r+1}x^{r-1}$$

$$+ \ldots + x^n$$

for positive integral values of n.

Solution: Step 1. The formula is true for n = 1.

Step 2. Assume the formula is true for n = k. Then

$$(a+x)^k = a^k + ka^{k-1}x + \frac{k(k-1)}{2!}a^{k-2}x^2 + \ldots + \frac{k(k-1)\ldots(k-r+2)}{(r-1)!}a^{k-r+1}x^{r-1}$$

$$+ \ldots + x^k .$$

Multiply both sides by a+x. The multiplication on the right may be written

$$a^{k+1} + ka^k x + \frac{k(k-1)}{2!}a^{k-1}x^2 + \ldots + \frac{k(k-1)\ldots(k-r+2)}{(r-1)!}a^{k-r+2}x^{r-1}$$

$$+ \ldots + ax^k$$

$$+ a^k x + ka^{k-1}x^2 + \ldots + \frac{k(k-1)\ldots(k-r+3)}{(r-2)!}a^{k-r+2}x^{r-1} + \ldots + x^{k+1} .$$

Since

$$\frac{k(k-1)\ldots(k-r+2)}{(r-1)!}a^{k-r+2}x^{r-1} + \frac{k(k-1)\ldots(k-r+3)}{(r-2)!}a^{k-r+2}x^{r-1}$$

$$= \frac{k(k-1)\ldots(k-r+3)}{(r-2)!}a^{k-r+2}x^{r-1}\left\{\frac{k-r+2}{r-1} + 1\right\} = \frac{(k+1)k(k-1)\ldots(k-r+3)}{(r-1)!}a^{k-r+2}x^{r-1},$$

the produc may be written

$$(a+x)^{k+1} = a^{k+1} + (k+1)a^k x + \ldots + \frac{(k+1)k(k-1)\ldots(k-r+3)}{(r-1)!}a^{k-r+2}x^{r-1}$$

$$+ \ldots + x^{k+1}$$

which is the binomial formula with n replaced by k+1.

Hence if the formula is true for n = k, it is true for n = k + 1. But the formula holds for n = 1; hence it holds for n = 1 + 1 = 2, and so on. Thus the formula is true for all positive integral values of n.

CHAPTER 19

THE BINOMIAL THEOREM

Basic Attacks and Strategies for Solving Problems in this Chapter. See pages 340 to 353 for step-by-step solutions to problems.

The *factorial* operator is the exclamation point, !, and it is placed after an integer variable or number to indicate the factorial function, which is defined as:

$$n! = n \cdot (n - 1) \cdot (n - 2) \ldots \cdot 2 \cdot 1.$$

From this definition it follows that

$$n! = n \cdot [(n - 1)!] = n(n - 1) \cdot [(n - 2)!]$$

etc. By general agreement, 0! is defined to be 1.

The *binomial theorem* or the *binomial formula* is a formula to expand a power of the binomial expression $a + b$, that is, it is a formula that enables someone to expand $(a + b)^n$.

Descriptions of the binomial theorem frequently use a shorthand notation for the coefficients of each term consisting of two numbers over each other enclosed in parentheses, e.g., $\binom{n}{k}$. This number is defined in terms of factorials, i.e., $\binom{n}{k}$ is defined to be equal to

$$\frac{n!}{k!(n - k)!}.$$

The notation $\binom{n}{k}$ is often called the binomial coefficient "n choose k." Note that, after simplification, $\binom{n}{k}$ also equals

$$\frac{n(n-1)(n-2)\ldots(n-k+1)}{k!}.$$

This last formula can also be used in cases when the power n is not integral, but rather a fraction. This notation also appears in the study of combinations and is sometimes written C_k^n.

Note that, in general, the binomial theorem is as follows:

$$(a+b)^n = \binom{n}{0}a^n b^0 + \binom{n}{1}a^{n-1}b^1 + \binom{n}{2}a^{n-2}b^2 + \ldots +$$

$$\binom{n}{r}a^{n-r}b^r + \ldots + \binom{n}{n-1}a^1 b^{n-1} + \binom{n}{n}a^0 b^n.$$

In general, for any n and r,

$$\binom{n}{0} = \binom{n}{n} = 1, \quad \binom{n}{1} = \binom{n}{n-1} = n, \quad \text{and} \quad \binom{n}{r} = \binom{n}{n-r}.$$

Applying the binomial theorem to $(a+b)^3$, we obtain

$$\binom{3}{0}a^3 b^0 + \binom{3}{1}a^2 b^1 + \binom{3}{2}a^1 b^2 + \binom{3}{3}a^0 b^3$$

which equals

$$a^3 + 3a^2 b + 3ab^2 + b^3.$$

The binomial expansion can sometimes be used to simplify (and factor) polynomials. For example, one might notice that

$$8x^6 - 12x^4 + 6x^2 - 1$$

can be rewritten as

$$(2x^2)^3 + 3(2x^2)^2(-1)^1 + 3(2x^2)(-1)^2 + (-1)^3$$

which is merely $(2x^2 - 1)^3$.

The values of the binomial coefficients, $\binom{n}{k}$, are often arranged in the shape of a triangle, called Pascal's Triangle. The top row (called row zero) consists only of 1, which is the sole coefficient (constant) in the expression $(a+b)^0$. The values in row n consist of the coefficients of the expression $(a+b)^n$. In each row, the two extreme values are always 1 and the second and next-to-last values in row n are always n. Because of the triangular shape of the arrangement of the coefficients (each row changing between an even and odd number of values), an arbitrary element is not under any element in the previous row, but rather under an element two rows above it. It can also be seen that each element is the sum of the two nearest elements (surrounding it) in the previous row.

```
          1
       1     1
     1    2    1
   1    3    3    1
```

Step-by-Step Solutions to Problems in this Chapter, "The Binomial Theorem"

● **PROBLEM** 19-1

Simplify the following numbers.

(a) $\dfrac{8!}{11!}$ (b) $\dfrac{5! - 8!}{4! - 7!}$

Solution:

(a)   Note $n! = n \cdot (n-1) \cdot (n-2) \cdot (n-3) \cdot \ldots \cdot 1$;

    also $n! = n \cdot (n-1)!$ or $n \cdot (n-1) \cdot (n-2)!$, etc.

    Thus, $\dfrac{8!}{11!} = \dfrac{\cancel{8!}}{11 \cdot 10 \cdot 9 \cdot \cancel{8!}} = \dfrac{1}{11 \cdot 10 \cdot 9} = \dfrac{1}{990}$.

(b)   Similarly,

$$\frac{5! - 8!}{4! - 7!} = \frac{5 \cdot 4! - 8 \cdot 7 \cdot 6 \cdot 5 \cdot 4!}{4! - 7 \cdot 6 \cdot 5 \cdot 4!}$$

Factoring 4!, $= \dfrac{\cancel{4!}\,[5 - (8 \cdot 7 \cdot 6 \cdot 5)]}{\cancel{4!}\,[1 - (7 \cdot 6 \cdot 5)]}$

$$= \frac{5 - 1680}{1 - 210}$$

$$= \frac{-1675}{-209}$$

$$= \frac{1675}{209} .$$

● **PROBLEM** 19-2

Find the value of $\dfrac{5!\,6!}{4!\,7!}$.

Solution: Apply the definition of factorial: If n is

340

any positive integer, the symbol n! is the product of the integers from 1 up to and including n.

Also if r and n are both positive integers and r is less than n, then $n! = n \cdot (n - 1) \ldots (r + 2)(r + 1) \, r!$. Use these two ideas to expand each factorial.

$$n! = 1 \cdot 2 \cdot 3 \ldots n$$

$5! = (4!)(5)$ and $7! = (6!)(7)$ Substituting the values of 5! and 7! we have

$$\frac{5! \, 6!}{4! \, 7!} = \frac{(4!)(5)(6!)}{(4!)(6!)(7)}$$. Dividing the common factors

4! and 6!

$$= \frac{5}{7} \, .$$

If n and r are positive integers, and $r < n$, show that
$$n! = r!(r + 1)(r + 2) \cdot \ldots \cdot n.$$

Solution: By definition of factorial,
$$r! = 1 \cdot 2 \cdot \ldots \cdot r \ .$$
Then,
$$r!(r+1)(r+2) \cdot \ldots \cdot n = (1 \cdot 2 \cdot \ldots \cdot r)(r+1)(r+2) \cdot \ldots \cdot n$$
$$r!(r+1)(r+2) \cdot \ldots \cdot n = 1 \cdot 2 \cdot \ldots \cdot n \qquad (1)$$
Again, by definition of factorial,
$$n! = 1 \cdot 2 \cdot \ldots \cdot n \ .$$
Hence, equation (1) becomes:
$$r!(r+1)(r+2) \cdot \ldots \cdot n = n!$$
or
$$n! = r!(r+1)(r+2) \cdot \ldots \cdot n \ .$$

Find the expansion of $(a - 2x)^7$.

Solution: Use the binomial formula:

$$(u + v)^n = u^n + nu^{n-1}v + \frac{n(n-1)}{2} u^{n-2}v^2$$
$$+ \frac{n(n-1)(n-2)}{2 \cdot 3} u^{n-3}v^3 + \ldots + v^n$$

and substitute a for u and (-2x) for v and 7 for n to obtain:

$$(a-2x)^7 = [a+(-2x)]^7$$

$$= a^7 + 7a^6(-2x) + \frac{7 \cdot \overset{3}{\cancel{6}}}{\cancel{2}} a^5(-2x)^2 + \frac{7 \cdot \overset{3}{\cancel{6}} \cdot 5}{\cancel{2} \cdot \cancel{3}} a^4(-2x)^3$$

$$+ \frac{7 \cdot \cancel{6} \cdot 5 \cdot \cancel{4}}{\cancel{2} \cdot \cancel{3} \cdot \cancel{4}} a^3(-2x)^4 + \frac{7 \cdot \cancel{6} \cdot 5 \cdot \cancel{4} \cdot \overset{3}{\cancel{3}}}{\cancel{2} \cdot \cancel{3} \cdot \cancel{4} \cdot \cancel{5}} a^2(-2x)^5$$

$$+ \frac{7 \cdot \cancel{6} \cdot \cancel{5} \cdot \cancel{4} \cdot \cancel{3} \cdot \cancel{2}}{\cancel{2} \cdot \cancel{3} \cdot \cancel{4} \cdot \cancel{5} \cdot \cancel{6}} a^1(-2x)^6 + \frac{\cancel{7} \cdot \cancel{6} \cdot \cancel{5} \cdot \cancel{4} \cdot \cancel{3} \cdot \cancel{2} \cdot 1}{\cancel{2} \cdot \cancel{3} \cdot \cancel{4} \cdot \cancel{5} \cdot \cancel{6} \cdot \cancel{7}} a^0(-2x)^7$$

$$(a - 2x)^7 = a^7 - 14a^6x + 84a^5x^2 - 280a^4x^3 + 560a^3x^4$$

$$- 672a^2x^5 + 448a\, x^6 - 128x^7.$$

Expand $(x + 2y)^5$.

Solution: Apply the binomial theorem. If n is a positive integer, then

$$(a + b)^n = \binom{n}{0}a^n b^0 + \binom{n}{1}a^{n-1}b + \binom{n}{2}a^{n-2}b^2 + \cdots + \binom{n}{r}a^{n-r}b^r$$

$$+ \cdots + \binom{n}{n}b^n.$$

Note that $\binom{n}{r} = \frac{n!}{r!\,(n-r)!}$ and that $0! = 1$. Then, we obtain:

$$(x + 2y)^5 = \binom{5}{0}x^5(2y)^0 + \binom{5}{1}x^4(2y)^1 + \binom{5}{2}x^3(2y)^2$$

$$+ \binom{5}{3}x^2(2y)^3 + \binom{5}{4}x^1(2y)^4 + \binom{5}{5}x^0(2y)^5$$

$$= \frac{5!}{0!5!}\, x^5 + \frac{5!}{1!4!}\, x^4 2y + \frac{5!}{2!3!}x^3\left(4y^2\right)$$

$$+ \frac{5!}{3!2!}x^2\left(8y^3\right) + \frac{5!}{4!1!}x\left(16y^4\right) + \frac{5!}{5!0!}1\left(32y^5\right)$$

$$= x^5 + \frac{5 \cdot \cancel{4!}}{\cancel{4!}}x^4 2y + \frac{5 \cdot \overset{2}{\cancel{4}} \cdot \cancel{3!}}{\cancel{2} \cdot 1 \cdot \cancel{3!}}\underset{1}{}x^3\left(4y^2\right)$$

$$+ \frac{5 \cdot \overset{2}{\cancel{4}} \cdot \cancel{3!}}{\cancel{3!} \cdot \cancel{2} \cdot 1}\underset{1}{}x^2\left(8y^3\right) + \frac{5 \cdot \cancel{4!}}{\cancel{4!}1!}x\left(16y^4\right) + \frac{5!}{\cancel{5!}0!}\left(32y^5\right)$$

$$= x^5 + 10x^4y + 40x^3y^2 + 80x^2y^3 + 80xy^4 + 32y^5.$$

● **PROBLEM 19-6**

Expand $\left(3p^2 - 2q^{\frac{1}{2}}\right)^4$ by means of the binomial theorem.

Solution: We apply the binomial theorem: If n is a positive integer, then

$$(a + b)^n = \binom{n}{0} a^n b^0 + \binom{n}{1} a^{n-1}b + \binom{n}{2} a^{n-2}b^2 + \ldots + \binom{n}{r} a^{n-r}b^r + \ldots + \binom{n}{n} a^0 b^n$$

where $\binom{n}{r} = \dfrac{n!}{r!(n-r)!}$.

We identify $3p^2$ with a, $-2q^{\frac{1}{2}}$ with b, and n with 4. The binomial theorem then gives us

$$\left(3p^2 - 2q^{\frac{1}{2}}\right)^4 = \binom{4}{0}\left(3p^2\right)^4\left(-2q^{\frac{1}{2}}\right)^0 + \binom{4}{1}\left(3p^2\right)^3\left(-2q^{\frac{1}{2}}\right) + \binom{4}{2}\left(3p^2\right)^2\left(-2q^{\frac{1}{2}}\right)^2$$

$$+ \binom{4}{3}\left(3p^2\right)^1\left(-2q^{\frac{1}{2}}\right)^3 + \binom{4}{4}\left(3p^2\right)^0\left(-2q^{\frac{1}{2}}\right)^4$$

$$= \frac{4!}{0!(4)!} \, 3^4 p^8 + \frac{4!}{1!3!} \, 3^3 p^6\left(-2q^{\frac{1}{2}}\right) + \frac{4!}{2!2!} \, 3^2 p^4 (-2)^2 q$$

$$+ \frac{4!}{3!1!} \, 3p^2(-2)^3 q^{3/2} + \frac{4!}{4!0!}(-2)^4 q^2$$

$$= 81p^8 + 4 \cdot \frac{3!}{3!} \, 27p^6\left(-2q^{\frac{1}{2}}\right) + \frac{4 \cdot 3 \cdot 2!}{2! \, 2!} \, 9p^4 4q$$

$$+ \frac{4 \cdot 3!}{3! \cdot 1!} \, 3p^2(-8)q^{3/2} + 16q^2$$

$$= 81p^8 - 216p^6 q^{\frac{1}{2}} + 216p^4 q - 96p^2 q^{3/2} + 16q^2 \, .$$

● **PROBLEM 19-7**

Give the expansion of $\left[r^2 - \dfrac{1}{s}\right]^5$.

Solution: Write the given expression as the sum of two terms raised to the 5th power:

$$\left[r^2 - \frac{1}{s}\right]^5 = \left[r^2 + \left(-\frac{1}{s}\right)\right]^5 \tag{1}$$

The Binomial Theorem can be used to expand the expression on the right side of equation (1). The Binomial Theorem is stated as:

$$(a+b)^n = a^n + na^{n-1}b + \frac{n(n-1)}{1 \cdot 2}a^{n-2}b^2 + \frac{n(n-1)(n-2)}{1 \cdot 2 \cdot 3}a^{n-3}b^3$$

$+ \ldots + nab^{n-1} + b^n$, where a and b are any two numbers.

Let $a = r^2$, $b = -\frac{1}{s}$, and $n = 5$. Then, using the Binomial Theorem:

$$\left(r^2 - \frac{1}{s}\right)^5 = \left[r^2 + \left(-\frac{1}{s}\right)\right]^5$$

$$= \left(r^2\right)^5 + 5\left(r^2\right)^{5-1}\left(-\frac{1}{s}\right) + \frac{5(5-1)}{1 \cdot 2}\left(r^2\right)^{5-2}\left(-\frac{1}{s}\right)^2$$

$$+ \frac{5(5-1)(5-2)}{1 \cdot 2 \cdot 3}\left(r^2\right)^{5-3}\left(-\frac{1}{s}\right)^3$$

$$+ \frac{5(5-1)(5-2)(5-3)}{1 \cdot 2 \cdot 3 \cdot 4}\left(r^2\right)^{5-4}\left(-\frac{1}{s}\right)^4$$

$$+ \frac{5(5-1)(5-2)(5-3)(5-4)}{1 \cdot 2 \cdot 3 \cdot 4 \cdot 5}\left(r^2\right)^{5-5}\left(-\frac{1}{s}\right)^5$$

$$= r^{10} - \frac{5\left(r^2\right)^4}{s} + \frac{5(\overset{2}{\cancel{4}})}{1 \cdot \underset{1}{\cancel{2}}}\left(r^2\right)^3\left[\frac{1}{s^2}\right] - \frac{5(\overset{2}{\cancel{4}})(\cancel{3})}{1 \cdot \underset{1}{\cancel{2}} \cdot \cancel{3}}\left(r^2\right)^2\left[\frac{1}{s^3}\right]$$

$$+ \frac{5(\cancel{4})(\cancel{3})(\cancel{2})}{1 \cdot \cancel{2} \cdot \cancel{3} \cdot \cancel{4}}\left(r^2\right)^1\left[\frac{1}{s^4}\right] - \frac{\cancel{5}(\cancel{4})(\cancel{3})(\cancel{2})(\cancel{1})}{\cancel{1} \cdot \cancel{2} \cdot \cancel{3} \cdot \cancel{4} \cdot \cancel{5}}\left(r^2\right)^0\left[\frac{1}{s^5}\right]$$

$$= r^{10} - \frac{5r^8}{s} + \frac{10r^6}{s^2} - \frac{10r^4}{s^3} + \frac{5r^2}{s^4} - (1)(1)\left[\frac{1}{s^5}\right]$$

$$\left(r^2 - \frac{1}{s}\right)^5 = r^{10} - \frac{5r^8}{s} + \frac{10r^6}{s^2} - \frac{10r^4}{s^3} + \frac{5r^2}{s^4} - \frac{1}{s^5}$$

● PROBLEM 19-8

Find the fifth term of $\left(2 + 2x^3\right)^{17}$.

Solution: Use the Binomial Theorem which states that

$$(c+b)^n = \frac{1}{0!} c^n + \frac{n}{1!} c^{n-1}b + \frac{n(n-1)}{2!} c^{n-2}b^2 + \ldots + ncb^{n-1} + b^n .$$

Replacing c by a and b by $2x^3$:

$$\left(a+2x^3\right)^{17} = \frac{1}{0!} a^{17} + \frac{17}{1!} a^{16}\left(2x^3\right)^1 + \frac{17 \cdot 16}{2!} a^{15}\left(2x^3\right)^2$$

$$+ \frac{17 \cdot 16 \cdot 15}{3!} a^{14}\left(2x^3\right)^3 + \frac{17 \cdot 16 \cdot 15 \cdot 14}{4!} a^{13}\left(2x^3\right)^4$$

$$+ \ldots$$

The fifth term of this expansion is:

$$\frac{17 \cdot 16 \cdot 15 \cdot 14}{4!} a^{13}(2x^3)^4 = \frac{17 \cdot \cancel{16} \cdot \cancel{15} \cdot \cancel{14}}{\cancel{4} \cdot \cancel{3} \cdot \cancel{2} \cdot 1} a^{13}(2^4)(x^3)^4$$

$$= \frac{17 \cdot 4 \cdot 5 \cdot 7}{1} a^{13} 16x^{12}$$

$$= 38,080 \, a^{13} \, x^{12}$$

Factor

$$8x^6y^3 + 12x^4y^2 + 6x^2y + 1$$

Solution: (1) $8x^6y^3 + 12x^4y^2 + 6x^2y + 1$

Look for the largest common factor:

L.C.F. $= 2x^2y$

Set $a = 2x^2y$

$b = 1$

Then we get

$$a^3 + 3a^2 + 3a + b^3$$

which is equal to $a^3 + 3ab^2 + 3a^2b + b^3$ when $b = 1$. Therefore, the factor is

$$(a + b)^3 = (2x^2y + 1)^3$$

Find the 5th term of the expansion of $(3y - 4w)^8$.

Solution: Use the binomial formula:

$$(u+v)^n = u^n + nu^{n-1}v + \frac{n(n-1)}{2}u^{n-2}v^2 + \frac{n(n-1)(n-2)}{2 \cdot 3}u^{n-2}v^3 + \ldots + v^n .$$

Applying the formula for the rth term where $r \leq n+1$; the rth term is:

$\frac{n(n-1)(n-2)\ldots(n-r+2)}{1 \cdot 2 \cdot 3 \ldots (r-1)} u^{n-r+1}v^{r-1}$. Let $u = 3y$, $v = -4w$, $n = 8$, $r = 5$.

Then $n - r + 1 = 8 - 5 + 1 = 4$

$r - 1 = 5 - 1 = 4$

Therefore $u^{n-r+1}v^{r-1} = u^4v^4$

$r - 1 = 4$

$n - r + 2 = 8 - 5 + 2 = 5$

So:　　　　　$n(n-1)(n-2)\dots(5) = 8\cdot7\cdot6\cdot5$

Thus the coefficient is:

$$\frac{8\cdot7\cdot6\cdot5}{1\cdot2\cdot8\cdot4}$$

$$u^4 = (3y)^4$$

$$v^4 = (-4w)^4$$

Therefore:　　　$(3y - 4w)^8 = \frac{8\cdot7\cdot6\cdot5}{4\cdot3\cdot2\cdot1}(3y)^4(-4w)^4$

$$= \frac{\overset{2}{8}(7)(\cancel{6})(5)}{\cancel{4}(\cancel{3})(\cancel{2})}\left(81y^4\right)\left(256w^4\right)$$

$$= 70\left(81y^4\right)\left(256w^4\right)$$

$$= 1{,}451{,}520y^4w^4$$

Find the term involving y^5 in the expansion of $\left(2x^2 + y\right)^{10}$.

Solution: The formula for the binomial expansion is:

$$(a + b)^n = a^n + na^{n-1}b + \frac{n(n-1)}{1\cdot2}a^{n-2}b^2 + \frac{n(n-1)(n-2)}{1\cdot2\cdot3}a^{n-3}b^3 + \dots + nab^{n-1} + b^n.$$

The rth term of the expansion of $(a + b)^n$ is

$$\text{rth term} = \frac{n(n-1)(n-2)\dots(n-r+2)}{(r-1)!}a^{n-r+1}b^{r-1}$$

In this example,

$$b^{r-1} = y^5$$

$$r-1 = 5$$

$$r = 6 \text{ and } n = 10$$

Thus,

$$\text{6th term} = \frac{10\cdot9\cdot8\cdot7\cdot6}{5!}\left(2x^2\right)^5y^5$$

$$= \frac{\overset{2}{10}\cdot\overset{3}{9}\cdot\overset{1}{\cancel{8}}\overset{1}{}\cdot7\cdot6}{\underset{1}{\cancel{5}}\cdot\underset{1}{\cancel{4}}\cdot\underset{1}{\cancel{3}}\cdot\underset{1}{\cancel{2}}\cdot1}\,32x^{10}y^5$$

$$= 8064\,x^{10}y^5$$

Find the constant term in the expansion of $\left(2x^2 + \frac{1}{x}\right)^9$.

Solution: The rth term in the expansion of $(a+b)^n$ is given by:

$$\frac{n(n-1)(n-2)\dots(n-r+2)}{1\cdot2\cdot3\cdots(r-1)}a^{n-r+1}b^{r-1}$$

Replacing a by $2x^2$ and b by $1/x$ in this formula, the rth term in the expansion of $\left(2x^2 + 1/x\right)^9$ is given by:

$$\frac{9(8)(7)\ldots(9-r+2)}{(1)(2)(3)\ldots(r-1)}\left(2x^2\right)^{9-r+1}\left(\frac{1}{x}\right)^{r-1} = \frac{9(8)(7)\ldots(11-r)}{(1)(2)(3)\ldots(r-1)}\left(2x^2\right)^{10-r}\left(\frac{1}{x}\right)^{r-1}$$

Then the rth term in the expansion will contain the factors $\left(2x^2\right)^{10-r}$ and $\left(\frac{1}{x}\right)^{r-1}$. Hence, as far as powers of x are concerned, the rth term will involve

$$\left(x^2\right)^{10-r}\left(\frac{1}{x}\right)^{r-1} \quad \text{or} \quad \left(x^2\right)^{10-r}\frac{(1)^{r-1}}{x^{r-1}} \quad \text{or} \quad \frac{x^{20-2r}}{x^{r-1}}$$

$$\text{or} \quad x^{20-2r-(r-1)}$$

$$\text{or} \quad x^{20-2r-r+1}$$

$$\text{or} \quad x^{21-3r}$$

The desired constant term is free of x; that is, the constant term has a factor of x^0 since $kx^0 = k(1) = k$, where k is the constant. Hence,

$$21 - 3r = 0$$
$$21 = 3r$$
$$\frac{21}{3} = r$$
$$7 = r$$

The rth term or seventh term can be found by using the Binomial Theorem which states that:

$$(a+b)^n = \frac{1}{0!}a^n + \frac{n}{1!}a^{n-1}b + \frac{n(n-1)}{2!}a^{n-2}b^2 + \ldots + nab^{n-1} + b^n .$$

Replacing a by $2x^2$ and b by $1/x$, $\left(2x^2 + 1/x\right)^9$ can be expanded as:

$$\left(2x^2 + \frac{1}{x}\right)^9 = \frac{1}{0!}\left(2x^2\right)^9 + \frac{9}{1!}\left(2x^2\right)^8\left(\frac{1}{x}\right) + \frac{9(8)}{2!}\left(2x^2\right)^7\left(\frac{1}{x}\right)^2$$

$$+ \frac{9(8)(7)}{3!}\left(2x^2\right)^6\left(\frac{1}{x}\right)^3 + \frac{9(8)(7)(6)}{4!}\left(2x^2\right)^5\left(\frac{1}{x}\right)^4$$

$$+ \frac{9(8)(7)(6)(5)}{5!}\left(2x^2\right)^4\left(\frac{1}{x}\right)^5 + \frac{9(8)(7)(6)(5)(4)}{6!}\left(2x^2\right)^3\left(\frac{1}{x}\right)^6$$

$$+ \frac{9(8)(7)(6)(5)(4)(3)}{7!}\left(2x^2\right)^2\left(\frac{1}{x}\right)^7$$

$$+ \frac{9(8)(7)(6)(5)(4)(3)(2)}{8!}\left(2x^2\right)\left(\frac{1}{x}\right)^8$$

$$+ \frac{9(8)(7)(6)(5)(4)(3)(2)(1)}{9!}\left(2x^2\right)^0\left(\frac{1}{x}\right)^9$$

Hence, the rth term or 7th term of this expansion is:

$$\frac{9(8)(7)(6)(5)(4)}{6!}\left(2x^2\right)^3\left(\frac{1}{x}\right)^6$$

$$= \frac{9 \cdot 8 \cdot 7 \cdot 6 \cdot 5 \cdot 4}{6 \cdot 5 \cdot 4 \cdot 3 \cdot 2 \cdot 1}\left(2x^2\right)^3\left(\frac{1}{x}\right)^6$$

$$= 3 \cdot 4 \cdot 7 \left(2x^2\right)^3\left(\frac{1}{x}\right)^6$$

$$= 84\left(2^3\right)\left(x^2\right)^3\left(\frac{1^6}{x^6}\right)$$

$$= 84(8)x^6 \frac{1}{x^6}$$

$$= 672 .$$

Hence, 672 is the desired constant term in the expansion of $\left(2x^2 + \frac{1}{x}\right)^9$.

Find the term involving x^3yz^2 in the expansion $(x + 2y - 3z)^6$.

Solution: Use the binomial formula: $a^n + na^{n-1}b + \frac{n(n-1)}{1\cdot2} \times$

$a^{n-2}b^2 + \frac{n(n-1)(n-2)}{1\cdot2\cdot3} a^{n-3}b^3 + \ldots + nab^{n-1} + b^n = (a+b)^n$

and associate 2y - 3z and substitute it for b in the formula, x for a, and 6 for n:

$[x + (2y - 3z)]^6 = x^6 + 6x^5(2y - 3z)^1 + \frac{6\cdot5}{1\cdot2} x^4(2y - 3z)^2$

$+ \frac{6\cdot5\cdot4}{1\cdot2\cdot3} x^3(2y - 3z)^3 + \frac{6\cdot5\cdot4\cdot3}{1\cdot2\cdot3\cdot4} x^2(2y - 3z)^4$

$+ \frac{6\cdot5\cdot4\cdot3\cdot2}{1\cdot2\cdot3\cdot4\cdot5} x(2y - 3z)^5 + (2y - 3z)^6.$

The term that involves x^3yz^2 is: $\frac{6\cdot5\cdot4}{1\cdot2\cdot3} x^3(2y - 3z)^3$ in

which $(2y - 3z)^3$ must be expanded.

$(2y - 3z)^3 = (2y - 3z)(2y - 3z)^2 = (2y - 3z)(4y^2 - 12yz + 9z^2)$

$= 8y^3 - 36zy^2 + 54yz^2 - 27z^3$

When $(2y - 3z)^3$ is multiplied by $\frac{6\cdot5\cdot4}{1\cdot2\cdot3} x^3$, the final term is

$\frac{6\cdot5\cdot4}{1\cdot2\cdot3} x^3\left[8y^3 - 36zy^2 + 54yz^2 - 27z^3\right].$

Distributing, notice that the term of interest involves

$\frac{6\cdot5\cdot4}{1\cdot2\cdot3} x^3\left(54yz^2\right) = 1080x^3yz^2.$

Find the value of $\left(a + \sqrt{a^2 - 1}\right)^7 + \left(a - \sqrt{a^2 - 1}\right)^7.$

Solution: Use the binomial theorem to expand the first term:

$$(a + b)^n = a^n + na^{n-1}b + \frac{n(n-1)}{1 \cdot 2} a^{n-2}b^2 + \frac{n(n-1)(n-2)}{1 \cdot 2 \cdot 3} a^{n-3}b^3$$
$$+ \ldots + b^n.$$

Now $\left(a + \sqrt{a^2 - 1}\right)^7 = a^7 + 7a^6\left(\sqrt{a^2 - 1}\right)^1 + \frac{7 \cdot 6}{1 \cdot 2} (a)^5\left(\sqrt{a^2 - 1}\right)^2$

$$+ \frac{7 \; 6 \; 5}{1 \; 2 \; 3} a^4\left(\sqrt{a^2 - 1}\right)^3 + \frac{7 \cdot 6 \cdot 5 \cdot 4}{1 \cdot 2 \cdot 3 \cdot 4} a^3\left(\sqrt{a^2 - 1}\right)^4$$

$$+ \frac{7 \cdot 6 \cdot 5 \cdot 4 \cdot 3}{1 \cdot 2 \cdot 3 \cdot 4 \cdot 5} a^2\left(\sqrt{a^2 - 1}\right)^5 + 7a\left(\sqrt{a^2 - 1}\right)^6 + \left(\sqrt{a^2 - 1}\right)^7$$

Expand $\left(a - \sqrt{a^2 - 1}\right)^7$ in the same fashion.

$$\left(a - \sqrt{a^2 - 1}\right)^7 = a^7 + 7a^6\left(-\sqrt{a^2 - 1}\right)^1 + \frac{7 \cdot 6}{1 \cdot 2} (a)^5\left(-\sqrt{a^2 - 1}\right)^2$$

$$+ \frac{7 \cdot 6 \cdot 5}{1 \cdot 2 \cdot 3} a^4\left(-\sqrt{a^2 - 1}\right)^3 + \frac{7 \cdot 6 \cdot 5 \cdot 4}{1 \cdot 2 \cdot 3 \cdot 4} a^3\left(-\sqrt{a^2 - 1}\right)^4$$

$$+ \frac{7 \cdot 6 \cdot 5 \cdot 4 \cdot 3}{1 \cdot 2 \cdot 3 \cdot 4 \cdot 5} a^2\left(-\sqrt{a^2 - 1}\right)^5 + 7a\left(-\sqrt{a^2 - 1}\right)^6 + \left(-\sqrt{a^2 - 1}\right)^7.$$

In this expansion, the odd powers of $-\sqrt{a^2 - 1}$ will cancel the corresponding terms in the previous expansion.

$$\left(a + \sqrt{a^2 - 1}\right)^7 + \left(a - \sqrt{a^2 - 1}\right)^7 = a^7 + 7a^6\left(\sqrt{a^2 - 1}\right)^1$$

$$+ \frac{7 \cdot \overset{3}{\cancel{6}}}{1 \cdot \underset{1}{\cancel{2}}} (a)^5\left(\sqrt{a^2 - 1}\right)^2 + \frac{7 \cdot 6 \cdot 5}{1 \cdot 2 \cdot 3} a^4\left(\sqrt{a^2 - 1}\right)^3$$

$$+ \frac{7 \cdot \overset{2}{\cancel{6}} \cdot 5 \cdot \cancel{4}}{1 \cdot \cancel{2} \cdot \cancel{3} \cdot \cancel{4}} a^3\left(\sqrt{a^2 - 1}\right)^4 + \frac{7 \cdot 6 \cdot 5 \cdot 4 \cdot 3}{1 \cdot 2 \cdot 3 \cdot 4 \cdot 5} a^2\left(\sqrt{a^2 - 1}\right)^5$$

$$+ 7a\left(\sqrt{a^2 - 1}\right)^6 + \left(\sqrt{a^2 - 1}\right)^7 + a^7 - 7a^6\left(\sqrt{a^2 - 1}\right)$$

$$+ \frac{7 \cdot \overset{3}{\cancel{6}}}{1 \cdot \underset{1}{\cancel{2}}} (a)^5\left(\sqrt{a^2 - 1}\right)^2 - \frac{7 \cdot 6 \cdot 5}{1 \cdot 2 \cdot 3} a^4\left(\sqrt{a^2 - 1}\right)^3 + \frac{7 \cdot \overset{2}{\cancel{6}} \cdot 5 \cdot 4}{1 \cdot \cancel{2} \cdot 3 \cdot 4} a^3\left(\sqrt{a^2 - 1}\right)^4$$

$$- \frac{7 \cdot 6 \cdot 5 \cdot 4 \cdot 3}{1 \cdot 2 \cdot 3 \cdot 4 \cdot 5} a^2\left(\sqrt{a^2 - 1}\right)^5 + 7a\left(\sqrt{a^2 - 1}\right)^6 - \left(\sqrt{a^2 - 1}\right)^7$$

$$= a^7 + 21a^5(a^2 - 1) + 35a^3(a^2 - 1)^2 + 7a(a^2 - 1)^3 + a^7 + 21a^5(a^2 - 1)$$

$$+ 35a^3(a^2 - 1)^2 + 7a(a^2 - 1)^3$$

$$= 2a^7 + 42a^5(a^2 - 1) + 70a^3(a^2 - 1)^2 + 14a(a^2 - 1)^3$$

$$= 2\left[a^7 + 21a^5(a^2 - 1) + 35a^3(a^2 - 1)^2 + 7a(a^2 - 1)^3\right]$$

$$= 2\left[a^7 + 21a^7 - 21a^5 + 35a^3(a^4 - 2a^2 + 1) + 7a(a^2 - 1)^2(a^2 - 1)\right]$$

$$= 2\left[22a^7 - 21a^5 + 35a^7 - 70a^5 + 35a^3 + 7a\left(a^4 - 2a^2 + 1\right)\left(a^2 - 1\right)\right]$$

$$= 2\left[57a^7 - 91a^5 + 35a^3 + 7a\left(a^6 - 2a^4 + a^2 - a^4 + 2a^2 - 1\right)\right]$$

$$= 2a\left[57a^6 - 91a^4 + 35a^2 + 7\left(a^6 - 2a^4 + a^2 - a^4 + 2a^2 - 1\right)\right]$$

$$= 2a\left[57a^6 - 91a^4 + 35a^2 + 7a^6 - 14a^4 + 7a^2 - 7a^4 + 14a^2 - 7\right]$$

$$= 2a\left[64a^6 - 112a^4 + 56a^2 - 7\right]$$

● **PROBLEM** 19-15

Compute the approximate value of $(1.01)^5$

<u>Solution:</u> The Binomial Theorem can be used to find the approximate value of $(1.01)^5$. The Binomial Theorem is stated as:

$$(a + b)^n = a^n + na^{n-1}b + \frac{n(n-1)}{1 \cdot 2}a^{n-2}b^2 + \frac{n(n-1)(n-2}{1 \cdot 2 \cdot 3} \times$$

$$a^{n-3}b^3 + \cdots + nab^{n-1} + b^n,$$

where a and b are any two numbers. In order to use this theorem, express $(1.01)^5$ as the sum of two numbers raised to the fifth power. Then,

$$(1.01)^5 = (1 + .01)^5$$

Now, let a = 1, b = .01, and n = 5 in the Binomial Theorem. Calculating the first four terms of this theorem with these substitutions:

$$(1.01)^5 = (1 + .01)^5$$

$$= (1)^5 + 5(1)^{5-1}(.01) + \frac{5(5 - 1)}{1 \cdot 2} \times$$

$$(1)^{5-2}(.01)^2 + \frac{5(5 - 1)(5 - 2)}{1 \cdot 2 \cdot 3} \times$$

$$(1)^{5-3}(.01)^3 + \cdots$$

$$= 1 + 5(1)^4(.01) + \frac{5(\cancel{4})^2}{\cancel{2}_1}(1)^3(.01)^2$$

350

$$+ \frac{5(4)(\cancel{3})}{\cancel{6}\;2}\;(1)^2\;(.01)^3 + \cdots$$

$$= 1 + 5(1)(.01) + 10(1)(.0001) + 10(1)$$

$$(.000001) + \cdots$$

$$= 1 + 0.05 + 0.001 + 0.00001$$

$$= 1.05101$$

Use the binomial formula with $n = 1/3$ to find an approximation to $\sqrt[3]{28}$.

<u>Solution</u>: To apply the binomial formula, try to express $\sqrt[3]{28}$ as the sum of two numbers raised to a power. Note that the formula is simplified if one of the numbers is one.

$$\sqrt[3]{28} = (28)^{(1/3)} = (27+1)^{1/3} .$$

We can write the expansion of $(x+y)^{1/3}$ to four terms and later substitute for x and y. We write out the binomial expansion to four terms when $n = 1/3$.

$$(x+y)^{1/3} = x^{1/3} + \frac{1}{3} x^{(1/3)-1} y + \frac{\left(\frac{1}{3}\right)\left(\frac{1}{3}-1\right)}{1\cdot 2} x^{(1/3)-2} y^2$$

$$+ \frac{\left(\frac{1}{3}\right)\left(\frac{1}{3}-1\right)\left(\frac{1}{3}-2\right)}{1\cdot 2\cdot 3} x^{(1/3)-3} y^3 + \cdots$$

$$= x^{1/3} + \frac{1}{3} x^{-2/3} y + \frac{\left(\frac{1}{3}\right)\left(\frac{-2}{3}\right)}{2} x^{-5/3} y^2$$

$$+ \frac{\left(\frac{1}{3}\right)\left(\frac{-2}{3}\right)\left(\frac{-5}{3}\right)}{1\cdot 2\cdot 3} x^{-8/3} y^3 + \cdots$$

$$= x^{1/3} + \frac{1}{3} x^{-2/3} y + \left(\frac{-2}{9}\right)\left(\frac{1}{2}\right) x^{-5/3} y^2$$

$$+ \frac{\overset{5}{\cancel{10}} \cdot 1}{3\cdot 3\cdot 3\cdot 1\cdot \underset{1}{\cancel{2}}\cdot 3} x^{-8/3} y^3 + \cdots$$

$$(x+y)^{1/3} = x^{1/3} + \frac{1}{3} x^{-2/3} y - \frac{1}{9} x^{-5/3} y^2$$

$$+ \frac{5}{81} x^{-8/3} y^3 + \cdots \qquad (1)$$

In this case, n is fractional. We obtain an infinite series and we can expand it for the first few terms if $|x| < |y|$. $x = 27$ and $y = 1$, and $|y| < |x|$, i.e., $|1| < |27|$. Therefore, using equation (1) with $x = 27$, $y = 1$ and $n = 1/3$ (writing only the first four terms):

351

$$\sqrt[3]{28} = (28)^{1/3}$$
$$= (27 + 1)^{1/3}$$
$$= 27^{1/3} + 1/3\left(27^{-2/3}\right)(1) - \frac{1}{9}\left(27^{-5/3}\right)\left(1^2\right)$$
$$+ \frac{5}{81}\left(27^{-8/3}\right)\left(1^3\right)$$
$$= 3 + \frac{1}{3}\left(\frac{1}{9}\right) - \frac{1}{9}\left(\frac{1}{243}\right) + \frac{5}{81}\left(\frac{1}{6,561}\right)$$
$$= 3 + 0.037037 - 0.000457 + 0.000009$$
$$= 3.036589$$

● **PROBLEM 19-17**

Expand

(1) $\left[\dfrac{1}{\sqrt{2x}} + \dfrac{1}{\sqrt{x}}\right]^4$

(2) $(x + 4y^2)^{-5}$

and simplify if possible.

Solution: (1) $\left[\dfrac{1}{\sqrt{2x}} + \dfrac{1}{\sqrt{x}}\right]^4 = \left[(2x)^{-\frac{1}{2}} + x^{-\frac{1}{2}}\right]^4$

which can be expanded using the form

$$(a+x)^n = a^n + na^{n-1}x + \frac{n(n-1)}{2!}a^{n-2}x^2 + \frac{n(n-1)(n-2)a^{n-3}x^3}{3!}$$

$$+ \ldots + \frac{n(n-1)(n-2)\ldots(n-r+2)}{(r-1)!}a^{n-r+1}x^{r-1} + \ldots + x^n$$

Therefore,

$$\left[(2x)^{-\frac{1}{2}} + x^{-\frac{1}{2}}\right]^4 = \left[(2x)^{-\frac{1}{2}}\right]^4 + 4\left[(2x)^{-\frac{1}{2}}\right]^3\left[x^{-\frac{1}{2}}\right] + \frac{4[3]}{2!}\left[(2x)^{-\frac{1}{2}}\right]^2\left[x^{-\frac{1}{2}}\right]^2$$

$$+ \frac{4\cdot3\cdot2}{3!}\left[(2x)^{-\frac{1}{2}}\right]\left[x^{-\frac{1}{2}}\right]^3 + \left[x^{-\frac{1}{2}}\right]^4$$

$$= \frac{1}{4x^2} + 4\left[(2x)^{-\frac{3}{2}}\right](x^{-\frac{1}{2}}) + 6\left[(2x)^{-1}\right]\left[x^{-1}\right]$$

$$+ 4\left[(2x)^{-\frac{1}{2}}\right](x^{-\frac{3}{2}}) + x^{-2}$$

$$= \frac{1}{4x^2} + \frac{\sqrt{2}}{x^2} + \frac{3}{x^2} + \frac{2\sqrt{2}}{x^2} + \frac{1}{x^2}$$

$$= x^{-2}\left[\frac{1}{4} + \sqrt{2} + 3 + 2\sqrt{2} + 1\right]$$

$$= x^{-2}\left[3\sqrt{2} + 4\frac{1}{4}\right]$$

(2) $(x + 4y^2)^{-5} = x^{-5} + (-5)x^{(-5)-1}(4y^2) + \dfrac{(-5)[(-5)-1]}{2!} x^{(-5)-2}(4y^2)^2$

$$+ \dfrac{(-5)[(-5)-1][(-5)-2]}{3!} x^{(-5)-3}(4y^2)^3 + \ldots$$

$$= x^{-5} - 20x^{-6}y^2 + 240x^{-7}y^4 - 2240x^{-8}y^6 + \ldots$$

For the expansion of $(a + b)^n$, where n is a real number but not a positive integer, the binomial theorem is still valid provided that $|a| > |b|$. However, the expansion of $(a + b)^n$ (for n is a real number but not a positive integer and $|a| > |b|$) is an endless succession of terms as seen in part (2) of this problem.

● **PROBLEM** 19-18

Express the coefficients of $(a+b)^0$, $(a+b)^1 \ldots (a+b)^9$ as Pascal's Triangle.

Solution:

$(a+b)^0$										1									
$(a+b)^1$									1		1								
$(a+b)^2$								1		2		1							
$(a+b)^3$							1		3		3		1						
$(a+b)^4$						1		4		6		4		1					
$(a+b)^5$					1		5		10		10		5		1				
$(a+b)^6$				1		6		15		20		15		6		1			
$(a+b)^7$			1		7		21		35		35		21		7		1		
$(a+b)^8$		1		8		28		56		70		56		28		8		1	
$(a+b)^9$	1		9		36		84		126		126		84		36		9		1

CHAPTER 20

LOGARITHMS AND EXPONENTIALS

> **Basic Attacks and Strategies for Solving Problems in this Chapter. See pages 354 to 401 for step-by-step solutions to problems.**

Logarithms (abbreviated "logs") can be considered to be the inverse function of *exponentials*. In other words, the equation $b^y = x$ and the equation $\log_b x = y$ indicate the same relationship between the three quantities b, y, and x. The answer (i.e., output) from finding the value of the exponential expression b^y is x, and if one wishes to go from this answer to the exponent of b which gave it, one uses the inverse function, the logarithm. Since the two equations express the same relationship between the three quantities, given one equation, one can rewrite the relationship using the alternate format.

Rules governing logarithms are derived from rules governing exponents (powers) (*cf.* Chapter 6). For example, let

$$a^m = b \quad \text{and} \quad a^n = c.$$

Then

$$\log_a b = m \quad \text{and} \quad \log_a c = n$$

(from the definition of logarithm given in the previous paragraph). Since

$$b \cdot c = a^m a^n = a^{m+n}$$

it follows that

$$\log_a(b \cdot c) = m + n = \log_a b + \log_a c.$$

Other rules can be proven in a similar way by referring to rules governing powers. The following are some of the major rules:

- $\log_b x + \log_b y = \log_b(x \cdot y)$

- $\log_b x - \log_b y = \log_b {}^x/_y$

- $\log_b b^x = x$

- $\log_b (x^n) = n(\log_b x)$

- $\log_b 1 = 0$

- $\log_b \,^1/_x = -\log_b x$

Although the base of the logarithm can be any number, there are three bases in common use: base 10 (called *common* logarithms), base 2 (often used in computer analysis), and base e (called *natural* or *Naperian* logarithms). Naperian logarithms are named after John Napier who is credited with discovering them. The number e is an irrational number which begins 2.718281828459045.... Sometimes natural logarithms are indicated ln and base 2 logarithms are indicated lg.

Tables of logarithms usually only give values for numbers between 1 and 10. Since $\log_{10} 1 = 0$ and $\log_{10} 10 = 1$, the values are between 0.0 and 1.0. These values are called the *mantissa* of the number. When the logarithm of a number less than 1 or greater than 10 is desired, the number is written in scientific (i.e., exponential) notation format (*cf.* Chapter 1), with a multiplier of 10 raised to some power. The power of 10 is called the *characteristic* and becomes the integer value added to the mantissa. For example, 2,000 can be written as 2×10^3. We know that $\log_{10} 2 = .30103$. Thus, using the rules for logarithms

$$\log_{10} 2000 = \log_{10} 2 + \log_{10} 10^3 = .30103 + 3 = 3.30103.$$

If a number is smaller than 1, the characteristic will be negative. In order to keep the mantissa positive, a negative characteristic is normally expressed by an overbar or by putting the negative characteristic *after* the mantissa. For example,

$$\log_{10} 0.2 = .30103 - 1 = -0.69897.$$

The *antilog* of a number is defined by reference to the definition of the logarithm. $\log_b x = a$ is the equivalent of $x = \text{antilog}_b a$. By reference to the definition of a logarithm, it is also true that $\text{antilog}_b a = b^a$.

The major usefulness of logarithms is related to the rules given above. The rules enable a person to translate multiplication into addition, and exponentiation into multiplication.

For example, given

$$y = \ln e^{3x},$$

we can rewrite the right side as $3x \cdot \ln e$, but since

$$\ln e = \ln e^1 = 1,$$

the equation reduces to $y = 3x$. If one knows the value of logarithms of certain primes, one can compute the logarithms of other numbers. For example, given that

$$\log_{10} 2 = .30103 \quad \text{and} \quad \log_{10} 3 = .47712$$

and rewriting 12 as $2^2 \cdot 3$ we can compute $\log_{10} 12$ as follows:

$$\log_{10} 12 = \log_{10} (2^2 \cdot 3) = 2 \log_{10} 2 + \log_{10} 3 = 2 \cdot (.30103) + .47712$$

$$= .60206 + .47712 = 1.07918.$$

Complicated numerical expressions can be computed using logs by first taking the log of the expression, transforming it into a simpler expression by using the rules given above, performing the calculations and finally taking the antilog of the result.

Since tables of logarithms normally give values for numbers with three or four significant digits, to obtain a logarithm for a number with more digits, *linear interpolation* is frequently used. We assume (for the sake of simplicity) that between any two values in a table, the progression of values is linear. Thus, we can make an educated guess to obtain a value between two values in a table. For example,

$$\log_{10} 5.600 = .74819 \quad \text{and} \quad \log_{10} 5.601 = .74827.$$

Now 5.6003 can be considered to be .3 of the way between 5.600 and 5.601 (considering this a unit jump). The difference between the corresponding logs is

$$.74827 - .74819 = .00008$$

and .3 of that is .000024. We truncate this to .00002 (five places as with the other log values) and add it to the $\log_{10} 5.600$ to obtain .74821.

Linear interpolation can also be used to find the antilog of a number.

The *colog* of a number is defined as follows:

$$\text{colog } n = \log {}^1\!/_n = - \log n.$$

Equations that have a variable in the exponent can often be solved by taking the log (with an appropriate base) of both sides of the equation. For example, given $2^{3x} = 8$, we can obtain $\log_2 2^{3x} = \log_2 8$ which can be rewritten as $3x = 3$ giving $x = 1$.

Step-by-Step Solutions to
Problems in this Chapter,
"Logarithms and Exponentials"

● **PROBLEM 20-1**

Find the following by inspection

(1) $\ln 50067400 - \ln 500674 - 2\ln 10$.

(2) $\ln e^{(5x^2+x)}$ where $x = 1$.

Solution: Before we solve either of these problems, a few rules about the natural log which are helpful are:

$$\ln a - \ln b = \ln \frac{a}{b}$$

$$\ln a + \ln b = \ln a \cdot b$$

$$\ln e^x = x$$

$$\ln a^n = n \ln a$$

(1) $\ln 50067400 - \ln 500674 - 2\ln 10$

$= \ln \dfrac{50067400}{500674} - 2\ln 10 = \ln 100 - 2\ln 10$

$= \ln 100 - \ln 10^2 = \ln 100 - \ln 100 = 0$

(2) $\ln e^{(5x^2+x)} = 5x^2+x = 5(1)^2+1 = 6$

● **PROBLEM 20-2**

Calculate the value of y at $x = 1$ for the equation

$$y = 2e^{\ln(x^3+2x^2+x+4)}$$

Solution: $y = 2e^{\ln(x^3+2x^2+x+4)}$

From the fact that $e^{\ln x} = x$, we get

$$y = 2(x^3+2x^2+x+4)$$
$$= 2x^3+4x^2+2x+8$$
$$= 2 + 4 + 2 + 8$$
$$= 16$$

● **PROBLEM 20-3**

Show that
$$a^{4\log_a b} = b^4.$$

Solution: Let $x = 4\log_a b$

$x = \log_a b^4$ Then, using the fact that

$m = \log_r c \Rightarrow r^m = c$, we get

$a^x = b^4$

Now, replacing x by $4\log_a b$, we get

$$a^{4\log_a b} = b^4.$$

● **PROBLEM 20-4**

Write $5^3 = 125$ in logarithmic form.

Solution: The statement $b^y = x$ is equivalent to the statement $\log_b x = y$, where b is the base and y is the exponent. The latter form is the logarithmic form. Thus $5^3 = 125$ in logarithmic form is $\log_5 125 = 3$, where the base is 5 and the logarithm of 125 is 3.

● **PROBLEM 20-5**

Find $\log_{10} 100$.

Solution: The following solution presents 2 methods for solving the given problem.

Method I.
   The statement $\log_{10} 100 = x$ is equivalent to $10^x = 100$.
   Since $10^2 = 100$, $\log_{10} 100 = 2$.

   Method II. Note that $100 = 10 \times 10$; thus
$\log_{10} 100 = \log_{10} (10 \times 10)$. Recall: $\log_x (a \times b)$
$= \log_x a + \log_x b$, therefore

   $$\log_{10} (10 \times 10) = \log_{10} 10 + \log_{10} 10$$

   $$= \quad 1 \quad + \quad 1$$

   $$= 2.$$

● **PROBLEM 20-6**

Write $\frac{1}{2} = \log_9 3$ in exponential form.

Solution:
   The statement $1/2 = \log_9 3$ in exponential form is
$9^{\frac{1}{2}} = 3$, where the base is 9 and the exponent is $\frac{1}{2}$.

● **PROBLEM 20-7**

If $\log_3 N = 2$, find N.

Solution:
   Since $\log_3 N = 2$ is equivalent to the equation $3^2 = N$, we obtain $N = 9$.

● **PROBLEM 20-8**

Find the value of x if $\log_4 64 = x$.

Solution:
   The exponential equivalent of
      $\log_4 64 = x$ is $4^x = 64$.

Since,

$$4^3 = 4 \cdot 4 \cdot 4 = 64$$

$$\log_4 64 = 3.$$

That is,

$$x = 3.$$

Express the logarithm of 7 to the base 3 in terms of common logarithms.

Solution:

If $\log_3 7 = x$, then $3^x = 7$. Take the logarithm of both sides:

$$\log 3^x = \log 7.$$

By the law of the logarithm of a power of a positive number which states that $\log a^n = n \log a$, $\log 3^x = x \log 3$. Hence, $x \log 3 = \log 7$. Divide both sides of this equation by $\log 3$:

$$\frac{x \, \log \, 3}{\log \, 3} = \frac{\log \, 7}{\log \, 3}.$$

Therefore, $x = \frac{\log \, 7}{\log \, 3} = \log_3 7$ is the logarithm of 7 to the base 3 expressed in terms of common logarithms.

Express the logarithm of $\dfrac{\sqrt{a^3}}{c^5 b^2}$ in terms of log a, log b and log c.

Solution: We apply the following properties of logarithms:

$$\log_b (P \cdot Q) = \log_b P + \log_b Q$$

$$\log_b (P/Q) = \log_b P - \log_b Q$$

$$\log_b (P^n) = n \log_b P$$

$$\log_b (\sqrt[n]{P}) = \frac{1}{n} \log_b P$$

Therefore,

$$\log \frac{\sqrt{a^3}}{c^5 b^2} = \log \frac{a^{3/2}}{c^5 b^2}$$

357

$$= \log a^{3/2} - \log\left(c^5 b^2\right)$$
$$= 3/2 \log a - \left(\log c^5 + \log b^2\right)$$
$$= 3/2 \log a - \log c^5 - \log b^2$$
$$= 3/2 \log a - 5 \log c - 2 \log b .$$

● **PROBLEM** 20-11

If $\log_{10} 3 = .4771$ and $\log_{10} 4 = .6021$, find $\log_{10} 12$.

Solution: Since $12 = 3 \times 4$,

$$\log_{10} 12 = \log_{10} (3)(4).$$

Since $\log_b (xy) = \log_b x + \log_b y$

$$\log_{10} (3 \times 4) = \log_{10} 3 + \log_{10} 4$$
$$= .4771 + .6021$$
$$= 1.0792.$$

Thus $\log_{10} 12 = 1.0792$.

● **PROBLEM** 20-12

Given that $\log_{10} 2 = 0.3010$ and $\log_{10} 3 = 0.4771$, find $\log_{10}\sqrt{6}$.

Solution: $\sqrt{6} = 6^{\frac{1}{2}}$, thus $\log_{10}\sqrt{6} = \log_{10} 6^{\frac{1}{2}}$. Since $\log_b x^y = y \log_b x$, $\log_{10} 6^{\frac{1}{2}} = \frac{1}{2} \log_{10} 6$. Therefore $\log_{10}\sqrt{6} = \frac{1}{2} \log_{10} 6$. $6 = 3 \cdot 2$, hence $\frac{1}{2} \log_{10} 6 = \frac{1}{2} \log_{10}(3 \cdot 2)$. Recall $\log_{10}(a \cdot b) = \log_{10} a + \log_{10} b$. Thus $\frac{1}{2} \log_{10}(3 \cdot 2) = \frac{1}{2}\left(\log_{10} 3 + \log_{10} 2\right)$. Replace our values for $\log_{10} 3$ and $\log_{10} 2$,

$$= \frac{1}{2}(0.4771 + 0.3010)$$
$$= \frac{1}{2}(0.7781)$$
$$\approx 0.3890$$

Therefore $\log_{10}\sqrt{6} = 0.3890$.

● **PROBLEM** 20-13

If $\log_4 7 = n$, find $\log_4 \frac{1}{7}$.

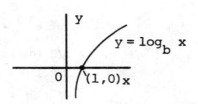

Solution: Apply the following property of logarithms $\log_b \frac{x}{y} = \log_b x - \log_b y$. Then,

$$\log_4 \frac{1}{7} = \log_4 1 - \log_4 7.$$

We note that if $x = 1$, $\log_b x = 0$. See the sketch. Thus $\log_4 1 = 0$, and if $\log_4 7 = n$, then $\log_4 \frac{1}{7} = 0 - n = -n$.

● **PROBLEM 20-14**

Find $\log_{10}(10^2 \cdot 10^{-3} \cdot 10^5)$.

Solution: Recall $\log_x (a \cdot b \cdot c) = \log_x a + \log_x b + \log_x c$. Thus

$$\log_{10}(10^2 \cdot 10^{-3} \cdot 10^5) = \log_{10}10^2 + \log_{10}10^{-3} + \log_{10}10^5$$

Recall $\log_b b^x = x$, since $b^x = b^x$; therefore, $\log_{10}10^2 + \log_{10}10^{-3} + \log_{10}10^5 = 2 + (-3) + 5 = 4$. Thus $\log_{10}(10^2 \cdot 10^{-3} \cdot 10^5) = 4$. Another method of finding $\log_{10}(10^2 \cdot 10^{-3} \cdot 10^5)$ is to note $10^2 \cdot 10^{-3} \cdot 10^5 = 10^{2+(-3)+5} = 10^4$ (because $a^x \cdot a^y \cdot a^z = a^{x+y+z}$). Thus $\log_{10}(10^2 \cdot 10^{-3} \cdot 10^5) = \log_{10}10^4 = 4$.

● **PROBLEM 20-15**

Find the values of the following logarithims:

    a) $\log_{10}10$ b) $\log_{10}100$ c) $\log_{10}1$

    d) $\log_{10}0.1$ e) $\log_{10}0.01$

Solution: The logarithmic expression $N = \log_b x$ is equivalent to $b^N = x$. Hence,

    a) Let $N_1 = \log_{10}10$. Then the logarithmic expression $N_1 = \log_{10}10$ is equivalent to $10^{N_1} = 10$. Since

$10^1 = 10$, $N_1 = 1$. Therefore, $N_1 = 1 = \log_{10} 10$.

b) Let $N_2 = \log_{10} 100$. Then the logarithmic expression $N_2 = \log_{10} 100$ is equivalent to $10^{N_2} = 100$. Since $10^2 = 100$, $N_2 = 2$. Therefore, $N_2 = 2 = \log_{10} 100$.

c) Let $N_3 = \log_{10} 1$. Then the logarithmic expression $N_3 = \log_{10} 1$ is equivalent to $10^{N_3} = 1$. Since $10^0 = 1$, $N_3 = 0$. Therefore, $N_3 = 0 = \log_{10} 1$.

d) Let $N_4 = \log_{10} 0.1 = \log_{10} \frac{1}{10}$. Then the logarithmic expression $N_4 = \log_1 0.1 = \log_{10} \frac{1}{10}$ is equivalent to $10^{N_4} = \frac{1}{10}$. Since $10^{-1} = \frac{1}{10^1} = \frac{1}{10}$, $N_4 = -1$. Therefore, $N_4 = -1 = \log_{10} 0.1$.

e) Let $N_5 = \log_{10} 0.01 = \log_{10} \frac{1}{100}$. Then the logarithmic expression $N_5 = \log_{10} 0.01 = \log_{10} \frac{1}{100}$ is equivalent to $10^{N_5} = \frac{1}{100}$. Since $10^{-2} = \frac{1}{10^2} = \frac{1}{100}$, $N_5 = -2$. Therefore, $N_5 = -2 = \log_{10} 0.01$.

● **PROBLEM 20-16**

Find the base b for which $\log_b 16 = \log_6 36$.

Solution: The statement $y = \log_b a$ is equivalent to the statement $b^y = a$. Thus, $x = \log_6 36$ is equivalent to $6^x = 36$. $6^2 = 36$, thus $x = 2$ and $\log_6 36 = 2$. Replacing $\log_6 36$ by 2 we obtain $\log_b 16 = 2$, or equivalently $b^2 = 16$. Thus $b = \sqrt{16} = 4$.

● **PROBLEM 20-17**

Evaluate $\log_{10} \sqrt[3]{7}$.

Solution: Since $\sqrt[a]{x} = x^{1/a}$, $\sqrt[3]{7} = 7^{\frac{1}{3}}$, and

$$\log_{10} \sqrt[3]{7} = \log_{10} 7^{\frac{1}{3}}$$

Recall the property of logarithms: $\log_b x^a = a \log_b x$.

Thus, $\log_{10} 7^{\frac{1}{3}} = \frac{1}{3} \log_{10} 7$.

From the table of common logarithms we find that $\log_{10} 7 = .8451$, thus

$$\frac{1}{3} \log_{10} 7 = \frac{1}{3} (.8451) = .2817$$

Therefore, $\log_{10} \sqrt[3]{7} = .2817$.

• **PROBLEM 20-18**

Find $\log \left(4 \frac{2}{7} \right)$.

Solution: $4 \frac{2}{7} = \frac{30}{7}$, thus $\log \left(4 \frac{2}{7} \right) = \log \frac{30}{7}$.

Since $\log_b \frac{x}{y} = \log_b x - \log_b y$,

$$\log \frac{30}{7} = \log 30 - \log 7.$$

Reducing 30 to prime factors,

$$\log \frac{30}{7} = \log (2 \times 3 \times 5) - \log 7$$

Recalling $\log_b xyz = \log_b x + \log_b y + \log_b z$,

$$\log \frac{30}{7} = \log 2 + \log 3 + \log 5 - \log 7.$$

From a log table we find

$$\log 2 = .3010$$
$$\log 3 = .4771$$
$$\log 5 = .6990$$
$$\log 7 = .8451$$

Hence, $\log \left(4 \frac{2}{7} \right) = .3010 + .4771 + .6990 - .8451 = .6320$.

• **PROBLEM 20-19**

Evaluate $\dfrac{\log_{10} 12}{\log_{10} 5}$.

361

Solution: First calculate $\log_{10}12$.

$$\log_{10}12 = \log_{10}(1.2 \times 10)$$

By the law of logarithms which states that $\log_b(x \cdot y) = \log_b x + \log_b y$,

$$\log_{10}12 = \log_{10}(1.2 \times 10) = \log_{10}1.2 + \log_{10}10$$

$$= 0.0792 + 1$$

$$= 1.0792$$

The $\log_{10}1.2$ was obtained from a table of common logarithms, base 10. Also, $\log_{10}5 = 0.6990$. This value was also obtained from a table of common logarithms, base 10.

$$\frac{\log_{10}12}{\log_{10}5} = \frac{1.0792}{0.6990} = 1.544.$$

● **PROBLEM 20-20**

Find the logarithm of 30,700.

Solution: First express 30,700 in scientific notation. $30,700 = 3.07 \times 10^4$. 4 is the characteristic. To find the mantissa, see a table of common logarithms of numbers The mantissa is 4871. Thus $\log 30,700 = 4 + .4871 = 4.4871$.

● **PROBLEM 20-21**

Find $\log 0.0364$.

Solution: $0.0364 = 3.64 \times 10^{-2}$. Therefore, the characteristic, the power of 10, is -2. From a table of logarithms, the mantissa for 3.64 is 0.5611. Therefore, $\log 0.0364 = -2 + 0.5611 = -1.4389$.

● **PROBLEM 20-22**

Find $\text{Antilog}_{10}\ 0.8762 - 2$.

Solution: Let $N = \text{Antilog}_{10}\ 0.8762 - 2$. The following relationship between log and antilog exists: $\log_{10}x = a$ is the equivalent of $x = \text{antilog}_{10}a$. Therefore,

$$\log_{10}N = 0.8762 - 2.$$

The characteristic is -2. The mantissa is 0.8762. The number that corresponds to this mantissa is 7.52. This number is found from a table of common logarithms, base 10. Therefore,

$$N = 7.52 \times 10^{-2}$$

$$= 7.52 \times \left(\frac{1}{10^2}\right)$$

$$= 7.52 \times \left(\frac{1}{100}\right)$$

$$= 7.52(.01)$$

$$N = 0.0752 .$$

Therefore, $N = \text{Antilog}_{10}\ 0.8762 - 2 = 0.0752$.

● PROBLEM 20-23

Find $\sqrt[5]{20}$.

Solution: Let $N = \sqrt[5]{20} = 20^{1/5}$. Then, taking the logarithm of both sides:

$$\log N = \log 20^{1/5};$$

and since $\log a^x = x \log a$, $= 1/5 \log 20$. Using a table of logs of numbers, we find that the mantissa for 20 is .3010; and the characteristic is 1, since for a number greater than 1 (in this case 20) the characteristic is positive and one less than the number of digits before the decimal. Thus, we have:

$$\log N = \frac{1.3010}{5} = .2602.$$

To find N look up .2602 in a Table of Mantissas of Common Logarithms. We find the closest number is 182. Since we have 0.2602, which has a characteristic of zero, then there is one digit to the left of the decimal point. Thus, we adjust the decimal point and N = 1.82. Thus,

$$N = \sqrt[5]{20} = 1.82.$$

● PROBLEM 20-24

Evaluate

$$\frac{542.3\sqrt{0.1383}}{32.72}$$ using logarithms.

Solution: Let x denote the above expression. Then,

$$x = \frac{542.3\sqrt{0.1383}}{32.72} .$$

Take the logarithms of both sides to obtain:

$$\log x = \log \frac{542.3(0.1383)^{1/2}}{32.72}$$

Apply the following properties of logarithms:

$$\log\left(\frac{a}{b}\right) = \log a - \log b$$

$$\log(a \cdot b) = \log a + \log b$$

$$\log a^n = n \log a$$

Then,

$$\log x = \log 542.3(0.1383)^{1/2} - \log 32.72$$
$$= \log 542.3 + \log(0.1383)^{1/2} - \log 32.72$$
$$= \log 542.3 + \frac{1}{2} \log(0.1383) - \log 32.72$$

To find the characteristic and mantissa of each number, express each number in powers of 10. The exponent is the corresponding characteristic

$$542.3 = 5.423 \times 10^2$$
$$0.1383 = 1.383 \times 10^{-1}$$
$$32.72 = 3.272 \times 10^1$$

Note that these numbers are expressed as numbers between 1 and 10, multiplied by a power of ten. Look up the corresponding number in a table of common logarithms of numbers. This is the mantissa. Then,

$$\log 542.3 = 2.7343$$

$$\log 0.1383 = -1.1408.$$

The form 19.1408 - 20 is more convenient for computation.

$$\log 32.72 = 1.5148$$

Hence

$$\tfrac{1}{2} \log 0.1383 = \tfrac{1}{2}(19.1408 - 20) = 9.5704 - 10,$$
$$\log 542.3 = 2.7342$$
$$\tfrac{1}{2} \log 0.1383 + \log 542.3 \qquad = 9.5704{-}10 + 2.7342 = 2.3046$$
$$\log 32.72 = 1.5148$$
$$\log 542.3 + \tfrac{1}{2} \log(0.1383) - \log 32.72 \qquad = 2.3046 - 1.5148 = 0.7898$$
$$\log x \qquad = 0.7898$$

Look up the mantissa and we find the number is 6163. To adjust the decimal point, we note that the characteristic is 0. Since the characteristic is positive, it is one less than the number of digits to the left of the decimal point. Here, there is one digit to the left of the decimal point. Thus, the number is 6.163, and

$$x = 6.163.$$

Therefore,

$$\frac{542.3 \sqrt{0.1383}}{32.72} = 6.163.$$

● PROBLEM 20-25

Evaluate $\sqrt{\dfrac{x^3 + 1}{x^3 - 1}}$ where $x = 1.47$.

Solution: $\log x = \log 1.47 = 0.1673$. This value can be found in a table of logarithms. By the law of the logarithm of a power of a positive number which states that

$$\log a^n = n \log a, \quad \log x^3 = 3 \log x = 3(0.1673)$$

$$= 0.5019.$$

Hence, x^3 = antilog 0.5019 = 3.18. This value is obtained from a table of logarithms by noting that the number that corresponds to the mantissa 0.5019 is approximately 3.18. Then $x^3 + 1 = 4.18$ and $x^3 - 1 = 2.18$. Let

$$N = \sqrt{\frac{x^3 + 1}{x^3 - 1}} = \sqrt{\frac{4.18}{2.18}} .$$

Take the logarithm of both sides of the above equation.

$$\log N = \log\sqrt{\frac{4.18}{2.18}} = \log\left(\frac{4.18}{2.18}\right)^{\frac{1}{2}}.$$

By the law of the logarithm of a power of a positive number which states that

$$\log a^n = n \log a, \quad \log\left(\frac{4.18}{2.18}\right)^{\frac{1}{2}} = \frac{1}{2} \log\left(\frac{4.18}{2.18}\right).$$

By the law of the logarithm of a quotient which states that $\log \frac{a}{b} = \log a - \log b$, $\log\left(\frac{4.18}{2.18}\right) = \log 4.18 - \log 2.18$. Hence,

$$\log N = \log\left(\frac{4.18}{2.18}\right)^{\frac{1}{2}} = \frac{1}{2} \log\left(\frac{4.18}{2.18}\right) = \frac{1}{2}\left(\log 4.18 - \log 2.18\right)$$

$$= \frac{1}{2}(0.6212 - 0.3385).$$

Note that the values for the two logs were found in a table of logarithms. Therefore,

$$\log N = \frac{1}{2}(0.6212 - 0.3385) = \frac{1}{2}(0.2827) = 0.1414.$$

$$N = \text{antilog } 0.1414 = 1.38.$$

Note that in a table of logarithms, the number that corresponds to the mantissa 0.1414 is approximately 1.38.

● **PROBLEM 20-26**

Use linear interpolation to find log 5.723.

Solution: Since 5.723 is .3 of the way from 5.72 to 5.73, we argue that log 5.723 is approximately .3 of the way from log 5.72 to log 5.73.

This is the basic idea involved in linear interpolation. We obtain log 5.72 and log 5.73 from a table of common logarithms, and find the mantissas to be 7574 and 7582, respectively. We now use interpolation to find the mantissa for 5.723.

        Note: Observe that 5.73 - 5.72 = 0.1, and
5.723 - 5.72 = .003; but we can rewrite these as 1 and
.3 by shifting the decimal two places.

$$\frac{.3}{1} = \frac{x}{8}$$

$$x = 2.4 \overset{\sim}{\sim} 2$$

        Thus, the mantissa of the given number is
7574 + 2 = 7576. Since the number is 5.723, and there is
one digit before the decimal point, we know that the
characteristic is one less than the number of digits, or
one less than one, or 0. Thus,

$$\log 5.723 = .7576.$$

● PROBLEM 20-27

Find log 0.7056.

Solution: First determine the characteristic by realiz-
ing that it will be one more than the number of zeros to
the right of the decimal point, with a negative sign
because the number is less than one. Thus, the characteristic is - 1. To compute the mantissa notice that the
number 7056 lies between 7050 and 7060, so that its
log will occur between the logs of those numbers. Interpolating we obtain

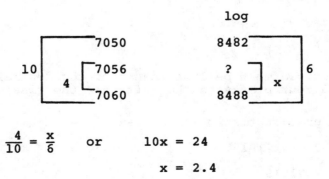

$$\frac{4}{10} = \frac{x}{6} \quad \text{or} \quad 10x = 24$$

$$x = 2.4$$

Now subtract this value from the higher mantissa

$$
\begin{array}{r}
8488 \\
2.4 \\
\hline
8485.6
\end{array}
$$

The mantissa is always less than one so the decimal point must be moved four places to the left.

Now - 1 can be written as 9 - 10 for convenience, so our final answer becomes

log 0.7056 = 9.84856-10

● **PROBLEM** 20-28

Find log 513.06

Solution: First determine the characteristic according to the rule that the characteristic is one less than the number of digits to the left of the decimal point. In this case it is two. Now find the mantissa by checking the mantissa for the number 51300 which is 7101 and the mantissa for 51400 which is 7110; the mantissa for 51306 will lie between these two mantissas. Now a proportion can be set up to determine the actual mantissa for the number.

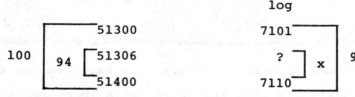

$\frac{94}{100} = \frac{x}{9}$, or cross-multiplying to obtain

(.94)9 = x = 8.46

Now subtract this value from the higher mantissa to obtain:

7110.00

−   8.46

7101.54

Since the mantissa must be less than one, the decimal must be moved four places to the left, and the final answer is

log 513.06 = characteristic + mantissa

= 2 + 0.71015

= 2.71015

Using a table of logarithms, determine the logarithm of 14.57.

Solution: This logarithm can be found by interpolation.

$$
\begin{array}{ccc}
x & & \log x \\
\end{array}
$$

$$
.010 \left[.007 \left[\begin{array}{c} 1.450 \\ 1.457 \\ 1.460 \end{array} \right. \quad \quad \begin{array}{c} .1614 \\ \log 1.457 \\ .1644 \end{array} \left] d \right] \right. .0030
$$

Set up the following proportion:

$$
\frac{.007}{.010} = \frac{d}{.0030}
$$

Cross multiply to obtain

$$
d = (.0030)\left(\frac{.007}{.010}\right)
$$

$$
= (3 \times 10^{-3})\left(\frac{7 \times 10^{-3}}{1 \times 10^{-2}}\right) = \frac{21 \times 10^{-6}}{1 \times 10^{-2}} = 21 \times 10^{-4}
$$

$$
= (21)(0.0001)
$$

$$
d = 0.0021
$$

Hence, $\log 1.457 = 0.1614 + 0.0021$

$$
= 0.1635
$$

Note $14.57 = 1.457 \times 10^{1}$. The characteristic is the exponent of 10, thus the characteristic is 1, and $\log 14.57 = 1 + 0.1635 = 1.1635$.

Find Antilog$_{10}$ 1.4850.

Solution: By definition, Antilog$_{10}$ $a = N$ is equivalent to $\log_{10} N = a$. Let Antilog$_{10}$ 1.4850 = N. Hence, Antilog$_{10}$ 1.4850 = N is equivalent to $\log_{10} N = 1.4850$. The characteristic is 1. The mantissa is 0.4850.

Therefore, the number that corresponds to this mantissa will be multiplied by 10^{1} or 10. The mantissas which appear in a table of common logarithms and are closest to the mantissa 0.4850 are 0.4843 and 0.4857. The number that corresponds to the mantissa 0.4850 will be found by interpolation.

$$
\begin{array}{cc}
\text{Number} & \text{Logarithms} \\
\end{array}
$$

$$
.01 \left[d \left[\begin{array}{c} 3.05 \\ x \\ 3.06 \end{array} \right. \quad \quad \begin{array}{c} 0.4843 \\ 0.4850 \\ 0.4857 \end{array} \left] .0007 \right] \right. .0014
$$

Set up the following proportion.

$$\frac{d}{.01} = \frac{.0007}{.0014}$$

cross-multiplying, $.0014d = (.01)(.0007)$, or $d = .01\left(\frac{.0007}{.0014}\right)$

$$= \left(1 \times 10^{-2}\right)\left(\frac{7 \times 10^{-4}}{1.4 \times 10^{-3}}\right)$$

$$= \frac{7 \times 10^{-6}}{1.4 \times 10^{-3}} = \frac{7}{1.4} \times \frac{10^{-6}}{10^{-3}} = 5 \times 10^{-6-(-3)}$$

$$= 5 \times 10^{-3} = 5 \times .001 = .005$$

Hence, $d = 0.005$

$$x = d + 3.05$$

$$= 0.005 + 3.050$$

$$= 3.055$$

Hence, $N = \text{Antilog}_{10} 1.4850 = 3.055 \times 10$

$$= 30.550$$

$$= 30.55$$

Therefore $\text{Antilog}_{10} 1.4850 = 30.55$.

● **PROBLEM** 20-31

Determine the value of x such that $10^x = 3.142$.

Solution: The statement $10^x = 3.142$ is equivalent by definition to $\log_{10} 3.142 = x$. Thus we must find $\log 3.142$,
using the following interpolation:

$$.01\left(.002 \left(\begin{array}{c|c} \text{Number} & \text{Log} \\ \hline 3.140 & .4969 \\ 3.142 & \\ 3.150 & .4983 \end{array}\right) x\right).0014$$

We set up the proportion,

$$\frac{.002}{.01} = \frac{x}{.0014}$$

Cross multiply to obtain,

$$.01x = .0000028$$

$$x = .00028$$

$$x \approx .0003$$

Thus $\log 3.142 = .4969 + .0003 = 0.4972$
Therefore $x = \log_{10} 3.142 = 0.4972$

Find the value of $(2.154)^5$.

Solution: We will use logs, in solving the given problem.

Let $x = (2.154)^5$. Now take the log of both sides:

$\log x = \log(2.154)^5$; and now by the rule $\log a^b = b \log a$ we obtain

$$\log x = 5 \log(2.154).$$

Log(2.154) is now found by interpolation, using a table of common logs. Notice 2.154 occurs between 2.150 and 2.160 which have recorded logs.

$$.010 \left[.006 \begin{bmatrix} N = 2.150 \\ N = 2.154 \\ N = 2.160 \end{bmatrix} \right. \qquad \begin{array}{l} \log x = .3324 \\ ? \\ \log x = .3345 \end{array} \left. \right] d \quad .0021$$

Now set up the proportion and use scientific notation.

$$\frac{.006}{.010} = \frac{d}{.0021}$$

$$\frac{6 \times 10^{-3}}{1 \times 10^{-2}} = \frac{d}{2.1 \times 10^{-3}}$$

$$\left(2.1 \times 10^{-3}\right)\left(\frac{6 \times 10^{-3}}{1 \times 10^{-2}}\right) = \left(2.1 \times 10^{-3}\right)\left(\frac{d}{2.1 \times 10^{-3}}\right)$$

$$\frac{12.6 \times 10^{-6}}{1 \times 10^{-2}} = d$$

$$12.6 \times 10^{-4} = d$$

$$(12.6)(0.0001) = d$$

$$0.00126 = d$$

or

$$d \approx 0.0013$$

Hence, $\log 2.154 = .3345 - d = .3345 - .0013$

$$\log 2.154 = .3332$$

Therefore,

$$\log x = 5 \log(2.154)$$

$$= 5(.3332)$$

$$\log x = 1.6660$$

The characteristic is 1. Therefore, the number that corresponds to the mantissa 0.6660 will be multiplied by 10^1 or 10. Using a table of common logarithms, the number that approximately corresponds to the mantissa is 4.63. Then,

$$x = (4.63)10$$

or $\qquad (2.154)^5 = x = 46.3$

Hence, $(2.154)^5 = 46.3$

Find $\sqrt[4]{36.91}$

Solution: We can use the rule for the log of a number raised to a power to find the solution.

$$\log a^b = b \log a$$

or $\log (36.91)^{\frac{1}{4}} = (\frac{1}{4}) \log 36.91$

log 36.91 lies between log 36.90 and log 37.00 Therefore, we set up the proportion:

$$
10 \begin{bmatrix} 36.90 & 5670 \\ 36.91 & ? \\ 37.00 & 5682 \end{bmatrix} x \bigg] 12
$$

$\frac{9}{10} = \frac{x}{12}$ or cross multiply to obtain

$(.9)(12) = x,$ $x = 10.8$

Now subtract 10.8 from 5682

$$
\begin{array}{r}
5682 \\
-\quad 10.8 \\
\hline
\end{array}
$$

5671.2. Since the mantissa must be less than one, move the decimal pint four places to the left i.e. (.56712). Now that the mantissa has been determined the characteristic is determined by examining the number of decimal places to the left of the decimal point of the original number. The characteristic is one less than this number. In this case the characteristic is 1. This process is used for characteristic determination because the number is greater than one.

$(\frac{1}{4}) \log 36.91 = \frac{1}{4} (1.56712) = .39178$

Now the answer is obtained by finding the number whose log is .39178. Again consult the table of common logarithms.

$$
180 \begin{bmatrix} & .39090 = \log 2.460 \\ 92 \begin{bmatrix} .39178 \\ .39270 = \log 2.470 \end{bmatrix} x & \end{bmatrix} 10
$$

Now set up a proportion.

$\frac{92}{180} = \frac{23}{45} = \frac{x}{10}$ or cross multiply to obtain

$$45x = 230 \quad \text{or} \quad x = \frac{230}{45},$$

$$x = 5.1$$

Now subtract 5.1 from 2470 to obtain

$$
\begin{array}{r}
2470.0 \\
-\quad 5.1 \\
\hline
2464.9
\end{array}
$$

Now convert to the answer by placing the decimal point between the two and four. (Note: from the above interpolation we knew the answer must be the integer 2 plus a rational part.)

$$\sqrt[4]{36.91} = 2.4649$$

● **PROBLEM** 20-34

Find $\text{Antilog}_{10} 0.5579 - 1$.

__Solution:__ By definition, $\text{Antilog}_{10} a = N$ is equivalent to $\log_{10} N = a$. Let $N = \text{Antilog}_{10} 0.5579 - 1$. Therefore, $\text{Antilog}_{10} 0.5579 - 1 = N$ is equivalent to $\log_{10} N = 0.5579 - 1$. The characteristic is -1. Therefore, the number that corresponds to the mantissa 0.5579 will be multiplied by 10^{-1}. The number that corresponds to the mantissa 0.5579 must be found through interpolation since only the mantissas 0.5575 and 0.5587 appear in a table of four-place logarithms, base 10.

	Number		log		
	3.61		0.5575		
.01 d	x		0.5579	.0004	.0012
	3.62		0.5587		

Set up the following proportion.

$$\frac{d}{.01} = \frac{.0004}{.0012}$$

$$d = .01\left(\frac{.0004}{.0012}\right)$$

$$= (1 \times 10^{-2})\left(\frac{4 \times 10^{-4}}{1.2 \times 10^{-3}}\right)$$

$$= \frac{4 \times 10^{-6}}{1.2 \times 10^{-3}}$$

$$= 3.3 \times 10^{-3}$$

$$= 3.3(.001)$$

$$d = 0.0033$$

Therefore, $x = 3.6100 + d$
$= 3.6100 + 0.0033$
$= 3.6133$

Hence, $\text{Antilog}_{10} 0.5579 - 1 = N = 10^{-1}(3.6133)$
$= 0.36133.$
$N \approx 0.3613.$

● **PROBLEM** 20-35

Find the product 5.06×71.32 by using logs and antilogs.

<u>Solution:</u> By definition, antilog $a = N$ is equivalent to log $N = a$. Now substitute the value for a in the antilog expression. Therefore, antilog $a = N$ becomes antilog(log N) $= N$. Let $N = 5.06 \times 71.32$; then, $5.06 \times 71.32 = $ antilog[log(5.06×71.32)]. Since $5.06 \times 71.32 = 5.06 \times 7.132 \times 10$, we write:

$$5.06 \times 71.32 = \text{antilog}[\log(5.06 \times 7.132 \times 10)] \qquad (1)$$

Evaluating the expression in the brackets:

$$\log(5.06 \times 7.132 \times 10) = \log 5.06 + \log 7.132 + \log 10.$$

This is true because of the following law of exponents:

$$\log abc = \log a + \log b + \log c.$$

Using a table of common logarithms to find the value of log 5.06 and noting that log 10 = 1,

$$\log(5.06 \times 7.132 \times 10) = 0.7042 + (\log 7.132) + 1$$
$$= 1.7042 + \log 7.132 \qquad (2)$$

We now evaluate log 7.132. The numbers that appear in a table of common logarithms which are closest to the number 7.132 are 7.13 and 7.14. The mantissa that corresponds to the number 7.132 will be found by interpolation.

		Number		Logarithm	
		7.13		0.8531	
.01	.002	7.132		x	d .0006
		7.14		0.8537	

Now, setting up the following proportion:

$$\frac{d}{.0006} = \frac{.002}{.01}$$

Cross-multiplying, $d = .0006\left(\frac{.002}{.01}\right)$

$$= \left(6 \times 10^{-4}\right)\left(\frac{2 \times 10^{-3}}{1 \times 10^{-2}}\right) = \frac{12 \times 10^{-4+(-3)}}{1 \times 10^{-2}}$$

$$= \frac{12 \times 10^{-7}}{1 \times 10^{-2}} = 12 \times 10^{-7-(-2)}$$

$$= 12 \times 10^{-5}$$

$$= \left(1.2 \times 10^1\right) \times 10^{-5} = 1.2 \times 10^{1+(-5)}$$
$$= 1.2 \times 10^{-4}$$
$$= 1.2 \times 0.0001$$
$$= 0.00012$$
$$\approx 0.0001$$

Hence, $\log 7.132 = x = 0.8531 + 0.0001$
$$= 0.8532.$$

Therefore, equation (2) becomes:
$$\log(5.06 \times 7.132 \times 10) = 1.7042 + 0.8532$$
$$= 2.5574$$

Equation (1) becomes:
$$5.06 \times 71.32 = \text{antilog}[2.5574] = M \qquad (3)$$

By definition, antilog[2.5574] is equivalent to log M = 2.5574. The characteristic is 2. The mantissa is 0.5574. The number that corresponds to this mantissa will be multiplied to 10^2 or 100. The mantissas which appear in a table of logarithms and are closest to the mantissa 0.5574 are 0.5563 and 0.5575. The number that corresponds to the mantissa 0.5574 can be found by interpolation.

	Number	Logarithm	
d	3.60	0.5563	.0011
	x	0.5574	
	3.61	0.5575	.0012

.01 on left, .0012 on right

Now setting up the following proportion:
$$\frac{d}{.01} = \frac{.0011}{.0012}$$

Cross-multiplying,
$$d = .01\left(\frac{.0011}{.0012}\right)$$
$$= \left(1 \times 10^{-2}\right)\left(\frac{1.1 \times 10^{-3}}{1.2 \times 10^{-3}}\right) = \frac{1.1 \times 10^{-2+(-3)}}{1.2 \times 10^{-3}}$$
$$= \frac{1.1 \times 10^{-5}}{1.2 \times 10^{-3}} = \frac{1.1}{1.2} \times 10^{-5-(-3)}$$
$$= 0.917 \times 10^{-2}$$
$$= 0.917 \times 0.01$$
$$= 0.00917$$
$$\approx 0.009$$

Hence, $x = 3.60 + 0.009$
$$= 3.600 + 0.009$$
$$= 3.609$$

Therefore, $M = 3.609 \times 10^2 = 360.9$. Equation (3) becomes:
$$5.06 \times 71.32 = M = 360.9$$

Calculate $\dfrac{50.73}{2.42}$, using logs and antilogs.

Solution: By definition, antilog $a = N$ is equivalent to $\log N = a$. Now substitute the value for a in the antilog expression. Therefore, antilog $a = N$ becomes antilog$(\log N) = N$. Let

$$N = \frac{50.73}{2.42} \; ;$$

then

$$\frac{50.73}{2.42} = \text{antilog}\left[\log\frac{50.73}{2.42}\right].$$

Since

$$\frac{50.73}{2.42} = \frac{5.073 \times 10}{2.42} \; ,$$

we write:

$$\frac{50.73}{2.42} = \text{antilog}\left[\log\frac{5.073 \times 10}{2.42}\right] \qquad (1)$$

Evaluating the expression in the brackets:

$$\log\frac{5.073 \times 10}{2.42} = (\log 5.073 + \log 10) - \log 2.42.$$

This statement is true by the two following laws of exponents:

$$\log ab = \log a + \log b$$

$$\log\frac{c}{d} = \log c - \log d$$

Noting that $\log 10 = 1$, and using a table of common logarithms to obtain $\log 2.42$, we obtain:

$$\log\frac{5.073 \times 10}{2.42} = (\log 5.073 + 1) - 0.3838,$$

or

$$\log\frac{5.073 \times 10}{2.42} = (\log 5.073) + 1 - 0.3838 \qquad (2)$$

Now, to find $\log 5.073$, we use interpolation.

Number		Logarithm	
5.07		0.7050	
.003 [5.073		x	] d .0009
5.08		0.7059	

.01

Set up the following proportion:

$$\frac{d}{.0009} = \frac{.003}{.01}$$

Cross-multiplying, $d = .0009\left(\dfrac{.003}{.01}\right)$

$$= \left(9 \times 10^{-4}\right)\left(\frac{3 \times 10^{-3}}{1 \times 10^{-2}}\right) = \frac{27 \times 10^{-4+(-3)}}{1 \times 10^{-2}}$$

$$= \frac{27 \times 10^{-7}}{1 \times 10^{-2}} = \frac{27}{1} \times 10^{-7-(-2)} = 27 \times 10^{-7+2}$$

$$= 27 \times 10^{-5}$$

$$= 2.7 \times 10^1 \times 10^{-5}$$
$$= 2.7 \times 10^{-4}$$
$$d = 0.00027 \approx 0.0003$$

Hence, $x = 0.7050 + d = 0.7050 + 0.0003$. Therefore,
$$\log 5.073 = x = 0.7053.$$

Rewriting equation (2):

$$\log \frac{5.073 \times 10}{2.42} = 0.7053 + 1 - 0.3838$$
$$= 1.7053 - 0.3838$$
$$= 1.3215$$

Therefore, equation (1) becomes:

$$\frac{50.73}{2.42} = \text{antilog}[1.3215] = M \qquad (3)$$

By definition, antilog[1.3215] is equivalent to log M = 1.3215. The characteristic is 1. The mantissa is 0.3215. The number that corresponds to this mantissa will be multiplied by 10^1 or 10. The mantissas which appear in a log table and are closest to the mantissa 0.3215 are 0.3201 and 0.3222. The number that corresponds to the mantissa 0.3215 will be found by interpolation.

$$.01 \begin{bmatrix} d \begin{bmatrix} 2.09 \\ x \\ 2.10 \end{bmatrix} & \begin{array}{c} 0.3201 \\ 0.3215 \end{array} \Big] .0014 \quad 0.0021 \\ 0.3222 \end{bmatrix}$$

Number		Logarithm

Now, set up the following proportion.

$$\frac{d}{.01} = \frac{.0014}{.0021}$$

Cross-multiplying, $d = (.01)\left(\frac{.0014}{.0021}\right)$

$$= \left(1 \times 10^{-2}\right)\left(\frac{1.4 \times 10^{-3}}{2.1 \times 10^{-3}}\right) = \frac{1.4 \times 10^{-2-3}}{2.1 \times 10^{-3}}$$

$$= \frac{1.4 \times 10^{-5}}{2.1 \times 10^{-3}} = \frac{1.4}{2.1} \times 10^{-5-(-3)}$$

$$= 0.67 \times 10^{-2}$$
$$= 0.67 \times 0.01$$
$$d = 0.0067 \approx 0.007$$

Hence, $x = 2.09 + d$
$$= 2.09 + 0.007$$
$$= 2.090 + 0.007$$
$$= 2.097$$

Therefore, $M = 2.097 \times 10 = 20.97$. Hence, equation (3) becomes:

$$\frac{50.73}{2.42} = \text{antilog}[1.3215]$$

$$= M$$

$$= 20.97 \ .$$

If $\operatorname{colog} a = b$, find $\log a$.

Solution: We will solve this problem for the general case, and then apply it to our specific case.

$$\underline{\operatorname{colog} N} = \log \frac{1}{N} = \log 1 - \log N$$

$$= 0 - \log N = -\log N$$

$$\operatorname{colog} a = b = -\log a$$

Therefore, $\log a = -b$

Write the following equations in logarithmic form.
(a) $3^4 = 81$ (b) $10^0 = 1$ (c) $M^k = 5$ (d) $5^k = M$

Solution: The expression $b^y = x$ is equivalent to the logarithmic expression $\log_b x = y$. Hence,

a) $3^4 = 81$ is equivalent to the logarithmic expression
$\log_3 81 = 4$

b) $10^0 = 1$ is equivalent to the logarithmic expression
$\log_{10} 1 = 0$

c) $M^k = 5$ is equivalent to the logarithmic expression
$\log_M 5 = k$

d) $5^k = M$ is equivalent to the logarithmic expression
$\log_5 M = k$.

The graph of an exponential function f contains the point $(2,9)$. What is the base of f?

Solution: Since f is an exponential function, we know that $f(x) = b^x$, where b, the base, is a positive number that we are to determine. An exponential function f may also be written as $y = f(x) = b^x$.

Since the exponential function f contains the point $(2,9)$,

$$9 = f(2) = b^2 \quad \text{or} \quad b^2 = 9$$
$$\sqrt{b^2} = \sqrt{9}$$
$$b = 3 .$$

Note that only the positive square root was taken, since for the base b, a positive number, is desired.

● **PROBLEM 20-40**

If f is the logarithmic function with base 4, find $f(4)$, $f\left(\frac{1}{4}\right)$, and $f(8)$.

Solution: Since f is the logarithmic function with base 4, then $y = f(x) = \log_4 x$. The values $f(4)$, $f\left(\frac{1}{4}\right)$, and $f(8)$ can be found by replacing x by 4, $\frac{1}{4}$, and 8 in the logarithmic function $y = f(x) = \log_4 x$. Hence, $f(4) = \log_4$. Let $N_1 = f(4) = \log_4 4$. By definition, $\log_x a = N$ is equivalent to $x^N = a$. Therefore, $N_1 = \log_4 4$ is equivalent to $4^{N_1} = 4$. Since $4^1 = 4$, $N_1 = 1$. Then, $N_1 = 1 = f(4)$.

For the second value $f\left(\frac{1}{4}\right)$, $f\left(\frac{1}{4}\right) = \log_4 \frac{1}{4}$. Let $N_2 = f\left(\frac{1}{4}\right) = \log_4 \frac{1}{4}$. Hence, $N_2 = \log_4 \frac{1}{4}$ is equivalent to $4^{N_2} = \frac{1}{4}$. Since $4^{-1} = \frac{1}{4^1} = \frac{1}{4}$, $N_2 = -1$. Then, $N_2 = -1 = f\left(\frac{1}{4}\right)$.

For the third value $f(8)$, $f(8) = \log_4 8$. Let $N_3 = f(8) = \log_4 8$. Hence, $N_3 = \log_4 8$ is equivalent to $4^{N_3} = 8$. Since $4 = 2^2$, $N_3 = \log_4 8$ is equivalent to $\left(2^2\right)^{N_3} = 8$ or $2^{2N_3} = 8$. Since $2^3 = 8$, $2N_3 = 3$. Dividing both sides of the equation $2N_3 = 3$ by 2:

$$\frac{2N_3}{2} = \frac{3}{2} \quad \text{or} \quad N_3 = \frac{3}{2} .$$

Then, $N_3 = \frac{3}{2} = f(8)$.

Solve the equation $\log_3(x^2 - 8x) = 2$.

Solution: The expression $\log_b a = y$ is equivalent to $b^y = a$. Hence, $\log_3(x^2 - 8x) = 2$ is equivalent to $3^2 = x^2 - 8x$. Therefore,

$$3^2 = x^2 - 8x$$

or $$9 = x^2 - 8x .$$

Subtract 9 both sides of this equation:

$$9 - 9 = x^2 - 8x - 9$$

$$0 = x^2 - 8x - 9.$$

Factoring this equation:

$$0 = (x - 9)(x + 1).$$

Whenever the product $ab = 0$ where a and b are any two numbers, either $a = 0$ or $b = 0$. Hence, either

$$x - 9 = 0 \quad \text{or} \quad x + 1 = 0$$

$$x = 9 \quad \text{or} \quad x = -1.$$

Solve for x in the equation $7^{2x-1} - 5^{3x} = 0$.

Solution: Writing the equation as $7^{2x-1} = 5^{3x}$, and equating logarithms of both members, we have

$$\log 7^{2x-1} = \log 5^{3x}$$

Recall $\log x^y = y \log x$, thus,

$$(2x - 1)\log 7 = 3x \log 5$$

Looking up $\log 7$ and $\log 5$ in our log table and substituting,

$$(2x - 1)(0.8451) = 3x(0.6990).$$

Hence

$$1.6902x - 0.8451 = 2.097x, \quad 0.4068x = -0.8451$$

and

$$x = -2.077.$$

Solve $8x^{\frac{3}{2n}} - 8x^{-\frac{3}{2n}} = 63$.

Solution: Multiply by $x^{\frac{3}{2n}}$ and transpose; thus

$$8x^{\frac{3}{2n}}\left(x^{\frac{3}{2n}}\right) - 8x^{-\frac{3}{2n}}\left(x^{\frac{3}{2n}}\right) = 63x^{\frac{3}{2n}}$$

$$8x^{\frac{3}{2n}+\frac{3}{2n}} - 8x^{-\frac{3}{2n}+\frac{3}{2n}} = 63x^{\frac{3}{2n}}$$

$$8x^{\frac{6}{2n}} - 8x^0 = 63x^{\frac{3}{n}}$$

$$8x^{\frac{3}{n}} - 8 \cdot 1 - 63x^{\frac{3}{2n}} = 0$$

$$8x^{\frac{3}{n}} - 63x^{\frac{3}{2n}} - 8 = 0.$$

Factor, $\left(x^{\frac{3}{2n}} - 8\right)\left(8x^{\frac{3}{2n}} + 1\right) = 0.$

Whenever a product of two numbers $ab = 0$ either $a = 0$ or $b = 0$. Hence,

$$x^{\frac{3}{2n}} - 8 = 0 \qquad \text{or} \qquad 8x^{\frac{3}{2n}} + 1 = 0$$

$$8x^{\frac{3}{2n}} = -1$$

$$x^{\frac{3}{2n}} = 8 \qquad\qquad x^{\frac{3}{2n}} = -\frac{1}{8}$$

$$x^{\frac{3}{2n}\cdot\frac{2n}{3}} = 8^{\frac{2n}{3}} \qquad\qquad x^{\frac{3}{2n}\cdot\frac{2n}{3}} = \left(\frac{-1}{8}\right)^{\frac{2n}{3}}$$

$$x = \left(2^3\right)^{\frac{2n}{3}} \qquad \text{or} \qquad x = \left(\frac{-1}{2^3}\right)^{\frac{2n}{3}}$$

$$x = \frac{(-1)^{\frac{2n}{3}}}{\left(2^3\right)^{\frac{2n}{3}}}$$

Note that $(-1)^{\frac{2n}{3}} = \sqrt[3]{(-1)^{2n}} = \sqrt[3]{1} = 1$

and $\left(2^3\right)^{\frac{2n}{3}} = \left[2^{3\left(\frac{1}{3}\right)}\right]^{2n} = 2^{2n}$. Therefore, $x = 2^{2n}$ or $x = \frac{1}{2^{2n}}$.

● PROBLEM 20-44

Express y in terms of x if
$$\log_b y = 2x + \log_b x .$$

<u>Solution:</u> Transposing $\log_b x$, we have

$$\log_b y - \log_b x = 2x,$$

A property of logarithms is that the logarithm of the quotient of two positive numbers S and T is equal to the difference of the logarithms of the numbers; that is,

$$\log_b \frac{S}{T} = \log_b S - \log_b T \ .$$

Therefore,

$$\log_b \frac{y}{x} = 2x \ .$$

Now use the definition of logarithm: The logarithm of N to the base b is $x = \log_b N$; and $b^x = N$ is an equivalent statement. Then,

$$2x = \log_b \frac{y}{x} \ \text{ is equivalent to}$$

$$b^{2x} = \frac{y}{x}$$

Solving for y we obtain:

$$y = x \cdot b^{2x}$$

● **PROBLEM 20-45**

Find x from the equation $a^x \cdot c^{-2x} = b^{3x+1}$.

<u>Solution:</u> An exponential equation is an equation involving one or more unknowns in an exponent. This can be solved by means of logarithms.

$$\log\left(a^x \cdot c^{-2x}\right) = \log b^{3x+1} \ .$$

Apply the following properties of logarithms. If P and Q are positive numbers, then:

(a) $\log_b (P \cdot Q) = \log_b P + \log_b Q$

(b) $\log_b\left(P^n\right) = n \log_b P$

Then, in this problem, we have:

$$\log a^x + \log c^{-2x} = \log b^{3x+1}$$

$$x \log a - 2x \log c = (3x + 1) \log b$$

$$x \log a - 2x \log c = 3x \log b + \log b$$

$$x \log a - 2x \log c - 3x \log b = \log b$$

$$x (\log a - 2 \log c - 3 \log b) = \log b$$

Hence,

$$x = \frac{\log b}{\log a - 2 \log c - 3 \log b} \ .$$

Solve the equation 2 log x - log 10x = 0.

Solution: We can use a fundamental property of logarithms to simplify the left-hand side of this equation.

The logarithm of the product of two or more positive numbers is equal to the sum of the logarithms of the several numbers. If P, Q, and R are positive numbers, then $\log (P \cdot Q \cdot R) = \log P + \log Q + \log R$.

$$2 \log x - \log 10x = 2 \log x - (\log 10 + \log x)$$

$$= 2 \log x - \log 10 - \log x$$

$$= \log x - \log 10.$$

But log 10 means that base 10 raised to what power = 10, or $10^? = 10$; and $10^1 = 10$. Thus, log 10 = 1, and the equation becomes: log x - 1 = 0.

Rewriting this equation:

$$\log x - 1 = 0$$

$$\log x = 1$$

Since the problem is in base 10, log x = 1 can be rewritten as,

$$10^1 = x. \text{ Thus } x = 10.$$

Solve $\log_2 (x - 1) + \log_2 (x + 1) = 3$.

Solution: Applying a property of logarithms, $\log_b x + \log_b y = \log_b xy$, to

$$\log_2 (x - 1) + \log_2 (x + 1) = 3$$

we get $\log_2 [(x - 1)(x + 1)] = 3$. $\log_b x = y$ is equivalent to $b^y = x$ by definition, thus $\log_2 [(x - 1)(x + 1)] = 3$ is equivalent to

$$(x - 1)(x + 1) = 2^3 = 8$$

$$x^2 = -1 = 8$$

$$x^2 - 9 = 0$$

$$0 = x^2 - 9 = x^2 - 3^2.$$

Thus we apply the formula for the difference of two squares, $a^2 - b^2 = (a + b)(a - b)$, replacing a by x and b by 3 and obtain $0 = x^2 - 3^2 = (x + 3)(x - 3)$. Whenever the product of two numbers $ab = 0$, either $a = 0$ or $b = 0$. Thus

$$(x + 3)(x - 3) = 0 \text{ means either}$$

$$x + 3 = 0 \text{ or } x - 3 = 0$$

and $\qquad\qquad x = -3 \text{ or } \quad x = 3.$

Therefore, $\{3, -3\}$ is the possible solution set, but we must check each in the given equation. This is necessary because we have not defined the logarithm of a negative number and, consequently, must rule out any value of x which would require the use of the logarithm of a negative number.

Check: Replacing x by 3 in our original equation

$$\log_2(x - 1) + \log_2(x + 1) = 3$$

$$\log_2(3 - 1) + \log_2(3 + 1) = 3$$

$$\log_2 2 + \log_2 4 = 3$$

$$1 + 2 = 3 \text{ since } 2^1 = 2 \text{ and } 2^2 = 4$$

$$3 = 3.$$

Replacing x by (-3) in our original equation

$$\log_2(x - 1) + \log_2(x + 1) = 3$$

$$\log_2(-3 - 1) + \log_2(-3 + 1) = 3$$

$$\log_2(-4) + \log_2(-2) = 3.$$

$x = -3$ cannot be accepted as a root because we have not defined the **logarithm** of a negative number. Thus our solution set is $\{3\}$.

● **PROBLEM 20-48**

Solve the equation $\log_{10}(x^2 + 3x) + \log_{10}5x = 1 + \log_{10}2x$.

Solution: We first subtract $\log_{10}2x$ from both sides of our equation so as to have the right-hand side free of logarithmic expressions and obtain

$$\log_{10}(x^2 + 3x) + \log_{10}5x - \log_{10}2x = 1$$

By the law of logarithms which states that $\log_b \frac{x}{y} = \log_b x - \log_b y$, $\log_{10}5x - \log_{10}2x = \log_{10}\frac{5x}{2x}$. Also, by the law of exponents which states that $\log_b(x \cdot y) = \log_b x + \log_b y$,

$$\log_{10}\left(x^2 + 3x\right) + \log_{10}5x - \log_{10}2x = \log_{10}\left(x^2 + 3x\right) + \log_{10}\frac{5x}{2x}$$

$$= \log_{10}\left(x^2 + 3x\right)\left(\frac{5x}{2x}\right)$$

$$= \log_{10}\frac{5x\left(x^2 + 3x\right)}{2x} = 1$$

Hence, $\log_{10}\dfrac{5x\left(x^2 + 3x\right)}{2x} = 1$ or $\log_{10}\dfrac{5\left(x^2 + 3x\right)}{2} = 1$. The expression $\log_{b}a = y$ is equivalent to $b^y = a$. Therefore, $\log_{10}\dfrac{5\left(x^2 + 3x\right)}{2} = 1$ is equivalent to $10^1 = \dfrac{5\left(x^2 + 3x\right)}{2}$. Hence, distributing:

$$10 = \frac{5x^2 + 15x}{2}$$

Multiply both sides of this equation by 2.

$$2(10) = 2\left(\frac{5x^2 + 15x}{2}\right)$$

$$20 = 5x^2 + 15x$$

Subtract 20 from both sides of this equation:

$$20 - 20 = 5x^2 + 15x - 20$$

$$0 = 5x^2 + 15x - 20$$

Factor out the common factor of 5 from the right side of this equation:

$$0 = 5\left(x^2 + 3x - 4\right)$$

Divide both sides of this equation by 5.

$$\frac{0}{5} = \frac{5\left(x^2 + 3x - 4\right)}{5}$$

$$0 = x^2 + 3x - 4$$

Factoring the right side of this equation:

$$0 = (x + 4)(x - 1)$$

Whenever the product $ab = 0$ where a and b are any two numbers, either $a = 0$ or $b = 0$. Therefore,

$$x + 4 = 0 \quad \text{or} \quad x - 1 = 0$$

$$x = -4 \quad \text{or} \quad x = 1$$

To check if these two values are indeed solutions, replace x by -4 and 1 in the original equation:
When $x = -4$,

$$\log_{10}\left((-4)^2 + 3(-4)\right) + \log_{10}5(-4) = 1 + \log_{10}2(-4)$$

$$\log_{10}(16 - 12) + \log_{10}(-20) = 1 + \log_{10}(-8)$$

$$\log_{10}4 + \log_{10}(-20) = 1 + \log_{10}(-8).$$

The number -4 is not a solution because the logarithm of a negative number does not exist. When $x = 1$,

$$\log_{10}\left((1)^2 + 3(1)\right) + \log_{10}5(1) = 1 + \log_{10}2(1)$$

$$\log_{10}(1+3) + \log_{10}5 = 1 + \log_{10}2$$

$$\log_{10}4 + \log_{10}5 = 1 + \log_{10}2$$

$$0.6021 + 0.6990 = 1.0000 + 0.3010$$
$$1.3011 \approx 1.3010.$$

Therefore, $x = 1$ is the only solution.

● PROBLEM 20-49

Solve the equation $x^{\log x} = 100x$.

Solution: By taking logarithms of both sides of the equation, we obtain the equivalent equation

$$\log \left(x^{\log x} \right) = \log 100x.$$

But, by the law of exponents which states $\log x^p = p \log x$,

$$\log \left(x^{\log x} \right) = (\log x)(\log x) = (\log x)^2.$$

Also, by another law of exponents which states

$$\log (x \cdot y) = \log x + \log y,$$

$$\log 100x = \log 100 + \log x.$$

Now, since $\log 100$ can be equivalently written as $\log_{10} 100$, then $\log_{10} 100 = x$ or $10^x = 100$; and we can replace $\log 100$ by 2 ($10^2 = 100$).

Thus, we have: $2 + \log x$.

We can therefore write our equation as

$$(\log x)^2 = 2 + \log x,$$

and so it is equivalent to the equation

$$(\log x)^2 - \log x - 2, \quad \text{and factoring:}$$

$$= (\log x - 2)(\log x + 1) = 0.$$

Now, $\quad \{(\log x - 2)(\log x + 1) = 0\}$

$$= \{x \mid \log x = 2 \text{ or } \log x = -1\}$$

$$= \{\log x = 2\} \cup \{\log x = -1\}.$$

Recall that when no base is expressed it is assumed to be 10. Thus, the equation $\log x = 2$ means $10^2 = x$, or $x = 100$; and $\log x = -1$ means $10^{-1} = x$, or $x = \frac{1}{10}$. Thus,

$$\{100\} \cup \left\{ \frac{1}{10} \right\} = \left\{ 100, \frac{1}{10} \right\},$$

and this is the set of numbers that solves the given equation.

Solve the equation $27^{x^2+1} = 243$.

Solution: We seek all numbers which satisfy the equation. If x is such a number, then

$$27^{x^2+1} = 243$$

Then, taking logarithms to the base 3 of both sides we have

$$\log_3 27^{x^2+1} = \log_3 243$$

Since $\log_b x^r = r \log_b x$, it follows that

$$(x^2 + 1)\log_3 27 = \log_3 243.$$

Note that the expression $\log_b x = y$ is equivalent to $b^y = x$. Hence, $\log_3 27 = N$ is equivalent to $3^N = 27$. Therefore, $N = 3$ and $\log_3 27 = 3$. Also, $\log_3 243 = M$ is equivalent to $3^M = 243$. Therefore, $M = 5$ and $\log_3 243 = 5$. Hence,

$$(x^2 + 1)3 = 5$$

or, by the commutative property of multiplication,

$$3(x^2 + 1) = 5.$$

Divide both sides of the equation by 3.

$$\frac{3(x^2 + 1)}{3} = \frac{5}{3}$$

$$x^2 + 1 = \frac{5}{3}$$

Subtract 1 from both sides of the equation.

$$x^2 + 1 - 1 = \frac{5}{3} - 1$$

$$x^2 = \frac{5}{3} - 1 = \frac{5}{3} - \frac{3}{3} = \frac{2}{3}$$

Therefore, $x = \pm \sqrt{\frac{2}{3}}$, i.e., $x = \sqrt{\frac{2}{3}}$ or $x = -\sqrt{\frac{2}{3}}$.

To check that each of these numbers satisfies the given equation, substitute each number for x in the given equation. Substituting $\sqrt{\frac{2}{3}}$ for x:

$$(27)^{(\sqrt{2/3})^2 + 1} = (27)^{(2/3)+1} = 27^{5/3} = \sqrt[3]{27^5} = \left(\sqrt[3]{27}\right)^5$$

$$= (3)^5$$

$$= 243 \checkmark$$

Substituting $-\sqrt{\frac{2}{3}}$ for x:

$$(27)^{(-\sqrt{2/3})^2+1} = (27)^{(2/3)+1} = 27^{5/3} = \sqrt[3]{27^5} = \left(\sqrt[3]{27}\right)^5$$

$$= (3)^5$$

$$= 243 \checkmark$$

Solve $2^{x+1} = 7^{x+2}$.

Solution: Take the logarithm of each side of the equation.
$\log 2^{x+1} = \log 7^{x+2}$. By the law of the logarithm of a power of a positive number which states that

$\log a^n = n \log a$, $\log 2^{x+1} = (x+1)\log 2$ and $\log 7^{x+2} = (x+2)\log 7$. Therefore: $(x+1)\log 2 = (x+2)\log 7$. Distributing, $x \log 2 + \log 2 = x \log 7 + 2 \log 7$. Subtract $x \log 7$ from both sides to obtain:

$x \log 2 + \log 2 - x \log 7 = x \log 7 + 2 \log 7 - x \log 7$

or $x \log 2 - x \log 7 + \log 2 = 2 \log 7$. Now, subtract $\log 2$ from both sides to obtain:

$x \log 2 - x \log 7 + \log 2 - \log 2 = 2 \log 7 - \log 2$

or $x \log 2 - x \log 7 = 2 \log 7 - \log 2$. Factoring out the common factor x from the left side:

$x(\log 2 - \log 7) = 2 \log 7 - \log 2$.

Dividing both sides by $\log 2 - \log 7$:

$$\frac{x (\log 2 - \log 7)}{\log 2 - \log 7} = \frac{2 \log 7 - \log 2}{\log 2 - \log 7}$$

$$x = \frac{2 \log 7 - \log 2}{\log 2 - \log 7}.$$

By the law of the logarithm of a power of a positive number which states that $\log a^n = n \log a$, $2 \log 7 = \log 7^2 = \log 49$. Therefore,

$$x = \frac{\log 49 - \log 2}{\log 2 - \log 7}.$$

By the law of the logarithm of a quotient which states that $\log \frac{a}{b} = \log a - \log b$, $\log 49 - \log 2 = \log \frac{49}{2}$ and $\log 2 - \log 7 = \log \frac{2}{7}$. Therefore,

$$x = \frac{\log 49 - \log 2}{\log 2 - \log 7} = \frac{\log \frac{49}{2}}{\log \frac{2}{7}}.$$

Solutions may be left in logarithmic form, as above. If a decimal approximation is desired, the final expression can be evaluated by means of a table of logarithms.

Solve for x: $\log(x + 1) + \log x = 1.3010$.

<u>Solution:</u> $\log(x + 1) + \log x = 1.3010$

Recall $\log a + \log b = \log(ab)$; thus $\log(x + 1) + \log x = \log(x + 1)x$.
Hence $\log (x + 1)(x) = 1.3010$. Take the antilog of each side,

$$\text{antilog}[\log(x + 1)(x)] = \text{antilog } 1.3010.$$

Since the antilog of $\log(x + 1)(x)$ is $(x + 1)(x)$,

$$x(x + 1) = \text{antilog } 1.3010$$

Evaluate antilog 1.3010. The antilog of 1.3010 is the number whose log is 1.3010. The characteristic is 1; hence to obtain the number multiply the antilog of the mantissa by 10'. The antilog of .3010 is 2.0, therefore our number is $2.0 \times 10' = 20$.

$$x^2 + x = 20$$

Add (-20) to both sides, $\qquad x^2 + x - 20 = 0$

Factor, $\qquad\qquad\qquad (x + 5)(x - 4) = 0$

Whenever a product of numbers $ab = 0$ either $a = 0$ or $b = 0$ thus $(x + 5)(x - 4) = 0$ means either $x + 5 = 0$ or $x - 4 = 0$, that is,

$$x = -5 \quad \text{or} \quad x = 4 .$$

The domain of the logarithmic function is the positive real numbers, thus the value -5 must be excluded, and our solution is $x = 4$.

Check: Replace x by 4 in our original equation,

$$\log(x + 1) + \log x = 1.3010$$
$$\log(4 + 1) + \log 4 = 1.3010$$
$$\log 5 + \log 4 = 1.3010$$
$$\log(5 \cdot 4) = 1.3010$$
$$\log 20 = 1.3010$$
$$\log(10 \cdot 2) = 1.3010$$
$$\log 10 + \log 2 = 1.3010$$
$$1 + .3010 = 1.3010$$
$$1.3010 = 1.3010$$

Find the inverse of the function

$$y = \ln\left(1 + \sqrt{1 - e^2 x^4}\right) - 2 \ln x - 1. \qquad (1)$$

<u>Solution:</u> Transfer all the natural logarithm functions to one side with the variables and constants on the other side. Thus,

$$y + 1 = \ln\left(1 + \sqrt{1 - e^2 x^4}\right) - 2 \ln x$$

Recall $a \log b = \log b^a$ thus $2\ln x = \ln x^2$ and

$$y + 1 = \ln\left(1 + \sqrt{1 - e^2 x^4}\right) - \ln x^2$$

Since $\ln a - \ln b = \ln \frac{a}{b}$, $\ln\left(1 + \sqrt{1 - e^2 x^4}\right) - \ln x^2 = \ln \frac{1 + \sqrt{1 - e^2 x^4}}{x^2}$

Thus,
$$y + 1 = \ln \frac{1 + \sqrt{1 - e^2 x^4}}{x^2},$$

The inverse function of the logarithmic function is the exponential function. $\ln u = v$ and $e^v = u$. Here

$$u = \frac{1 + \sqrt{1 - e^2 x^4}}{x^2}, \quad v = y + 1$$

$$e^{y+1} = \frac{1 + \sqrt{1 - e^2 x^4}}{x^2}.$$

Then, multiply by x^2 and subtract 1 from both sides to obtain:
$$x^2 e^{y+1} - 1 = \sqrt{1 - e^2 x^4}.$$

Rationalizing this equation, we get by squaring:
$$x^4 e^{2y+2} - 2x^2 e^{y+1} + 1 = 1 - e^2 x^4.$$

Now x must be positive in order that the term involving $\ln x$, in equation (1), have meaning. $\ln x$ exists for $x > 0$. In particular, x cannot be zero; hence, after subtracting 1 from each member of the last equation, we may further simplify by dividing by ex^2.

$$x^4 e^{2y+2} - 2x^2 e^{y+1} = -e^2 x^4 \quad \text{(subtracting 1 from each side)}$$

$$x^2 e^{2y+1} - 2e^y = -ex^2, \quad \left(\text{dividing by } ex^2\right)$$

$$x^2 e^{2y+1} + ex^2 = 2e^y$$

$$ex^2\left(e^{2y} + 1\right) = 2e^y, \quad \left(\text{factor out } ex^2\right)$$

Solve for x^2
$$x^2 = \frac{2e^y}{e\left(e^{2y} + 1\right)} = \frac{2}{e\left(e^y + e^{-y}\right)}.$$

Remembering that x must be positive, we extract only the positive square root to obtain

$$x = \sqrt{\frac{2}{e\left(e^y + e^{-y}\right)}}. \tag{2}$$

This is the required inverse function.

The form of relation (2) would seem to indicate that, if $x = x_1$ is the value corresponding to $y = y_1$, then $y = -y_1$ will also yield $x = x_1$. However, equation (1) shows that y cannot be replaced by $-y$ without creating thereby a different functional relation. In fact, y as given by (1) will be non-negative for every real value of x in its permissible range, $0 < x \leq e^{-\frac{1}{2}}$. $x > 0$ because $\ln 0$ does not exist. If $x \neq e^{-\frac{1}{2}}$, for example $x = e$, the

$$\ln\left(1 + \sqrt{1 - e^2 e^4}\right)$$

does not exist. The functional relation $y = \ln\left(1 - \sqrt{1 - e^2 x^4}\right) - 2 \ln x - 1$ yields only non-positive values of y and likewise leads to the same inverse relation (2). If you select an x from the domain $0 < x \leq e^{-\frac{1}{2}}$, for example $x = e^{-\frac{1}{2}}$, and substitute it in $y = \ln\left(1 - \sqrt{1 - e^2 x^4}\right) - 2 \ln x - 1$, then $y = \ln\left(1 - \sqrt{1 - e^2 e^{-4/2}}\right) - 2 \ln e^{-\frac{1}{2}} - 1$

$$= \ln\left(1 - \sqrt{1 - e^0}\right) - 2\left(-\tfrac{1}{2}\right) - 1$$

$$y = \ln(1) + 1 - 1 = 0.$$

Therefore, y is a non-positive value.

Our example thus illustrates the fact that all conclusions drawn from a deduced inverse function must be checked against the original relation.

● PROBLEM 20-54

Solve for x in the equation
$$3251 = 2184(1.02)^x.$$

Solution: Taking the logarithm of each member of the given equation we obtain
$$\log 3251 = \log\left[2184(1.02)^x\right]$$

Since $\log(ab) = \log a + \log b$,
$$\log 3251 = \log 2184 + \log(1.02)^x$$

since $\log a^b = b \log a$,
$$\log 3251 = \log 2184 + x \log 1.02.$$

Adding $(-\log 2184)$ to both sides,
$$x \log 1.02 = \log 3251 - \log 2184,$$

dividing both sides by $\log 1.02$,
$$x = \frac{\log 3251 - \log 2184}{\log 1.02}$$

Solving for our logarithms:
$$3,251 = 3.251 \times 10^3$$

Thus the characteristic is 3 and we interpolate to find log 3.251:

Number	log
3.250	5119
.01 .001 3.251	? x 13
3.260	5132

390

we set up the following proportion:

$$\frac{.001}{.01} = = \frac{x}{13}$$

Cross multiply to obtain, $.013 = .01x$

$$13 = 10x$$

$$x = \frac{13}{10} = 1.3$$

Thus $\log 3.251 = 5119 + 1.3 \approx .5120$, and

$\log 3.251 = 3 + .5120 = 3.5120$

$2,184 = 2.184 \times 10^3$

Thus the characteristic is 3 and we interpolate to find $\log 2.184$:

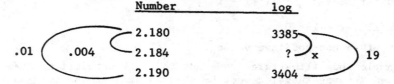

Number	log
2.180	3385
2.184	?
2.190	3404

We set up the following proportion:

$$\frac{.004}{.01} = \frac{x}{19}$$

cross multiply to obtain,

$$.076 = .01x$$
$$76 = 10x$$
$$x = 7.6$$

Thus $\log 2.184 = 3385 + 7.6 \approx 3393$ and

$\log 2.184 = 3 + .3393 = 3.3393$

Last, $\log 1.02$ can be found in a table of logarithms,

$\log 1.02 = 0.0086$

Substitute these values into our equation for x:

$$x = \frac{3.5120 - 3.3393}{0.0086}$$

$$= \frac{0.1727}{0.0086}$$

$$= 20.08$$

● PROBLEM 20-55

Solve the inequality $(.3)^x < \frac{4}{3}$.

Solution: This is an exponential inequality which involves one or more unknowns in an exponent, and is solved by means of logarithms. If $M < N$, then

$\log_b M < \log_b N$, and conversely. Thus $(.3)^x < \frac{4}{3}$ if,

391

and only if, $\log (.3)^x < \log \frac{4}{3}$. We can then use the fact that:

$\log a^b = b \log a$ and,

$\log \frac{a}{b} = \log a - \log b$. Thus,

from $\log (.3)^x < \log \frac{4}{3}$ we have:

$x \log (.3) < \log 4 - \log 3$,

Express log (.3). The characteristic of the common logarithm of any positive number smaller than 1 is negative and is obtained by adding one more than the number of zeros between the decimal point and the first digit. The mantissa is obtained by looking it up in a table of common logarithms. For .3, the characteristic is - 1 since it is less than one and there are no zeros between the decimal point and the first digit. Its mantissa is .4771. Thus log .3 = (.4771 - 1). The log 4 and log 3 can be obtained from a table of mantissas of common logarithms.

Thus, solving the inequality we obtain:

$x (.4771 - 1) < .6021 - .4771$,

$- .5229x < .1250$,

$x > - .239$.

Therefore, $\{x > - .239\}$ is the solution set of our given inequality.

● **PROBLEM** 20-56

Solve $\quad\quad\quad 5^{x+y} = 100 \quad\quad\quad\quad\quad\quad (1)$
and $\quad\quad\quad\quad 2^{2x-y} = 10 \quad\quad\quad\quad\quad\quad (2)$
for x and y.

Solution: If we equate the common logarithms of the members of each of (1) and of (2), we get

$$\log 5^{x+y} = \log 100$$
$$\log 2^{2x-y} = \log 10.$$

Recalling that $\log_b a^y = y \log_b a$,

$\log 5^{x+y} = (x + y) \log 5$ and $\log 2^{2x-y} = (2x - y) \log 2$.

We also know $\log_{10} 100 = 2$ $\left(\text{since } 10^2 = 100\right)$ and

$$\log_{10} 10 = 1 \left(\text{since } 10^1 = 10\right).$$

Substituting these values into equations (1) and (2) we obtain

$$(x + y)\log 5 = 2 \qquad\qquad (3)$$

$$(2x - y)\log 2 = 1. \qquad\qquad (4)$$

If we solve these equations for x + y and 2x - y we obtain

$$x + y = \frac{2}{\log 5}$$

$$2x - y = \frac{1}{\log 2}.$$

We observe in a log table that log 5 = .6990 and log 2 = .3010. Thus,

$$x + y = \frac{2}{\log 5} = \frac{2}{.6990} = 2.86 \qquad\qquad (5)$$

$$2x - y = \frac{1}{\log 2} = \frac{1}{.3010} = 3.32. \qquad\qquad (6)$$

Adding equations (5) and (6) we obtain

$$
\begin{array}{r}
x + y = 2.86 \\
2x - y = 3.32 \\
\hline
3x = 6.18 \\
x = 2.06.
\end{array}
$$

Substituting 2.06 for x in (5) and solving for y gives

$$2.06 + y = 2.86$$

$$y = 2.86 - 2.06$$

$$y = .80.$$

Therefore the solution set of this system is {(x,y)} = {(2.06, .80)}.

● **PROBLEM** 20-57

Under favorable conditions a single cell of the bacterium Escherichia coli divides into two about every 20 minutes. If this same rate of division is maintained for 10 hours, how many organisms will be produced from a single cell?

Solution: The 10-hour interval may be divided into 30 periods of 20 minutes each. This is because for each hour there are three twenty minute periods. Thus in 10 hours, there are 3 × 10 = 30 periods. At the end of the first period there are 2 bacteria. These two then divide and at the end of the second period there are $2 \cdot 2 = 2^2$ bacteria. Now, each bacterium (four in total) divide in two. At the end of the third period there are $8 = 2 \cdot 2 \cdot 2 = 2^3$ bacteria. Then each of these divide in two to form $16 = 2^4$ bacteria at the end of the fourth period.

A pattern has been set up where at the end of the nth period, there are 2^n bacteria. At the end of the 30th period there will be 2^{30} bacteria, which we will call N, the number we were seeking. It is easier to compute 2^{30} with logarithms.

$$N = 2^{30}$$

Taking logarithms: $\log N = \log 2^{30}$

Then since $\log x^y = y \log x,$

$$\log N = 30 \log 2$$

See a table of logarithms for log 2.

$$\log N = 30(0.3010) = 9.0300$$

The characteristic is 9 and its mantissa is 0.0300. Expressing this as a number, we have

$$10^{9.0300} = 10^{0.0300} \times 10^9 .$$

See a table of common mantissas of logarithms for the number corresponding to 0.0300. Note that the characteristic of 0.0300 which is zero, will be one less than the number of digits to the left of the decimal point of the corresponding number . Therefore, there is one digit to the left of the decimal point. Thus, the number is 1.072. Then we see that

$$N = 1.072 \times 10^9 = 1,072,000,000,$$

so a single cell is potentially capable of producing about a billion organisms in a 10-hour period.

● **PROBLEM 20-58**

From the given graph find as well as you can (a) $\log_e 1.5$, (b) $\log_e .5$, (c) the number x for which $\log_e x = 1.5$, and (d) the value of e.

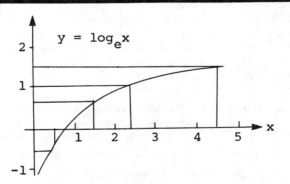

Solution: The smooth curve drawn is $y = \log_e x$. For

(a) $\log_e 1.5$ and (b) $\log_e .5$, find the corresponding x-

values, x = 1.5 and x = .5, and move along these vertical lines until you reach the curve, $y = \log_e x$. Then find

the y-values from the corresponding projections onto the

y - axis. We find that (a) $\log_e 1.5 = .4$, and (b)
$\log_e .5 = -.7$. For (c), we are given the ordinate. Move
along the horizontal line, $y = 1.5$, up to the curve of
$\log_e x$, and then find its abscissa, which is 4.5. Thus,
(c) $x = 4.5$. The number e satisfies the equation $\log_e x$
$= 1$, and from the figure, it appears that $e = 2.7$
(actually, to 5 decimal places, $e = 2.71828$).

● **PROBLEM 20-59**

Construct the graph of $y = \log_2 x$.

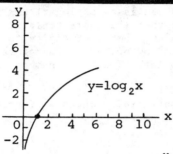

Solution: The equations $u = \log_b v$ and $v = b^u$ are equivalent.
Therefore, the relation $y = \log_2 x$ is equivalent to $x = 2^y$. Hence we
assume values of y and compute the corresponding values of x, getting
the table:

x:	$\frac{1}{8}$	$\frac{1}{4}$	$\frac{1}{2}$	1	2	4	8
y:	-3	-2	-1	0	1	2	3

For example, if $y = -3$, then $x = 2^y = 2^{-3} = \frac{1}{2^3} = \frac{1}{8}$.

The points corresponding to these values are plotted on the coordinate
system in the figure. The smooth curve joining these points
is the desired graph of $y = \log_2 x$. It should be noted that the graph
lies entirely to the right of the y-axis. The graph of $y = \log_b x$ for
any $b > 1$ will be similar to that in the figure. Some of the properties of this function which can be noted from the graph are:

    I. $\log_b x$ is not defined for negative values of x or zero.

   II. $\log_b 1 = 0$.

  III. If $x > 1$, then $\log_b x > 0$.

  IV. If $0 < x < 1$, then $\log_b x < 0$.

(a) Graph the functions

     (1) $y = \ln x$

     (2) $y = e^x$

(b) State the domain and range for (1) and (2).

Solution:

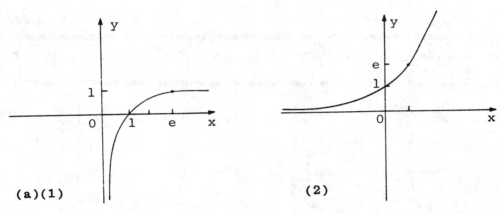

(a)(1)   (2)

(b)(1)   the domain for $y = \ln x$: all $x > 0$

         range for $y = \ln x$: all real numbers

   (2)   the domain for $y = e^x$: all real numbers

         range for $y = e^x$: all $y > 0$

Construct the graph of $y = 3^x$.

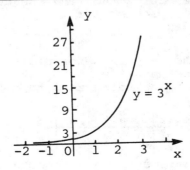

Solution: Assume values of x and compute the corresponding values of y by substituting into $y = 3^x$, obtaining the following table of values:

x:	-3	-2	-1	0	1	2	3
y:	$\frac{1}{27}$	$\frac{1}{9}$	$\frac{1}{3}$	1	3	9	27

The points corresponding to these pairs of values are plotted on the coordinate system of the figure and these points are joined by a smooth curve, which is the desired graph of the function. Note that the values of y are all positive. Furthermore, if x < 0, then y increases to a small extent as x does. If x > 0, y increases at a more rapid rate.

• **PROBLEM 20-62**

Graph the following functions: (A) $y = 2^x$, (B) $y = 4^x$, (C) 4^{-x}, (D) $y = 3 \cdot 2^x$.

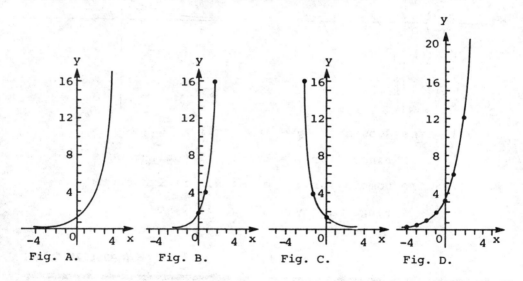

Fig. A.   Fig. B.   Fig. C.   Fig. D.

<u>(A) Solution:</u> When graphing a function y = f(x), set up a table consisting of two columns: one for x and one for y. Choose values for x and find the corresponding value for y. In this problem if:

$x = -4$, then $y = 2^x = 2^{-4} = \frac{1}{2^4} = \frac{1}{16}$.

Similarly, find other y values for different values of x. It is best to choose negative and positive values of x centering around and including zero to determine the nature of the graph.

x	-4	-3	-2	-1	0	1	2	3	4
y	$\frac{1}{16}$	$\frac{1}{8}$	$\frac{1}{4}$	$\frac{1}{2}$	1	2	4	8	16

Plot these points and draw a smooth curve through them. This is the graph of the exponential function $y = 2^x$.

397

Note that from the table and graph constructed, as you increase x by 1 each time moving from x = - 4 to x = 0, the y values increase slightly. However, when you move through the positive values of x, the change in y is much greater for each unit change in x.

See Figure A.

(B) Solution: Construct a table in the same manner as problem A. The table of values for the integers - 3 to 3 can be determined to be:

x	-3	-2	-1	0	1	2	3
y	$\frac{1}{64}$	$\frac{1}{16}$	$\frac{1}{4}$	1	4	16	64

Then plot these points and draw a smooth curve.

Figure B is the graph of $y = 4^x$ although it is not practical to plot the points corresponding to x = - 3 or x = 3 on this coordinate system.

When this curve is compared to the graph of $y = 2^x$ (Figure A), we see that the general shape is the same. Both curves pass through the point (0,1), that is, both have a y-intercept of 1. If we consider the curves to the left of the y-axes, we see that the curve of Figure B approaches the x-axis faster than the curve of Figure A. If we consider the same negative value on both curves, the y-value in Figure B is smaller than in Figure A. If x = - 3, then $y = 2^{-3} = \frac{1}{8}$ for Figure A and $y = 4^{-3} = \frac{1}{64}$ for Figure B. Thus (the) point $\left(- 3, \frac{1}{64}\right)$ is closer to the x-axis than $\left(- 3, \frac{1}{8}\right)$.

(C) Solution: Obtain a table of ordered pairs as in Examples A and B. Plot the points and draw the smooth curve by connecting them.

$$f(- 3) = 4^{-(-3)} = 4^3 = 64$$

$$f(- 2) = 4^{-(-2)} = 4^2 = 16$$

$$f(- 1) = 4^{-(-1)} = 4^1 = 4$$

$$f(0) = 4^0 = 1$$

$$f(1) = 4^{-1} = \frac{1}{4}$$

$$f(2) = 4^{-2} = \frac{1}{4^2} = \frac{1}{16}$$

$$f(3) = 4^{-3} = \frac{1}{4^3} = \frac{1}{64}$$

x	-3	-2	-1	0	1	2	3
y	64	16	4	1	$\frac{1}{4}$	$\frac{1}{16}$	$\frac{1}{64}$

Figure C is the graph of $y = 4^{-x}$.

398

The graph of $y = 4^x$ of the Figure B and the graph of $y = 4^{-x}$ of Figure C are mirror images of each other.

<u>(D) Solution:</u> In Example A we determined the values of 2^x for x an integer and $-4 \leq x \leq 4$. The values of y for this function then must be three times the corresponding values of y of Example A.

x	-4	-3	-2	-1	0	1	2	3	4
y	$\frac{3}{16}$	$\frac{3}{8}$	$\frac{3}{4}$	$\frac{3}{2}$	3	6	12	24	48

The graph of this function is shown in Figure D.

From these four examples we can see some of the features of the graph of $y = ab^x$, a > 0, and b > 0. The y-intercept of the function is a: If a > 1, the curve will approach the x-axis to the left of the y-axis and the y value increases as the x value increases. The graph will be in quadrants I and II.

The y-intercept of Examples A, B, and C is 1 since a = 1. It is true for Examples A and B that the curve approaches the x-axis as x becomes more negative and the y-value increases as x increases. However when a = 1 in Example C, the reverse occurs. As x decreases, y increases and the curve approaches the x-axis as x becomes more positive.

● **PROBLEM 20-63**

Plot the functions $y = 4^x$, $y = \log_4 x$, and $x = 4^y$ on the same graph.

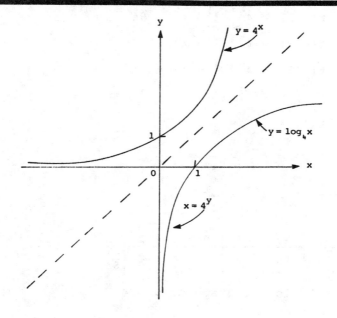

399

Solution:

Note that $x = 4^y$ is the same as $y = \log_4 x$ hence they have the same graph. Also, note that $y_1 = f_1(x) = \log_4 x$ is the inverse function of $y_2 = f_2(x) = 4^x$, and vice versa. Finally, observe that the graph of y_1 is the reflection of y_2 through the line $y = x$, and vice versa.

● PROBLEM 20-64

Solve for x and y, given

  (1) $3^{2x} \cdot 27^{y^2} + 9 = 90$ (1)

  (2) $2x + y^2 + 4 = 0$ (2)

Solution: From (1) one has

$$3^{2x} \cdot 27^{y^2} + 9 = 90$$

$$3^{2x} \cdot 3^{3y^2} + 9 = 90$$

$$3^{2x} \cdot 3^{3y^2} = 90 - 9 = 81 = 3^4$$

$$2x + 3y^2 - 4 = 0 \qquad\qquad (3)$$

With

$$2x + y^2 + 4 = 0$$

Simultaneous solution yields

$$2x + 3y^2 - 4 = 0$$

$$2x + y^2 + 4 = 0$$

$$2y^2 - 8 = 0$$

$$y^2 = 4$$

$$y = \pm 2$$

Solving for x by replacing y in (2) or (3) by ± 2, one obtains from (2)

$$2x + y^2 + 4 = 0$$

$$2x + 8 = 0$$

$$x = -4$$

Therefore, the solutions are $\begin{cases} x = -4 \\ y = 2 \end{cases}$ and $\begin{cases} x = -4 \\ y = -2 \end{cases}$

Check:

$$3^{2(-4)} \cdot 27^{(\pm 2)^2} + 9 = 3^{-8} \cdot (3^3)^4 + 3^2$$

400

$$= 3^{-8}3^{12} + 3^2 = 3^{12-8}+3^2 = 3^4 + 3^2$$

$$= 81 + 9 = 90$$

$$2 \times (-4) + (\pm 2)^2 + 4$$

$$= -8 + 4 + 4 = 0$$

● **PROBLEM** 20-65

Given $f(x) = (5a)^x$, find the inverse of $f(x)$, that is $f^{-1}(x)$.

<u>Solution:</u> Before we can solve this problem, we must know that if $f(x) = a^x$, then $f^{-1}(x) = \log_a x$. Therefore, when $f(x) = (5a)^x$,

$$f^{-1}(x) = \log_{5a} x$$

CHAPTER 21

PERMUTATIONS

Basic Attacks and Strategies for Solving Problems in this Chapter. See pages 402 to 408 for step-by-step solutions to problems.

A *permutation* refers to an arrangement of a group of objects in some definite order. If we have a pool of n different objects and we want to know how many different arrangements of k of those objects are possible, we refer to the "permutation of n different objects taken k at a time" and write

$$_nP_k, \quad p(n, k), \quad \text{or} \quad P(n, k).$$

The value of

$$P(n,k) \quad \text{is} \quad \frac{n!}{(n-k)!}.$$

Computing permutations is based on the *Fundamental Principle of Counting*, which states that if one thing can be done in m different ways, and a second thing can be done in n different ways, then the number of ways of performing these two acts in succession is $m \cdot n$.

Using these principles we can see that, given 4 objects (call them A, B, C, and D), if we take two of them at a time, there are 12 different permutations that result. This is true from the formula

$$P(4,2) = \frac{4!}{(4-2)!} = \frac{24}{2} = 12$$

and also from the Fundamental Principle, since there are 4 different first choices and 3 different second choices (after one was chosen). The 12 permutations are:

$$AB, AC, AD, BA, BC, BD, CA, CB, CD, DA, DB, DC.$$

Note that AB and BA, though they contain the same elements, are not the same permuation, since the *order* of the elements is different.

If we are given 3 objects (call them A, B, and C), and wanted to know the different permutations of all of them together, we would want the value of $P(3, 3) = 3! = 6$. The permutations are

$$ABC, ACB, BAC, BCA, CAB, CBA.$$

● **PROBLEM 21-1**

Find $_9P_4$.

Solution: Using the general formula for permutations of b different things taken a at a time, $_bP_a = \dfrac{b!}{(b-a)!}$, we substitute 9 for b and 4 for a. Hence $_9P_4 = \dfrac{9!}{(9-4)!} = \dfrac{9!}{5!}$. Evaluating our factorials, we obtain:

$$_9P_4 = \frac{9\cdot 8\cdot 7\cdot 6\cdot (5\cdot 4\cdot 3\cdot 2\cdot 1)}{(5\cdot 4\cdot 3\cdot 2\cdot 1)}$$

cancelling 5! in the numerator and denominator:

$$_9P_4 = 9\cdot 8\cdot 7\cdot 6$$
$$= 3,024 \quad.$$

● **PROBLEM 21-2**

Calculate the number of permutations of the letters a,b,c,d taken two at a time.

Solution: The first of the two letters may be taken in 4 ways (a,b,c, or d). The second letter may therefore be selected from the remaining three letters in 3 ways. By the fundamental principle the total number of ways of selecting two letters is equal to the product of the number of ways of selecting each letter, hence

$$p(4,2) = 4\cdot 3 = 12.$$

The list of these permutations is:

```
ab   ba   ca   da

ac   bc   cb   db

ad   bd   cd   dc.
```

● **PROBLEM** 21-3

Calculate the number of permutations of the letters
a,b,c,d taken four at a time.

<u>Solution:</u> The number of permutations of the four letters
taken four at a time equals the number of ways the four
letters can be arranged or ordered. Consider four places
to be filled by the four letters. The first place can be
filled in four ways choosing from the four letters. The
second place may be filled in three ways selecting one of
the three remaining letters. The third place may be filled
in two ways with one of the two still remaining. The fourth
place is filled one way with the last letter. By the
fundamental principle, the total number of ways of ordering
the letters equals the product of the number of ways of
filling each ordered place, or $4 \cdot 3 \cdot 2 \cdot 1 = 24 = P(4,4) = 4!$
(read 'four factorial').

In general, for n objects taken r at a time,

$$P(n,r) = n(n-1)(n-2)...(n-r+1) = \frac{n!}{(n-r)!} \quad (r < n).$$

For the special case where r = n,

$$P(n,n) = n(n-1)(n-2)...(3)(2)(1) = n!,$$

since (n-r)! = 0! which = 1 by definition.

● **PROBLEM** 21-4

How many permutations of two letters each can be formed from
the letters a,b,c,d,e? Actually write these permutations.

<u>Solution:</u> We recall the general formula for the number of
permutations of n different things taken r at a time
$_nP_r = n!/(n-r)!$. The number of permutations of 2 letters
that can be formed from the 5 given letters is $_5P_2$.

$$_5P_2 = \frac{5!}{(5-2)!} = \frac{5!}{3!} = \frac{5 \cdot 4 \cdot \cancel{3!}}{\cancel{3!}} = 20$$

Thus, the 20 permutations are:

```
ab    ac    ad    ae
ba    bc    bd    be
ca    cb    cd    ce
```

da   db   dc   de
ea   eb   ec   ed

● **PROBLEM** 21-5

> In how many ways may 3 books be placed next to each other
> on a shelf?

Solution: We construct a pattern of 3 boxes to represent
the places where the 3 books are to be placed next to each
other on the shelf:

Since there are 3 books, the first place may be filled in 3
ways. There are then 2 books left, so that the second place
may be filled in 2 ways. There is only 1 book left to fill
the last place. Hence our boxes take the following form:

3	2	1

The Fundamental Principle of Counting states that if one
thing can be done in a different ways and, when it is done
in any one of these ways, a second thing can be done in b
different ways, and a third thing can be done in c ways, ...
then all the things in succession can be done in a×b×c ...
different ways. Thus the books can be arranged in 3·2·1 = 6
ways.

This can also be seen using the following approach.
Since the arrangement of books on the shelf is important,
this is a permutations problem. Recalling the general
formula for the number of permutations of n things taken r
at a time, $_nP_r = n!/(n-r)!$, we replace n by 3 and r by 3
to obtain

$$_3P_3 = \frac{3!}{(3-3)!} = \frac{3!}{0!} = \frac{3\cdot2\cdot1}{1} = 6$$

● **PROBLEM** 21-6

> Candidates for 3 different political offices are to be
> chosen from a list of 10 people. In how many ways may
> this be done?

Solution: There are 10 choices for the first office, and to
go with each choice there are 9 choices for the second office,
and to go with each of these there are 8 choices for the
third office. The Fundamental Principle of Counting states
that if one thing can be done in a different ways, a second
thing can be done in b different ways, and a third thing
can be done in c different ways ..., then all the things in
succession can be done in a×b×c ... different ways. Hence,

there are $10 \cdot 9 \cdot 8 = 720$ ways of choosing the officers.

This can also be seen using the following approach. Since the arrangement of candidates on each slate is important (each arrangement represents people running for different political offices), this is a permutations problem. Recalling the general formula for the number of permutations of n things taken r at a time, $_nP_r = n!/(n-r)!$, we replace n by 10 and r by 3 to obtain

$$_{10}P_3 = \frac{10!}{(10-3)!} = \frac{10!}{7!} = \frac{10 \cdot 9 \cdot 8 \cdot \cancel{7}}{\cancel{7}!} = 720$$

● **PROBLEM** 21-7

A club wishes to select a president, vice-president and treasurer from five members. How many possible slates of officers are there if no person can hold more than one office?

Solution: There are five choices for president, and to go with each choice there are four choices for vice-president, and to go with each president-vice-president choice there are three choices for treasurer. The Fundamental Principle of Counting states that if one thing can be done in a different ways, a second thing can be done in b different ways, and a third thing can then be done in c ways ...,then all the things in succession can be done in a×b×c ... different ways. Hence, there are $5 \cdot 4 \cdot 3 = 60$ choices.
This can also be seen using the following approach. Since the arrangement of people on each slate is important, this is a permutations problem. Recalling the general formula for the number of permutations of n things taken r at a time, $_nP_r = n!/(n-r)!$, we replace n by 5 and r by 3 to obtain

$$_5P_3 = \frac{5!}{(5-3)!} = \frac{5 \cdot 4 \cdot 3 \cdot 2 \cdot 1}{2 \cdot 1} = 60$$

● **PROBLEM** 21-8

How many telephone numbers of four different digits each can be made from the digits 0,1,2,3,4,5,6,7,8,9?

Solution: A different arrangement of the same four digits produces a different telephone number. Since we are concerned with the order in which the digits appear, we are dealing with permutations.

There are ten digits to choose from and four different ones are to be chosen at a time. The general formula for the number of permutations of n things taken r at a time is

$$P(n,r) = \frac{n!}{(n-r)!} .$$

Here n = 10, r = 4, and the desired number is

$$P(10,4) = \frac{10!}{(10-4)!} = \frac{10!}{6!} = \frac{10 \cdot 9 \cdot 8 \cdot 7 \cdot 6!}{6!}$$

$$= 5040$$

Thus 5040 telephone numbers of four digits each can be made from the 10 digits.

In how many ways can the letters in the word "Monday" be arranged?

Solution: The word Monday contains 6 different letters. Since different letter arrangements yield different "words", we seek the number of permutations of 6 different objects taken 6 at a time.

Recall the general formula for the number of permutations of n things taken r at a time:

$$_nP_r = \frac{n!}{(n-r)!} \quad . \quad \text{Thus, } _6P_6 = \frac{6!}{(6-6)!} = \frac{6!}{0!} \quad .$$

Since 0! = 1 by definition,

$$_6P_6 = \frac{6!}{1} = 6 \cdot 5 \cdot 4 \cdot 3 \cdot 2 \cdot 1 = 720.$$

Thus, the letters in the word "Monday" may be arranged in 720 ways. We can arrive at the same conclusion using the fundamental theorem of counting, which states that for a given sequence of n events $E_1, E_2, \ldots E_n$, if for each i, E_i can occur m_i ways,

then the total number of distinct ways the event may take place is $m_1 \cdot m_2 \cdot m_3 \cdot \ldots \cdot m_n$.

Thus, the first of the 6 letters may be chosen 6 ways
        the second of the 6 letters may be chosen 5 ways
        the third of the 6 letters may be chosen 4 ways
        the fourth of the 6 letters may be chosen 3 ways
        the fifth of the 6 letters may be chosen 2 ways
        the sixth of the 6 letters may be chosen 1 way

Hence the total number of ways the letters may be arranged is 6 × 5 × 4 × 3 × 2 × 1 = 720.

Find the number of permutations of the seven letters of the word "algebra."

Solution: A permutation is an ordered arrangement of a set of objects. For example, if you are given 4 letters a,b,c,d and you choose two at a time, some permutations you can obtain are: ab, ac, ad, ba, bc, bd, ca, cb.

For n things, we can arrange the first object in n different ways, the second in n-1 different ways, the third can be done in n-2 different ways, etc. Thus the n objects can be arranged in order in

$$n! = n \cdot n-1 \cdot n-2 \ldots 1 \text{ ways}$$

Temporarily place subscripts, 1 and 2, on the a's to distinguish them, so that we now have $7! = 5040$ possible permutations of the seven distinct objects. Of these 5040 arrangements, half will contain the a's in the order a_1, a_2 and the other half will contain them in the order a_2, a_1. If we assume the two a's are indistinct, then we apply the following theorem. The number P of distinct permutations of n objects taken at a time, of which n_1 are alike, n_2 are alike of another kind,. . . ,n_k are alike of still another kind, with $n_1 + n_2 + .. + n_k = n$ is $P = \dfrac{n!}{n_1! \; n_2! \; \ldots \; n_k!}$ Then, here in this example, the 2 a's are alike so

$$P = \frac{7!}{2!} = 2520 \text{ permutations of the letters of}$$

the word algebra, when the a's are indistinguishable.

● **PROBLEM 21-11**

In how many ways may a party of four women and four men be seated at a round table if the women and men are to occupy alternate seats?

Solution: If we consider the seats indistinguishable, then this is a problem in circular permutations, as opposed to linear permutations. In the standard linear permutation approach each chair is distinguishable from the others. Thus, if a woman is seated first, she may be chosen 4 ways, then a man seated next to her may be chosen 4 ways, the next woman can be chosen 3 ways and the man next to her can be chosen in 3 ways ... Our diagram to the linear approach shows the number of ways each seat can be occupied.

4	4	3	3	2	2	1	1

By the Fundamental Principle of Counting there are thus $4 \cdot 4 \cdot 3 \cdot 3 \cdot 2 \cdot 2 \cdot 1 \cdot 1 = 576$ ways to seat the people.

However, if the seats are indistinguishable then so long as each person has the same two people on each side, the seating arrangement is considered the same. Thus we may suppose one person, say a woman is seated in a part-icular place, and then arrange the remaining three women and four men relative to her. Because of the alternate seating scheme, there are three possible places for the remaining three women, so that there are $3! = 6$ ways of seating them. There are four possible places for the four men, whence there are $4! = 24$ ways in which the men may be seated. Hence the total number of arrangements is $6 \cdot 24 = 144$. In general, the formula for circular per-mutations of n things and n other things which are alter-nating is $(n - 1)! n!$. In our case we have

$$(4 - 1)! 4! = 3! 4! = 3 \cdot 2 \cdot 4 \cdot 3 \cdot 2 = 144.$$

Prove this identity: $P(n,n-1) = P(n,n)$.

Solution: The general formula for the number of permutations of x objects taken r at a time is

$$P(x,r) = \frac{x!}{(x-r)!}$$

Thus, evaluating the left side of the given identity, we obtain

$$P(n,n-1) = \frac{n!}{[n-(n-1)]!} = \frac{n!}{(n-n+1)!} = \frac{n!}{1!} = \frac{n!}{1} = n!$$

Evaluating the right side of the identity we obtain

$$P(n,n) = \frac{n!}{(n-n)!} = \frac{n!}{0!}$$

$0! = 1$, by definition, hence $\frac{n!}{0!} = \frac{n!}{1} = n!$

Thus $P(n,n-1) = n! = P(n,n)$. Therefore, by the transitive property (If $a = b$ and $b = c$, $a = c$),

$$P(n,n-1) = P(n,n).$$

CHAPTER 22

COMBINATIONS

Basic Attacks and Strategies for Solving Problems in this Chapter. See pages 409 to 421 for step-by-step solutions to problems.

A *combination* refers to a collection of objects in which the order or arrangement is not considered. If we have a pool of n different objects and we want to know how many different collections of k of those objects are possible, we refer to the "combinations of n different objects taken k at a time" and write

$$_nC_k \quad \text{or} \quad C(n, k).$$

The value of

$$C(n,k) \text{ is } \frac{n!}{k!(n-k)!},$$

which is sometimes written $\binom{n}{k}$ (*cf.* Chapter 19).

There are always fewer combinations than permutations. For example, a collection of three objects (i.e., a combination) can be arranged in six different ways (i.e., the permutations). Thus, if we have m possible permutations of three objects (from some larger pool), there will be $^m/_{3!}$ combinations.

In dealing with problems in which one must count occurrences, one must first determine whether the problem asks for a simple application of the fundamental principle of counting, or for computing a permutation or a combination.

- In determining the number of configurations that two coins may have (i.e., which coin has heads or tails up), one is merely applying the fundamental principle. Since there are two possibilities for the first coin and two for the second, there are $2 \times 2 = 4$ total configurations possible.

- In determining the number of 2 person committees possible from a pool of 5 people, order is unimportant. Thus, this is a combinations problem. We want the value of

$$C(5,2) = \frac{5!}{3!2!} = \frac{5 \cdot 4}{2!} = 10.$$

- In determining the number of ways that two people out of a group of 5 people can hold the separate offices of president and vice-president, order is important (i.e., A being president and B being vice-president is a different situation than if B were president and A were vice-president). Thus, this is a permutations problem. We want the value of

$$P(5, 2) = {}^{5!}/_{3!} = 5 \cdot 4 = 20.$$

Some questions ask information about groups of different sizes, for example, groups of *at least three* members. In these cases, one must compute the successive values of combinations or permutations starting with 3 members and incrementing to the maximum possible. For example, asking how many committees are possible that have at least 3 people (out of a pool of 5) is equivalent to asking how many 3 person committees and 4 person committees and 5 person committees are possible. Since, in this type of question, the order of the group is not important, it is a combination question, rather than a permutation question.

In both combination and permutation problems, one normally assumes distinct objects which cannot be reused in a specific group. Problems that deal with letters of the alphabet sometimes do not assume this restriction, since the same letter can appear more than once in a word (i.e., a group). In these types of problems, one must merely apply the fundamental principle of counting carefully. For example, if we wish to know the maximum number of three letter words that begin and end with a consonant and have a vowel (a, e, i, o, or u) as the middle letter, we would have $21 \cdot 5 \cdot 21 = 2,205$ since there are 21 consonants as the first or last letters and 5 vowels as the middle letter. The number 21 is used twice, since it is permitted to have the same consonant to begin and to end a word (e.g., *pop, mom*). If we demanded that no letter be used twice, we would have $21 \cdot 5 \cdot 20 = 2,100$ possibilities.

Step-by-Step Solutions to Problems in this Chapter, "Combinations"

● **PROBLEM 22-1**

Find the value of C(n,0).

__Solution:__ Starting with the formula for combinations:

$$C(n,r) = \frac{n!}{(n - r)!\,r!}$$

and substituting r = 0, we have

$$C(n,0) = \frac{n!}{(n - 0)!\,0!}$$

Hence,

$$C(n,0) = \frac{n!}{n!\,0!}$$

Recall that $0! = 1$ by definition. Then

$$C(n,0) = \frac{n!}{n!}$$

$$= 1 .$$

● **PROBLEM 22-2**

Find $_9C_4$.

__Solution:__ By definition, combinations of b different things taken a at a time, $_bC_a = \dfrac{b!}{a!\,(b-a)!}$; hence by substitution, $_9C_4 = \dfrac{9!}{4!\,(9-4)!}$.

$$= \frac{9!}{4!\,5!}$$

$$= \frac{9.8.7.6.(5.4.3.2.1)}{4.3.2.1.(5.4.3.2.1)}$$

Cancelling $5! = (5.4.3.2.1)$ out of numerator and denominator, we multiply to obtain:

$$_9C_4 = \frac{3024}{24} = 126 \quad .$$

How many different five-card hands can be obtained from a fify-two card deck?

Solution: Since the order of the cards is unimportant, we are dealing with a combination problem, as opposed to a permutation problem. Recall our general formula for combinations $_bC_a = \frac{b!}{a!(b-a)!}$.

We want to know how many $\underline{5}$ card hands can be obtained from a $\underline{52}$ card deck, hence $a = 5$, $b = 52$. Substituting 5 for a and 52 for b in our general formula we obtain:

$$_{52}C_5 = \frac{52!}{5!(52-5)!} = \frac{52!}{5!(47)!} \quad .$$

Recall that $n! = n \cdot (n-1). = n \cdot (n-1) \cdot (n-2). = n \cdot (n-1)(n-2)(n-3) \ldots$

hence $52! = \frac{52 \cdot 51 \cdot 50 \cdot 49 \cdot 48 \cdot (47).}{5.(47).}$

Cancelling $(47)!$ from numerator and denominator, and evaluating $5!$:

$$\frac{52 \cdot 51 \cdot 50 \cdot 49 \cdot 48}{5 \cdot 4 \cdot 3 \cdot 2 \cdot 1} \quad .$$

Performing the necessary multiplications and divisions:

$$= 2,598,960 \quad .$$

Thus 2,598,960 different five-card hands can be obtained from a fifty-two card deck.

In how many different ways may a pair of dice fall?

Solution: The Fundamental Principle of Counting states that if one thing can be done in a different ways, and when it is done in any one of these ways, a second thing can be done in b different ways, then both things in succession can be done in $a \times b$ different ways. A die has 6 sides, thus it may land in any of six ways. Since each die may land in 6 ways, by the Fundamental Principle both die may fall in $6 \times 6 = 36$ ways. We can verify this result by enumerating all the possible ordered pairs of dice throws:

1,1	1,2	1,3	1,4	1,5	1,6
2,1	2,2	2,3	2,4	2,5	2,6
3,1	3,2	3,3	3,4	3,5	3,6
4,1	4,2	4,3	4,4	4,5	4,6
5,1	5,2	5,3	5,4	5,5	5,6
6,1	6,2	6,3	6,4	6,5	6,6

● **PROBLEM 22-5**

In how many ways can we select a committee of 3 from a group of 10 people?

Solution: The arrangement or order of people chosen is unimportant. Thus this is a combinations problem. Recalling the general formula for the number of combinations of n different things taken r at a time

$$C(n,r) = \frac{n!}{r!(n-r)!} \quad ,$$

the number of committees of 3 from a group of 10 people is

$$C(10,3) = \frac{10!}{3!(10-3)!} = \frac{10!}{3!7!} = \frac{10 \cdot \overset{3}{9} \cdot \overset{4}{8} \cdot 7!}{3 \cdot 2 \cdot 1 \cdot 7!} = 120.$$

● **PROBLEM 22-6**

How many committees of four members each can be formed from a group of seven persons?

Solution: This is a problem in combinations, rather than permutations, since the order is of no consequence. Thus, a committee consisting of Smith, Jones, Young, and Robinson is the same as a committee consisting of Smith, Robinson, Jones, and Young. The number of combinations of n different objects taken r at a time is equal to:

$$\frac{n(n-1)\ldots(n-r+1)}{1 \cdot 2 \cdots r}.$$

In this example, n = 7, r = 4, therefore

$$c(7,4) = \frac{7 \cdot 6 \cdot 5 \cdot 4}{1 \cdot 2 \cdot 3 \cdot 4} = 35.$$

Alternately: The first member may be selected from the seven persons in 7 ways. The second member may be selected from the remaining six people in 6 ways. The third member may be selected in 5 ways from the remaining five people. The fourth member may be selected from the remaining four people in 4 ways. By the fundamental principle the total number of ways of picking the four members is equal to the product of the number of ways of picking each member, or $7 \cdot 6 \cdot 5 \cdot 4$ ways. This is a permutation of 7 people selected 4 at a time. To account for the number of ways in which the same four-person committee is selected, but in a different order, divide $7 \cdot 6 \cdot 5 \cdot 4$ by the number of ways in which the same committee of four can be selected. This equals a permutation of 4 people selected 4 at a time. This, by the fundamental principle, applied as above, equals $4 \cdot 3 \cdot 2 \cdot 1$. Then

$$\frac{7 \cdot 6 \cdot 5 \cdot 4}{4 \cdot 3 \cdot 2 \cdot 1} = 35.$$

● **PROBLEM 22-7**

How many baseball teams of nine members can be chosen from among twelve boys, without regard to the position played by each member?

Solution: Since there is no regard to position, this is a combinations problem (if order or arrangement had been important it would have been a permutations problem). The general formula for the number of combinations of n things taken r at a time is

$$C(n,r) = \frac{n!}{r!(n-r)!} \cdot$$

We have to find the number of combinations of 12 things taken 9 at a time. Hence we have

$$C(12,9) = \frac{12!}{9!(12-9)!} = \frac{12!}{9!3!} = \frac{12 \cdot 11 \cdot 10 \cdot 9!}{3 \cdot 2 \cdot 1 \cdot 9!} = 220$$

Therefore, there are 220 possible teams.

● **PROBLEM 22-8**

A manufacturer produces 7 different items. He packages assortments of equal parts of 3 different items. How many different assortments can be packaged?

Solution: Since we are not concerned with the order of the items, we are dealing with combinations. Thus the number of assortments is the number of combinations of 7 items taken 3 at a time. Recall the general formula for the number of combinations of n items taken r at a time,

$$C(n,r) = \frac{n!}{r!(n-r)!}$$

$$C(7,3) = \frac{7!}{3!(7-3)!}$$

412

$$= \frac{7!}{3!4!}$$

$$= \frac{7 \cdot \cancel{6} \cdot 5 \cdot \cancel{4}!}{\cancel{3} \cdot \cancel{2} \cdot \cancel{4}!}$$

$$= 35$$

Thus, 35 different assortments can be packaged.

● PROBLEM 22-9

A Sunday school class of 12 members is to be seated on seven chairs and a bench that accommodates five persons. In how many ways can the bench be occupied?

Solution: If we are concerned with the order of people on the bench (so that we consider the same five people sitting in different arrangements as distinct ways), then this is a permutations problem. Recalling the general formula for the number of permutations of n elements taken r at a time

$$p(n,r) = \frac{n!}{(n-r)!}$$

we find the number of permutations of 12 elements taken 5 at a time, or p(12,5). Thus

$$p(12,5) = \frac{12!}{(12-5)!} = \frac{12!}{7!} = 12 \cdot 11 \cdot 10 \cdot 9 \cdot 8 = 95,040$$

If we are not concerned with the order of the people on the bench this becomes a combinations problem. Recalling the general formula for the number of combinations of n elements taken r at a time

$$c(n,r) = \frac{n!}{r!(n-r)!}$$

we find the number of combinations of 12 elements taken 5 at a time, or c(12,5).

$$c(12,5) = \frac{12!}{5!(12-5)!} = \frac{12!}{5!7!} = \frac{12 \cdot 11 \cdot 10 \cdot 9 \cdot 8 \cdot 7!}{5 \cdot 4 \cdot 3 \cdot 2 \cdot 7!} = 792.$$

● PROBLEM 22-10

How many different sums of money can be obtained by choosing two coins from a box containing a penny, a nickel, a dime, a quarter, and a half dollar?

Solution: The order makes no difference here, since a selection of a penny and a dime is the same as a selection of a dime and a penny, insofar as a sum of money is concerned. This is a case of combinations, then, rather than permutations. Then the number of combinations of n different objects taken r at a time is equal to:

$$\frac{n(n-1)\ldots(n-r+1)}{1 \cdot 2 \cdots r} \; .$$

In this example, n = 5, r = 2, therefore

$$C(5,2) = \frac{5 \cdot 4}{1 \cdot 2} = 10.$$

As in the problem of selecting four committee members from a group of seven people, a distinct two coins can be selected from five coins in

$$\frac{5 \cdot 4}{1 \cdot 2} = 10 \text{ ways (applying the fundamental principle).}$$

● **PROBLEM 22-11**

How many "words" each consisting of two vowels and three consonants, can be formed from the letters of the word "integral"?

Solution: To find the number of ways to choose vowels or consonants from letters, we use combinations. The number of combinations of n different objects taken r at a time is defined to be

$$C(n,r) = \frac{n!}{r!(n-r)!} .$$

Then, we first select the two vowels to be used, from among the three vowels in integral; this can be done in C(3,2) = 3 ways. Next, we select the three consonants from the five in integral; this yields C(5,3) = 10 possible choices. To find the number of ordered arrangements of 5 letters selected five at a time, we need to find the number of permutations of choosing r from n objects. Symbolically, it is P(n,r) which is defined to be

$$P(n,r) = \frac{n!}{(n-r)!}$$

We permute the five chosen letters in all possible ways, of which there are P(5,5) = 5! = 120 arrangements. Finally, to find the total number of words which can be formed, we apply the Fundamental Counting Principle which states that if one event can be performed in m ways, another one in n ways, and another in k ways, then the total number of ways in which all events can occur is m × n × k ways. Hence the total number of possible words is, by the fundamental principle

$$C(3,2)C(5,3)P(5,5) = 3 \cdot 10 \cdot 120 = 3600.$$

● **PROBLEM 22-12**

From 12 books in how many ways can a selection of 5 be made, (1) when one specified book is always included, (2) when one specified book is always excluded?

Solution: Here the formula for combinations is appropriate: the number of combinations of n things taken r at a time:

$$C(n,r) = {}_nC_r = \frac{n!}{r!(n-r)!}$$

where n = 11, and r = 4.

(1) Since the specified book is to be included in every selection, we have only to choose 4 out of the remaining 11.
Hence the number of ways = $^{11}C_4$

$$^{11}C_4 = \frac{11!}{4!(11-4)!}$$

$$= \frac{11!}{4!7!}$$

$$= \frac{11 \cdot 10 \cdot 9 \cdot 8 \cdot 7!}{4 \cdot 3 \cdot 2 \cdot 1 \cdot 7!}$$

$$= \frac{11 \times 10 \times 9 \times 8}{1 \times 2 \times 3 \times 4}$$

$$= 330.$$

(2) Since the specified book is always to be excluded, we have
to select the 5 books out of the remaining 11.

Hence the number of ways = $^{11}C_5$

$$^{11}C_5 = \frac{11!}{5!(11-5)!}$$

$$= \frac{11!}{5!6!}$$

$$= \frac{11 \cdot 10 \cdot 9 \cdot 8 \cdot 7 \cdot 6!}{5 \cdot 4 \cdot 3 \cdot 2 \cdot 1 \cdot 6!}$$

$$= \frac{11 \times 10 \times 9 \times 8 \times 7}{1 \times 2 \times 3 \times 4 \times 5}$$

$$= 462$$

● **PROBLEM** 22-13

How many groups can be formed from ten objects taking at
least three at a time?

Solution: The number of combinations of n objects
taken r at a time is $C(n,r) = n!/r!(n - r)!$. Thus, the
number of groups that can be formed from 10 objects taking
three at a time is $C(10, 3)$,

from 10 objects taking 4 at a time is $C(10,4)$,

from 10 objects taking 5 at a time is $C(10,5)$,

from 10 objects taking 6 at a time is $C(10,6)$,

.
.
.

from 10 objects taking 10 at a time is $C(10, 10)$.

Therefore, the number of groups that can be formed

from 10 objects taking at least three at a time is
C(10, 3) + C(10, 4) + C(10, 5) + C(10, 6) + C(10, 7) +
C(10, 8) + C(10, 9) + C(10, 10),

$$\frac{10!}{3!7!} + \frac{10!}{4!6!} + \frac{10!}{5!5!} + \frac{10!}{6!4!} + \frac{10!}{7!3!} + \frac{10!}{8!2!} + \frac{10!}{9!1!} + \frac{10!}{10!0!} =$$

$$\frac{\overset{3\cdot 4}{10\cdot 9\cdot 8\cdot 7!}}{3\cdot 2\cdot 7!} + \frac{\overset{5\cdot 3\cdot 2}{10\cdot 9\cdot 8\cdot 7\cdot 6!}}{4\cdot 3\cdot 2\cdot 6!} + \frac{\overset{2\cdot 3\cdot 2\ \cdot\ 3}{10\cdot 9\cdot 8\cdot 7\cdot 6\cdot 5!}}{5\cdot 4\cdot 3\cdot 2\cdot 5!} + \frac{10\cdot 9\cdot 8\cdot 7\cdot 6!}{4\cdot 3\cdot 2\cdot 6!}$$

$$+ \frac{10\cdot 9\cdot 8\cdot 7!}{3\cdot 2\cdot 7!} + \frac{10\cdot 9\cdot 8!}{2\cdot 8!} + \frac{10\cdot 9!}{1\cdot 9!} + \frac{10!}{10!\cdot 1}$$

= 120 + 210 + 252 + 210 + 120 + 45 + 10 + 1

= 968.

● **PROBLEM 22-14**

> A boy has in his pocket a penny, a nickel, a dime, and a quarter. How many different sums of money can he take out if he removes one or more coins?

<u>Solution:</u> In this problem, we are not considering order; that is, we are not concerned whether we choose a penny first and a nickel second or vice versa. (It is still the same arrangement.) Thus, we are considering combinations, not permutations. We consider the following cases to solve this problem:

a)   the boy removes one coin
b)   the boy removes two coins
c)   the boy removes three coins
d)   the boy removes four coins

Now, a combination of n things taken r at a time is:
$$C(n,r) = \frac{n!}{r!(n-r)!} .$$

Thus if for a) the boy removes one coin, then we want to find the number of combinations of 4 coins taken one at a time. Similarly for b), c), and d). The total number of combinations of 4 things taken 1,2,3, or 4 at a time is

$$C(4,1) + C(4,2) + C(4,3) + C(4,4) = \frac{4!}{1!3!} + \frac{4!}{2!2!} + \frac{4!}{3!1!} + \frac{4!}{4!0!}$$

$$= \frac{4\cdot 3!}{1\cdot 3!} + \frac{4\cdot 3\cdot 2!}{2\cdot 1\cdot 2!} + \frac{4\cdot 3!}{3!\cdot 1} + \frac{4!}{4!\cdot 1}$$

$$= 4 + \frac{12}{2} + 4 + 1$$

$$= 4 + 6 + 4 + 1$$

$$= 15.$$

From 10 men and 6 women, how many committees of 5 people can be chosen:

(a) If each committee is to have exactly 3 men?
(b) If each committee is to have at least 3 men?

Solution:
(a) The order in which the people on the committee are chosen is unimportant, thus this is a problem involving combinations. The general formula for the number of combinations of n different things taken r at a time is $C(n,r) = n!/r!(n-r)!$. Thus, the number of ways to choose 3 men from 10 men is $C(10,3)$

$$= \frac{10!}{3!(10-3)!} = \frac{10!}{3!7!} = \frac{10 \cdot \overset{3}{\cancel{9}} \cdot \overset{4}{\cancel{8}} \cdot \cancel{7!}}{\cancel{3} \cdot \cancel{2} \cdot 1 \cdot \cancel{7!}} = 120.$$

The number of ways to choose 2 women from 6 women is $C(6,2)$

$$= \frac{6!}{2!(6-2)!} = \frac{6!}{2!4!} = \frac{\overset{3}{\cancel{6}} \cdot 5 \cdot \cancel{4!}}{\cancel{2} \cdot 1 \cdot \cancel{4!}} = 15.$$

The Fundamental Principle of Counting states that if the first of two independent acts can be performed in a ways, and if the second act can be performed in b ways, then the number of ways of performing the two acts in the order stated is ab. Thus by the fundamental principle, the number of ways to choose the committee is $C(10,3) \cdot C(6,2) = 120 \cdot 5 = 1,800$.

(b) If the committee is to contain at least 3 men, the possibilities are 3 men and 2 women, 4 men and 1 woman, 5 men and no women.

We have just shown that the number of committees consisting of 3 men and 2 women is 1,800. The number of committees containing 4 men and 1 woman is

$$C(10,4) \cdot C(6,1) = \frac{10!}{4!(10-4)!} \cdot \frac{6!}{1!(6-1)!} = \frac{10!}{4!6!} \cdot \frac{6!}{1!5!}$$

$$= \frac{10 \cdot \overset{3}{\cancel{9}} \cdot \overset{\cancel{8}}{8} \cdot 7 \cdot \cancel{6!}}{\cancel{4} \cdot \cancel{3} \cdot \cancel{2} \cdot 1 \cdot \cancel{6!}} \cdot \frac{6 \cdot \cancel{5!}}{1 \cdot \cancel{5!}} = 210 \cdot 6 = 1,260 .$$

The number of committees consisting of 5 men is

$$C(10,5) = \frac{10!}{5!(10-5)!} = \frac{10!}{5!5!} = \frac{\overset{2}{\cancel{10}} \cdot \overset{3}{\cancel{9}} \cdot \overset{2}{\cancel{8}} \cdot 7 \cdot \overset{3}{\cancel{6}} \cdot \cancel{5!}}{\cancel{5} \cdot \cancel{4} \cdot \cancel{3} \cdot \cancel{2} \cdot 1 \cdot \cancel{5!}} = 252$$

The probability that any of several mutually exclusive events will occur is the sum of the probabilities of the separate events.

Hence the number of committees containing at least 3 men is

$$1,800 + 1,260 + 252 = 3,312.$$

Two ordinary dice are rolled. In how many different ways
can they fall? How many of these ways will give a sum of
nine?

Solution: The first die can fall in any one of six
positions, and for each of these positions the second
die can also fall in six positions; so there are
6 x 6 = 36 ways that the dice can fall. This is shown
in the accompanying construction of possible pairs for
the faces of the two dice.

1,1	1,2	1,3	1,4	1,5	1,6
2,1	2,2	2,3	2,4	2,5	2,6
3,1	3,2	3,3	3,4	3,5	3,6
4,1	4,2	4,3	4,4	4,5	4,6
5,1	5,2	5,3	5,4	5,5	5,6
6,1	6,2	6,3	6,4	6,5	6,6

Of these 36 ways, there are four ways of obtaining a sum
of nine: (6,3), (5,4), (4,5), (3,6) (circled in the figure).

How many four-letter "words" (we use "word" to mean any
sequence of letters) which begin and end with a vowel may
be formed from the letters a, e, i, p, q.
    (a)   if no repetitions are allowed?
    (b)   if repetitions are allowed?

Solution: (a) We construct a pattern of four boxes to
represent the four letters of the word to be formed:

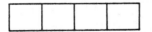

    Since the word is to begin with a vowel, the first
letter may be chosen in 3 ways (since there are three
vowels in the given letters):

3			

    Since the word must end with a vowel, and repetition of
letters is not allowed, after the first place has been
chosen only 2 letters remain from which to choose the
letter in last place:

3			2

Having chosen first and last letters, thus using up two letters of the original five, 3 letters are left from which to choose the letter in second position:

| 3 | 3 | | 2 |

Finally, 3 letters having been used, there remain 2 from which to choose the letter in third position:

| 3 | 3 | 2 | 2 |

The Fundamental Principle of Counting states that if one thing can be done in a different ways and, when it is done in any one of these ways, a second thing can be done in b different ways, and a third thing then can be done in c ways, ... then all the things in succession can be done in a×b×c··· different ways. Thus if no repetitions are allowed, then 3×3×2×2 = 36 words can be formed.

(b) Now we consider the case where repetitions are allowed. Once again we construct a pattern of four boxes to represent the four letters of the word to be formed:

| | | | |

Since the word is to begin with a vowel, the first letter may be chosen in 3 ways:

| 3 | | | |

Since the word must end with a vowel, and repetition is allowed, the last letter may also be chosen in 3 ways:

| 3 | | | 3 |

Since there are no specifications regarding the second and third letters, and repetition is allowed, the second and third letters may both be chosen in 5 ways. Thus we arrive at the following array:

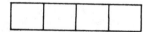

| 3 | 5 | 5 | 3 |

Using the Fundamental Principle of Counting, there are thus 3×5×5×3 = 225 words which may be formed if repetitions are allowed.

● **PROBLEM 22-18**

If in a series of license plates the letters I and O are not used and if four successive zeroes can not be used as the four digits, how many different license plates can be made?

<u>Solution:</u> A license plate consists of two letters and four digits. Of the 26 letters of the alphabet, since I and O are not used, there are now 24 choices for each of the two letters. Since four successive zeroes can not be used as the four digits, the digits would form the numbers from 0001 to 9999; so there are 9999 choices for the four digits. The fundamental principle states that if one thing can be done in m different ways and, when it is done in any one of these ways, a second thing can be done in n different ways, and if a third thing can then be done in p ways, ... then the successive things can be done in mnp... different ways. Therefore,

$$\text{Number of license plates} = (24)(24)(9999)$$

$$= 5,759,424$$

A baseball manager after determining his starting players now must determine his batting order. If the pitcher is to bat last, how many different ways can the manager turn in his batting order?

<u>Solution:</u> Since the pitcher is to bat last only the order of the other eight players must be determined. The fundamental principle states that if one thing can be done in m different ways and, when it is done in any one of these ways, a second thing can be done in n different ways, and if after it has been done in any one of these ways, a third thing can be done in p different ways, . . .,the several things can be done in m·n·p·...different ways. Using the fundamental principle, he has 8 choices for the lead-off man. Once this choice is made, he has 7 choices for the second batter, and so on.

$$\text{Number of batting orders} = 8(7)(6)(5)(4)(3)(2)(1)$$

$$= 8!$$

$$= 40,320$$

Two cards are to be drawn in order from a pack of 4 cards (say, an ace, king, queen, and jack), the drawn card not being replaced before the second card is drawn. How many different drawings are possible?

<u>Solution:</u> The first card can be drawn in four ways, and then the second card in three ways. By the fundamental principle of counting, which states that if the first of two independent acts can be performed in x ways, and if the

second act can be performed in y ways, then the number of
ways of performing the two acts, in the order stated, is
xy, there are 4 · 3 = 12 different drawings. (We regard
here ace first, king second as a different drawing from
king first, ace second.)

   If the first card were to be replaced before the second is drawn in this example, then the answer would be
4 · 4 = 16 different drawings.

CHAPTER 23

PROBABILITY

> **Basic Attacks and Strategies for Solving Problems in this Chapter. See pages 422 to 441 for step-by-step solutions to problems.**

The *probability* that a specified event will occur, often indicated as $P(event)$, is the ratio of the number of times the specified event can occur to the number of times the specified event does and does not occur.

Since the probability is a ratio, and the numerator of the ratio (i.e., the number of times the specified event can occur) is always less than the denominator (i.e., the number of times the specified event does and does not occur), the value of the probability of any event is always less than or equal to one.

For example, the probability that a coin will have a head up after being tossed is $1/2$, since there is only one head on a coin and two possible sides (that is, the *head* side and the *tail* [non-head] side).

Given three red balls and seven black balls in a bag, the probability of picking one of the red balls out of the bag is $3/10$ since there are a total of three red balls possible to pick, and ten (3 red and 7 non-red) balls total in the bag from which to pick.

The probability that either one of two mutually exclusive events, say E_1 and E_2, will occur is merely the sum of the probabilities of the separate events, i.e.,

$$P(E_1 \text{ or } E_2) = P(E_1) + P(E_2).$$

Thus, the probability that either a 3 or a 5 will be the top face after rolling a die is

$$\frac{1}{6} + \frac{1}{6} = \frac{1}{3}.$$

The probability that both of two mutually independent events, say E_1 and E_2, will occur is the product of the probabilities of the separate events, i.e.,

$$P(E_1 \text{ and } E_2) = P(E_1) \cdot P(E_2).$$

Thus, the probability that in two successive rolls of a die a 3 will be the top face followed by a 5 is

$$\frac{1}{6} \cdot \frac{1}{6} = \frac{1}{36}.$$

The probability that two dependent events, say E_1 and E_2, will occur in the order proposed is the product of the probability of the first event and the probability of the second event given the first.

Thus, given a bag with three white balls and four black balls, and given that any ball picked is not replaced into the bag, the probability that if we pick two balls out of the bag, one after another, both will be white is computed as follows. The probability that the first ball is white is

$$\frac{3}{3+4} = \frac{3}{7}.$$

The probability that the second ball is white given that the first ball was white is

$$\frac{2}{2+4} = \frac{2}{6} = \frac{1}{3}.$$

Thus, the desired probability is

$$\frac{3}{7} \cdot \frac{1}{3} = \frac{1}{7}.$$

When dealing with repeated events, the following theorem is useful:

> If p is the probability that an event will occur in a single trial, and q is the probability that this event will fail in this trial, then
>
> $$_nC_r \, p^r \, q^{q-r}$$
>
> is the probability that this event will happen exactly r times in n trials.

Conditional probability refers to computing the probability that event E_2 occurs on the condition that event E_1 has occurred. The probability that E_2 occurs given E_1 is

$$P(E_2 \,|\, E_1) = \frac{P(E_1 \text{ and } E_2)}{P(E_1)}.$$

The numerator of the fraction is the probability that both events occur. This probability does not take into account the fact that one of the two events occurred and we are now trying to determine the likelihood of the other.

As an example, we are given (as event E_1) that after 3 tosses of a coin tails were up (at least) twice. We wish to compute the probability that (as E_2) all three were tails. (Note that in this example, E_2 demands that E_1 occur.) $P(E_1$ and $E_2)$ is the probability that "tails were up twice (after three coin tosses)" and also "tails were up three times." In other words, we merely calculate the probability that all three tosses were tails. The total number of combinations of heads and tails for three tosses is $2^3 = 8$. The total number of events in which tails occurs three times for three tosses is only 1. Thus, $P(E_1$ and $E_2)$ is $1/8$. $P(E_1)$ is the probability that in three tosses, tails were up (at least) twice. The only possibilities are T, T, H or T, H, T or H, T, T or T, T, T. As before, the total number of combinations of heads and tails for three tosses is 8. Thus,

$$P(E_1) = \frac{4}{8} = \frac{1}{2}.$$

Thus,

$$P(E_2 \mid E_1) = \frac{1/8}{1/2} = \frac{2}{8} = \frac{1}{4}.$$

Step-by-Step Solutions to Problems in this Chapter, "Probability"

● PROBLEM 23-1

What is the probability of throwing a "six" with a single die?

Solution: The die may land in any of 6 ways:

$$1, \quad 2, \quad 3, \quad 4, \quad 5, \quad 6$$

The probability of throwing a six,

$$P(6) = \frac{\text{number of ways to get a six}}{\text{number of ways the die may land}}$$

Thus $P(6) = \frac{1}{6}$.

● PROBLEM 23-2

A box contains 7 red, 5 white, and 4 black balls. What is the probability of your drawing at random one red ball? One black ball?

Solution: There are 7 + 5 + 4 = 16 balls in the box. The probability of drawing one red ball,

$$P(R) = \frac{\text{number of possible ways of drawing a red ball}}{\text{number of ways of drawing any ball}}$$

$$P(R) = \frac{7}{16}.$$

Similarly, the probability of drawing one black ball

$$P(B) = \frac{\text{number of possible ways of drawing a black ball}}{\text{number of ways of drawing any ball}}$$

Thus,

$$P(B) = \frac{4}{16} = \frac{1}{4}$$

● **PROBLEM 23-3**

In a single throw of a single die, find the probability of obtaining either a 2 or a 5.

Solution: In a single throw, the die may land in any of 6 ways:

1 2 3 4 5 6.

The probability of obtaining a 2,

$$P(2) = \frac{\text{number of ways of obtaining a 2}}{\text{numbers of ways the die may land}}$$

$$P(2) = \frac{1}{6}.$$

Similarly, the probability of obtaining a 5,

$$P(5) = \frac{\text{number of ways of obtaining a 5}}{\text{number of ways the die may land}}$$

$$P(5) = \frac{1}{6}.$$

The probability that either one of two mutually exclusive events will occur is the sum of the probabilities of the separate events. Thus the probability of obtaining either a 2 or a 5, P(2) or P(5), is

$$P(2) + P(5) = \frac{1}{6} + \frac{1}{6} = \frac{2}{6} = \frac{1}{3}$$

● **PROBLEM 23-4**

Suppose that we have a bag containing two red balls, three white balls, and six blue balls. What is the probability of obtaining a red or a white ball on one withdrawal?

○ red ball

○ white ball  Fig. A.

● blue ball

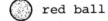
$$U = \left\{ \begin{array}{l} \text{Red,Red, White,White,White, Blue,Blue,Blue,Blue,Blue,Blue} \\ \underbrace{}_{\text{2 Red}} \quad \underbrace{}_{\text{3 White}} \quad \underbrace{}_{\text{6 Blue}} \end{array} \right\}$$

Fig. B.

423

Solution: The bag is shown in figure A.

There are two favorable possibilities: drawing a red ball and drawing a white ball. The universal set of equally likely possibilities contains eleven elements; the number of red balls is two, and the number of white balls is three. Refer to figure B.

Hence the probability of drawing a red ball is $\frac{2}{11}$ and the probability of drawing a white ball is $\frac{3}{11}$.

In the set notation, the probability of A, P(A), or the probability of B, P(B), is P(A) + P(B) therefore the probability of drawing a red, or a white ball is the probability of drawing a red ball plus the probability of drawing a white ball

$$P(red) = \frac{2}{11} \qquad\qquad P(white) = \frac{3}{11}$$

$$P(red \text{ or } white) = \frac{2}{11} + \frac{3}{11} = \frac{5}{11}.$$

Note: In the above example the probability of drawing a blue ball would be $\frac{6}{11}$. Therefore the sum of the probability of a red ball, the probability of a white ball, and the probability of a blue ball is $\frac{2}{11} + \frac{3}{11} + \frac{6}{11} = 1$.

If there are no possible results that are considered favorable, then the probability P(F) is obviously 0. If every result is considered favorable, then P(F) = 1. Hence the probability P(F) of a favorable result F always satisfies the inequality

$$0 \leq P(F) \leq 1.$$

● **PROBLEM 23-5**

An urn contains 6 white, 4 black, and 2 red balls. In a single draw, find the probability of drawing: (a) a red ball; (b) a black ball; (c) either a white or a black ball. Assume all outcomes equally likely.

Solution: The urn contains 6 white balls, 4 black balls, and 2 red balls, or a total of 12 balls.

(a) The probability of drawing a red ball,

$$P(R) = \frac{\text{number of ways of drawing a red ball}}{\text{number of ways of selecting a ball}}$$

$$P(R) = \frac{2}{12} = \frac{1}{6}.$$

(b) The probability of drawing a black ball,

$$P(B) = \frac{\text{number of ways of drawing a black ball}}{\text{number of ways of selecting a ball}}$$

$$P(B) = \frac{4}{12} = \frac{1}{3}.$$

(c) The probability that either one of two mutually exclusive events will occur is the sum of the probabilities of the separate events. Thus the probability of drawing either a white [P(W)] or a black ball [P(B)] is P(W) + P(B).

$$P(W) = \frac{\text{number of ways of drawing a white ball}}{\text{number of ways of selecting a ball}}$$

$$= \frac{6}{12} = \frac{1}{2}.$$

$P(B) = \frac{1}{3}$ [shown in part (b)].

Thus, $P(W \text{ or } B) = P(W) + P(B) = \frac{6}{12} + \frac{4}{12}$

$$= \frac{10}{12}$$

$$= \frac{5}{6}.$$

● **PROBLEM 23-6**

Determine the probability of getting 6 or 7 in a toss of two dice.

Solution: Let A = the event that a 6 is obtained in a toss of two dice

B = the event that a 7 is obtained in a toss of two dice.

Then, the probability of getting 6 or 7 in a toss of two dice is

$$P(A \text{ or } B) = P(A \cup B).$$

The union symbol "$\cup$" means that A and/or B can occur. Now $P(A \cup B) = P(A) + P(B)$ if A and B are mutually exclusive. Two or more events are said to be mutually exclusive if the

425

occurrence of any one of them excludes the occurrence of the others. In this case, we cannot obtain a six and a seven in a single toss of two dice. Thus, A and B are mutually exclusive.

To calculate P(A) and P(B), use the following table.

Note: There are 36 different tosses of two dice.

A = a 6 is obtained in a toss of two dice

$$= \Big\{ (1,5), \ (2,4), \ (3,3), \ (4,2), \ (5,1) \Big\}$$

B = a 7 is obtained in a toss of two dice

$$= \Big\{ (1,6), \ (2,5), \ (3,4), \ (4,3), \ (5,2), \ (6,1) \Big\}.$$

$$P(A) = \frac{\text{number of ways to obtain a 6 in a toss of two dice}}{\text{number of ways to toss two dice}}$$

$$= \frac{5}{36}$$

$$P(B) = \frac{\text{number of ways to obtain a 7 in a toss of two dice}}{\text{number of ways to toss two dice}}$$

$$= \frac{6}{36} = \frac{1}{6}.$$

Therefore, $P(A \cup B) = P(A) + P(B) = \frac{5}{36} + \frac{6}{36} = \frac{11}{36}.$

● **PROBLEM 23-7**

A penny is to be tossed 3 times. What is the probability there will be 2 heads and 1 tail?

Solution: We start this problem by constructing a set of all possible outcomes:

We can have heads on all 3 tosses: (HHH)
head on first 2 tosses, tail on the third: (HHT)   (1)
head on first toss, tail on next two: (HTT)
                    •   (HTH)   (2)
                    •   (THH)   (3)
                    •   (THT)
                                                 (TTH)
                                                 (TTT)

Hence there are eight possible outcomes (2 possibilities on first toss x 2 on second x 2 on third = 2 x 2 x 2 = 8).

We assume that these outcomes are all equally likely and assign the probability 1/8 to each. Now we look for the set of outcomes that produce 2 heads and 1 tail. We see there are 3 such outcomes out of the 8 possibilities

426

(numbered (1), (2), (3) in our listing). Hence the
probability of 2 heads and 1 tail is 3/8.

Find the probability of throwing two sixes in one toss of a pair
of dice.

Solution: To find the probability of throwing two sixes in one toss of
a pair of dice, first we express it symbolically.

P(throwing two sixes in one toss of a pair of dice) =

P(throwing a six in one toss of a die) X P(throwing a six in one toss
of a die).
This is true because the event of tossing a die is independent of tossing another die. That is, the occurrence of one event has no effect
upon the occurrence or non-occurrence of the other. Now,

P(throwing a six in one toss) =

$$\frac{\text{number of ways to obtain a six}}{\text{number of ways to obtain any face value of a die}} = \frac{1}{6}.$$

Hence, the probability of obtaining two sixes is $\left(\frac{1}{6}\right)\left(\frac{1}{6}\right) = \frac{1}{36}$.

A bag contains 4 black and 5 blue marbles. A marble is
drawn and then replaced, after which a second marble is
drawn. What is the probability that the first is black
and second blue?

Solution: Let C = event that the first marble drawn is
              black.
          D = event that the second marble drawn is
              blue.

The probability that the first is black and the second
is blue can be expressed symbolically:

P(C and D) = P(CD).

We can apply the following theorem. If two events A and
B, are independent, then the probability that A and B will
occur is,

P(A and B) = P(AB) = P(A) · P(B).

Note that two or more events are said to be independent if
the occurrence of one event has no effect upon the occurrence or non-occurrence of the other. In this case the
occurrence of choosing a black marble has no effect on
the selection of a blue marble and vice versa; since, when
a marble is drawn it is then replaced before the next

marble is drawn. Therefore, C and D are two independent events.

$$P(CD) = P(C) \cdot P(D)$$

$$P(C) = \frac{\text{number of ways to choose a black marble}}{\text{number of ways to choose a marble}}$$

$$= \frac{4}{9}.$$

$$P(D) = \frac{\text{number of ways to choose a blue marble}}{\text{number of ways to choose a marble}}$$

$$= \frac{5}{9}.$$

$$P(CD) = P(C) \cdot P(D) = \frac{4}{9} \cdot \frac{5}{9} = \frac{20}{81}.$$

● **PROBLEM 23-10**

A box contains 4 black marbles, 3 red marbles, and 2 white marbles. What is the probability that a black marble, then a red marble, then a white marble is drawn without replacement?

Solution: Here we have three dependent events. There is a total of 9 marbles from which to draw. We assume on the first draw we will get a black marble. Since the probability of drawing a black marble is the

$$\frac{\text{number of ways of drawing a black marble}}{\text{number of ways of drawing 1 out of (4+3+2) marbles}},$$

$$P(A) = \frac{4}{4 + 3 + 2} = \frac{4}{9}$$

There are now 8 marbles left in the box.

On the second draw we get a red marble. Since the probability of drawing a red marble is
$$\frac{\text{number of ways of drawing a red marble}}{\text{number of ways of drawing 1 out of the 8 remaining marbles}},$$

$$P(B) = \frac{3}{8}$$

There are now 7 marbles remaining in the box.

On the last draw we get a white marble. Since the probability of drawing a white marble is

$$\frac{\text{number of ways of drawing a white marble}}{\text{number of ways of drawing 1 out of the 7 remaining marbles}},$$

$$P(C) = \frac{2}{7}$$

428

When dealing with two or more dependent events, if P_1 is the probability of a first event, P_2 the probability that, after the first has happened, the second will occur, P_3 the probability that, after the first and second have happened, the third will occur, etc., then the probability that all events will happen in the given order is the product $P_1 \cdot P_2 \cdot P_3 \ldots$

Thus, $P(A \cap B \cap C) = P(A) \cdot P(B) \cdot P(C)$

$$= \frac{4}{9} \cdot \frac{3}{8} \cdot \frac{2}{7}$$

$$= \frac{1}{21} .$$

● **PROBLEM** 23-11

There is a box containing 5 white balls, 4 black balls, and 7 red balls. If two balls are drawn one at a time from the box and neither is replaced, find the probability that
(a)   both balls will be white.
(b)   the first ball will be white and the second red.
(c)   if a third ball is drawn, find the probability that the three balls will be drawn in the order white, black, red.

Solution: This problem involves dependent events. Two or more events are said to be dependent if the occurrence of one event has an effect upon the occurrence or non-occurrence of the other. If you are drawing objects without replacement, the second draw is dependent on the occurrence of the first draw. We apply the following theorem for this type of problem. If the probability of occurrence of one event is p and the probability of the occurrence of a second event is q, then the probability that both events will happen in the order stated is pq.

(a) To find the probability that both balls will be white, we express it symbolically.

p (both balls will be white) =
p (first ball will be white and the second ball will be white) =
p (first ball will be white) p(second ball will be white) =

$$= \left(\frac{\text{number of ways to choose a white ball}}{\text{number of ways to choose a ball}} \right) \left(\frac{\text{number of ways to choose a second white ball after removal of the first white ball}}{\text{number of ways to choose a ball after removal of the first ball}} \right)$$

$$= \frac{\overset{1}{\cancel{5}}}{\underset{4}{\cancel{16}}} \cdot \frac{\overset{1}{\cancel{4}}}{\underset{3}{\cancel{15}}} = \frac{1}{12}$$

(b) p (first ball will be white and the second red)

  = p (first ball will be white) p(the second ball will be red)

429

$$= \left(\frac{\text{number of ways to choose a white ball}}{\text{number of ways to choose a ball}} \right) \left(\frac{\text{number of ways to choose a red ball}}{\text{number of ways to choose a ball after the removal of the first}} \right)$$

$$= \frac{\overset{1}{\cancel{5}}}{16} \cdot \frac{7}{\underset{3}{\cancel{15}}} = \frac{7}{48}$$

(c) p (three balls drawn in the order white, black, red)

= p (first ball is white) p(second ball is black) p(third ball is red)

$$= \left(\frac{\text{number of ways to choose that the first ball is white}}{\text{number of ways to choose the first ball}} \right) \left(\frac{\text{number of ways to choose that second one is black}}{\text{number of ways to choose the second one}} \right)$$

$$\left(\frac{\text{number of ways to choose that the third one is red}}{\text{number of ways to choose the third one}} \right)$$

$$= \underset{4}{\frac{\overset{1}{\cancel{5}}}{\cancel{16}}} \quad \underset{3}{\frac{\overset{1}{\cancel{4}}}{\cancel{15}}} \quad \underset{2}{\frac{\overset{1}{\cancel{7}}}{\cancel{14}}} = \frac{1}{24}$$

● PROBLEM 23-12

What is the chance of throwing a number greater than 4 with an ordinary die whose faces are numbered from 1 to 6?

Solution: If an event can happen in s ways and fail to happen in f ways, and if all these ways (s + f) are assumed to be equally likely, then the probability (p) that the event will happen is

$$p = \frac{s}{s + f} = \frac{\text{(successful ways)}}{\text{(total ways)}}$$

In our case there are 6 possible ways in which the die can fall (1,2,3,4,5, or 6). Of these, two are favorable to the event required, 5 or 6, therefore the required chance $= \frac{2}{6} = \frac{1}{3}$.

● PROBLEM 23-13

What is the probability that the sum 11 will appear in a single throw of 2 dice?

Solution: There are 6 ways the first die may be tossed and 6 ways the second die may be tossed. The Fundamental Principle of Counting states that if the first of two independent acts can be performed in x ways, and if the second act can be performed in y ways, then the number of ways of performing the two acts, in the order stated, is xy. Thus there are 6 X 6 = 36 ways that two dice can be thrown (see accompanying figure).

1,1	1,2	1,3	1,4	1,5	1,6
2,1	2,2	2,3	2,4	2,5	2,6
3,1	3,2	3,3	3,4	3,5	3,6
4,1	4,2	4,3	4,4	4,5	4,6
5,1	5,2	5,3	5,4	5,5	5,6
6,1	6,2	6,3	6,4	6,5	6,6

The number of possible ways that an 11 will appear are circled in the figure. Let us call this set A. Thus,

$$A = \{(5,6),(6,5)\}.$$

The probability that an 11 will appear,

$$p(11) = \frac{\text{number of possible ways of obtaining an 11}}{\text{number of ways that 2 dice can be thrown}}$$

Therefore

$$p(11) = \frac{2}{36} = \frac{1}{18} .$$

● PROBLEM 23-14

What is the probability of making a 7 in one throw of a pair of dice?

Solution: There are 6 X 6 = 36 ways that two dice can be thrown, as shown in the accompanying figure.

1,1	1,2	1,3	1,4	1,5	1,6
2,1	2,2	2,3	2,4	2,5	2,6
3,1	3,2	3,3	3,4	3,5	3,6
4,1	4,2	4,3	4,4	4,5	4,6
5,1	5,2	5,3	5,4	5,5	5,6
6,1	6,2	6,3	6,4	6,5	6,6

The number of possible ways that a 7 will appear are circled in the figure. Let us call this set B. Thus,

$$B = \{(1,6),(2,5),(3,4),(4,3),(5,2),(6,1)\}.$$

The probability that a 7 will appear,

$$p(7) = \frac{\text{number of possible ways of obtaining a 7}}{\text{number of ways that 2 dice can be thrown}}$$

$$p(7) = \frac{6}{36} = \frac{1}{6} .$$

● PROBLEM 23-15

If two dice are cast, what is the probability the sum will be less than 5?

431

Solution: If A, B, and C are mutually exclusive events, that is, their intersection is the null set, then $P(A \cup B \cup C) = P(A) + P(B) + P(C)$. Since the obtaining of sums of 2, 3, and 4 are mutually exclusive events, the probability of obtaining a sum less than 5 is the sum of the probabilities of obtaining a sum of 2, 3, and 4. To obtain the sum of 2 with 2 die, we have the following possibilities: (1,1).

Similarly for the sum of 3, we have: (1,2) and (2,1). For the sum of 4, we obtain: (1,3), (3,1), and (2,2).

Thus P_1 = probability of obtaining a sum of 2

$\quad = \dfrac{\text{number of ways to obtain a sum of 2}}{\text{number of ways to throw 2 dice}}$

$\quad = \dfrac{1}{36}$

P_2 = probability of obtaining a sum of 3

$\quad = \dfrac{\text{number of ways to obtain a sum of 3}}{\text{number of ways to throw 2 dice}}$

$\quad = \dfrac{2}{36} = \dfrac{1}{18}.$

P_3 = probability of obtaining a sum of 4

$\quad = \dfrac{\text{number of ways to obtain a sum of 4}}{\text{number of ways to throw 2 dice}}$

$\quad = \dfrac{3}{36} = \dfrac{1}{12}.$

The probability of obtaining a sum less than 5 is

$$P_1 + P_2 + P_3 = \dfrac{1}{36} + \dfrac{1}{18} + \dfrac{1}{12}.$$

$$= \dfrac{1}{36} + \dfrac{2}{36} + \dfrac{3}{36} = \dfrac{6}{36} = \dfrac{1}{6}$$

● **PROBLEM 23-16**

Find the probability that when a pair of dice are thrown, the sum of the two up faces is greater than 7 or the same number appears on each face.

Solution: The sample space consists of 36 equally likely outcomes as shown in the accompanying figure. Those outcomes that give a sum greater than 7 are

$G = \{(6,2),(6,3),(6,4),(6,5),(6,6),(5,3),(5,4),(5,5),$
$\quad (5,6),(4,4),(4,5),(4,6),(3,5),(3,6),(2,6)\}$

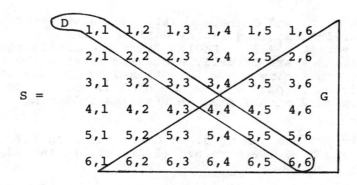

$$S = \begin{matrix} 1,1 & 1,2 & 1,3 & 1,4 & 1,5 & 1,6 \\ 2,1 & 2,2 & 2,3 & 2,4 & 2,5 & 2,6 \\ 3,1 & 3,2 & 3,3 & 3,4 & 3,5 & 3,6 \\ 4,1 & 4,2 & 4,3 & 4,4 & 4,5 & 4,6 \\ 5,1 & 5,2 & 5,3 & 5,4 & 5,5 & 5,6 \\ 6,1 & 6,2 & 6,3 & 6,4 & 6,5 & 6,6 \end{matrix}$$

Then P(G) =

= $\dfrac{\text{number of ways the two faces will be greater than 7}}{\text{total number of ways that the dice may fall}}$

= $\dfrac{15}{36}$.

Those outcomes where each die is the same are

D = {(1,1),(2,2),(3,3),(4,4),(5,5),(6,6)}

Then P(D) =

= $\dfrac{\text{number of ways that the same number can appear on each face}}{\text{total number of ways that the dice may fall}}$

= $\dfrac{6}{36}$.

The probability of G or D is P(G∪D). Recall that

P(G∪D) = P(G) + P(D) - P(G∩D).

But G ∩ D = {(4,4),(5,5),(6,6)}.

Hence, P(G ∩ D) = $\dfrac{3}{36}$

and P(G ∪ D) = $\dfrac{15}{36} + \dfrac{6}{36} - \dfrac{3}{36} = \dfrac{18}{36} = \dfrac{1}{2}$.

● **PROBLEM 23-17**

In a single throw of a pair of dice, find the probability of obtaining a total of 4 or less.

Solution: Each die may land in 6 ways. By the Fundamental Principle of Counting the pair of dice may thus land in 6 X 6 = 36 ways:

```
1,1    1,2    1,3    1,4    1,5    1,6

2,1    2,2    2,3    2,4    2,5    2,6

3,1    3,2    3,3    3,4    3,5    3,6

4,1    4,2    4,3    4,4    4,5    4,6

5,1    5,2    5,3    5,4    5,5    5,6

6,1    6,2    6,3    6,4    6,5    6,6
```

Let us call the possible outcomes which are circled above set A. Then the elements of set A, A = { (1,1), (1,2), (1,3), (2,1), (2,2), (3,1)} are all the possible ways of obtaining four or less.

The probability of obtaining 4 or less,

$$P[(x,y) \le 4] = \frac{\text{number of ways of obtaining 4 or less}}{\text{number of ways the dice may land}} \overset{\left(\substack{\text{number of} \\ \text{elements} \\ \text{in set A}}\right)}{}$$

$$= \frac{6}{36} = \frac{1}{6}.$$

● **PROBLEM 23-18**

The probability that A wins a certain game is $\frac{2}{3}$. If A plays 5 games, what is the probability that A will win (a) exactly 3 games? (b) at least 3 games?

Solution: We shall apply the following theorem. If p is the probability that an event will happen in a single trial and q is the probability that this event will fail in this trial, then $_nC_r \, p^r q^{n-r}$ is the probability that this event will happen exactly r times in n trials. $_nC_r$, the number of combinations of n different objects taken r at a time, is

$$_nC_r = \frac{n!}{r!(n-r)!}.$$

Note that p + q = 1.

(a) We are given the probability of a success, p, which is winning a game: $p = \frac{2}{3}$. Therefore from p + q = 1, $q = 1 - p = 1 - \frac{2}{3} = \frac{1}{3}$. The number of ways of winning 3 games out of 5 is

$$_5C_3 = \binom{5}{3} = \frac{5!}{3!2!} = \frac{5 \cdot \overset{2}{\cancel{4}} \cdot \cancel{3!}}{\cancel{3!} \cdot \underset{1}{\cancel{2}} \cdot 1} = 10.$$

Thus, the probability of A winning 3 games is

$$_nC_r \; p^r q^{n-r} = \; _5C_3\left(\frac{2}{3}\right)^3\left(\frac{1}{3}\right)^2 = 10 \; \frac{2}{3}\cdot\frac{2}{3}\cdot\frac{2}{3}\cdot\frac{1}{3}\cdot\frac{1}{3} = \frac{80}{243}$$

(b) To win at least 3 games A must win either exactly 3 or exactly 4 or all 5 games. In order that A will win at least 3 games, we must calculate the probability that A will win three games, four games, and five games.

$$P = \; _5C_3\left(\frac{2}{3}\right)^3\left(\frac{1}{3}\right)^2 + \; _5C_4\left(\frac{2}{3}\right)^4\left(\frac{1}{3}\right)^1 + \; _5C_5\left(\frac{2}{3}\right)^5\left(\frac{1}{3}\right)^0.$$

$$= \frac{5!}{3!2!}\left(\frac{2}{3}\right)^3\left(\frac{1}{3}\right)^2 + \frac{5!}{4!1!}\left(\frac{2}{3}\right)^4\left(\frac{1}{3}\right)^1 + \frac{5!}{5!0!}\left(\frac{2}{3}\right)^5$$

$$= \frac{5\cdot\cancel{4}^2\cdot\cancel{3}}{\cancel{3!}\;\cancel{2}\cdot 1}\frac{8}{243} + \frac{5\cdot\cancel{4!}}{\cancel{4!}\;1!}\frac{16}{243} + \frac{32}{243}$$

$$= 10\cdot\frac{8}{243} + 5\cdot\frac{16}{243} + \frac{32}{243} = \frac{192}{243} = \frac{64}{81}.$$

● **PROBLEM 23-19**

Find the chance of throwing at least one ace in a single throw with two dice.

$$S \; = \;
\begin{matrix}
1,1 & 1,2 & 1,3 & 1,4 & 1,5 & 1,6 \\
2,1 & 2,2 & 2,3 & 2,4 & 2,5 & 2,6 \\
3,1 & 3,2 & 3,3 & 3,4 & 3,5 & 3,6 \\
4,1 & 4,2 & 4,3 & 4,4 & 4,5 & 4,6 \\
5,1 & 5,2 & 5,3 & 5,4 & 5,5 & 5,6 \\
6,1 & 6,2 & 6,3 & 6,4 & 6,5 & 6,6
\end{matrix}$$

Solution: The chance of throwing at least one ace is the chance of throwing an ace on either the first die, or the second die , or both.

The sample space, S, consists of 6 × 6 = 36 equally likely outcomes (see figure). Those outcomes that give an ace on the first die are shown in the first row of the figure. Let us call this row A. Thus,

A = {(1,1),(1,2),(1,3),(1,4),(1,5),(1,6)}.

Then $P(A)$ = $\dfrac{\text{number of ways of getting ace on first die}}{\text{total number of ways that the dice may fall}}$

$$= \frac{6}{36}$$

Those outcomes that give an ace on the second die are shown in the first column of the figure. Let us call this column B. Thus,

B = {(1,1),(2,1),(3,1),(4,1),(5,1),(6,1)}.

Then P(B) = $\dfrac{\text{number of ways of getting ace on second die}}{\text{total number of ways that the dice may fall}}$

$= \dfrac{6}{36}$

The probability of an ace on the first die or on the second die is P(A) or P(B), P(A∪B). Recall that

P(A∪B) = P(A) + P(B) - P(A∩B).

Now, A∩B = (1,1);
then P(A∩B)

$= \dfrac{\text{number of ways of getting ace on both dice}}{\text{total number of ways that the dice may fall}} = \dfrac{1}{36}$

Therefore, P(A∪B) = $\dfrac{6}{36} + \dfrac{6}{36} - \dfrac{1}{36} = \dfrac{11}{36}$.

Thus, the chance of throwing at least one ace in a single throw of 2 dice is 11/36.

● PROBLEM 23-20

A die is tossed five times. What is the probability that an ace will appear: (a) at least twice; (b) at least once?

Solution: This is a problem involving repeated trials of an experiment. The experiment is "tossing a die five times". Apply the following theorem: If p is the probability that an event will happen in a single trial and q is the probability that this event will fail in this trial, then

$$_nC_r p^r q^{n-r}$$

is the probability that this event will happen exactly r times in n trials.

(a) To find the probability that an ace will occur at least twice, find the probability that it will occur twice, or three times, or four times, or five times. The sum (the word "or" implies addition in set notation) of these probabilities will be the probability that an ace will happen at least twice. p = probability that an ace will occur in a given trial

$= \dfrac{\text{number of ways to obtain an ace}}{\text{number of ways to obtain any face of a die}}$

$= \dfrac{1}{6}$

An experiment can only succeed or fail, hence the probability of success, p, plus the probability of failure, q, is one; p+q = 1. Then q = 1-p = 1 - 1/6 = 5/6. Therefore, using $_nC_r p^r q^{n-r}$, p (at least two aces) =

$$_5C_2 (1/6)^2(5/6)^3 + {_5C_3}(1/6)^3(5/6)^2$$

$$+ {}_5C_4 (1/6)^4 (5/6)^1 + {}_5C_5 (1/6)^5 (5/6)^0$$

${}_nC_r$ is a symbol for a combination of n things, r at a time, where r objects are chosen from n objects.

$$_nC_r = \frac{n!}{r! \ (n-r)!}$$

Apply this formula. Then,

$$_5C_2(1/6)^2(5/6)^3 + {}_5C_3(1/6)^3(5/6)^2$$
$$+ {}_5C_4(1/6)^4(5/6)^1 + {}_5C_5(1/6)^5(5/6)^0$$

$$= \frac{5!}{2! 3!}\left(\frac{125}{6^5}\right) + \frac{5!}{2! 3!}\left(\frac{25}{6^5}\right) + \frac{5!}{4! 1!}\left(\frac{5}{6^5}\right) + \frac{5!}{5! 0!}\left(\frac{1}{6^5}\right)$$

$$= \frac{5 \cdot 4 \cdot 3!}{2 \cdot 1 \cdot 3!}\left(\frac{125}{6^5}\right) + \frac{5 \cdot 4 \cdot 3!}{2 \cdot 1 \cdot 3!}\left(\frac{25}{6^5}\right) + \frac{5 \cdot 4!}{4! 1!}\left(\frac{5}{6^5}\right) + \frac{1}{6^5}$$

$$= 10\left(\frac{125}{6^5}\right) + 10\left(\frac{25}{6^5}\right) + 5\left(\frac{5}{6^5}\right) + \frac{1}{6^5}$$

$$= \frac{1250 + 250 + 25 + 1}{6^5} = \frac{1526}{7776} = \frac{763}{3888}$$

Therefore, the probability that an ace will appear at least twice is $\frac{763}{3888}$.

(b) An ace can be obtained at least once by tossing one ace, 2 aces, 3 aces,..., or 5 aces. Hence, the probability of obtaining at least one ace is the sum of the individual probabilities of obtaining one, two, three,..., up to five aces. Apply the same method as in part (a).

$$p(\text{at least one ace}) = {}_5C_1(1/6)^1(5/6)^4 + {}_5C_2(1/6)^2(5/6)^3$$
$$+ {}_5C_3(1/6)^3(5/6)^2 + {}_5C_4(1/6)^4(5/6)^1$$
$$+ {}_5C_5(1/6)^5(5/6)^0$$

$$= \frac{5!}{1! 4!}\left(\frac{625}{6^5}\right) + \frac{5!}{2! 3!}\left(\frac{125}{6^5}\right) + \frac{5!}{3! 2!}\left(\frac{25}{6^5}\right) + \frac{5!}{4! 1!}\left(\frac{5}{6^5}\right)$$

$$+ \frac{5!}{5! 0!}\left(\frac{1}{6^5}\right) = \frac{5 \cdot 4!}{1 \cdot 4!}\left(\frac{625}{6^5}\right) + \frac{5 \cdot 4 \cdot 3!}{2 \cdot 1 \cdot 3!}\left(\frac{125}{6^5}\right) + \frac{5 \cdot 4 \cdot 3!}{3! 2 \cdot 1}\left(\frac{25}{6^5}\right)$$

$$+ \frac{5 \cdot 4!}{4! \cdot 1}\left(\frac{5}{6^5}\right) + \left(\frac{1}{6^5}\right) = 5\left(\frac{625}{6^5}\right) + 10\left(\frac{125}{6^5}\right) + 10\left(\frac{25}{6^5}\right)$$

$$+ 5\left(\frac{5}{6^5}\right) + \frac{1}{6^5}$$

$$= \frac{3125 + 1250 + 250 + 25 + 1}{6^5} = \frac{4651}{7776}$$

An alternate, shorter method, is to calculate the probability of failure, (obtaining no aces) and subtract this from one. This is true because $q+p = 1$, hence $q = 1-p$.

p (at least one ace) = 1 - p (no aces)

$$p \text{ (no aces)} = {}_5C_0(1/6)^0(5/6)^5 = \frac{5!}{0!5!}\left(\frac{5^5}{6^5}\right)$$
$$= \frac{3125}{7776}$$

Thus,

$$p \text{ (at least one ace)} = 1 - p \text{ (no aces)}$$
$$= 1 - \frac{3125}{7776} = \frac{4651}{7776}$$

Therefore, the probability that an ace appears at least once is

$$\frac{4651}{7776}\,.$$

● **PROBLEM** 23-21

A coin is tossed 3 times, and 2 heads and 1 tail fall. What is the probability that the first toss was heads?

Solution: This problem is one of conditional probability. Given two events, P_1 and P_2, the probability that event P_2 will occur on the condition that we have event P_1 is

$$P(P_2/P_1) = \frac{P(P_1 \text{ and } P_2)}{P(P_1)} = \frac{P(P_1 P_2)}{P(P_1)}$$

Define

P_1: 2 heads and 1 tail fall,

P_2: the first toss is heads.

$P(P_1) = \dfrac{\text{number of ways to obtain 2 heads and 1 tail}}{\text{number of possibilities resulting from 3 tosses}}$

$= \Big(\{H,H,T\},\{H,T,H\},\{T,H,H\}\Big)/\Big(\{H,H,H\},\{H,H,T\},\{H,T,T\},\{H,T,H\},$
$\{T,T,H\},\{T,H,T\},\{T,H,H\},\{T,T,T\}\Big)$

$= 3/8$

$P(P_1 P_2) = P(\text{2 heads and 1 tail and the first toss is heads})$

$= \dfrac{\text{number of ways to obtain } P_1 \text{ and } P_2}{\text{number of possibilities resulting from 3 tosses}}$

$= \dfrac{(\{H,H,T\},\ \{H,T,H\})}{8} = 2/8 = 1/4$

$P(P_2/P_1) = \dfrac{P(P_1 P_2)}{P(P_1)} = \dfrac{1/4}{3/8} = 2/3$

438

A coin is tossed 3 times. Find the probability that all
3 are heads,
   (a) if it is known that the first is heads,
   (b) if it is known that the first 2 are heads,
   (c) if it is known that 2 of them are heads.

<u>Solution:</u> This problem is one of conditional probability.
If we have two events, A and B, the probability of event
A given that event B has occurred is

$$P(A/B) = \frac{P(AB)}{P(B)}.$$

   (a) We are asked to find the probability that all
three tosses are heads given that the first toss is heads.
The first event is A and the second is B.

P(A/B) = probability that all three tosses are heads
given that the first toss is heads

$$= \frac{\text{the number of ways that all three tosses are heads given that the first toss is a head}}{\text{the number of possibilities resulting from 3 tosses}}$$

$$= \frac{\{H,H,H\}}{\{\{H,H,H\}, \{H,H,T\}, \{H,T,H\}, \{H,T,T\}, \{T,T,T\}, \{T,T,H\}, \{T,H,T\}, \{T,H,H\}}$$

$$= \frac{1}{8}.$$

P(B) = P(first toss is a head)

$$= \frac{\text{the number of ways to obtain a head on the first toss}}{\text{the number of ways to obtain a head or a tail on the first of 3 tosses}}$$

$$= \frac{\{H,H,H\}, \{H,H,T\}, \{H,T,H\}, \{H,T,T\}}{8}$$

$$= \frac{4}{8}$$

$$= \frac{1}{2}.$$

$$P(A/B) = \frac{P(AB)}{P(B)} = \frac{\frac{1}{8}}{\frac{1}{2}} = \frac{1}{8} \quad \frac{2}{1} = \frac{1}{4}.$$

To see what happens, in detail, we note that if the first
toss is heads, the logical possibilities are HHH, HHT,
HTH, HTT. There is only one of these for which the second
and third are heads. Hence,

$$P(A/B) = \frac{1}{4}.$$

(b)   The problem here is to find the probability that all 3 tosses are heads given that the first two tosses are heads.

$P(A/B)$ = the probability that all three tosses are heads given that the first two are heads

= $\dfrac{\text{the number of ways to obtain 3 heads given that the first two tosses are heads}}{\text{the number of possibilities resulting from 3 tosses}}$

$$= \frac{1}{8}.$$

$P(B)$ = the probability that the first two are heads

= $\dfrac{\text{number of ways to obtain heads on the first two tosses}}{\text{number of possibilities resulting from three tosses}}$

$$= \frac{\{H,H,H\}, \ \{H,H,T\}}{8} = \frac{2}{8} = \frac{1}{4}.$$

$$P(A/B) = \frac{P(AB)}{P(B)} = \frac{\frac{1}{8}}{\frac{1}{4}} = \frac{4}{8} = \frac{1}{2}.$$

(c)   In this last part, we are asked to find the probability that all 3 are heads on the condition that any 2 of them are heads.
Define:

A = the event that all three are heads

B = the event that two of them are heads

$P(A/B)$ = the probability that all three tosses are heads knowing that two of them are heads

$$= \frac{1}{8}.$$

$P(B)$ = the probability that two tosses are heads

= $\dfrac{\text{number of ways to obtain at least two heads out of three tosses}}{\text{number of possibilities resulting from 3 tosses}}$

$$= \frac{\{H,H,T\}, \ \{H,H,H\}, \ \{H,T,H\}, \ \{T,H,H\}}{8}$$

$$= \frac{4}{8}$$

$$= \frac{1}{2}.$$

$$P(A/B) = \frac{P(AB)}{P(B)} = \frac{\frac{1}{8}}{\frac{1}{2}} = \frac{2}{8} = \frac{1}{4}.$$

CHAPTER 24

VECTORS, MATRICES AND DETERMINANTS

> **Basic Attacks and Strategies for Solving Problems in this Chapter. See pages 442 to 489 for step-by-step solutions to problems.**

A *vector* is a quantity that has both *direction* and *magnitude* (i.e., size or length). To distinguish them from vectors, numbers associated with points on the real number line are also called *scalars*. Vectors are designated in various ways. Often, vectors are indicated by bold typeface, such as the two standard direction vectors, **i** (the unit-length vector in the positive *x*-direction) and **j** (the unit-length vector in the positive *y*-direction). In some situations, vectors are indicated by an overbar arrow, such as $\bar{a}$. Other times, vectors are indicated by an overbar arrow over two letters (designating the origin point and terminus point of the vectors), such as $\overrightarrow{AB}$. Sometimes vectors are indicated by a column of numbers (one for each dimension of the vector space), enclosed by parentheses, such as $\begin{pmatrix} 1 \\ 0 \end{pmatrix}$ (which is normally considered equal to **i**). It should be noted that mathematical vectors do not have location, merely length and direction. Thus, the vector $\bar{a}$ from (0, 0) to (1, 1) and the vector $\bar{b}$ from (1, 0) to (2, 1) are equal since they are the same length ($\sqrt{2}$) and point in the same direction (along the 45° line above the horizontal *x*-axis). One can calculate a vector between two points by subtracting the corresponding coordinates of the first point from the second point. Thus,

$$\bar{a} \text{ is } \begin{pmatrix} 1 \\ 1 \end{pmatrix} \text{ and } \bar{b} \text{ is also } \begin{pmatrix} 1 \\ 1 \end{pmatrix}.$$

For simplicity, the column vectors are often written as row vectors, e.g., (1, 1).

The length or magnitude of a vector can be computed using the Pythagorean theorem, that is, by taking the square root of the sum of the squares of the component directions. For example, the length of the vector (3, 4) is 5, of the vector (1, 1) is $\sqrt{2}$, of the vector (1,1,2) is $\sqrt{6}$, and of the vector (0,0,1,0) is 1.

The vector with a value of 1 is the jth component of zeroes for all other components is usually designated as $\mathbf{e}_j$ and called the *unit vector in the jth direction*. These unit vectors are assumed to have the correct number of elements in them that correspond to the situation in which they are used.

Vectors can be used to analyze problems involving forces exerted in various directions (for example, finding the resultant [combined] force). Conversely, given a force in a specified direction, one can find the component forces in other directions.

In problems like these, the vector represents the force by the length of the vector corresponding to the magnitude of the force and the direction of the vector corresponding to the direction of the force. Often the force is indicated merely by the direction and the magnitude. Sometimes it is easier to solve such problems by decomposing each force into its horizontal and vertical directions.

Finding the component forces in the horizontal and vertical directions (i.e., the coordinate axes) is simply a matter of assuming that the vector is a hypotenuse of a right triangle and finding the length of the other two sides using the sine and cosine functions (*cf.* Chapter 28). Computing resultant forces is simply a matter of adding two or more vectors together. Sometimes it may be convenient to use the Law of Cosines when the vectors are such that one has two sides of a triangle (and the angle between them) and needs the third side (opposite the angle). The Law of Cosines states:

$$c^2 = a^2 + b^2 - 2ab \cos \gamma,$$

in which γ is the angle between sides a and b (i.e., opposite side c).

A *matrix* is a collection of values stored in a rectangular arrangement having m rows and n columns (i.e., an $m \times n$ matrix) (and surrounded by large parentheses). One can also consider a matrix to be a collection (i.e., a row vector) of n (column) vectors each of length m (or a column vector of m row vectors each of length n). Pre-multiplying a matrix A by $\mathbf{e}_i$ results in the ith row of A. An element in a matrix is located by the row number and column number (with row number first). Thus a_{ij} indicates the element in row i and column j. A matrix A is often indicated by an arbitrary element in parentheses, e.g., (a_{ij}). These are two special matrices of variable sizes. The identity matrix, I, is a square matrix (i.e., the same number of rows as columns) with the diagonal elements (i.e., a_{ii}) all equal to 1 and every other element being zero. The zero matrix, 0, has the appropriate dimensions for the situation in which it is needed, and all elements are zero.

Matrices can be added *provided the corresponding column and row dimensions are the same*. The elements in the sum matrix are merely the sum of the corresponding elements in the components matrices. That is, given $m \times n$ matrices, A, B, and C, $C = A + B$ if

$$c_{ij} = a_{ij} + b_{ij}.$$

This definition can be shown to be consistent with the definition of vector addition and viewing matrices as vectors of vectors. In a similar way, if d is a scalar, one can compute the scalar multiple of a matrix A to obtain dA by multiplying every element of A by d. That is, if

$$A = (a_{ij}), \quad \text{then} \quad dA = (da_{ij}).$$

Matrices can be multiplied *provided that the dimensions are compatible*. If A is an $m \times n$ matrix and B is an $n \times p$ matrix, then the product $C = AB$ can be computed and is of size $m \times p$. Note that unless $p = m$, it is impossible to compute the product BA, however. The product matrix is obtained as follows: to obtain the element c_{ij} in $C = AB$, one multiplies the elements of the ith row of A by the corresponding elements in the jth column of B and adds these products together. In other words,

$$c_{ij} = \sum_{k=1}^{n} a_{ik} b_{kj}$$

for all possible values of i and j. As in the case of matrix addition, it is possible to consider matrix multiplication from the perspective of a matrix as a collection of vectors.

The product of a matrix and a column vector follows from the definition of the product of two matrices. The vector is considered to be a matrix of dimension $n \times 1$ and the standard definition of a product of two matrices is used.

A matrix can be subdivided ("partitioned") into *blocks* by introducing horizontal lines between certain rows and vertical lines between certain columns. Each of these blocks can be considered as elements of a smaller matrix. It is possible to perform matrix operations (e.g., multiplication) by using the appropriate block of one matrix and the appropriate (and compatible) block of a second matrix. The products of these blocks are then evaluated by the standard rules of matrix multiplication.

Matrices can be classified according to the distribution of non-zero elements. A *triangular matrix* is a square ($n \times n$) matrix whose non-zero elements lie all either on the diagonal and above the diagonal or on the diagonal and below the diagonal. An *upper triangular matrix* is a triangular matrix all of whose non-zero elements lie on or above the diagonal. A *lower triangular matrix* is a triangular matrix all of whose non-zero elements lie on or below the diagonal.

Given matrix $A = (a_{ij})$ or size $m \times n$, the *transpose* of matrix A,

$$A^T = (a_{ij}^T),$$

is the $n \times m$ matrix such that $a_{ij}^T = a_{ji}$. One can consider the transpose of A to be the matrix obtained by taking row i of A and turning it into column i of A^T.

A determinant is a function that assigns a value to an $n \times n$ square matrix A. The value is formally defined as the sum of all signed products of all possible permutations of n elements of A, taken one from each column and row. The sign is determined by ordering the elements, either by row or column subscript, and then examining the permutation of the other subscripts. If the total number of *inversions*, k, is even, the sign is positive (i.e., $(-1)^k$), but if the total number is odd, the sign is negative (i.e., $(-1)^k$). An inversion is a reversal of the order of numbers. For example, 1 2 3 has no inversions (since the numbers are all in proper order). On the other hand, 3 2 1 has 3 inversions (relative to 3, both 2 and 1 are out of order, thus giving 2 inversions and relative to 2, 1 is out of order, thus giving 1 more inversion).

In general, determinants are evaluated using rules other than the definition in the previous paragraph. For a 1×1 matrix, the determinant value equals the value of the element. For a 2×2 matrix A, the determinant value equals $a_{11}a_{22} - a_{21}a_{12}$, that is, the product of the elements on the major diagonal minus the product of the elements on the cross diagonal.

Higher order determinants are frequently evaluated by using a technique known as *expanding by minors*. A *minor* of an element a_{ij} in an $n \times n$ matrix A is the determinant of the $(n-1) \times (n-1)$ matrix obtained by removing row i and column j from A. The *cofactor* of an element a_{ij} is $(-1)^{i+j} \times$ minor of a_{ij}. To find the value of a determinant A, one chooses either a column or a row, and sums the products of each element times its cofactor.

If a row or column filled with numerous zeroes is chosen for the expansion, the computations are simplified. It can thus be deduced that if A is a triangular matrix, then the determinant of A is merely the product of the elements on the main diagonal.

If each element in any row or column of a matrix is zero, the value of the determinant is zero.

The determinant of a matrix equals the determinant of its transpose.

If two rows (or columns) of a matrix are identical, the value of the determinant is zero.

If B is derived from A by interchanging two adjacent rows or columns, then $\det(B) = -\det(A)$.

If A and B are the same except for elements of column i, and if C is derived from A and B by duplicating the identical columns and adding the two different column i's together, the

$$\det(A) + \det(B) = \det(C).$$

The value of a determinant remains unchanged if a multiple of one row (column) is added to another row (column).

If each element of any row (column) is multiplied by a constant d, the value of the determinant is multiplied by d.

The *adjoint* of a matrix A is the transpose of the matrix of cofactors of the elements of A. The *inverse* of a square matrix A, denoted A^{-1}, is the unique matrix A^{-1} such that

$$A^{-1}A = A \cdot A^{-1} = I.$$

Not every matrix has an inverse. A square matrix A has an inverse if and only if $\det(A) \neq 0$. In these cases,

$$A^{-1} = \frac{1}{\det(A)} \operatorname{adj}(A).$$

Matrices that have inverses are said to be *invertible* or *nonsingular*. If the determinant of a matrix is zero, the matrix has no inverse and is said to be *singular*.

The inverse of a matrix A can be obtained by this procedure. An identity matrix of the same as A is placed immediately to the right of A. Elementary row operations are performed on both matrices together so that the A is slowly transformed into the identity matrix. When the original matrix A becomes the identity, the matrix that began as the identity has become A^{-1}.

There are three types of elementary row operations.

1. Adding or subtracting a multiple of some row to another row.

2. Multiplying all elements of a row by some constant.

3. Interchanging two rows.

There are certain special forms of matrices. A is said to be in *echelon form* if (1) the first non-zero entry in any row is in a column to the left of the first non-zero entry in the next row, (2) the very first non-zero entry is one, and (3) any row of all zeroes lies below all rows that have some non-zero entries. A is said to be in *reduced echelon form* if (1) the first non-zero entry in *each* row is one, and (2) each column containing such a non-zero item has zeroes for all other entries.

Step-by-Step Solutions to Problems in this Chapter, "Vectors, Matrices and Determinants"

Which of the following vectors are equal to $\overrightarrow{MN}$ if
M = (2, 1) and N = (3, - 4)?
(a) $\overrightarrow{AB}$, where A = (1, - 1) and B = (2, 3)
(b) $\overrightarrow{CD}$, where C = (- 4, 5) and D = (- 3, 10)
(c) $\overrightarrow{EF}$, where E = (3, - 2) and F = (4, - 7).

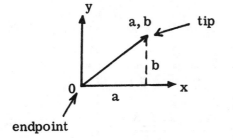

Figure A: (a-0, b-0) represents
        the vector.

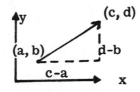

Figure B: (c-a, d-b) represents the
        vector.

<u>Solution:</u> With each ordered pair in the plane there can be associated a vector from the origin to that point.

The vector is determined by subtracting the co-ordinates of the endpoint from the corresponding co-ordinates of the tip. As for $\overrightarrow{MN}$, the tip is the point corresponding to the second letter of the alphabetical notation, N, while the endpoint is the point corresponding to the first, M. In this problem the vectors are of a general nature wherein their endpoints do not lie at the origin.

We first find the ordered pair which represents $\overrightarrow{MN}$.

$$\overrightarrow{MN} = (3 - 2, -4 - 1) = (1, -5)$$

Now, we find the ordered pair representing each vector.

(a) $\overrightarrow{AB} = (2 - 1, 3 - (-1)) = (1, 4)$

(b) $\overrightarrow{CD} = ((-3) - (-4), 10 - 5) = (1, 5)$

(c) $\overrightarrow{EF} = (4 - 3, -7 - (-2)) = (1, -5)$

Only $\overrightarrow{EF}$ and $\overrightarrow{MN}$ are equal.

● **PROBLEM 24-2**

A force of 315 lbs. is acting at an angle of 67° with the horizontal. What are its horizontal and vertical components?

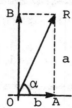

Solution: Construct the figure shown.
    OR = vector force = c.
    b = OA = horizontal component.
    a = OB = vertical component.

In △OAR: c = 315; α = 67°.		
$\frac{a}{c} = \sin \alpha$, or $a = c \sin \alpha.$	log c = 2.49831 log sin α = 9.96403 − 10 log a = 2.46234	a = 289.96 lbs.
$\frac{b}{c} = \cos \alpha$, or $b = c \cos \alpha.$	log c = 2.49831 log cos α = 9.59188 − 10 log b = 2.09019	b = 123.08 lbs.

● **PROBLEM 24-3**

Two forces of 50 lbs. and 30 lbs. have an included angle of 60°. Find the magnitude and direction of their resultant.

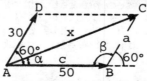

Solution: Construct the parallelogram and label it as in the figure. Since $\overline{AD}$ is parallel to $\overline{BC}$ we have $\angle ABC = \beta = 180° - 60° = 120°$.

By the law of cosines:
$$x^2 = c^2 + a^2 - 2\,ac\cos\beta$$
$$= 2500 + 900 - 2(50)(30)(-\tfrac{1}{2})$$
$$= 2500 + 900 + 1500 = 4900.$$
$$x = 70 \text{ lbs.}$$

$$\cos\alpha = \frac{x^2 + c^2 - a^2}{2xc} = \frac{4900 + 2500 - 900}{2(70)(50)} = \frac{13}{14} = .9286.$$

$$\alpha = 21°47'.$$

● **PROBLEM 24-4**

Two forces act simultaneously on a body free to move. One force of 112 lbs. is acting due east, while the other of 88 lbs. is acting due north. Find the magnitude and direction of their resultant.

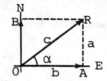

<u>Solution:</u> Construct the figure shown.

OA = b = 112 lbs.
OB = 88 lbs. = RA = a.

In △OAR: a = 88; b = 112.		
$\dfrac{a}{b} = \tan\alpha$.	log a = 1.94448 log b = 2.04922 log tan α = 9.89526 − 10	α = 38°9'25''
$\dfrac{a}{c} = \sin\alpha$, or $c = \dfrac{a}{\sin\alpha}$	log a = 11.94448 − 10 log sin α = 9.79086 − 10 log c = 2.15362	c = 142.44
Therefore the resultant is 142.44 lbs. and its direction is 38°9'25'' north of east.		

● **PROBLEM 24-5**

Find the force required to prevent a 500-pound object from sliding down a 40° incline, disregarding friction.

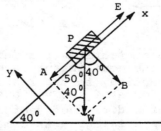

<u>Solution:</u> The weight of the object acts as a 500-pound force vertically downward. The component of the force parallel to the plane tends to force the object down

444

the incline. The component perpendicular to the plane
tends to force the object against the plane. The required force is parallel to the plane, equal in magnitude and opposite in direction to the parallel component. Observing the figure, the triangle WPB is
similar to the triangle made by the inclined plane
and the ground. Angle WPB therefore equals 40°. It
follows then that angle APW equals 50°, as it is
complementary to angle WPB. By super posing the coordinate axes with the x axis parallel to the inclined
plane, observe that the component of the force to be
determined lies parallel to the inclined plane and can
be found by multiplying the magnitude of weight by
the cos of 50°. Here cos 50° can be calculated using
the rule cos θ = adjacent/hypotenuse for right
triangles. The adjacent side is $\overrightarrow{AP}$ and the hypotenuse
is $\overrightarrow{PW}$. Therefore cos 50° = AP/PW or (PW) cos 50° = AP,
where Ap and PW represent the magnitudes of the force
vectors.

$\overrightarrow{PW}$ = 500, $\angle$APW = 90° - 40° = 50°

$\cos \angle APW = \dfrac{AP}{PW}$, where AP = $|\overrightarrow{AP}|$

AP = PW cos $\angle$APW

AP = 500 cos 50°

AP = 500(0.6428) = 321
The required force $\overrightarrow{PE}$ is 321 pounds.

● **PROBLEM** 24-6

Find the magnitude and direction of the force necessary
to counteract the effect of a force of 60 pounds and a
force of 40 pounds that act on a point at an angle of
60° with each other.

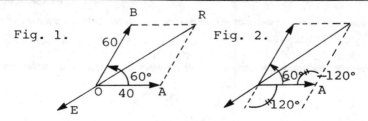

Fig. 1.   Fig. 2.

Solution: The problem requires finding the equilibrant
of the two forces. The equilibrant is equal in magnitude,
but opposite in direction, to the resultant force. We
find the resultant force of two vectors by using the
parallelogram law of addition of vectors. Two vectors
are drawn and the parallelogram is completed. The resultant is drawn connecting the two opposite vertices
(see figure 1).We indicate the magnitude of a vector as
the corresponding segment. In the parallelogram, resultant

$\vec{OR}$ may be found by solving triangle OAR.

$$OA = BR = 40$$
$$AR = OB = 60$$
$$\angle A = 180° - 60° = 120° \text{(see fig.2)}$$

By the Law of Cosines:

$$(OR)^2 = (OA)^2 + (AR)^2 - [2(OA)(AR) \cos 120°]$$
$$(OR)^2 = (40)^2 + (60)^2 - [(2)(40)(60)(-\tfrac{1}{2})]$$
$$(OR)^2 = 7,600$$
$$OR = 87$$

From the Law of Sines,

$$\sin \angle AOR = \frac{60 \sin 120°}{87} = \frac{60(0.8660)}{87} = 0.5972$$

$$\angle AOR = 37° \quad \text{(to the nearest degree)}$$

$$\angle AOE = 180° - 37° = 143°, \text{ and the required}$$

force is a force of 87 pounds, 143° from the 40-pound
force in the opposite direction from the 60-pound
force.

● **PROBLEM 24-7**

Prove that if $A = (a_{ij})$ and $B = (b_{ij})$ are $m \times n$ matrices over a field K and c is an element of K, then $A + B = (c_{ij})$ where $c_{ij} = a_{ij} + b_{ij}$, $i = 1,2,\ldots,m$, $j = 1,2,\ldots,n$ and $cA = (d_{ij})$ where $d_{ij} = ca_{ij}$, $i = 1,2,\ldots,m$, $j = 1,2,\ldots,n$.

Solution: Let e_j represent the vector $(0,\ldots,0,1,0,\ldots,0)$ where the 1 is in the ith place. Letting $C = (c_{ij})$, then the ith row of C is given by $e_i \cdot C = (c_{i1}, c_{i2}, \ldots, c_{in})$; $i = 1,2,\ldots,m$. On the other hand,

$$e_i(A+B) = e_iA + e_iB = (a_{i1}, a_{i2}, \ldots, a_{in}) + (b_{i1}, b_{i2}, \ldots, b_{in})$$

Adding corresponding elements in the ith row,

$$e_i(A+B) = (a_{i1} + b_{i1}, a_{i2} + b_{i2}, \ldots, a_{in} + b_{in}) .$$

But, by definition, $c_{ij} = a_{ij} + b_{ij}$. Therefore, by substitution, $e_i(A+B) = (c_{i1}, c_{i2}, \ldots, c_{in}) = e_iC$.

This is true for all $i = 1,\ldots,m$. Therefore, $A + B = C$.

(b)   Let $D = (d_{ij})$ then,

$$e_i \cdot D = (d_{i1}, d_{i2}, \ldots, d_{in}) , \quad i = 1,2,\ldots,m .$$

On the other hand,

$$e_i(cA) = c(e_iA) = c(a_{i1}, a_{i2}, \ldots, a_{in}), \quad i = 1,2,\ldots,m$$
$$= (ca_{i1}, ca_{i2}, \ldots, ca_{in}) .$$

But, by definition, $d_{ij} = ca_{ij}$. Therefore, $e_i(cA) = (d_{i1}, \ldots, d_{in})$ by substitution; i.e., $e_i(cA) = e_iD$. This holds for all $i = 1, \ldots, m$. Therefore, $cA = D$.

● **PROBLEM** 24-8

If $A = \begin{bmatrix} 2 & -2 & 4 \\ -1 & 1 & 1 \end{bmatrix}$ and $B = \begin{bmatrix} 0 & 1 & -3 \\ 1 & 3 & 1 \end{bmatrix}$, find $2A + B$.

<u>Solution</u>: For an m×n matrix, $A = (a_{ij})$, we know $cA = (ca_{ij})$. Hence,

$$2A = 2\begin{bmatrix} 2 & -2 & 4 \\ -1 & 1 & 1 \end{bmatrix} = \begin{bmatrix} 2 \cdot 2 & 2 \cdot (-2) & 2 \cdot 4 \\ 2 \cdot (-1) & 2 \cdot 1 & 2 \cdot 1 \end{bmatrix} = \begin{bmatrix} 4 & -4 & 8 \\ -2 & 2 & 2 \end{bmatrix}.$$

For two m×n matrices, $A = (\alpha_{ij})$ and $B = (\beta_{ij})$, the ith row of the matrix $A + B$ is given by $e_i \cdot (A+B) = (\alpha_{i1} + \beta_{i1}, \ldots, \alpha_{in} + \beta_{in})$. Thus,

$$2A + B = \begin{bmatrix} 4 & -4 & 8 \\ -2 & 2 & 2 \end{bmatrix} + \begin{bmatrix} 0 & 1 & -3 \\ 1 & 3 & 1 \end{bmatrix} = \begin{bmatrix} 4+0 & -4+1 & 8-3 \\ -2+1 & 2+3 & 2+1 \end{bmatrix}$$

$$2A + B = \begin{bmatrix} 4 & -3 & 5 \\ -1 & 5 & 3 \end{bmatrix}.$$

● **PROBLEM** 24-9

Show that

a) $A + B = B + A$ where
$$A = \begin{bmatrix} 3 & 1 & 1 \\ 2 & -1 & 1 \end{bmatrix}; \quad B = \begin{bmatrix} 4 & 2 & -1 \\ 0 & 0 & 2 \end{bmatrix}.$$

b) $(A+B) + C = A + (B+C)$ where
$$A = \begin{bmatrix} -2 & 6 \\ 2 & 1 \end{bmatrix}, \quad B = \begin{bmatrix} 2 & 1 \\ 0 & 3 \end{bmatrix} \text{ and } C = \begin{bmatrix} -1 & 0 \\ 7 & 2 \end{bmatrix}.$$

c)   If A and the zero matrix (0_{ij}) have the same size, then $A + 0 = A$ where
$$A = \begin{bmatrix} 2 & 1 \\ 1 & 2 \end{bmatrix}.$$

d) $A + (-A) = 0$ where
$$A = \begin{bmatrix} 2 & 1 \\ 1 & 2 \end{bmatrix}.$$

e) $(ab)A = a(bA)$ where $a = -5$, $b = 3$ and
$$A = \begin{bmatrix} 6 & -1 & 0 \\ 1 & 2 & 1 \end{bmatrix}.$$

f) Find B if 2A - 3B + C = 0 where

$$A = \begin{bmatrix} -1 & 3 \\ 0 & 0 \end{bmatrix} \text{ and } C = \begin{bmatrix} -2 & -1 \\ -1 & 1 \end{bmatrix}.$$

Solution: a) By the definition of matrix addition,

$$A + B = \begin{bmatrix} 3 & 1 & 1 \\ 2 & -1 & 1 \end{bmatrix} + \begin{bmatrix} 4 & 2 & -1 \\ 0 & 0 & 2 \end{bmatrix}$$

$$= \begin{bmatrix} 3+4 & 1+2 & 1+(-1) \\ 2+0 & -1+0 & 1+2 \end{bmatrix}$$

$$= \begin{bmatrix} 7 & 3 & 0 \\ 2 & -1 & 3 \end{bmatrix}$$

and

$$B + A = \begin{bmatrix} 4 & 2 & -1 \\ 0 & 0 & 2 \end{bmatrix} + \begin{bmatrix} 3 & 1 & 1 \\ 2 & -1 & 1 \end{bmatrix}$$

$$= \begin{bmatrix} 4+3 & 2+1 & -1+1 \\ 0+2 & 0+(-1) & 2+1 \end{bmatrix} + \begin{bmatrix} 7 & 3 & 0 \\ 2 & -1 & 3 \end{bmatrix}$$

Thus, A + B = B + A .

b)
$$A + B = \begin{bmatrix} -2 & 6 \\ 2 & 1 \end{bmatrix} + \begin{bmatrix} 2 & 1 \\ 0 & 3 \end{bmatrix}$$

$$= \begin{bmatrix} -2+2 & 6+1 \\ 2+0 & 1+3 \end{bmatrix} = \begin{bmatrix} 0 & 7 \\ 2 & 4 \end{bmatrix}$$

and

$$(A+B) + C = \begin{bmatrix} 0 & 7 \\ 2 & 4 \end{bmatrix} + \begin{bmatrix} -1 & 0 \\ 7 & 2 \end{bmatrix} = \begin{bmatrix} 0+(-1) & 7+0 \\ 2+7 & 4+2 \end{bmatrix} = \begin{bmatrix} -1 & 7 \\ 9 & 6 \end{bmatrix}.$$

$$B + C = \begin{bmatrix} 2 & 1 \\ 0 & 3 \end{bmatrix} + \begin{bmatrix} -1 & 0 \\ 7 & 2 \end{bmatrix} = \begin{bmatrix} 2+(-1) & 1+0 \\ 0+7 & 3+2 \end{bmatrix} = \begin{bmatrix} 1 & 1 \\ 7 & 5 \end{bmatrix}$$

and

$$A + (B+C) = \begin{bmatrix} -2 & 6 \\ 2 & 1 \end{bmatrix} + \begin{bmatrix} 1 & 1 \\ 7 & 5 \end{bmatrix} = \begin{bmatrix} -2+1 & 6+1 \\ 2+7 & 1+5 \end{bmatrix} = \begin{bmatrix} -1 & 7 \\ 9 & 6 \end{bmatrix}.$$

Thus, (A+B) + C = A + (B+C) .

c) An m×n matrix all of whose elements are zeros is called a zero matrix and is usually denoted by $_m 0_n$.

$$A = \begin{bmatrix} 2 & 1 \\ 1 & 2 \end{bmatrix} \qquad 0 = \begin{bmatrix} 0 & 0 \\ 0 & 0 \end{bmatrix}.$$

Thus,

$$A + 0 = \begin{bmatrix} 2 & 1 \\ 1 & 2 \end{bmatrix} + \begin{bmatrix} 0 & 0 \\ 0 & 0 \end{bmatrix} = \begin{bmatrix} 2+0 & 1+0 \\ 1+0 & 2+0 \end{bmatrix} = \begin{bmatrix} 2 & 1 \\ 1 & 2 \end{bmatrix}.$$

Hence, A + 0 = A .

d)
$$-A = -1 \cdot \begin{bmatrix} 2 & 1 \\ 1 & 2 \end{bmatrix} = \begin{bmatrix} -1 \cdot 2 & -1 \cdot 1 \\ -1 \cdot 1 & -1 \cdot 2 \end{bmatrix} = \begin{bmatrix} -2 & -1 \\ -1 & -2 \end{bmatrix}.$$

Thus,

$$A + (-A) = \begin{bmatrix} 2 & 1 \\ 1 & 2 \end{bmatrix} + \begin{bmatrix} -2 & -1 \\ -1 & -2 \end{bmatrix} = \begin{bmatrix} 2+(-2) & 1+(-1) \\ 1+(-1) & 2+(-2) \end{bmatrix} = \begin{bmatrix} 0 & 0 \\ 0 & 0 \end{bmatrix}$$

Therefore, $A + (-A) = 0$.

e)   If $A = \begin{bmatrix} a_1 & b_1 \\ c_1 & d_1 \end{bmatrix}$ and a is any scalar from a field, aA is

defined by
$$aA = \begin{bmatrix} aa_1 & ab_1 \\ ac_1 & ad_1 \end{bmatrix} .$$

So, $bA = 3 \begin{bmatrix} 6 & -1 & 0 \\ 1 & 2 & 1 \end{bmatrix} = \begin{bmatrix} 3 \cdot 6 & 3 \cdot (-1) & 3 \cdot 0 \\ 3 \cdot 1 & 3 \cdot 2 & 3 \cdot 1 \end{bmatrix}$

$$= \begin{bmatrix} 18 & -3 & 0 \\ 3 & 6 & 3 \end{bmatrix}$$

and
$$a(bA) = -5 \begin{bmatrix} 18 & -3 & 0 \\ 3 & 6 & 3 \end{bmatrix} = \begin{bmatrix} -90 & 15 & 0 \\ -15 & -30 & -15 \end{bmatrix}$$

$$(ab)A = ((-5)(3)) \begin{bmatrix} 6 & -1 & 0 \\ 1 & 2 & 1 \end{bmatrix} = -15 \begin{bmatrix} 6 & -1 & 0 \\ 1 & 2 & 1 \end{bmatrix} = \begin{bmatrix} -90 & 15 & 0 \\ -15 & -30 & -15 \end{bmatrix}$$

Thus, $(ab)A = a(bA)$.

f) $2A - 3B + C = 2A + C - 3B = 0$ since matrix addition is commutative.

Now, add $3B$ to both sides of the equation,
$$2A + C - 3B = 0 ,$$
to obtain $2A + C - 3B + 3B = 0 + 3B$.   (1)

Using the laws we exemplified in parts a) through d), (1) becomes
$2A + C = 3B$. Now,
$$\tfrac{1}{3}(2A + C) = \tfrac{1}{3}(3B)$$
which implies $\qquad B = \tfrac{1}{3}(2A + C)$.

$$2A + C = \begin{bmatrix} 2(-1) & 2(3) \\ 2(0) & 2(0) \end{bmatrix} + \begin{bmatrix} -2 & -1 \\ -1 & 1 \end{bmatrix} = \begin{bmatrix} -4 & 5 \\ -1 & 1 \end{bmatrix} .$$
Thus,
$$\tfrac{1}{3}(2A + C) = \tfrac{1}{3}\begin{bmatrix} -4 & 5 \\ -1 & 1 \end{bmatrix} = \begin{bmatrix} -4/3 & 5/3 \\ -1/3 & 1/3 \end{bmatrix} .$$

● **PROBLEM 24-10**

a) If $A = (a_{ij})$ is a $p \times q$ matrix and $B = (b_{ij})$ is a $q \times r$ matrix, prove AB is the $p \times r$ matrix (c_{ij}) where
$$c_{ij} = \sum_{k=1}^{q} a_{ik} b_{kj} , \qquad \begin{array}{l} i = 1,2,\ldots,p \\ j = 1,2,\ldots,r \end{array} .$$

b) If $A = \begin{bmatrix} 2 & 1 & 1 \\ -1 & 2 & 3 \\ 1 & 0 & 1 \end{bmatrix}$ and $B = \begin{bmatrix} 2 & 1 \\ -1 & 1 \\ 2 & -1 \end{bmatrix}$, find AB .

Solution: Let e_i represent $(0,\ldots,0,1,0,\ldots,0)$ with the 1 in the ith place. Let $C = (c_{ij})$. Consequently, the ith row of C is given by $e_i \cdot (C) = (c_{i1}, \ldots, c_{ir}) = \sum_{j=1}^{r} c_{ij} \cdot e_j$, $i = 1, \ldots, p$. On the other hand, since

$$e_i A = \sum_{k=1}^{r} a_{ik} e_k \quad \text{and} \quad e_k B = \sum_{j=1}^{r} b_{kj} e_j \ ,$$

$$e_i (AB) = (e_i A)B = \sum_{k=1}^{r} (a_{ik} e_k)B = \sum_{k=1}^{r} a_{ik} \left(\sum_{j=1}^{r} b_{kj} e_j \right)$$

$$= \sum_{j=1}^{r} \left(\sum_{k=1}^{r} a_{ik} b_k \right) e_j \ , \quad i = 1, 2, \ldots, p \ .$$

But, by definition, $\sum_{k=1}^{r} a_{ik} b_{kj} = c_{ij}$. Thus, by substitution,

$$e_i (A \cdot B) = \sum_{j=1}^{r} c_{ij} e_j = e_i C \ .$$

This holds for all $i = 1, \ldots, p$. Therefore, $AB = C$.

$$AB = \begin{bmatrix} 2 & 1 & 1 \\ -1 & 2 & 3 \\ 1 & 0 & 1 \end{bmatrix} \begin{bmatrix} 2 & 1 \\ -1 & 1 \\ 2 & -1 \end{bmatrix}$$

$$= \begin{bmatrix} (2 \cdot 2 + 1 \cdot (-1) + 1 \cdot 2) & (2 \cdot 1 + 1 \cdot 1 + 1 \cdot (-1)) \\ (-1 \cdot 2 + 2 \cdot (-1) + 3 \cdot 2) & (-1 \cdot 1 + 2 \cdot 1 + 3 \cdot (-1)) \\ (1 \cdot 2 + 0 \cdot (-1) + 1 \cdot 2) & (1 \cdot 1 + 0 \cdot 1 + 1 \cdot (-1)) \end{bmatrix}$$

$$= \begin{bmatrix} 4-1+2 & 2+1-1 \\ -2-2+6 & -1+2-3 \\ 2+0+2 & 1+0-1 \end{bmatrix} = \begin{bmatrix} 5 & 2 \\ 2 & -2 \\ 4 & 0 \end{bmatrix}$$

● PROBLEM 24-11

If $A = \begin{bmatrix} 1 & 2 \\ -1 & 3 \end{bmatrix}$ and $B = \begin{bmatrix} 2 & 1 \\ 0 & 1 \end{bmatrix}$, show $AB \ne BA$.

Solution: A is 2×2 and B is 2×2; the product AB is a 2×2 matrix.

$$AB = \begin{bmatrix} 1 & 2 \\ -1 & 3 \end{bmatrix} \begin{bmatrix} 2 & 1 \\ 0 & 1 \end{bmatrix} = \begin{bmatrix} 1 \cdot 2 + 2 \cdot 0 & 1 \cdot 1 + 2 \cdot 1 \\ -1 \cdot 2 + 3 \cdot 0 & -1 \cdot 1 + 3 \cdot 1 \end{bmatrix} = \begin{bmatrix} 2+0 & 1+2 \\ -2+0 & -1+3 \end{bmatrix}$$

$$= \begin{bmatrix} 2 & 3 \\ -2 & 2 \end{bmatrix} \ .$$

Now, $BA = \begin{bmatrix} 2 & 1 \\ 0 & 1 \end{bmatrix} \begin{bmatrix} 1 & 2 \\ -1 & 3 \end{bmatrix} = \begin{bmatrix} 2 \cdot 1 + 1 \cdot (-1) & 2 \cdot 2 + 1 \cdot 3 \\ 0 \cdot 1 + 1 \cdot (-1) & 0 \cdot 2 + 1 \cdot 3 \end{bmatrix}$

$$= \begin{bmatrix} 2-1 & 4+3 \\ 0-1 & 0+3 \end{bmatrix} = \begin{bmatrix} 1 & 7 \\ -1 & 3 \end{bmatrix}$$

Thus, $AB \neq BA$.

a) Suppose $A = \begin{bmatrix} 1 & 3 \\ 2 & -1 \end{bmatrix}$ and $B = \begin{bmatrix} 2 & 0 & -4 \\ 3 & -2 & 6 \end{bmatrix}$. Find i) AB and
ii) BA .

b) Suppose $A = [2,1]$ and $B = \begin{bmatrix} 1 & -2 & 0 \\ 4 & 5 & -3 \end{bmatrix}$. Find i) AB, and
ii) BA.

Solution: Suppose $A = (a_{ij})$ and $B = (b_{jk})$ are matrices such that the number of columns of A equals the number of rows of B; also suppose A is an $m \times n$ matrix and B is an $n \times s$ matrix.

$$A = \begin{bmatrix} a_{11} & \cdots & a_{1n} \\ \cdot & & \cdot \\ \cdot & & \cdot \\ \cdot & & \cdot \\ a_{m1} & \cdots & a_{mn} \end{bmatrix}$$

$$B = \begin{bmatrix} b_{11} & \cdots & b_{1s} \\ \cdot & & \cdot \\ \cdot & & \cdot \\ \cdot & & \cdot \\ b_{n1} & \cdots & b_{ns} \end{bmatrix} .$$

Then the product AB is the $m \times s$ matrix whose ik-element is
$\sum_{j=1}^{n} a_{ij} b_{jk} = a_{i1} b_{ik} + a_{i2} b_{2k} + \ldots + a_{in} b_{nk}$. If $A_1, \ldots, A_m$ are the
row vectors of the matrix A, and if $B^1, \ldots, B^s$ are the column vectors of the matrix B, then the ik-element of the product AB is equal to $A_1, \ldots, B^k$. Thus,

$$\begin{matrix} A_1 \cdot B^1 & \cdots & A_1 \cdot B^s \\ \cdot & & \cdot \\ \cdot & & \cdot \\ \cdot & & \cdot \\ A_m B^1 & \cdots & A_m \cdot B^s \end{matrix}$$

a) i) $A = \begin{bmatrix} 1 & 3 \\ 2 & -1 \end{bmatrix}$ and $B = \begin{bmatrix} 2 & 0 & -4 \\ 3 & -2 & 6 \end{bmatrix}$. Here, A is 2×2 and

451

B is 2×3. The product AB is a 2×3 matrix. To obtain the components in the first row of AB, multiply the first row $(1,3)$ of A by the columns $\begin{bmatrix} 2 \\ 3 \end{bmatrix}$, $\begin{bmatrix} 0 \\ -2 \end{bmatrix}$ and $\begin{bmatrix} -4 \\ 6 \end{bmatrix}$ of B, respectively.

$$\begin{bmatrix} 1 & 3 \\ 2 & -1 \end{bmatrix} \begin{bmatrix} 2 & 0 & -4 \\ 3 & -2 & 6 \end{bmatrix} = [1\cdot2+3\cdot3 \quad 1\cdot0+3\cdot(-2) \quad 1\cdot(-4)+3\cdot6]$$

$$= [2+0 \quad 0-6 \quad -4+18] = [11 \quad -6 \quad 14]$$

To obtain the components in the second row of AB, multiply the second row $(2,01)$ of A by the columns of B, respectively.

$$\begin{bmatrix} 1 & 3 \\ 2 & -1 \end{bmatrix} \begin{bmatrix} 2 & 0 & -4 \\ 3 & -2 & 6 \end{bmatrix} = [(2\cdot2+(-1)\cdot3) \quad (2\cdot0+(-1)\cdot(-2)) \quad (2\cdot(-4)+(-1)\cdot6)]$$

$$= [4-3 \quad 0+2 \quad -8-6] = [1 \quad 2 \quad -14].$$

Thus,

$$AB = \begin{bmatrix} 11 & -6 & 14 \\ 1 & 2 & -14 \end{bmatrix}.$$

ii) Here B is 2×3 and A is 2×2. Since the number of columns of B is not equal to the number of rows of A, the product BA is not defined.

b) i) Since A is a 1×2 and B is a 2×3, the product AB is a 1×3 matrix.

$$AB = \begin{bmatrix} 2,1 \end{bmatrix} \begin{bmatrix} 1 & -2 & 0 \\ 4 & 5 & -3 \end{bmatrix} = \begin{bmatrix} (2\cdot1+1\cdot4, \quad 2\cdot(-2)+1\cdot5, \quad 2\cdot0+1\cdot(-3) \end{bmatrix}$$

$$= \begin{bmatrix} 6 & 1 & -3 \end{bmatrix}$$

ii) In this case, B is 2×3 and A is 1×2. Since the number of columns of B is not equal to the number of rows of A, the product BA is not defined.

● **PROBLEM 24-13**

Find $A(B+C)$ and $AB + AC$ if $A = \begin{bmatrix} 2 & 2 & 3 \\ 3 & -1 & 2 \end{bmatrix}$, $B = \begin{bmatrix} 1 & 0 \\ 2 & 2 \\ 3 & -1 \end{bmatrix}$ and $C = \begin{bmatrix} -1 & 2 \\ 1 & 0 \\ 2 & -2 \end{bmatrix}$.

Solution: $B + C = \begin{bmatrix} 1 & 0 \\ 2 & 2 \\ 3 & -1 \end{bmatrix} + \begin{bmatrix} -1 & 2 \\ 1 & 0 \\ 2 & -2 \end{bmatrix}$

$$\begin{bmatrix} 1+(-1) & 0+2 \\ 2+1 & 2+0 \\ 3+2 & -1+(-2) \end{bmatrix} = \begin{bmatrix} 0 & 2 \\ 3 & 2 \\ 5 & -3 \end{bmatrix}$$

then,

$$A(B+C) = \begin{bmatrix} 2 & 2 & 3 \\ 3 & -1 & 2 \end{bmatrix} \begin{bmatrix} 0 & 2 \\ 3 & 2 \\ 5 & -3 \end{bmatrix}$$

452

$$= \begin{bmatrix} 2 \cdot 0 + 2 \cdot 3 + 3 \cdot 5 & 2 \cdot 2 + 2 \cdot 2 + 3 \cdot (-3) \\ 3 \cdot 0 + (-1) \cdot 3 + 2 \cdot 5 & 3 \cdot 2 + (-1) \cdot 2 + 2 \cdot (-3) \end{bmatrix}.$$

$$A(B+C) = \begin{bmatrix} 0+6+15 & 4+4-9 \\ 0-3+10 & 6-2-6 \end{bmatrix} = \begin{bmatrix} 21 & -1 \\ 7 & -2 \end{bmatrix}.$$

$$AB = \begin{bmatrix} 2 & 2 & 3 \\ 3 & -1 & 2 \end{bmatrix} \begin{bmatrix} 1 & 0 \\ 2 & 2 \\ 3 & -1 \end{bmatrix}$$

$$= \begin{bmatrix} 2 \cdot 1 + 2 \cdot 2 + 3 \cdot 3 & 2 \cdot 0 + 2 \cdot 2 + 3 \cdot (-1) \\ 3 \cdot 1 + (-1) \cdot 2 + 2 \cdot 3 & 3 \cdot 0 + (-1) \cdot 2 + 2 \cdot (-1) \end{bmatrix}$$

$$= \begin{bmatrix} 2+4+9 & 0+4+(-3) \\ 3-2+6 & 0-2-2 \end{bmatrix} = \begin{bmatrix} 15 & 1 \\ 7 & -4 \end{bmatrix}.$$

$$AC = \begin{bmatrix} 2 & 2 & 3 \\ 3 & -1 & 2 \end{bmatrix} \begin{bmatrix} -1 & 2 \\ 1 & 0 \\ 2 & -2 \end{bmatrix}$$

$$= \begin{bmatrix} 2 \cdot (-1) + 2 \cdot 1 + 3 \cdot 2 & 2 \cdot 2 + 2 \cdot 0 + 3 \cdot (-2) \\ 3 \cdot (-1) + (-1) \cdot 1 + 2 \cdot 2 & 3 \cdot 2 + (-1) \cdot 0 + 2 \cdot (-2) \end{bmatrix}$$

$$AC = \begin{bmatrix} -2+2+6 & 4+0-6 \\ -3-1+4 & 6+0-4 \end{bmatrix} = \begin{bmatrix} 6 & -2 \\ 0 & 2 \end{bmatrix}.$$

Then,
$$AB + AC = \begin{bmatrix} 15 & 1 \\ 7 & -4 \end{bmatrix} + \begin{bmatrix} 6 & -2 \\ 0 & 2 \end{bmatrix} = \begin{bmatrix} 15+6 & 1+(-2) \\ 7+0 & -4+2 \end{bmatrix}$$

$$= \begin{bmatrix} 21 & -1 \\ 7 & -2 \end{bmatrix}$$

Remark: $A(B+C) = \begin{bmatrix} 21 & -1 \\ 7 & -2 \end{bmatrix}$ and $AB + AC = \begin{bmatrix} 21 & -1 \\ 7 & -2 \end{bmatrix}$.

Thus, $A(B+C) = AB + AC$. This is called the left distributive law.

● PROBLEM 24-14

Find (i) A^2 (ii) A^3 (iii) A^4 when $A = \begin{bmatrix} 1 & 2 \\ -1 & 1 \end{bmatrix}$.

<u>Solution</u>: $A^2 = AA = \begin{bmatrix} 1 & 2 \\ -1 & 1 \end{bmatrix} \begin{bmatrix} 1 & 2 \\ -1 & 1 \end{bmatrix} = \begin{bmatrix} 1-2 & 2+2 \\ -1-1 & -2+1 \end{bmatrix} = \begin{bmatrix} -1 & 4 \\ -2 & -1 \end{bmatrix}.$

$A^3 = AAA = A^2A = \begin{bmatrix} -1 & 4 \\ -2 & -1 \end{bmatrix} \begin{bmatrix} 1 & 2 \\ -1 & 1 \end{bmatrix} = \begin{bmatrix} -1-4 & -2+4 \\ -2+1 & -4-1 \end{bmatrix} = \begin{bmatrix} -5 & 2 \\ -1 & -5 \end{bmatrix}.$

The usual laws for exponents are $A^m A^n = A^{m+n}$ and $(A^m)^n = A^{mn}$. Thus,
$A^4 = A^3A$ or, $A^4 = (A^2)^2 = A^2A^2$.

$$A^4 = A^3A = \begin{bmatrix} -5 & 2 \\ -1 & -5 \end{bmatrix} \begin{bmatrix} 1 & 2 \\ -1 & 1 \end{bmatrix} = \begin{bmatrix} -5-2 & -10+2 \\ -1+5 & -2-5 \end{bmatrix} = \begin{bmatrix} -7 & -8 \\ 4 & -7 \end{bmatrix}.$$

Observe that

$$A^4 = A^2A^2 = \begin{bmatrix} -1 & 4 \\ -2 & -1 \end{bmatrix} \begin{bmatrix} -1 & 4 \\ -2 & -1 \end{bmatrix} = \begin{bmatrix} 1-8 & -4-4 \\ 2+2 & -8+1 \end{bmatrix}$$

$$= \begin{bmatrix} -7 & -8 \\ 4 & -7 \end{bmatrix}$$

● **PROBLEM** 24-15

a)   Show that:

   (i) $A0 = 0$ (ii) $0A = 0$ (iii) $AI = A$ (iv) $IA = A$

where 0 and I denote the zero and identity matrices respectively, and

$$A = \begin{bmatrix} 2 & 1 & 3 \\ 4 & -1 & -1 \end{bmatrix}$$

b)   Give examples of the following rules: (i)   if A has a row of zeros, the same row of AB consists of zeros. (ii)   if B has a column of zeros, the same column of AB consists of zeros.

Solution: a)   The mxn matrix whose entries are all zero is called the zero matrix and is denoted by 0 . The nxn matrix with 1's on the diagonal and 0's elsewhere, denoted by I, is called the unit or identity matrix; e.g., in R^3 ,

$$I = \begin{bmatrix} 1 & 0 & 0 \\ 0 & 1 & 0 \\ 0 & 0 & 1 \end{bmatrix}$$

(i)

$$A = \begin{bmatrix} 2 & 1 & 3 \\ 4 & -1 & -7 \end{bmatrix}$$

$$0 = \begin{bmatrix} 0 & 0 & 0 & 0 \\ 0 & 0 & 0 & 0 \\ 0 & 0 & 0 & 0 \end{bmatrix}$$

$$A0 = \begin{bmatrix} 2 & 1 & 3 \\ 4 & -1 & -7 \end{bmatrix} \begin{bmatrix} 0 & 0 & 0 & 0 \\ 0 & 0 & 0 & 0 \\ 0 & 0 & 0 & 0 \end{bmatrix}$$

$$= \begin{bmatrix} 0+0+0 & 0+0+0 & 0+0+0 & 0+0+0 \\ 0+0+0 & 0+0+0 & 0+0+0 & 0+0+0 \end{bmatrix} = \begin{bmatrix} 0 & 0 & 0 & 0 \\ 0 & 0 & 0 & 0 \\ 0 & 0 & 0 & 0 \end{bmatrix} .$$

(ii)   Likewise, $0A = \begin{bmatrix} 0 & 0 \\ 0 & 0 \end{bmatrix} \begin{bmatrix} 2 & 1 & 3 \\ 4 & -1 & -7 \end{bmatrix}$

$$= \begin{bmatrix} 0+0 & 0+0 & 0+0 \\ 0+0 & 0+0 & 0+0 \end{bmatrix} = \begin{bmatrix} 0 & 0 & 0 \\ 0 & 0 & 0 \end{bmatrix}$$

(iii)

$$I = \begin{bmatrix} 1 & 0 & 0 \\ 0 & 1 & 0 \\ 0 & 0 & 1 \end{bmatrix}$$

454

$$AI = \begin{bmatrix} 2 & 1 & 3 \\ 4 & -1 & -7 \end{bmatrix} \begin{bmatrix} 1 & 0 & 0 \\ 0 & 1 & 0 \\ 0 & 0 & 1 \end{bmatrix}$$

$$= \begin{bmatrix} 2+0+0 & 0+1+0 & 0+0+3 \\ 4+0+0 & 0-1+0 & 0+0-7 \end{bmatrix}$$

$$= \begin{bmatrix} 2 & 1 & 3 \\ 4 & -1 & -7 \end{bmatrix}.$$

Thus, AI = A.

(iv)

$$I = \begin{bmatrix} 1 & 0 \\ 0 & 1 \end{bmatrix}$$

$$IA = \begin{bmatrix} 1 & 0 \\ 0 & 1 \end{bmatrix} \begin{bmatrix} 2 & 1 & 3 \\ 4 & -1 & -7 \end{bmatrix}$$

$$= \begin{bmatrix} 2+0 & 1+0 & 3+0 \\ 0+4 & 0+(-1) & 0+(-7) \end{bmatrix}$$

$$= \begin{bmatrix} 2 & 1 & 3 \\ 4 & -1 & -7 \end{bmatrix}.$$

Thus, IA = A.
Observe that the zero and identity matrices must have the appropriate size in order for the products to be defined. For example,

$$\begin{bmatrix} 0 & 0 & 0 \\ 0 & 0 & 0 \end{bmatrix} \begin{bmatrix} 2 & 1 & 3 \\ 4 & -1 & -7 \end{bmatrix}$$

is not defined.

b) Let $A = \begin{bmatrix} -2 & -3 & 1 \\ 0 & 0 & 0 \\ 1 & -1 & 0 \end{bmatrix}$ and $B = \begin{bmatrix} 6 & -2 & 1 \\ 3 & 1 & 2 \\ -1 & 1 & 1 \end{bmatrix}$.

Then,

$$AB = \begin{bmatrix} -2 & -3 & 1 \\ 0 & 0 & 0 \\ 1 & -1 & 0 \end{bmatrix} \begin{bmatrix} 6 & -2 & 1 \\ 3 & 1 & 2 \\ -1 & 1 & 1 \end{bmatrix}$$

$$AB = \begin{bmatrix} -12-9-1 & 4-3+1 & -2-6+1 \\ 0+0+0 & 0+0+0 & 0+0+0 \\ 6-3+0 & -2-1+0 & 1-2+0 \end{bmatrix}$$

$$= \begin{bmatrix} -22 & 2 & -7 \\ 0 & 0 & 0 \\ 3 & -3 & -1 \end{bmatrix}.$$

(ii) Let $A = \begin{bmatrix} 4 & 1 & 0 \\ -1 & 2 & 3 \\ 1 & 0 & 1 \end{bmatrix}$ and $B = \begin{bmatrix} 1 & 0 & 1 \\ -2 & 0 & 1 \\ 1 & 0 & 1 \end{bmatrix}$.

Then,

$$AB = \begin{bmatrix} 4 & 1 & 0 \\ -1 & 2 & 3 \\ 1 & 0 & 1 \end{bmatrix} \begin{bmatrix} 1 & 0 & 1 \\ -2 & 0 & 1 \\ 1 & 0 & 1 \end{bmatrix}$$

$$= \begin{bmatrix} 4-2+0 & 0+0+0 & 4+1+0 \\ -1-4+3 & 0+0+0 & -1+2+3 \\ 1+0+1 & 0+0+0 & 1+0+1 \end{bmatrix}$$

$$= \begin{bmatrix} 2 & 0 & 5 \\ -2 & 0 & 4 \\ 2 & 0 & 2 \end{bmatrix}.$$

Compute AB using block multiplication where

$$A = \begin{bmatrix} 1 & 2 & | & 1 \\ 3 & 4 & | & 0 \\ - & - & - & - \\ 0 & 0 & | & 2 \end{bmatrix} \quad \text{and} \quad B = \begin{bmatrix} 1 & 2 & 3 & | & 1 \\ 4 & 5 & 6 & | & 1 \\ - & - & - & - & - \\ 0 & 0 & 0 & | & 1 \end{bmatrix}.$$

Solution: Using a system of horizontal and vertical lines, one can partition a matrix A into smaller "submatrices" of A. The matrix A is then called a block matrix. A given matrix may be divided into blocks in different ways. For example,

$$\begin{bmatrix} 1 & -2 & 0 & 1 \\ 2 & 3 & 5 & 7 \\ 3 & 1 & 4 & 5 \end{bmatrix} = \begin{bmatrix} 1 & -2 & | & 0 & 1 \\ 2 & 3 & | & 5 & 7 \\ - & - & - & - & - \\ 3 & 1 & | & 4 & 5 \end{bmatrix}$$

$$= \begin{bmatrix} 1 & -2 & 0 & | & 1 \\ - & - & - & - & - \\ 2 & 3 & 5 & | & 7 \\ - & - & - & - & - \\ 3 & 1 & 4 & | & 5 \end{bmatrix}$$

$$A = \begin{bmatrix} 1 & 2 & | & 1 \\ 3 & 4 & | & 0 \\ - & - & - & - \\ 0 & 0 & | & 2 \end{bmatrix}.$$

Let $E \begin{bmatrix} 1 & 2 \\ 3 & 4 \end{bmatrix}$, $F = \begin{bmatrix} 1 \\ 0 \end{bmatrix}$ and $G[2]$ then,

$$A = \begin{bmatrix} E & | & F \\ - & - & - \\ 0 & | & G \end{bmatrix}$$

$$B = \begin{bmatrix} 1 & 2 & 3 & | & 1 \\ 4 & 5 & 6 & | & 1 \\ - & - & - & - & - \\ 0 & 0 & 0 & | & 1 \end{bmatrix}. \quad \text{Let} \quad R = \begin{bmatrix} 1 & 2 & 3 \\ 4 & 5 & 6 \end{bmatrix} \quad S = \begin{bmatrix} 1 \\ 1 \end{bmatrix} \quad T = [1].$$

Then,

$$B = \begin{bmatrix} R & | & S \\ - & - & - \\ 0 & | & T \end{bmatrix}.$$

After partitioning the matrices into block matrices, multiplication of the matrices is the usual matrix multiplication with each entire block considered as a unit entry of the matrix. If two matrices can be multiplied, then they can be multiplied as block matrices if they are each partitioned into blocks similarly; that is, into an equal number of blocks so that corresponding blocks have the same size. Suppose

$$A = \begin{bmatrix} A_1 & | & A_2 \\ - & - & - \\ A_3 & | & A_4 \end{bmatrix} \quad \text{and} \quad B = \begin{bmatrix} B_1 & | & B_2 \\ - & - & - \\ B_3 & | & B_4 \end{bmatrix}$$

where A_1 and B_1, A_2 and B_2, A_3 and B_3, A_4 and B_4 are the same sizes, respectively. Then AB is given by

$$A_1B_1 + A_2B_3 \qquad A_1B_2 + A_2B_4$$

$$A_3B_1 + A_3B_3 \qquad A_3B_2 + A_4B_4$$

In the problem,

$$AB = \begin{bmatrix} E & F \\ 0 & G \end{bmatrix} \begin{bmatrix} R & S \\ 0 & T \end{bmatrix} = \begin{bmatrix} ER+F\cdot 0 & ES+FT \\ 0R+G\cdot 0 & 0S+GT \end{bmatrix}$$

$$= \begin{bmatrix} ER & ES+FT \\ 0 & GT \end{bmatrix} =$$

$$= \begin{bmatrix} \begin{bmatrix} 1 & 2 \\ 3 & 4 \end{bmatrix}\begin{bmatrix} 1 & 2 & 3 \\ 4 & 5 & 6 \end{bmatrix} & \begin{bmatrix} 1 & 2 \\ 3 & 4 \end{bmatrix}\begin{bmatrix} 1 \\ 1 \end{bmatrix} + \begin{bmatrix} 1 \\ 0 \end{bmatrix}[1] \\ 0 & [2]\,[1] \end{bmatrix}$$

$$= \begin{bmatrix} \begin{bmatrix} 1+8 & 2+10 & 3+12 \\ 3+16 & 6+20 & 9+24 \end{bmatrix} & \begin{bmatrix} 1+2 \\ 3+4 \end{bmatrix} + \begin{bmatrix} 1 \\ 0 \end{bmatrix} \\ [0 \quad 0 \quad 0] & [2] \end{bmatrix}$$

$$= \begin{bmatrix} 9 & 12 & 15 & 4 \\ 19 & 26 & 33 & 7 \\ 0 & 0 & 0 & 2 \end{bmatrix}$$

● **PROBLEM 24-17**

Define (1) An upper triangular matrix.
   (2) A lower triangular matrix.
   (3) A properly triangular matrix.
Give examples.

Solution: A triangular matrix is an $n \times n$ matrix whose non-zero elements lie on the diagonal and all are either above or below the diagonal.
Definition: (1) An upper triangular matrix is an $n \times n$ matrix all of whose non-zero entries lie on its diagonal and above. Example:

$$\begin{bmatrix} 1 & 2 & 3 \\ 0 & 0 & 0 \\ 0 & 0 & 6 \end{bmatrix}$$

(2) A lower triangular matrix is an $n \times n$ matrix all of whose non-zero entries lie on its diagonal and below. Example:

$$\begin{bmatrix} 1 & 0 & 0 \\ 2 & 0 & 0 \\ 3 & 0 & 6 \end{bmatrix} .$$

(3) A properly triangular matrix is an $n \times n$ matrix whose diagonal
entries are all zero. Example:

$$\begin{bmatrix} 0 & 1 & 2 \\ 0 & 0 & 3 \\ 0 & 0 & 0 \end{bmatrix}$$

It should be noted that the zero matrix and all diagonal matrices are
triangular matrices.

● PROBLEM 24-18

Define the transpose of a matrix. Find the transpose of the following
matrices:

$$A = \begin{bmatrix} 4 & -2 & 3 \\ 0 & 5 & -2 \end{bmatrix} \quad B = \begin{bmatrix} 6 & 2 & -4 \\ 3 & -1 & 2 \\ 0 & 4 & 3 \end{bmatrix}$$

$$C = \begin{bmatrix} 5 & 4 \\ 3 & 2 \\ 2 & -3 \end{bmatrix} \quad D = \begin{bmatrix} 3 & -5 & 1 \end{bmatrix} \quad E = \begin{bmatrix} 2 \\ -1 \\ 3 \end{bmatrix}$$

Solution: Definition: If $A = [a_{ij}]$ is an $m \times n$ matrix, then the
$n \times m$ matrix $A^t = [a^t_{ij}]$ where

$$a^t_{ij} = a_{ji} \quad [1 \le i \le m . \quad 1 \le j \le n]$$

is called the transpose of A. Thus, the transpose of A is obtained
by interchanging the rows and columns of A.

$$A = \begin{bmatrix} 4 & -2 & 3 \\ 0 & 5 & -2 \end{bmatrix} .$$

Then,

$$A^t = \begin{bmatrix} 4 & 0 \\ -2 & 5 \\ 3 & -2 \end{bmatrix}$$

$$B = \begin{bmatrix} 6 & 2 & -4 \\ 3 & -1 & 2 \\ 0 & 4 & 3 \end{bmatrix} .$$

$$B^t = \begin{bmatrix} 6 & 3 & 0 \\ 2 & -1 & 4 \\ -4 & 2 & 3 \end{bmatrix} .$$

$$C = \begin{bmatrix} 5 & 4 \\ -3 & 2 \\ 2 & -3 \end{bmatrix} ; \quad \text{thus,} \quad C^t = \begin{bmatrix} 5 & -3 & 2 \\ 4 & 2 & -3 \end{bmatrix} .$$

$$D = \begin{bmatrix} 3 & -5 & 1 \end{bmatrix} .$$

Then,

$$D^t = \begin{bmatrix} 3 \\ -5 \\ 1 \end{bmatrix} .$$

$$E = \begin{bmatrix} 2 \\ -1 \\ 3 \end{bmatrix} ; \quad \text{hence,} \quad E^t = \begin{bmatrix} 2 & -1 & 3 \end{bmatrix} .$$

458

Define det A and find the determinant of the following matrices:

(a) $[a_{11}]$ (b) $\begin{bmatrix} a_{11} & a_{12} \\ a_{21} & a_{22} \end{bmatrix}$ (c) $\begin{bmatrix} 0 & 0 & 0 \\ 0 & 0 & 0 \\ 0 & 0 & 0 \end{bmatrix}$

(d) $\begin{bmatrix} a_{11} & a_{12} & a_{13} \\ a_{21} & a_{22} & a_{23} \\ a_{31} & a_{32} & a_{33} \end{bmatrix}$

Solution: Determinants are formally defined as:

$$Det(A) = |A| = \sum_{\sigma} sgn(\sigma) \prod_{j=1}^{n} a_{\sigma(j)j}$$

$$= \sum_{\sigma} sgn(\sigma) \, a_{\sigma(1)1} \, a_{\sigma(2)2} \cdots a_{\sigma(n)n}$$

where summation extends over the $n!$, different permutations, σ, of the n symbols $1, 2, \ldots, n$ and $sgn(\sigma) = +1$, σ even
-1, σ odd .

$|A|$ is also known as an $n \times n$ determinant or a determinant of order n.

(a) det A = a_{11}

(b) det A = $\begin{vmatrix} a_{11} & a_{12} \\ a_{21} & a_{22} \end{vmatrix} = a_{11}a_{22} - a_{21}a_{12}$

(c) det A = $|0| = 0$

(d) det A = $\begin{vmatrix} a_{11} & a_{12} & a_{13} \\ a_{21} & a_{22} & a_{23} \\ a_{31} & a_{32} & a_{33} \end{vmatrix}$

The permutations of S_3 and their signs are:

Permutation	Sign	Permutation	Sign
1 2 3	+	2 1 3	-
1 3 2	-	3 1 2	+
2 3 1	+	3 2 1	-

Then, det A = $a_{11}a_{22}a_{33} - a_{11}a_{32}a_{23} + a_{21}a_{32}a_{13} - a_{21}a_{12}a_{33} + a_{31}a_{12}a_{23}$

$- a_{31}a_{22}a_{13}$.

a) Find the determinant of an arbitrary 3×3 matrix.
b) Find det A where:

$$A = \begin{bmatrix} -5 & 0 & 2 \\ 6 & 1 & 2 \\ 2 & 3 & 1 \end{bmatrix}$$

Solution: Let

$$A = \begin{bmatrix} b_{11} & b_{12} & b_{13} \\ b_{21} & b_{22} & b_{23} \\ b_{31} & b_{32} & b_{33} \end{bmatrix} .$$

$$\det A = \begin{vmatrix} b_{11} & b_{12} & b_{13} \\ b_{21} & b_{22} & b_{23} \\ b_{31} & b_{32} & b_{33} \end{vmatrix} .$$

Expand the above determinant by minors, using the first column.

$$\det A = +b_{11} \begin{vmatrix} b_{22} & b_{23} \\ b_{32} & b_{33} \end{vmatrix} - b_{21} \begin{vmatrix} b_{12} & b_{13} \\ b_{32} & b_{33} \end{vmatrix}$$

$$+ b_{31} \begin{vmatrix} b_{12} & b_{13} \\ b_{22} & b_{23} \end{vmatrix}$$

$$\det A = b_{11}(b_{22}b_{33} - b_{32}b_{23}) - b_{21}(b_{12}b_{33} - b_{32}b_{13}) + b_{31}(b_{12}b_{23} - b_{22}b_{13}).$$

Now expand the determinant by minors, using the second row:

$$\det A = -b_{21} \begin{vmatrix} b_{12} & b_{13} \\ b_{32} & b_{33} \end{vmatrix} + b_{22} \begin{vmatrix} b_{11} & b_{13} \\ b_{31} & b_{33} \end{vmatrix}$$

$$- b_{23} \begin{vmatrix} b_{11} & b_{12} \\ b_{31} & b_{32} \end{vmatrix} .$$

$$\det A = -b_{21}(b_{12}b_{33} - b_{32}b_{13}) + b_{22}(b_{11}b_{33} - b_{31}b_{13}) - b_{23}(b_{11}b_{32} - b_{31}b_{12})$$

$$= b_{22}b_{11}b_{33} - b_{22}b_{31}b_{13} - b_{23}b_{11}b_{32} + b_{23}b_{31}b_{12} - b_{21}(b_{12}b_{33} - b_{32}b_{13})$$

$$= b_{11}(b_{22}b_{33} - b_{32}b_{23}) - b_{21}(b_{12}b_{33} - b_{32}b_{13}) + b_{31}(b_{12}b_{23} - b_{22}b_{13})$$

Clearly, this is the same as the first answer. Note, also, that det A can be rearranged algebraically until it can be written as:

$$\det A = b_{11}b_{22}b_{33} + b_{12}b_{23}b_{31} + b_{13}b_{32}b_{21}$$

460

$$- [b_{13}b_{22}b_{31} + b_{23}b_{32}b_{11} + b_{33}b_{21}b_{12}] \ .$$

It is easy to remember this result (det of a 3x3 matrix) using the following mnemonic device: Figure 1

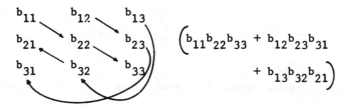

Figure 2:

$$\left(b_{13}b_{22}b_{31} + b_{23}b_{32}b_{11} + b_{33}b_{21}b_{12} \right)$$

$$\det A = b_{11}b_{22}b_{33} + b_{12}b_{23}b_{31} + b_{13}b_{32}b_{21}$$

$$- [b_{13}b_{22}b_{31} + b_{23}b_{32}b_{11} + b_{33}b_{21}b_{12}] \ .$$

This makes taking 3x3 determinants simpler.

b) Expand the determinant by minors, using the first column.

$$\det A = \ -5 \begin{vmatrix} 1 & 3 \\ 3 & 1 \end{vmatrix} - 6 \begin{vmatrix} 0 & 2 \\ 3 & 1 \end{vmatrix} + 2 \begin{vmatrix} 0 & 2 \\ 1 & 2 \end{vmatrix}$$

$$= \ -5(1-6) - 6(0-6) + 2(0-2) = +25 + 36 - 4 = 57.$$

● **PROBLEM 24-21**

Find the determinant of the following matrix:

$$A \ = \ \begin{bmatrix} 2 & 0 & 3 & 0 \\ 2 & 1 & 1 & 2 \\ 3 & -1 & 1 & -2 \\ 2 & 1 & -2 & 1 \end{bmatrix}$$

Solution: Use the method of expansion by minors.

$$A \ = \ \begin{bmatrix} 2 & 0 & 3 & 0 \\ 2 & 1 & 1 & 2 \\ 3 & -1 & 1 & -2 \\ 2 & 1 & -2 & 1 \end{bmatrix}$$

Expanding along the first row:

461

$$\det A = 2 \begin{vmatrix} 1 & 1 & 2 \\ -1 & 1 & -2 \\ 1 & -2 & 1 \end{vmatrix} + 3 \begin{vmatrix} 2 & 1 & 2 \\ 3 & -1 & -2 \\ 2 & 1 & 1 \end{vmatrix}$$

Note that the minors, whose multiplying factors were zero, have been eliminated. This illustrates the general principle that, when evaluating determinants, expansion along the row (or column) containing the most zeros is the optimal procedure.

Add the second row to the first row for each of the 3 by 3 determinants:

$$\det A = 2 \begin{vmatrix} 0 & 2 & 0 \\ -1 & 1 & -2 \\ 1 & -2 & 1 \end{vmatrix} + 3 \begin{vmatrix} 5 & 0 & 0 \\ 3 & -1 & -2 \\ 2 & 1 & 1 \end{vmatrix}$$

Now expand the above determinants by minors using the first row.

$$\det A = 2(-2) \begin{vmatrix} -1 & -2 \\ 1 & 1 \end{vmatrix} + 3(5) \begin{vmatrix} -1 & -2 \\ 1 & 1 \end{vmatrix}$$

$$= (-4)(-1+2) + 15(-1+2)$$

$$= -4+15 = 11 \ .$$

● **PROBLEM 24-22**

Find the determinant of the matrix A where:

$$A = \begin{bmatrix} 2 & 7 & -3 & 8 & 3 \\ 0 & -3 & 7 & 5 & 1 \\ 0 & 0 & 6 & 7 & 6 \\ 0 & 0 & 0 & 9 & 8 \\ 0 & 0 & 0 & 0 & 4 \end{bmatrix}$$

Solution: A is an upper triangular matrix. As we know, if A is an n×m triangular matrix (upper or lower), then det A is the product of the entries on the main diagonal.
Hence,

$$\det A = (2) \cdot (-3) \cdot (6) \cdot (9) \cdot (4) = -1296 \ .$$

● **PROBLEM 24-23**

Evaluate det A where:

$$A = \begin{bmatrix} 0 & 1 & 5 \\ 3 & -6 & 9 \\ 2 & 6 & 1 \end{bmatrix}$$

Solution: Interchange the first and second rows of matrix A, obtaining matrix

$$B = \begin{bmatrix} 3 & -6 & 9 \\ 0 & 1 & 5 \\ 2 & 6 & 1 \end{bmatrix} \quad ;$$

and by the properties of the function,

$$\det A = -\det B .$$

$$= - \det \begin{bmatrix} 3 & -6 & 9 \\ 0 & 1 & 5 \\ 2 & 6 & 1 \end{bmatrix}$$

or

$$\det A = -3 \det \begin{bmatrix} 1 & -2 & 3 \\ 0 & 1 & 5 \\ 2 & 6 & 1 \end{bmatrix} .$$

A common factor of 3 from the first row of the matrix B was taken out. Add -2 times the first row to the third row. The value of the determinant of A will remain the same.
Thus,

$$\det A = -3 \det \begin{bmatrix} 1 & -2 & 3 \\ 0 & 1 & 5 \\ 0 & 10 & -5 \end{bmatrix} .$$

Add -10 times the second row to the third row.
Thus,

$$\det A = -3 \det \begin{bmatrix} 1 & -2 & 3 \\ 0 & 1 & 5 \\ 0 & 0 & -55 \end{bmatrix} .$$

As we know, the determinant of a triangular matrix is equal to the product of the diagonal elements.
Thus,

$$\det A = (-3) \cdot (1) \cdot (1) \cdot (-55) = 165 .$$

● **PROBLEM 24-24**

Find the adjoint A of the following matrices:

(a) $\quad A = \begin{bmatrix} a_{11} & a_{12} \\ a_{21} & a_{22} \end{bmatrix}$ (b) $\quad A = \begin{bmatrix} 1 & 0 & 5 \\ 2 & 1 & 0 \\ 0 & 4 & 0 \end{bmatrix}$

(c) $\quad A = \begin{bmatrix} \lambda_1 & \cdots & 0 \\ & \ddots & \\ 0 & & \lambda_n \end{bmatrix} , \quad \lambda_i \neq 0 , \quad i = 1, 2, \ldots, n$

Solution:a)The transpose of the matrix of cofactors of the elements, a_{ij} of A, is called the adjoint of A.

$$A = \begin{bmatrix} a_{11} & a_{12} \\ a_{21} & a_{22} \end{bmatrix} .$$

The cofactors of the four elements are:

$$A_{11} = a_{22} , \quad A_{12} = -a_{21}$$
$$A_{21} = -a_{12}, \quad A_{22} = a_{11} .$$

The matrix of the cofactors is:

$$\begin{bmatrix} a_{22} & -a_{21} \\ -a_{12} & a_{11} \end{bmatrix}$$

Thus,

$$\text{adj } A = \begin{bmatrix} a_{22} & -a_{12} \\ -a_{21} & a_{11} \end{bmatrix} .$$

(b)

$$A = \begin{bmatrix} 1 & 0 & 5 \\ 2 & 1 & 0 \\ 0 & 4 & 0 \end{bmatrix}$$

The cofactors of the nine elements are:

$$A_{11} = + \begin{vmatrix} 1 & 0 \\ 4 & 0 \end{vmatrix} = 1(0-0) = 0$$

$$A_{12} = - \begin{vmatrix} 2 & 0 \\ 0 & 0 \end{vmatrix} = 0$$

$$A_{13} = + \begin{vmatrix} 2 & 1 \\ 0 & 4 \end{vmatrix} = (8-0) = +8$$

$$A_{21} = - \begin{vmatrix} 0 & 5 \\ 4 & 0 \end{vmatrix} = -(0-20) = +20$$

$$A_{22} = + \begin{vmatrix} 1 & 5 \\ 0 & 0 \end{vmatrix} = 0$$

$$A_{23} = - \begin{vmatrix} 1 & 0 \\ 0 & 4 \end{vmatrix} = -(4-0) = -4$$

$$A_{31} = + \begin{vmatrix} 0 & 5 \\ 1 & 0 \end{vmatrix} = (0-5) = -5$$

$$A_{32} = - \begin{vmatrix} 1 & 5 \\ 2 & 0 \end{vmatrix} = -(0-10) = +10$$

$$A_{33} = + \begin{vmatrix} 1 & 0 \\ 2 & 1 \end{vmatrix} = (1-0) = 1$$

The matrix of the cofactors is:

$$\begin{bmatrix} 0 & 0 & 8 \\ 20 & 0 & -4 \\ -5 & +10 & 1 \end{bmatrix}$$

Thus,

$$\text{adj } A = \begin{bmatrix} 0 & 20 & -5 \\ 0 & 0 & +10 \\ 8 & -4 & 1 \end{bmatrix}$$

(c)

$$A = \begin{bmatrix} \lambda_1 & & & 0 \\ & \ddots & & \\ & & \ddots & \\ 0 & & \cdot & \lambda_n \end{bmatrix}$$

Here, the matrix A is a diagonal matrix. We know,

$$A^{-1} = \frac{1}{\det A} \; (\text{adj } A)$$

or

$$\text{adj } A = |A| \; A^{-1}.$$

We also know that the inverse of a diagonal matrix is simply made up of the inverses of its non-diagonal elements. Now,

$$A = \begin{bmatrix} \lambda_1 & & 0 \\ & \ddots & \\ 0 & \cdot & \lambda_n \end{bmatrix}.$$

Then,

$$A^{-1} = \begin{bmatrix} \lambda_1^{-1} & 0 & \ldots & 0 \\ 0 & \lambda_2^{-1} & & \vdots \\ \vdots & & \ddots & \\ 0 & & \cdot & \lambda_n^{-1} \end{bmatrix},$$

providing $\lambda_i \neq 0$ $(i = 1, \ldots, n)$. Hence,

$$\text{adj } A = |A| \begin{bmatrix} \lambda_1^{-1} & 0 & \ldots & 0 \\ 0 & \lambda_2^{-1} & \ldots & 0 \\ \vdots & & \ddots & \\ 0 & 0 & \cdot & \lambda_n^{-1} \end{bmatrix}$$

Since $|A| = \lambda_1 \lambda_2 \cdots \lambda_n = \prod_{i=1}^{n} \lambda_i$,

$$\text{adj } A = \prod_{i=1}^{n} \lambda_i \begin{bmatrix} \dfrac{1}{\lambda_1} & 0 & \ldots & 0 \\ 0 & \dfrac{1}{\lambda_2} & & 0 \\ \vdots & & \ddots & \vdots \\ 0 & & \ldots & \dfrac{1}{\lambda_n} \end{bmatrix}$$

Given:

$$A = \begin{bmatrix} 3 & 1 & 2 \\ 0 & 1 & 1 \\ -1 & 1 & 0 \end{bmatrix}.$$

Show that $(adj A) \cdot A = (det A)I$ where I is the identity matrix.

Solution: It is known that the classical adjoint, or adj A, is the transpose of the matrix of cofactors of the elements a_{ij} of A. The cofactors of the nine elements of the given matrix A are:

$$A_{11} = + \begin{vmatrix} 1 & 1 \\ 1 & 0 \end{vmatrix} = (0-1) = -1$$

$$A_{12} = - \begin{vmatrix} 0 & 1 \\ -1 & 0 \end{vmatrix} = -(0+1) = -1$$

$$A_{13} = + \begin{vmatrix} 0 & 1 \\ -1 & 1 \end{vmatrix} = (0+1) = 1$$

$$A_{21} = - \begin{vmatrix} 1 & 2 \\ 1 & 0 \end{vmatrix} = -(0-2) = +2$$

$$A_{22} = + \begin{vmatrix} 3 & 2 \\ -1 & 0 \end{vmatrix} = (0+2) = 2$$

$$A_{23} = - \begin{vmatrix} 3 & 1 \\ -1 & 1 \end{vmatrix} = -(3+1) = -4$$

$$A_{31} = + \begin{vmatrix} 1 & 2 \\ 1 & 1 \end{vmatrix} = (1-2) = -1$$

$$A_{32} = - \begin{vmatrix} 3 & 2 \\ 0 & 1 \end{vmatrix} = -(3-0) = -3$$

$$A_{33} = + \begin{vmatrix} 3 & 1 \\ 0 & 1 \end{vmatrix} = (3-0) = 3 .$$

Then the matrix of the cofactors is:

$$\begin{bmatrix} -1 & -1 & 1 \\ 2 & 2 & -4 \\ -1 & -3 & 3 \end{bmatrix} .$$

Hence,

$$adj A = \begin{bmatrix} -1 & 2 & -1 \\ -1 & 2 & -3 \\ 1 & -4 & 3 \end{bmatrix} .$$

$$(adj A) \cdot A = \begin{bmatrix} -1 & 2 & -1 \\ -1 & 2 & -3 \\ 1 & -4 & 3 \end{bmatrix} \begin{bmatrix} 3 & 1 & 2 \\ 0 & 1 & 1 \\ -1 & 1 & 0 \end{bmatrix}$$

$$= \begin{bmatrix} -3+0+1 & -1+2-1 & -2+2+0 \\ -3+0+3 & -1+2-3 & -2+2+0 \\ 3+0-3 & 1-4+3 & 2-4+0 \end{bmatrix}$$

$$= \begin{bmatrix} -2 & 0 & 0 \\ 0 & -2 & 0 \\ 0 & 0 & -2 \end{bmatrix} .$$

$$\text{adj } A \cdot = -2 \begin{bmatrix} 1 & 0 & 0 \\ 0 & 1 & 0 \\ 0 & 0 & 1 \end{bmatrix} = -2I$$

$$\det A = \begin{vmatrix} 3 & 1 & 2 \\ 0 & 1 & 1 \\ -1 & 1 & 0 \end{vmatrix} .$$

$$\det A = 3 \begin{vmatrix} 1 & 1 \\ 1 & 0 \end{vmatrix} - 0 \begin{vmatrix} 1 & 2 \\ 1 & 0 \end{vmatrix} - 1 \begin{vmatrix} 1 & 2 \\ 1 & 1 \end{vmatrix}$$

$$= 3(0-1) -1 (1-2)$$

$$= -3 + 1 = -2 .$$

Hence,

$$\text{adj } A \cdot A = -2I = (\det A)I .$$

● **PROBLEM 24-26**

Compute the determinants of each of the following matrices and find which of the matrices are invertible.

(a) $\begin{bmatrix} 3 & 1 & 2 \\ 1 & 0 & 6 \\ -1 & 1 & 1 \end{bmatrix}$ (b) $\begin{bmatrix} -1 & 1 & 3 \\ 2 & 1 & 1 \\ 4 & 2 & 2 \end{bmatrix}$ (c) $\begin{bmatrix} 2 & 1 & 1 \\ 0 & 0 & 0 \\ 4 & 3 & 1 \end{bmatrix}$

<u>Solution</u>: We can evaluate determinants by using the basic properties of the determinant function.

Properties of Determinants:
(1) If each element in a row (or column) is zero, the value of the determinant is zero.
(2) If two rows (or columns) of a determinant are identical, the value of the determinant is zero.
(3) The determinant of a matrix A and its transpose A^t are equal: $|A| = |A^t|$.
(4) The matrix A has an inverse if and only if $\det A \neq 0$.

$$\det A = \begin{vmatrix} 3 & 1 & 2 \\ 1 & 0 & 6 \\ -1 & 1 & 1 \end{vmatrix}$$

$$= 3 \begin{vmatrix} 0 & 6 \\ 1 & 1 \end{vmatrix} - 1 \begin{vmatrix} 1 & 6 \\ -1 & 1 \end{vmatrix} + 2 \begin{vmatrix} 1 & 0 \\ -1 & 1 \end{vmatrix}$$

$$= 3(0-6) - 1(1+6) + 2(1-0)$$

$$= -18-7+2 = -23 .$$

467

Since det A = -23 ≠ 0, this matrix is invertible.

(b)

$$A = \begin{bmatrix} -1 & 1 & 3 \\ 2 & 1 & 1 \\ 4 & 2 & 2 \end{bmatrix}$$

Here det A = 0, since the third row is a multiple of the second row.
Since det A = 0, the matrix is not invertible.

(c)

$$A = \begin{bmatrix} 2 & 1 & 1 \\ 0 & 0 & 0 \\ 4 & 3 & 1 \end{bmatrix}$$

$$\det A = \begin{vmatrix} 2 & 1 & 1 \\ 0 & 0 & 0 \\ 4 & 3 & 1 \end{vmatrix} = 0.$$

Here, each element in the second row is zero, therefore, the value of
the determinant is zero. Since det A = 0, the matrix is not invertible.

Given:

$$A = \begin{bmatrix} 2 & -1 & 1 \\ 4 & 1 & -3 \\ 2 & -1 & 3 \end{bmatrix},$$

Evaluate det A and det A^{-1}. What is the relation between det A and
det A^{-1} ?

Solution: First find det A.

$$\det A = \begin{vmatrix} 2 & -1 & 1 \\ 4 & 1 & -3 \\ 2 & -1 & 3 \end{vmatrix}$$

Add the third row to the second row:

$$\det A = \begin{vmatrix} 2 & -1 & 1 \\ 6 & 0 & 0 \\ 2 & -1 & 3 \end{vmatrix}$$

Expand the determinant by minors, using the second row.

$$\det A = -6 \begin{vmatrix} -1 & 1 \\ -1 & 3 \end{vmatrix} + 0 \begin{vmatrix} 2 & 1 \\ 2 & 3 \end{vmatrix} - 0 \begin{vmatrix} 2 & -1 \\ 2 & -1 \end{vmatrix}$$

$$= -6(-3+1) = 12.$$

Thus, det A = 12. Next, find A^{-1}. We know,

$$A^{-1} = \frac{1}{\det A} \cdot \text{adj } A,$$

where adj A is the transpose of the matrix of cofactors. Then, the
cofactors are:

$$A_{11} = + \begin{vmatrix} 1 & -3 \\ -1 & 3 \end{vmatrix} = (3-3) = 0$$

$$A_{12} = - \begin{vmatrix} 4 & -3 \\ 2 & 3 \end{vmatrix} = -(12+6) = -18$$

$$A_{13} = + \begin{vmatrix} 4 & 1 \\ 2 & -1 \end{vmatrix} = +(-4-2) = -6$$

$$A_{21} = - \begin{vmatrix} -1 & 1 \\ -1 & 3 \end{vmatrix} = -(-3+1) = 2$$

$$A_{22} = + \begin{vmatrix} 2 & 1 \\ 2 & 3 \end{vmatrix} = (6-2) = 4$$

$$A_{23} = - \begin{vmatrix} 2 & -1 \\ 2 & -1 \end{vmatrix} = -(-2+2) = 0$$

$$A_{31} = + \begin{vmatrix} -1 & 1 \\ 1 & -3 \end{vmatrix} = (3-1) = 2$$

$$A_{32} = - \begin{vmatrix} 2 & 1 \\ 4 & -3 \end{vmatrix} = -(-6-4) = 10$$

$$A_{33} = + \begin{vmatrix} 2 & -1 \\ 4 & 1 \end{vmatrix} = (2+4) = 6 .$$

The matrix of the cofactors is:

$$\begin{bmatrix} 0 & -18 & -6 \\ 2 & 4 & 0 \\ 2 & 10 & 6 \end{bmatrix} .$$

Hence,

$$adj\ A = \begin{bmatrix} 0 & 2 & 2 \\ -18 & 4 & 10 \\ -6 & 0 & 6 \end{bmatrix} .$$

Then,

$$A^{-1} = \frac{1}{det\ A}\ (adj\ A)$$

$$= \frac{1}{12} \begin{bmatrix} 0 & 2 & 2 \\ -18 & 4 & 10 \\ -6 & 0 & 6 \end{bmatrix}$$

$$= \begin{bmatrix} 0 & 2/12 & 2/12 \\ -18/12 & 4/12 & 10/12 \\ -6/12 & 0 & 6/12 \end{bmatrix} = \begin{bmatrix} 0 & 1/6 & 1/6 \\ -3/2 & 1/3 & 5/6 \\ -1/2 & 0 & 1/2 \end{bmatrix} .$$

Now, find $det\ A^{-1}$:

$$det\ A^{-1} = \begin{vmatrix} 0 & 1/6 & 1/6 \\ -3/2 & 1/3 & 5/6 \\ -1/2 & 0 & 1/2 \end{vmatrix} .$$

Add the third column to the first column.

$$det\ A^{-1} = \begin{vmatrix} 1/6 & 1/6 & 1/6 \\ -4/6 & 1/3 & 5/6 \\ 0 & 0 & 1/2 \end{vmatrix} .$$

Expand the determinant by minors, using the third row:

$$\det A^{-1} = 0 \begin{vmatrix} 1/6 & 1/6 \\ 1/3 & 5/6 \end{vmatrix} - 0 \begin{vmatrix} 1/6 & 1/6 \\ -4/6 & 5/6 \end{vmatrix} + 1/2 \begin{vmatrix} 1/6 & 1/6 \\ -4/6 & 1/3 \end{vmatrix}$$

$$= 1/2(1/18 + 4/36)$$

$$= 1/2(1/6) = 1/12 .$$

Thus, $\det A = 12$, while $\det A^{-1} = 1/12$. That is,

$$\det A^{-1} = \frac{1}{\det A} .$$

We emphasize that A has an inverse if and only if $\det A \neq 0$.

● PROBLEM 24-28

Find the value of $\begin{vmatrix} 67 & 19 & 21 \\ 39 & 13 & 14 \\ 81 & 24 & 26 \end{vmatrix}$

Solution: Our aim in this problem is to break down the given determinant into one that is easier to evaluate. We can therefore rewrite our determinant as:

$$\begin{vmatrix} 67 & 19 & 21 \\ 39 & 13 & 14 \\ 81 & 24 & 26 \end{vmatrix} = \begin{vmatrix} 10 + 57 & 19 & 21 \\ 0 + 39 & 13 & 14 \\ 9 + 72 & 24 & 26 \end{vmatrix} .$$

Now we can make use of one of the well-known properties of determinants; that is, if each element of a column of a determinant is expressed as the sum of two terms, the determinant can be expressed as the sum of two determinants. Thus,

$$\begin{vmatrix} 10 + 57 & 19 & 21 \\ 0 + 39 & 13 & 14 \\ 9 + 72 & 24 & 26 \end{vmatrix} = \begin{vmatrix} 10 & 19 & 21 \\ 0 & 13 & 14 \\ 9 & 24 & 26 \end{vmatrix} + \begin{vmatrix} 57 & 19 & 21 \\ 39 & 13 & 14 \\ 72 & 24 & 26 \end{vmatrix} .$$

The determinant can again be simplified further. Let us examine the second determinant in the above sum. Remember that multiplying each element in a column of a determinant by a number and adding that product to the corresponding elements in another column does not change the value of the determinant. Therefore, we can perform this on the determinant using - 3 as the number, and adding the product of - 3 and the elements of column two to the corresponding elements of column one. Thus, we obtain:

$$\begin{vmatrix} 57 & 19 & 21 \\ 39 & 13 & 14 \\ 72 & 24 & 26 \end{vmatrix} = \begin{vmatrix} 57 + (- 3)(19) & 19 & 21 \\ 39 + (- 3)(13) & 13 & 14 \\ 72 + (- 3)(24) & 24 & 26 \end{vmatrix}$$

$$= \begin{vmatrix} 0 & 19 & 21 \\ 0 & 13 & 14 \\ 0 & 24 & 26 \end{vmatrix} .$$

Now, since each element in a column of a determinant is zero, the value of the determinant is zero. Thus, the value of the second determinant in the above sum is zero, and we have:

$$\begin{vmatrix} 67 & 19 & 21 \\ 39 & 13 & 14 \\ 81 & 24 & 26 \end{vmatrix} = \begin{vmatrix} 10 & 19 & 21 \\ 0 & 13 & 14 \\ 9 & 24 & 26 \end{vmatrix} .$$

But, this can be rewritten as:

$$\begin{vmatrix} 10 & 19 & 19+2 \\ 0 & 13 & 13+1 \\ 9 & 24 & 24+2 \end{vmatrix} = \begin{vmatrix} 10 & 19 & 19 \\ 0 & 13 & 13 \\ 9 & 24 & 24 \end{vmatrix} + \begin{vmatrix} 10 & 19 & 2 \\ 0 & 13 & 1 \\ 9 & 24 & 2 \end{vmatrix} .$$

If two columns of a determinant have the same elements, then its value is zero. Thus the first determinant in the above sum is zero, and we are left with:

$$\begin{vmatrix} 10 & 19 & 2 \\ 0 & 13 & 1 \\ 9 & 24 & 2 \end{vmatrix} .$$

We now use minors to determine the value of the determinant. Let us choose column one, and call its elements a_1, a_2, a_3. Then their corresponding minors are A_1, A_2, A_3. We form the products a_1A_1, a_2A_2, a_3A_3. Since a_1 is in the first row and the first column, and $1 + 1 = 2$, which is even, the sign of a_1A_1 is positive. Similarly, the sign of a_2A_2 is negative, and that of a_3A_3 is positive. Thus, we have:

$a_1A_1 - a_2A_2 + a_3A_3$, and substituting we obtain:

$10A_1 - 0A_2 + 9A_3$. The second term vanishes. We find the minors A_1 and A_3 by eliminating from the determinant the row and column that a_1 and a_3 are found in. Thus,

$$A_1 = \begin{vmatrix} 13 & 1 \\ 24 & 2 \end{vmatrix}, \qquad A_3 = \begin{vmatrix} 19 & 2 \\ 13 & 1 \end{vmatrix}$$

and
$$\begin{vmatrix} 67 & 19 & 21 \\ 39 & 13 & 14 \\ 81 & 24 & 26 \end{vmatrix} = \begin{vmatrix} 10 & 19 & 2 \\ 0 & 13 & 1 \\ 9 & 24 & 2 \end{vmatrix}$$

$$= 10 \begin{vmatrix} 13 & 1 \\ 24 & 2 \end{vmatrix} + 9 \begin{vmatrix} 19 & 2 \\ 13 & 1 \end{vmatrix}.$$

Now, these two determinants are easily evaluated. The first,

$$\begin{vmatrix} 13 & 1 \\ 24 & 2 \end{vmatrix} = (13)(2) - (1)(24) = 26 - 24 = 2, \text{ and the second}$$

$$\begin{vmatrix} 19 & 2 \\ 13 & 1 \end{vmatrix} = (19)(1) - (2)(13) = 19 - 26 = -7.$$

Thus,
$$\begin{vmatrix} 67 & 19 & 21 \\ 39 & 13 & 14 \\ 81 & 24 & 26 \end{vmatrix} = 10(2) + 9(-7) = 20 - 63 = -43.$$

Another way to approach this problem is to use the expansion scheme for determinants of third order. Using this method we rewrite the given determinant as follows:

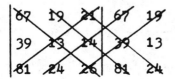

We multiply the elements falling on the same diagonal, thus obtaining six terms. The three terms on the lines sloping downward from left to right have a positive value, and the three on the lines sloping downward from right to left have a negative value. Upon expanding we obtain:

$$(67)(13)(26) + (19)(14)(81) + (21)(39)(24) - (21)(13)(81)$$
$$- (67)(14)(24) - (19)(39)(26).$$

Performing the indicated operations should give us the same value obtained using our previous method, that is -43.

The advantage of the first method is that it does not involve a long multiplication process.

Find the inverse M^{-1} of the matrix

$$M = \begin{bmatrix} 1 & 3 & 4 \\ -2 & 4 & 5 \\ 3 & 1 & 6 \end{bmatrix}$$

and verify that $MM^{-1} = I$.

Solution: We find the determinant of the matrix M using the method of minors and cofactors, and expanding in the first row, with the following scheme giving the sign of each minor (in this case each element of the first row).

$$\begin{bmatrix} + & - & + \\ - & + & - \\ + & - & + \end{bmatrix}$$

$$D = 1 \begin{vmatrix} 4 & 5 \\ 1 & 6 \end{vmatrix} -3 \begin{vmatrix} -2 & 5 \\ 3 & 6 \end{vmatrix} +4 \begin{vmatrix} -2 & 4 \\ 3 & 1 \end{vmatrix}$$

$$= 1(24-5) -3(-12-15) +4(-2-12) = 19 + 81 - 56 = 44$$

Furthermore, each entry A_{ij} consists of the cofactor (the determinant of the two by two matrix of entries resulting when any combination of row i and column j is excluded) of each element of the original matrix.

$$A_{11} = \begin{vmatrix} 4 & 5 \\ 1 & 6 \end{vmatrix} = 19 \quad A_{12} = - \begin{vmatrix} -2 & 5 \\ 3 & 6 \end{vmatrix} = 27 \text{ and } A_{13} = \begin{vmatrix} -2 & 4 \\ 3 & 1 \end{vmatrix} = -14$$

Similarly, $A_{21} = -14$, $A_{22} = -6$, $A_{23} = 8$, $A_{31} = -1$, $A_{32} = -13$, and $A_{33} = 10$.

Hence, by the definition of the multiplicative inverse, the inverse matrix M^{-1} becomes

$$\frac{1}{\text{determinant of } M} \text{ [matrix A]}$$

where A is the matrix of cofactors.

$$M^{-1} = \frac{1}{44} \begin{bmatrix} 19 & -14 & -1 \\ 27 & -6 & -13 \\ -14 & 8 & 10 \end{bmatrix}$$

Therefore,

$$M^{-1}M = \frac{1}{44} \begin{bmatrix} 19 & -14 & -1 \\ 27 & -6 & -13 \\ -14 & 8 & 10 \end{bmatrix} \begin{bmatrix} 1 & 3 & 4 \\ -2 & 4 & 5 \\ 3 & 1 & 6 \end{bmatrix}$$

$$= \frac{1}{44} \begin{bmatrix} 19 + 28 - 3 & 57 - 56 - 1 & 76 - 70 - 6 \\ 27 + 12 - 39 & 81 - 24 - 13 & 108 - 30 - 78 \\ -14 - 16 + 30 & -42 + 32 + 10 & -56 + 40 + 60 \end{bmatrix}$$

$$= \frac{1}{44} \begin{bmatrix} 44 & 0 & 0 \\ 0 & 44 & 0 \\ 0 & 0 & 44 \end{bmatrix} = \begin{bmatrix} 1 & 0 & 0 \\ 0 & 1 & 0 \\ 0 & 0 & 1 \end{bmatrix}$$

Show that $\begin{vmatrix} b + c & a - b & a \\ c + a & b - c & b \\ a + b & c - a & c \end{vmatrix} = 3abc - a^3 - b^3 - c^3.$

Solution: The following is a known property of determinants:

$\begin{vmatrix} a_1 + \bar{a}_1 & b_1 & c_1 \\ a_2 + \bar{a}_2 & b_2 & c_2 \\ a_3 + \bar{a}_3 & b_3 & c_3 \end{vmatrix} =$

$\begin{vmatrix} a_1 & b_1 & c_1 \\ a_2 & b_2 & c_2 \\ a_3 & b_3 & c_3 \end{vmatrix} + \begin{vmatrix} \bar{a}_1 & b_1 & c_1 \\ \bar{a}_2 & b_2 & b_2 \\ \bar{a}_3 & b_3 & c_3 \end{vmatrix}.$ Notice that in our given

determinant there are two columns with their elements expressed as the sum of two terms. We will first apply the above property to the first column. We thus obtain:

$\begin{vmatrix} b + c & a - b & a \\ c + a & b - c & b \\ a + b & c - a & c \end{vmatrix} = \begin{vmatrix} b & a - b & a \\ c & b - c & b \\ a & c - a & c \end{vmatrix} + \begin{vmatrix} c & a - b & a \\ a & b - c & b \\ b & c - a & c \end{vmatrix}.$

Now, applying this property to the second column of both determinants on the right side of the equal sign we obtain:

$\begin{vmatrix} b & a & a \\ c & b & b \\ a & c & c \end{vmatrix} + \begin{vmatrix} b & -b & a \\ c & -c & b \\ a & -a & c \end{vmatrix} + \begin{vmatrix} c & a & a \\ a & b & b \\ b & c & c \end{vmatrix} + \begin{vmatrix} c & -b & a \\ a & -c & b \\ b & -a & c \end{vmatrix}.$

But, if each element in a column of a determinant is multiplied by a number p, in this case p = - 1, then the value of the determinant is multiplied by p. That is,

$\begin{vmatrix} a_1 & pb_1 & c_1 \\ a_2 & pb_2 & c_2 \\ a_3 & pb_3 & c_3 \end{vmatrix} = p \begin{vmatrix} a_1 & b_1 & c_1 \\ a_2 & b_2 & c_2 \\ a_3 & b_3 & c_3 \end{vmatrix}.$

Thus, our above determinants become,

$$= \begin{vmatrix} b & a & a \\ c & b & b \\ a & c & c \end{vmatrix} - \begin{vmatrix} b & b & a \\ c & c & b \\ a & a & c \end{vmatrix} + \begin{vmatrix} c & a & a \\ a & b & b \\ b & c & c \end{vmatrix} - \begin{vmatrix} c & b & a \\ a & c & b \\ b & a & c \end{vmatrix}.$$

Recall that when two columns of a determinant are identical, the value of the determinant is zero. Thus, the first three determinants vanish and we are

left with $- \begin{vmatrix} c & b & a \\ a & c & b \\ b & a & c \end{vmatrix}.$

To evaluate this third order determinant we employ the following method: rewrite the first two columns of the determinant next to the third column, obtaining:

$$- \begin{vmatrix} c & b & a \\ a & c & b \\ b & a & c \end{vmatrix} \begin{matrix} c & b \\ a & c \\ b & a \end{matrix} \; .$$

Draw three diagonal lines sloping downward from left to right, each of which encompasses three elements of the determinant. Do this also from right to left.

The diagram now looks like: -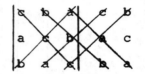

We now form the products of the elements in each of the six diagonals, preceding each of the terms in the left to right diagonals by a positive sign, and each of the terms in the right to left diagonals by a negative sign. The sum of the six products is the required expansion of the determinant. Thus, we obtain:

$$- (c \cdot c \cdot c + b \cdot b \cdot b + a \cdot a \cdot a - acb - cba - bac) =$$

$$- c^3 - b^3 - a^3 + 3abc = 3abc - a^3 - b^3 - c^3.$$

● PROBLEM 24-31

Find the inverse of the matrix A where

$$A = \begin{bmatrix} 1 & 1 & 1 & 1 \\ 0 & 1 & 1 & 1 \\ 0 & 0 & 1 & 1 \\ 0 & 0 & 0 & 1 \end{bmatrix}$$

Show that the inverse of a diagonal matrix is obtained by inverting the diagonal entries.

Solution:

$$[A : I] = \begin{bmatrix} 1 & 1 & 1 & 1 & \vdots & 1 & 0 & 0 & 0 \\ 0 & 1 & 1 & 1 & \vdots & 0 & 1 & 0 & 0 \\ 0 & 0 & 1 & 1 & \vdots & 0 & 0 & 1 & 0 \\ 0 & 0 & 0 & 1 & \vdots & 0 & 0 & 0 & 1 \end{bmatrix}$$

Subtract the second row from the first row:

$$\begin{bmatrix} 1 & 0 & 0 & 0 & \vdots & 1 & -1 & 0 & 0 \\ 0 & 1 & 1 & 1 & \vdots & 0 & 1 & 0 & 0 \\ 0 & 0 & 1 & 1 & \vdots & 0 & 0 & 1 & 0 \\ 0 & 0 & 0 & 1 & \vdots & 0 & 0 & 0 & 1 \end{bmatrix}$$

Subtract the third row from the second row, and the fourth row from the third row:

$$\begin{bmatrix} 1 & 0 & 0 & 0 & \vdots & 1 & -1 & 0 & 0 \\ 0 & 1 & 0 & 0 & \vdots & 0 & 1 & -1 & 0 \\ 0 & 0 & 1 & 0 & \vdots & 0 & 0 & 1 & -1 \\ 0 & 0 & 0 & 1 & \vdots & 0 & 0 & 0 & 1 \end{bmatrix}$$

Hence

$$A^{-1} = \begin{bmatrix} 1 & -1 & 0 & 0 \\ 0 & 1 & -1 & 0 \\ 0 & 0 & 1 & -1 \\ 0 & 0 & 0 & 1 \end{bmatrix}$$

A diagonal matrix is a square matrix whose non-diagonal entries are all zero. Let A be a diagonal matrix whose diagonal entries are all non-zero, and let

$$A = \begin{bmatrix} a_{11} & 0 & \cdots & & & 0 \\ 0 & a_{22} & \cdots & & & 0 \\ \cdot & & & & & \cdot \\ \cdot & & 0 & a_{kk} & & \cdot \\ \cdot & & & & & \cdot \\ 0 & \cdots & & & & a_{nn} \end{bmatrix}$$

with $a_{ii} \neq 0$, $i = 1,...,n$.

Now apply the procedure for finding the inverse at a matrix.

Then

$$[A : I] = \begin{bmatrix} a_{11} & & 0 & \vdots & 1 & 0 & \cdots & 0 \\ 0 & a_{22} & 0 & \vdots & 0 & 1 & \cdots & 0 \\ \cdot & & & \vdots & \cdot & & & \\ \cdot & & & \vdots & \cdot & & & \\ 0 & \cdots & a_{nn} & \vdots & 0 & \cdots & & 1 \end{bmatrix}$$

Multiply the first row by $\frac{1}{a_{11}}$, the second row by $\frac{1}{a_{22}}$ . . . and the nth row by $\frac{1}{a_{nn}}$, to obtain

$$[I : B] = \begin{bmatrix} 1 & 0 & . & . & . & . & . & 0 & | & 1/a_{11} & 0 & . & . & . & . & . & 0 \\ 0 & 1 & . & . & . & . & 0 & | & 0 & 1/a_{22} & . & . & . & . & 0 \\ . & & & & & & & | & . & & & & & \\ . & & & & & & & | & . & & & & & \\ . & & & & & & & | & . & & & & & \\ 0 & . & . & . & . & . & . & 1 & | & 0 & . & . & . & . & . & . & . & . & 1/a_{nn} \end{bmatrix}$$

Hence

$$A^{-1} = \begin{bmatrix} 1/a_{11} & 0 & . & . & . & . & 0 \\ 0 & 1/a_{22} & . & . & 0 \\ . & & & & . \\ . & & & & . \\ . & & & & . \\ 0 & . & . & . & . & . & . & 1/a_{nn} \end{bmatrix}$$

Thus the inverse of a diagonal matrix is obtained by inverting the diagonal entries.

Observe that if one of the diagonal entries is zero, the matrix is not invertible. For example,

$$\begin{bmatrix} 1 & 0 & 0 \\ 0 & 0 & 0 \\ 0 & 0 & 3 \end{bmatrix}$$

is not invertible.

● PROBLEM 24-32

Find the inverses of the following matrices.

(1)

$$A = \begin{bmatrix} 3 & 1 \\ -1 & 6 \end{bmatrix}$$

(2)

$$A = \begin{bmatrix} 1 & -7 & -14 \\ 2 & 1 & -1 \\ 1 & 3 & 4 \end{bmatrix}$$

(3)

$$A = \begin{bmatrix} 3 & 1 & 0 \\ 1 & -1 & 2 \\ 1 & 1 & 1 \end{bmatrix} .$$

<u>Solution:</u>   The method of solution is the same in all three cases, namely, forming the block matrix $[A : I]$ where I is the $n \times n$ identity matrix, and using elementary row operations to reduce it to $[I : A^{-1}]$.

(1) $A = \begin{bmatrix} 3 & 1 \\ -1 & 6 \end{bmatrix}$.

Now $[A : I] = \begin{bmatrix} 3 & 1 & \vdots & 1 & 0 \\ -1 & 6 & \vdots & 0 & 1 \end{bmatrix}$.

Multiply the first row by 6:

$$\begin{bmatrix} 18 & 6 & \vdots & 6 & 0 \\ -1 & 6 & \vdots & 0 & 1 \end{bmatrix}$$

Subtract the second row from the first row:

$$\begin{bmatrix} 19 & 0 & \vdots & 6 & -1 \\ -1 & 6 & \vdots & 0 & 1 \end{bmatrix}$$

Multiply the second row by 19:

$$\begin{bmatrix} 19 & 0 & \vdots & 6 & -1 \\ -19 & 114 & \vdots & 0 & 19 \end{bmatrix}$$

Add the first row to the second row:

$$\begin{bmatrix} 19 & 0 & \vdots & 6 & -1 \\ 0 & 114 & \vdots & 6 & 18 \end{bmatrix}$$

Divide the first and second rows by 19:

$$\begin{bmatrix} 1 & 0 & \vdots & 6/19 & -1/19 \\ 0 & 6 & \vdots & 6/19 & 18/19 \end{bmatrix}$$

Divide the second row by 6:

$$\begin{bmatrix} 1 & 0 & \vdots & 6/19 & -1/19 \\ 0 & 1 & \vdots & 1/19 & 3/19 \end{bmatrix}$$

Therefore
$$A^{-1} = \begin{bmatrix} 6/19 & -1/19 \\ 1/19 & 3/19 \end{bmatrix}$$

(2)

$$A = \begin{bmatrix} 1 & -7 & -14 \\ 2 & 1 & -1 \\ 1 & 3 & 4 \end{bmatrix}$$

$$[A : I] = \begin{bmatrix} 1 & -7 & -14 & \vdots & 1 & 0 & 0 \\ 2 & 1 & -1 & \vdots & 0 & 1 & 0 \\ 1 & 3 & 4 & \vdots & 0 & 0 & 1 \end{bmatrix}$$

478

Subtract the first row from the third row:

$$\begin{bmatrix} 1 & -7 & -14 & \vdots & 1 & 0 & 0 \\ 2 & 1 & -1 & \vdots & 0 & 1 & 0 \\ 0 & 10 & 18 & \vdots & -1 & 0 & 1 \end{bmatrix}$$

Divide the third row by 2:

$$\begin{bmatrix} 1 & -7 & -14 & \vdots & 1 & 0 & 0 \\ 2 & 1 & -1 & \vdots & 0 & 1 & 0 \\ 0 & 5 & 9 & \vdots & -1/2 & 0 & 1/2 \end{bmatrix}$$

Add -2 times the first row to the second row:

$$\begin{bmatrix} 1 & -7 & 9 & \vdots & 1 & 0 & 0 \\ 0 & 15 & 27 & \vdots & -2 & 1 & 0 \\ 0 & 5 & 9 & \vdots & -1/2 & 0 & 1/2 \end{bmatrix}$$

Divide the second row by 3:

$$\begin{bmatrix} 1 & -7 & -14 & \vdots & 1 & 0 & 0 \\ 0 & 5 & 9 & \vdots & -2/3 & 1/3 & 0 \\ 0 & 5 & 9 & \vdots & -1/2 & 0 & 1/2 \end{bmatrix}$$

Subtract the second row from the third row:

$$\begin{bmatrix} 1 & -7 & -14 & \vdots & 1 & 0 & 0 \\ 0 & 5 & 9 & \vdots & -2/3 & 1/3 & 0 \\ 0 & 0 & 0 & \vdots & 1/6 & -1/3 & 1/2 \end{bmatrix}$$

At this point A is row equivalent to

$$F = \begin{bmatrix} 1 & -7 & -14 \\ 0 & 5 & 9 \\ 0 & 0 & 0 \end{bmatrix}$$

The matrix A is singular and therefore A does not have an inverse.

(3)

$$A = \begin{bmatrix} 3 & 1 & 0 \\ 1 & -1 & 2 \\ 1 & 1 & 1 \end{bmatrix}$$

$$[A : I] = \begin{bmatrix} 3 & 1 & 0 & \vdots & 1 & 0 & 0 \\ 1 & -1 & 2 & \vdots & 0 & 1 & 0 \\ 1 & 1 & 1 & \vdots & 0 & 0 & 1 \end{bmatrix}$$

Interchange the first and third rows:

$$\begin{bmatrix} 1 & 1 & 1 & 0 & 0 & 1 \\ 1 & -1 & 2 & 0 & 1 & 0 \\ 3 & 1 & 0 & 1 & 0 & 0 \end{bmatrix}$$

Subtract the **first row** from the second row and add −3 times the first row to the third row:

$$\begin{bmatrix} 1 & 1 & 1 & 0 & 0 & 1 \\ 0 & -2 & 1 & 0 & 1 & -1 \\ 0 & -2 & -3 & 1 & 0 & -3 \end{bmatrix}$$

Divide the second row by −2:

$$\begin{bmatrix} 1 & 1 & 1 & 0 & 0 & 1 \\ 0 & 1 & -1/2 & 0 & -1/2 & 1/2 \\ 0 & -2 & -3 & 1 & 0 & -3 \end{bmatrix}$$

Subtract the second row from the first row:

$$\begin{bmatrix} 1 & 0 & 3/2 & 0 & 1/2 & 1/2 \\ 0 & 1 & -1/2 & 0 & -1/2 & 1/2 \\ 0 & -2 & -3 & 1 & 0 & -3 \end{bmatrix}$$

Add 2 times the second row to the third row:

$$\begin{bmatrix} 1 & 0 & 3/2 & 0 & 1/2 & 1/2 \\ 0 & 1 & -1/2 & 0 & -1/2 & 1/2 \\ 0 & 0 & -4 & 1 & -1 & -2 \end{bmatrix}$$

Divide the third row by −4:

$$\begin{bmatrix} 1 & 0 & 3/2 & 0 & 1/2 & 1/2 \\ 0 & 1 & -1/2 & 0 & -1/2 & 1/2 \\ 0 & 0 & 1 & -1/4 & 1/4 & +2/4 \end{bmatrix}$$

Add −3/2 times the third row to the first row and add 1/2 times the third row to the second row:

$$\begin{bmatrix} 1 & 0 & 0 & 3/8 & +1/8 & -2/8 \\ 0 & 1 & 0 & -1/8 & -3/8 & 6/8 \\ 0 & 0 & 1 & -1/4 & 1/4 & +2/4 \end{bmatrix}$$

Thus

$$A^{-1} = \begin{bmatrix} 3/8 & 1/8 & -2/8 \\ -1/8 & -3/8 & 6/8 \\ -1/4 & 1/4 & 2/4 \end{bmatrix} .$$

Let
$$A = \begin{bmatrix} 1 & 2 \\ 3 & 4 \end{bmatrix}$$

Find the inverse of A directly by solving for the entries of the matrix B which satisfies the equation

$$A \cdot B = I ,$$

where A·B is matrix multiplication.

Solution: The problem asks us to solve for the entries a,b,c, and d of the matrix

$$B = \begin{bmatrix} a & b \\ c & d \end{bmatrix} ,$$

given that A·B = I .

Since A·B = I, we have

$$\begin{bmatrix} 1 & 2 \\ 3 & 4 \end{bmatrix} \cdot \begin{bmatrix} a & b \\ c & d \end{bmatrix} = \begin{bmatrix} 1 & 0 \\ 0 & 1 \end{bmatrix}$$

After performing the multiplication of the matrices, we obtain

$$\begin{bmatrix} a+2c & b+2d \\ 3a+4c & 3b+4d \end{bmatrix} = \begin{bmatrix} 1 & 0 \\ 0 & 1 \end{bmatrix} .$$

Recall that two matrices are equal if, and only, if, their corresponding entries are equal. Thus from the last equation we may conclude that a+2c = 1, b+2d = 0, 3a+4c = 0, 3b+4d = 1.

From these four equations, we can obtain two sets of linear equations from which we can solve for each of a,b,c,d. That is, we have the set:

$$a + 2c = 1$$
$$3a + 4c = 0$$

whose solutions are a = -2 and c = 3/2, and the set

$$b + 2d = 0$$
$$3b + 4d = 1$$

whose solutions are b = 1 and d = -1/2 . Hence,

$$B = \begin{bmatrix} a & b \\ c & d \end{bmatrix} = \begin{bmatrix} -2 & 1 \\ 3/2 & -1/2 \end{bmatrix} .$$

Since B satisfies AB = I, B = A^{-1} . Hence,

$$A^{-1} = \begin{bmatrix} -2 & 1 \\ 3/2 & -1/2 \end{bmatrix} .$$

The method used consisted of obtaining sets of linear equations whose unique solutions yielded the required inverse.

Let

$$A = \begin{bmatrix} 1 & 2 & 3 \\ 1 & 3 & 2 \\ 1 & 1 & 5 \end{bmatrix}$$

Show how we obtain the inverse of A by reducing the matrix $[A : I]$ to a matrix of the form $[I : B]$.

Solution: Let e_1, e_2, e_3 be the row vectors of the identity matrix, i.e., $e_1 = (1,0,0)$; $e_2 = (0,1,0)$; $e_3 = (0,0,1)$. Then

$$e_1 A = (1,0,0) \cdot \begin{bmatrix} 1 & 2 & 3 \\ 1 & 3 & 2 \\ 1 & 1 & 5 \end{bmatrix}$$

$$= (1,2,3) = e_1 + 2e_2 + 3e_3$$

$$e_2 A = (0,1,0) \begin{bmatrix} 1 & 2 & 3 \\ 1 & 3 & 2 \\ 1 & 1 & 5 \end{bmatrix}$$

$$= (1,3,2) = e_1 + 3e_2 + 2e_3$$

and

$$e_3 A = (0,0,1) \begin{bmatrix} 1 & 2 & 3 \\ 1 & 3 & 2 \\ 1 & 1 & 5 \end{bmatrix}$$

$$= (1,1,5) = e_1 + e_2 + 5e_3 \quad .$$

So we have

$$e_1 + 2e_2 + 3e_3 = e_1 A$$

$$e_1 + 3e_2 + 2e_3 = e_2 A$$

$$e_1 + e_3 + 5e_3 = e_3 A$$

This is equivalent to

$$e_1 + 2e_2 + 3e_3 = e_1 A + 0e_2 A + 0e_3 A$$

$$e_1 + 3e_2 + 2e_3 = 0e_1 A + e_2 A + 0e_3 A$$

$$e_1 + e_2 + 5e_3 = 0e_1 A + 0e_2 A + e_3 A \quad .$$

The coefficients of the above system correspond to the coefficients of the matrix $(A : I)$.

By applying transformations to the equations that correspond to row operations on $(A : I)$, we can transform the system of equations to the system

$$e_1 = 13e_1 A - 7e_2 A - 5e_3 A$$

$$e_2 = -3e_1 A + 2e_2 A + e_3 A$$

$$e_3 = -2e_1 A + e_2 A + e_3 A \quad .$$

This is equivalent to

$$e_1 + 0e_2 + 0e_3 = 13e_1A - 7e_2A - 5e_3A$$

$$0e_1 + e_2 + 0e_3 = -3e_1A + 2e_2A + e_3A$$

$$0e_1 + 0e_2 + e_3 = -2e_1A + e_2A + e_3A \quad .$$

By comparing coefficients, the above equations correspond to the matrix $(I : B)$, and

$$B = \begin{bmatrix} 13 & -7 & -5 \\ -3 & 2 & 1 \\ -2 & 1 & 1 \end{bmatrix} .$$

Multiplying both sides of the equations by A^{-1}, we have

$$e_1A^{-1} = 13e_1 - 7e_2 - 5e_3$$

$$e_2A^{-1} = -3e_1 + 2e_2 + e_3$$

$$e_3A^{-1} = -2e_1 + e_2 + e_3$$

which shows that

$$A^{-1} = \begin{bmatrix} 13 & -7 & -5 \\ -3 & 2 & 1 \\ -2 & 1 & 1 \end{bmatrix} .$$

So $A^{-1} = B$ where B is the matrix obtained by reduction of $(A : I)$.

Let us go through the steps in the transformation of the equations which correspond to row operations on $(A : I)$. Now

$$\begin{array}{l} e_1 + 2e_2 + 3e_3 \\ e_1 + 3e_2 + 2e_3 \longrightarrow \\ e_1 + e_2 + 5e_3 \end{array} \begin{bmatrix} 1 & 2 & 3 & \vdots & 1 & 0 & 0 \\ 1 & 3 & 2 & \vdots & 0 & 1 & 0 \\ 1 & 1 & 5 & \vdots & 0 & 0 & 1 \end{bmatrix}$$

Consider the following three types of operations for transforming linear systems:

T_{ij} : Interchange the ith and jth equations.

$T_i(c)$: Replace the ith equation with c times the ith equation, where $c \neq 0$.

$T_{ij}(c)$: Replace the ith equation with the sum of the ith equation and c times the jth equation ($j \neq i$).

Ther, $[T_{21}(-1), T_{31}(-1)]$:

$$e_1 + 2e_2 + 3e_3 = e_1A + 0e_2A + 0e_3A$$
$$0e_1 + e_2 - e_3 = -e_1A + e_2A + 0e_3A$$
$$0e_1 - e_2 + 2e_3 = -e_1A + 0e_2A + e_3A$$

$$\longrightarrow \begin{bmatrix} 1 & 2 & 3 & | & 1 & 0 & 0 \\ 0 & 1 & -1 & | & -1 & 1 & 0 \\ 0 & -1 & 2 & | & -1 & 0 & 1 \end{bmatrix}$$

$[T_{32}(1)] =$

$$e_1 + 2e_2 + 3e_3 = e_1A + 0e_2 + 0e_3A$$
$$0e_1 + e_2 - e_3 = -e_1A + e_2A + 0e_3A$$
$$0e_1 + 0e_2 + e_3 = -2e_1A + e_2A + e_3A$$

$$\longrightarrow \begin{bmatrix} 1 & 2 & 3 & | & 1 & 0 & 0 \\ 0 & 1 & -1 & | & -1 & 1 & 0 \\ 0 & 0 & 1 & | & -2 & 1 & 1 \end{bmatrix}$$

$[T_{13}(-3), T_{23}(1)] =$

$$e_1 + 2e_2 + 0e_3 = 7e_1A - 3e_2A - 3e_3A$$
$$0e_1 + e_2 + 0e_3 = -3e_1A + 2e_2A + e_3A$$
$$0e_1 + 0e_2 + e_3 = -2e_1A + e_2A + e_3A$$

$$\longrightarrow \begin{bmatrix} 1 & 2 & 0 & | & 7 & -3 & -3 \\ 0 & 1 & 0 & | & -3 & 2 & 1 \\ 0 & 0 & 1 & | & -2 & 1 & 1 \end{bmatrix}$$

$[T_{12}(-2)] =$

$$e_1 + 0e_2 + 0e_3 = 13e_1A - 7e_2A - 5e_3A$$
$$0e_1 + e_2 + 0e_3 = -3e_1A + 2e_2A + e_3A$$
$$0e_1 + 0e_2 + e_3 = -2e_1A + e_2A + e_3A$$

$$\longrightarrow \begin{bmatrix} 1 & 0 & 0 & | & 13 & -7 & -5 \\ 0 & 1 & 0 & | & -3 & 2 & 1 \\ 0 & 0 & 1 & | & -2 & 1 & 1 \end{bmatrix}$$

Thus

$$A^{-1} = \begin{bmatrix} 13 & -7 & -5 \\ -3 & 2 & 1 \\ -2 & 1 & 1 \end{bmatrix} .$$

● **PROBLEM 24-35**

Use the classical adjoint to find A^{-1} where

$$A = \begin{bmatrix} 1 & 0 & -1 \\ 0 & 2 & 2 \\ 1 & 1 & -1 \end{bmatrix}$$

<u>Solution:</u> Recall some definitions: If $A = (a_{ij})$, then a co-factor of an entry a_{ij} is denoted A_{ij} and is given by $(-1)^{i+j}$ times the determinant of the $(n-1) \times (n-1)$ minor matrix obtained from A by deleting its ith row and jth column.

By the matrix of cofactors, we mean the matrix

$$C = \begin{bmatrix} A_{11} & \cdots\cdots\cdots & A_{1n} \\ \cdot & \cdots\cdots\cdots & \cdot \\ \cdot & & \cdot \\ \cdot & & \cdot \\ A_{n1} & \cdots\cdots\cdots & A_{nn} \end{bmatrix} \quad .$$

Then the adjoint of A is C^T, i.e.,

$$\text{adj } A = \begin{bmatrix} A_{11} & A_{21} & \cdots\cdot & A_{n1} \\ A_{12} & A_{22} & \cdots\cdot & \\ \cdot & \cdot & & \cdot \\ \cdot & \cdot & & \cdot \\ \cdot & \cdot & & \cdot \\ A_{1n} & A_{2n} & \cdots\cdot & A_{nn} \end{bmatrix}$$

Recall that A^{-1} exists if and only if det $A = |A| \neq 0$. The rule for obtaining A^{-1} is then

$$A^{-1} = \frac{1}{|A|} [\text{adj } A]$$

where $|A|$ = determinant of the $n \times n$ square matrix. Let us first compute the determinant of matrix A

$$A = \begin{bmatrix} 1 & 0 & -1 \\ 0 & 2 & 2 \\ 1 & 1 & -1 \end{bmatrix}$$

$$|A| = \begin{vmatrix} 1 & 0 & -1 \\ 0 & 2 & 2 \\ 1 & 1 & -1 \end{vmatrix}$$

$$= 1 \begin{vmatrix} 2 & 2 \\ 1 & -1 \end{vmatrix} - 0 \begin{vmatrix} 0 & 2 \\ 1 & -1 \end{vmatrix} + (-1) \begin{vmatrix} 0 & 2 \\ 1 & 1 \end{vmatrix}$$

$$= 1(-2-2) - 0(0-2) - 1(0-2)$$

$$= -4-0+2 = -2 \quad .$$

We find that $|A| \neq 0$. Therefore A^{-1} exists. The classical adjoint of A is found by replacing each element of A by its cofactor and taking the transpose of the resulting matrix.

Let us now compute the cofactors of the entries of A.

$$A = \begin{bmatrix} 1 & 0 & -1 \\ 0 & 2 & 2 \\ 1 & 1 & -1 \end{bmatrix}$$

To find A_{11}, we delete the first row and first column of A to obtain the matrix

$$\begin{bmatrix} 2 & 2 \\ 1 & -1 \end{bmatrix} \quad .$$

The cofactor A_{11} is then $(-1)^{1+1}$ times the determinant of the

We find the cofactors of the remaining elements of A by the same method. The cofactors of the nine elements of A are

$$A_{11} = + \begin{vmatrix} 2 & 2 \\ 1 & -1 \end{vmatrix}, \quad A_{12} = - \begin{vmatrix} 0 & 2 \\ 1 & -1 \end{vmatrix}, \quad A_{13} = + \begin{vmatrix} 0 & 2 \\ 1 & 1 \end{vmatrix}$$

$$= (-2-2) \qquad\qquad = -(0-2) \qquad\qquad = (0-2)$$

$$= -4 \qquad\qquad\quad = 2 \qquad\qquad\qquad = -2$$

$$A_{21} = - \begin{vmatrix} 0 & -1 \\ 1 & -1 \end{vmatrix}, \quad A_{22} = + \begin{vmatrix} 1 & -1 \\ 1 & -1 \end{vmatrix}, \quad A_{23} = - \begin{vmatrix} 1 & 0 \\ 1 & 1 \end{vmatrix}$$

$$= -(0+1) \qquad\qquad = (-1+1) \qquad\qquad = -(1-0)$$

$$= -1 \qquad\qquad\quad = 0 \qquad\qquad\qquad = -1$$

$$A_{31} = + \begin{vmatrix} 0 & -1 \\ 2 & 2 \end{vmatrix}, \quad A_{32} = - \begin{vmatrix} 1 & -1 \\ 0 & 2 \end{vmatrix}, \quad A_{33} = + \begin{vmatrix} 1 & 0 \\ 0 & 2 \end{vmatrix}$$

$$= (0+2) \qquad\qquad = -(2-0) \qquad\qquad = (2-0)$$

$$= 2 \qquad\qquad\qquad = -2 \qquad\qquad\qquad = 2$$

above matrix, i.e.,

$$A_{11} = (1)^2 \cdot \begin{vmatrix} 2 & 2 \\ 1 & -1 \end{vmatrix} = \begin{vmatrix} 2 & 2 \\ 1 & -1 \end{vmatrix}$$

$$= (-2-2) = -4 .$$

The matrix of cofactors C is given by

$$C = \begin{bmatrix} -4 & 2 & -2 \\ -1 & 0 & -1 \\ 2 & -2 & 2 \end{bmatrix} .$$

We form the transpose of the matrix of cofactors to obtain the classical adjoint of A:

$$\text{Adj } A = \begin{bmatrix} -4 & -1 & 2 \\ 2 & 0 & -2 \\ -2 & -1 & 2 \end{bmatrix}$$

Now,

$$A^{-1} = \frac{1}{|A|} [\text{adj } A]$$

$$= -\frac{1}{2} \begin{bmatrix} -4 & -1 & 2 \\ 2 & 0 & -2 \\ -2 & -1 & 2 \end{bmatrix} ,$$

So

$$A^{-1} = \begin{bmatrix} 2 & 1/2 & -1 \\ -1 & 0 & 1 \\ 1 & 1/2 & -1 \end{bmatrix} .$$

It is easy to check the computation by verifying that

$$AA^{-1} = I$$

$$\begin{bmatrix} 1 & 0 & -1 \\ 0 & 2 & 2 \\ 1 & 1 & -1 \end{bmatrix} \begin{bmatrix} 2 & 1/2 & -1 \\ -1 & 0 & 1 \\ 1 & 1/2 & -1 \end{bmatrix}$$

$$= \begin{bmatrix} 2 + 0 - 1 & 1/2 + 0 - 1/2 & -1 + 0 + 1 \\ 0 - 2 + 2 & 0 + 0 + 1 & 0 + 2 - 2 \\ 2 - 1 - 1 & 1/2 + 0 - 1/2 & -1 + 1 + 1 \end{bmatrix}$$

$$= \begin{bmatrix} 1 & 0 & 0 \\ 0 & 1 & 0 \\ 0 & 0 & 1 \end{bmatrix} = I .$$

● **PROBLEM 24-36**

Define elementary row operations and give an example.

Solution: The three elementary row operations on a matrix A are:
1. Interchange the i-th and the j-th row of A.
2. Add the i-th row of A to the j-th row of A, i ≠ j.
3. Multiply the i-th row of A by a non-zero scalar k. Let

$$A = \begin{bmatrix} 1 & 6 & 3 & 4 \\ 1 & 2 & 1 & 1 \\ -1 & 2 & 1 & 2 \end{bmatrix} .$$

We perform the following row operations on matrix A.
(1) Interchange the first and the second rows

$$\begin{bmatrix} 1 & 2 & 1 & 1 \\ 1 & 6 & 3 & 4 \\ -1 & 2 & 1 & 2 \end{bmatrix} .$$

(2) Add the first row to the third row and -1 times the first row to the second row. Adding the first row to the third row,

$$\begin{bmatrix} 1 & 2 & 1 & 1 \\ 1 & 6 & 3 & 4 \\ -1+1 & 2+2 & 1+1 & 2+1 \end{bmatrix} = \begin{bmatrix} 1 & 2 & 1 & 1 \\ 1 & 6 & 3 & 4 \\ 0 & 4 & 2 & 3 \end{bmatrix} ;$$

adding -1 times the first row to the second,

$$\begin{bmatrix} 1 & 2 & 1 & 1 \\ (-1\cdot1)+1 & (-1\cdot2)+6 & (-1\cdot1)+3 & (-1\cdot1)+4 \\ 0 & 4 & 2 & 3 \end{bmatrix}$$

$$= \begin{bmatrix} 1 & 2 & 1 & 1 \\ 0 & 4 & 2 & 3 \\ 0 & 4 & 2 & 3 \end{bmatrix} \cdot$$

Add -1 times the second row to the third row.

$$\begin{bmatrix} 1 & 2 & 1 & 1 \\ 0 & 4 & 2 & 3 \\ 0 & 0 & 0 & 0 \end{bmatrix}$$

Divide the second row by 4.

$$\begin{bmatrix} 1 & 2 & 1 & 1 \\ 0 & 1 & \frac{1}{2} & 3/4 \\ 0 & 0 & 0 & 0 \end{bmatrix}$$

Add -2 times the second row to the first row.

$$\begin{bmatrix} (-2\cdot 0)+1 & (-2\cdot 1)+2 & (-2\cdot \frac{1}{2})+1 & (-2\cdot 3/4)+1 \\ 0 & 1 & \frac{1}{2} & 3/4 \\ 0 & 0 & 0 & 0 \end{bmatrix}$$

$$= \begin{bmatrix} 1 & 0 & 0 & -\frac{1}{2} \\ 0 & 1 & \frac{1}{2} & 3/4 \\ 0 & 0 & 0 & 0 \end{bmatrix} \cdot$$

Note that this matrix is in row-reduced echelon form. The elementary row operations can be applied to reduce a matrix to echelon form, and this technique is used in solving systems of linear equations.

● PROBLEM 24-37

Given

$$A = \begin{bmatrix} 1 & -2 & 3 & -1 \\ 2 & -1 & 2 & 2 \\ 3 & 1 & 2 & 3 \end{bmatrix} ,$$

(i) Reduce A to echelon form.
(ii) Reduce A to row reduced echelon form.

Solution: To obtain the echelon form, the first non-zero entry of any row must be contained in a column to the left of the first non-zero entry in the next row. Also, the first non-zero entry must be a 1, and any row of all zeros lies below all the rows that have non-zero entries. In reduced echelon form, the column containing the first non-zero entry of the ith row is $\vec{e}_i$.

(i) To reduce A to echelon form, apply the following row operations:

add -2 times the first row to the second row and -3 times the first
row to the third row.

$$
\begin{bmatrix}
1 & -2 & 3 & -1 \\
0 & 3 & -4 & 4 \\
0 & 7 & -7 & 6
\end{bmatrix} .
$$

Multiply the second row by 7 and the third row by 3.

$$
\begin{bmatrix}
1 & -2 & 3 & -1 \\
0 & 21 & -28 & 28 \\
0 & 21 & -21 & 18
\end{bmatrix} .
$$

Then, add -1 times the second row to the third row, to obtain

$$
\begin{bmatrix}
1 & -2 & 3 & -1 \\
0 & 3 & -4 & 4 \\
0 & 0 & 7 & -10
\end{bmatrix} .
$$

Divide the second row by 3 and the third row by 7 to obtain the echelon
form.

$$
\begin{bmatrix}
1 & -2 & 3 & -1 \\
0 & 1 & -4/3 & 4/3 \\
0 & 0 & 1 & -10/7
\end{bmatrix} .
$$

(ii)   To obtain the reduced echelon form, add 2 times the second row to
the first row.

$$
\begin{bmatrix}
1 & 0 & 1/3 & 5/3 \\
0 & 1 & -4/3 & 4/3 \\
0 & 0 & 1 & -10/7
\end{bmatrix} .
$$

Add -1/3 times the third row to the first row and 4/3 times the third
row to the second row, resulting in the row reduced echelon form.

$$
\begin{bmatrix}
1 & 0 & 0 & 15/7 \\
0 & 1 & 0 & -4/7 \\
0 & 0 & 1 & -10/7
\end{bmatrix} .
$$

CHAPTER 25

DETERMINANTS, MATRICES AND SYSTEMS OF EQUATIONS

> **Basic Attacks and Strategies for Solving Problems in this Chapter. See pages 490 to 516 for step-by-step solutions to problems.**

Given a system of linear equations (*cf.* Chapter 11), one can associate with it a *system matrix* composed of the coefficients of the variables, a variable vector composed of the variables, and a constant vector composed of the constant terms of the equations. Thus, given a system, such as

$$4x + 2y = 0$$

$$x - 2y = 4$$

it can be rewritten as a matrix and vector equation $A\vec{x} = \vec{b}$, such as

$$\begin{pmatrix} 4 & 2 \\ 1 & -2 \end{pmatrix} \begin{pmatrix} x \\ y \end{pmatrix} = \begin{pmatrix} 0 \\ 4 \end{pmatrix}.$$

A linear system can be solved using the determinant of the system matrix A by *Cramer's rule*. Cramer's rule uses auxiliary matrices A_i that are created by substituting column i of A with the constant vector $\vec{b}$. Cramer's rule says that the solution is given by

$$x_i = \frac{\det(A_i)}{\det(A)}$$

for all possible values of i. Note that Cramer's rule can be applied only if $\det(A) \neq 0$, i.e., if A is nonsingular.

A system of linear equations is said to be *homogeneous* if the constant vector $\vec{b}$ is the zero vector. A homogeneous system has either (a) the unique trivial solution $\vec{x}$ equaling the zero vector, or (b) an infinite number of non-trivial solutions in addition to the trivial solution. The unique solution occurs when $\det(A) \neq 0$.

If a vector $\vec{z}$ can be shown to be the sum of multiples of two other vectors,

$\vec{x}$ and $\vec{y}$, (i.e., $\vec{z} = a\vec{x} + b\vec{y}$ for some a and b) then $\vec{z}$ is said to be a *linear combination* of $\vec{x}$ and $\vec{y}$. If a set of vectors is such that no vector is a linear combination of other vectors, that set is said to be *linearly independent*. For example, the unit direction vectors e_i are all linearly independent.

Given a set of vectors V, the *space spanned by V* is the set of all vectors that are linear combinations of vectors in V. The *dimensions* of a space is the minimum number of vectors needed to span that space. The *column rank* of a matrix A is the dimension of the space spanned by the columns of A, considered as vectors. The *rank* of a matrix A equals the column rank of A.

It may occur that an $n \times n$ matrix is of a rank less than n (as when certain columns or rows are duplicates or multiples of other columns or rows). In this case, $\det(A) = 0$. If the rank of A equals the dimension n, then A is said to be of *full rank*.

A matrix A may be said to be *equivalent* to another matrix B, if by using the elementary row operations (*cf.* Chapter 24) and elementary column operations (parallel to the row operations), A can be transformed into B.

In general, if a system has fewer equations than variables, the system does not have a unique solution, and at least one of the solutions is nontrivial (i.e., some variables have values other than zero). In general, the system matrix can be reduced to echelon form (*cf.* Chapter 24) and converted back to another (simpler) linear system. Some of the variables may then be seen to have unique values. Some variables may be dependent on others. At this point, one can choose an arbitrary value for certain variables and let the others be determined based on those values.

Systems of equations can also be solved using the following method of *augmenting the system matrix*. One creates a new $n \times (n + 1)$ matrix C from system matrix A and constant vector b, by adding b to A as the $n + 1^{st}$ column. One then performs the elementary row operations on C (*cf.* Chapter 24) that are needed to convert C to echelon form. If it turns out that the echelon form of C has a row whose first non-zero entry is in the last column, then the original system has no solution.

The *Gauss-Jordan elimination method* uses the augmented system matrix and reduces it to echelon form. Next, a new system is created and the system is rewritten by solving for the leading variable. At this point, one substitutes the constant values of variables into those equations that utilize them, and if there are any variables with indeterminant values (i.e., no equations corresponding to certain variables), those variables can be assigned an arbitrary value.

● **PROBLEM** 25-1

Solve the following linear equations by using Cramer's Rule:

$$-2x_1 + 3x_2 - x_3 = 1$$
$$x_1 + 2x_2 - x_3 = 4$$
$$-2x_1 - x_2 + x_3 = -3 .$$

<u>Solution</u>: Consider a system of n linear equations in n unknowns:

$$a_{11}x_1 + a_{12}x_2 + \ldots + a_{1n}x_n = b_1$$

$$a_{21}x_1 + a_{22}x_2 + \ldots + a_{2n}x_n = b_2$$

$$. \quad . \quad . \quad . \quad . \quad . \quad . \quad . \quad . \quad . \quad . \quad .$$

$$a_{n1}x_1 + a_{n2}x_2 + \ldots + a_{nn}x_n = b_n .$$

Write the above equations in matrix notation.

$$\begin{bmatrix} a_{11} & a_{12} & \cdots & a_{1n} \\ a_{21} & a_{22} & \cdots & a_{2n} \\ \vdots & & & \vdots \\ a_{n1} & a_{n2} & \cdots & a_{nn} \end{bmatrix} \begin{bmatrix} x_1 \\ x_2 \\ \vdots \\ x_n \end{bmatrix} = \begin{bmatrix} b_1 \\ b_2 \\ \vdots \\ b_n \end{bmatrix}$$

or, $AX = B$.

Let A be an $n \times n$ matrix over the field F such that $\det A \neq 0$. If $b_1, b_2, \ldots, b_n$ are any scalars in F, the unique solution of the system of equations $AX = B$ is given by:

$$x_i = \frac{\det A_i}{\det A} \qquad i = 1, 2, \ldots, n ,$$

where A_i is the $n \times n$ matrix obtained from A by replacing the ith column of A by the column vector

$$\begin{bmatrix} b_1 \\ b_2 \\ \cdot \\ \cdot \\ \cdot \\ b_n \end{bmatrix}$$

The above theorem is known as "Cramer's Rule" for solving systems of linear equations. Cramer's Rule applies only to systems of n linear equations in n unknowns with non-zero determinants.

Consider the given linear system:

$$-2x_1 + 3x_2 - x_3 = 1$$
$$x_1 + 2x_2 - x_3 = 4$$
$$-2x_1 - x_2 + x_3 = -3$$

or,

$$\begin{bmatrix} -2 & 3 & -1 \\ 1 & 2 & --1 \\ -2 & -1 & 1 \end{bmatrix} \quad \begin{bmatrix} x_1 \\ x_2 \\ x_3 \end{bmatrix} = \begin{bmatrix} 1 \\ 4 \\ 3 \end{bmatrix}$$

$$A \qquad\qquad X \;=\; B$$

$$\text{Det } A = \begin{vmatrix} -2 & 3 & -1 \\ 1 & 2 & -1 \\ -2 & -1 & 1 \end{vmatrix}$$

$$\text{Det } A = -2 \begin{vmatrix} 2 & -1 \\ -1 & 1 \end{vmatrix} - 3 \begin{vmatrix} 1 & -1 \\ -2 & 1 \end{vmatrix} - 1 \begin{vmatrix} 1 & 2 \\ -2 & -1 \end{vmatrix}$$

$$= -2(2-1) - 3(1-2) - (-1+4)$$

$$= -2 + 3 - 3 = -2 \; .$$

Since $\det A \neq 0$, the system has a unique solution. Now,

$$x_1 = \frac{\det A_1}{\det A}, \quad x_2 = \frac{\det A_2}{\det A}, \quad x_3 = \frac{\det A_3}{\det A} \; .$$

$\text{Det } A_1$ is the determinant of the matrix obtained by replacing the 1st column of A by the column of B. Thus,

$$\det A_1 = \begin{vmatrix} 1 & 3 & -1 \\ 4 & 2 & -1 \\ 3 & -1 & 1 \end{vmatrix}$$

$$= -4 \; .$$

Then,

$$x_1 = \frac{-4}{-2} = 2 \; .$$

$$x_2 = \frac{\begin{vmatrix} -2 & 1 & -1 \\ 1 & 4 & -1 \\ -2 & -3 & 1 \end{vmatrix}}{|A|} = \frac{-6}{-2} = 3 \; .$$

$$x_3 = \frac{\det A_3}{\det A} = \frac{\begin{vmatrix} -2 & 3 & 1 \\ 1 & 2 & 4 \\ -2 & -1 & -3 \end{vmatrix}}{-2} = \frac{-8}{-2} = 4 .$$

Thus,

$$x_1 = 2 , \quad x_2 = 3 , \quad x_3 = 4 ,$$

is the unique solution to the given system.

● **PROBLEM 25-2**

Solve the following homogeneous equations:

$$x_1 + 2x_2 + x_3 = 0$$

$$x_2 - 3x_3 = 0$$

$$-x_1 + x_2 - x_3 = 0 .$$

Solution: A homogeneous system of linear equations has either a) the unique trivial solution $x_1 = x_2 = \ldots = x_n = 0$ or b) an infinite number of non-trivial solutions, plus the trivial solution.

First, write the above equations in matrix notation:

$$\begin{bmatrix} 1 & 2 & 1 \\ 0 & 1 & -3 \\ -1 & 1 & -1 \end{bmatrix} \begin{bmatrix} x_1 \\ x_2 \\ x_3 \end{bmatrix} = \begin{bmatrix} 0 \\ 0 \\ 0 \end{bmatrix}$$

or, $AX = 0$.

$$\det A = 1 \begin{vmatrix} 1 & -3 \\ 1 & -1 \end{vmatrix} - 2 \begin{vmatrix} 0 & -3 \\ -1 & -1 \end{vmatrix} + 1 \begin{vmatrix} 0 & 1 \\ -1 & 1 \end{vmatrix}$$

$$= \left[(-1+3) - 2(0-3) + (0+1) \right]$$

$$= 2 + 6 + 1$$

$$= 9 .$$

Hence, $\det A \neq 0$, and according to Cramer's Rule, the above system has a unique solution.

Therefore, $X = 0$, i.e., $x_1 = x_2 = x_3 = 0$ since the homogeneous system $AX = 0$ has a non-zero solution if and only if $\det A = 0$.

● **PROBLEM 25-3**

Solve the system of linear equations:

$$3x + 2y + 4z = 1$$

$$2x - y + z = 0$$

$$x + 2y + 3z = 1 .$$

Solution: Use Cramer's Rule to solve this system. Write the above equations in matrix form:

$$\begin{bmatrix} 3 & 2 & 4 \\ 2 & -1 & 1 \\ 1 & 2 & 3 \end{bmatrix} \begin{bmatrix} x \\ y \\ z \end{bmatrix} = \begin{bmatrix} 1 \\ 0 \\ 1 \end{bmatrix}.$$

Then

$$A = \begin{bmatrix} 3 & 2 & 4 \\ 2 & -1 & 1 \\ 1 & 2 & 3 \end{bmatrix}, \quad B = \begin{bmatrix} 1 \\ 0 \\ 1 \end{bmatrix}.$$

First, check that $\det A \neq 0$.

$$\det A = \begin{vmatrix} 3 & 2 & 4 \\ 2 & -1 & 1 \\ 1 & 2 & 3 \end{vmatrix}$$

$$\det A = 3 \begin{vmatrix} -1 & 1 \\ 2 & 3 \end{vmatrix} - 2 \begin{vmatrix} 2 & 1 \\ 1 & 3 \end{vmatrix} + 4 \begin{vmatrix} 2 & -1 \\ 1 & 2 \end{vmatrix}$$

$$= 3(-3-2) - 2(6-1) + 4(4+1)$$
$$= -15 - 10 + 20 = -5.$$

Since $\det A \neq 0$, the system has a unique solution. Then,

$$x = \frac{\det A_1}{\det A}, \quad y = \frac{\det A_2}{\det A}, \quad z = \frac{\det A_3}{\det A}.$$

$\det A_1$ is the determinant of the matrix obtained by replacing the first column of A by the column vector B.
Thus,

$$\det A_1 = \begin{vmatrix} 1 & 2 & 4 \\ 0 & -1 & 1 \\ 1 & 2 & 3 \end{vmatrix}.$$

Expand the determinant by minors, using the first column.

$$\det A_1 = 1 \begin{vmatrix} -1 & 1 \\ 2 & 3 \end{vmatrix} + 1 \begin{vmatrix} 2 & 4 \\ -1 & 1 \end{vmatrix}$$

$$= 1(-3-2) + 1(2+4)$$

$$= -5 + 6 = +1.$$

Now, we have $\det A = -5$.

Thus, $x = \dfrac{\det A_1}{\det A} = \dfrac{1}{-5} = -\dfrac{1}{5}$.

$$y = \frac{\det A_2}{\det A} = \frac{\begin{vmatrix} 3 & 1 & 4 \\ 2 & 0 & 1 \\ 1 & 1 & 3 \end{vmatrix}}{-5}.$$

Now, expand $\det A_2$ along the second row:

$$\begin{vmatrix} 3 & 1 & 4 \\ 2 & 0 & 1 \\ 1 & 1 & 3 \end{vmatrix} = -2 \begin{vmatrix} 1 & 4 \\ 1 & 3 \end{vmatrix} - 1 \begin{vmatrix} 3 & 1 \\ 1 & 1 \end{vmatrix}$$

$$= -2(3-4) - 1(3-1)$$
$$= 2 - 2 = 0$$

$$y = \frac{0}{-5} = 0 .$$

$$z = \frac{\det A_3}{\det A} = \frac{\begin{vmatrix} 3 & 2 & 1 \\ 2 & -1 & 0 \\ 1 & 2 & 1 \end{vmatrix}}{-5}$$

Expand determinant A_3 by minors, using the third column.

$$\begin{vmatrix} 3 & 2 & 1 \\ 2 & -1 & 0 \\ 1 & 2 & 1 \end{vmatrix} = + 1 \begin{vmatrix} 2 & -1 \\ 1 & 2 \end{vmatrix} + 1 \begin{vmatrix} 3 & 2 \\ 2 & -1 \end{vmatrix}$$

$$= (4+1) + (-3-4)$$
$$= 5 - 7 = -2 .$$

Then,

$$z = \frac{-2}{-5} = \frac{2}{5} .$$

Thus $x = -1/5$, $y = 0$, $z = 2/5$.

● **PROBLEM 25-4**

Let the homogeneous linear system $AX = B$ be given by

$$\begin{bmatrix} 1 & 2 & 0 \\ 0 & 1 & 3 \\ 2 & 1 & 3 \end{bmatrix} \begin{bmatrix} x_1 \\ x_2 \\ x_3 \end{bmatrix} = \begin{bmatrix} 0 \\ 0 \\ 0 \end{bmatrix} . \qquad (1)$$

Show that A has only the trivial solution, $(x_1, x_2, x_3) = (0,0,0)$.

Solution: The system (1) will have only the trivial solution if A^{-1} exists, for then

$$\begin{bmatrix} x_1 \\ x_2 \\ x_3 \end{bmatrix} = \begin{bmatrix} 0 \\ 0 \\ 0 \end{bmatrix} \begin{bmatrix} 1 & 2 & 0 \\ 0 & 1 & 3 \\ 2 & 1 & 3 \end{bmatrix}^{-1} = \begin{bmatrix} 0 \\ 0 \\ 0 \end{bmatrix} .$$

Thus, we must show that A is non-singular. But an $n \times n$ matrix is non-singular if and only if rank $A = n$. To show this, use the folowing reasoning: A has full rank (i.e., rank equal to n) if and only if it is equivalent to the $n \times n$ identity matrix, I_n . But I_n is non-singular. Hence, A is non-singular. Conversely, if an $n \times n$ matrix is non-singular, it is equivalent to an identity matrix, I_n which has rank equal to n. Hence, rank $A = n$. We must find the rank of the matrix in (1). By applying elementary row operations it can be seen that

$$\begin{bmatrix} 1 & 2 & 0 \\ 0 & 1 & 3 \\ 2 & 1 & 3 \end{bmatrix}$$

is equivalent to

$$\begin{bmatrix} 1 & 0 & -6 \\ 0 & 1 & 3 \\ 0 & 0 & 1 \end{bmatrix},$$

i.e., rank A = 3 . Hence, A is non-singular and the system (1) has only the trivial solution.

The above problem is an illustration of the theorem below. A necessary and sufficient condition for a system of n homogeneous linear equations, AX = 0 , to have only the trivial solution, X = 0 , is that rank A = n .

● PROBLEM 25-5

Consider the system of equations

$$\begin{bmatrix} 2 & 1 & 3 \\ 1 & -2 & 2 \\ 0 & 1 & 3 \end{bmatrix} \begin{bmatrix} x_1 \\ x_2 \\ x_3 \end{bmatrix} = \begin{bmatrix} 1 \\ 2 \\ 3 \end{bmatrix}.$$

Show that the system has a solution without actually computing a solution.

Solution: Let AX = B be a system of m linear equations in n unknowns, where A = $[a_{ij}]$ is an m Xn matrix, X is an n-dimensional column vector and B is an m-dimensional column vector. Thus, the system has the form

$$a_{11}x_1 + a_{12}x_2 + \ldots + a_{1n}x_n = b_1$$

$$a_{21}x_1 + a_{22}x_2 + \ldots + a_{2n}x_n = b_2$$

$$\cdot \qquad \cdot \qquad \qquad \cdot \qquad \cdot$$

$$\cdot \qquad \cdot \qquad \qquad \cdot \qquad \cdot \qquad (1)$$

$$\cdot \qquad \cdot \qquad \qquad \cdot \qquad \cdot$$

$$a_{m1}x_1 + a_{n2}x_2 + \ldots + a_{mn}x_n = b_m ,$$

The system (1) may also be written as the vector equation

$$x_1 \begin{bmatrix} a_{11} \\ a_{21} \\ \cdot \\ \cdot \\ \cdot \\ a_{m1} \end{bmatrix} + x_2 \begin{bmatrix} a_{12} \\ a_{22} \\ \cdot \\ \cdot \\ \cdot \\ a_{m2} \end{bmatrix} + \ldots + x_n \begin{bmatrix} a_{1n} \\ a_{2n} \\ \cdot \\ \cdot \\ \cdot \\ a_{mn} \end{bmatrix} = \begin{bmatrix} b_1 \\ b_2 \\ \cdot \\ \cdot \\ \cdot \\ b_m \end{bmatrix}$$

Thus, AX = B has a solution when B is a linear combination of the columns of A; that is, if B belongs to the column space of A. But the dimension of the column space of A is the rank of A. Since B is in this space, the rank of A is also the rank of the augmented

495

matrix $[A\,|\,B]$. Thus, a necessary and sufficient condition for $AX = B$ to have a solution is that rank $A = \text{rank}[A\,|\,B]$.

Applying elementary rwo operations to the given matrix, it can be seen that

$$\begin{bmatrix} 2 & 1 & 3 \\ 1 & -2 & 2 \\ 0 & 1 & 3 \end{bmatrix}$$

is equivalent to

$$\begin{bmatrix} 1 & 0 & 8 \\ 0 & 1 & 3 \\ 0 & 0 & -16 \end{bmatrix} \tag{2}$$

Since the columns of the matrix (2) form a basis for R^3, rank $A = 3$. Next, form the augmented matrix

$$\begin{bmatrix} 2 & 1 & 3 & | & 1 \\ 1 & -2 & 2 & | & 2 \\ 0 & 1 & 3 & | & 3 \end{bmatrix}$$

and reduce to echelon form:

$$\begin{bmatrix} 1 & 0 & 0 & | & -1 \\ 0 & 1 & 0 & | & -3/8 \\ 0 & 0 & 1 & | & 9/8 \end{bmatrix}$$

Thus, rank $[A\,|\,B] = 3$ and the given system of equations has a solution.

● **PROBLEM 25-6**

Solve the following system of equations by forming the matrix of coefficients and reducing it to echelon form.

$$\begin{aligned} 3x + 2y - z &= 0 \\ x - y + 2z &= 0 \\ x + y - 6z &= 0 \end{aligned} \tag{1}$$

Solution: The most general linear system of m equations in n unknowns (variables) $x_1, x_2, \ldots, x_n$ is of the form

$$\begin{aligned} a_{11}x_1 + a_{12}x_2 + \ldots + a_{1n}x_n &= c_1 \\ a_{21}x_1 + a_{22}x_2 + \ldots + a_{2n}x_n &= c_2 \\ \cdots \cdots \cdots \cdots \cdots \cdots \cdots \\ a_{m1}x_1 + a_{m2}x_2 + \ldots + a_{mn}x_n &= c_m, \end{aligned} \tag{2}$$

where $a_{11}, a_{12}, \ldots, a_{1n}$, $a_{21}, a_{22}, \ldots, a_{2n}$, $a_{m1}, a_{m2}, \ldots, a_{mn}$ are called the coefficients of the system (2), and $c_1, c_2, \ldots, c_m$ are called the constants of the system (2).

If $c_1 = c_2 = \ldots = c_m = 0$, a system of the form (2) is called a homogeneous system. Let $A = (a_{ij})$ be a matrix of size $m \times n$ where the entries a_{ij} are the same as the coefficients of the system given by (2). Then we say that A is the matrix of coefficients of (2). In order to solve the homogeneous system

$$a_{11}x_1 + a_{12}x_2 + \ldots + a_{1n}x_n = 0$$
$$a_{21}x_1 + a_{22}x_2 + \ldots + a_{2n}x_n = 0 \qquad \text{(3)}$$
$$\cdot$$
$$\cdot$$
$$\cdot$$
$$a_{m1}x_1 + a_{m2}x_2 + \ldots + a_{mn}x_n = 0 \quad ,$$

first form the matrix of coefficients $A = (a_{ij})$. Next, obtain a row-echelon matrix $R = (r_{ij})$ by the method of elementary row operations on the matrix A. Since the solutions to (3) are the same as the solutions to

$$r_{11}x_1 + r_{12}x_2 + \ldots + r_{1n}x_n = 0$$
$$r_{21}x_1 + r_{22}x_2 + \ldots + r_{2n}x_n = 0$$
$$\cdot \qquad \text{(4)}$$
$$\cdot$$
$$\cdot$$
$$r_{m1}x_1 + r_{m2}x_2 + \ldots + r_{mn}x_n = 0$$

where the coefficients r_{ij} are identically the entries of R (i.e., R is the coefficient matrix of the homogeneous system (4)) we solve the original system (3) by solving the reduced system (4). Now solve the given homogeneous system (1).

The matrix of coefficients of the system (1) is

$$\begin{bmatrix} 3 & 2 & -1 \\ 1 & -1 & 2 \\ 1 & 1 & -6 \end{bmatrix} \, .$$

Reduce the above matrix to echelon form. Add -3 times the second row to the first row, and add -1 times the second row to the third row:

$$\begin{bmatrix} 0 & 5 & -7 \\ 1 & -1 & 2 \\ 0 & 2 & -8 \end{bmatrix}$$

Divide the third row by 2.

$$\begin{bmatrix} 0 & 5 & -7 \\ 1 & -1 & 2 \\ 0 & 1 & -4 \end{bmatrix}$$

Add -5 times the third row to the first row, and add the third row to the second row.

$$\begin{bmatrix} 0 & 0 & 13 \\ 1 & 0 & -2 \\ 0 & 1 & -4 \end{bmatrix}$$

497

Divide row one by 13; then add 2 times the resulting row one to row two. Next, add 4 times the resulting row one to row three:

$$\begin{bmatrix} 0 & 0 & 1 \\ 1 & 0 & 0 \\ 0 & 1 & 0 \end{bmatrix}$$

Interchange rows one and two, then rows two and three:

$$\begin{bmatrix} 1 & 0 & 0 \\ 0 & 1 & 0 \\ 0 & 0 & 1 \end{bmatrix}$$

This matrix is reduced and gives the system

$$x = 0$$
$$y = 0$$
$$z = 0 \quad .$$

Thus, the unique solution to the original system is $x = y = z = 0$, the solution that is called the trivial solution.

● **PROBLEM 25-7**

Let $A = \begin{bmatrix} 2 & 1 & 4 \\ 3 & 0 & 1 \\ 2 & -1 & 1 \end{bmatrix}$ be the coefficient matrix of a homogeneous system in x,y, and z. Solve this system to illustrate that a homogeneous system of 3 equations in the unknowns, x,y,z has a unique solution.

Solution: The method of solving a homogeneous system is reduction of the coefficient matrix to echelon form. To do this, perform the following row operations on matrix A:
Divide the first row by 2

$$\begin{bmatrix} 1 & 1/2 & 2 \\ 3 & 0 & 1 \\ 2 & -1 & 1 \end{bmatrix} \quad .$$

Add -3 times the first row to the second row and -2 times the first row to the third row

$$\begin{bmatrix} 1 & 1/2 & 2 \\ 0 & -3/2 & -5 \\ 0 & -2 & -3 \end{bmatrix} \quad .$$

Add -4/3 times the second row to the third row,

$$\begin{bmatrix} 1 & 1/2 & 2 \\ 0 & -3/2 & -5 \\ 0 & 0 & 11/3 \end{bmatrix} \quad .$$

Multiply the third row by 3/11. Next, add -2 times the third row to

498

the first row and 5 times the third row to the second row.

$$\begin{bmatrix} 1 & 1/2 & 0 \\ 0 & -3/2 & 0 \\ 0 & 0 & 1 \end{bmatrix} .$$

Multiply the second row by $-2/3$. Finally, add $-1/2$ times the second row to the first row.

$$\begin{bmatrix} 1 & 0 & 0 \\ 0 & 1 & 0 \\ 0 & 0 & 1 \end{bmatrix} .$$

The above matrix is an identity matrix and the corresponding system of equations is

$$x = 0$$
$$y = 0$$
$$z = 0 .$$

Thus, the system has a unique solution and it is the trivial solution. In general, a homogeneous system of n equations in n unknowns has a unique solution if and only if the coefficient matrix can be reduced to the n by n identity matrix. In this case, the solution is the trivial solution, and the dimension of the solution space is zero. We can look at this in the following way:

Since the coefficient matrix A can be reduced by row operations to the identity matrix, the row vectors of A are linearly independent. Therefore, there cannot exist non-zero real numbers x_1, x_2, x_3 such that

$$a_{j1}x_1 + a_{j2}x_2 + a_{j3}x_3 = 0$$

$$= a_{k1}x_1 + a_{k2}x_2 + a_{k3}x_3 ,$$

where $j \ne k$, $j,k = 1,2,3$; i.e., there is no non-trivial solution.

● **PROBLEM** 25-8

Solve the following system of equations:

$$x + 3y = 0$$
$$2x + 6y + 4z = 0 \qquad (1)$$

Solution: To solve the given system of equations, first form the matrix of the coefficients. Then reduce this matrix to echelon form. The matrix of the coefficients is

$$\begin{bmatrix} 1 & 3 & 0 \\ 2 & 6 & 4 \end{bmatrix} .$$

Now add -2 times the first row to the second row

$$\begin{bmatrix} 1 & 3 & 0 \\ 0 & 0 & 4 \end{bmatrix} .$$

Divide the second row by 4

$$\begin{bmatrix} 1 & 3 & 0 \\ 0 & 0 & 1 \end{bmatrix} .$$

The above is the matrix of coefficients for
$$x + 3y = 0$$
$$z = 0 \ . \tag{2}$$

This system is easy to solve. We have $z = 0$ and can assign y any value. Then compute x from (2). This gives a solution to (1).

Solve the following homogeneous system of linear equations.
$$2x_1 + 2x_2 - x_3 + x_5 = 0$$
$$-x_1 - x_2 + 2x_3 - 3x_4 + x_5 = 0$$
$$x_1 + x_3 - 2x_3 - x_5 = 0 \tag{1}$$
$$x_3 + x_4 + x_5 = 0$$

Solution: The system (1) has five unknowns but only four equations. We know that a homogeneous system of linear equations with more unknowns than equations has a non-zero (non-trivial) solution. Now, to solve the system (1), form the matrix of the coefficients. Then reduce this matrix to reduced row-echelon form. The coefficient matrix is

$$A = \begin{bmatrix} 2 & 2 & -1 & 0 & 1 \\ -1 & -1 & 2 & -3 & 1 \\ 1 & 1 & -2 & 0 & -1 \\ 0 & 0 & 1 & 1 & 1 \end{bmatrix} \ . \tag{1}$$

Add the fourth row to the first row and the third row to the second row.

$$\begin{bmatrix} 2 & 2 & 0 & 1 & 2 \\ 0 & 0 & 0 & -3 & 0 \\ 1 & 1 & -2 & 0 & -1 \\ 0 & 0 & 1 & 1 & 1 \end{bmatrix} \ . \tag{2}$$

Divide the second row by -3. Then add -1 times the second row to the first row and to the fourth row.

$$\begin{bmatrix} 2 & 2 & 0 & 0 & 2 \\ 0 & 0 & 0 & 1 & 0 \\ 1 & 1 & -2 & 0 & -1 \\ 0 & 0 & 1 & 0 & 1 \end{bmatrix} \tag{3}$$

Divide the first row by 2. Then add -1 times the first row to the third row.

$$\begin{bmatrix} 1 & 1 & 0 & 0 & 1 \\ 0 & 0 & 0 & 1 & 0 \\ 0 & 0 & -2 & 0 & -2 \\ 0 & 0 & 1 & 0 & 1 \end{bmatrix} \ . \tag{4}$$

Add 2 times the fourth row to the third row

$$\begin{bmatrix} 1 & 1 & 0 & 0 & 1 \\ 0 & 0 & 0 & 1 & 0 \\ 0 & 0 & 0 & 0 & 0 \\ 0 & 0 & 1 & 0 & 1 \end{bmatrix}$$ (5)

Interchange the second and fourth rows. Next, interchange the third and fourth rows.

$$\begin{bmatrix} 1 & 1 & 0 & 0 & 1 \\ 0 & 0 & 1 & 0 & 1 \\ 0 & 0 & 0 & 1 & 0 \\ 0 & 0 & 0 & 0 & 0 \end{bmatrix}$$ (6)

This matrix is in row reduced echelon form. The corresponding system of equations is

$$x_1 + x_2 \qquad + x_5 = 0$$
$$x_3 + x_5 = 0$$
$$x_4 = 0 .$$

Solving for the leading variables yields

$$x_1 = -x_2 - x_5$$
$$x_3 = -x_5$$
$$x_4 = 0 .$$

The solution set is, therefore, given by
$$x_1 = -s - t, \ x_2 = s, \ x_3 = -t, \ x_4 = 0, \ x_5 = t.$$

That is, we have chosen x_2 and x_5 to be free variables and, hence, set them equal to the parameters s and t, respectively. The dependent variables are then x_1 and x_3, and their dependence on s and t are given by the reduced form of the system. Thus, any solution vector is of the form $(-s-t, s, -t, 0, t)$. Recall that a basis for the subspace of vectors of this form (i.e., the solution space) would be the vectors

(1) $(-0-1,0,-1,0,1) = (-1,0,-1,0,1)$

(2) $(-1-0,1,-0,0,0) = (-1,1,0,0,0,)$

● **PROBLEM 25-10**

[A] Show that each of the following systems has a non-zero solution:

   (a) $x + 2y - 3z + w = 0$
              $x - 3y + z - 2w = 0$ (1)
           $2x + y - 3z + 5w = 0$

   (b) $x + y - z = 0$
           $2x - 3y + z = 0$ (2)
           $x - 4y + 2z = 0$

[B] Show that following system has a unique solution:

       $x + y - z = 0$
       $2x + 4y - z = 0$ (3)
       $3x + 2y + 2z = 0 .$

Solution: [A] For a homogeneous system of linear equations, exactly one of the following is true.
(i) The system has only the trivial solution.
(ii) The system has infinitely many non-trivial solutions.

Any homogeneous linear system that has fewer linearly independent equations than unknowns always has a non-zero solution.

(a) Now, system (1) has a non-zero solution since there are four unknowns but only three equations.

(b) The coefficient matrix for the system (2) is

$$\begin{bmatrix} 1 & 1 & -1 \\ 2 & -3 & 1 \\ 1 & -4 & 2 \end{bmatrix} .$$

Add -2 times the first row to the second row and -1 times the first row to the third row

$$\begin{bmatrix} 1 & 1 & -1 \\ 0 & -5 & 3 \\ 0 & -5 & 3 \end{bmatrix} .$$

Add -1 times the second row to the third row

$$\begin{bmatrix} 1 & 1 & -1 \\ 0 & -5 & 3 \\ 0 & 0 & 0 \end{bmatrix} .$$

This is the matrix of coefficients of the system

$$\begin{aligned} x + y - z &= 0 \\ - 5y + 3z &= 0 \\ 0 &= 0 \end{aligned} .$$

The system has a non-zero solution since we obtained only two equations in the three unknowns when we reduced the system to echelon form. For example, let z = 5; then y = 3 and x = 2 solves the system.

[B] The matrix of coefficients for system (3) is

$$\begin{bmatrix} 1 & 1 & -1 \\ 2 & 4 & -1 \\ 3 & 2 & 2 \end{bmatrix} .$$

Reduce it to echelon form. Add -2 times the first row to the second row and -3 times the first row to the third row

$$\begin{bmatrix} 1 & 1 & -1 \\ 0 & 2 & 1 \\ 0 & -1 & 5 \end{bmatrix} .$$

Multiply the third row by 2. Then, add the second row to the third row

$$\begin{bmatrix} 1 & 1 & -1 \\ 0 & 2 & 1 \\ 0 & 0 & 11 \end{bmatrix} .$$

The corresponding system of equations is

$$\begin{aligned} x + y - z &= 0 \\ 2y + z &= 0 \\ 11z &= 0 \end{aligned} .$$

Solving this system yields $x = y = z = 0$. In general, a system of homogeneous equations in n unknowns has a zero solution if the corresponding reduced matrix has exactly n rows with non-zero entries.

● **PROBLEM 25-11**

By forming the augmented matrix and row reducing, determine the solutions of the following system

$$
\begin{aligned}
2x - y + 3z &= 4 \\
3x \quad\;\; + 2z &= 5 \\
-2x + y + 4z &= 6 .
\end{aligned}
\tag{1}
$$

Solution: The system of equations

$$
\begin{aligned}
a_{11}x_1 + a_{12}x_2 + \cdots + a_{1n}x_n &= c_1 \\
a_{21}x_1 + a_{22}x_2 + \cdots + a_{2n}x_n &= c_2 \\
&\;\;\vdots \\
a_{m1}x_1 + a_{m2}x_2 + \cdots + a_{mn}x_n &= c_m
\end{aligned}
\tag{2}
$$

is called a non-homogeneous linear system if the constants $c_1, c_2, \ldots, c_m$ are not all zero.

We form the $m \times (n+1)$ matrix A' defined by

$$
A' = \begin{bmatrix}
a_{11} & a_{12} & \cdots & a_{1n} & c_1 \\
a_{21} & a_{22} & \cdots & a_{2n} & c_2 \\
\vdots & & & & \vdots \\
a_{m1} & a_{m2} & \cdots & a_{mn} & c_m
\end{bmatrix} .
$$

This matrix is called the augmented matrix of the system (2). The first n columns of A' consist of the coefficient matrix of (2), and the last column of A' consists of the corresponding constants.

To solve the non-homogeneous linear system, form the augmented matrix A'. Apply row operations to A' to reduce it to echelon form. Now, the augmented matrix of the system (1) is

$$
\begin{bmatrix}
2 & -1 & 3 & 4 \\
3 & 0 & 2 & 5 \\
-2 & 1 & 4 & 6
\end{bmatrix} .
$$

Add the first row to the third row

$$
\begin{bmatrix}
2 & -1 & 3 & 4 \\
3 & 0 & 2 & 5 \\
0 & 0 & 7 & 10
\end{bmatrix} .
$$

This is the augmented matrix of

$$
\begin{aligned}
2x - y + 3z &= 4 \\
3x \quad\;\; + 2z &= 5 \\
7z &= 10 .
\end{aligned}
$$

503

The system has been sufficiently simplified now so that the solution can be found.

From the last equation we have z = 10/7. Substituting this value into the second equation and solving for x gives x = 5/7. Substituting x = 5/7 and z = 10/7 into the first equation and solving for y yields y = 12/7 . The solution to system (1) is, therefore,

$$x = 5/7 \ , \ y = 12/7 \ , \ z = 10/7 \ .$$

Note: We could have further reduced the matrix to row-reduced echelon form and solved the system directly from the reduced matrix. That is, by adding -2/3 times the second row to the first row, we have

$$\begin{bmatrix} 0 & -1 & 5/3 & \vdots & 2/3 \\ 3 & 0 & 2 & \vdots & 5 \\ 0 & 0 & 7 & \vdots & 10 \end{bmatrix} .$$

Multiplying the second row by 1/3 and the first row by -1 and interchanging the two, then multiplying the third row by 1/7 results in

$$\begin{bmatrix} 1 & 0 & 2/3 & 5/3 \\ 0 & 1 & -5/3 & -2/3 \\ 0 & 0 & 1 & 10/7 \end{bmatrix} .$$

Then, adding 5/3 times the third row to the second and -2/3 times the third row to the first gives

$$\begin{bmatrix} 1 & 0 & 0 & 5/7 \\ 0 & 1 & 0 & 12/7 \\ 0 & 0 & 1 & 10/7 \end{bmatrix} .$$

The solution to a non-homogeneous system found in the above manner is called the particular solution. The non-homogeneous system will be satisfied by any sum of the particular solution and a solution to the corresponding homogeneous system. In this case, the only solution to the homogeneous system is the trivial solution. Therefore, the only solution to the non-homogeneous problem is the particular solution.

● **PROBLEM 25-12**

Show that the following non-homogeneous system of linear equations has no solution

$$\begin{array}{rcrcrcl} x & + & 2y & - & 3z & = & -1 \\ 3x & - & y & + & 2z & = & 7 \\ 5x & + & 3y & - & 4z & = & 2 \end{array} \qquad (1)$$

Solution: The system (1) has no solution if the echelon form of the augmented matrix has a row whose first non-zero entry k appears in the last column. This corresponds in equation form to the statement 0 = k, which shows that all the equations of the system cannot be satisfied simultaneously. The augmented matrix for the system (1) is

$$\begin{bmatrix} 1 & 2 & -3 & \vdots & -1 \\ 3 & -1 & 2 & \vdots & 7 \\ 5 & 3 & -4 & \vdots & 2 \end{bmatrix}$$

Now, apply row operations to this matrix to reduce it to echelon form.

Add -3 times the first row to the second row and -5 times the first row to the third row

$$\begin{bmatrix} 1 & 2 & -3 & | & -1 \\ 0 & -7 & 11 & | & 10 \\ 0 & -7 & 11 & | & 7 \end{bmatrix} .$$

Add -1 times the second row to the third row.

$$\begin{bmatrix} 1 & 2 & -3 & | & -1 \\ 0 & -7 & 11 & | & 10 \\ 0 & 0 & 0 & | & -3 \end{bmatrix} .$$

Thus, this matrix has a row in which the first non-zero entry appears in the last column. Therefore, the system (1) has no solution.

● **PROBLEM 25-13**

Solve the following system by Gauss-Jordan elimination

$$\begin{aligned} x_1 + 3x_2 - 2x_3 + 2x_5 &= 0 \\ 2x_1 + 6x_2 - 5x_3 - 2x_4 + 4x_5 - 3x_6 &= -1 \\ 5x_3 + 10x_4 + 15x_6 &= 5 \\ 2x_1 + 6x_2 + 8x_4 + 4x_5 + 18x_6 &= 6 . \end{aligned}$$

Solution: The augmented matrix for the system is

$$\begin{bmatrix} 1 & 3 & -2 & 0 & 2 & 0 & | & 0 \\ 2 & 6 & -5 & -2 & 4 & -3 & | & -1 \\ 0 & 0 & 5 & 10 & 0 & 15 & | & 5 \\ 2 & 6 & 0 & 8 & 4 & 18 & | & 6 \end{bmatrix} .$$

Reduce this matrix to row-reduced echelon form. Add -2 times the first row to the second and fourth rows

$$\begin{bmatrix} 1 & 3 & -2 & 0 & 2 & 0 & | & 0 \\ 0 & 0 & -1 & -2 & 0 & -3 & | & -1 \\ 0 & 0 & 5 & 10 & 0 & 15 & | & 5 \\ 0 & 0 & 4 & 8 & 0 & 18 & | & 6 \end{bmatrix} .$$

Add 5 times the second row to the third row and 4 times the second row to the fourth row.

$$\begin{bmatrix} 1 & 3 & -2 & 0 & 2 & 0 & | & 0 \\ 0 & 0 & -1 & -2 & 0 & -3 & | & -1 \\ 0 & 0 & 0 & 0 & 0 & 0 & | & 0 \\ 0 & 0 & 0 & 0 & 0 & 6 & | & 2 \end{bmatrix} .$$

Multiply the second row by -1 and the fourth row by 1/6. Then, interchange the third and fourth rows.

$$\begin{bmatrix} 1 & 3 & -2 & 0 & 2 & 0 & | & 0 \\ 0 & 0 & 1 & 2 & 0 & 3 & | & 1 \\ 0 & 0 & 0 & 0 & 0 & 1 & | & 1/3 \\ 0 & 0 & 0 & 0 & 0 & 0 & | & 0 \end{bmatrix}.$$

Add -3 times the third row to the second row. Then add 2 times the second row to the first row.

$$\begin{bmatrix} 1 & 3 & 0 & 4 & 2 & 0 & | & 0 \\ 0 & 0 & 1 & 2 & 0 & 0 & | & 0 \\ 0 & 0 & 0 & 0 & 0 & 1 & | & 1/3 \\ 0 & 0 & 0 & 0 & 0 & 0 & | & 0 \end{bmatrix}.$$

Now the corresponding system of equations is

$$x_1 + 3x_2 + 4x_4 + 2x_5 = 0$$
$$x_3 + 2x_4 = 0$$
$$x_6 = 1/3 .$$

Then, solving for the leading variables results in

$$x_1 = -3x_2 - 4x_4 - 2x_5$$
$$x_3 = -2x_4$$
$$x_6 = 1/3 .$$

If we assign x_2, x_4 and x_5 the arbitrary values r,s, and t, respectively, the solution set is given by the formulas,

$$x_1 = -3r - 4s - 2t, \quad x_2 = r, \quad x_3 = -2s, \quad x_4 = s$$
$$x_5 = t, \quad x_6 = 1/3 .$$

● PROBLEM 25-14

Show that the following system has more than one solution.

$$\begin{array}{rcrcrcl} 3x & - & y & + & 7z & = & 0 \\ 2x & - & y & + & 4z & = & \frac{1}{2} \\ x & - & y & + & z & = & 1 \\ 6x & - & 4y & + & 10z & = & 3 . \end{array} \qquad (1)$$

Solution: The augmented matrix for the system (1) is

$$\begin{bmatrix} 3 & -1 & 7 & | & 0 \\ 2 & -1 & 4 & | & \frac{1}{2} \\ 1 & -1 & 1 & | & 1 \\ 6 & -4 & 10 & | & 3 \end{bmatrix}.$$

Reduce it to row reduced echelon form. Add -3 times the third row to the first row, add -2 times the third row to the second row and add -6 times the third row to the fourth row

$$\begin{bmatrix} 0 & 2 & 4 & -3 \\ 0 & 1 & 2 & -3/2 \\ 1 & -1 & 1 & 1 \\ 0 & 2 & 4 & -3 \end{bmatrix}.$$

Add -1 times the first row to the fourth row; add the second row to the third row

$$\begin{bmatrix} 0 & 2 & 4 & -3 \\ 0 & 1 & 2 & -3/2 \\ 1 & 0 & 3 & -1/2 \\ 0 & 0 & 0 & 0 \end{bmatrix}.$$

Add -2 times the second row to the first row

$$\begin{bmatrix} 0 & 0 & 0 & 0 \\ 0 & 1 & 2 & -3/2 \\ 1 & 0 & 3 & -1/2 \\ 0 & 0 & 0 & 0 \end{bmatrix}.$$

Interchanging row one and row three gives

$$\begin{bmatrix} 1 & 0 & 3 & -1/2 \\ 0 & 1 & 2 & -3/2 \\ 0 & 0 & 0 & 0 \\ 0 & 0 & 0 & 0 \end{bmatrix}.$$

This is the augmented matrix for the system

$$\begin{aligned} x + 3z &= -\tfrac{1}{2} \\ y + 2z &= -3/2 \\ 0 &= 0 \\ 0 &= 0 . \end{aligned}$$

We can write the above system as

$$\begin{aligned} x &= -3z - \tfrac{1}{2} \\ y &= -2z - 3/2 . \end{aligned}$$

Since z can be assigned any arbitrary value, there are infinitely many solutions, one for each value of z. Thus, the system (1) has more than one solution.

In general, let R be a row reduced echelon form of the augmented matrix of the given system. Let r be the number of non-zero rows of this R and n be the number of unknowns of the system. If n > r, then the system has more than one particular solution.

When n = r, the system may or may not have a solution, and there cannot be more than one particular solution. That is, if there is no row whose only non-zero entry is in the last column, then the particular solution exists and is unique.

Suppose that the augmented matrix for a system of linear equations has been reduced by row operations to the given reduced row echelon form. Solve the system.

(a) $\begin{bmatrix} 1 & 0 & 0 & | & 5 \\ 0 & 1 & 0 & | & -2 \\ 0 & 0 & 1 & | & 4 \end{bmatrix}$ (b) $\begin{bmatrix} 1 & 0 & 0 & 4 & | & -1 \\ 0 & 1 & 0 & 2 & | & 6 \\ 0 & 0 & 1 & 3 & | & 2 \end{bmatrix}$

(c) $\begin{bmatrix} 1 & 6 & 0 & 0 & 4 & | & -2 \\ 0 & 0 & 1 & 0 & 3 & | & 1 \\ 0 & 0 & 0 & 1 & 5 & | & 2 \\ 0 & 0 & 0 & 0 & 0 & | & 0 \end{bmatrix}$ (d) $\begin{bmatrix} 1 & 0 & 0 & | & 0 \\ 0 & 1 & 0 & | & 0 \\ 0 & 0 & 0 & | & 1 \end{bmatrix}$

Solution: (a) If the number of non-zero rows, r, of the reduced matrix is equal to the number of unknowns, n, and there is no inconsistent equation   0 = k, then the system has a unique particular solution.

(a) The corresponding system of equations is

$$x_1 \qquad\qquad = 5$$
$$\qquad x_2 \qquad = -2$$
$$\qquad\qquad x_3 \quad = \quad 4 \ .$$

Therefore, the solution to the system is

$$x_1 = 5, \ x_2 = -2 \ \text{and} \ x_3 = 4 \ .$$

(b) Since n > r, we have more than one solution. The corresponding system of equations is

$$x_1 \qquad\qquad + 4x_4 = -1$$
$$\qquad x_2 \qquad + 2x_4 = 6$$
$$\qquad\qquad x_3 + 3x_4 = 2 \ .$$

The above system can be written as

$$x_1 = -1 - 4x_4$$
$$x_2 = 6 - 2x_4 \qquad\qquad (1)$$
$$x_3 = 2 - 3x_4 \ .$$

Assign x_4 any value, and then compute x_1, x_2 and x_3 from (1). Thus, we have many solutions, one for each value of x_4 .

(c)   n > r; therefore, the system has many solutions. The corresponding system of equations is

$$x_1 + 6x_2 \qquad\qquad + 4x_5 = -2$$
$$\qquad\qquad x_3 \qquad + 3x_5 = 1$$
$$\qquad\qquad\qquad x_4 + 5x_5 = 2 \ .$$

The above equations can be written as

$$x_1 = -2 - 4x_5 - 6x_2$$
$$x_3 = 1 - 3x_5$$
$$x_4 = 2 - 5x_5 .$$

Since x_5 can be assigned an arbitrary value, t, and x_2 can be assigned an arbitrary value, s, there are infinitely many solutions. The solution set is given by the formula

$$x_1 = -2 - 4t - 6s, \quad x_2 = s, \quad x_3 = 1 - 3t ,$$
$$x_4 = 2 - 5t, \quad x_5 = t.$$

(d) This system has no solution since the row-reduced echelon form has a row in which the first non-zero entry is in the last column.

Find the necessary and sufficient conditions for the existence of a solution to the following system.

$$
\begin{aligned}
x + y + 2z &= a_1 \\
-2x \qquad - z &= a_2 \qquad\qquad (1)\\
x + 3y + 5z &= a_3 .
\end{aligned}
$$

Solution: The procedure for solving nonhomogeneous systems is to form the augmented matrix of the system and reduce this matrix to a row-echelon matrix. Suppose, during this reduction, we obtain a matrix in which there is a row where the first non-zero entry appears in the last column. This matrix is the augmented matrix of a system with no solution.

If we obtain a reduced matrix with the property that no row has its first non-zero entry in the last column then the system has a solution. Now, the augmented matrix of the system (1) is

$$
\left[
\begin{array}{ccc|c}
1 & 1 & 2 & a_1 \\
-2 & 0 & -1 & a_2 \\
1 & 3 & 5 & a_3
\end{array}
\right] .
$$

Add 2 times the first row to the second row and -1 times the first row to the third row

$$
\left[
\begin{array}{ccc|c}
1 & 1 & 2 & a_1 \\
0 & 2 & 3 & a_2+2a_1 \\
0 & 2 & 3 & a_3-a_1
\end{array}
\right] .
$$

Add -1 times the second row to the third row

$$
\left[
\begin{array}{ccc|c}
1 & 1 & 2 & a_1 \\
0 & 2 & 3 & a_2+2a_1 \\
0 & 0 & 0 & a_3-a_2-3a_1
\end{array}
\right] .
$$

Suppose $a_3 - a_2 - 3a_1 \neq 0$. Then the reduced matrix has a row in which the first non-zero entry appears in the last column. Thus, the system has no solution. Therefore, a necessary condition for a solution to exist is that $a_3 - a_2 - 3a_1$ must be equal to zero.

Let $a_3 - a_2 - 3a_1 = 0$. Then, we have

$$\begin{bmatrix} 1 & 1 & 2 & | & a_1 \\ 0 & 2 & 3 & | & a_2+2a_1 \\ 0 & 0 & 0 & | & 0 \end{bmatrix}$$

which is the augmented matrix for the system

$$\begin{aligned} x + y + 2z &= a_1 \\ 2y + 3z &= a_2 + 2a_1 \end{aligned} .$$

The above system can be written as

$$\begin{aligned} x + \tfrac{1}{2}z &= -a_2/2 \\ y + 3/2z &= a_1 + a_2/2 \end{aligned} .$$

Assign to z any arbitrary value, then compute x and y from the above equations to obtain a solution to system (1). Thus, we can conclude that $a_3 - a_2 - 3a_1 = 0$ is a necessary and sufficient condition for the existence of a solution to a system (1).

● **PROBLEM** 25-17

Determine the values of a so that the following system of equations has: (a) no solution, (b) more than one solution, (c) a unique solution.

$$\begin{aligned} x + y - z &= 1 \\ 2x + 3y + az &= 3 \\ x + ay + 3z &= 2 \end{aligned}$$

Solution: First form the augmented matrix for the system and then reduce it to echelon form.

The augmented matrix is

$$\begin{bmatrix} 1 & 1 & -1 & | & 1 \\ 2 & 3 & a & | & 3 \\ 1 & a & 3 & | & 2 \end{bmatrix} .$$

Add -2 times the first row to the second row and -1 times the first row to the third row.

$$\begin{bmatrix} 1 & 1 & -1 & | & 1 \\ 0 & 1 & a+2 & | & 1 \\ 0 & a-1 & 4 & | & 1 \end{bmatrix} .$$

Add $-(a-1)$ times the second row to the third row

$$\begin{bmatrix} 1 & 1 & -1 & | & 1 \\ 0 & 1 & a+2 & | & 1 \\ 0 & 0 & 4-(a+2)(a-1) & | & 1-(a-1) \end{bmatrix} .$$

The above matrix can be written as

$$\begin{bmatrix} 1 & 1 & -1 & \vdots & 1 \\ 0 & 1 & a+2 & \vdots & 1 \\ 0 & 0 & (3+a)(2-a) & \vdots & (2-a) \end{bmatrix} \qquad (1)$$

since $4-(a+2)(a-1) = 4-(a^2+a-2) = 6 - a - a^2 = (3+a)(2-a)$. Suppose $(3+a) = 0$. Then $(2-a) \neq 0$, and we have a reduced matrix with the property that a row has its first non-zero entry in the last column. In this case, the system has no solution. Thus, the system has a solution only if $(3+a) \neq 0$, that is, if $a \neq -3$. If $a = 3$, then the last row of the matrix (1) becomes $[0\ 0\ 0\ 5]$, and the system has no solution. Suppose $a = -2$. Then the last row of the matrix (1) is $[0\ 0\ 0\ 0]$.

In this case, the system has an infinite number of solutions. We can summarize our results as follows: If
(a) $a = -3$, then the system has no solution.
(b) if $a = 2$, the system has more than one solution.
(c) if $a \neq 2$ and $a \neq -3$, the system has a unique solution.

● **PROBLEM 25-18**

Solve the following system

$$\begin{aligned} x_1 - 2x_2 - 3x_3 &= 3 \\ 2x_1 - x_2 - 4x_3 &= 7 \\ 3x_1 - 3x_2 - 5x_3 &= 8 \ . \end{aligned} \qquad (1)$$

<u>Solution</u>: The matrix of coefficients for the system (1) is

$$A = \begin{bmatrix} 1 & -2 & -3 \\ 2 & -1 & -4 \\ 3 & -3 & -5 \end{bmatrix} \ .$$

The system (1) may be written in matrix form as

$$\begin{bmatrix} 1 & -2 & -3 \\ 2 & -1 & -4 \\ 3 & -3 & -5 \end{bmatrix} \begin{bmatrix} x_1 \\ x_2 \\ x_3 \end{bmatrix} = \begin{bmatrix} 3 \\ 7 \\ 8 \end{bmatrix} \ . \qquad (2)$$

Let

$$X = \begin{bmatrix} x_1 \\ x_2 \\ x_3 \end{bmatrix}, \qquad b = \begin{bmatrix} 3 \\ 7 \\ 8 \end{bmatrix} \ .$$

Then equation (2) is written

$$\vec{AX} = \vec{b} \ . \qquad (3)$$

A solution vector $\vec{x}$ can be found by multiplying both sides of equation (3) by A^{-1}. Then we have

$$A^{-1}AX = A^{-1}b \ ,$$

but $A^{-1}A = I$. Hence,

$$IX = A^{-1}b$$

or

$$X = A^{-1}b \ . \qquad (4)$$

511

Thus, the solutions of a system of linear equations can be obtained by finding the inverse matrix of the coefficient matrix of the system and then solving equation (4). To find A^{-1}, first form the matrix $[A : I]$, and reduce this matrix, by applying row operations, to the form $[I : B]$. Then, $B = A^{-1}$. Now,

$$[A : I] = \begin{bmatrix} 1 & -2 & -3 & | & 1 & 0 & 0 \\ 2 & -1 & -4 & | & 0 & 1 & 0 \\ 3 & -3 & -5 & | & 0 & 0 & 1 \end{bmatrix} .$$

Add -2 times the first row to the second row and -3 times the first row to the third row

$$\begin{bmatrix} 1 & -2 & -3 & | & 1 & 0 & 0 \\ 0 & 3 & 2 & | & -2 & 1 & 0 \\ 0 & 3 & 4 & | & -3 & 0 & 1 \end{bmatrix} .$$

Now add -1 times the second row to the third row

$$\begin{bmatrix} 1 & -2 & -3 & | & 1 & 0 & 0 \\ 0 & 3 & 2 & | & -2 & 1 & 0 \\ 0 & 0 & 2 & | & -1 & -1 & 0 \end{bmatrix} .$$

Divide the third row by 2

$$\begin{bmatrix} 1 & -2 & -3 & | & 1 & 0 & 0 \\ 0 & 3 & 2 & | & -2 & 1 & 0 \\ 0 & 0 & 1 & | & -\frac{1}{2} & -\frac{1}{2} & \frac{1}{2} \end{bmatrix} .$$

Add -2 times the third row to the second row and 3 times the third row to the first row

$$\begin{bmatrix} 1 & -2 & 0 & | & -\frac{1}{2} & -3/2 & 3/2 \\ 0 & 3 & 0 & | & -1 & 2 & -1 \\ 0 & 0 & 1 & | & -\frac{1}{2} & -\frac{1}{2} & \frac{1}{2} \end{bmatrix} .$$

Divide the second row by 3; then add 2 times the resulting second row to the first row.

$$\begin{bmatrix} 1 & 0 & 0 & | & -7/6 & -1/6 & 5/6 \\ 0 & 1 & 0 & | & -1/3 & 2/3 & -1/3 \\ 0 & 0 & 1 & | & -\frac{1}{2} & -\frac{1}{2} & \frac{1}{2} \end{bmatrix} .$$

Thus,

$$A^{-1} = \begin{bmatrix} -7/6 & -1/6 & 5/6 \\ -1/3 & 2/3 & -1/3 \\ -\frac{1}{2} & -\frac{1}{2} & \frac{1}{2} \end{bmatrix} .$$

Then equation (4) becomes

$$\begin{bmatrix} x_1 \\ x_2 \\ x_3 \end{bmatrix} = \begin{bmatrix} -7/6 & -1/6 & 5/6 \\ -1/3 & 2/3 & -1/3 \\ -\frac{1}{2} & -\frac{1}{2} & \frac{1}{2} \end{bmatrix} \begin{bmatrix} 3 \\ 7 \\ 8 \end{bmatrix} .$$

Multiplying, we have

$$
\begin{bmatrix} x_1 \\ x_2 \\ x_3 \end{bmatrix} = \begin{bmatrix} -21/6 - 7/6 + 40/6 \\ -1 + 14/3 - 8/3 \\ -3/2 - 7/2 + 4 \end{bmatrix} = \begin{bmatrix} 2 \\ 1 \\ -1 \end{bmatrix} .
$$

Thus,

$$
x_1 = 2, \ x_2 = 1, \ x_3 = -1 .
$$

It is interesting to note that the calculation of the inverse matrix is closely related to the solution of simultaneous equations. Indeed, the two processes are essentially the same. To show this, we can use the method of successive elimination. Eliminating x_1 from the second and third equations,

$$
\begin{aligned}
x_1 - 2x_2 - 3x_3 &= 3 \\
3x_2 + 2x_3 &= 1 \\
3x_2 + 4x_3 &= -1 .
\end{aligned}
$$

Then, eliminating x_2 from the third equation,

$$
\begin{aligned}
x_1 - 2x_2 - 3x_3 &= 3 \\
3x_2 + 2x_3 &= 1 \\
2x_3 &= -2 .
\end{aligned}
$$

We obtain

$$
\begin{aligned}
x_1 - 2x_2 - 3x_3 &= 3 \\
3x_2 + 2x_3 &= 1 \\
x_3 &= -1 ,
\end{aligned}
$$

$$
\begin{aligned}
x_1 - 2x_2 &= 0 \\
3x_2 &= 3 \\
x_3 &= -1
\end{aligned}
$$

and, finally,

$$
\begin{aligned}
x_1 &= 2 \\
x_2 &= 1 \\
x_3 &= -1 .
\end{aligned}
$$

It can be seen that the solution to the system $A\vec{x} = \vec{b}$, calculated directly, is the same solution we obtained by calculating A^{-1} and finding $A^{-1}\vec{b}$. Observe that the inverse matrix can also be found by using the formula

$$
A^{-1} = \frac{1}{\det A} \operatorname{adj} A .
$$

Consider the following nonhomogeneous system of linear equations.

$$2x + y - 3z = 1 \qquad (1)$$
$$3x + 2y - 2z = 2$$
$$x + y + z = 1 \ .$$

Show that (i) any two solutions to the system (1) differ by a vector which is a solution to the homogeneous system

$$2x + y - 3z = 0$$
$$3x + 2y - 2z = 0 \qquad (2)$$
$$x + y + z = 0 \ .$$

(ii) the sum of a solution to (1) and a solution to (2) gives a solution to (1).

Solution: Consider a system of linear equations

$$a_{11}x_1 + a_{12}x_2 + \ldots + a_{1n}x_n = b_1$$
$$a_{21}x_1 + a_{22}x_2 + \ldots + a_{2n}x_n = b_2 \qquad (3)$$
$$\cdots \cdots \cdots \cdots \cdots \cdots$$
$$a_{m1}x_1 + a_{m2}x_2 + \ldots + a_{mn}x_n = b_m \ .$$

We say that a vector $v = (c_1, c_2, \ldots, c_n)$ of R^n is a solution vector of (3) if $x_1 = c_1$, $x_2 = c_2, \ldots, x_n = c_n$ is a solution of (3). Now the augmented matrix for system (1) is

$$\begin{bmatrix} 2 & 1 & -3 & | & 1 \\ 3 & 2 & -2 & | & 2 \\ 1 & 1 & 1 & | & 1 \end{bmatrix} \ .$$

Reduction of the above matrix gives

$$\begin{bmatrix} 1 & 0 & -4 & | & 0 \\ 0 & 1 & 5 & | & 1 \\ 0 & 0 & 0 & | & 0 \end{bmatrix} \ .$$

The corresponding system of equations is

$$x - 4z = 0$$
$$y + 5z = 1 \ .$$

There is more than one solution. Let $z = 1$; then $y = -4$ and $x = 4$. Let $z = 0$; then $y = 1$ and $x = 0$. Thus, $(4, -4, 1)$ and $(0, 1, 0)$ are vector solutions to system (1). The coefficient matrix of the system (2) is

$$\begin{bmatrix} 2 & 1 & -3 \\ 3 & 2 & -2 \\ 1 & 1 & 1 \end{bmatrix} \ .$$

Reduction of this matrix gives

$$\begin{bmatrix} 1 & 0 & -4 \\ 0 & 1 & 5 \\ 0 & 0 & 0 \end{bmatrix}$$

The corresponding system of equations is

$$x - 4z = 0$$
$$y + 5z = 0 .$$

Let $z = 1$; then $y = -5$ and $x = 4$. Now the difference of two solutions of system (1), $(4,-4,1) - (0,1,0) = (4,-5,1)$, is a solution to the homogeneous system (1). Add $(4,-4,1)$ to $(4,-5,1)$. This gives $(8,-9,2)$ which is also a solution to system (1).

(It is easy to check that $(8,-9,2)$ satisfies $x - 4z = 0$ and $y + 5z = 1$.)

Generalize these results as follows: To solve the nonhomogeneous system, first find one solution to this system and all solutions to its associated homogeneous system. Then any other solution to the given nonhomogeneous system is obtained by adding the particular solution of the nonhomogeneous system to the general solution of the homogeneous system.

● PROBLEM 25-20

If the method of Gauss elimination corresponds in its final form to an echelon matrix, what is the matrix analogue of the Gauss-Jordan method for solving linear systems of equations? Explain by example.

Solution: Consider the equations:

$$x_1 + x_2 + 2x_3 = 9$$
$$2x_1 + 4x_2 - 3x_3 = 1 \tag{1}$$
$$3x_1 + 6x_2 - 5x_3 = 0 .$$

The augmented coefficient matrix of (1) is:

$$\begin{bmatrix} 1 & 1 & -2 & | & 9 \\ 2 & 4 & -3 & | & 1 \\ 3 & 6 & -5 & | & 0 \end{bmatrix} \tag{2}$$

Recall that in the method of Gauss elimination we successively eliminate unknowns from successive equations. Thus, for example, we can use the first equation of system (1) to eliminate x_1 from the remaining two equations. This amounts to applying elementary row operations on (2) to reduce the second and third elements of the first column to zero. This gives

$$\begin{bmatrix} 1 & 1 & 2 & | & 9 \\ 0 & 2 & -7 & | & -17 \\ 0 & 3 & -11 & | & -27 \end{bmatrix} \tag{3}$$

In equation form this is:

$$x_1 + x_2 + 2x_3 = 9$$
$$2x_2 - 7x_3 = -17 \tag{4}$$
$$3x_2 - 11x_3 = -27 .$$

515

According to the method of Gauss, we would now proceed to use the second equation of (4) to eliminate x_2 from the third equation. But, by Gauss-Jordan, we use the second equation of (4) to eliminate x_2 from the other two equations. Multiply the second row of (3) by .5 to obtain

$$\begin{bmatrix} 1 & 1 & 2 & \vdots & 9 \\ 0 & 1 & -3.5 & \vdots & -8.5 \\ 0 & 3 & -11 & \vdots & -27 \end{bmatrix} \quad . \tag{5}$$

Now use elementary row operations on (5) to reduce the first and third elements of the second column to zero. This results in

$$\begin{bmatrix} 1 & 0 & 1.5 & \vdots & 17.5 \\ 0 & 1 & -3.5 & \vdots & -8.5 \\ 0 & 0 & -.5 & \vdots & -1.5 \end{bmatrix} \quad . \tag{6}$$

In equation form this is:

$$\begin{aligned} x_1 + 1.5x_3 &= 17.5 \\ x_2 - 3.5x_3 &= -8.5 \\ - .5x_3 &= -1.5 \quad . \end{aligned} \tag{7}$$

The final step of the Gauss-Jordan method is elimination of x_3 from both the first and second equations of (7). We obtain:

$$\begin{aligned} x_1 &= 1 \\ x_2 &= 2 \\ x_3 &= 3 \end{aligned}$$

The coefficient matrix of this system is:

$$\begin{bmatrix} 1 & 0 & 0 & \vdots & 1 \\ 0 & 1 & 0 & \vdots & 2 \\ 0 & 0 & 1 & \vdots & 3 \end{bmatrix} \quad .$$

Thus, it can be seen that the Gauss-Jordan method corresponds to applying elementary row operations to reduce a matrix to row reduced echelon form.

CHAPTER 26

PARTIAL FRACTIONS

> **Basic Attacks and Strategies for Solving Problems in this Chapter. See pages 517 to 538 for step-by-step solutions to problems.**

We know that

$$\frac{5}{6} = \frac{1}{2} + \frac{1}{3},$$

and that

$$\frac{4}{9} = \frac{1}{3} + \frac{1}{3^2}.$$

The process of converting a fraction with a denominator that can be factored to a sum of fractions with denominators that are factors of the original denominator is called *decomposition into partial fractions*.

To find the appropriate *numerators* of the partial fractions, the standard technique is to solve a linear system. The linear system is obtained by equating the original fraction to a sum of the partial fractions with factors of the original denominator as denominators and variables as numerators. For example, we write

$$\frac{5}{6} = \frac{A}{2} + \frac{B}{3}.$$

The fractions on the right side are then combined to produce

$$\frac{3A + 2B}{6}.$$

It follows that $5 = 3A + 2B$. In this case, there are no other constraints that produce another equation (and thus force a unique solution), and so there are several possibilities for values of A and B. One possible set of values is $A = B = 1$. Another set of values is $A = 5$ and $B = -5$.

Before finding the values of the numerators, one must first determine the various *denominators* for the partial fractions.

The first step in decomposing a fraction with polynomials in the denominator and numerator into partial fractions is to factor the denominator into the prime factors, i.e., into linear terms and irreducible (unfactorable) quadratics. In general, there should be as many partial fractions as the degree of the original denominator. These are various possible scenarios for combinations of linear and quadratic terms as factors.

- All the factors may be linear terms and none of these terms is raised to any power. As an example, suppose the denominator of the original fractions is $x^2 - 2x - 3$. This can be factored into $x + 1$ and $x - 3$. In a case like this, for each factor, one creates a new partial fraction having a variable as the numerator and the linear factor as a denominator and sums all the partial fractions. For example, we decompose

$$\frac{1}{x^2 - 2x - 3} \text{ into } \frac{A}{x+1} + \frac{B}{x-3}.$$

Note that since the original denominator was of degree 2, it is decomposed into two partial fractions.

- All the factors may be linear terms, but one (or more) of these terms is raised to some power. As an example, suppose the denominator of the original fraction is

$$x^3 + 5x^2 + 7x + 3,$$

which can be factored into $(x + 1)^2$ and $x - 3$. In this case, the factor raised to a power, e.g., $(x - r)^n$, corresponds to multiple fractions, one for each power, and the denominators of these fractions are merely all possible powers of the linear factor, i.e.,

$$x - r, (x - r)^2, (x - r)^3, ..., (x - r)^n.$$

For example, we decompose

$$\frac{1}{x^3 + 5x^2 + 7x + 3} \text{ into } \frac{A}{x+1} + \frac{B}{(x+1)^2} + \frac{C}{x-3}.$$

Note that since the original denominator was of degree 3, it is decomposed into three partial fractions.

- If one (or more) of the factors is an irreducible quadratic (such as $x^2 + 1$), this still corresponds to *two* partial fractions, but one of these fractions has an x in the *numerator*. For example, we decompose

$$\frac{1}{(x-1)(x^2+1)} \text{ into } \frac{A}{x-1} + \frac{B}{x^2+1} + \frac{Cx}{x^2+1}.$$

- If one (or more) of the factors is an irreducible quadratic raised to a power, one combines the procedures outlined in the previous two cases. Each power of the factor must be represented, and, for each power, two fractions are included, one of which has an x in its numerator. For example, we decompose

$$\frac{1}{(x-1)(x^2+1)^2} \text{ into } \frac{A}{x-1} + \frac{B}{x^2+1} + \frac{Cx}{x^2+1} + \frac{D}{(x^2+1)^2} + \frac{Ex}{(x^2+1)^2}.$$

Note that the original denominator was of degree 5 and that it has been decomposed into five partial fractions.

- If the original fraction is such that the degree of the polynomial in the numerator is greater than (or equal to) the degree of the denominator, one should first divide. This division produces a polynomial added to a fraction of polynomials. In this fraction the degree of the numerator will be less than the degree of the denominator (*cf.* Chapter 4 for the description of long division of two polynomials).

The factors of the denominator should be known in order for the decomposition to be performed. In theory, any polynomial with real coefficients can be expressed as a product of real linear and quadratic factors. In practice, however, an arbitrary polynomial may be difficult to factor.

Once the original fraction has been decomposed into partial fractions, it remains to find the values of the temporary variables used as numerators, A, B, etc. This is done by (1) combining the partial fractions into one fraction, (2) setting corresponding coefficients of the powers of x equal to each other, thereby creating a linear system of the temporary variables A, B, etc., and finally (3) solving the linear system.

For example (as was shown above),

$$\frac{1}{x^2-2x-3} = \frac{A}{x+1} + \frac{B}{x-3}.$$

The right side can be combined to produce

$$\frac{A(x-3)+B(x+1)}{(x+1)(x-3)}$$

which equals

$$\frac{Ax-3A+Bx+B}{(x+1)(x-3)}$$

which, in turn, equals

$$\frac{(A+B)x+(-3A+B)}{(x+1)(x-3)}.$$

The numerators can now be set equal to each other and then the corresponding coefficients of powers of x equated. We obtain

$$1 = (A + B)x + (- 3A + B).$$

Since there are no powers of x on the left, the coefficient is zero and we obtain the equation $0 = A + B$. Setting the constant terms equal, we obtain the equation $1 = - 3A + B$. This gives a 2×2 linear system which can be solved by the methods presented in Chapters 11 or 25. The solution is $A = - 1/4$ and $b = 1/4$.

If there are only linear factors, the values of the temporary variables can be found using an alternate method associated with Oliver Heaviside. Instead of rewriting the numerator of the right side based on the coefficients of x, one leaves it in terms of the coefficients of the temporary variables, A, B, etc. In the example of the previous paragraph, we would obtain

$$1 = A(x - 3) + B(x + 1).$$

The root of the linear term that was the denominator of A (i.e., the term now multiplying B) is $- 1$. Letting $x = - 1$ in this equation results in

$$1 = A(- 1 - 3) + 0 = A \cdot (- 4)$$

implying that $A = - 1/4$. The root of the linear term that was the denominator of B (i.e., the term now multiplying A) is 3. Letting $x = 3$ in this equation results in

$$1 = 0 + B(3 + 1) = B \cdot 4$$

implying that $B = 1/4$.

Step-by-Step Solutions to
Problems in this Chapter,
"Partial Fractions"

● **PROBLEM 26-1**

Decompose $\dfrac{x^3+6x^2+3x+6}{x^3+2x^2}$ into partial fractions.

Solution: To decompose the given equation into partial fractions, we first find the factors of the denominator. The denominator, $x^3 + 2x^2$, can be factored as $x^2(x + 2)$. We then have three terms: $\dfrac{A}{x}$, $\dfrac{B}{x^2}$, and $\dfrac{C}{(x + 2)}$. Since the denominator is of the same power as the numerator, we can divide the numerator by the denominator, and write the expression as:

$$1 + \frac{4x^2 + 3x + 6}{x^3 + 2x^2}.$$

We now find three numbers: A, B, and C, such that:

$$1 + \frac{4x^2 + 3x + 6}{x^3 + 2x^2} = \frac{A}{x} + \frac{B}{x^2} + \frac{C}{x + 2} + 1.$$

Considering the 1 as a constant, we can ignore it momentarily and solve for the numerical values of A, B and C by multiplying both sides of the equation by $x^3 + 2x^2$. We find,

$$4x^2 + 3x + 6 = Ax(x+2) + B(x+2) + Cx^2.$$

We can combine like powers of x to find

$$4x^2 + 3x + 6 = Ax^2 + 2Ax + Bx + 2B + Cx^2.$$

$$4x^2 + 3x + 6 = (A+C)x^2 + (2A+B)x + 2B.$$

Now we equate like powers of x to find A, B, and C. We find,

$$A + C = 4.$$
$$2A + B = 3.$$

$$2B = 6.$$

Therefore, $\quad\quad\quad B = 3, \; A = 0, \; C = 4.$

$$\frac{x^3+6x^2+3x+6}{x^3+2x^2} = 1 + \frac{4x^2+3x+6}{x^3+2x^2}$$

$$= 1 + \frac{0}{x} + \frac{3}{x^2} + \frac{4}{x+2}$$

$$= 1 + \frac{3}{x^2} + \frac{4}{x+2} \; .$$

● **PROBLEM 26-2**

Decompose the following into partial fractions.

$$\frac{2x^2 + 5x - 1}{x^3 + x^2 - 2x}$$

Solution:

Since $x^3 + x^2 - 2x = x(x - 1)(x + 2)$,

the denominator is a product of distinct linear factors, and we try to find A_1, A_2, and A_3 such that

$$\frac{2x^2 + 5x - 1}{x^3 + x^2 - 2x} = \frac{A_1}{x} + \frac{A_2}{x - 1} + \frac{A_3}{x + 2}$$

$$= \frac{A_1(x - 1)(x + 2) + A_2(x)(x + 2) + A_3(x)(x - 1)}{x^3 + x^2 - 2x}$$

Setting the two numerators equal, we obtain:

$$2x^2 + 5x - 1 = A_1(x - 1)(x + 2) + A_2x(x + 2)$$
$$+ A_3x(x - 1).$$

Multiplying out and collecting like powers of x yields:

$$(A_1 + A_2 + A_3)x^2 + (A_1 + 2A_2 - A_3)x - 2A_1$$

$$= 2x^2 + 5x - 1.$$

Equating coefficients of like powers of x,

518

we have the equations:

$$A_1 + A_2 + A_3 = 2$$

$$A_1 + 2A_2 - A_3 = 5$$

$$- 2A_1 = - 1.$$

Solving these equations we find that

$$A_1 = \frac{1}{2} \ , \quad A_2 = 2 \ \text{ and } \ A_3 = - \frac{1}{2} \ .$$

Therefore, we have:

$$\frac{2x^2 + 5x - 1}{x^3 + x^2 - 2x}$$

$$= \frac{1}{2x} + \frac{2}{x-1} + \frac{-1}{2(x+2)}$$

$$= \frac{1}{2x} + \frac{2}{x-1} + \frac{1}{2x+4}$$

● PROBLEM 26-3

Decompose

$$\frac{3x + 6}{x^3 + 2x^2 - 3x}$$

into partial fractions.

<u>Solution</u>: Factoring the denominator, we can write:

$$\frac{3x + 6}{x^3 + 2x^2 - 3x} = \frac{3x + 6}{x(x - 1)(x + 3)}$$

$$= \frac{A}{x} + \frac{B}{(x - 1)} + \frac{C}{(x + 3)}$$

519

$$= \frac{A(x - 1)(x + 3) + B(x)(x + 3) + C(x)(x - 1)}{x(x - 1)(x + 3)}.$$

Since the denominators of the fractions are equal, the numerators must also be the same, and we write:

$$3x + 6 = A(x - 1)(x + 3) + Bx(x + 3)$$

$$+ Cx(x - 1)$$

$$= A(x^2 + 2x - 3) + Bx^2 + 3Bx + Cx^2 - Cx$$

$$= x^2(A + B + C) + x(2A + 3B - C) - 3A.$$

Equating coefficients of like powers of x, we have:

$$A + B + C = 0$$

$$2A + 3B - C = 3$$

$$- 3A = 6,$$

therefore $\qquad A = - 2.$

Solving for B and C by substitution, we find that $B = \frac{9}{4}$ and $C = - \frac{1}{4}$. Therefore, we have $A = -2$, $B = \frac{9}{4}$, and $C = - \frac{1}{4}$. Now, substituting the values of A, B, and C into

$$\frac{3x + 6}{x^3 + 2x^2 - 3x} = \frac{A}{x} + \frac{B}{x-1} + \frac{C}{x+3}$$

resulting in

$$\frac{3x + 6}{x^3 + 2x^2 - 3x} = \frac{-2}{x} + \frac{\frac{9}{4}}{x-1} + \frac{-\frac{1}{4}}{x+3}$$

$$= - \frac{2}{x} + \frac{9}{4(x-1)} - \frac{1}{4(x+3)}$$

$$= - \frac{2}{x} + \frac{9}{4x-4} + \frac{1}{4x+12}.$$

Decompose

$$\frac{3x^2 + 2x - 2}{x^3 - 1}$$

into partial fractions.

Solution:

The denominator can be factored into the product

$$x^3 - 1 = (x - 1)(x^2 + x + 1),$$

and we write;

$$\frac{3x^2 + 2x - 2}{x^3 - 1} = \frac{A}{x - 1} + \frac{Bx + C}{x^2 + x + 1}$$

$$= \frac{A(x^2 + x + 1) + (Bx + C)(x - 1)}{(x - 1)(x^2 + x + 1)} .$$

Setting the numerators of the above fractions equal, we have:

$$3x^2 + 2x - 2 = A(x^2 + x + 1) + (Bx + C)(x - 1)$$

Now we multiply out and collect like powers of x. We obtain:

$$3x^2 + 2x - 2 = (A + B)x^2 + (A - B + C)x + (A - C).$$

Equating coefficients of like powers of x, we obtain:

$$A + B = 3$$
$$A - B + C = 2$$
$$A - C = -2.$$

Solving for A, B, and C, we find A = 1, B = 2, and C = 3. Therefore we have:

$$\frac{3x^2 + 2x - 2}{x^3 - 1} = \frac{1}{x - 1} + \frac{2x + 3}{x^2 + x + 1}$$

Decompose

$$\frac{x^2+8}{x^3-7x+6}$$

into partial fractions. (1)

Solution: First, notice that the denominator may be factored,

$$x^3-7x+6 = (x-1)(x-2)(x+3) \qquad (2)$$

so that

$$\frac{x^2+8}{x^3-7x+6} = \frac{x^2+8}{(x-1)(x-2)(x+3)} \qquad (3)$$

We may now write

$$\frac{x^2+8}{(x-1)(x-2)(x+3)} = \frac{A}{(x-1)} + \frac{B}{(x-2)} + \frac{C}{(x+3)} \qquad (4)$$

Now, multiplying both sides of eq.(4) by $(x-1)(x-2)(x+3)$ results in

$$x^2+8 = A(x-2)(x+3) + B(x-1)(x+3) + C(x-1)(x-2) \qquad (5)$$

It is easy to see that we could evaluate each of the constants A,B andC by simply plugging in certain values of x.

For x=1, we have
$$1^2+8 = A(1-2)(1+3) \quad \text{or} \quad A=-\tfrac{9}{4} \ .$$

For x=2, we obtain
$$2^2+8 = B(2-1)(2+3) \quad \text{or} \quad B=\tfrac{12}{5} \ .$$

Lastly, for x=-3, we obtain
$$(-3)^2+8 = C(-3-1)(-3-2) \quad \text{or} \quad C=\tfrac{17}{20}$$

Thus, expression (1) is decomposed into

$$\frac{x^2 + 8}{x^3 - 7x + 6} = \frac{-9/4}{x-1} + \frac{12/5}{x-2} + \frac{17/20}{x+3} = -\frac{9}{4(x-1)} + \frac{12}{5(x-2)} + \frac{17}{20(x+3)}.$$

(6)

● **PROBLEM 26-6**

Decompose

$$\frac{x^3 + 5x^2 + 2x - 4}{(x^4 - 1)}$$

into partial fractions.

Solution: The denominator $(x^4 - 1)$ of the given expression can be factored into:

$$(x - 1)(x + 1)(x^2 + 1).$$

Then

$$\frac{x^3 + 5x^2 + 2x - 4}{(x^4 - 1)} = \frac{A}{x - 1} + \frac{B}{x + 1}$$

$$+ \frac{Cx + D}{x^2 + 1}.$$

When the denominator contains a term in x^2, an x term must appear in the numerator in addition to a constant.

Now,

$$\frac{x^3 + 5x^2 + 2x - 4}{(x^4 - 1)}$$

$$= \frac{A(x+1)(x^2+1) + B(x-1)(x^2+1) + (Cx+D)(x-1)(x+1)}{(x - 1)(x + 1)(x^2 + 1)}$$

Since the denominators are the same, the numerators are also equal and we have:

$$x^3 + 5x^2 + 2x - 4 = A(x + 1)(x^2 + 1)$$

$$+ B(x - 1)(x^2 + 1) + (Cx + D)(x - 1)(x + 1) =$$

523

$$(Ax^3 + Ax^2 + Ax + A + Bx^3 - Bx^2 + Bx - B + Cx^3$$

$$+ Dx^2 - Cx - D) = (A + B + C)x^3 + (A - B + D)x^2$$

$$+ (A + B - C)x + (A - B - D), \quad \text{collecting terms.}$$

We have:

$$x^3 + 5x^2 + 2x - 4 = (A + B + C)x^3$$

$$+ (A - B + D)x^2 + (A + B-C)x + (A - B - D).$$

Equating coefficients of like powers of x, we obtain:

$$A + B + C = 1$$
$$A - B + D = 5$$
$$A + B - C = 2$$
$$A - B - D = -4.$$

Solving these equations simultaneously, we obtain:

$$A = 1, \quad B = \frac{1}{2}, \quad C = -\frac{1}{2} \quad \text{and} \quad D = 4\frac{1}{2}.$$

Then, substituting the values of A, B, C and D into the equation

$$\frac{x^3 + 5x^2 + 2x - 4}{x^4 - 1}$$

$$= \frac{A}{x - 1} + \frac{B}{x + 1} + \frac{Cx + D}{x^2 + 1}$$

One obtains

$$\frac{x^3 + 5x^2 + 2x - 4}{x^4 - 1}$$

$$= \frac{1}{x - 1} + \frac{\frac{1}{2}}{x + 1} + \frac{-\frac{1}{2}x + 4\frac{1}{2}}{x^2 + 1}.$$

$$\frac{x^3 + 5x^2 + 2x - 4}{x^4 - 1}$$

$$= \frac{1}{x - 1} + \frac{1}{2(x + 1)} + \frac{-\frac{1}{2}x + \frac{9}{2}}{x^2 + 1}$$

$$= \frac{1}{x - 1} + \frac{1}{2(x + 1)} + \frac{\frac{-1 \cdot x + 9}{2}}{x^2 + 1}$$

$$= \frac{1}{x - 1} + \frac{1}{2(x + 1)} + \frac{-x + 9}{2(x^2 + 1)}$$

$$= \frac{1}{x - 1} + \frac{1}{2(x + 1)} + \frac{9 - x}{2(x^2 + 1)}$$

$$= \frac{1}{x - 1} + \frac{1}{2(x + 1)} + \frac{9}{2(x^2 + 1)} - \frac{x}{2(x^2 + 1)}$$

$$= \frac{1}{x - 1} + \frac{1}{2x + 2} + \frac{9}{2x^2 + 2} - \frac{x}{2x^2 + 2} \ .$$

● **PROBLEM 26-7**

Decompose

$$\frac{x^2 + 4x}{(x - 2)^2 (x^2 + 4)}$$

into partial fractions.

Solution: We write:

$$\frac{x^2 + 4x}{(x - 2)^2 (x^2 + 4)} = \frac{A}{(x - 2)} + \frac{B}{(x - 2)^2}$$

525

$$+ \frac{Cx + D}{(x^2 + 4)}$$

$$= \frac{A(x-2)(x^2+4)+B(x^2+4)+(Cx+D)(x-2)^2}{(x-2)^2(x^2+4)} ,$$

where the rules for square and repeated factors have been followed. Equating the numerators (denominators are equal) we have:

$$x^2 + 4x = A(x - 2)(x^2 + 4) + B(x^2 + 4) + (Cx + D)$$

$$(x - 2)^2.$$

Multiplying out and collecting like powers of x we obtain:

$$(A + C)x^3 + (-2A + B - 4C + D)x^2 + (4A + 4C - 4D)x$$

$$+ (-8A + 4B + 4D) = x^2 + 4x.$$

Now we equate the coefficients of like powers of x, obtaining the equations:

$$A + C = 0$$
$$-2A + B - 4C + D = 1$$
$$4A + 4C - 4D = 4$$
$$-8A + 4B + 4D = 0.$$

Solving these equations, we find the values of A, B, C and D to be: $A = \frac{1}{4}$, $B = \frac{3}{2}$, $C = -\frac{1}{4}$ and $D = -1$.

By substitution:

$$\frac{x^2 + 4x}{(x - 2)^2(x^2 + 4)} = \frac{\frac{1}{4}}{(x - 2)} + \frac{\frac{3}{2}}{(x - 2)^2}$$

$$- \frac{\frac{1}{4}x - 1}{(x^2 + 4)}.$$

Decompose

$$\frac{x^3 + 5x - 4}{x^2 - x - 2} \tag{1}$$

into partial fractions.

Solution: It is important to note that the degree of the numerator is higher than the degree of the denominator. Since our usual procedure for determining partial fractions requires that the degree of the numerator be less than the degree of the denominator, we must first divide the numerator by the denominator. Doing this, we obtain

$$
x^2-x-2 \overline{\smash{\big)}\,
\begin{array}{l}
x+1+\dfrac{8x-2}{x^2-x-2} \\[2mm]
x^3+5x-4 \\
-(x^3-x^2-2x) \\
\hline
x^2+7x-4 \\
-(x^2-\ x-2) \\
\hline
8x-2
\end{array}}
\tag{2}
$$

Thus,

$$\frac{x^3+5x-4}{x^2-\ x-2} = x+1+\frac{8x-2}{x^2-x-2} \tag{3}$$

We will now go through our usual procedure to determine the partial fraction decomposition of

$$\frac{8x-2}{x^2-x-2} = \frac{8x-2}{(x-2)(x+1)} \tag{4}$$

We have,

$$\frac{8x-2}{(x-2)(x+1)} = \frac{A}{x-2} + \frac{B}{x+1} \tag{5}$$

Multiplying both sides of eq. (5) by (x-2)(x+1) results in

$$8x-2 = A(x+1) + B(x-2)$$

Letting x=-1, we obtain

$$8(-1)-2 = B(-1-2) \text{ or } B = \frac{10}{3}.$$

For x=2, we have

$$8(2)-2 = A(2+1) \text{ or } A = \frac{14}{3}.$$

Thus,

$$\frac{8x-2}{x^2-x-2} = \frac{\frac{14}{3}}{x-2} + \frac{\frac{10}{3}}{x+1} \tag{6}$$

Substituting eq. (6) into eq. (3) gives us the required decomposition.

$$\frac{x^3+5x-4}{x^2-x-2} = x+1 + \frac{14}{3(x-2)} + \frac{10}{3(x+1)} \tag{7}$$

● **PROBLEM 26-9**

Decompose

$$\frac{x^4 - x^3 + 2x^2 - x + 2}{(x-1)(x^2+2)^2}$$

into partial fractions.

Solution: First, decompose the given expression as follows:

$$\frac{x^4 - x^3 + 2x^2 - x + 2}{(x-1)(x^2+2)^2} = \frac{A}{x-1}$$

$$+ \frac{Bx + C}{x^2 + 2} + \frac{Dx + E}{(x^2+2)^2}$$

$$= \frac{A(x^2+2)^2 + (Bx+C)(x-1)(x^2+2) + (Dx+E)(x-1)}{(x-1)(x^2+2)^2}.$$

We can now equate the numerators of the first and last fractions, obtaining:

$$x^4 - x^3 + 2x^2 - x + 2 = A(x^2 + 2)^2 + (Bx + C)$$

$$(x - 1)(x^2 + 2) + (Dx + E)(x - 1).$$

Multiplying out and collecting like powers of x, we obtain:

$$(A + B)x^4 + (C - B)x^3 + (4A - C + 2B + D)x^2$$

$$+ (2C - 2B + E - D)x + (4A - 2C - E)$$

528

$$= x^4 - x^3 + 2x^2 - x + 2.$$

Now equate the coefficients of like powers of x. Doing this, we have:

$$A + B = 1, \quad C - B = -1,$$

$$4A - C + 2B + D = 2,$$

$$2C - 2B + E - D = -1,$$

$$4A - 2C - E = 2.$$

Solving for A, B, C, D, and E, we find that

$$A = \frac{1}{3}, \quad B = \frac{2}{3}, \quad C = -\frac{1}{3}, \quad D = -1, \quad E = 0.$$

Substituting these values, we have:

$$\frac{x^4 - x^3 + 2x^2 - x + 2}{(x-1)(x^2+2)^2}$$

$$= \frac{1}{3(x-1)} + \frac{2x-1}{3(x^2+2)} - \frac{x}{(x^2+2)^2}.$$

● **PROBLEM** 26-10

Decompose $\frac{2}{x^2-x-2}$ into partial fractions.

Solution: First, let us factor out the denominator, $(x^2-2x+2) = (x+1)(x-2)$. We have,

$$\frac{2}{x^2-x-2} = \frac{2}{(x+1)(x-2)} = \frac{A}{x+1} + \frac{B}{x-2}$$

Multiplying through by (x+1)(x-2) results in

$$2 = A(x-2) + B(x+1)$$

Multiplying out and collecting like powers of x, we have:

$$2 = (A+B)x - 2A+B$$

Equating coefficients of like powers of x, we obtain:

$$A+B = 0$$
$$-2A+B = 2$$

Solving for A and B, we find

$$A = -2/3 \text{ and } B = 2/3.$$

Therefore,

$$\frac{2}{x^2-2x-2} = \frac{-{}^2/{}_3}{x+1} + \frac{{}^2/{}_3}{x-2} = -\frac{2}{3(x+1)} + \frac{2}{3(x-2)}.$$

Decompose

$$\frac{18 + 11x - x^2}{(x - 1)(x + 1)(x^2 + 3x + 3)}$$

into partial fractions.

Solution: Recognizing the presence of a quadratic factor, we use the rule for that case and write:

$$\frac{18 + 11x - x^2}{(x - 1)(x + 1)(x^2 + 3x + 3)} = \frac{A}{x - 1} + \frac{B}{x + 1}$$
$$+ \frac{Cx + D}{x^2 + 3x + 3}$$

$$= \frac{A(x+1)(x^2+3x+3)+B(x-1)(x^2+3x+3)+(Cx+D)(x-1)(x+1)}{(x - 1)(x + 1)(x^2 + 3x + 3)}.$$

Because the denominators are the same, the two numerators must be equal. We now write:

$$A(x + 1)(x^2 + 3x + 3) + B(x - 1)(x^2 + 3x + 3)$$
$$+ (Cx + D)(x - 1)(x + 1)$$
$$= 18 + 11x - x^2.$$

To evaluate A, B, C, and D we multiply out and collect like powers of x. We obtain:

$$(A + B + C)x^3 + (4A + 2B + D)x^2 + (6A - C)x$$
$$+ (3A - 3B - D) = 18 + 11x - x^2.$$

The two polynomials that are the coefficients of like powers of x must be equal. Therefore,

$$A + B + C = 0.$$

$$4A + 2B + D = -1,$$

$$6A - C = 11,$$

$$3A - 3B - D = 18.$$

We now solve these equations simultaneously for A, B, C, and D. Adding the first and third, and the second and fourth,

$$7A + B = 11,$$

$$7A - B = 17.$$

From these two equations we find that B = -3 and A = 2. Substituting these values above, we obtain C = 1 and D = -3. Therefore,

$$\frac{18 + 11x - x^2}{(x - 1)(x + 1)(x^2 + 3x + 3)}$$

$$= \frac{2}{x - 1} - \frac{3}{x + 1} + \frac{x - 3}{x^2 + 3x + 3}.$$

● **PROBLEM 26-12**

Obtain the partial fraction decomposition of

$$\frac{x^3 - 9x^2 + 10x - 3}{x^2 - 1} \qquad (1)$$

Solution: Before we proceed with the usual procedure of computing the partial fractions of the above expression (1), let us note that the degree of the numerator
$$f(x) = x^3 - 9x^2 + 10x - 3$$
is not less than the degree of the denominator $g(x) = x^2 - 1$, and therefore, we must first actually divide $f(x)$ by $g(x)$. We have

$$
\begin{array}{r}
x - 9 + \dfrac{11x - 12}{x^2 - 1} \\[4pt]
x^2 - 1 \overline{\smash{)}\ x^3 - 9x^2 + 10x - 3} \\
\underline{x^3 - x} \\
-9x^2 + 11x \\
\underline{-9x^2 + 9} \\
11x - 12
\end{array}
\qquad (2)
$$

531

Thus,

$$\frac{x^3-9x^2+10x-3}{x^2-1} = x-9 + \frac{11x-12}{x^2-1} \tag{3}$$

Now, we wish to find the partial fraction decomposition for the term

$$\frac{11x-12}{x^2-1} \tag{4}$$

Since the degree of the numerator is less than the degree of the denominator, we may proceed with the usual procedure for calculating partial fractions.

We may write

$$\frac{11x-12}{x^2-1} = \frac{11x-12}{(x+1)(x-1)} = \frac{A}{(x+1)} + \frac{B}{(x-1)} \tag{5}$$

Multiplying eq.(5) by $(x+1)(x-1)$ results in

$$11x-12 = A(x-1) + B(x+1) \tag{6}$$

If we now let $x=1$, then we obtain

$$11-12 = 2B \text{ or } B = -\tfrac{1}{2}$$

Letting $x=-1$, we find

$$-11-12 = -2A, \text{ or } A = {}^{23}/_2 .$$

Thus, the decomposition for $\frac{11x-12}{x^2-1}$ is

$$\frac{11x-12}{x^2-1} = \frac{{}^{23}/_2}{x+1} + \frac{-{}^{1}/_2}{x-1} . \tag{7}$$

Substituting eq. (7) into eq. (3) gives us the final result.

$$\frac{x^3-9x^2+10x-3}{x^2-1} = x-9 + \frac{23}{2(x+1)} - \frac{1}{2(x-1)} . \tag{8}$$

Decompose

$$\frac{x^3 + 5x^2 + 2x - 4}{x(x^2 + 4)^2}$$

into partial fractions.

Solution:

$$\frac{x^3 + 5x^2 + 2x - 4}{x(x^2 + 4)^2} = \frac{A}{x} + \frac{Bx + C}{(x^2 + 4)}$$

$$+ \frac{Dx + E}{(x^2 + 4)^2}$$

$$= \frac{A(x^2+4)^2 + (Bx+C)(x^2+4)(x) + (Dx+E)(x)}{x(x^2 + 4)^2} .$$

Setting the numerators equal, we obtain:

$$x^3 + 5x^2 + 2x - 4 = A(x^2 + 4)^2 + (Bx + C)(x^2 + 4)$$

$$(x) + (Dx + E)(x).$$

Multiplying out and collecting like powers of x, we obtain:

$$(A + B)x^4 + Cx^3 + (8A + 4B + D)x^2 + (4C + E)x + 16A$$

$$= x^3 + 5x^2 + 2x - 4.$$

Now we equate the coefficients of like powers of x. This gives the following equations:

$$A + B = 0$$
$$C = 1$$
$$8A + 4B + D = 5$$
$$4C + E = 2$$
$$16A = -4$$

Solving these equations, we find:

$$A = -\frac{1}{4}, \quad B = \frac{1}{4}, \quad C = 1, \quad D = 6, \quad E = -2.$$

Substituting, we have:

$$\frac{x^3 + 5x^2 + 2x - 4}{x(x^2 + 4)^2} = \frac{-\frac{1}{4}}{x} + \frac{\frac{1}{4}x + 1}{(x^2 + 4)}$$

$$+ \frac{6x - 2}{(x^2 + 4)^2}$$

$$= -\frac{1}{4x} + \frac{x + 4}{4(x^2 + 4)} + \frac{6x - 2}{(x^2 + 4)^2} \quad .$$

● **PROBLEM** 26-14

Decompose $\dfrac{y^2}{(y^2 + 1)^3}$ into partial fractions.

Solution:

$$\frac{y^2}{(y^2 + 1)^3} = \frac{2y\,A_1 + A_2}{y^2 + 1} + \frac{2y\,A_3 + A_4}{(y^2 + 1)^2}$$

$$+ \frac{2y\,A_5 + A_6}{(y^2 + 1)^3}$$

$$= \frac{(2yA_1 + A_2)(y^2+1)^2 + (2yA_3+A_4)(y^2+1) + 2yA_5 + A_6}{(y^2 + 1)^3}$$

Canceling out the denominator, and expanding, yields:

$$y^2 = 2A_1 y^5 + 4A_1 y^3 + 2A_1 y + A_2 y^4 + 2A_2 y^2 + A_2$$

$$+ 2A_3 y^3 + 2A_3 y + A_4 y^2 + A_4 + 2A_5 y + A_6$$

Equating like coefficients,

y^5 yields: $2A_1 = 0$.

y^4 : $A_2 = 0$.

y^3 : $4A_1 + 2A_3 = 0$.

y^2 : $2A_2 + A_4 = 1$.

y : $2A_1 + 2A_3 + 2A_5 = 0$.

1 : $A_2 + A_4 + A_6 = 0$.

We have:

$A_1 = A_3 = A_5 = 0$, $A_2 = 0$, $A_4 = 1$, and $A_6 = -1$.

Thus,

$$\frac{y^2}{(y^2 + 1)^3} = \frac{1}{(y^2 + 1)^2} + \frac{-1}{(y^2 + 1)^3}.$$

● **PROBLEM 26-15**

Find the partial fraction decomposition of

$$\frac{3x^3 - 12x^2 + 21x - 3}{(x+1)(x-2)^3} \tag{1}$$

Solution: First, let us note that for the (x+1) factor of the denominator, there corresponds a partial fraction of the form A/(x+1). Similarly, for the $(x-2)^3$ factor of the denominator, there correspond three partial fractions of the form B/(x-2), $C/(x-2)^2$, and $D/(x-3)^3$. Therefore, the decomposition of the given expression has the form

$$\frac{3x^3 - 12x^2 + 21x - 3}{(x+1)(x-2)^3} = \frac{A}{x+1} + \frac{B}{x-2} + \frac{C}{(x-2)^2} + \frac{D}{(x-2)^3} \tag{2}$$

Now, multiplying both sides by $(x+1)(x-2)^3$ results in

$$3x^3 - 12x^2 + 21x - 3 = A(x-2)^3 + B(x+1)(x-2)^2 + C(x+1)(x-2) + D(x+1) \tag{3}$$

At this point, two of the unknown coefficients may be determined easily. If we let x=2 in eq. (3), then we obtain

$$24 - 48 + 42 - 3 = 3D \quad \text{or} \quad D = {}^{15}/_3 = 5. \tag{4}$$

Setting x=-1 gives us

$$-3 - 12 - 21 - 3 = -27A \quad \text{or} \quad A = {}^{39}/_{27} = {}^{13}/_9. \tag{5}$$

Now, the remaining constants may be found by comparing the coefficients. Indeed, if the right side of eq. (3) is expanded and the like powers of x are collected, we find that the coefficient of the x^3 term is A+B. This must, of course, be equal to the coefficient of the x^3 term on the left of eq. (3), that is,

$$A + B = 3$$

Now, since $A = {}^{13}/_9$ as found earlier, we see that $B = 3 - {}^{13}/_9 = {}^{14}/_9$.

Lastly, by letting x=0 in eq.(3), we obtain

$$-3 = -8A+4B-2C+D. \quad \text{Solving for C, we have } C = {}^4/_3.$$

Therefore, the partial fraction decomposition is

$$\frac{3x^3-12x^2+21x-3}{(x+1)(x-2)^3} = \frac{13}{9(x+1)} + \frac{14}{9(x-2)} + \frac{4}{3(x-2)^2} + \frac{5}{(x-2)^3}$$

● **PROBLEM 26-16**

Decompose the expression

$$\frac{x^2}{x^3-2x^2-5x+6} \qquad (1)$$

into partial fractions.

Solution: First, let us note that the denominator of expression (1) may be factored into

$$x^3-2x^2-5x+6 = (x-3)(x+2)(x-1) \qquad (2)$$

So that

$$\frac{x^2}{x^3-2x^2-5x+6} = \frac{x^2}{(x-3)(x+2)(x-1)} \qquad (3)$$

Now, following the usual procedure for calculating partial fractions, we write

$$\frac{x^2}{(x-3)(x+2)(x-1)} = \frac{A}{x-3} + \frac{B}{x+2} + \frac{C}{x-1} \qquad (4)$$

536

Next, multiplying eq. (4) by $(x-3)(x+2)(x-1)$ results in

$$x^2 = A(x+2)(x-1) + B(x-3)(x-1) + C(x-3)(x+2) \qquad (5)$$

We can now determine each of the constants A,B and C by simply plugging in appropriate values of x in eq. (5).

For x=1, we obtain

$$(1)^2 = C(1-3)(1+2) \text{ so that } C = -\frac{1}{6}.$$

For x=2, eq. (5) gives us

$$4 = B(2-3)(2-1) \text{ so that } B=-4.$$

For x=3, we find

$$(3)^2 = A(3+2)(3-1) \text{ so that } A = {}^9/_{10}$$

Thus, substituting the values of A,B and C into eq. (4) gives

$$\frac{x^2}{(x-3)(x+2)(x-1)} = \frac{{}^9/_{10}}{x-3} + \frac{-4}{x-2} + \frac{-{}^1/_6}{x-1}$$

$$= \frac{9}{10(x-3)} - \frac{4}{x-2} - \frac{1}{6(x-1)}.$$

● PROBLEM 26-17

Decompose

$$\frac{1}{(x+1)(x^4+2x^3+3x^2+2x+1)}$$

into partial fractions.

Solution: The denominator of the given expression has the factored form $(x+1)(x^2+x+1)^2$. The decomposition of the given expression is:

$$\frac{1}{(x+1)(x^4+2x^3+3x^2+2x+1)} = \frac{1}{(x+1)(x^2+x+1)^2}$$

$$= \frac{A_1}{x+1} + \frac{A_2x+A_3}{x^2+x+1} + \frac{A_4x+A_5}{(x^2+x+1)^2}$$

Multiplying both sides of the equation above by $(x+1)(x^2+x+1)^2$ gives us:

537

$$1 = A_1 (x^2+x+1)^2 + (A_2x+A_3)(x+1)(x^2+x+1) + (A_4x+A_5)(x+1) \qquad (1)$$

Letting $x = -1$, we have

$$1 = A_1 ((-1)^2+(-1)+1)^2 + 0 + 0,$$
$$A_1 = 1$$

To find the other constants, expanding eq. (1), collecting like powers of x, and equating the coefficients, we have:

$$1 = (A_1+A_2)x^4 + (2A_1+2A_2+A_3)x^3$$
$$+ (3A_1+2A_2+2A_3+A_4)x^2$$
$$+ (2A_1+A_2+2A_3+A_4+A_5)x$$
$$+ (A_1+A_3+A_5)$$

$$\begin{cases} A_1 = 1 \\ A_1+A_2 = 0 \\ 2A_1+2A_2+A_3 = 0 \\ 3A_1+2A_2+2A_3+A_4 = 0 \\ 2A_1+A_2+2A_3+A_4+A_5 = 0 \\ A_1+A_3+A_5 = 1 \end{cases}$$

Solving the system of equations above, one has $A_2 = -1$, $A_3 = 0$, $A_4 = -1$, and $A_5 = 0$.

Therefore,

$$\frac{1}{(x+1)(x^4+2x^3+3x^2+2x+1)} = \frac{1}{(x+1)(x^2+x+1)^2}$$

$$= \frac{A_1}{x+1} + \frac{A_2x+A_3}{x^2+x+1} + \frac{A_4x+A_5}{(x^2+x+1)^2}$$

$$= \frac{1}{x+1} + \frac{(-1)x+0}{x^2+x+1} + \frac{(-1)x+0}{(x^2+x+1)^2}$$

$$= \frac{1}{x+1} - \frac{x}{x^2+x+1} - \frac{x}{(x^2+x+1)^2} .$$

CHAPTER 27

SERIES

Basic Attacks and Strategies for Solving Problems in this Chapter. See pages 539 to 568 for step-by-step solutions to problems.

The capital Greek letter sigma, Σ, is the standard symbol used in mathematics to indicate a repeated *sum* and is often called the *summation symbol*. It is used with some integer variable, called the *index*, to abbreviate a long expression in which the terms are essentially the same. The summation symbol normally includes the index below it along with a starting value, and a value above it that indicates the final value for the index. An expression following the summation symbol gives the general expression for a term which will be part of the repeated sum. For example,

$$\sum_{i=1}^{5} 2(i^2)$$

is a shorthand for the repeated sum

$$2(1^2) + 2(2^2) + 2(3^2) + 2(4^2) + 2(5^2).$$

Each term in this sum is the same except for the number being raised to a power, and that number progresses by one's from one to five. A general sum is sometimes expressed as

$$\sum_{i=1}^{n} a_i \quad \text{or even} \quad \sum a_i.$$

A *series* is an infinite sequence (*cf.* Chapter 17) of sums with the same general term. The nth item in the sequence has n terms in its summation. As the number of the term in the series increases, we may ask the question of whether the value of the term (i.e., of the corresponding sum) will also increase toward infinity, or whether it will "level off" and achieve a plateau (i.e., a mathematical "limit"). If the sums which make up the items in the series grow toward infinity, the series is said to *diverge*. If, however, the sums reach a plateau and do not grow past a given number, the series is said to *converge*.

A well-known *convergent* series is

$$\sum_{i=1}^{n} \frac{1}{i^2} = 1 + \frac{1}{4} + \frac{1}{9} + \frac{1}{16} + \ldots + \frac{1}{n^2}.$$

A well-known *divergent* series is

$$\sum_{i=1}^{n} \frac{1}{i} = 1 + \frac{1}{2} + \frac{1}{3} + \frac{1}{4} + \ldots + \frac{1}{n}.$$

In general, the series

$$\sum_{i=1}^{n} \frac{1}{i^p}$$

converges if $p > 1$ and diverges if $p \leq 1$.

One standard test for convergence or divergence is called the *comparison test*, in which a general item in a series is compared to a corresponding general item in another series known to be convergent or divergent. If the series is (item by item) less than a convergent series, it also is convergent. If, however, the series is (item by item) greater than a divergent series, it also is divergent.

Another standard test for convergence or divergence is called the *ratio test*. This may be applied to any series in which each term of the sum is positive. One forms the ratio of two successive terms of the sum, i.e., $\frac{a_{n+1}}{a_n}$, and determines what the value will be as n continues to grow larger (i.e., as the "limit of n approaches infinity"). If the value of this ratio approaches some number less than 1, the series converges. If the value of this ratio approaches some number greater than 1, the series diverges. If the value of this ratio approaches 1, the series may either converge or diverge, and so another test must be performed. (We give some rules to help evaluate the ratio in simple cases: (1) if the ratio can be algebraically transformed to an expression in which the numerator is a constant and the denominator contains the index variable n, this ratio will approach the value of zero as n gets larger; (2) if the ratio can be algebraically transformed to an expression in which the denominator is a constant and the numerator contains the index variable n, this ratio will grow and grow as n gets larger.)

When these tests fail, other tests can be used, such as the *root test*. Consult a standard calculus book for the rules.

An *alternating series* is one in which two consecutive terms in the sum have different signs. An alternating series converges if (a) the magnitudes of the terms (eventually) start decreasing, i.e., for all $n > k$ for some k,

$$|a_{k+1}| < |a_k|$$

and also (b) a_n approaches zero as n gets larger.

A series

$$\sum_{i=1}^{n} a_i$$

is said to *converge absolutely* if

$$\sum_{i=1}^{n} |a_i|$$

converges. When some of the terms of a series

$$\sum_{i=1}^{n} a_i$$

are positive and others are negative, the series converges if

$$\sum_{i=1}^{n} |a_i|$$

converges. If the series of absolute values diverges, the original series may still converge, in which case, we say that it *converges conditionally.*

A series is called a *power series* if it is of the form

$$\sum_{i=0}^{n} a_i (x - x_0)^i,$$

where x_0 is a constant (perhaps zero) and x is a variable. The *radius of convergence* is the number R such that if $|x| < R$ the series converges absolutely, and if $|x| > R$ the series diverges. When $|x| = R$ the series may or may not converge at either or both endpoints. Techniques for finding the radius of convergence of series may be found in standard calculus texts.

● **PROBLEM** 27-1

Find the numerical value of the following:

a) $\displaystyle\sum_{j=1}^{7} (2j + 1)$ b) $\displaystyle\sum_{j=1}^{21} (3j - 2)$.

Solution: If $A(r)$ is some mathematical expression and n is a positive integer, then the symbol $\displaystyle\sum_{r=0}^{n} A(r)$ means "Successively replace the letter r in the expression $A(r)$ with the numbers $0,1,2,...,n$ and add up the terms. The symbol Σ is the Greek letter sigma and is a shorthand way to denote "the sum". It avoids having to write the sum $A(0) + A(1) + A(2) + ... + A(n)$.

a) For a) successively replace j by $1,...,7$ and add up the terms.

$$\sum_{j=1}^{7} (2j+1) = \left(2(1)+1\right) + \left(2(2)+1\right) + \left(2(3)+1\right) + \left(2(4)+1\right) + \left(2(5)+1\right)$$
$$+ \left(2(6)+1\right) + \left(2(7)+1\right)$$
$$= (2+1) + (4+1) + (6+1) + (8+1) + (10+1) + (12+1) + (14+1)$$
$$= 3 + 5 + 7 + 9 + 11 + 13 + 15$$
$$= 63 .$$

b) For b) successively replace j by $1,2,3,...,21$ and add up the terms.

$$\sum_{j=1}^{21} (3j-2) = \left(3(1)-2\right) + \left(3(2)-2\right) + \left(3(3)-2\right) + \left(3(4)-2\right) + \left(3(5)-2\right)$$
$$+ \left(3(6)-2\right) + \left(3(7)-2\right) + \left(3(8)-2\right) + \left(3(9)-2\right)$$
$$+ \left(3(10)-2\right) + \left(3(11)-2\right) + \left(3(12)-2\right) + \left(3(13)-2\right)$$
$$+ \left(3(14)-2\right) + \left(3(15)-2\right) + \left(3(16)-2\right) + \left(3(17)-2\right)$$
$$+ \left(3(18)-2\right) + \left(3(19)-2\right) + \left(3(20)-2\right) + \left(3(21)-2\right)$$
$$= (3-2) + (6-2) + (9-2) + (12-2) + (15-2) + (18-2) + (21-2)$$
$$+ (24-2) + (27-2) + (30-2) + (33-2) + (36-2)$$
$$+ (39-2) + (42-2) + (45-2) + (48-2) + (51-2)$$

$$+ (54-2) + (57-2) + (60-2) + (63-2)$$

$$= 1 + 4 + 7 + 10 + 13 + 16 + 19 + 22 + 25 + 28 + 31 + 34$$

$$+ 37 + 40 + 43 + 46 + 49 + 52 + 55 + 58 + 61$$

$$= 651 \ .$$

● PROBLEM 27-2

Establish the convergence or divergence of the series:

$$\frac{1}{1 + \sqrt{1}} + \frac{1}{1 + \sqrt{2}} + \frac{1}{1 + \sqrt{3}} + \frac{1}{1 + \sqrt{4}} + \ \ldots \ .$$

Solution: To establish the convergence or divergence of the given series we first determine the nth term of the series. By studying the law of formation of the terms of the series we find the nth term to be $\frac{1}{1 + \sqrt{n}}$. To determine whether this series is convergent or divergent we use the comparison test. We choose $\frac{1}{n}$, which is a known divergent series since it is a p-series, $\frac{1}{n^p}$, with p = 1. If we can show $\frac{1}{1 + \sqrt{n}} > \frac{1}{n}$, then $\frac{1}{1 + \sqrt{n}}$ is divergent. But we can see this is true, since $1 + \sqrt{n} < n$ for n > 1. Therefore the given series is divergent.

● PROBLEM 27-3

Establish the convergence or divergence of the series:

$$\sin \frac{\pi}{2} + \frac{1}{4} \sin \frac{\pi}{4} + \frac{1}{9} \sin \frac{\pi}{6} + \frac{1}{16} \sin \frac{\pi}{8} + \ \ldots \ .$$

Solution: To establish the convergence or divergence of the given series, we first determine the nth term of the series. By studying the law of formation of the terms of the series, we find the nth term to be: $\frac{1}{n^2} \sin \frac{\pi}{2n}$. To determine whether this series is convergent or divergent, we use the comparison test. We choose $\frac{1}{n^2}$, which is a known convergent series, since it is a p-series, $\frac{1}{n^p}$, with p = 2. If we can show $\frac{1}{n^2} \sin \frac{\pi}{2n} < \frac{1}{n^2}$, then

$\frac{1}{n^2} \sin \frac{\pi}{2n}$ is convergent. But we can see this is true since $\sin \frac{\pi}{2n}$ is less than 1 for $n > 1$. Therefore, the given series is convergent.

Test the series:

$$1 + \frac{2!}{2^2} + \frac{3!}{3^3} + \frac{4!}{4^4} + \ldots$$

by means of the ratio test. If this test fails, use another test.

Solution: To make use of the ratio test, we find the nth term of the given series, and the (n+1)th term. If we let the first term, $1 = u_1$, then $\frac{2!}{2^2} = u_2$, $\frac{3!}{3^3} = u_3$, etc., up to $u_n + u_{n+1}$. We examine the terms of the series to find the law of formation, from which we conclude:

$$u_n = \frac{n!}{n^n} \quad \text{and,} \quad u_{n+1} = \frac{(n+1)!}{(n+1)^{n+1}}.$$

Forming the ratio $\frac{u_{n+1}}{u_n}$ we obtain:

$$\frac{(n+1)!}{(n+1)^{n+1}} \times \frac{n^n}{n!}$$

$$\frac{(n+1)(n!)}{(n+1)^n (n+1)} \times \frac{n^n}{n!} = \frac{n^n}{(n+1)^n}.$$

Now, we find $\lim\limits_{n \to \infty} \left| \frac{n^n}{(n+1)^n} \right|$. This can be rewritten as:

$$\lim_{n \to \infty} \frac{n^n}{\left[n\left(1 + \frac{1}{n}\right) \right]^n} = \lim_{n \to \infty} \frac{n^n}{n^n \cdot \left(1 + \frac{1}{n}\right)^n} = \lim_{n \to \infty} \frac{1}{\left(1 + \frac{1}{n}\right)^n}.$$

We now use the definition:

$$e = \lim_{x \to 0} (1 + x)^{\frac{1}{x}}.$$

If we let $x = \frac{1}{n}$ in this definition, we have:

$$\lim_{\frac{1}{n} \to 0} \left(1 + \frac{1}{n}\right)^{\frac{1}{\frac{1}{n}}} = \lim_{n \to \infty} \left(1 + \frac{1}{n}\right)^n, \quad \text{which is what we have}$$

above. Therefore, $\lim\limits_{n\to\infty} \dfrac{1}{\left(1 + \frac{1}{n}\right)^n} = \dfrac{1}{e}$. Since $e \approx 2.7$,

$\dfrac{1}{e} \approx \dfrac{1}{2.7}$ which is less than 1.

Hence, by the ratio test, the given series in convergent.

● PROBLEM 27-5

Test the series:

$$1 - \frac{3^2}{2^2} + \frac{3^4}{2^2 \cdot 4^2} + - \frac{3^6}{2^2 \cdot 4^2 \cdot 6^2} + \ldots$$

by means of the ratio test. If this test fails, use another test.

__Solution:__ To make use of the ratio test, we find the nth term of the given series, and the (n+1)th term. If we let the first term, $1 = u_1$, then

$\dfrac{3^2}{2^2} = u_2$, $\dfrac{3^4}{2^2 \cdot 4^2} = u_3$, etc. up to $u_n \pm u_{n+1}$. We

examine the terms of the series to find the law of formation, from which we conclude:

$$u_n = \frac{3^{2n-2}}{2^2 \cdot 4^2 \cdot \ldots (2n-2)^2},$$

and

$$u_{n+1} = \frac{3^{2(n+1)-2}}{2^2 \cdot 4^2 \cdot \ldots (2n-2)^2 [2(n+1)-2]^2}$$

$$= \frac{3^{2n}}{2^2 \cdot 4^2 \cdot \ldots (2n-2)^2 (2n)^2}.$$

Forming the ratio $\dfrac{u_{n+1}}{u_n}$, we obtain:

$$\frac{3^{2n}}{2^2 \cdot 4^2 \cdot \ldots (2n-2)^2 (2n)^2} \times \frac{2^2 \cdot 4^2 \cdot \ldots (2n-2)^2}{3^{2n-2}}$$

$$= \frac{3^{2n}}{(2n)^2 \times 3^{2n-2}} = \frac{3^{2n-(2n-2)}}{(2n)^2} = \frac{3^2}{4n^2}.$$

Now, we find:

$$\lim_{n\to\infty}\left|\frac{3^2}{4n^2}\right| = 0.$$

Since $\lim_{n\to\infty}\left|\frac{u_{n+1}}{u_n}\right| = 0$ and $0 < 1$, the given series converges.

● **PROBLEM 27-6**

Determine whether the following series are convergent for the given values of a, p, and r.

(1) $a + ar + ar^2 + \ldots + ar^{n-1} + \ldots$

where a = 5   r = 9

(2) $\frac{1}{1^p} + \frac{1}{2^p} + \frac{1}{3^p} + \ldots + \frac{1}{n^p} + \ldots$

where p = 5.

Solution: (1) $a + ar + ar^2 + \ldots + ar^{n-1} + \ldots$

represents the geometric series.

If $|r| < 1$ then the series is convergent.

If $|r| \geq 1$ then the series is divergent.

In this case, r=9, $|r| \geq 1$.   Therefore, the series diverges.

(2) $\frac{1}{1^p} + \frac{1}{2^p} + \frac{1}{3^p} \ldots$ represents the p-series.   If p > 1, the

series converges.

If $p \leq 1$, the series diverges.   In this case, p = 5, so p > 1.   Therefore, the series converges.

● **PROBLEM 27-7**

Determine whether the following are convergent.

(a) $\frac{9^{n-4}}{3^{2n}}$

(b) $\frac{1}{2n^2}$

(c) $\frac{2n}{(n+1)}$

Solution: (a) $\dfrac{9^{n-4}}{3^{2n}} = \dfrac{3^{2(n-4)}}{3^n} = 3^{2n-8-2n} = 3^{-8}$

Yes, this series converges to $\dfrac{1}{3^8}$.

(b) $\dfrac{1}{2n^2} = \dfrac{\frac{1}{n^2}}{2} = 0$

Yes, this series converges to 0.

(c) $\dfrac{2n}{n+1} = \dfrac{2}{1+\frac{1}{n}} = \dfrac{2}{1+0} = 2$

Yes, this series converges to 2.

● PROBLEM 27-8

Establish the convergence or divergence of the series:

$$\frac{1}{1 + \sqrt{1}} + \frac{1}{1 + \sqrt{2}} + \frac{1}{1 + \sqrt{3}} + \frac{1}{1 + \sqrt{4}} + \dots .$$

Solution: To establish the convergence or divergence of the given series we first determine the nth term of the series. By studying the law of formation of the terms of the series we find the nth term to be $\dfrac{1}{1 + \sqrt{n}}$. To determine whether this series is convergent or divergent we use the comparison test. We choose $\dfrac{1}{n}$, which is a known divergent series since it is a p-series, $\dfrac{1}{n^p}$, with p = 1. If we can show $\dfrac{1}{1 + \sqrt{n}} > \dfrac{1}{n}$, then $\dfrac{1}{1 + \sqrt{n}}$ is divergent. But we can see this is true, since $1 + \sqrt{n} < n$ for n > 1. Therefore the given series is divergent.

● PROBLEM 27-9

Test the alternating series:

$$\frac{1 + \sqrt{2}}{2} - \frac{1 + \sqrt{3}}{4} + \frac{1 + \sqrt{4}}{6} - \frac{1 + \sqrt{5}}{8} + \dots$$

for convergence.

Solution: An alternating series is convergent if (a) the terms, after a certain nth term, decrease numerically, i.e., $u_{n+1} < u_n$, and (b) the general term approaches 0 as n becomes infinite. Therefore, we determine the nth term of the given alternating series. By discovering the law of formation, we find that the general term is $\pm \dfrac{1 + \sqrt{n+1}}{2n}$. Therefore, the preceding term is $\pm \dfrac{1 + \sqrt{n}}{2(n-1)}$. To satisfy condition (a) stated above, we must show that:

$$\frac{1 + \sqrt{n+1}}{2n} < \frac{1 + \sqrt{n}}{2(n-1)}.$$

Obtaining a common denominator for both these terms,

$$\frac{1}{2} \cdot \frac{\left(1 + \sqrt{n+1}\right)(n-1)}{n(n-1)} < \frac{\left(1 + \sqrt{n}\right)(n)}{n(n-1)} \cdot \frac{1}{2}.$$

Since the denominators are the same, to prove condition (a) we must show,

$$1 + \sqrt{n+1}(n-1) < \left(1 + \sqrt{n}\right)(n),$$

which is obvious, since subtracting 1 from n has a greater effect than adding 1 to $\sqrt{n}$. Since $u_{n+1} < u_n$, we have the first condition for convergence.

Now we must show that

$$\lim_{n \to \infty} \frac{1 + \sqrt{n}}{2n - 2} = 0.$$

We find that $\lim\limits_{n \to \infty} \dfrac{1 + \sqrt{n}}{2n - 2} = \dfrac{\infty}{\infty}$, which is an indeterminate form. We therefore apply L'Hospital's Rule, obtaining:

$$\lim_{n \to \infty} \frac{\frac{1}{2}n^{-1/2}}{2} = \lim_{n \to \infty} \frac{1}{4\sqrt{n}} = 0.$$

Since both conditions hold, the given alternating series is convergent.

● **PROBLEM 27-10**

Write the (n+1)th term of the series whose nth term is

(a) $\dfrac{2^n}{3^{n-2}}$

(b) $\dfrac{(-1)^{n-1}}{n^2 + 4}$

Solution: (a) $\dfrac{2^{n+1}}{3^{(n+1)-2}} = \dfrac{2^{n+1}}{3^{n-1}}$

(b) $\dfrac{(-1)^{(n+1)-1}}{(n+1)^2+4} = \dfrac{(-1)^n}{n^2+2n+5}$

● **PROBLEM** 27-11

Write the series represented by each of the following:

(1) $\displaystyle\sum_{n=0}^{4} \dfrac{2^n}{4^n+n}$

(2) $\displaystyle\sum_{n=0}^{\infty} \dfrac{(-1)^{n+1}}{2a+n}$

Solution: (a) $\displaystyle\sum_{n=0}^{4} \dfrac{2^n}{4^n+n} = 1 + \dfrac{2}{5} + \dfrac{2}{9} + \dfrac{8}{67} + \dfrac{4}{65}$

(b) $\displaystyle\sum_{n=4}^{\infty} \dfrac{(-1)^{n+1}}{2a+n} = -\dfrac{1}{2a+4} + \dfrac{1}{2a+5} - \dfrac{1}{2a+6} \cdots$

● **PROBLEM** 27-12

Find all values of x for which the series:

$1 + \dfrac{1}{3}x - \dfrac{1\cdot2}{3\cdot6}x^2 + \dfrac{1\cdot2\cdot5}{3\cdot6\cdot9}x^3 - \dfrac{1\cdot2\cdot5\cdot8}{3\cdot6\cdot9\cdot12}x^4 + \cdots$

converges.

Solution: To find the values for which the given series converges we use the ratio test. Upon examining the terms of the series we find the nth term to be:

$$\pm \dfrac{1\cdot2\cdot5\cdot\ldots(3n-7)}{3\cdot6\cdot9\cdot\ldots(3n-3)}x^{n-1}.$$

Then the (n+1)th. term is:

$$\pm \dfrac{1\cdot2\cdot5\cdot\ldots(3n-7)(3n-4)}{3\cdot6\cdot9\cdot\ldots(3n-3)(3n)}x^{n}.$$

Forming the ratio $\dfrac{u_{n+1}}{u_n}$, we have:

546

$$\frac{1 \cdot 2 \cdot 5 \cdot \ldots (3n-7)(3n-4)x^n}{3 \cdot 6 \cdot 9 \cdot \ldots (3n-3)(3n)} \times \frac{3 \cdot 6 \cdot 9 \cdot \ldots (3n-3)}{1 \cdot 2 \cdot 5 \cdot \ldots (3n-7)x^{n-1}}$$

$$= \frac{(3n-4)x^n}{(3n)x^{n-1}} = \frac{(3n-4)x}{3n} = \frac{3nx - 4x}{3n}.$$

We now find,

$$\lim_{n \to \infty} \left| \frac{3nx - 4x}{3n} \right|.$$

Dividing both numerator and denominator by n, we obtain:

$$\lim_{n \to \infty} \left| \frac{3x - \frac{4x}{n}}{3} \right| = \left| \frac{3x}{3} \right| = |x|.$$

The given series is convergent when $|x| < 1$, and divergent when $|x| > 1$. Therefore, we conclude that the series converges when $1 > x > -1$. But we must still test the endpoints of the interval, since the ratio test gives no information when $x = 1$. We find that both endpoints are not included in the interval of convergence since, when $x = \pm 1$, $\lim_{n \to \infty}$ of the

nth term is not 0. We find:

$$\lim_{n \to \infty} \frac{1 \cdot 2 \cdot 5 \cdot \ldots (3n-7)}{3 \cdot 6 \cdot 9 \cdot \ldots (3n-3)} = \frac{\infty}{\infty}.$$

Using Gauss's test (an advanced test) we find that the given series converges for

$$1 \geq x \geq -1.$$

● **PROBLEM 27-13**

From the series for $(1+x)^{-1}$, obtain the series for $\ln(1+x)$.

Solution: First, we determine the series for $(1+x)^{-1}$. To do this we find f(x), f(0), f'(x), f'(0), f"(x), f"(0), etc. We find:

$$f(x) = (1 + x)^{-1} \qquad f(0) = 1$$

$$f'(x) = -(1+x)^{-2} \qquad f'(0) = -1$$

$$f''(x) = 2(1+x)^{-3} \qquad f''(0) = 2$$

$$f'''(x) = -6(1+x)^{-4} \qquad f'''(0) = -6$$

$$f^4(x) = 24(1+x)^{-5} \qquad f^4(0) = 24$$

$$f^5(x) = -120(1+x)^{-6} \qquad f^5(0) = -120.$$

We develop the series as follows:

$$f(x) = f(0) + f'(0)x + \frac{f''(0)}{2!}x^2 + \frac{f''(0)}{3!}x^3 + \frac{f^4(0)}{4!}x^4$$

$$+ \frac{f^5(0)}{5!}x^5 + \ldots .$$

By substitution:

$$(1+x)^{-1} = 1 - x + \frac{2}{2!}x^2 - \frac{6}{3!}x^3 + \frac{24}{4!}x^4 - \frac{120}{5!}x^5 + \ldots$$

$$= 1 - x + x^2 - x^3 + x^4 - x^5 + \ldots \quad \pm x^n$$
$$\overset{+}{\underset{-}{}} \ldots .$$

To obtain the series for $\ln(1 + x)$, we find

$$\int_0^x \frac{dx}{1 + x}.$$

$$\int_0^x \frac{dx}{1+x} = \int_0^x \left(1 - x + x^2 - x^3 + \ldots \quad \pm x^n + \ldots\right)dx$$

$$= \left[\ln(1+x)\right]_0^x$$

$$= x - \frac{x^2}{2} + \frac{x^3}{3} - \frac{x^4}{4} + \frac{x^5}{5} + \ldots$$
$$\pm \frac{x^n}{n} \quad \pm \frac{x^{n+1}}{n+1} \quad \ldots .$$

Therefore,

$$\ln(1+x) = x - \frac{x^2}{2} + \frac{x^3}{3} - \frac{x^4}{4} + \ldots + \frac{(-1)^{n+1} x^n}{n} + \ldots .$$

Show that:

a)

$$\lim_{n \to \infty} \frac{n^4 + n^3 - 1}{(n^2 + 2)(n^2 - n - 1)} = 1$$

b)

$$\lim_{n \to \infty} \left(1 + \frac{C}{n^2}\right)^n = 1 \ ,$$

C a constant.

Solution: By definition,

$$\lim_{n \to \infty} p_n = p$$

if for all $\varepsilon > 0$, there exists N such that $n \geq N$ implies that

$$|p_n - p| < \varepsilon \ .$$

a)

$$\lim_{n \to \infty} \frac{n^4 + n^3 - 1}{(n^2 + 2)(n^2 - n - 1)}$$

$$= \lim_{n \to \infty} \frac{n^4(1 + 1/n - 1/n^4)}{n^2(1 + 2/n^2)n^2(1 - 1/n - 1/n^2)}$$

$$= \lim_{n \to \infty} \frac{(1 + 1/n - 1/n^4)}{(1 + 2/n^2)(1 - 1/n - 1/n^2)}$$

$$= \frac{1 + 0 - 0}{(1 + 0)(1 - 0 - 0)} = 1$$

since

$$\lim_{n \to \infty} \frac{1}{n^p} = 0 \text{ for } p > 0 \ .$$

b) We want to show that

$$\lim_{n \to \infty} \left(1 + \frac{C}{n^2}\right)^n = 1 \ .$$

However, from the definition of a convergent sequence, this is equivalent to showing that

$$\left|\left(1 + \frac{C}{n^2}\right)^n - 1\right| < \varepsilon$$

where $\varepsilon > 0$, and $n \geq N$.

To do this we make use of the binomial theorem,

$$(a+b)^m = \sum_{r=0}^{m} \binom{m}{r} a^{m-r} b^r \ ,$$

where

$$\binom{m}{r} = \frac{m!}{r!\,(m-r)!} \ .$$

Hence,

$$\left| \left(1 + \frac{c}{n^2}\right)^n - 1 \right| = \left| \sum_{r=0}^{n} \binom{n}{r} \left(\frac{c}{n^2}\right)^r - 1 \right|$$

$$= \left| \sum_{r=1}^{n} \binom{n}{r} \left(\frac{c}{n^2}\right)^r \right|$$

$$\leq \sum_{r=1}^{n} \frac{n^r}{r!} \left(\frac{|c|}{n^2}\right)^r \qquad \text{since } \binom{n}{r} \leq \frac{n^r}{r!}$$

$$\leq \sum_{r=1}^{n} \left(\frac{|c|}{n}\right)^r \qquad \text{since } \frac{n^r}{r!\,n^r} < 1 \ .$$

But

$$\sum_{r=1}^{n} \left(\frac{|c|}{n}\right)^r = \frac{1 - \left(\frac{|c|}{n}\right)^{n+1}}{1 - \frac{c}{n}} - 1 = \frac{\frac{|c|}{n} - \left(\frac{|c|}{n}\right)^{n+1}}{1 - \frac{c}{n}}$$

$$= \frac{\frac{|c|}{n}\left(1 - \left(\frac{|c|}{n}\right)^{n}\right)}{1 - \frac{c}{n}}$$

$$< \frac{\frac{|c|}{n}}{1 - \frac{c}{n}}$$

$$\left(\text{if } n > |c| \text{ since } 1 - \left(\frac{|c|}{n}\right)^{n} < 1\right).$$

Therefore,

$$\left| \left(1 + \frac{c}{n^2}\right)^n - 1 \right| < \frac{\frac{|c|}{n}}{1 - \frac{c}{n}} = \frac{|c|}{n - |c|}$$

which $\to 0$ as $n \to \infty$.
Thus,

$$\lim_{n \to \infty} \left(1 + \frac{c}{n^2}\right)^n = 1 \ .$$

Evaluate the following limits:

(1) $\lim\limits_{n\to\infty} \dfrac{2n^2+4}{n^2+3n}$

(2) $\lim\limits_{n\to\infty} \dfrac{1}{n}$

(3) $\lim\limits_{n\to\infty} \dfrac{(n-4)!}{n!}$

Solution: (1) $\lim\limits_{n\to\infty} \dfrac{2n^2+4}{n^2+3n} = \lim\limits_{n\to\infty} \dfrac{2+\dfrac{4}{n^2}}{1+\dfrac{3}{n}}$

$$= \dfrac{\lim\limits_{n\to\infty} 2 + \lim\limits_{n\to\infty} \dfrac{4}{n^2}}{\lim\limits_{n\to\infty} 1 + \lim\limits_{n\to\infty} \dfrac{3}{n}} = \dfrac{2+0}{1+0} = 2$$

(2) $\lim\limits_{n\to\infty} \dfrac{1}{n} = 0$

(3) $\lim\limits_{n\to\infty} \dfrac{(n-4)!}{n!} = \dfrac{(n-4)(n-5)(n-6)\dots}{n(n-1)(n-2)(n-3)(n-4)\dots}$

$$\lim\limits_{n\to\infty} \dfrac{1}{n(n-1)(n-2)(n-3)} = 0$$

Find

$$\lim\limits_{x\to\infty} (x\sqrt{x^2+1} - x^2).$$

Solution: We can find this limit by three different methods. For the first method, let

$$\lim\limits_{x\to\infty} (x\sqrt{x^2+1} - x^2) = \lim\limits_{x\to\infty} (x\sqrt{x^2+1} - x^2)\left(\dfrac{x\sqrt{x^2+1} + x^2}{x\sqrt{x^2+1} + x^2}\right)$$

$$= \lim\limits_{x\to\infty} \dfrac{x^2}{x\sqrt{x^2+1} + x^2}$$

$$= \lim_{x \to \infty} \frac{1}{\left(\sqrt{1 + \dfrac{1}{x^2}} + 1 \right)} = \frac{1}{2}$$

since

$$\lim_{n \to \infty} \frac{1}{n^2} = 0 .$$

For the second method, we use the following theorem:
Let
$$f(x), g(x) \in C^{n+1} \quad \text{for } a \le x \le b .$$

In addition, let $f^{(k)}(a) = g^{(k)}(a) = 0$ for $k = 0, 1, \ldots , n$
and let $g^{(n+1)}(a) \neq 0$. Then,

$$\lim_{x \to a^+} \frac{f(x)}{g(x)} = \frac{f^{(n+1)}(a)}{g^{(n+1)}(a)} .$$

To be able to apply this theorem to the given problem, we must replace x by $\dfrac{1}{y}$ and let y approach zero. Then

$$\lim_{x \to \infty} (x\sqrt{x^2 + 1} - x^2) = \lim_{y \to 0} \left(\frac{1}{y} \sqrt{\frac{1}{y^2} + 1} - \frac{1}{y^2} \right) \qquad (1)$$

$$= \lim_{y \to 0} \left(\frac{\sqrt{y^2 + 1} - 1}{y^2} \right) .$$

Now, let $f(y) = (y^2 + 1)^{1/2} - 1$ and $g(y) = y^2$

Then

$$f'(y) = y(y^2 + 1)^{-1/2} , \quad g'(y) = 2y$$

$$f''(y) = (y^2 + 1)^{-1/2} - y^2(y^2 + 1)^{-3/2} \qquad g''(y) = 2$$

Here, $n = 1$ since

$$f(0), \; g(0), \; f'(0), \; g'(0)$$

each equal zero. Hence,

$$\lim_{y \to 0} \frac{f(y)}{g(y)} = \frac{f''(a)}{g''(a)} \quad \text{where } a \text{ is equal to } 0$$

hence,

$$\lim_{y \to 0} \frac{f(y)}{g(y)} = \frac{f''(0)}{g''(0)} = \frac{1}{2} .$$

Thus, by (1),

552

$$\lim_{x \to \infty} (x\sqrt{x^2 + 1} - x^2) = \frac{1}{2}.$$

For the third method expand the function in powers of $1/x$. Hence,

$$\lim_{x \to \infty} x\sqrt{x^2 + 1} - x^2 = \lim_{x \to \infty} x^2 \left[\left(1 + \frac{1}{x^2}\right)^{\frac{1}{2}} - 1 \right]$$

(by the binomial theorem)

$$= \lim_{x \to \infty} x^2 \left[\frac{1}{2x^2} - \frac{1}{8x^4} + \ldots \right]$$

$$= \lim_{x \to \infty} \left[\frac{1}{2} - \frac{1}{8x^2} + \ldots \right] = \frac{1}{2}$$

Since $\lim\limits_{n \to \infty} \dfrac{1}{n^p} = 0$ for $p > 0$.

and

$$\lim_{n \to \infty} (S_{n_1} + S_{n_2} + \ldots) = S_1 + S_2 + \ldots$$

given that

$$\lim_{n \to \infty} S_{n_i} = S_i.$$

● **PROBLEM 27-17**

Find the limit of the sequence defined by

$$x_1 = \frac{2}{3}$$

and

$$x_{n+1} = \frac{(x_n + 1)}{(2x_n + 1)}.$$

Solution: We write the first four terms of the sequence

$$\left\{ \frac{2}{3}, \frac{5}{7}, \frac{12}{17}, \frac{29}{41}, \ldots \right\}.$$

Note that we pass from a/b to

$$\frac{\frac{a}{b} + 1}{\frac{2a}{b} + 1} = (a+b)/(2a+b)$$

from one term of the sequence to the next. To find the limit of this sequence, apply Banach's fixed point theorem, which states for the case of one variable:

Let S be a closed nonempty subset of R. Let f be a contraction mapping on S; f maps S into S such that for some k, $0 < k < 1$, and all x and y in S,

$$|f(x) - f(y)| \leq k |x - y| . \tag{1}$$

Then, there is one and only one point x in S for which $f(x) = x$. In addition, if $x_1 \in S$ and $x_{n+1} = f(x_n)$ for all n, then $x_n \to x$ as $n \to \infty$.

To apply this theorem, first note that for $x \geq 0$,

$$\frac{1}{2} \leq f(x) = \frac{x + 1}{2x + 1} \leq 1$$

because

$$1 - \frac{x + 1}{2x + 1} = \frac{x}{2x + 1} \geq 0 .$$

Therefore, with

$$S = \left[\frac{1}{2} ; 1 \right] ,$$

f maps S to S and is a contraction. To prove this, note that

$$f(x) - f(y) = f'(z)(x - y)$$

where $x < z < y$, by the Mean Value Theorem. Hence,

$$|f(x) - f(y)| = |f'(z)| |x-y| \leq k|x-y|$$

if $|f'(z)| \leq k$. Therefore, to prove that f is a contraction mapping it suffices to show that $|f'(x)| \leq k < 1$ where $x \in S$. Since

$$|f'(x)| = \frac{1}{(2x+1)^2} \leq \frac{1}{4} ,$$

(1) is satisfied. Consequently, for $x, y \in S$ we have

$$|f(x) - f(y)| \leq \frac{1}{4} |x-y|$$

Therefore, by the theorem, $x_n \to x$, where

$$x = f(x) = \frac{x+1}{2x+1} \quad \text{and} \quad \frac{1}{2} \leq x \leq 1 .$$

554

Hence, since

$$x = \frac{x+1}{2x+1} \ ,$$

we have

$$2x^2 + x = x + 1$$

or

$$x^2 = \frac{1}{2} \Longrightarrow x = \left(\frac{1}{2}\right)^{\frac{1}{2}} \quad \left(\frac{1}{2} \leq x \leq 1\right) .$$

Thus,

$$\lim_{n \to \infty} x_n = x = \left(\frac{1}{2}\right)^{\frac{1}{2}} .$$

● **PROBLEM** 27-18

Let $\{a_n\}$ be the sequence

$$a_n = \left(1 + \frac{1}{n}\right)^n .$$

Show that a_n is convergent.

Solution: To solve this problem, we make use of the binomial theorem,

$$(1+b)^n = 1 + nb + \frac{n(n-1)}{1 \cdot 2} b^2 + \frac{n(n-1)(n-2)}{1 \cdot 2 \cdot 3} b^3 + \ldots + b^n . \qquad (1)$$

Note that the coefficient of b^k $(1 \leq k \leq n)$ in this expansion is

$$\frac{n(n-1)(n-2) \ldots (n-k+1)}{k!} .$$

Substituting $\frac{1}{n}$ in place of b in (1) yields

$$\left(1 + \frac{1}{n}\right)^n = 1 + n\left(\frac{1}{n}\right) + \frac{n(n-1)}{1 \cdot 2}\left(\frac{1}{n}\right)^2 + \frac{n(n-1)(n-2)}{1 \cdot 2 \cdot 3}\left(\frac{1}{n}\right)^3$$

$$+ \ldots + \left(\frac{1}{n}\right)^n$$

Now, the expression on the right has n+1 terms, where $(k+1)^{th}$ term is

$$\frac{n(n-1)(n-2) \ldots (n-k+1)}{k! n^k} = \frac{1}{k!} \cdot \left(\frac{n-1}{n}\right)\left(\frac{n-2}{n}\right) \ldots$$

$$\left(\frac{n-k+1}{n}\right)$$

$$= \frac{1}{k!} \cdot (1) \cdot \left(1 - \frac{1}{n}\right)\left(1 - \frac{2}{n}\right) \cdots \left(1 - \frac{k-1}{n}\right) \ .$$

Hence,

$$\left(1 + \frac{1}{n}\right)^n = 1 + 1 + \frac{1 - \frac{1}{n}}{2!} + \frac{\left(1 - \frac{1}{n}\right)\left(1 - \frac{2}{n}\right)}{3!} + \cdots$$

$$+ \frac{\left(1 - \frac{1}{n}\right) \cdots \left(1 - \frac{n-1}{n}\right)}{n!} \ . \tag{2}$$

Now, assume that we follow the same procedure again; however, this time we replace n with n+1 to form a corresponding formula for the expression

$$\left(1 + \frac{1}{n+1}\right)^{n+1} \ .$$

Note that on the right hand side of equation (2), each of the numerators after the first two terms increases if n is replaced by n+1. In addition, the total number of terms on the right is increased from n+1 to n+2. Therefore,

$$\left(1 + \frac{1}{n}\right)^n < \left(1 + \frac{1}{n+1}\right)^{n+1} \ .$$

Hence, the sequence is monotonically increasing.

Furthermore, from (2) we have

$$\left(1 + \frac{1}{n}\right)^n < 1 + 1 + \frac{1}{2!} + \frac{1}{3!} + \cdots + \frac{1}{n!} \ .$$

However, from a previous problem we have

$$1 + \frac{1}{2!} + \frac{1}{3!} + \cdots + \frac{1}{n!} < 2 \ .$$

Therefore,

$$\left(1 + \frac{1}{n}\right)^n < 1 + 1 + \frac{1}{2!} + \cdots + \frac{1}{n!} < 1 + 2 < 3 \ .$$

Thus, the sequence is bounded above. Consequently, since $\{a_n\}$ is a monotonically increasing sequence that is bounded above, it has a limit and is therefore convergent. Note that since

$$\lim_{n \to \infty} \left(1 + \frac{1}{n}\right)^n = e \ ,$$

the exponential function, this limit is denoted by e.

Determine if the series

$$\frac{1}{2} + \frac{1}{3} + \frac{1}{2^2} + \frac{1}{3^2} + \frac{1}{2^3} + \frac{1}{3^3} + \ldots.$$

is convergent or divergent.

Solution: The series can be rewritten as the sum of the sequence of numbers given by

$$a_n = \begin{cases} \dfrac{1}{2^{(n+1)/2}} & \text{if n is odd (n > 0)} \\[4mm] \dfrac{1}{3^{n/2}} & \text{if n is even (n > 0)} \end{cases}$$

Now the ratio test states: If $a_k > 0$ and $\lim\limits_{k \to \infty} \dfrac{a_{k+1}}{a_k} = \ell < 1$,

then
$$\sum_{k=1}^{\infty} a_k$$

converges. Similarly, if

$$\lim_{k \to \infty} \frac{a_{k+1}}{a_k} = \ell \ (1 < \ell \leq \infty) \qquad \text{then}$$

$$\sum_{k=1}^{\infty} a_k \quad \text{diverges.}$$

If $\ell = 1$, the test fails. Therefore applying this test gives:

If a_n is odd,

$$\lim_{n \to \infty} \frac{a_{n+1}}{a_n}$$

$$= \lim_{n \to \infty} \frac{\dfrac{1}{3^{n/2}}}{\dfrac{1}{2^{(n+1)/2}}} = \lim_{n \to \infty} \frac{2^{(n+1)/2}}{3^{n/2}} = \lim_{n \to \infty} \left(\frac{2}{3}\right)^{n/2} 2^{\frac{1}{2}} = 0 \ .$$

If a_n is even,

$$\lim_{n \to \infty} \frac{a_{n+1}}{a_n} = \lim_{n \to \infty} \frac{\frac{1}{2^{(n+1)/2}}}{\frac{1}{3^{n/2}}} = \lim_{n \to \infty} \left(\frac{3}{2}\right)^{n/2} 2^{\frac{1}{2}}$$

and no limit exists.

Hence, the ratio test gives two different values, one < 1 and the other > 1, therefore the test fails to determine if the series is convergent. Thus, another test, known as the root test is now applied. This test states:
Let

$$\sum_{k=1}^{\infty} a_k$$

be a series of nonnegative terms, and let

$$\lim_{n \to \infty} \left(\sqrt[n]{a_n}\right) = S, \text{ where } 0 \le S \le \infty \quad . \quad \text{If:}$$

1) $0 \le S < 1$, the series converges

2) $1 < S \le \infty$, the series diverges

3) $S = 1$, the series may converge or diverge.

Applying this test yields, if a_n is odd

$$\lim_{n \to \infty} \sqrt[n]{a_n} = \lim_{n \to \infty} \sqrt[n]{\frac{1}{2^{(n+1)/2}}} = \lim_{n \to \infty} \sqrt[n]{\frac{1}{2^{n/2}}} \sqrt[n]{\frac{1}{2^{\frac{1}{2}}}}$$

$$= \lim_{n \to \infty} \frac{1}{\sqrt{2}} \frac{1}{2^{1/2n}} = \frac{1}{\sqrt{2}} < 1 \quad .$$

If a_n is even

$$\lim_{n \to \infty} \sqrt[n]{a_n} = \lim_{n \to \infty} \sqrt[n]{\frac{1}{3^{n/2}}} = \frac{1}{\sqrt{3}} < 1 \quad .$$

Thus, since for both cases, the

$$\lim_{n \to \infty} \sqrt[n]{a_n} < 1 \ ,$$

the series converges.

Determine if the following series are absolutely convergent, conditionally convergent or divergent.

a)

$$\sum_{n=1}^{\infty} \frac{(-1)^{n+1}}{n}$$

b)

$$\sum_{n=2}^{\infty} (-1)^n \left(\frac{n}{1 + n^2}\right)^n$$

c)

$$\sum_{n=1}^{\infty} \frac{(-1)^n 2^n}{n!}$$

<u>Solution:</u> a) To determine if the series

$$\sum_{n=1}^{\infty} \frac{(-1)^{n+1}}{n}$$

is convergent or divergent, the following test called
the alternating series test is used. This test states:

An alternating series

$$a_1 - a_2 + a_3 - a_4 + \ldots = \sum_{n=1}^{\infty} (-1)^{n+1} a_n \,, \quad a_n > 0 \,,$$

converges if the following two conditions are satisfied:

    i) its terms are decreasing in absolute value:

$$|a_{n+1}| \leq |a_n| \text{ for } n = 1, 2, \ldots$$

    ii)

$$\lim_{n \to \infty} a_n = 0$$

For this series, the terms are decreasing in absolute
value since

$$1 > \frac{1}{2} > \frac{1}{3} \ldots \quad .$$

Also, the nth term approaches zero so

$$\lim_{n \to \infty} a_n = 0 \; .$$

Hence, the series converges. Next, if $\Sigma|a_n|$ converges also, then the series Σa_n is absolutely convergent. But the series of absolute values is the harmonic series

$$\sum_{n=1}^{\infty} \frac{1}{n}$$

which is known to diverge. Hence,

$$\sum_{n=1}^{\infty} |a_n| \quad \text{does not converge and the series} \quad \sum_{n=1}^{\infty} \frac{(-1)^{n+1}}{n}$$

is conditionally convergent.

b) For the series

$$\sum_{n=2}^{\infty} (-1)^n \left[\frac{n}{n^2 + 1}\right]^n$$

use the root test, which states: Let a series

$$\sum_{n=1}^{\infty} a_n$$

be given and let

$$\lim_{n \to \infty} \sqrt[n]{|a_n|} = R$$

Then if $R < 1$, the series is absolutely convergent. If $R > 1$, the series diverges. If $R = 1$, the test fails.

For this series

$$\lim_{n \to \infty} \sqrt[n]{|a_n|} = \lim_{n \to \infty} \sqrt[n]{\left(\frac{n}{1+n^2}\right)^n} = \lim_{n \to \infty} \frac{n}{1+n^2} = \lim_{n \to \infty} \frac{1}{\frac{1}{n} + n} = 0$$

Therefore the series converges absolutely.

c) For the series

$$\sum_{n=1}^{\infty} \frac{(-1)^n\, 2^n}{n!} \quad,$$

use the ratio test. This states that if $a_n \neq 0$ for $n = 1, 2, \ldots$ and

$$\lim_{n \to \infty} \left| \frac{a_{n+1}}{a_n} \right| = L$$

then if $L < 1$,

$$\sum_{n=1}^{\infty} a_n$$

is absolutely convergent, if $L = 1$, the test fails, if $L > 1$,

$$\sum_{n=1}^{\infty} a_n \qquad \text{is divergent.}$$

Here

$$\lim_{n \to \infty} \left| \frac{a_{n+1}}{a_n} \right| = \lim_{n \to \infty} \frac{2^{n+1}}{(n+1)!} \cdot \frac{n!}{2^n} = \lim_{n \to \infty} \frac{2^{n+1} \, 2^{-n} \, (1 \cdot 2 \cdot 3 \, \ldots \, n)}{1 \cdot 2 \cdot 3 \cdot n \cdot (n+1)}$$

$$= \lim_{n \to \infty} \frac{2}{n+1} = 0 \; .$$

Hence $L = 0$ and the series converges absolutely.

● **PROBLEM** 27-21

a) Define the general form, radius of convergence, and interval of convergence of a power series.

b) Given the power series $\Sigma a_n x^n$, show that the series converges, if given that:

(i)
$$R = 1/L \quad \text{where} \quad L = \lim_{n \to \infty} \left| \frac{a_{n+1}}{a_n} \right|$$

exists and $|x| < R$ or

(ii)
$$R = 1/\alpha \quad \text{where} \quad \alpha = \lim_{n \to \infty} \sup \sqrt[n]{|a_n|} \quad \text{and} \quad |x| < R.$$

Solution: By definition the general form of a power series in powers of $x - x_0$, where x_0 is fixed and x is variable, is

$$\sum_{n=0}^{\infty} a_n (x-x_0)^n = a_0 + a_1 (x-x_0) + a_2 (x-x_0)^2 + \ldots$$

$$+ a_n(x-x_0)^n + \ldots . \tag{1}$$

However, in the study of power series it is convenient to look at a series of the form,

$$a_0 + a_1x + a_2x^2 + \ldots = \sum_{n=0}^{\infty} a_n x^n ;$$

this is a power series in x (i.e., $x_0 = 0$). The reason for this is that equation (1) can be reduced to the above form by the substitution t = x - a.

Furthermore, in general a power series converges absolutely for $|x| < R$ and diverges for $|x| > R$, where the constant R is called the radius of convergence of the series. However the series may or may not converge at the end points x = ±R. That is, it may converge at both, at just one, or at neither. In addition, the interval of convergence of the series is the interval -R < x < R, with the possible inclusion of the endpoints.

b) For the power series $\Sigma a_n x^n$ if we are given that

(i)
$$R = \frac{1}{L}$$

where

$$L = \lim_{n \to \infty} \left| \frac{a_{n+1}}{a_n} \right|$$

exists, for this case let $U_n = a_n x^n$. Then

$$\lim_{n \to \infty} \left| \frac{U_{n+1}}{U_n} \right| = \lim_{n \to \infty} \left| \frac{a_{n+1}}{a_n} \right| |x| = L|x| .$$

However, by the ratio test, the series ΣU_n converges absolutely if

$$\lim_{n \to \infty} \left| \frac{U_{n+1}}{U_n} \right| < 1 ,$$

and diverges if

$$\lim_{n \to \infty} \left| \frac{U_{n+1}}{U_n} \right| > 1 .$$

Consequently, for convergence

$$L|x| = \frac{|x|}{R} < 1 .$$

Therefore, the power series converges absolutely if $|x| < R$, diverges if $|x| > R$. Also if L = 0 we have R = ∞ , and if L = ∞ then R = 0.

(ii) For this case, again let $U_n = a_n x^n$. Then

$$\lim_{n \to \infty} \sup \sqrt[n]{|U_n|} = \lim_{n \to \infty} \sup \sqrt[n]{|a_n x^n|}$$

$$= |x| \lim_{n \to \infty} \sup \sqrt[n]{|a_n|}$$

$$= |x| \alpha = \frac{|x|}{R}$$

However, by the root test, the series ΣU_n converges absolutely if

$$\lim_{n \to \infty} \sup \sqrt[n]{|U_n|} < 1$$

and diverges if

$$\lim_{n \to \infty} \sup \sqrt[n]{|U_n|} > 1 .$$

Therefore, for convergence of the power series

$$\frac{|x|}{R} < 1 \quad \text{or} \quad |x| < R \ ;$$

if $|x| > R$, the power series diverges. In addition, if $\alpha = 0$, $R = + \infty$; if $\alpha = + \infty$, $R = 0$.

We make note that the second method of finding R is more powerful than the first. That is the limit,

$$\lim_{n \to \infty} \left| \frac{a_{n+1}}{a_n} \right|$$

does not exist for certain power series.

Find the radius of convergence of the following power series:

a)
$$\sum_{n=1}^{\infty} \frac{x^n}{n}$$

b)
$$\sum_{n=1}^{\infty} \frac{(2n)!}{(n!)^2} x^n$$

c)
$$\sum_{n=1}^{\infty} \frac{(3n)!}{(n!)^2} x^n$$

Solution: a)

$$\sum_{n=1}^{\infty} \frac{x^n}{n} = x + \frac{x^2}{2} + \frac{x^3}{3} + \cdots .$$

Here we let

$$U_n = \frac{x^n}{n}$$

so that

$$\lim_{n \to \infty} \left| \frac{U_{n+1}}{U_n} \right| = \lim_{n \to \infty} \left| \frac{x^{n+1}}{n+1} \cdot \frac{n}{x^n} \right| = \lim_{n \to \infty} |x| \frac{n}{n+1} = |x| .$$

Therefore, R = 1 and the series converges for -1 < x < 1. Note that the series may or may not converge at the end points x = ±1. That is, it may converge at both x = ±1, at just one, or neither. For example, given the endpoints ±1, we have at x = 1 the series,

$$\sum_{n=1}^{\infty} \frac{1}{n} ,$$

which is the divergent harmonic series. For x = -1 we have the alternating series,

$$\sum_{n=1}^{\infty} \frac{(-1)^n}{n}$$

where

$$\left| \frac{(-1)^{n+1}}{n+1} \right| < \left| \frac{(-1)^n}{n} \right|$$

and

$$\lim_{n \to \infty} \frac{(-1)^n}{n} = 0$$

Hence this series is convergent. Thus, $R = 1$ and the series converges absolutely for $-1 \le x < 1$.

b)

$$\sum_{n=1}^{\infty} \frac{(2n)!}{(n!)^2} x^n = \frac{2!}{(1)^2} x + \frac{4!}{(2!)^2} x^2 + \frac{6!}{(3!)^2} x^3 + \dots \; .$$

Here we have

$$a_n = \frac{(2n)!}{(n!)^2} \quad \text{and} \quad U_n = \frac{(2n)!}{(n!)^2} x^n$$

$$\lim_{n \to \infty} \left| \frac{U_{n+1}}{U_n} \right| = \lim_{n \to \infty} \left| \frac{(2n+2)!}{[(n+1)!]^2} x^{n+1} \cdot \frac{(n!)^2}{(2n)! x^n} \right| =$$

$$= \lim_{n \to \infty} |x| \frac{(2n+2)(2n+1)}{(n+1)^2} = 4|x| \; .$$

Therefore $R = \frac{1}{4}$ since we want $4|x| < 1$.

c)

$$\sum_{n=1}^{\infty} \frac{(3n)!}{(n!)^2} x^n = 3! x + \frac{6!}{(2!)^2} x^2 + \frac{9!}{(3!)^2} x^3 + \dots \; .$$

Here,

$$a_n = \frac{(3n)!}{(n!)^2} \quad \text{and} \quad U_n = \frac{(3n)!}{(n!)^2} x^n$$

$$\lim_{n \to \infty} \left| \frac{U_{n+1}}{U_n} \right| = \lim_{n \to \infty} \left| \frac{(3n+3)! \; x^{n+1}}{[(n+1)!]^2} \cdot \frac{(n!)^2}{(3n)! \; x^n} \right|$$

$$= \lim_{n \to \infty} x \frac{(3n+3)(3n+2)(3n+1)}{(n+1)^2} = \infty$$

Therefore, the series does not converge for any value of x except $x = 0$. This means that

$$R = 0 \left(\text{i.e., } R = \frac{1}{\infty} \right) \; .$$

Considering all possibilities, find the radius of convergence of the series

$$\sum \frac{(pn)!}{(n!)^q} x^n \, ,$$

where p is a positive integer and q > 0.

<u>Solution</u>: To find R for the given series, three cases must be considered. They are the cases (i) p = q , (ii) p > q , (iii) p < q . However, for each case we will need the ratio test. Hence, in general,

$$\lim_{n\to\infty} \left| \frac{U_{n+1}}{U_n} \right| = \lim_{n\to\infty} \left| \frac{(pn+p)! \; x^{n+1}}{[(n+1)!]^q} \cdot \frac{(n!)^q}{(pn)! x^n} \right|$$

$$= |x| \lim_{n\to\infty} \frac{(pn+p)(pn+p-1) \ldots (pn+p-(p-1))}{(n+1)^q}$$

$$= |x| \lim_{n\to\infty} \frac{p(n+1)p\left(n+1 - \frac{1}{p}\right) \ldots p\left(n+1 - \frac{p-1}{p}\right)}{(n+1)^q} \, .$$

Now since there are p number of terms in the numerator (this because $\frac{(n+x)!}{n!}$ has, after cancellations, x terms in the numerator), we have

$$\lim_{n\to\infty} \left| \frac{U_{n+1}}{U_n} \right| = |x| \; p^p \lim_{n\to\infty} \frac{(n+1)\left(n+1 - \frac{1}{p}\right) \ldots \left(n + \frac{1}{p}\right)}{(n+1)^q} \quad (1)$$

Case (1): Assume p = q , then

$$\lim_{n\to\infty} \left| \frac{U_{n+1}}{U_n} \right| = |x| \; p^p \lim_{n\to\infty} \frac{(n+1)\left(n+1 - \frac{1}{p}\right) \ldots \left(n + \frac{1}{p}\right)}{(n+1)^p}$$

$$= x \; p^p \lim_{n\to\infty} \frac{(n+1)}{(n+1)} \frac{\left(n+1 - \frac{1}{p}\right)}{(n+1)} \ldots \frac{\left(n + \frac{1}{p}\right)}{(n+1)}$$

$$= |x| \; p^p \lim_{n\to\infty} \frac{(n+1)}{(n+1)} \lim_{n\to\infty} \frac{\left(n+1 - \frac{1}{p}\right)}{(n+1)} \ldots \lim_{n\to\infty} \frac{\left(n + \frac{1}{p}\right)}{(n+1)} \quad (2)$$

since

$$\lim_{n\to\infty} (a_1 \cdot a_2 \cdot a_3 \cdot \ldots a_k) = \lim_{n\to\infty} a_1 \lim_{n\to\infty} a_2 \ldots \lim_{n\to\infty} a_k \, .$$

Therefore,

$$\lim_{n\to\infty}\left|\frac{U_{n+1}}{U_n}\right| = |x|\, p^p$$

since each limit in (2) equals 1. Consequently, the series converges if $|x| < p^{-p}$ therefore $R = p^{-p}$.

Case (ii): Assume $p > q$, then from (1),

$$\lim_{n\to\infty}\left|\frac{U_{n+1}}{U_n}\right| = |x|\, p^p \lim_{n\to\infty}\frac{(n+1)\left(n+1-\frac{1}{p}\right)\cdots\left(n+\frac{1}{p}\right)}{(n+1)^q} \tag{3}$$

$$= |x|\, p^p \lim_{n\to\infty}\frac{(n+1)}{(n+1)}\frac{\left(n+1-\frac{1}{p}\right)}{(n+1)}\cdots\frac{\left(n+1-\frac{q-1}{p}\right)\left(n+1-\frac{q}{p}\right)}{(n+1)}$$

$$\times \cdots \times \left(n+\frac{1}{p}\right)$$

$$= |x|\, p^p \lim_{n\to\infty}\frac{(n+1)}{(n+1)}\lim_{n\to\infty}\frac{\left(n+1-\frac{1}{p}\right)}{(n+1)}\cdots\lim_{n\to\infty}\frac{\left(n+1-\frac{q-1}{p}\right)}{(n+1)}$$

$$\lim_{n\to\infty}\left(n+1-\frac{q}{p}\right)\cdots\left(n+\frac{1}{p}\right)\ .$$

Now the first q limits each equal 1, however, the last limit

$$\lim_{n\to\infty}\left(n+1-\frac{q}{p}\right)\cdots\left(n+\frac{1}{p}\right) = \infty\ ,$$

hence,

$$\lim_{n\to\infty}\left|\frac{U_{n+1}}{U_n}\right| = \infty\ ,\ \text{therefore } R = 0\ .$$

Case (iii): Assume $p < q$, then from (3),

$$\lim_{n\to\infty}\left|\frac{U_{n+1}}{U_n}\right| = |x|\, p^p \lim_{n\to\infty}\frac{(n+1)}{(n+1)}\frac{\left(n+1-\frac{1}{p}\right)}{(n+1)}\cdots\frac{\left(n+\frac{1}{p}\right)}{(n+1)}\frac{1}{(n+1)^{q-p}}$$

$$\lim_{n\to\infty}\left|\frac{U_{n+1}}{U_n}\right| = |x|\, p^p \lim_{n\to\infty}\frac{1}{(n+1)^{q-p}} = |x|\, p^p \cdot 0 = 0$$

Therefore, for this case $R = \infty$.

Summarizing, we have

$$R = \begin{cases} p^{-p} & \text{for } p = q \\ 0 & \text{for } p > q \\ \infty & \text{for } p < q \end{cases}$$

SECTION 2 — PLANE TRIGONOMETRY

CHAPTER 28

ANGLES AND ARCS

> **Basic Attacks and Strategies for Solving Problems in this Chapter. See pages 569 to 577 for step-by-step solutions to problems.**

An angle can be measured either in terms of *degrees* or *radians*. Both are related to a full circle which has a measure of 360° or 2π radians (note that 2π is the circumference of a circle with a radius equal to one). To convert between degree measure and radian measure, one merely writes an equation of two fractions with each denominator being the measure of a full circle (or half circle) according to each different measure, and each numerator being the corresponding measure of the angle. For example, to convert 45° to radians, we write

$$\frac{45}{360} = \frac{r}{2\pi}.$$

Solving for *r*, we get

$$r = \frac{\pi}{4}.$$

A section of the circumference of a circle corresponds to a certain angle measured at the center of a circle (by drawing a pie shaped region from the circumference to the center). Thus, we can relate an angle to a portion of a "walk" around the circumference of a circle. In this sense, 360° can be interpreted to mean that one has walked completely around a circle and is back at the starting point. In that sense, 0° has the same endpoint as 360° and these angles are said to be *coterminal*. Angles between 0° and 360° are called primary and other angles can be found to be coterminal with primary angles by subtracting (or adding, if they are negative) multiples of 360° or 2π radians.

Dividing an angle by the measure of a full circle gives the fraction of the circle it measures. This fraction can be used to compute the area of the sector the angle slices or the length of the arc on the circumference of a circle it marks. One computes the appropriate area of a circle ($\pi \cdot r^2$) or circumference ($2\pi r$) and then multiplies by the fraction of the circle that the angle measures. It is also possible to convert from the length of an arc or the area of the sector to the associated central angle, given the radius of the circle.

The Pythagorean theorem (i.e.,

$$a^2 + b^2 = c^2,$$

where c is the length of the hypotenuse and a and b are the lengths of the other two sides of a triangle) can be used to confirm that a given angle is indeed a right (90°) angle.

Step-by-Step Solutions to Problems in this Chapter, "Angles and Arcs"

● PROBLEM 28-1

Complete the following table:

Width of θ in radians	1	2	3	4	5	6	7	8	9
	0	$\frac{1}{6}\pi$	$\frac{1}{4}\pi$		$\frac{1}{2}\pi$	$\frac{2}{3}\pi$			π
Width of θ in degrees	$0°$			$60°$			$135°$	$150°$	

Solution: If an angle θ is A degrees wide and also t radians wide, then the numbers A and t are related by the equation:

$$\frac{A}{180°} = \frac{t}{\pi} \qquad (1)$$

Thus, equation (1) can be used to complete the table. For column (2):

$$\frac{A}{180°} = \frac{1/6\ \pi}{\pi}$$

$$\frac{A}{180°} = 1/6$$

Multiplying both sides by $180°$,

$$180°\left(\frac{A}{180°}\right) = 180°(1/6)$$

$$A = 30°$$

For column (3):

$$\frac{A}{180°} = \frac{\cancel{\frac{1}{4}}\,\cancel{\pi}}{\cancel{\pi}}$$

$$\frac{A}{180°} = \frac{1}{4}$$

Multiplying both sides by 180°,

$$180°\left(\frac{A}{180°}\right) = 180°\,(\tfrac{1}{4})$$

$$A = 45°$$

For column (4):

$$\frac{60°}{180°} = \frac{t}{\pi}$$

$$\frac{1}{3} = \frac{t}{\pi}$$

Multiplying both sides by π,

$$\pi(1/3) = \cancel{\pi}(t/\cancel{\pi})$$
$$1/3\ \pi = t$$

For column (5):

$$\frac{A}{180°} = \frac{\cancel{\frac{1}{2}}\,\cancel{\pi}}{\cancel{\pi}}$$

$$\frac{A}{180°} = \frac{1}{2}$$

Multiplying both sides by 180°,

$$180°\left(\frac{A}{180°}\right) = 180°\,(\tfrac{1}{2})$$

$$A = 90°$$

For column (6):

$$\frac{A}{180°} = \frac{2/3\ \cancel{\pi}}{\cancel{\pi}}$$

$$\frac{A}{180°} = 2/3$$

Multiplying both sides by 180°,

$$180°\left(\frac{A}{180°}\right) = 180°\,(2/3)$$
$$A = 120°$$

For column (7):

$$\frac{135°}{180°} = \frac{t}{\pi}$$

$$\frac{27}{36} = \frac{t}{\pi}$$

$$\frac{3}{4} = \frac{t}{\pi}$$

Multiplying both sides by π,

$$\pi(3/4) = \cancel{\pi}(t/\cancel{\pi})$$

$$3/4 \ \pi = t$$

For column (8):

$$\frac{150^\circ}{180^\circ} = \frac{t}{\pi}$$

$$\frac{50}{60} = \frac{t}{\pi}$$

$$\frac{5}{6} = \frac{t}{\pi}$$

Multiplying both sides by π , $\pi\left(\frac{5}{6}\right) = \cancel{\pi}\left(\frac{t}{\cancel{\pi}}\right).$

$$\frac{5}{6} \ \pi = t$$

For column (9):

$$\frac{A}{180^\circ} = \frac{\cancel{\pi}}{\cancel{\pi}}$$

$$\frac{A}{180^\circ} = 1$$

Multiplying both sides by 180°,

$$\cancel{180}^\circ\left(\frac{A}{\cancel{180}^\circ}\right) = 180^\circ(1)$$

$$A = 180^\circ$$

All of the computed values are now put into the table as follows:

	1	2	3	4	5	6	7	8	9
Width of θ in radians	0	$\frac{1}{6}\pi$	$\frac{1}{4}\pi$	$\frac{1}{3}\pi$	$\frac{1}{2}\pi$	$\frac{2}{3}\pi$	$\frac{3}{4}\pi$	$\frac{5}{6}\pi$	π
Width of θ in degrees	0°	30°	45°	60°	90°	120°	135°	150°	180°

● PROBLEM 28-2

What primary angle is coterminal with the angle of - 743°?

Solution:
$$- 743° = \alpha - n \cdot 360° \qquad (1)$$

$$- 743° = \alpha - 3 \cdot 360° = \alpha - 1080° \quad (2)$$

Multiply both sides of equation (2) by - 1.

$$- 1(- 743°) = - 1(\alpha - 1080°)$$

$$743° = -\alpha + 1080°$$

$$743° = 1080° - \alpha \qquad (3)$$

Note that the positive integer value chosen for n
results in an angle (in equation (3)) which is larger
than but closest to the angle of 743°. Also,

$$0° \le \alpha \le 360°.$$

From equation (3),

$$\alpha = 1080° - 743° = 337°.$$

● **PROBLEM 28-3**

What primary angle is coterminal with the angle of 1243°?

Solution: Using the formula, $\beta = \alpha + n \cdot 360°$, obtain
$1243° = \alpha + n \cdot 360°$. In this formula choose the largest
positive integer for n which, when multiplied by 360,
is closest to but smaller than the given angle. Also, α
is an angle between 0° and 360°, that is

$$0° \le \alpha \le 360°.$$

Since $3 \cdot 360° = 1080°$ and $4 \cdot 360° = 1440°$, n = 3, and
$1243° = \alpha + 1080°$ or $\alpha = 1243° - 1080° = 163°$. Thus, the
angles 1243° and 163° are coterminal.

● **PROBLEM 28-4**

What primary angle is coterminal with the angle of $5\frac{1}{4}\pi$ radians?

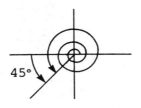

45°

Solution: The figure illustrates the angle of $5\frac{1}{4}\pi$ radians.
Note that the angle of $5\frac{1}{4}\pi$ radians has a reference angle of 45°.
However, we seek a primary angle (an angle between 0° and 360°)
which is coterminal with $5\frac{1}{4}\pi$; that is, which has the same terminal
side as an angle of $5\frac{1}{4}\pi$ radians. Also, since a primary angle is a
positive angle, its initial side is the positive x-axis and the
angle revolves in the counter-clockwise direction. Therefore, a
primary angle with the same terminal side as an angle of $5\frac{1}{4}\pi$ radians
is $(180° + 45°) = 225° = (\pi + \pi/4)$ radians = $5/4 \pi$ radians.

If θ is an angle of 30°, what is its width in radians?

<u>Solution:</u> If an angle θ is A degrees wide and also t radians wide, then the numbers A and t are related by the equation:

$$\frac{A}{180°} = \frac{t}{\pi} \qquad (1)$$

If the given angle of 30° is t radians wide, then by using equation (1):

$$\frac{30°}{180°} = \frac{t}{\pi}$$

$$\frac{1}{6} = \frac{t}{\pi}$$

Multiplying both sides by π,

$$\pi\left(\frac{1}{6}\right) = \pi\left(\frac{t}{\pi}\right)$$

$$\frac{1}{6}\pi = t$$

Hence, the angle's width in radians is $\frac{1}{6}\pi$.

Find the area of a sector in which the measure of the central angle is 60° and the radius of the circle is 2.

<u>Solution:</u> Note the diagram.

The central angle, 60° is $\frac{1}{6} \times (360°)$; that is, 60° = $\frac{1}{6} \times$ (circumference of circle). Therefore, one-sixth of the area of the circle is covered. Using the fact that the area of the total circumference of the circle is πr^2, or A = πr^2 : the area covered by the sector given in this problem is

$$A = \frac{1}{6}\pi r^2 = \frac{1}{6}\pi(2)^2 = \frac{1}{6}\pi(4) = \frac{4\pi}{6}$$

$$= \frac{2\pi}{3}.$$

Find the length of the minor arc of a circle of radius 1 whose central angle has a measure of 90°.

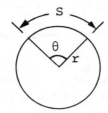

Solution: For problems dealing with arc length of a circle, the following information is helpful. On a circle of radius r, a central angle of θ radians intercepts an arc of length

$$s = r\theta;$$

that is, arc length = radius x central angle in radians. (See Figure). Therefore, in this problem:

$$\text{length of minor arc} = s = (1)(90°)$$
$$= (1)\left(\frac{\pi}{2} \text{ radians}\right)$$
$$= \frac{\pi}{2} \text{ radians}$$

We could have obtained this answer by noting that a central angle of 90° yields an arc equal to ¼ the circumference, since

$$90° = \tfrac{1}{4}(360°) = \tfrac{1}{4} \times (\text{circumference of circle})$$
$$= \tfrac{1}{4}(2\pi \text{ radians})$$
$$= \frac{2\pi}{4} \text{ radians}$$
$$= \frac{\pi}{2} \text{ radians}$$

● PROBLEM 28-8

What length of arc is subtended by a central angle of 75° on a circle 13.7 inches in radius?

Solution: Let θ denote a central angle in a circle of radius r, and let s be the length of the intercepted arc, measured in the same units as the radius. Then if θ is an angle measured in radians, the length of the arc, s, is s = rθ.

We must express the given angle in radians. Since 2π radians = 360° then 1° = 2π/360 radians = π/180 radians. Thus, 75° = 75 · π/180 radians = 5/12 π radians.

Substituting this value into the relation s = rθ, we have

$$s = 13.7 \left[\frac{5}{12}\pi\right] \text{ inches} = (5.71 \ \pi) \text{ inches}$$

$$= (5.71)(3.14) \text{ inches}$$

$$\approx 17.9 \text{ inches.}$$

● **PROBLEM** 28-9

If the central angle of a circle of radius 5 in. is 30 , what is the length of the intercepted arc, and, what is the area of the sector?

Solution: To find the length of the intercepted arc, use the formula

(1) $s = r\theta$, where s is the length of the arc of a circle whose radius is r and whose central angle θ is expressed in radians. We are given the radius and the angle which is measured in degrees. We first convert the given angle, 30˚, to radian measure; we have

$$\frac{\theta}{2\pi} = \frac{30˚}{360˚} \quad ,$$

$$\theta = \frac{\pi}{6} \quad .$$

Hence equation (1) gives

$$s = r\theta = 5 \times \frac{\pi}{6} = 2.618 \text{ in., approx.,}$$

for the length of arc. To find the area of the Sector, A, apply the formula

(2) $\qquad\qquad\qquad A = \frac{1}{2} r^2 \theta \ .$

Thus equation (2) yields for the area

$$A = \frac{1}{2} \times 5^2 \times \frac{\pi}{6} = 6.545 \text{ in.}^2 \text{ approx.}$$

● **PROBLEM** 28-10

In a circle whose radius is 8 inches, find the number of degrees contained in the central angle whose arc length is 2π inches.

Solution: The measure of a central angle is equal to the measure of the arc it intercepts.

The ratio of arc length to circumference, in linear units, will be equal to the ratio of arc length to circumference as measured in degrees.

If n = the number of degrees in the arc 2π inches long,

Then $\qquad\qquad \dfrac{\text{length of arc}}{\text{circumference}} = \dfrac{n}{360°}$

By substitution, $\qquad \dfrac{2\pi \text{in.}}{2\pi(8 \text{ in.})} = \dfrac{n}{360°}$

$$\frac{1}{8} = \frac{n}{360°}$$

$$360° = 8n$$

$$n = 45°$$

Therefore, the central angle contains 45°.

The legs of a right triangle are 3 feet and 4 feet in length. What is the length of the hypotenuse of the triangle?

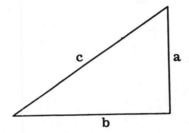

Solution: Apply the Pythagorean Theorem, which states that the sum of the squares of the lengths of two legs, a and b, of a right triangle is equal to the square of the length of the hypotenuse, c: $a^2 + b^2 = c^2$.

We are given that a = 3, b = 4, and substituting in the above formula,

$$3^2 + 4^2 = c^2$$

$$9 + 16 = c^2$$

$$25 = c^2$$

$$5 = c$$

Therefore, the hypotenuse of the triangle is 5 feet.

The lengths of the sides of a triangle are 8, 15, and 17. Show that the triangle is a right triangle.

Solution: This problem requires the use of the converse of the Pythagorean Theorem, which states that if the square of the length of one side of a triangle is equal to the sum of the squares of the lengths of the other two sides, then the triangle is a right triangle.

Let x = 17, the longest side of the triangle and let y = 8 and z = 15.

Then, $x^2 = (17)^2 = 289$

and $y^2 + z^2 = (8)^2 + (15)^2 = 64 + 225 = 289.$

Since $17^2 = 8^2 + 15^2$, the triangle is a right triangle.

CHAPTER 29

DEFINITIONS OF THE TRIGONOMETRIC FUNCTIONS

> **Basic Attacks and Strategies for Solving Problems in this Chapter. See pages 578 to 588 for step-by-step solutions to problems.**

The six trigonometric functions are defined in terms of the sides of a right triangle.

The names of the six functions are sine (abbreviated *sin*), cosine (*cos*), tangent (*tan*), cotangent (*cot*), secant (*sec*), and cosecant (*csc*). They are defined as follows:

$$\sin \alpha = \frac{\text{opposite}}{\text{hypotenuse}}$$

$$\cos \alpha = \frac{\text{adjacent}}{\text{hypotenuse}}$$

$$\tan \alpha = \frac{\text{opposite}}{\text{adjacent}}$$

$$\cot \alpha = \frac{1}{\tan \alpha} = \frac{\text{adjacent}}{\text{opposite}}$$

$$\sec \alpha = \frac{1}{\cos \alpha} = \frac{\text{hypotenuse}}{\text{adjacent}}$$

$$\csc \alpha = \frac{1}{\sin \alpha} = \frac{\text{hypotenuse}}{\text{opposite}}$$

As can be seen, three functions are merely reciprocals of the three other functions, but can also be expressed in terms of ratios of certain sides of a triangle. It is also common to express tan α as $\frac{\sin \alpha}{\cos \alpha}$.

If one depicts angles looking at a circle centered at the origin in the Cartesian plane, the point at which the circle intersects the positive x-axis is the location of the zero angle and positive angles are measured from the positive x-axis counter-clockwise. Angles between 0° and 90° are located in the quadrant where both x and y are positive, and this area is called quadrant I. Angles between 90° and 180° are located in quadrant II, where x is negative but y is positive. Angles be-tween 180° and 270° are located in quadrant III, where both x and y are negative. Angles between 270° and 360° are in quadrant IV, where x is positive but y is negative. Negative angles are measured clockwise from the positive x-axis and in most cases are rewritten in terms of a corresponding, coterminal positive angle before any calculations are performed. Similarly, angles greater than 360° are reduced to a coterminal angle less than 360°.

Depending on the quadrant, the values of the various trigonometric functions are positive or negative. The sign of the function can always be determined by first drawing the proper triangle within the unit circle. The adjacent side is measured along the positive or negative x-axis and the opposite side is drawn from a point on the circle to the x-axis. For example, consider an angle α between 90° and 180°. Then the triangle drawn in quadrant II actually has a central angle (180 − α) at the origin, and the "opposite" side is opposite this angle. If the adjacent side is along the negative x-axis, its value is negative. If the opposite side is below the x-axis, its value is also negative. The hypotenuse is always considered posi-tive.

In quadrant I, all trigonometric functions are positive.

In quadrant II, sine is positive, but cosine and tangent are negative.

In quadrant III, tangent is positive, but sine and cosine are negative.

In quadrant IV, cosine is positive, but sine and tangent are negative.

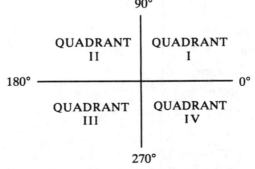

The signs of cotangent, secant and cosecant can be found by looking at the associated functions of which they are reciprocals. In fact, the sign of tangent can be determined by inspection of the signs of sine and cosine.

Given any two sides of a right triangle, the third side can always be found using the Pythagorean theorem (*cf.* Chapter 28). Once the three sides of a triangle have been determined, all six trigonometric functions can be computed. Thus, given any one trigonometric function, one can determine the other five trigonometric functions. (Given one trigonometric function, one can label two appropriate sides of a right triangle with arbitrary lengths, as long as the ratios give the proper value for the given trigonometric function. For example, given that sin α = .8 = $^4/_5$, one can arbitrarily label a triangle so that the opposite side has length 4 and the hypotenuse has length 5.) The signs of the functions may have to be determined by other means, however.

From plane geometry, it is known that similar triangles have equal corresponding angles and proportional sides. Using this fact, we can use trigonometric functions and known lengths of sides of smaller triangles to compute the sides of larger triangles. For example, given a distance of 40 meters from the base of a tree, and given that the angle of the line of sight to the top of the tree with the horizontal is 26.565°, we can determine the height of the tree. The height (opposite) over the distance from the tree (adjacent) gives the tangent of the angle. The value of tan 26.565° is 0.500. Thus, the height of the tree is half of the distance to the tree, that is, the height is 20 meters.

● **PROBLEM** 29-1

Construct a table to indicate the signs of all the trig-onometric functions for all four quadrants.

Solution:

Quadrant	sinα	cosα	tanα	cotα	secα	cscα
I	+	+	+	+	+	+
II	+	−	−	−	−	+
III	−	−	+	+	−	−
IV	−	+	−	−	+	−

● **PROBLEM** 29-2

P is a point on the terminal side of an angle θ. Find the values of the trigonometric functions of θ (smallest positive angle in standard position) if the point P is (1) (4,0), (2), (0,1), (3), (−1,0), and (4) (0,−9).

Solution: (1) θ = 0°

(2) θ = 90°

(3) θ = 180°

(4) θ = 270°

The values of the trigonometric functions for each value of θ are given in the table below.

θ	sinθ	cosθ	tanθ	cotθ	secθ	cscθ
0⁰	0	1	0	±∞	1	±∞
90⁰	1	0	±∞	0	±∞	1
180⁰	0	-1	0	±∞	-1	±∞
270⁰	-1	0	±∞	0	±∞	-1

● **PROBLEM** 29-3

Given the right triangle with a = 3, b = 4, and c = 5, find the values of the trigonometric functions of α.

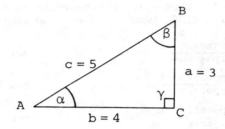

Solution: In the accompanying figure, a is the side opposite angle α, b is the side opposite angle β, and c is the side opposite angle γ. The values of the trigonometric functions of α are:

$$\cos \alpha = \frac{\text{adjacent side}}{\text{hypotenuse}}, \quad \sin \alpha = \frac{\text{opposite side}}{\text{hypotenuse}},$$

$$\tan \alpha = \frac{\text{opposite side}}{\text{adjacent side}}, \quad \cot \alpha = \frac{1}{\tan \alpha},$$

$$\sec \alpha = \frac{1}{\cos \alpha}, \quad \text{and} \quad \csc \alpha = \frac{1}{\sin \alpha}.$$

Therefore: $\cos \alpha = \frac{4}{5}, \qquad \sin \alpha = \frac{3}{5}$,

$\tan \alpha = \frac{3}{4}, \qquad \cot \alpha = \frac{1}{3/4} = \frac{4}{3}$,

$\sec \alpha = \frac{1}{4/5} = \frac{5}{4}, \csc \alpha = \frac{1}{3/5} = \frac{5}{3}$.

● **PROBLEM** 29-4

Calculate the values of the six trigonometric functions at the point $\frac{1}{3}\pi$.

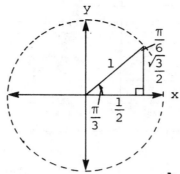

Solution: To find the trigonometric point $P(\frac{1}{3}\pi)$, proceed around the unit circle in a counterclockwise direction, since $\frac{\pi}{3}$ is a positive angle. Recall that $\sin 60°$ i.e., $\sin(\pi/3) = \sqrt{3}/2$. Now, using the Pythagorean theorem and the fact that the hypotenuse is unity because it is a unit circle we can compute the third side, which we find to be $1/2$ (see figure). Therefore, the coordinates of the trigonometric point $P(\frac{1}{3}\pi)$ are $(1/2, 1/2\sqrt{3})$. Hence, we apply the following equations:

$$\cos \theta = \frac{\text{adjacent side}}{\text{hypotenuse}} \qquad \sec \theta = \frac{1}{\cos \theta} = \frac{\text{hypotenuse}}{\text{adjacent side}}$$

$$\sin \theta = \frac{\text{opposite side}}{\text{hypotenuse}} \qquad \csc \theta = \frac{1}{\sin \theta} = \frac{\text{hypotenuse}}{\text{opposite side}}$$

$$\tan \theta = \frac{\text{opposite side}}{\text{adjacent side}} \qquad \cot \theta = \frac{\cos \theta}{\sin \theta} = \frac{\text{adjacent side}}{\text{opposite side}}$$

Thus,

$$\cos \frac{1}{3}\pi = \frac{1}{2}, \qquad \sec \frac{1}{3}\pi = 2,$$

$$\sin \frac{1}{3}\pi = \frac{1}{2}\sqrt{3}, \qquad \csc \frac{1}{3}\pi = 2/\sqrt{3} = 2/\sqrt{3} \cdot \sqrt{3}/\sqrt{3}$$
$$= 2/3 \sqrt{3}$$

$$\tan \frac{1}{3}\pi = \sqrt{3}, \qquad \cot \frac{1}{3}\pi = 1/\sqrt{3} = 1/\sqrt{3} \cdot \sqrt{3}/\sqrt{3}$$
$$= 1/3 \sqrt{3}$$

● **PROBLEM 29-5**

Find the values of the trigonometric functions of an angle of 300°.

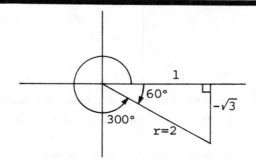

Solution: An angle of 300° is a fourth quadrant angle and its reference angle is an angle of 60°. In the fourth quadrant the sine, tangent, cotangent and cosecant functions are negative. This yields

$$\sin 300° = \sin 60° = -\frac{\sqrt{3}}{2},$$

$$\cos 300° = \cos 60° = \frac{1}{2},$$

$$\tan 300° = \tan 60° = -\sqrt{3},$$

$$\cot 300° = \cot 60° = -\frac{\sqrt{3}}{3},$$

$$\sec 300° = \sec 60° = \frac{1}{\cos 60°} = \frac{1}{\frac{1}{2}} = 2, \text{ and}$$

$$\csc 300° = \csc 60° = \frac{1}{\sin 60°} = \frac{1}{-\sqrt{3}/2} = -\frac{2}{\sqrt{3}}$$

$$= -\frac{2\sqrt{3}}{\sqrt{3}\sqrt{3}} = -\frac{2\sqrt{3}}{3}.$$

● **PROBLEM 29-6**

Given that tan θ = 2 and cos θ is negative, find the other functions of θ.

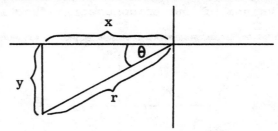

Solution: Since cos θ is negative, θ must be a second or third quadrant angle. In the second quadrant, the tangent function is negative. Hence, θ must be a third quadrant angle.

In the figure, the trigonometric functions have the following values:

$$\sin \theta = \frac{\text{opposite side}}{\text{hypotenuse}} = \frac{y}{r},$$

$$\cos \theta = \frac{\text{adjacent side}}{\text{hypotenuse}} = \frac{x}{r},$$

$$\tan \theta = \frac{\text{opposite side}}{\text{adjacent side}} = \frac{y}{x},$$

$$\cot \theta = \frac{1}{\tan \theta} = \frac{x}{y},$$

$$\sec \theta = \frac{1}{\cos \theta} = \frac{1}{x/r} = \frac{r}{x}, \qquad \text{and}$$

581

$$\csc \theta = \frac{1}{\sin \theta} = \frac{1}{y/r} = \frac{r}{y} \ .$$

Also, from the figure, $r^2 = x^2 + y^2$ (from the Pythagorean Theorem), or $r = \sqrt{x^2 + y^2}$. Therefore, in this problem,

$$\tan \theta = 2 = \frac{-2}{-1} = \frac{y}{x} \ .$$

Hence, $y = -2$ and $x = -1$. Also, in this problem,

$$r^2 = x^2 + y^2 = (-1)^2 + (-2)^2 = 1 + 4$$

or $r^2 = 5$ or $r = \sqrt{5}$. Therefore,

$$\sin \theta = \frac{-2}{\sqrt{5}} = -\frac{2\sqrt{5}}{5} \ , \qquad \cos \theta = \frac{-1}{\sqrt{5}} = -\frac{\sqrt{5}}{5} \ ,$$

$$\tan \theta = 2, \qquad \cot \theta = \frac{-1}{-2} = \frac{1}{2} \ ,$$

$$\sec \theta = \frac{\sqrt{5}}{-1} = -\sqrt{5}, \quad \text{and} \quad \csc \theta = \frac{\sqrt{5}}{-2} = -\frac{\sqrt{5}}{2} \ .$$

● **PROBLEM 29-7**

Find the values of the trigonometric functions of an angle of - 510°.

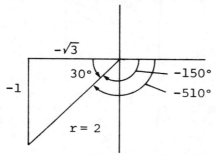

Solution: We see that angles of - 510° and - 150° are coterminal angles having the same values for the trigonometric functions. The reference angle of an angle of - 150° is an angle of 30° and - 150° is a third quadrant angle. In the third quadrant, the tangent and cotangent functions are positive while the other four functions are negative. This yields

$$\sin (-510°) = \sin 30° = -\frac{1}{2} \ ,$$

$$\cos (-510°) = \cos 30° = -\frac{\sqrt{3}}{2} \ ,$$

$$\tan (-510°) = \tan 30° = \frac{\sqrt{3}}{3} \ ,$$

$$\cot(-510°) = \cot 30° = \frac{1}{\tan 30°} = \frac{1}{\sqrt{3}/3} = \frac{3}{\sqrt{3}}$$

$$= \frac{3\sqrt{3}}{\sqrt{3}\sqrt{3}} = \frac{3\sqrt{3}}{3} = \sqrt{3},$$

$$\sec(-510°) = \sec 30° = \frac{1}{\cos 30°} = \frac{1}{-\sqrt{3}/2} = -\frac{2}{\sqrt{3}}$$

$$= -\frac{2\sqrt{3}}{\sqrt{3}\sqrt{3}} = -\frac{2\sqrt{3}}{3},$$

$$\csc(-510°) = \csc 30° = \frac{1}{\sin 30°} = \frac{1}{-\frac{1}{2}} = -2.$$

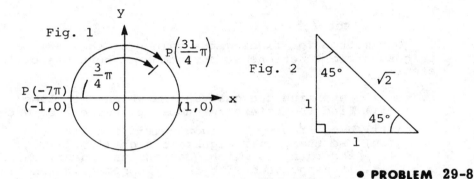

Fig. 1

Fig. 2

● **PROBLEM** 29-8

Find sec $\left[-\frac{31}{4}\pi\right]$.

<u>Solution:</u> We obtain the point P $\left[-\frac{31}{4}\pi\right]$ by proceeding

$\frac{31}{4}\pi = 7\pi + \frac{3}{4}\pi$ units around the unit circle from the

point (1, 0) in a clockwise direction. Movement in a clockwise direction will make the value of the angle negative.
Now, $7\pi = 1260°$; and starting at point (1, 0), and moving
in a clockwise direction, when we return to the point (1,0)
we have covered 360°. Doing this three times gives us an
angle of 1080°, again leaving us at (1, 0). Proceeding
another 180° takes us to (- 1, 0), where we have now
covered - 7π, or 1260°. Since $\frac{3}{4}\pi = 135°$, we proceed from

(- 1,0) to (0, 1) (covering 90°) and then another 45°;
thus we have proceeded $-\frac{31}{4}\pi$. (see Figure 1). The point

is in the first quadrant. Since each quadrant is $\frac{1}{2}\pi$,

and our desired point is $\frac{1}{4}\pi$ into the 1st quadrant, we

use the reference number $\frac{1}{4}\pi$ (or the reference angle 45°).

Hence,

$$\sec \left(- \frac{31}{4}\, \pi \right) = \sec \frac{1}{4}\, \pi = \sqrt{2}.$$

Now, recall that $\sec x = \dfrac{1}{\cos x}$, and

$$\cos x = \frac{\text{adjacent}}{\text{hypotenuse}} \; ; \quad \text{therefore,}$$

$$\sec x = \frac{\text{hypotenuse}}{\text{adjacent}} \; .$$

Observing the 45-45 right triangle, (see Figure 2)

$$\cos 45° = \frac{1}{\sqrt{2}} \; . \text{ Thus,}$$

$$\sec 45° = \sqrt{2}.$$

Our method for finding the values of the trigonometric functions at any number t consists of four steps:

(i) Determine the quadrant in which $P(t)$ lies;
(ii) Find the reference number t_1 associated with t;

(iii) Use a table of trigonometric functions;
(iv) Use the proper algebraic sign (+ or -), according to which quadrant the point, which is associated with the trigonometric function, lies in.

● **PROBLEM 29-9**

From a point 5 ft. above the horizontal ground, and 30 ft. from the trunk of a tree, the line of sight to the top of the tree is measured as 52° with the horizontal. Find the height of the tree.

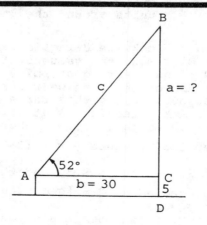

Solution: See the accompanying figure. A is taken at the observer's eye, B is the top of the tree, and the required height is BD = a + 5 ft. One relation connecting the known angle A = 52°, the known distance b = 30 ft., and the unknown length a is

$$\tan A = \frac{\text{opposite side}}{\text{adjacent side}} = \frac{a}{b}$$

We therefore have, with the aid of trigonometric tables,

$$a = b \tan A = 30 \tan 52° .$$

We find in a table of trigonometric functions that $\tan 52°$ is 1.280. Then

$$a = 30 \times 1.280 = 38.4 \text{ ft.;}$$

and the height of the tree is

$$BD = a + 5 = 43.4 \text{ ft.}$$

This result may be checked by first finding c from the relation $\cos A =$

$$\cos A = \frac{\text{adjacent side}}{\text{hypotenuse}} = \frac{b}{c}$$

and then getting a from $\sin A = \dfrac{\text{opposite side}}{\text{hypotenuse}} = \dfrac{a}{c}$

$$\cos A = \frac{b}{c}$$

$$\cos 52° = \frac{30}{c}$$

$$c = \frac{30}{\cos 52°} = \frac{30}{.6157} \approx 48.7$$

$$\sin A = \frac{a}{c}$$

$$\sin 52° = \frac{a}{48.7}$$

$$a = 48.7 (\sin 52°) = 48.7(.7880)$$

$$a \approx 38.4$$

● **PROBLEM** 29-10

A ship is 67 mi west and 40 mi north of a port. What is the distance and bearing of the ship from the port? (See Figure).

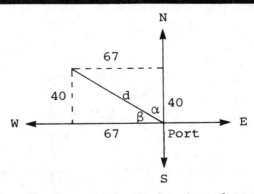

<u>Solution:</u> In order to specify the bearing of some traveling vehicle, the angle of the direction in which the vehicle is traveling must be discovered. To do this we superimpose our direction upon the coordinate axes by extending a line from the origin to the point (x,y) where x is the distance traveled east or west, and y is the distance traveled north or south. The distance in the direction of travel is determined using the Pythagorean theorem: $d^2 = x^2 + y^2$. The angle

that the direction makes with the east/west line is determined using the form

$$\tan \beta = \frac{y}{x} = \frac{\text{side opposite } \beta}{\text{side adjacent } \beta} \ .$$

The angle that the direction makes with a north/south line is determined using the form

$$\cot \beta = \frac{\text{side adjacent } \beta}{\text{side opposite } \beta} = \frac{x}{y}$$

or

$$\tan \alpha = \frac{\text{side opposite } \alpha}{\text{side adjacent } \alpha} = \frac{x}{y}$$

$$\tan \alpha = \frac{67}{40} = 1.6750 = \tan 59°9.7'$$

Hence

$$\alpha = 59°9.7'$$

and the bearing (azimuth) of the ship is N59° 9.7'W.

$$d = \sqrt{(67)^2 + (40)^2} = \sqrt{4489 + 1600} = \sqrt{6089}$$

$$d = 78.032$$

● **PROBLEM** 29-11

(1)   Show that tan α =bc in Fig. 1.

(2)   Determine which segment in Fig. 1 has measure equal to sec α .

(3)   Determine which segment in Fig. 2 has length equal to cot α.

(4)   Determine which segment in Fig. 2 has measure equal to csc α.

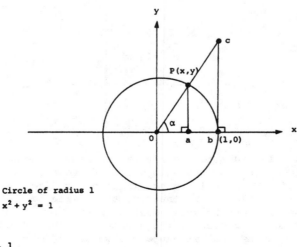

Circle of radius 1
$x^2 + y^2 = 1$

Fig. 1

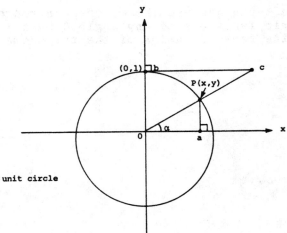

unit circle

Fig. 2

__Solution:__ (1) By definition, $\tan \alpha = \dfrac{aP}{ao}$.

From Fig. 1 ⟨ Poa = ⟨ cob = α and ⟨ Pao = ⟨ cbo = 90^0.

By A.A. = A.A., one has Δ aoP ~ Δ boc.

Hence $\dfrac{ao}{bo} = \dfrac{aP}{bc} = \dfrac{oP}{oc}$, which gives

$$\dfrac{aP}{ao} = \dfrac{bc}{bo} = \tan \alpha .$$

Since bo = 1, $\tan \alpha = \dfrac{bc}{bo} = bc.$

(2) By definition, $\sec \alpha = \dfrac{oP}{ao}$.

From part (1), one has $\dfrac{ao}{bo} = \dfrac{oP}{oc}$,

which gives $\dfrac{oP}{ao} = \dfrac{oc}{bo} = \dfrac{oc}{1} = oc = \sec \alpha$, i.e. $\sec \alpha = oc.$

(3) From Fig. 2, one has

⟨ Pao = ⟨ obc = 90^0, thus bc//ao. Then

⟨ Poa = ⟨ ocb = α. Therefore, Δ bco ~ Δ aoP.

Since Δ bco ~ Δ aoP, $\dfrac{ao}{bc} = \dfrac{aP}{bo} = \dfrac{oP}{co}$.

$$\cot \alpha = \dfrac{ao}{aP} = \dfrac{bc}{bo} = \dfrac{bc}{1} = bc$$

(4) $\csc \alpha = \dfrac{oP}{aP}$ Using the result from part (3), one obtains

$$\dfrac{oP}{aP} = \dfrac{co}{bo} . \quad \csc \alpha = \dfrac{oP}{aP} = \dfrac{co}{bo} = \dfrac{co}{1} = co.$$

By the same technique used here, one can represent all the trigonometric functions of any angle θ by a line segment, i.e. the line representation of the trigonometric functions.

CHAPTER 30

TABLES AND LOGARITHMS OF TRIGONOMETRIC FUNCTIONS

> **Basic Attacks and Strategies for Solving Problems in this Chapter. See pages 589 to 592 for step-by-step solutions to problems.**

A degree (as a measure of angle size) can be subdivided into 60 *minutes* (which are indicated by an apostrophe). Most tables of trigonometric values indicate angles by means of degrees and minutes, so a fractional angle (e.g., 30.5°) must first be translated into degree and minute measures (i.e., 30°30′).

$$30.5° = \text{thirty degrees and a half}$$
$$= 30° + 0.5°$$
$$= 30° + {}^{30°}\!/_{60°}$$
$$= 30° + 30 \text{ minutes}$$
$$= 30° + 30′$$
$$= 30°30′$$

Values for angles not exactly listed in tables can be obtained by linear interpolation (as was explained in Chapter 20 in the discussion of tables of logarithms).

Books of tables sometimes also include *tables of logarithms of trigonometric functions*, eliminating the necessity of having to first look up the value of a trigonometric function and then look up the logarithm of that value. The use of such tables has declined since the advent of the pocket calculator.

Step-by-Step Solutions to Problems in this Chapter, "Tables and Logarithms of the Trigonometric Functions"

● **PROBLEM** 30-1

If tan θ = 3.8436, find θ.

Solution: Looking through a table of trigonometric functions under the vertical column marked tan, it is found that the angle θ = 75°35' corresponds to the number 3.8436.

● **PROBLEM** 30-2

Find cos 37°12'.

Solution: See a table of natural trigonometric functions, which is constructed in terms of multiples of ten seconds. The cosine of 37°12' lies between 37°10' and 37°20'. Therefore, we must interpolate. The cosine decreases as the angle increases, so we form our proportion as follows, where

x = the cosine of the angle 37°12'
d = the difference between the cos 37°10' and cos 37°12':

$$10' \begin{bmatrix} \begin{array}{l} \cos 37°10' = 0.7969 \\ 2' \begin{bmatrix} \cos 37°12' = x \end{bmatrix} d \\ \cos 37°20' = 0.7951 \end{array} \end{bmatrix} -0.0018$$

$$\frac{2}{10} = \frac{d}{-0.0018}$$

Cross multiply to obtain:

$$10d = 2(-0.0018)$$

$$d = .2(-0.0018)$$
$$= -0.00036$$
$$d \approx -0.0004$$

Thus,
$$x = 0.7969 - 0.0004$$
$$= 0.7965$$

Since the cosine is positive in the first quadrant,

$$\cos 37°12' = 0.7965$$

Remember that results obtained by interpolation are approximations. You should not use an answer that is more accurate than the original data, in this case, four significant digits.

● PROBLEM 30-3

Find the value of tan 38°46' by use of interpolation.

Solution: Since 38°46' is between 38°40' and 38°50', we assume that tan 38°36' is between tan 38°40' and tan 38°50'. In fact, since 38°46' is six-tenths of the way from 38°40' toward 38°50', we assume that tan 38°46' is six-tenths of the way from tan 38°40' = .8002 toward tan 38°50' = .8050. Using these assumptions we perform the following interpolation:

$$10'\left[\begin{array}{c} 6'\left[\begin{array}{c} \tan 38°40' = .8002 \\ \tan 38°46' = \quad ? \end{array}\right]c \\ \tan 38°50 = .8050 \end{array}\right].0048$$

Set up the proportion $\dfrac{c}{.0048} = \dfrac{6}{10}$

$$10c = 6(.0048)$$

$$c = \frac{6}{10}(.0048) = .0029$$

$$\tan 38°46' = .8002 + .0029 = .8031$$

Therefore, c was added because tan θ increases from

$$θ = 38°40' \text{ to } θ = 38°50'.$$

● PROBLEM 30-4

Find θ if sin θ = .6212, and $-\dfrac{\pi}{2} \le θ \le \dfrac{\pi}{2}$.

Solution: Since 0.6212 is not found in the sine table, we proceed by finding the two numbers closest to .6212, one greater and the other less than it, and interpolating.

$$10' \begin{bmatrix} c \begin{bmatrix} -\sin 38°20' &= .6302 \\ -\sin \theta &= .6212 \\ -\sin 38°30' &= .6225 \end{bmatrix} .0010 \end{bmatrix} .0023$$

We set up the proportion $\dfrac{c}{10'} = \dfrac{.0010}{.0023}$.

$$c = \frac{.0010}{.0023}(10') = \frac{10}{23}(10')$$

$\theta = 4'$ to the nearest minute.

Thus, $\theta = 38°20' + 4'$

$= 38°24'$.

● **PROBLEM 30-5**

.Find log sin 36° 41´.

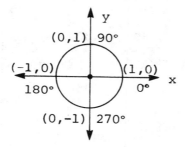

__Solution:__   From a table of logarithms of trigonometric functions we find log sin 36°40´ and log sin 36°50´. Then, by the process of interpolation we find: log sin 36°41´ = 9.77626 - 10. Since all values of the sine function for acute angles are in the range of 0 < sin x ≤ 1, the characteristic is negative. (Recall that for a number less than one, the characteristic is negative.) The range of sine can be seen by inspecting the accompanying figure. Sine is given by the y coordinate; cos is given by the x coordinate. Observe that y value varies from 0 to 1, as the angle varies from 0° to 90°.

● **PROBLEM 30-6**

Find log cos 49°13.6´.

__Solution:__   First consult a table of logarithms of trigono-metric functions.

Notice that 49°13.6´ lies between 49°10´ and 49°20´, so that the log of 49°13.6´ will occur between the logs of 49°10´ and 49°20´ and can be determined by inter-polation.

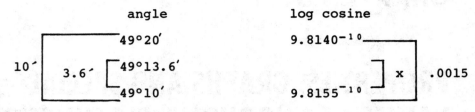

Now set up the proportion

$$\frac{3.6}{10} = \frac{x}{.0015} \quad \text{or} \quad 10x = .00540$$

$$x = .000540$$

Since cos decreases on the interval $0 < \theta < \pi/2$, subtract x from the log cosine of $49°1\overline{0}'$.

$$\begin{array}{r} 9.8155 \\ - \quad .000540 \\ \hline 9.814960 \end{array}$$

Thus, log cosine $49°13.6'$ is $9.81496-10$.

CHAPTER 31

PROPERTIES, GRAPHS AND SPECIAL VALUES OF TRIGONOMETRIC FUNCTIONS

Basic Attacks and Strategies for Solving Problems in this Chapter. See pages 593 to 603 for step-by-step solutions to problems.

Two angles are called complementary if their measures sum to 90°.

Since trigonometric functions are defined in terms of the sides of a right triangle, if α and β are the two acute angles, then $\alpha + \beta = 90°$, i.e., they are complementary angles. By the definition of the trigonometric functions, it can be seen that $\sin \alpha = \cos \beta$, that is, $\sin \alpha = \cos (90° - \alpha)$.

Certain angles play key roles in the study of trigonometric functions. In particular, there are the angles of 0°, 30°, 45°, 60°, and 90°. Because of the rules for complementary angles, and the ability to derive any trigonometric function from one value, one need only remember a (relatively) few values. For example, it is good to remember that

$$\sin 0° = 0, \quad \sin 30° = \frac{1}{2}, \quad \text{and} \quad \tan 45° = 1.$$

The values of all other trigonometric functions and the values of trigonometric functions of 60° and 90° can be derived from these.

Because of similar triangles, the values of trigonometric functions for angles in quadrants II, III and IV can be related to values of trigonometric functions in quadrant I. One need to make sure that the sign of the function is correct. The following rules hold:

$$\sin (\alpha + 90°) = \cos \alpha$$

$$\cos (\alpha + 90°) = - \sin \alpha.$$

Other rules applying to tangent, secant, cotangent and cosecant can be derived from these two rules.

Trigonometric functions are said to be *periodic* because their values repeat according to a regular pattern. For example, $\sin \alpha = \sin(2\pi + \alpha)$. Since the repetitions occur at intervals of 2π, sine is said to have a *period* of 2π. The *amplitude* is half of the maximum height minus the minimum depth, which in the case of the sine function turns out to be 1. The inclusion of a constant addition term to the angle shifts the values of the function, e.g., sin ($\alpha + {}^\pi/_4$). This constant term is called the *phase angle*.

The graphs of the six standard trigonometric functions may be found in any trigonometric or calculus book.

Step-by-Step Solutions to Problems in this Chapter, "Properties, Graphs and Special Values of Trigonometric Functions"

● PROBLEM 31-1

Find the values of all the trigonometric functions for θ equal to

(1) 0^0

(2) 30^0

(3) 45^0

(4) 60^0

(5) 90^0

Solution:

θ	sinθ	cosθ	tanθ	cotθ	secθ	cscθ
0^0	0	1	0	$\pm\infty$	1	$\pm\infty$
30^0	$\frac{1}{2}$	$\frac{\sqrt{3}}{2}$	$\frac{\sqrt{3}}{3}$	$\sqrt{3}$	$\frac{2\sqrt{3}}{3}$	2
45^0	$\frac{\sqrt{2}}{2}$	$\frac{\sqrt{2}}{2}$	1	1	$\sqrt{2}$	$\sqrt{2}$
60^0	$\frac{\sqrt{3}}{2}$	$\frac{1}{2}$	$\sqrt{3}$	$\frac{\sqrt{3}}{3}$	2	$\frac{2}{3}\sqrt{3}$
90^0	1	0	$\pm\infty$	0	$\pm\infty$	1

● PROBLEM 31-2

Find the exact values of the trigonometric functions for

$$\theta = 120^0.$$

Solution: (1) $\sin 120^0 = \sin(90^0+30^0) = \cos 30^0$

$$= \frac{\sqrt{3}}{2}$$

$\cos 120^0 = \cos(90^0+30^0) = -\sin 30^0$

$$= -\frac{1}{2}$$

$\tan 120^0 = \tan(90^0+30^0) = -\cot 30^0$

$$= -\sqrt{3}$$

$\cot 120^\circ = \cot(90^\circ+30^\circ) = -\tan 30^\circ$

$$= -\frac{\sqrt{3}}{3}$$

$\sec 120^0 = \sec(90^0+30^0) = -\csc 30^0$

$$= -2$$

$\csc 120^0 = \csc(90^0+30^0) = \sec 30^0$

$$= \frac{2\sqrt{3}}{3}$$

One can also use any of the following formulas

(a) $\sin 120^0 = \sin(180^0 - 60^0) = \sin 60^0$

$$= \frac{\sqrt{3}}{2}$$

(b) $\sin 120^0 = \sin(180^0 + (-60^0)) = -\sin(-60^0) = -(-\sin 60^0)$

$$= \frac{\sqrt{3}}{2}$$

(c) $\sin 120^0 = \sin(90^0 - (-30^0)) = \cos(-30^0) = \cos 30^0$

$$= \frac{\sqrt{3}}{2}$$

Similarly, for the other trigonometric functions.

● **PROBLEM 31-3**

Given $\triangle ABC$, $\angle BAC = 60^0$ as shown in the figure. Calculate

(1) $\csc(\alpha+\beta)$

(2) $\cot\left(\frac{\alpha+\beta}{2}\right)$

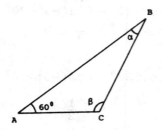

Solution: (1) Observe that

$$\alpha + \beta + \angle BAC = \alpha + \beta + 60^0 = 180^0.$$ Therefore, one has

$$\sin(\alpha+\beta) = \sin(180^0-60^0) = \sin 60^0 = \frac{\sqrt{3}}{2}.$$ Thus,

$$\csc(\alpha+\beta) = \frac{1}{\sin(\alpha+\beta)} = \frac{1}{\frac{\sqrt{3}}{2}} = \frac{2\sqrt{3}}{3}.$$

or $\csc(\alpha+\beta) = \csc(180^0-60^0) = \csc 60^0$

$$= \frac{2}{3}\sqrt{3}.$$

(2) $$\cot\left[\frac{1}{2}(\alpha+\beta)\right] = \cot\left[\frac{180^0-60^0}{2}\right]$$

$$= \cot(90^0-30^0) = \tan 30^0$$

$$= \frac{\sqrt{3}}{3}$$

● **PROBLEM 31-4**

Find all values of θ such that sec θ = 2.

Solution: First find all values of θ, $0^0 \le \theta < 360^0$, such that sec θ = 2. They are θ = 60^0 and θ = 300^0.

To obtain all values of θ for which sec θ = 2, add $360^0 \cdot n$ (or 2nπ) where n = 0,±1,±2,±3.....

Hence, all the values of θ for which sec θ = 2 are given by

$$\theta = 360^0 \cdot n + 60^0 \quad \text{and} \quad \theta = 360^0 \cdot n + 300^0$$

(or $\theta = 2n\pi + \frac{\pi}{3}$ and $\theta = 2n\pi + \frac{5}{3}\pi$), where n is any integer.

Graph   (1) $y = \sin x$

         (2) $y = 4 \sin x$

         (3) $y = \sin 4x$

         (4) $y = \sin(x + \frac{\pi}{4})$

         (5) $y = A \sin(Bx+C) + D$

where A, B, C, D are real constants.

<u>Solution</u>:   (1)

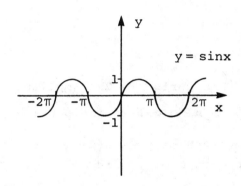

(1)

$y = \sin x$ is a periodic function with period $T = 2\pi$. Its amplitude is

$$\frac{1}{2}\left[(y_{max}) - (y_{min})\right] = \frac{1}{2}\left[1 - (-1)\right] = 1.$$

The graph is symmetrical with respect to the origin.

(2) The graph of $y = 4 \sin x$ is the same as $y = \sin x$ except that the amplitude is

$$\frac{1}{2}\left[(y_{max}) - (y_{min})\right] = \frac{1}{2}\left[4 - (-4)\right] = 4.$$

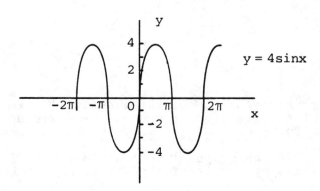

(2)

(3) y = sin 4x has an amplitude 1, but a period of

$$T = \frac{2\pi}{4} = \frac{\pi}{2} .$$

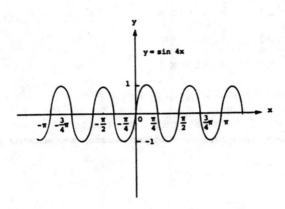

(3)

(4) y = sin(x + $\frac{\pi}{4}$) has period T = 2π, amplitude 1, and phase

shift of $\frac{\pi}{4}$ ($\frac{\pi}{4}$ is also called the phase angle).

Note that the graph of y = sin(x + $\frac{\pi}{4}$) can be obtained by
simply shifting the graph of y = sin x by $\frac{\pi}{4}$ to the left of
origin.

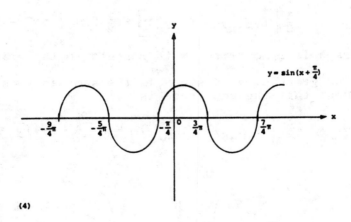

(4)

(5) For y = A sin(Bx+C) + D, the constant |A| is the ampli-
tude of the function, the constant B decides the period
of the function, $T = \frac{2\pi}{|B|}$: C is the phase angle, and D
will shift the graph of y = A sin(Bx+C) up (or down) along
the y-axis by D units for positive (or negative) D.

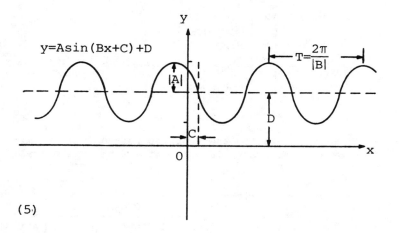

$y = A\sin(Bx + C) + D$

$|A|$

$T = \dfrac{2\pi}{|B|}$

D

C

0

y

x

(5.)

● **PROBLEM 31-6**

Sketch three periods of the graph y = 3 cos 2x.

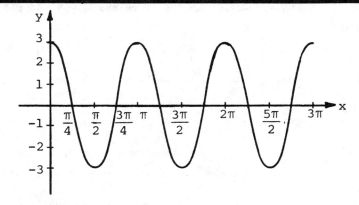

Solution: The coefficient of the function is 3, which means that the maximum and minimum values are 3 and - 3, respectively. The period of the cosine function is the co-efficient of x multiplied by $\frac{\pi}{2}$ radians. Therefore, the period of the cosine function given in this problem is

$$2 \left[\frac{\pi}{2} \text{ radians} \right] = \pi \text{ radians}$$

and with this knowledge, we sketch the curve as in the Figure.

● **PROBLEM 31-7**

Graph y = csc x, 0 ≤ x ≤ 2π.

598

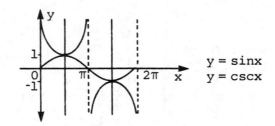

$y = \sin x$

$y = \csc x$

<u>Solution:</u> To plot points for the function cosecant of x, first find
the y-values of the reciprocal function, the sine of the angle x.

x	0	$\frac{\pi}{2}$	π	$\frac{3\pi}{2}$	2π
sin x	0	1	0	-1	0
csc x	not defined	1	not defined	-1	not defined

Since the sine and cosecant are reciprocals, we state the following
conclusions based on properties of real numbers.

(1)   For $0 \le x \le \frac{\pi}{2}$, $0 \le \sin x \le 1$ and $\csc x \ge 1$.

In fact, as sin x increases, csc x decreases. For example:

$\sin 0^{o} = \sin 0 = 0$ $\qquad\qquad$ $\csc 0 = $ undefined

$\sin 30^{o} = \sin \frac{\pi}{6} = \frac{1}{2} = .5000$ $\qquad$ $\csc \frac{\pi}{6} = 2$

$\sin 45^{o} = \sin \frac{\pi}{4} = \frac{1}{\sqrt{2}} = \frac{\sqrt{2}}{2} \approx \frac{1.414}{2} = .7070$ $\quad$ $\csc \frac{\pi}{4} = \sqrt{2} \approx 1.414$

$\sin 90^{o} = \sin \frac{\pi}{2} = 1.000$ $\qquad\qquad$ $\csc \frac{\pi}{2} = 1$

(2)   For $\frac{\pi}{2} \le x \le \pi$, sin x decreases from 1 to 0. Hence, csc x will
increase from 1 to very large values. We can observe this from specific
examples:

$\sin \frac{\pi}{2} = 1$ $\qquad\qquad\qquad\qquad$ $\csc \frac{\pi}{2} = 1$

$\sin \frac{2}{3}\pi = \frac{\sqrt{3}}{2} \approx .87$ $\qquad\qquad$ $\csc \frac{2}{3}\pi \approx 1.15$

$\sin \frac{3}{4}\pi \approx .707$ $\qquad\qquad\quad$ $\csc \frac{3}{4}\pi \approx 1.4$

$\sin \frac{5}{6}\pi = .500$ $\qquad\qquad\quad$ $\csc \frac{5}{6}\pi = 2$

$\sin \pi = 0$ $\qquad\qquad\qquad\quad$ $\csc \pi = $ undefined

(3)   For $\pi \le x \le \frac{3\pi}{2}$, sin x decreases from 0 to -1. Hence, csc x will
be increasing and will increase from very large negative values to -1.

(4)   For $\frac{3\pi}{2} \le x \le 2\pi$, $-1 \le \sin x \le 0$, and the graph will be increasing.
Hence, csc x will decrease from -1 to very large negative values.
The student can verify conclusions (3) and (4) in a similar manner to
that used for (1) and (2), that is, choosing specific angles between
$\frac{\pi}{2}$ and $\frac{3\pi}{2}$, and then between $\frac{3\pi}{2}$ and 2π .

599

The graphs of both functions are shown in the accompanying figure. Note that the range of the sine function is $-1 \leq \sin x \leq 1$, but the range of the cosecant function is $\csc x \geq 1$ or $\csc x \leq -1$.

● **PROBLEM** 31-8

Graph   (1) $y = \cot x$

             (2) $y = \sec x$

             (3) $y = \csc x$

<u>Solution</u>:   (1) $y = \cot x = \dfrac{\cos x}{\sin x} = \dfrac{1}{\tan x}$

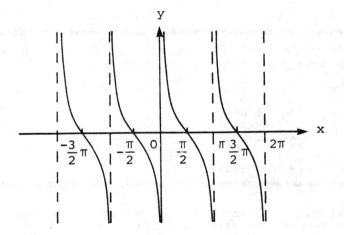

(2) $y = \sec x = \dfrac{1}{\cos x}$, $x \neq 2k\pi + \dfrac{\pi}{2}$, $k = 0,\ \pm1,\ \pm2, \ldots$

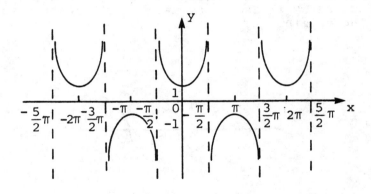

(3) $y = \csc x = \dfrac{1}{\sin x}$, $x \neq 2k\pi$, $k = 0, \pm 1, \pm 2, \ldots$.

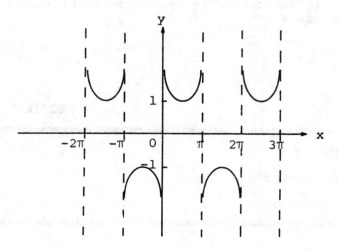

● PROBLEM 31-9

Classify the following functions into as many different classes of functions as possible.

    (1) $y = \sin x$

    (2) $y = \cos x$

    (3) $y = \tan x$

    (4) $y = \sec x$

<u>Solution</u>: (1) The graph of $y = \sin x$ is shown in Figure 1. It is an odd function (a function $f(x)$ is said to be an odd function if for all x in its domain $f(-x) = -f(x)$). It is periodic with period $T = 2\pi$ (a function $f(x)$ is said to be periodic with period T if there exists a positive number T such that $f(x+T) = f(x)$ and if T is the smallest positive number for which this relationship holds). In addition, the function is bounded by ±1 and is symmetric with respect to the y-axis and the origin.

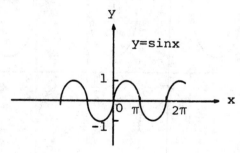

(2) $y = \cos x$ is an even function since $y = \cos x = \cos(-x)$; it is also a periodic function with period $T = 2\pi$. It is bounded by ± 1. The graph of $y = \cos x$ is shown in Fig. 2, and it is symmetric with respect to the y-axis and the origin.

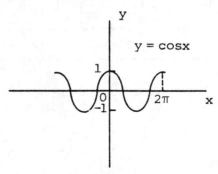

(3) $y = \tan x$ is periodic, $(T = 2\pi)$, odd, and unbounded.

(4) $y = \sec x$ is periodic $(T = 2\pi)$, even, and unbounded.

● **PROBLEM 31-10**

Graph   (1) $y = 2 \sin x + 4 \sin x$

          (2) $y = \sin x + \cos x$

Solution:

(1) To obtain the graph of $y = 2 \sin x + 4 \sin x$ we first draw the graphs of $y_1 = 2 \sin x$ and $y_2 + 4 \sin x$ as shown in Fig. 1. Next, draw the graph of $y = y_1 + y_2$ by adding corresponding ordinates as illustrated in Fig. 1. For instance, at $x = a_1$, the ordinate of y is $a_1 d_1$, which is the algebraic sum of the ordinates $a_1 b_1$ of y_1 and $a_1 c_1$ of y_2.

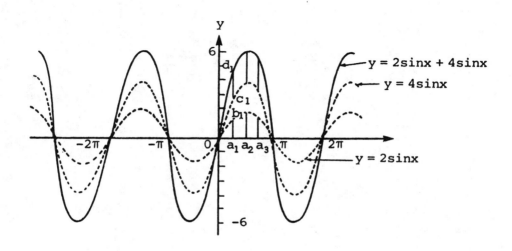

Another way of obtaining the graph of the given function is to construct a table as the following

x	....	$-\frac{\pi}{2}$	$-\frac{\pi}{3}$	$-\frac{\pi}{4}$	$-\frac{\pi}{6}$	0	$\frac{\pi}{6}$	$\frac{\pi}{4}$	$\frac{\pi}{3}$	$\frac{\pi}{2}$	....
y = 2sinx + 4sinx	....	- 6	$-3\sqrt{3}$	$-3\sqrt{2}$	-3	0	3	$3\sqrt{2}$	$3\sqrt{3}$	6	....

(2)   The graph of $y = \sin x + \cos x$ is obtained by using the methods described in part (1) of this problem; it is shown in Fig. 2.

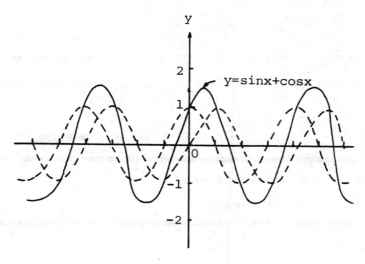

CHAPTER 32

TRIGONOMETRIC IDENTITIES AND FORMULAS

> **Basic Attacks and Strategies for Solving Problems in this Chapter. See pages 604 to 617 for step-by-step solutions to problems.**

The following are some of the standard and commonly used trigonometric identities and formulas.

$$\sin(\alpha + \beta) = \sin\alpha \cos\beta + \cos\alpha \sin\beta$$

$$\cos(\alpha + \beta) = \cos\alpha \cos\beta - \sin\alpha \sin\beta$$

$$\sin^2\alpha + \cos^2\alpha = 1$$

$$\sin(-\alpha) = -\sin\alpha$$

$$\cos(-\alpha) = \cos\alpha$$

$$\sin\alpha + \sin\beta = 2\sin\tfrac{1}{2}(\alpha + \beta)\cos\tfrac{1}{2}(\alpha - \beta)$$

$$\cos\alpha + \cos\beta = 2\cos\tfrac{1}{2}(\alpha + \beta)\cos\tfrac{1}{2}(\alpha - \beta)$$

$$\sin^2\alpha = \tfrac{1}{2}(1 - \cos 2\alpha)$$

$$\cos^2\alpha = \tfrac{1}{2}(1 + \cos 2\alpha)$$

From the first two identities it follows that:

$$\sin 2\alpha = 2\sin\alpha \cos\alpha$$

$$\cos 2\alpha = \cos^2\alpha - \sin^2\alpha.$$

Using the formulas for negative angles, one can obtain identities for the differences of angles from the formulas for sums. Given the formula for double angles, one can substitute $\alpha = \beta/2$ to obtain a formula for a half angle. Given the formulas for sine and cosine, one can obtain a similar formula for other trigonometric functions.

● PROBLEM 32-1

Derive the formulas for

    (1)   cot 2α in terms of cot α

    (2)   sin $\frac{1}{2}\alpha$ in terms of cos θ

    (3)   tan $\frac{1}{2}\alpha$ in terms of cos θ

<u>Solution</u>: (1)

$$\cot 2\alpha = \frac{\cos 2\alpha}{\sin 2\alpha} = \frac{\cos(\alpha+\alpha)}{\sin(\alpha+\alpha)}$$

$$= \frac{\cos\alpha \cos\alpha - \sin\alpha \sin\alpha}{\sin\alpha \cos\alpha + \cos\alpha \sin\alpha}$$

Dividing both numerator and denominator by $\sin^2\alpha$, one obtains

$$\cot 2\alpha = \frac{\frac{\cos^2\alpha}{\sin^2\alpha} - 1}{\frac{\cos\alpha}{\sin\alpha} + \frac{\cos\alpha}{\sin\alpha}} = \frac{\cot^2\alpha - 1}{2\cot\alpha}$$

(2) $\cos 2\theta = 1 - 2\sin^2\theta$. Let $\theta = \frac{\alpha}{2}$

$$\cos 2\left(\frac{\alpha}{2}\right) = \cos\alpha = 1 - 2\sin^2\frac{\alpha}{2}$$

$$\sin^2\frac{\alpha}{2} = \frac{1}{2}\left[1 - \cos\alpha\right]$$

$$\sin\frac{\alpha}{2} = \pm\sqrt{\frac{1-\cos\alpha}{2}}$$

(3) $\cos 2\theta = 2\cos^2\theta - 1$. Letting $\theta = \frac{\alpha}{2}$, one can

obtain

$$\cos \frac{\alpha}{2} = \pm \sqrt{\frac{\cos\alpha + 1}{2}}$$ by the same method as in part (2).

Now, $$\tan \frac{\alpha}{2} = \frac{\sin \frac{\alpha}{2}}{\cos \frac{\alpha}{2}} = \pm \sqrt{\frac{\frac{1-\cos\alpha}{2}}{\frac{1+\cos\alpha}{2}}} = \pm \sqrt{\frac{1-\cos\alpha}{1+\cos\alpha}}$$

● **PROBLEM 32-2**

(a) Prove $1 + \tan^2\theta = \sec^2\theta$.

(b) Given $\tan \theta = \frac{4}{3}$, find the values of other trigonometric functions of θ, $0° \leq \theta < 360°$.

<u>Solution</u>: (a) $1 + \tan^2\theta = 1 + \left[\dfrac{\sin\theta}{\cos\theta} \right]^2$

$$= 1 + \frac{\sin^2\theta}{\cos^2\theta} = \frac{\cos^2\theta + \sin^2\theta}{\cos^2\theta}$$

$$= \frac{1}{\cos^2\theta} = \sec^2\theta$$

(b) For $0° \leq \theta < 360°$, there are two angles for which $\tan\theta > 0$. θ is either a first or third quadrant angle.

First quadrant	Third quadrant

$$\cot\theta = \frac{1}{\tan\theta} = \frac{1}{\frac{4}{3}} = \frac{3}{4} \qquad\qquad \cot\theta = \frac{3}{4}$$

Use the result from part (a).

$$\sec \theta = \sqrt{1 + \tan^2\theta} = \sqrt{1 + \left(\frac{4}{3}\right)^2} \qquad \sec\theta = -\frac{5}{3}$$

$$= \frac{5}{3}$$

$$\cos \theta = \frac{1}{\sec\theta} = \frac{1}{\frac{5}{3}} = \frac{3}{5} \qquad\qquad \cos\theta = -\frac{3}{5}$$

$$\frac{\sin\theta}{\cos\theta} = \tan\theta \qquad\qquad\qquad \sin\theta = -\frac{4}{5}$$

$$\sin\theta = \tan\theta \, \cos\theta = \frac{4}{3} \times \frac{3}{5}$$

$$= \frac{4}{5}$$

$$\csc \theta = \frac{1}{\sin\theta} = \frac{5}{4} \qquad\qquad\qquad \csc\theta = -\frac{5}{4}$$

605

Find the exact value for

    (1) $\sin 37.5^0 \sin 7.5^0$

    (2) $\sin 52.5^0 \cos 7.5^0$

Solution: Use the formulas for the products of sines and cosines.

(1) $\sin 37.5^0 \sin 7.5^0$

$$= -\frac{1}{2}\left[\cos(37.5^0 + 7.5^0) - \cos(37.5^0 - 7.5^0)\right]$$

$$= -\frac{1}{2}\left[\cos 45^0 - \cos 30^0\right]$$

$$= -\frac{1}{2}\left(\frac{\sqrt{2}}{2} - \frac{\sqrt{3}}{2}\right) = -\left(\frac{\sqrt{2} - \sqrt{3}}{4}\right) = \frac{\sqrt{3} - \sqrt{2}}{4}$$

(2) $\sin 52.5^0 \cos 7.5^0$

$$= \frac{1}{2}\left[\sin(52.5^0 + 7.5^0) + \sin(52.5^0 - 7.5^0)\right]$$

$$= \frac{1}{2}(\sin 60^0 + \sin 45^0) = \frac{1}{2}\left(\frac{\sqrt{3}}{2} + \frac{\sqrt{2}}{2}\right)$$

$$= \frac{\sqrt{3} + \sqrt{2}}{4}$$

Find the exact value for

    (1) $\sin 75^0 - \sin 15^0$

    (2) $\sin^2 22.5^0$

    (3) $\tan^2 15^0$

Solution: (1) By use of the formula for the difference of sines, one obtains

$$\sin 75^0 - \sin 15^0 = 2 \cos \frac{1}{2}(75^0 + 15^0)\sin \frac{1}{2}(75^0 - 15^0)$$

$$= 2 \cos\left(\frac{90^0}{2}\right)\sin\left(\frac{60^0}{2}\right) = 2 \cos 45^0\sin 30^0$$

$$= 2\left(\frac{\sqrt{2}}{2}\right)\left(\frac{1}{2}\right) = \frac{2\sqrt{2}}{4} = \frac{\sqrt{2}}{2}$$

(2) $\cos 2\alpha = 1 - 2\sin^2\alpha$, so

$\sin^2\alpha = \dfrac{1 - \cos 2\alpha}{2}$. Then

$\sin^2 22.5^0 = \dfrac{1 - \cos 2 \times 22.5^0}{2}$

$= \dfrac{1}{2} - \dfrac{1}{2} \cos 45^0 = \dfrac{1}{2} - \dfrac{1}{2}\left(\dfrac{\sqrt{2}}{2}\right) = \dfrac{1}{2} - \dfrac{\sqrt{2}}{4}$

(3)   Using the double angle formula

$\cos 2\alpha = 2\cos^2\alpha - 1$,   one gets $\cos^2\alpha = \dfrac{1}{2}(\cos 2\alpha + 1)$.

Now, $\tan^2\alpha = \dfrac{\sin^2\alpha}{\cos^2\alpha} = \dfrac{\dfrac{1 - \cos 2\alpha}{2}}{\dfrac{1 + \cos 2\alpha}{2}} = \dfrac{1 - \cos 2\alpha}{1 + \cos 2\alpha}$

so $\tan^2 15^0 = \dfrac{1 - \cos 2 \times 15^0}{1 + \cos 2 \times 15^0} = \dfrac{1 - \cos 30^0}{1 + \cos 30^0}$.

$= \dfrac{1 - \dfrac{\sqrt{3}}{2}}{1 + \dfrac{\sqrt{3}}{2}} = \dfrac{2 - \sqrt{3}}{2 + \sqrt{3}}$

● **PROBLEM 32-5**

Simplify
$$\frac{\tfrac{1}{2}(\cos 3\alpha + \cos \alpha)}{\sin \alpha - \sin(-3\alpha)}$$

Solution:
$\dfrac{\tfrac{1}{2}(\cos 3\alpha + \cos \alpha)}{\sin \alpha - \sin(-3\alpha)} = \dfrac{\cos 3\alpha + \cos \alpha}{2(\sin \alpha + \sin 3\alpha)}$

By the sum of cosines formula

$\cos A + \cos B = 2\cos \tfrac{1}{2}(A+B)\cos \tfrac{1}{2}(A-B)$

and the sum of sines formula

$\sin A + \sin B = 2\sin \tfrac{1}{2}(A+B)\cos \tfrac{1}{2}(A-B)$, one gets

$= \dfrac{2\cos \tfrac{1}{2}(3\alpha+2)\cos \tfrac{1}{2}(3\alpha-\alpha)}{2(2\sin \tfrac{1}{2}(\alpha+3\alpha)\cos \tfrac{1}{2}(\alpha-3\alpha))}$

$= \left(\tfrac{1}{2}\right)\dfrac{\cos 2\alpha \cos \alpha}{\sin 2\alpha \cos(-\alpha)} = \left(\tfrac{1}{2}\right)\dfrac{\cos 2\alpha \cos \alpha}{\sin 2\alpha \cos \alpha}$

$= \left(\tfrac{1}{2}\right)\dfrac{\cos 2\alpha}{\sin 2\alpha} = \tfrac{1}{2}\cot 2\alpha$

Find the exact value for $\sin^4 15^0$.

Solution: $\sin^4 15^0 = (\sin^2 15^0)^2$

$$(\sin^2\alpha)^2 = \left(\frac{1 - \cos^2\alpha}{2}\right)^2$$

$$= \frac{1}{4}(1 - 2\cos 2\alpha + \cos^2 2\alpha)$$

$$= \frac{1}{4}\left(1 - 2\cos 2\alpha + \left[\frac{1}{2}(\cos 2(2\alpha)+1)\right]\right)$$

$$= \frac{1}{4} - \frac{1}{2}\cos 2\alpha + \frac{1}{8}\cos 4\alpha + \frac{1}{8}$$

$$= \frac{3}{8} - \frac{1}{2}\cos 2\alpha + \frac{1}{8}\cos 4\alpha$$

$$\sin^4 15^0 = (\sin^2 15^0)^2 = \frac{3}{8} - \frac{1}{2}\cos(2\times 15^0) + \frac{1}{8}\cos(4\times 15^0)$$

$$= \frac{3}{8} - \frac{1}{2}\cos 30^0 + \frac{1}{8}\cos 60^0$$

$$= \frac{3}{8} - \frac{1}{2}\cdot\frac{\sqrt{3}}{2} + \frac{1}{8}\times\frac{1}{2} = \frac{7}{16} - \frac{\sqrt{3}}{4}$$

Change $4 + (\tan\theta - \cot\theta)^2$ to $\sec^2\theta + \csc^2\theta$.

Solution: If we square the binomial in the first expression, we have

$$4 + (\tan\theta - \cot\theta)^2 = 4 + (\tan\theta - \cot\theta)(\tan\theta - \cot\theta)$$

$$= 4 + \tan^2\theta - 2\tan\theta\cot\theta + \cot^2\theta$$

Since $\cot\theta = \frac{1}{\tan\theta}$ the term $-2\tan\theta\cot\theta = -2\tan\theta\left(\frac{1}{\tan\theta}\right) = -2(1) = -2$.

Thus

$$4 + (\tan\theta - \cot\theta)^2 = 4 + \tan^2\theta - 2 + \cot^2\theta$$

$$= 2 + \tan^2\theta + \cot^2\theta$$

Since $2 = 1 + 1$,

$$= 1 + \tan^2\theta + 1 + \cot^2\theta$$

Recall $1 + \tan^2\theta = \sec^2\theta$ and $1 + \cot^2\theta = \csc^2\theta$. Replacing these values we obtain

$$4 + (\tan\theta - \cot\theta)^2 = \sec^2\theta + \csc^2\theta$$

Change $\tan\theta(\sin\theta + \cot\theta\cos\theta)$ to $\sec\theta$.

<u>Solution:</u> Distribute to obtain,

tan θ(sin θ + cot θ cos θ) = tan θ sin θ +
$$\text{tan } θ \text{ cot } θ \text{ cos } θ$$

Recall that cot θ = 1/tan θ, and replace cot θ by 1/tan θ:

$$= \text{tan } θ \text{ sin } θ + \text{tan } θ \text{ (1/tan } θ)\text{cos } θ$$

$$= \text{tan } θ \text{ sin } θ + \text{cos } θ$$

Since tan θ = sin θ/cos θ we may replace tan θ by sin θ/cos θ:

$$= \frac{\sin θ}{\cos θ} \sin θ + \cos θ$$

$$= \frac{\sin^2 θ}{\cos θ} + \cos θ.$$

To combine terms, we convert cos θ into a fraction whose denominator is cos θ, thus

$$= \frac{\sin^2 θ}{\cos θ} + \left(\frac{\cos θ}{\cos θ}\right) \cdot \cos θ. \quad \text{(Note that}$$

cos θ/cos θ equals one, so the equation is unaltered)

$$= \frac{\sin^2 θ}{\cos θ} + \frac{\cos^2 θ}{\cos θ}$$

$$= \frac{\sin^2 θ + \cos^2 θ}{\cos θ}.$$

Recall the identity sin² θ + cos² θ = 1; hence,

$$= \frac{1}{\cos θ}$$

$$= \sec θ.$$

● **PROBLEM 32-9**

Reduce the expression $\dfrac{\tan x - \cot x}{\tan x + \cot x}$ to one involving only sin x.

<u>Solution:</u> Since, by definition, tan x = $\dfrac{\sin x}{\cos x}$ and

$$\cot x = \frac{1}{\tan x} = \frac{1}{\sin x/\cos x} = \frac{\cos x}{\sin x},$$

$$\frac{\tan x - \cot x}{\tan x + \cot x} = \frac{\dfrac{\sin x}{\cos x} - \dfrac{\cos x}{\sin x}}{\dfrac{\sin x}{\cos x} + \dfrac{\cos x}{\sin x}}$$

$$= \frac{\dfrac{\sin x(\sin x)}{\sin x(\cos x)} - \dfrac{\cos x(\cos x)}{\cos x(\sin x)}}{\dfrac{\sin x(\sin x)}{\sin x(\cos x)} + \dfrac{\cos x(\cos x)}{\cos x(\sin x)}}$$

$$= \frac{\dfrac{\sin^2 x - \cos^2 x}{\sin x \cos x}}{\dfrac{\sin^2 x + \cos^2 x}{\sin x \cos x}}$$

$$= \frac{\sin^2 x - \cos^2 x}{\sin x \cos x} \times \frac{\sin x \cos x}{\sin^2 x + \cos^2 x}$$

$$= \frac{\sin^2 x - \cos^2 x}{\sin^2 x + \cos^2 x}$$

Since $\sin^2 x + \cos^2 x = 1$ or $\cos^2 x = 1 - \sin^2 x$,

$$\frac{\tan x - \cot x}{\tan x + \cot x} = \frac{\sin^2 x - \cos^2 x}{\sin^2 x + \cos^2 x} = \frac{\sin^2 x - \cos^2 x}{1}$$

$$= \sin^2 x - \cos^2 x$$

$$= \sin^2 x - \left(1 - \sin^2 x\right)$$

$$= \sin^2 x - 1 + \sin^2 x$$

$$= 2 \sin^2 x - 1.$$

● **PROBLEM 32-10**

Find sin 105° without the use of a trig. table.

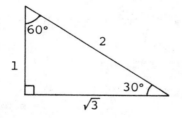 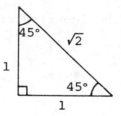

Solution: We note that $105° = 60° + 45°$ and find the sine of the sum of two angles.

$$\sin 105° = \sin(60° + 45°)$$

Using the formula for the sine of the sum of two numbers,

$$\sin(x+y) = \sin x \cos y + \cos x \sin y, \quad \sin(60° + 45°)$$
$$= \sin 60° \cos 45° + \cos 60° \sin 45° .$$

Now we must find the values of $\sin 60°$, $\cos 45°$, $\cos 60°$, and $\sin 45°$. Observing a 30-60 and 45-45 right triangle we note:

$$\sin = \frac{\text{opposite}}{\text{hypotenuse}} ; \quad \text{thus,} \quad \sin 60° = \frac{\sqrt{3}}{2}$$

$$\sin 45° = \frac{1}{\sqrt{2}} = \frac{\sqrt{2}}{2}$$

$$\cos = \frac{adjacent}{hypotenuse} \quad ; \quad thus, \quad \cos 45° = \frac{1}{\sqrt{2}} = \frac{\sqrt{2}}{2}$$

$$\cos 60° = \frac{1}{2}$$

Substituting, we obtain:

$$\frac{\sqrt{3}}{2} \cdot \frac{\sqrt{2}}{2} + \frac{1}{2} \cdot \frac{\sqrt{2}}{2}$$

Multiply the fractions (recall $\sqrt{a} \cdot \sqrt{b} = \sqrt{ab}$) to obtain $\frac{\sqrt{6}}{4} + \frac{\sqrt{2}}{4}$

$= \frac{\sqrt{6} + \sqrt{2}}{4}$. Therefore, $\sin 105° = \frac{\sqrt{6} + \sqrt{2}}{4}$.

● **PROBLEM** 32-11

Find $\cos \frac{1}{12} \pi$.

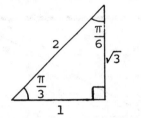 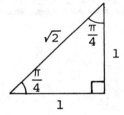

<u>Solution:</u> Express $\cos \frac{\pi}{12}$ in terms of angles whose values of the trigonometric functions are known.

$$\cos \frac{1}{12} \pi = \cos \left(\frac{4}{12} \pi - \frac{3}{12} \pi \right)$$

$$= \cos \left(\frac{1}{3} \pi - \frac{1}{4} \pi \right)$$

Now apply the difference formula for the cosine of two angles, α and β. $\cos (\alpha - \beta) = \cos \alpha \cos \beta + \sin \alpha \sin \beta$. In this example, $\alpha = \frac{1}{3} \pi$ and $\beta = \frac{1}{4} \pi$.

$$\cos \left(\frac{1}{3} \pi - \frac{1}{4} \pi \right) = \cos \frac{1}{3} \pi \cos \frac{1}{4} \pi$$

$$+ \sin \frac{1}{3} \pi \sin \frac{1}{4} \pi$$

See the accompanying diagrams to find the values of these angles. We find:

$$\cos \frac{\pi}{3} = \frac{1}{2}$$

$$\cos \frac{\pi}{4} = \frac{1}{\sqrt{2}}$$

$$\sin \frac{\pi}{3} = \frac{\sqrt{3}}{2}$$

$$\sin \frac{\pi}{4} = \frac{1}{\sqrt{2}}$$

Thus, $\cos \left(\frac{1}{3}\pi - \frac{1}{4}\pi\right) = \frac{1}{2} \cdot \frac{1}{\sqrt{2}}\left(\frac{\sqrt{2}}{\sqrt{2}}\right)$

$$+ \frac{\sqrt{3}}{2} \cdot \frac{1}{\sqrt{2}}\left(\frac{\sqrt{2}}{\sqrt{2}}\right) = \frac{\sqrt{2}}{4} + \frac{\sqrt{2}\,\sqrt{3}}{4}$$

$$= \frac{1}{4}\sqrt{2}\,(1 + \sqrt{3})$$

● **PROBLEM** 32-12

Find the sine and cosine of 75°.

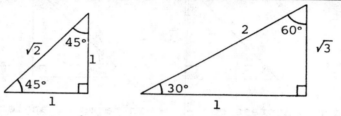

Solution: The angle 75° may be expressed as the sum or difference of two special angles, whose functions are known, in various ways: 75° = 45° + 30°, 75° = 120° - 45°, etc. If we choose the first of these, we use the sine of the sum of two angles,

$$\sin(\alpha + \beta) = \sin \alpha \cos \beta + \sin \beta \cos \alpha .$$

Here $\alpha = 45°$ and $\beta = 30°$. Then,

$$\sin 75° = \sin(45° + 30°) = \sin 45° \cos 30° + \cos 45° \sin 30°.$$

To find the sine and cosine of 45° and 30°, construct 45°, 45°, 90° and 30°, 60°, 90° triangles (see Figures). Since $\sin(\maltese) = \frac{\text{side opposite}(\maltese)}{\text{hypotenuse}}$, and $\cos(\maltese) = \frac{\text{side adjacent }(\maltese)}{\text{hypotenuse}}$; then

$$\sin 45° = \frac{1}{\sqrt{2}} = \frac{1}{\sqrt{2}}\left(\frac{\sqrt{2}}{\sqrt{2}}\right) = \frac{\sqrt{2}}{2} , \quad \sin 30° = \tfrac{1}{2} ,$$

and

$$\cos 45° = \frac{1}{\sqrt{2}} = \frac{1}{\sqrt{2}}\left(\frac{\sqrt{2}}{\sqrt{2}}\right) = \frac{\sqrt{2}}{2} , \quad \cos 30° = \frac{\sqrt{3}}{2} .$$

Substituting,

$$\sin 75° = \frac{\sqrt{2}}{2} \cdot \frac{\sqrt{3}}{2} + \frac{\sqrt{2}}{2} \cdot \frac{1}{2} = \frac{\sqrt{2}\sqrt{3}}{4} + \frac{\sqrt{2}}{4} = \tfrac{1}{4}\sqrt{6} + \tfrac{1}{4}\sqrt{2}$$

$$= \tfrac{1}{4}\left(\sqrt{6} + \sqrt{2}\right).$$

To find $\cos 75°$, apply the formula for the cosine of the sum of two angles,

$$\cos(\alpha + \beta) = \cos \alpha \cos \beta - \sin \alpha \sin \beta .$$

Here $\alpha = 45°$ and $\beta = 30°$. Thus,

$$\cos 75° = \cos(45° + 30°) = \cos 45° \cos 30° - \sin 45° \sin 30°.$$

$$= \frac{\sqrt{2}}{2} \cdot \frac{\sqrt{3}}{2} - \frac{\sqrt{2}}{2} \cdot \frac{1}{2} = \tfrac{1}{4}\left(\sqrt{6} - \sqrt{2}\right).$$

If we use the approximate values, $\sqrt{6} = 2.4494$, $\sqrt{2} = 1.4142$, we find

$$\sin 75° = 0.9659, \quad \cos 75° = 0.2588,$$

which check with the values given in the tables.

● **PROBLEM 32-13**

Find an expression for $\tan(u + v)$.

Solution: By definition of the tangent function, $\tan \theta = \dfrac{\sin \theta}{\cos \theta}$.

Then,

$$\tan(u+v) = \frac{\sin(u+v)}{\cos(u+v)} \qquad (1)$$

The addition formulas for the sine and cosine functions are:

$$\sin(\alpha+\beta) = \sin \alpha \cos \beta + \cos \alpha \sin \beta$$
$$\cos(\alpha+\beta) = \cos \alpha \cos \beta - \sin \alpha \sin \beta$$

Replacing α by u and β by v, and using these addition formulas, equation (1) becomes:

$$\tan(u+v) = \frac{\sin(u+v)}{\cos(u+v)} = \frac{\sin u \cos v + \cos u \sin v}{\cos u \cos v - \sin u \sin v}$$

If neither $\cos u = 0$ nor $\cos v = 0$, we can divide both the numerator and the denominator of this fraction by the product $\cos u \cos v$ to obtain a formula that involves only the tangent function:

$$\tan(u+v) = \frac{\dfrac{\sin u \cos v + \cos u \sin v}{\cos u \cos v}}{\dfrac{\cos u \cos v - \sin u \sin v}{\cos u \cos v}}$$

$$= \frac{\dfrac{\sin u \cos v}{\cos u \cos v} + \dfrac{\cos u \sin v}{\cos u \cos v}}{\dfrac{\cos u \cos v}{\cos u \cos v} - \dfrac{\sin u \sin v}{\cos u \cos v}}$$

$$= \frac{\dfrac{\sin u}{\cos u} + \dfrac{\sin v}{\cos v}}{1 - \dfrac{\sin u \sin v}{\cos u \cos v}}$$

$$= \frac{\dfrac{\sin u}{\cos u} + \dfrac{\sin v}{\cos v}}{1 - \dfrac{\sin u \sin v}{\cos u \cos v}}$$

$$tan(u+v) = \frac{tan\ u + tan\ v}{1 - tan\ u\ tan\ v} \ .$$

● **PROBLEM 32-14**

Find $\sin 15^{\circ}$, $\cos 15^{\circ}$, $\tan 15^{\circ}$, and $\cot 15^{\circ}$.

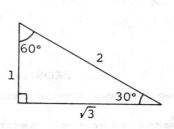
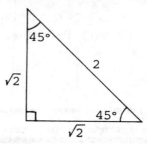

Solution: To find the values of these trigonometric functions, use the following subtraction formulas for the sine, cosine, tangent, and cotangent functions.

$$\sin(\alpha - \beta) = \sin \alpha \cos \beta - \cos \alpha \sin \beta$$

$$\cos(\alpha - \beta) = \cos \alpha \cos \beta + \sin \alpha \sin \beta$$

$$\tan(\alpha - \beta) = \frac{tan\ \alpha - tan\ \beta}{1 + tan\ \alpha\ tan\ \beta}$$

$$\cot(\alpha - \beta) = \frac{cot\ \alpha\ cot\ \beta + 1}{cot\ \beta - cot\ \alpha}$$

Recall that in a 30° - 60° right triangle:

$$\sin 30^{\circ} = \tfrac{1}{2} \ \text{and} \ \cos 30^{\circ} = \frac{\sqrt{3}}{2} \ ,$$

$$\sin 60^{\circ} = \frac{\sqrt{3}}{2} \ \text{and} \ \cos 60^{\circ} = \tfrac{1}{2} \ .$$

In a 45° - 45° right triangle:

$$\sin 45^{\circ} = \frac{\sqrt{2}}{2} \ \text{and} \ \cos 45^{\circ} = \frac{\sqrt{2}}{2} \ \text{(see figures)}$$

Now, substitute $\alpha = 45^{\circ}$ and $\beta = 30^{\circ}$ in the above formulas.

$$\sin 15^{\circ} = \sin(45^{\circ} - 30^{\circ}) = \sin 45^{\circ} \cos 30^{\circ} - \cos 45^{\circ} \sin 30^{\circ}$$

$$= \left(\frac{\sqrt{2}}{2}\right)\left(\frac{\sqrt{3}}{2}\right) - \left(\frac{\sqrt{2}}{2}\right)\left(\frac{1}{2}\right)$$

$$= \frac{\sqrt{2}\sqrt{3}}{4} - \frac{\sqrt{2}}{4} = \frac{\sqrt{2}(\sqrt{3} - 1)}{4}$$

$$\cos 15^{\circ} = \cos(45^{\circ} - 30^{\circ}) = \cos 45^{\circ} \cos 30^{\circ} + \sin 45^{\circ} \sin 30^{\circ}$$

$$= \left(\frac{\sqrt{2}}{2}\right)\left(\frac{\sqrt{3}}{2}\right) + \left(\frac{\sqrt{2}}{2}\right)\left(\frac{1}{2}\right)$$

$$= \frac{\sqrt{2}\sqrt{3}}{4} + \frac{\sqrt{2}}{4} = \frac{\sqrt{2}(\sqrt{3} + 1)}{4}$$

$$\tan 15^{\circ} = \frac{tan\ 45^{\circ} - tan\ 30^{\circ}}{1 + tan\ 45^{\circ}\ tan\ 30^{\circ}}$$

$$= \frac{1 - \frac{\sqrt{3}}{3}}{1 + 1\left(\frac{\sqrt{3}}{3}\right)} = \frac{1 - \frac{\sqrt{3}}{3}}{1 + \frac{\sqrt{3}}{3}}$$

Obtaining a common denominator of 3 for both the numerator and the denominator,

$$\tan 15^\circ = \frac{\frac{3}{3} - \frac{\sqrt{3}}{3}}{\frac{3}{3} + \frac{\sqrt{3}}{3}} = \frac{\frac{3 - \sqrt{3}}{3}}{\frac{3 + \sqrt{3}}{3}} = \frac{3 - \sqrt{3}}{\cancel{3}} \cdot \frac{\cancel{3}}{3 + \sqrt{3}} = \frac{3 - \sqrt{3}}{3 + \sqrt{3}}$$

$$\cot 15^\circ = \cot(45^\circ - 30^\circ)$$

$$= \frac{\cot 45^\circ \, \cot 30^\circ + 1}{\cot 30^\circ - \cot 45^\circ}$$

$$= \frac{(1)(\sqrt{3}) + 1}{\sqrt{3} - 1}$$

$$= \frac{\sqrt{3} + 1}{\sqrt{3} - 1}$$

● **PROBLEM 32-15**

Find the trigonometric functions of 195° using functions of 225° and 30°.

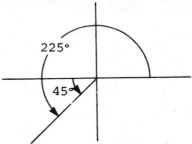

225°

45°

Solution: $195^\circ = 225^\circ - 30^\circ$. Hence, $\sin 195^\circ = \sin(225^\circ - 30^\circ)$. Using the subtraction formula for sines, $\sin(\alpha - \beta) = \sin \alpha \cos \beta - \cos \alpha \sin \beta$ where $\alpha = 225^\circ$ and $\beta = 30^\circ$, $\sin 195^\circ = \sin(225^\circ - 30^\circ) = \sin 225^\circ \cos 30^\circ - \cos 225^\circ \sin 30^\circ$. The reference angle for 225° is 45° ($225^\circ - 180^\circ = 45^\circ$), and the angle 225° is a third quadrant angle. See the figure.

In the third quadrant both the sine and cosine functions are negative. From a table of trigonometric functions, $\sin 225^\circ = \sin 45^\circ = -0.7071 = -\frac{\sqrt{2}}{2}$, $\cos 225^\circ = \cos 45^\circ = -0.7071 = -\frac{\sqrt{2}}{2}$, $\cos 30^\circ = 0.8660 = \frac{\sqrt{3}}{2}$ and $\sin 30^\circ = 0.5000 = \frac{1}{2}$. Therefore,

$$\sin 195^\circ = \sin(225^\circ - 30^\circ) = \left(-\frac{\sqrt{2}}{2}\right)\left(\frac{\sqrt{3}}{2}\right) - \left(-\frac{\sqrt{2}}{2}\right)\left(\frac{1}{2}\right)$$

$$= \frac{-\sqrt{2}\sqrt{3}}{4} + \frac{\sqrt{2}}{4}$$

$$= \frac{-\sqrt{2}\sqrt{3} + \sqrt{2}}{4}$$

$$= \frac{\sqrt{2}(-\sqrt{3} + 1)}{4}$$

Hence, $\sin 195^\circ = \frac{\sqrt{2}(1 - \sqrt{3})}{4}$.

● **PROBLEM 32-16**

Derive a formula for $\sin 3\alpha$ in terms of $\sin \alpha$.

<u>Solution:</u> We may regard 3α as $\alpha + 2\alpha$, and use the addition formula for the sine of two angles.

$$\sin(a + b) = \sin a \cos a + \cos a \sin b$$

where $a = \alpha$ and $b = 2\alpha$.

$$\sin 3\alpha = \sin(\alpha + 2\alpha) = \sin \alpha \cos 2\alpha + \cos \alpha \sin 2\alpha.$$

Now replace $\sin 2\alpha$ and $\cos 2\alpha$ by the expressions

$$\cos 2a = \cos^2 a - \sin^2 a$$

and

$$\sin 2a = 2 \sin a \cos a.$$

We find that

$$\sin 3\alpha = (\sin \alpha)(\cos^2\alpha - \sin^2\alpha) + (\cos \alpha)(2 \sin \alpha \cos \alpha)$$

$$= \sin \alpha \cos^2\alpha - \sin^3\alpha + 2 \sin \alpha \cos^2\alpha$$

$$= 3 \sin \alpha \cos^2\alpha - \sin^3\alpha.$$

Finally, since we wish a result involving only $\sin \alpha$, replace $\cos^2\alpha$ by $1 - \sin^2\alpha$; then

$$\sin 3\alpha = 3(\sin \alpha)(1 - \sin^2\alpha) - \sin^3\alpha$$

$$= 3 \sin \alpha - 3 \sin^3\alpha - \sin^3\alpha$$

$$= 3 \sin \alpha - 4 \sin^3\alpha.$$

This is the desired identity.

● **PROBLEM 32-17**

Derive a formula for $\cos 3\theta$ which involves only the cosine function.

<u>Solution:</u> $\cos 3\theta = \cos(2\theta + \theta)$
    Using the addition formula for the cosine function which states that $\cos(\alpha + \beta) = \cos \alpha \cos \beta - \sin \alpha \sin \beta$ and replacing α by 2θ and β by θ:

$$\cos 3\theta = \cos(2\theta + \theta) = \cos 2\theta \cos \theta - \sin 2\theta \sin \theta . \quad (1)$$

The double-angle formulas for $\cos 2\alpha$ and $\sin 2\alpha$ yield:

$$\cos 2\alpha = 2 \cos^2 \alpha - 1 \quad \text{and} \quad \sin 2\alpha = 2 \sin \alpha \cos \alpha .$$

Hence, replacing α by θ in both cases:

$$\cos 3\theta = \left(2 \cos^2 \theta - 1\right)\cos \theta - (2 \sin \theta \cos \theta)\sin \theta$$
$$= 2 \cos^3 \theta - \cos \theta - 2 \sin^2 \theta \cos \theta \quad (2)$$

Since $\sin^2 \theta = 1 - \cos^2 \theta$, equation (2) becomes:

$$\cos 3\theta = 2 \cos^3 \theta - \cos \theta - 2\left(1 - \cos^2 \theta\right)\cos \theta$$
$$= 2 \cos^3 \theta - \cos \theta - 2 \cos \theta + 2 \cos^3 \theta$$
$$\cos 3\theta = 4 \cos^3 \theta - 3 \cos \theta .$$

CHAPTER 33

SOLVING TRIANGLES

> **Basic Attacks and Strategies for Solving Problems in this Chapter. See pages 618 to 627 for step-by-step solutions to problems.**

Solving a triangle refers to finding all the angles and the lengths of all the sides, given some preliminary information.

If it is known that the triangle is a right triangle, then trigonometric functions or the Pythagorean theorem (*cf.* Chapter 28) can be used to find the length of the sides.

If it is not known whether the triangle is a right triangle, one can try to compute its altitude from a base side and see if, by chance, it matches a given side (implying that it must in fact be a right triangle).

If the triangle is known not to be a right triangle, the *Law of Cosines* can be used to find a side (or an angle). The law of cosines can be considered to be a version of the Pythagorean theorem that can be used for any triangle and any angle. It states that

$$c^2 = a^2 + b^2 - 2ab \cos \gamma,$$

where γ is the angle opposite side c (i.e., between sides a and b) (*cf.* Chapter 24). Since for a right triangle $r = 90°$, $\cos r = 0$ and we see that the law of cosines actually reduces to the Pythagorean theorem in this case. It is also possible to determine information using the *Law of Sines*, which states:

$$\frac{\sin \alpha}{a} = \frac{\sin \beta}{b} = \frac{\sin \gamma}{c},$$

where α is the angle opposite side a, β is opposite b, and γ is opposite c.

In solving triangles, one should first draw a triangle and label the sides and angles according to the information given. Then one can determine other information. If two angles are known, one should immediately compute the third angle (since all three angles of a triangle must always sum to 180°). It may be useful to draw associated triangles in order to determine if some side can be determined using a given angle and a trigonometric function.

Step-by-Step Solutions to Problems in this Chapter, "Solving Triangles"

● **PROBLEM 33-1**

Solve the oblique triangle ABC for side c, and the two unknown angles, where a = 20, b = 40, α = 30 ; and α is the angle between sides b and c.

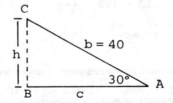

<u>Solution:</u> If we draw an altitude h from b to c, as in the accompanying diagram, we can find the length of this altitude by trigonometry.

$$\sin 30° = \frac{h}{b}$$

$$h = b \sin 30°$$

Since sin 30° = ½, and b = 40,

$$h = 40 \left(\tfrac{1}{2}\right) = 20 = \text{side a}$$

Thus, the triangle must have the altitude as one of its sides; therefore, we have a right triangle with angles 30°, 60°, and 90°. Sides of such right triangles are in proportion 1 : √3 : 2, and the lengths are therefore 20 : 20√3 : 40. Hence, c = 20√3, and the two unknown angles are 60° and 90°.

Given a = 8, c = 7, β = 135°, find b.

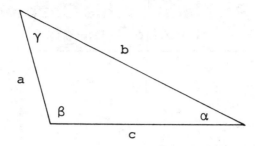

<u>Solution:</u> Use the law of cosines to find one side given two sides and an included angle.

$b^2 = a^2 + c^2 - 2ac \cos(\text{included angle})$ where the included angle is the angle between the two given sides.

$b^2 = a^2 + c^2 - 2ac \cos β$

$$b^2 = 64 + 49 - 2 \cdot 8 \cdot 7 \cos 135°$$
$$= 113 - 112 \cdot \left(\frac{\sqrt{2}}{2}\right) = 113 + 79.184$$
$$= 192.184$$
$$b = 13.863$$

Solve triangle ABC, given a = 30, b = 50, ∠C = 25°.

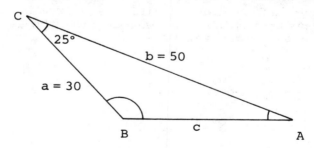

<u>Solution:</u> Two of the sides of △ABC, and their included angle are given. We wish to find the third side, c. Therefore use the law of cosines to find c.

$$c^2 = a^2 + b^2 - 2ab \cos C$$

$$c^2 = 30^2 + 50^2 - 2(30)(50) \cos 25°$$

$$c^2 = 900 + 2500 - 2(30)(50)(0.9063)$$

$$c^2 = 681.1$$

$$c = 26 \text{ (to two significant digits)}$$

Use the law of sines to find one of the remaining angles.

$$\frac{\sin A}{30} = \frac{\sin 25°}{26}$$

$$\sin A = \frac{30 \sin 25°}{26} = \frac{30(0.4226)}{26}$$

$$\sin A = 0.4876$$

$$\angle A = 29° \text{ (to the nearest degree)}$$

$\angle B$ can be found from $\angle A$ and $\angle C$.

$$\angle A + \angle B + \angle C = 180°$$

$$\angle B = 180° - \angle A - \angle C$$

$$= 180° - (\angle A + \angle C)$$

$$= 180° - (29° + 25°)$$

$$= 180° - (54°)$$

$$\angle B = 126°$$

● **PROBLEM 33-4**

In right triangle ABC, if a = 3, and b = 4, find c, α, β.

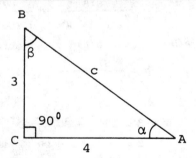

Solution: The figure can be drawn with accuracy in this case. (See the Figure).

The Pythagorean Theorem $c^2 = a^2 + b^2$ such that $c^2 = 3^2 + 4^2 = 25$ shows that c = 5. The ratio of the two known sides is, indeed, a value of a trigonometric function of one of the acute angles. In fact,

$$\tan \alpha = \frac{\text{side opposite } \angle\alpha}{\text{side adjacent } \angle\alpha} = \frac{3}{4}$$

Similarly for $\angle \beta$, $\tan \beta = \frac{4}{3}$

$$\cot \alpha = \frac{1}{\tan \alpha} = \frac{1}{\frac{\text{side opposite } \angle\alpha}{\text{side adjacent } \angle\alpha}} = \frac{\text{side adjacent } \angle\alpha}{\text{side opposite } \angle\alpha} = \frac{4}{3}$$

Similarly for $\measuredangle \beta$, $\cot \beta = \frac{3}{4}$. It is merely a matter of choosing the preferred one of these ratios. Let us choose $\tan \alpha = \frac{3}{4}$. Thus,

$$\tan \alpha = \frac{3}{4} = 0.75000$$
$$\alpha = 36°52.2'$$
$$\alpha + \beta + 90° = 180°$$
$$\alpha + \beta = 180° - 90° = 90°$$

Therefore,
$$\beta = 90° - \alpha = 90° - 36°52.2' = 53°7.8'$$

● **PROBLEM 33-5**

In triangle ABC, if $a = 675$, $\alpha = 48°36'$, find b, c, β.

<u>Solution:</u> If we wish to work with α, the functions of α involving a and one other side are

$$\sin \alpha = \frac{\text{opposite}}{\text{hypotenuse}}$$

$\sin \alpha = a/c$

$$\tan \alpha = \frac{\text{opposite}}{\text{adjacent}}$$

$\tan \alpha = a/b$, and

$$\cot \alpha = \frac{1}{\tan \alpha} = \frac{\text{adjacent}}{\text{opposite}}$$

$\cot \alpha = b/a$. Either of these ratios can be chosen. If the first is chosen, $\sin \alpha = a/c$ or $c \sin \alpha = a$ and $c = a/\sin \alpha$. Thus,

$$c = \frac{675}{\sin 48°36'} = \frac{675}{0.75011} = 899.85$$

Using $\tan \alpha = a/b$ or $b = a/\tan \alpha$, we obtain

$$b = \frac{675}{\tan 48°36'} = 595.08$$

To find β, $\alpha + \beta + 90° = 180°$
$$\alpha + \beta = 180° - 90° = 90°$$
$$\beta = 90° - \alpha = 90° - 48°36'$$
$$\beta = 41°24'$$

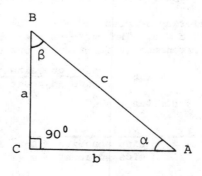

621

Find all the sides and angles of triangle ABC, given a = 137, c = 78.0,
< C = 23°0' .

Solution: Draw triangle ABC, filling in the given information. Thus

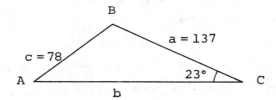

we divide the problem into 3 parts:

    (1)   find angle A.

    (2)   find angle B .

    (3)   find side b .

In order to find angle A we may use the law of sines, $\dfrac{\sin A}{\text{side } a} = \dfrac{\sin C}{\text{side } c}$,

because we are given side a = 137, side c = 78, and sin C = sin 23° ;

thus, $\dfrac{\sin A}{137} = \dfrac{\sin 23°0'}{78}$.

Using our Trig. table we find sin 23° = 0.3907 . Thus

$$\frac{\sin A}{137} = \frac{0.3907}{78}$$

Multiplying both sides by 137 we obtain

$$\sin A = \frac{137(0.3907)}{78}$$

$$\sin A = 0.6862 .$$

Using our Trig. table we find that < A = 43°20' . Our Trig. table
only gives values of sine between 0° and 90° . Since we are dealing
with an angle in a triangle, which can take on values greater than 90°
(recall there are 180° in a triangle) we must examine what happens
to the sine function in the second quadrant, that is between 90° and
180°

We use our trignometric identity $\sin \theta = \sin(180-\theta)$:

$$\sin 43°20' = \sin(180-43°20') = \sin 136°40'$$

Hence

$$< A = 43°20' \text{ } \underline{\text{or}} \text{ } 136°40'$$

Thus there are two solutions. We now proceed to the next part of our problem, finding angle B, and side b.

$$< A + < B < C = 180° \quad \text{(There are } 180° \text{ in a}$$
$$< A + < B + 23° = 180° \quad \text{triangle)}$$
$$< A + < B = 157°$$

If $< A = 43°20'$ then

$$43°20' + < B = 157°$$
$$< B = 157° - 43°20' = 113°40'$$

Since we now know $< B$, we may apply the law of sines:

$$\frac{\sin B}{\text{side } b} = \frac{\sin C}{\text{side } c}$$

to find side b:

$$\frac{\sin 113°40'}{b} = \frac{\sin 23°}{78}$$

Cross multiplying we obtain

$$b \sin 23° = 78 \sin 113°40'$$
$$b = \frac{78 \sin 113°40'}{\sin 23°}$$

Substituting in the values $\sin 113°40' = 0.9159$ and $\sin 23° = 0.3907$ we obtain

$$b = 78\left(\frac{0.9159}{0.3907}\right) = 183$$

Hence if we choose $< A = 43°20'$ then

$$< B = 113°40' \text{ and}$$

$$\text{side } b = 183$$

If, however, we choose $< A = 136°40'$, then since

$$< A + < B = 157°$$
$$136°40' + < B = 157°$$

and

$$< B = 157° - 136°40' = 20°20'$$

Applying the law of sines to find side b gives us:

$$\frac{\sin B}{\text{side } b} = \frac{\sin C}{\text{side } c}$$

$$\frac{\sin 20°20'}{b} = \frac{\sin 23°}{78}$$

Cross multiplying gives us

$$b \sin 23° = 78 \sin 20°20'$$

$$b = \frac{78 \sin 20°20'}{\sin 23°}$$

Substituting in $\sin 20°20' = 0.3475$ and $\sin 23° = 0.3907$ we obtain

$$b = \frac{78(0.3475)}{0.3907} = 69$$

Hence if we choose $< A = 136°40'$, then $< B = 20°20'$, and side $b = 69$.

● PROBLEM 33-7

Find all the sides and angles of triangle ABC, given $a = 43$, $b = 32$, $< B = 67°$.

Solution: Draw triangle ABC, filling in the given information. Thus

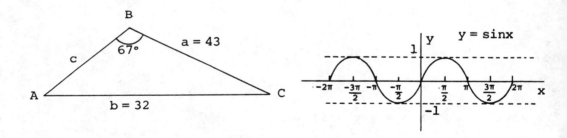

we divide the problem into 3 parts:

   (1)   find angle A

   (2)   find angle C

   (3)   find side c

In order to find angle A we may use the law of sines, $\frac{\sin A}{\text{side } a} = \frac{\sin B}{\text{side } b}$, because we are given side $a = 43$, side $b = 32$, and $\sin B = \sin 67°$. Thus, $\frac{\sin A}{43} = \frac{\sin 67}{32}$, multiplying both sides by 43,

$$\sin A = \frac{43 \sin 67°}{32}$$

In our trig table we find $\sin 67° = 0.9205$. Thus,

$$\sin A = \frac{43(0.9205)}{32} = 1.2369$$

We found sin A = 1.2369; however the sine function is only defined

on the interval [-1,1] — that is, it can only take on values between

-1 and 1, which can be seen from the graph y = sin x, as shown.

Thus there is no angle A such that sin A = 1.2369, and no triangle

can contain such an angle, hence triangle ABC is non-existent, and

we cannot solve for the other sides or angles.

● PROBLEM 33-8

In the accompanying figure, given b = 16.351, c = 11.189,
α = 42° 19.8'; find a,β,γ .

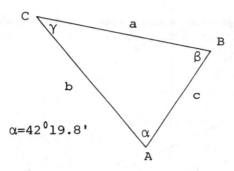

$$\tfrac{1}{2} \beta + \tfrac{1}{2} \gamma = 68°50.1´$$

$$\underline{\tfrac{1}{2} \beta - \tfrac{1}{2} \gamma = 25°50.0´}$$

$$\beta = 93°100.1´$$

Since 60´ = 1°, we have: β = 94°40.1´

We obtain the value for γ by substituting into the
equation, β + γ = 137°40.2´, and obtain: γ = 43°0.1´.

To obtain the value of side a, we use the law of
sines to form the proportion:

$$\frac{a}{\sin \alpha} = \frac{c}{\sin \gamma}$$

Since we know c, and can find sin γ and sin α, this law
enables us to compute the value of a. Solving for a,

$$a = \sin \alpha \left(\frac{c}{\sin \gamma} \right)$$

$$a = \sin 42°19.8´ \left(\frac{11.189}{\sin 43°0.1´} \right)$$

We must now obtain values for sin 42°19.8´ and
43°0.1´. Since the former is very close to 42°40´ and
the latter is very close to 43°, we do not need interpolation. (In fact, interpolating will give us the same

values as those for sin 42°20´ and 43°). Using a table we
find:

sin 42°20´ = .6734 $\simeq$ sin 42°19.8´

sin 43°= .6820 $\simeq$ sin 43°0.1´;

and substituting:

$$a = .6734 \left(\frac{11.189}{.6820}\right) = 11.0479 \simeq 11.048$$

Therefore,   a = 11.048

β = 94°40.1´

γ = 43°0.1´

● **PROBLEM 33-9**

Show that in any triangle ABC

$$\frac{a+b}{c} = \frac{\cos \frac{1}{2}(A-B)}{\sin \frac{1}{2}C} \quad \text{and}$$

$$\frac{a-b}{c} = \frac{\sin \frac{1}{2}(A-B)}{\cos \frac{1}{2}C}$$

<u>Solution</u>: By the law of sines, one has

$$\frac{a}{c} = \frac{\sin A}{\sin C} \quad \text{and} \quad \frac{b}{c} = \frac{\sin B}{\sin C}.$$

Adding these two equations together, one has

$$\frac{a}{c} + \frac{b}{c} = \frac{\sin A}{\sin C} + \frac{\sin B}{\sin C}$$

so that $\frac{a+b}{c} = \frac{\sin A + \sin B}{\sin C}$.

Using the formula for sum of sines, one obtains

$$\frac{a+b}{c} = \frac{2 \sin \frac{1}{2}(A+B) \cos \frac{1}{2}(A-B)}{\sin C}$$

sin C may be written as

$$\sin 2\left(\frac{C}{2}\right) = 2 \sin \frac{C}{2} \cos \frac{C}{2}.$$

Hence,

$$\frac{a+b}{c} = \frac{2 \sin\left(\frac{A+B}{2}\right) \cos\left(\frac{A-B}{2}\right)}{2 \sin \frac{C}{2} \cos \frac{C}{2}}$$

$$= \frac{\sin\left(\frac{A+B}{2}\right) \cos\left(\frac{A-B}{2}\right)}{\sin \frac{C}{2} \cos \frac{C}{2}}$$

Since $A + B + C = 180°$

$$A + B = 180° - C, \quad \frac{A+B}{2} = 90° - \frac{C}{2}$$

$$\sin\left(\frac{A+B}{2}\right) = \sin\left(90° - \frac{C}{2}\right) = \cos \frac{C}{2}$$

Therefore,

$$\frac{a+b}{c} = \frac{\sin\left(\frac{A+B}{2}\right) \cos\left(\frac{A-B}{2}\right)}{\sin \frac{C}{2} \cos \frac{C}{2}}$$

$$= \frac{\cos \frac{C}{2} \cos\left(\frac{A-B}{2}\right)}{\sin \frac{C}{2} \cos \frac{C}{2}}$$

$$= \frac{\cos\left(\frac{A-B}{2}\right)}{\sin \frac{C}{2}} = \frac{\cos \frac{1}{2}(A-B)}{\sin \frac{1}{2} c}$$

Similarly, one obtains

$$\frac{a}{c} = \frac{\sin A}{\sin C}, \quad \frac{b}{c} = \frac{\sin B}{\sin C}, \quad \frac{a-b}{c} = \frac{\sin A - \sin B}{\sin C}$$

$$\frac{a-b}{c} = \frac{2 \cos \frac{1}{2}(A+B) \sin \frac{1}{2}(A-B)}{2 \sin \frac{C}{2} \cos \frac{C}{2}} = \frac{\sin \frac{1}{2}(A-B)}{\cos \frac{C}{2}}$$

These two equations, which were just verified, are a pair of Mollweide's formulas. Mollweide's formulas are the following, for any triangle ABC:

$$\frac{a+b}{c} = \frac{\cos \frac{1}{2}(A-B)}{\sin \frac{C}{2}}, \quad \frac{a-b}{c} = \frac{\sin \frac{1}{2}(A-B)}{\cos \frac{C}{2}}$$

$$\frac{b+c}{a} = \frac{\cos \frac{1}{2}(B-C)}{\sin \frac{A}{2}}, \quad \frac{b-c}{a} = \frac{\sin \frac{1}{2}(B-C)}{\cos \frac{A}{2}}$$

$$\frac{c+a}{b} = \frac{\cos \frac{1}{2}(C-A)}{\sin \frac{A}{2}}, \quad \frac{c-a}{b} = \frac{\sin \frac{1}{2}(C-A)}{\cos \frac{B}{2}}$$

CHAPTER 34

INVERSE TRIGONOMETRIC FUNCTIONS

> **Basic Attacks and Strategies for Solving Problems in this Chapter. See pages 628 to 640 for step-by-step solutions to problems.**

Inverse trigonometric functions are inverse relationships (*cf.* Chapter 7) whose original function is a trigonometric function. Given an inverse trigonometric function, we input a numeric value and obtain an angle as output.

Inverse trigonometric functions are denoted in one of two ways: (a) one can use the standard exponent of –1 as would be used for any inverse function, or (b) one can prefix the function with the word *arc*. Thus, the inverse function of $a = \sin \alpha$ is $\alpha = \arcsin a = \sin^{-1} a$. The use of the notation "arcsin *a*" is often read "the *angle* (i.e, arc) whose sin is *a*," thereby helping someone remember that the output of an inverse trigonometric function is an *angle*.

Since the trigonometric functions are periodic, the inverse trigonometric functions are technically not functions as several angles may correspond to the same trigonometric value. To ensure that the inverse trigonometric functions are (in fact) functions when desired, the output values are usually restricted to certain *principal values*, either values between 0° and 360° or values between –180° and 180°. When designating inverse trigonometric functions as strict functions, one usually capitalizes the first letter, e.g., $\text{Sin}^{-1} a$ or Arcsin*a*.

Since an inverse trigonometric function results in an angle, one may use an inverse trigonometric function as an angle to compute the value of a trigonometric function. For example,

$$\sin\left(\cos^{-1}\frac{1}{2}\right) = \sin 60° = \frac{\sqrt{3}}{2}.$$

One need not ever compute the angle, though, since the argument to the inverse trigonometric function can be interpreted as two sides of a right triangle. Thus, one can compute the third side (using the Pythagorean theorem) and easily find the value for the desired trigonometric function by examining the sides of the triangle. This can be done even if the argument includes a variable.

When given equations involving inverse trigonometric functions whose arguments include variables, it is sometimes helpful to equate the inverse trigonometric functions to angle variables and rewrite the equations using these new angles. For example, given $\sin^{-1}(1 - x)$, one might write α and remember that $\sin \alpha = 1 - x$.

Graphs of the inverse trigonometric functions can be found in standard trigonometry and calculus books.

● **PROBLEM 34-1**

Calculate the following numbers.

(a) Arctan $\sqrt{3}$ (c) $\text{Tan}^{-1} 1.871$

(b) $\text{Tan}^{-1} .2027$

Solution: a) The expression $\tan y = x$ is equivalent to arctan $x = \tan^{-1} x = y$. Let the expression arctan $\sqrt{3} = y$. Hence, the expression arctan $\sqrt{3} = y$ is equivalent to $\tan y = \sqrt{3} = 1.7321$. In a table of trigonometric functions, the number y that corresponds to $\tan y = 1.7321$ is approximately 1.05.

b) Note that the expression $\tan^{-1} .2027 = $ arctan .2027 . Let the expression $\tan^{-1} .2027 = y$. Hence, the expression $\tan^{-1} .2027 = $ arctan .2027 $= y$ is equivalent to $\tan y = .2027$. In a table of trigonometric functions, the number y that corresponds to $\tan y = .2027$ is .20.

c) Note that the expression $\tan^{-1} 1.871 = $ arctan 1.871. Let the expression $\tan^{-1} 1.871 = y$. Hence, the expression $\tan^{-1} 1.871 = $ arctan $1.871 = y$ is equivalent to $\tan y = 1.871$. In a table of trigonometric functions, the number y that corresponds to $\tan y = 1.871$ is 1.08.

● **PROBLEM 34-2**

Find arcsin $\frac{1}{2}$.

Solution: Let $y = $ arcsin $\frac{1}{2}$; then $\sin y = \frac{1}{2}$. Since $\sin 30° = \frac{1}{2}$, we have y = 30°, 150°. Because the sine has a period of 360°, any angle which we can obtain by adding any integral multiple of 360° to 30° or 150° would also satisfy this equation. We thus get

$$\text{arcsin } \tfrac{1}{2} = \left\{ \begin{array}{l} 2n\pi + \tfrac{1}{6}\pi \\[2mm] 2n\pi + \tfrac{5}{6}\pi \end{array} \right. \text{ or } \left\{ \begin{array}{l} n(360°) + 30° \\[2mm] n(360°) + 150° \end{array} \right. ,$$

where n may be any integer, positive, negative, or zero.

● **PROBLEM 34-3**

In $\triangle ABC$, $A = \text{arc cos}\left(-\tfrac{\sqrt{3}}{2}\right)$. What is the value of A expressed in radians?

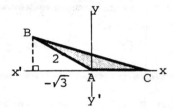

Solution: The expression "arc cos$\left(-\tfrac{\sqrt{3}}{2}\right)$" means "the angle whose cosine equals $-\tfrac{\sqrt{3}}{2}$." Angles whose cosine equals $-\tfrac{\sqrt{3}}{2}$ are 150°, 210°, -150°, and -210°.

Since the principal value of an arc cosine of an angle is the positive angle having the smallest numerical value of the angle, 150°, or $\tfrac{5\pi}{6}$, is the principal value of angle A.

● **PROBLEM 34-4**

Evaluate: (a) $\sin^{-1}\tfrac{\sqrt{3}}{2}$, (b) $\tan^{-1}\left(-\sqrt{3}\right)$.

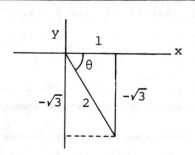

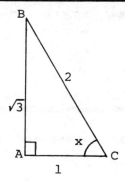

Solution: (a) Recall that inverse sines are angles. Thus we are looking for the angle whose sin is $\frac{\sqrt{3}}{2}$. $\sin^{-1}\frac{\sqrt{3}}{2} = x$ means $\sin x = \frac{\sqrt{3}}{2}$ where $\sin = \frac{\text{opposite}}{\text{hypotenuse}}$.

We note that triangle ABC is a 30-60 right triangle, and angle $x = 60°$. Since $\sin 60° = \frac{\sqrt{3}}{2}$,

$$\sin^{-1}\frac{\sqrt{3}}{2} = 60° .$$

(b) Recall that inverse tangents are angles. Thus we are looking for the angle whose tangent is $-\sqrt{3}$. $\text{Tan}^{-1}\left(-\sqrt{3}\right) = \theta$ means $\tan\theta = -\sqrt{3}$ where $\tan = \frac{\text{opposite}}{\text{adjacent}}$.

Since tangent is negative in the 4th quadrant, we draw our triangle there, and note it is a 30-60 right triangle, and angle $\theta = (-60°)$. Since $\tan(-60°) = \frac{-\sqrt{3}}{1}$, $\tan^{-1}\left(-\sqrt{3}\right) = -60°$.

Evaluate $\arctan\frac{5}{12} - \arccos\frac{3}{5}$.

Solution: Let $\alpha = \arctan\frac{5}{12}$, $\beta = \arccos\frac{3}{5}$. We are to evaluate the angle $\alpha - \beta$, where $-\frac{\pi}{2} < \alpha < \frac{\pi}{2}$ and $0 \leq \beta \leq \pi$. We first find some function of $\alpha - \beta$, say $\sin(\alpha - \beta)$. Since $\tan\alpha = \frac{5}{12}$, using the Pythagorean theorem and the fact that $\tan\alpha = \frac{\text{opposite}}{\text{adjacent}}$, we obtain hypotenuse = 13; thus $\sin\alpha = \frac{5}{13}$ and $\cos\alpha = \frac{12}{13}$. Also, since $\cos\beta = \frac{3}{5}$, $\sin\beta = \frac{4}{5}$. Then, using the formula for the sin of the difference of two angles, $\sin(\alpha - \beta) = \sin\alpha\cos\beta - \cos\alpha\sin\beta$

$$= \frac{5}{13} \cdot \frac{3}{5} - \frac{12}{13} \cdot \frac{4}{5}$$

$$= \frac{15}{65} - \frac{48}{65} = -\frac{33}{65} = -.5077$$

approximately. Now, observe that α is less than $\frac{\pi}{4}$ and β is greater than $\frac{\pi}{4}$ $\left(\text{this can be seen by observing that } \sin\alpha = \frac{5}{13} \text{ and } \sin\beta = \frac{4}{5}, \text{ and also observing the sin function from } 0 \text{ to } \frac{\pi}{2}\right)$. Hence $\alpha - \beta$ is a negative angle. Reference to tables gives us

$$\alpha - \beta = -\arcsin .5077 = -.5325 \text{ approximately.}$$

If principal values are used, then what does the ratio $\dfrac{\text{arc sin } \frac{1}{2}}{\text{arc tan } 1}$

equal?

<u>Solution:</u> The expression "arc sin ½" means "the angle whose sine equals ½"; similarly, "arc tan 1" means "the angle whose tangent equals 1."

Evaluate $\cos[\text{arc sin } (-1)]$.

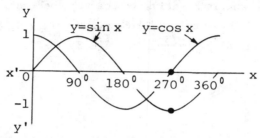

<u>Solution:</u> The expression "arc sin (-1)" means "the angle whose sine equals -1." Between $0°$ and $360°$, the only angle whose sine equals -1 is $270°$.

Hence, $\cos[\text{arc sin}(-1)] = \cos 270°$.

The value of $\cos 270° = 0$.

Note: In problems of this type, a sketch of $y = \sin x$ and $y = \cos x$ is very useful.

Evaluate $\cos\left(\sin^{-1} \frac{1}{2}\right)$.

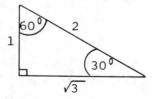

<u>Solution:</u> Inverse sine of $\frac{1}{2}$ is the angle whose sin is $\frac{1}{2}$. From our diagram of a 30° - 60° - 90° right triangle we observe $\sin 30° = \frac{1}{2}$ so $\sin^{-1} \frac{1}{2} = 30°$. Thus $\cos\left(\sin^{-1} \frac{1}{2}\right) =$

$\cos(30°)$. Consulting our diagram we see $\cos 30° = \frac{\sqrt{3}}{2}$. Therefore $\cos\left(\sin^{-1}\frac{1}{2}\right) = \frac{\sqrt{3}}{2}$.

● PROBLEM 34-9

Express in radical form the positive value of $\sin(\text{arc } \cos \frac{1}{2})$.

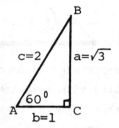

<u>Solution:</u> The expression "arc cos $\frac{1}{2}$" means "the angle whose cosine equals $\frac{1}{2}$." In the diagram, note that A is the angle whose cosine is $\frac{1}{2}$.

Since $60°$ is the angle whose cosine equals $\frac{1}{2}$, A = $60°$. The positive value of $\sin(\text{arc } \cos \frac{1}{2}) = \sin 60° = \frac{\sqrt{3}}{2}$.

● PROBLEM 34-10

Find $\sin\left(\text{Arc } \tan \frac{3}{4}\right)$.

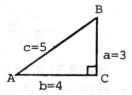

<u>Solution:</u> The expression "arc tan $\frac{3}{4}$" means "the angle whose tangent equals $\frac{3}{4}$." Spelled with a capital "A", "Arc tan" refers to the principal value of that angle, that is, its measure as a positive acute angle.

Draw a right triangle ABC with an acute angle A such that the leg adjacent to A is 4 and the leg opposite A is 3. Thus, $\tan A = \frac{3}{4}$ and the expression "Arc tan $\frac{3}{4}$" can be replaced by A. Hence, $\sin\left(\text{Arc } \tan \frac{3}{4}\right) = \sin A$.

Apply the Law of Pythagoras: if the legs of a right triangle are 3 and 4, its hypotenuse is 5. Hence, c = 5.

Therefore, $\sin A = \frac{3}{5}$, or $\sin\left(\text{Arc tan } \frac{3}{4}\right) = \frac{3}{5}$.

● **PROBLEM 34-11**

What is the value of $\tan\left(\text{Arc cos } \frac{\sqrt{2}}{2}\right)$?

<u>Solution:</u> The expression "arc cos $\frac{\sqrt{2}}{2}$ " means "the angle whose cosine is $\frac{\sqrt{2}}{2}$ " . Spelled with a capital "A," "Arc cos" refers to the principle value of that angle. If the cosine is positive, the principle value of an arc cosine is its measure as a positive acute angle.

Hence, Arc cos $\frac{\sqrt{2}}{2} = 45°$.

Therefore, $\tan\left(\text{Arc cos } \frac{\sqrt{2}}{2}\right) = \tan 45° = 1$.

● **PROBLEM 34-12**

Find sin arccos $\frac{4}{5}$, if arccos $\frac{4}{5}$ is in quadrant I.

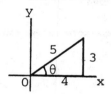

<u>Solution:</u> Let $\theta = \text{arccos } \frac{4}{5}$; then $\cos \theta = \frac{4}{5}$. We can then construct the triangle of the figure. From the triangle we get $\sin \theta = \frac{3}{5}$; therefore sin arccos $\frac{4}{5} = \frac{3}{5}$.

● **PROBLEM 34-13**

Find sin(Arctan x), where x may be any real number.

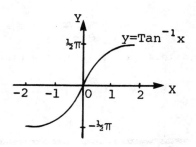

Solution: Let $t = $ Arctan x. Then $\tan t = x$, and $-\tfrac{1}{2}\pi < t < \tfrac{1}{2}\pi$. Also t and x have the same sign (see Figure). Now, we express $\sin t$ in terms of x as follows:

since $\tan t = x = \dfrac{\sin t}{\cos t}$, and

$$\sin t = \frac{\sin t}{\cos t} \cdot \cos t ,$$

we write

$$\sin t = x \cos t$$
$$\sin^2 t = x^2 \cos^2 t,$$

and by the identity

$$\sin^2 t + \cos^2 t = 1,$$
$$\sin^2 t = x^2 (1 - \sin^2 t).$$

Solving for $\sin^2 t$,

$$1 = \frac{x^2 (1 - \sin^2 t)}{\sin^2 t}$$

$$1 = \frac{x^2 - x^2 \sin^2 t}{\sin^2 t}$$

$$1 = \frac{x^2}{\sin^2 t} - \frac{x^2 \sin^2 t}{\sin^2 t}$$

$$1 = \frac{x^2}{\sin^2 t} - x^2$$

$$1 + x^2 = \frac{x^2}{\sin^2 t}$$

$$\sin^2 t (1 + x^2) = x^2$$
$$\sin^2 t = \frac{x^2}{1 + x^2} .$$

Hence,

$$\sin t = \frac{x}{\sqrt{1 + x^2}} \quad \text{or} \quad \frac{-x}{\sqrt{1 + x^2}} .$$

Since $-\tfrac{1}{2}\pi < t < \tfrac{1}{2}\pi$, $\sin t$ and t have the same sign. From above we know that t and x have the same sign. Thus, $\sin t$ and x must have the same sign. Now

$$\frac{x}{\sqrt{1 + x^2}}$$

has the same sign as x, while

634

$$\frac{-x}{\sqrt{1 + x^2}}$$

has the opposite sign; thus $\sin t = \sin(\text{Arctan } x) = \dfrac{x}{\sqrt{1 + x^2}}$.

● **PROBLEM 34-14**

Evaluate $\tan\left[\frac{1}{2} \arcsin\left(-\frac{8}{17}\right)\right]$.

Solution: Let $\theta = \arcsin\left(-\frac{8}{17}\right)$. Then $\sin \theta = -\frac{8}{17}$. Thus, we wish to evaluate $\tan\left(\frac{\theta}{2}\right)$. Using the half-angle formula for tan we obtain:

$$\tan\left(\frac{\theta}{2}\right) = \frac{1 - \cos \theta}{\sin \theta} .$$

Thus, we must find, in addition to $\sin \theta = -\frac{8}{17}$, the corresponding value of $\cos \theta$. Observe that when $\sin \theta$ is negative, θ must be a negative angle in the fourth quadrant, since we have the restriction on the inverse sin function $-\frac{\pi}{2} \le \theta \le \frac{\pi}{2}$. Thus, $\cos \theta$ will be positive (since cos is positive in the fourth quadrant), and using the identity $\sin^2\theta + \cos^2\theta = 1$ we obtain:

$$\cos \theta = \sqrt{1 - \sin^2\theta} = \sqrt{1 - \left(-\frac{8}{17}\right)^2} = \sqrt{1 - \frac{64}{289}}$$

$$= \sqrt{\frac{225}{289}} = \frac{15}{17} .$$

Therefore the desired value is

$$\tan\left(\frac{\theta}{2}\right) = \frac{1 - \frac{15}{17}}{-\frac{8}{17}} = \frac{2}{17} \cdot \left(-\frac{17}{8}\right) = -\frac{1}{4} ,$$

or

$$\tan\left[\frac{1}{2} \arcsin\left(-\frac{8}{17}\right)\right] = -\frac{1}{4} .$$

● **PROBLEM 34-15**

Find cos (Arccos x - Arccos 3y).

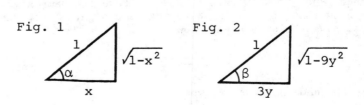

Fig. 1

Fig. 2

635

<u>Solution:</u> Let Arccos x = α and Arccos 3y = β. We can then construct the triangles of Figs. 1 and 2.

$$\cos(\alpha - \beta) = \cos\alpha\cos\beta + \sin\alpha\sin\beta$$
$$= (x)(3y) + (\sqrt{1 - x^2})(\sqrt{1 - 9y^2})$$
$$= 3xy + \sqrt{(1 - x^2)(1 - 9y^2)}.$$

Note: sin α and sin β are both positive since α and β must be between 0 and π; we thus take only the positive radical.

Find sin $\left(\sin^{-1} 1 + \sin^{-1} \frac{1}{2}\right)$.

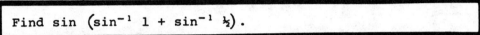

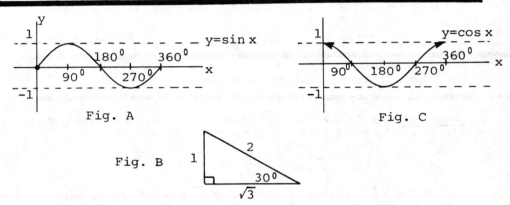

Fig. A   Fig. B   Fig. C

<u>Solution:</u> Recall that inverse sines are angles. Hence sin⁻¹ 1 is the angle whose sine is 1.

From the diagram of the sine function, we see that the angle whose sine is 1 is 90°.

Similarly, sin⁻¹ ½ is the angle whose sine is ½; that is, the angle whose opposite side is 1 and whose hypotenuse is 2.

We note this is a 30-60 right triangle and the angle whose sine is ½ is a 30° angle hence,

$$\sin\left(\sin^{-1} 1 + \sin^{-1} \tfrac{1}{2}\right) = \sin(90° + 30°)$$

Using the sum of the sines formula

sin (x + y) = sin x cos y + cos x sin y:

$$= \sin 90° \cos 30° + \cos 90° \sin 30°$$

Now we must find the values of sin 90°, cos 30°, cos 90°, and sin 30°.

We observed from our sine graph, sin 90° = 1. To

636

calculate cos 90° we can observe the graph of the cos function.

Thus we see cos 90° = 0.

To calculate cos 30° and sin 30° we look at a 30-60 right triangle:

$$\cos \theta = \frac{\text{adjacent side}}{\text{hypotenuse}} \text{ , hence } \cos 30° = \frac{\sqrt{3}}{2}$$

$$\sin \theta = \frac{\text{opposite side}}{\text{hypotenuse}} \text{ , hence } \sin 30° = \tfrac{1}{2}$$

$$= 1 \cdot \frac{\sqrt{3}}{2} + 0 \cdot \frac{1}{2} = \frac{\sqrt{3}}{2} .$$

● **PROBLEM 34-17**

Show that if $x > 0$, then Arctan x = Arccot $\frac{1}{x}$.

Solution: Let u = Arctan x. Then tan $u = x$, and $-\tfrac{1}{2}\pi < u < \tfrac{1}{2}\pi$. Since $x > 0$, $0 < u < \tfrac{1}{2}\pi$. Observe that

$$\frac{1}{x} = \frac{1}{\tan u} = \cot u ,$$

and $0 < u < \tfrac{1}{2}\pi$. Since $\frac{1}{x} = \cot u$ and $0 < u < \tfrac{1}{2}\pi$ we have $u = \text{Arccot}\left(\frac{1}{x}\right)$. But we have already stated that u = Arctan x. Hence Arctan x = Arccot $\frac{1}{x}$.

● **PROBLEM 34-18**

Show that Arcsin x + Arccos x = $\tfrac{1}{2}\pi$ for any number x such that $-1 \leq x \leq 1$.

Solution: Let u = Arcsin x. Then sin $u = x$. We restrict u such that $-\tfrac{1}{2}\pi \leq u \leq \tfrac{1}{2}\pi$ so that the inverse of the sin function is also a function. Now let $v = \tfrac{1}{2}\pi - u$; then $0 \leq v \leq \pi$, and cos $v = \cos(\tfrac{1}{2}\pi - u)$. Recall the formula for the cosine of the difference of two angles, $\cos(a - b) = \cos a \cos b + \sin a \sin b$. Using this we obtain:

$$\cos(\tfrac{1}{2}\pi - u) = \cos \tfrac{1}{2}\pi \cos u + \sin \tfrac{1}{2}\pi \sin u,$$

and since cos $\tfrac{1}{2}\pi = 0$ and sin $\tfrac{1}{2}\pi = 1$, we have:

$$\cos(\tfrac{1}{2}\pi - u) = \sin u = x.$$

Since $0 \leq v \leq \pi$, and cos $v = x$, we have v = Arccos x. Hence,

$$\text{Arcsin } x + \text{Arccos } x = u + v = u + (\tfrac{1}{2}\pi - u) = \tfrac{1}{2}\pi .$$

637

Solve the equation arcsin x + arccos (1 - x) = 0.

Solution: Let α = arcsin x, β = arccos(1 - x). Then we must solve the equation

$$\alpha + \beta = 0,$$

where sin α = x and cos β = 1 - x. Now, by use of the identity $\sin^2 a + \cos^2 a = 1$ we obtain:

$$\cos \alpha = \pm \sqrt{1 - \sin^2 \alpha} = \pm \sqrt{1 - x^2} \; ,$$

and

$$\sin \beta = \pm \sqrt{1 - \cos^2 \beta} = \pm \sqrt{1 - (1-x)^2} = \pm \sqrt{2x - x^2} \; .$$

We now make use of the formula for the sin of the sum of two angles which states,

$$\sin(\alpha + \beta) = \sin \alpha \cos \beta + \cos \alpha \sin \beta \; ;$$

and this equals 0 since $\alpha + \beta = 0$, and sin 0 = 0. Substituting the above values for sin α, cos β, cos α, sin β we have:

$$\sin \alpha \cos \beta + \cos \alpha \sin \beta = 0$$
$$x(1 - x) \pm \sqrt{1 - x^2} \sqrt{2x - x^2} = 0$$
$$x^2 (1 - x)^2 = (1 - x^2)(2x - x^2)$$
$$x^2 (1 - 2x + x^2) = (1 - x^2)(2x - x^2)$$

Observe that x = 0 satisfies this equation. Substituting this value in the given equation we obtain:

$$\text{arcsin } 0 + \text{arccos}(1 - 0) = 0.$$

Now, since arcsin 0 = 0 and arccos 1 = 0 we have:

$$0 + 0 = 0;$$
$$0 = 0$$

thus this value of x also satisfies the original equation. Removing the factor x from $x^2 (1 - 2x + x^2) = (1 - x^2)(2x - x^2)$ yields:

$$x^2 (1 - 2x + x^2) = (1 - x^2)(2 - x)x$$
$$x(1 - 2x + x^2) = (1 - x^2)(2 - x)$$
$$x - 2x^2 + x^3 = 2 - x - 2x^2 + x^3$$
$$2x = 2$$
$$x = 1$$

Thus, x = 1 is another possible solution to the original equation. But arcsin 1 = π/2 and arccos 0 = π/2 , thus

$$\text{arcsin } 1 + \text{arccos}(1 - 1) = \frac{\pi}{2} + \frac{\pi}{2} = \pi \neq 0.$$

Thus, x = 1 does not satisfy the given equation. (Notice that outside the restricted values for arccos we have arccos 0 = - π/2 also, which, together with arcsin 1 = π/2 , makes x = 1 a solution.)
 Hence x = 0 is the only solution of the given equation when only the restricted values are permitted (recall that for the inverse sine function,

$$- \frac{\pi}{2} \leq (\text{arcsin } x = \alpha) \leq \frac{\pi}{2}$$

and for the inverse cos function

$$0 \leq (\arccos(1 - x) = \beta) \leq \pi .$$

Graph   (1) $y = \sin^{-1} x$

          (2) $y = \cos^{-1} x$

          (3) $y = \tan^{-1} x$

          (4) $y = \cot^{-1} x$

          (5) $y = \sec^{-1} x$

          (6) $y = \csc^{-1} x$

Solution:

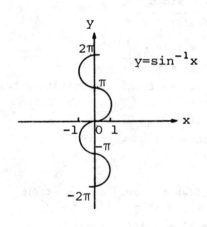

$y = \sin^{-1} x$

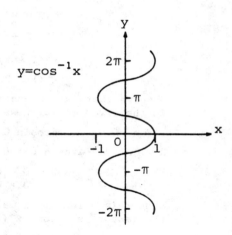

$y = \cos^{-1} x$

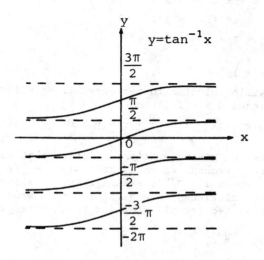

$y = \tan^{-1} x$

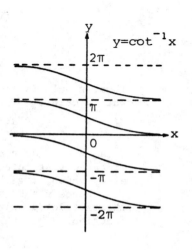

$y = \cot^{-1} x$

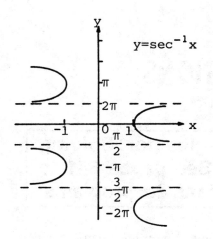

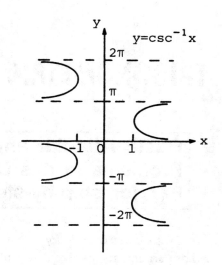

CHAPTER 35

TRIGONOMETRIC EQUATIONS

> **Basic Attacks and Strategies for Solving Problems in this Chapter. See pages 641 to 677 for step-by-step solutions to problems.**

Trigonometric equations, that is, equations involving trigonometric functions whose arguments involve a variable, are solved using the standard techniques for solving any equation.

Sometimes an equation is quite straightforward and can be transformed into an equation involving an inverse trigonometric function.

Sometimes, upon analysis, an equation presents an impossibility, e.g., $\sin x = 2$ cannot be solved since it is known that the absolute value of $\sin x$ never exceeds 1.

When the same trigonometric function appears in several places in an equation, it is sometimes helpful to substitute a different variable for the trigonometric function and solve an algebraic equation first. For example, substituting w for $\sin x$ in

$$6 \sin^2 x + \sin x - 1 = 0$$

gives

$$6w^2 + w - 1 = 0$$

which can be factored into

$$(3w - 1)(2w + 1) = 0.$$

One obtains the results that $w = \frac{1}{3}$ or $w = -\frac{1}{2}$ and then substitutes $\sin x$ for w in these two equations. This process holds even if the actual variable substitution does not take place.

Sometimes, one needs to transform one trigonometric function to another before proceeding. For example, in an equation containing both $\cos^2 x$ and $\sin x$, one could first convert $\cos^2 x$ to $\sin^2 x$ by using the Pythagorean identity

$$\sin^2 x + \cos^2 x = 1.$$

If an equation contains different multiples of the same angles, one should use the double-angle or half-angle formulas to convert all the trigonometric functions to functions using the same angle.

If an equation contains several different trigonometric functions, a good strategy is to convert all the functions to sines and cosines and then try a further conversion to one of the two functions.

When proving *identities,* one is trying to show equality for any value of the variable. In these cases, it is sometimes helpful to make use of other known identities.

Because the trigonometric functions are periodic, it is possible that an equation may have more than one solution, all of which are reasonably small. For example, $\sin 3x = \frac{1}{2}$ implies that $3x = 30°$ and therefore that $x = 10°$. However, it is also true that sine is positive for angles in quadrant II and for angles greater than $360°$. One could also choose $3x$ to be $390°$ or $150°$, in which cases, we obtain the solutions of $x = 130°$ or $x = 50°$.

Step-by-Step Solutions to Problems in this Chapter, "Trigonometric Equations"

● PROBLEM 35-1

Find the solution set for sin x = 0.

Solution: The solution set consists of all distinct values of the angle x which satisfy the equation. Note that x = 0 and x = π are both elements of the solution set. Similarly, any angle nπ, n = 0, ±1, ±2,..., satisfies the equality. Thus, the solution set is {x | x = nπ,n = 0, ±1, ±2,...}. This set is an infinite set.

● PROBLEM 35-2

Find the solution set of cos x = 3.

Solution: Since the values of the cosine function only range from - 1 to +1; that is, since $-1 \leq \cos x \leq 1$, there is no value of x which satisfies the equality cos x = 3. Therefore, the solution set is the empty set.

● PROBLEM 35-3

Find the solution set on [0,2π) for sin x = cos x.

Solution: Since the equation does not lend itself to factoring, we divide both sides by cos x, obtaining sin x/cos x = 1, or tan x = 1. The solution set of this new equation is {π/4, 5π/4}. However, in dividing the original equation by cos x, it was assumed that cos x ≠ 0 because division by 0 is not permitted. However, cos x may be equal to zero. When cos x = 0,

x = π/2 or x = 3π/2.   Checking the four solutions
X = π/4, π/2, 5π/4,   3π/2   in the original equation:

sin π/4 = 0.7071 = cos π/4 √

sin π/2 = 1 ≠ cos π/2 since cos π/2 = 0.

sin 5π/4 =-0.7071 = cos 5π/4 √

sin 3π/2 = -1 ≠ cos 3π/2   since cos 3π/2 = 0.

Therefore, the solution set of the original equation is:

{π/4, 5π/4}.

● **PROBLEM 35-4**

Find the solution set on $[0,2\pi]$ of the equation

$\sqrt{1 + \sin^2 x} = \sqrt{2} \sin x$.

Solution: Since the unknown quantity is involved in
the radicand, squaring of both sides to eliminate the
radical is suggested. Thus, we obtain $1 + \sin^2 x =$

$2 \sin^2 x$. Hence, $\sin^2 x - 1 = 0$,

or $\sin^2 x = 1$

$\sqrt{\sin^2 x} = \pm\sqrt{1}$

$\sin x = \pm 1$

When $\sin x = 1$ on $[0,2\pi]$, $x = \pi/2$. When $\sin x = -1$ on
$[0,2\pi]$, $x = 3\pi/2$.

The complete solution set seems to be $\{\pi/2, 3\pi/2\}$. Since
we squared both sides of the equation, we should try each
element in the original equation. When $x = \pi/2$, we obtain
$\sqrt{1+1} = \sqrt{2}\cdot 1$. When $x = 3\pi/2$, we obtain $\sqrt{1+1} = \sqrt{2}(-1)$. The
second element does not satisfy the original equation,
hence does not belong to the solution set. An extraneous
root was introduced by squaring the equation, it would
seem. Thus, the solution set is $\{\pi/2\}$.

● **PROBLEM 35-5**

Find the solution set on $[0,\pi]$ for the equation
tan x sin x - sin x - tan x + 1 = 0.

Solution: This equation can be factored to obtain
(sin x - 1)(tan x - 1) = 0.

The values of x satisfying this equation may be found by setting each factor equal to zero.

$$\sin x - 1 = 0 ; \quad \tan x - 1 = 0$$

or $\qquad \sin x = 1 ; \qquad \tan x = 1$

Keeping in mind that our solution set cannot contain values exceeding π or less than zero. We find that $x = \pi/2$, is our only acceptable solution for the first equation and $x = \pi/4$ is the only acceptable solution for the second, i.e.,

$$\sin(k \cdot \pi/2) = 1 \quad \text{where} \quad k = 1,3,5,\ldots \quad \text{but } k = 3,5,\ldots \quad \text{is}$$

unacceptable

$$\tan(k\pi/4) = \frac{\sin(k \cdot \pi/4)}{\cos(k \cdot \pi/4)} = \frac{\sqrt{2}/2}{\sqrt{2}/2}, k = 1,3,5,\ldots \quad \text{but}$$

$k = 3,5,\ldots$ is also unacceptable.

By substituting $\pi/2$ into $(\sin x - 1)(\tan x - 1) = 0$ we arrive at the undefined quantity $0 \times (1/0 - 1)$ so we must specify that $\pi/4$ is our only solution as its substitution leads to a valid identity.

● **PROBLEM 35-6**

Find all angles on $[0°,360°)$ which satisfy $\sin 2x - \sqrt{2} \sin x = 0$.

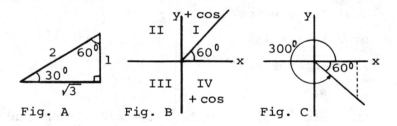

Fig. A   Fig. B   Fig. C

Solution: The equation contains mixed expressions and must be expressed as an equation involving only one multiple of the angle. It seems to be practical to express $\sin 2x$ in terms of the angle x; by using the double angle formula we can express $\sin 2x$ in terms of an angle multiplied by unity.

$$\sin(x + x) = \sin x \cos x + \sin x \cos x = 2 \sin x \cos x$$

$$\sin 2x - \sqrt{2} \sin x = 2 \sin x \cos x - \sqrt{2} \sin x = 0$$

$$\sin x(2 \cos x - \sqrt{2}) = 0$$

Setting each factor = 0,

$$\sin x = 0$$
$$x = 0, \pi$$
$$2 \cos x - \sqrt{2} = 0$$

$$\cos x = \frac{\sqrt{2}}{2}$$

$$x = \frac{\pi}{4}, \frac{7\pi}{4}$$

The solution set is $\left\{0, \frac{\pi}{4}, \pi, \frac{7\pi}{4}\right\}$

643

Solve $2 \sin^2\theta + 3 \cos \theta - 3 = 0$ for θ if $0 \le \theta < 360^\circ$.

Solution: The solution to the equation can be found by expressing the equation in terms of one trigonometric function. Here the convenient function is $\cos \theta$. Using the identity $\sin^2\theta + \cos^2\theta = 1$, we can eliminate $\sin^2\theta$ from the equation by substituting $1 - \cos^2\theta$:

$$2\left(1 - \cos^2\theta\right) + 3 \cos \theta - 3 = 0$$

distributing: $\quad 2 - 2 \cos^2\theta + 3 \cos \theta - 3 = 0$

adding: $\qquad -2\cos^2\theta + 3 \cos \theta - 1 = 0$

multiply by -1: $2 \cos^2\theta - 3 \cos \theta + 1 = 0$

factoring: $\qquad (2 \cos \theta - 1)(\cos \theta - 1) = 0$

Hence, $\qquad\qquad\qquad \cos \theta = \frac{1}{2}$ or $\cos \theta = 1$

Observe Figure A, of a 30-60 right triangle. From it we find $\cos 60^\circ = \dfrac{\text{adjacent side}}{\text{hypotenuse}} = \dfrac{1}{2}$.

The cosine is positive in quadrants I and IV; thus, θ can be 60° or 300°. (See Figures B and C). Then,

$$\theta = 60^\circ, 300^\circ \quad \text{or} \quad \theta = 0^\circ (\text{since } \cos 0^\circ = 1)$$

Substitution verifies all three. The solution set is

$$\{0^\circ, 60^\circ, 300^\circ\}$$

By removing the restriction that $0 \le \theta < 360^\circ$, the solution set is $\{0^\circ + 360^\circ k, \ 60^\circ + 360^\circ k, \ 300^\circ k\}$, k an integer.

Find the solution set of $\sin^2 \theta + \sin \theta = 0$.

Solution: Factoring the left side of the equation, obtain $\sin \theta (\sin \theta + 1) = 0$. Setting each factor equal to zero and solving for $\sin \theta$, obtain $\sin \theta = 0$ and $\sin \theta = -1$. For $\sin \theta = 0$, $\theta = 0$, π, ... and all integral multiples of π.

For $\sin \theta = -1$, note that the sign is negative and that the value of the sine is one. Thus, $\theta = 3\pi/2$ and all integral multiples of 2π plus $3\pi/2$. Therefore, from the first equation, the solution set contains the elements $n\pi$, $n = 0, \pm1, \ldots$. From the second, the solution set contains the elements $3\pi/2 + 2n\pi$, $n = 0, \pm1,\ldots$ The solution set of the original equation is then

$$\left\{\theta \mid \theta = n\pi, \text{ or } 3\pi/2 + 2n\pi, \ n = 0, \pm1, \pm2,\ldots\right\}$$

Find the solution set on $[0,2\pi]$ for the equation
$\sin x \cos x = \cos x$.

Solution: Dividing by cos x we obtain sin x = 1.
However, this operation assumes that cos x ≠ 0, since
division by 0 is not permitted. Hence we have that
cos x may also be equal to zero. Therefore, the solution
set consists of all values of x which satisfy the two
equations sin x = 1 and cos x = 0. When sin x = 1 on
$[0,2\pi]$ x = $\pi/2$. When cos x = 0 on $[0,2\pi]$, x = $\pi/2$ or
x = $3\pi/2$.

Note that one of the values of x obtained from the cosine
function (x = $\pi/2$) is the same as the value of x obtained
from the sine function. Therefore, the complete solution set
is {$\pi/2$, $3\pi/2$}.

Find the solution set on $[0,2\pi)$ for the equation $\cot^2\theta$
$+ (1 - \sqrt{3}) \cot \theta - \sqrt{3} = 0$.

Solution: Factoring, we obtain $(\cot \theta - \sqrt{3})(\cot \theta + 1) = 0$. Hence, $\cot \theta - \sqrt{3} = 0$ or $\cot \theta + 1 = 0$
$\cot \theta = \sqrt{3}$ or $\cot \theta = -1$.

The first equation gives the set {$\pi/6$, $7\pi/6$} while the
second gives {$3\pi/4$, $7\pi/4$}. Note that the solutions to
the first equation $\cot \theta = \sqrt{3}$ are angles which lie in the
first and third quadrants, since the cotangent function
is positive in these two quadrants. Also, the solutions
to the second equation $\cot \theta = -1$ are angles which lie
in the second and fourth quadrants, since the cotangent
function is negative in these quadrants. The complete
solution set on $[0,2\pi)$ is then {$\pi/6$, $7\pi/6$, $3\pi/4$, $7\pi/4$}.

Find the solution set on $[0,2\pi)$ of $2 \tan x + \sqrt{3} \sin x$
$\sec^2 x = 0$.

Solution: Use the following facts concerning trigonometric functions to rewrite the given equation:

$\tan x = \dfrac{\sin x}{\cos x}$, $\sec x = \dfrac{1}{\cos x}$.

Therefore, the given equation becomes:

$$2 \tan x + \sqrt{3} \sin x \sec^2 x = \frac{2 \sin x}{\cos x} + \frac{\sqrt{3} \sin x}{\cos^2 x} = 0 \qquad (1)$$

Multiply both sides of equation (1) by $\cos^2 x$. Hence,

equation (1) becomes: $\cos^2 x \left[\dfrac{2 \sin x}{\cos x} + \dfrac{\sqrt{3} \sin x}{\cos^2 x} \right] = \cos^2 x \,(0)$

$$\text{or } 2 \sin x \cos x + \sqrt{3} \sin x = 0$$

$$\text{or } \sin x \,(2 \cos x + \sqrt{3}) = 0$$

Therefore : $\sin x = 0$ or $2 \cos x + \sqrt{3} = 0$

$$2 \cos x = -\sqrt{3}$$

$$\cos x = -\frac{\sqrt{3}}{2}$$

When $\sin x = 0$ on $[0, 2\pi)$, $x = 0$ or $x = \pi$. When $\cos x = -\frac{\sqrt{3}}{2}$ on $[0, 2\pi)$, $x = 5\pi/6$ or $x = 7\pi/6$. Therefore, the complete solution set is $\{0, 5\pi/6, \pi, 7\pi/6\}$.

● **PROBLEM 35-12**

Solve for θ : $\sin \theta + 2\tan \theta = 0$, $0 \le \theta \le 2\pi$.

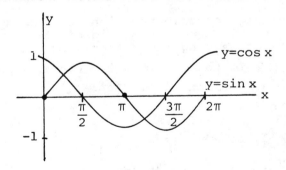

<u>Solution:</u> $\sin \theta + 2\tan \theta = 0$

Since $\tan \theta = \dfrac{\sin \theta}{\cos \theta}$, by substitution: $\sin \theta + 2 \dfrac{\sin \theta}{\cos \theta} = 0$

Multiplying both sides by $\cos \theta$ gives us $\cos \theta \left(\sin \theta + 2 \dfrac{\sin \theta}{\cos \theta} \right) = 0(\cos \theta)$

Distributing we obtain,

$$\sin \theta \cos \theta + 2\sin \theta = 0$$

Factoring out $\sin \theta$ gives us

646

$$\sin \theta (\cos \theta + 2) = 0$$

If $xy = 0$, either $x = 0$ or $y = 0$, hence either $\sin \theta = 0$ or

$\cos \theta + 2 = 0$. Subtracting 2 from each side of the latter gives us

$\cos \theta = -2$ Thus $\sin \theta = 0$ or $\cos \theta = -2$

On the given interval, $0 \leq \theta \leq 2\pi$, $\sin \theta = 0$ when $\theta = 0$ or
$\theta = \pi$; and $\cos \theta = -2$ for no angles of θ ($\cos$ is only defined on the interval $[-1,1]$).

Thus the solution set is $\{0, \pi\}$. Check these values. If we do not
restrict θ as we have done, then the solutions may be expressed as
$\theta = 0 + k\pi$, where k is any integer. That is, by adding any integral
multiple of π to the angle θ, we obtain an angle coterminal with
either zero or π. (By coterminal we mean the initial sides of the
angles lie on the positive branch of the x-axis and the terminal sides
coincide.)

● **PROBLEM** 35-13

Solve the equation

$$\sin^2 \theta + 2 \cos \theta - 1 = 0$$

for non-negative values of θ less than 2π.

Solution: Two trigonometric functions of the unknown θ
itself appear in this equation. Accordingly, we make use
of the identity connecting these functions, namely,
$\sin^2 \theta + \cos^2 \theta = 1$, to transform it into an equation involving only one function of θ. Replace
$\sin^2$ by $1 - \cos^2 \theta$.

$$\sin^2 \theta + 2 \cos \theta - 1 = 0$$

$$\cancel{1} - \cos^2 \theta + 2 \cos \theta - \cancel{1} = 0.$$

Factor out $\cos \theta$

$$\cos \theta (2 - \cos \theta) = 0.$$

Whenever a product of two numbers $ab = 0$, either $a = 0$ or
$b = 0$, hence $\cos \theta = 0$ or $2 - \cos \theta = 0$. Thus $\cos \theta = 0$
or $\cos \theta = 2$.

Now there are two angles in the range $0 \leq \theta < 2\pi$ for
which $\cos \theta = 0$ namely

$$\theta = \frac{\pi}{2}, \qquad \theta = \frac{3\pi}{2}.$$

But, since a cosine of an angle can never exceed unity, the relation cos θ = 2 does not yield a value of θ. Hence we have just two solutions, as given above. It is easy to check these solutions.

Check: $\sin^2 \theta + 2 \cos \theta - 1 = 0$

$(\sin \theta)^2 + 2 \cos \theta - 1 = 0$

For $\theta = \frac{\pi}{2}$

$(\sin \pi/2)^2 + 2 \cos \frac{\pi}{2} - 1 = 0$

$1 + 2 \cdot 0 - 1 = 0$
$0 = 0 \checkmark$

For $\theta = \frac{3}{2}\pi$

$(\sin 3/2\pi)^2 + 2 \cos \frac{3}{2}\pi - 1 = 0$

$(-1)^2 + 2 \cdot 0 - 1 = 0$
$0 = 0 \checkmark$

● **PROBLEM 35-14**

Solve the equation $\cos x + \sqrt{3} \sin x - 2 = 0$ for all x such that $0 \le x < 2\pi$.

Solution: One of the methods for solving this kind of equation is to put the equation into the following form:

$$\sin \theta \cos x + \cos \theta \sin x = \sin(\theta + x)$$

where the values of sin θ and cos θ are known (as the coefficients of cos x and sin x, respectively).

The given equation is rewritten as

$$\cos x + \sqrt{3} \sin x = 2$$

$$\frac{1}{2} \cos x + \frac{\sqrt{3}}{2} \sin x = 1$$

Hence $\sin \theta = \frac{1}{2}$, $\cos \theta = \frac{\sqrt{3}}{2}$

which gives $\theta = 30°$. Therefore,

$$\frac{1}{2} \cos x + \frac{\sqrt{3}}{2} \sin x = \sin 30° \cos x + \cos 30° \sin x$$

$$= \sin(30^\circ + x) = 1$$

$$30^\circ + x = 90^\circ$$

$$x = 60^\circ$$

For $0 \le x < 360^\circ$, $\sin(30^\circ + x) = 1$ has only one solu-tion for x, $x = 60^\circ$.

In general, any equation of the form

$$a \sin x \pm b \cos x = c$$

may be solved by the method used in this problem. The general procedure is the following: let $d = \sqrt{a^2 + b^2}$, then divide both sides of the equation by d as

$$\frac{a}{d} \sin x \pm \frac{b}{d} \cos x = \frac{c}{d} \text{ where } \sin \theta = \frac{b}{d}, \cos \theta = \frac{a}{d} \text{ and}$$

$$\sin^2 \theta + \cos^2 \theta = \left(\frac{b}{d}\right)^2 + \left(\frac{a}{d}\right)^2 = \frac{a^2 + b^2}{d^2}$$

$$= \frac{a^2 + b^2}{(\sqrt{a^2 + b^2})^2} = 1$$

as expected.

Thus, $\frac{a}{d} \sin x \pm \frac{b}{d} \cos x = \cos\theta\sin x \pm \sin\theta\cos x$

$$\theta \pm x = \sin^{-1}\left(\frac{c}{d}\right)$$

$$= \sin(\theta \pm x) = \frac{c}{d}$$

$$x = \pm \left[\sin^{-1}\left(\frac{c}{d}\right) - \theta \right]$$

● **PROBLEM 35-15**

Find the solution set of $2 \cos^2 x - 5 \cos x + 2 = 0$.

Solution: Factoring, we obtain $(\cos x - 2)(2 \cos x - 1) = 0$. Setting each factor equal to zero, we obtain $\cos x = 2$ and $\cos x = 1/2$. There is no value of x satisfying the first factor because the range of values for cos x is from -1 to +1; that is, $-1 \le \cos x \le 1$. Therefore, the solution set of the first factor is the empty set. For the second factor the solution set is $\{x \mid x = \pi/3 + 2n\pi$ or $5\pi/3 + 2n\pi, n = 0, \pm 1, \pm 2, \ldots\}$. Since the first solution set is the empty set, the second set is the complete solution set.

Determine all angles x, $0° \leq x < 360°$, such that $\sin 2x = -\frac{1}{2}$.

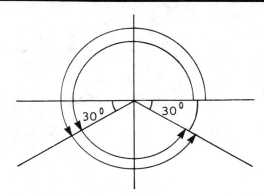

Solution: To determine all values of x such that $0° \leq x < 360°$ and $\sin 2x = -\frac{1}{2}$, we must determine all values of $2x$ such that

$$2 \cdot 0° \leq 2 \cdot x < 2 \cdot 360° \quad \text{and } \sin 2x = -\frac{1}{2} ,$$

or all values of $2x$ must be determined such that

$$0° \leq 2x < 720° \quad \text{and} \quad \sin 2x = -\frac{1}{2} .$$

Since the sine function is negative in only the third and fourth quadrants, the angle $2x$ may lie in only these two quadrants. Also, since the $\sin 30° = -\frac{1}{2}$ in the third and fourth quadrants, any angle with a reference angle of $30°$ will satisfy the equation $\sin 2x = -\frac{1}{2}$. The angles that satisfy this equation and which are in the range of $0° \leq 2x < 720°$ are $180° + 30° = 210°$, $360° - 30° = 330°$, $360° + 210° = 570°$, and $720° - 30° = 690°$ (see diagram). Therefore, $2x = 210°$ or $x = 105°$, $2x = 330°$ or $x = 165°$, $2x = 570°$ or $x = 285°$, and $2x = 690°$ or $x = 345°$. Hence, the solutions of the equation are $x = 105°$, $x = 165°$, $x = 285°$, and $x = 345°$. These solutions are checked by substituting each of them into the equation.

$$\sin 2(105°) = \sin 210° = \sin 30° = -\frac{1}{2} \checkmark$$
$$\sin 2(165°) = \sin 330° = \sin 30° = -\frac{1}{2} \checkmark$$
$$\sin 2(285°) = \sin 570° = \sin 30° = -\frac{1}{2} \checkmark$$
$$\sin 2(345°) = \sin 690° = \sin 30° = -\frac{1}{2} \checkmark .$$

Note that the angles $210°$, $333°$, $570°$, and $690°$ lie in either the third or fourth quadrant, in which the sine function is always negative.

Determine all values of x such that $0° \leq x < 360°$ and $\tan 2x = -1$.

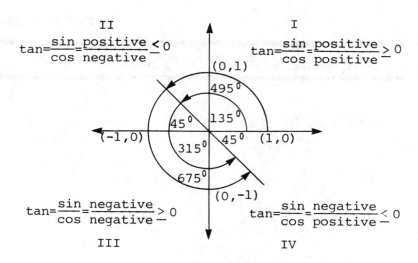

Solution: To determine all angles x such that 0° <
x < 360°, we must find all angles 2x such that 2 · ⁻0°
< 2 · x < 2 · 360° or 0° < 2x · 720°. Now determine in
what quadrants tan is negative, to conform to the e-
quation tan θ = − 1. (See Figure). Thus, the desired
angles lie in the second and fourth quadrants. For
tan θ = sin θ/cos θ = 1 to be true, either sin θ and
cos θ both equal 1, or sin θ = cos θ. Since sine and
cosine are never equal to 1 simultaneously, we seek the
angle for which sin and cos are equal. 45° satisfies
this relation:

$$\frac{\sin 45°}{\cos 45°} = \frac{\sqrt{2}/2}{\sqrt{2}/2} = 1.$$

We must pick angles of 45° in the 2nd and 4th quadrants.
They are 135° in the second quadrant, and 315° in the
fourth quadrant. This is through one revolution or 360°.
Through a second revolution (720°) the chosen angles
are 495° in the second quadrant and 675° in the fourth
(see Figure).

The angles are 135°, 315°, 495°, 675°. According to
the equation each angle is twice the required angle.
Therefore, the values of x which we are interested in are

$$\frac{135°}{2} = 67.5°$$

$$\frac{315°}{2} = 157.5°$$

$$\frac{495°}{2} = 247.5°$$

$$\frac{675°}{2} = 337.5°$$

Solve the equation

$$3 \tan \theta + \sec \theta + 1 = 0$$

for non-negative values of θ less than 2π, that is, $0 \leq \theta < 2\pi$.

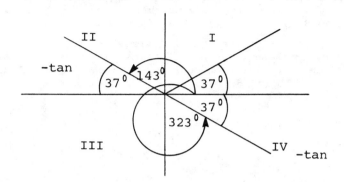

Solution: Here the two functions involved are connected by the identity $\sec^2\theta = 1 + \tan^2\theta$. This can be derived from $\sin^2\theta + \cos^2\theta = 1$, by dividing by $\cos^2\theta$.

$$\frac{\sin^2\theta}{\cos^2\theta} + \frac{\cos^2\theta}{\cos^2\theta} = \frac{1}{\cos^2\theta}$$

$$\tan^2\theta + 1 = \sec^2\theta$$

Since both functions appear to the first degree in the equation, the introduction of an irrationality is unavoidable. Therefore a redundancy may arise, and all solutions obtained must be checked.

We choose to eliminate $\sec \theta$; thus, solving for $\sec \theta$ in the given equation,

$$3 \tan \theta + \sec \theta + 1 = 0$$
$$3 \tan \theta + 1 = -\sec \theta$$

Solving for $\sec \theta$ in the identity $\sec^2\theta = \tan^2\theta + 1$, and substituting:

$$\sec \theta = \pm \sqrt{1 + \tan^2\theta}$$

$$3 \tan \theta + 1 = -\sec \theta = \pm \sqrt{1 + \tan^2\theta}$$

Squaring both sides:
$$(3 \tan \theta + 1)^2 = \left(\pm \sqrt{1 + \tan^2\theta} \right)^2$$

$$9 \tan^2\theta + 6 \tan \theta + 1 = 1 + \tan^2\theta$$

Subtract $\tan^2\theta + 1$ from both sides to obtain:

$$9 \tan^2\theta + 6 \tan \theta + 1 - (\tan^2\theta + 1) = 1 + \tan^2\theta$$
$$- (\tan^2\theta + 1)$$

$$8 \tan^2\theta + 6 \tan \theta = 0$$

Factor $2 \tan \theta$ from the left side:

$$(2 \tan \theta)(4 \tan \theta + 3) = 0. \quad \text{Thus,}$$

$$2 \tan \theta = 0, \quad \text{or} \quad 4 \tan \theta + 3 = 0$$

$$\tan \theta = 0, \quad \text{or} \quad \tan \theta = -3/4$$

Therefore, we must find the angles whose tangent is 0, and the angle whose tangent is (- 3/4) = -.75, where the angle is greater than or equal to 0 and less than 2π (or 360°).

The angle whose tangent is 0 is 0° and 180°. Thus, for $\tan \theta = 0$ we find $\theta = 0^\circ$, and $\theta = 180^\circ$.

Now, we must find θ such that $\tan \theta = -.75$. Referring to a table of trigonometric functions, we find $\tan 37^\circ \approx .75$. Thus, 37° is our reference angle. Since the tangent function is negative in the **second** and fourth quadrants, $\theta = 143^\circ, 323^\circ$ (see figure).

Therefore, the roots of the equation are $\theta = 0^\circ, 180^\circ, 143^\circ, 323^\circ$. But, before accepting these roots as solutions to the original equation, we substitute each root into the equation as a check for validity. Thus, when $\theta = 0^\circ$, $3 \tan \theta + \sec \theta + 1 = 0$ becomes,

$$3 \tan 0^\circ + \sec 0^\circ + 1 = 0, \text{ and}$$

since $\tan 0^\circ = 0$ and $\sec 0^\circ = \dfrac{1}{\cos 0^\circ} = 1$, we have:

$$3(0) + 1 + 1 \overset{?}{=} 0 \; ; \quad 2 \neq 0$$

Thus, $\theta = 0^\circ$ is an extraneous root.

when $\theta = 180^\circ$, we have
$3 \tan 180^\circ + \sec 180^\circ + 1 = 0$, since
$\tan 180^\circ = 0$, and $\sec 180^\circ = \cos 180^\circ = -1$
we have $\qquad 0 + (-1) + 1 = 0$
Therefore, $\theta = 180^\circ$ is a solution of the given equation.
When $\theta = 143^\circ$, $3 \tan \theta + \sec \theta + 1 = 0$ becomes,

$$3 \tan 143^\circ + \sec 143^\circ + 1 = 0; \text{ and}$$

since $\sec 143^\circ = -\sec 37^\circ$ (the sign is negative because sec is negative in the second quadrant), we have:

$$3 \tan 143^\circ - \sec 37^\circ + 1 = 0$$

From a table of trig functions we find that $\sec 37^\circ \approx 1.25$, and we found previously that $\tan 143^\circ \approx -.75$; thus:

$$3(- .75) - 1.25 + 1 \overset{?}{=} 0$$
$$-2.25 - .25 \overset{?}{=} 0$$
$$-2.5 \neq 0$$

Therefore, $\theta = 143^\circ$ is not a solution of the given equation,

When $\theta = 323^\circ$, we have: $3 \tan 323^\circ + \sec 323^\circ + 1 = 0$;
and since sec is positive in Quadrant IV, $\sec 323^\circ = \sec 37^\circ$. We have previously found that $\tan 323^\circ \approx - .75$. Thus,

$$3 \tan (323^\circ) + \sec 37^\circ + 1 \overset{?}{=} 0$$
$$3(- .75) + 1.25 + 1 = 0$$
$$-2.25 + 2.25 = 0$$
$$0 = 0$$

Therefore, $\theta = 323^\circ$ is a solution of the given equation.
Thus, the solutions of the given equation are

$$\theta = 180^\circ, 323^\circ \text{ for } 0 \leq \theta < 2\pi.$$

Solve the equation

$$2 \sin 2\theta + \cos 2\theta + 2 \sin \theta = 1$$

for non-negative values of θ less than 2π.

Solution: Here we have functions of both θ itself and 2θ. Hence we first transform so as to get an equivalent equation involving functions of θ only. We set $\sin 2\theta = 2 \sin \theta \cos \theta$ and $\cos 2\theta = 1 - 2 \sin^2 \theta$, the latter form being chosen so that the constant terms in the equation will cancel.

This leads to

$$2 \sin 2\theta + \cos 2\theta + 2 \sin \theta = 1$$

$$2(2 \sin \theta \cos \theta) + \left(1 - 2 \sin^2 \theta\right) + 2 \sin \theta = 1$$

$$4 \sin \theta \cos \theta + 1 - 2 \sin^2 \theta + 2 \sin \theta = 1$$

$$4 \sin \theta \cos \theta - 2 \sin^2 \theta + 2 \sin \theta = 0 \quad \text{(subtracting 1 from both sides)}$$

$$(2 \sin \theta)(2 \cos \theta - \sin \theta + 1) = 0 \quad \text{(Factoring out } 2 \sin \theta)$$

$$2(\sin \theta)(2 \cos \theta - \sin \theta + 1) = 0.$$

Whenever a product of two numbers $ab = 0$ either $a = 0$ or $b = 0$. Hence $2 \sin \theta = 0$ or $2 \cos \theta - \sin \theta + 1 = 0$. When the factor $\sin \theta$ is set equal to zero, the last equation is satisfied, whence we get

$$\theta = 0, \quad \theta = \pi.$$

It is easy to verify these solutions in the original equation.

Check:

$\theta = 0$

$$2 \sin 2(0) + \cos 2(0) + 2 \sin (0) = 1$$

$$2 \sin (0) + \cos (0) + 2 \sin (0) = 1$$

$$2 \cdot 0 \qquad 1 \quad + \quad 2 \cdot 0 \; = 1$$

$$1 \qquad\qquad = 1$$

$\theta = \pi$

$$2 \sin 2(\pi) + \cos 2(\pi) + 2 \sin \pi = 1$$

$$2 \sin 2\pi \quad + \cos 2\pi \quad + 2 \sin \pi = 1$$

$$2(0) \qquad + \qquad 1 \quad + 2(0) \quad = 1$$

$$1 \qquad\qquad\qquad = 1$$

To find possible solutions of the equation
$2 \cos \theta - \sin \theta + 1 = 0$, we transpose to get
$2 \cos \theta + 1 = \sin \theta$.
From the Pythagorean relation $\cos^2 \theta + \sin^2 \theta = 1$,
we obtain an expression for $\sin \theta$.

$$\sin^2 \theta = 1 - \cos^2 \theta$$

$$\sin \theta = \pm \sqrt{1 - \cos^2 \theta}.$$

Substitute this into the factor $2 \cos \theta - \sin \theta + 1 = 0$.

$$2 \cos \theta - \sin \theta + 1 = 2 \cos \theta \pm \sqrt{1 - \cos^2 \theta} + 1 = 0$$

Transpose $\pm \sqrt{1 - \cos^2 \theta}$,

$$2 \cos \theta + 1 = \pm \sqrt{1 - \cos^2 \theta}.$$

To solve this radical equation, we square both sides:

$$4 \cos^2 \theta + 4 \cos \theta + 1 = 1 - \cos^2 \theta.$$

Subtract 1 and add $\cos^2 \theta$ to both sides to obtain:

$$5 \cos^2 \theta + 4 \cos \theta = 0$$

$$\cos \theta (5 \cos \theta + 4) = 0.$$

Set both factors equal to zero.

$$\cos \theta = 0 \qquad 5 \cos \theta + 4 = 0$$

$$5 \cos \theta = -4$$

$$\cos \theta = 0 \qquad \cos \theta = -\frac{4}{5}.$$

Note that all values of θ obtained must be checked because
the process of rationalizing a radical equation may lead
to extraneous roots.
Corresponding to $\cos \theta = 0$, $\theta = \frac{\pi}{2}$ and $\theta = \frac{3}{2}\pi$. Substitute these values into the given equation,
$2 \cos \theta - \sin \theta + 1 = 0$.
Verify both values of θ:

For $\theta = \frac{\pi}{2}$.

$$2 \cos \theta - \sin \theta + 1 = 0$$

$$2 \cos \frac{\pi}{2} - \sin \frac{\pi}{2} + 1 = 0$$

$$2(0) - 1 + 1 = 0$$

$$0 = 0.$$

655

For $\theta = \frac{3\pi}{2}$

$$2 \cos \frac{3\pi}{2} - \sin \frac{3\pi}{2} + 1 \overset{?}{=} 0$$

$$- (-1) + 1 \neq 0$$

$$2 \neq 0.$$

Thus $\theta = \frac{3\pi}{2}$ is an extraneous root.

If $\cos \theta = -\frac{4}{5}$, θ may be in either the second or third quadrant; from tables we then get $\theta = 143° \ 8'$ and $\theta = 216° \ 52'$ (approx.). Only the latter of these is found to satisfy the given equation by similarly substituting these values of θ into $2 \cos \theta - \sin \theta + 1 = 0$. For $\theta = 143° \ 8'$, the reference angle is $36° \ 52'$ and it is in quadrant II.

$$2 \cos 143° \ 8' - \sin 143° \ 8' + 1 = 0$$

$$-2 \cos 36° \ 52' - \sin 36° \ 52' + 1 = 0.$$

Approximate $36° \ 52'$ by $37°$.

$$-2(.7986) - (.6018) + 1 = 0$$

$$-1.5972 \quad - \quad .6018 \quad + 1 = 0$$
$$-2.199 + 1 \quad \neq 0$$
$$-1.199 \quad \neq 0.$$

For $\theta = 216° \ 52'$, the reference angle is $36° \ 52'$ and θ is in quadrant III. We shall approximate by $37°$.

$$2 \cos 216° \ 52' - \sin 216° \ 52' + 1 = 0$$

$$-2 \cos 37° \quad - (-\sin 37°) \quad + 1 = 0$$

$$-2(.7986) \quad + (.6018) \quad + 1 = 0$$

$$-1.5972 \quad + 1.6018 \quad = 0$$

$$.0046 \quad \approx 0$$

Hence there are two more valid solutions:

$$\theta = \frac{\pi}{2}, \quad \theta = 216° \ 52'.$$

● **PROBLEM 35-20**

Find the solution set of $5\tan^2\alpha - 2\tan \alpha - 1 = 0$.

Solution: Use the quadratic formula $x = \dfrac{-b \pm \sqrt{b^2 - 4ac}}{2a}$,

656

to find $\tan \alpha$. For the given equation, a = 5, b = -2, c = -1, and we solve for x = $\tan \alpha$.

$$\tan \alpha = \frac{-(-2) \pm \sqrt{(-2)^2 - 4(5)(-1)}}{2(5)} = \frac{2 \pm \sqrt{4 + 20}}{10}$$

$$= \frac{2 \pm \sqrt{24}}{10}$$

$$= \frac{2 \pm \sqrt{4}\sqrt{6}}{10}$$

$$\tan \alpha = \frac{2 \pm 2\sqrt{6}}{10} = \frac{2 \pm 2(2.449)}{10}$$

$$\tan \alpha = \frac{2 \pm 4.8989}{10}$$

Therefore, $\tan \alpha = \dfrac{2 + 4.899}{10}$ and $\tan \alpha = \dfrac{2 - 4.899}{10}$

$$= \frac{6.899}{10} \qquad\qquad = \frac{-2.899}{10}$$

$$\tan \alpha = 0.6899 \quad (1) \qquad \tan \alpha = -0.2899 \quad (2)$$

For the $\tan \alpha$ = 0.6899, α lies in the first and third quadrants since $\tan \alpha > 0$ in these quadrants. For $\tan \alpha$ = 0.6899, α = 34° 36' (an angle in the first quadrant) and α = 214° 36' (an angle in the third quadrant). Note that the reference angle of 214° 36' is 34° 36', 214° 36' - 180° = 34° 36'. Hence, the solution set given by equation (1) is
$$\{\alpha \,|\, \alpha = 34°36' + n\pi, \; n = 0, \pm 1, \pm 2, \ldots\}.$$

For the $\tan \alpha$ = -0.2899, α lies in the second and fourth quadrants since $\tan \alpha < 0$ in these quadrants. For $\tan \alpha$ = -0.2899, α = 163°50'. Note that the angle 163°50' is a reference angle for 16°10', 180° - 163°50' = 16°10'. Hence, the solution set given by equation (2) is
$$\{\alpha \,|\, \alpha = 163°50' + n\pi, \; n = 0, \pm 1, \pm 2, \ldots\}.$$

Therefore, the entire solution set is
$$\{\alpha \,|\, \alpha = 34°36' + n\pi \;\text{ or }\; \alpha = 163°50' + n\pi, \; n = 0, \pm 1, \pm 2, \ldots\}.$$

● **PROBLEM 35-21**

Solve: $2 \cos 3x + 1 = 0$.

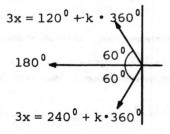

657

<u>Solution:</u> $2 \cos 3x + 1 = 0$

$$2 \cos 3x = -1$$

$$\cos 3x = -\frac{1}{2}.$$

The cosine is negative in quadrants II and III. The reference angle is 60° since $\cos 60° = \frac{1}{2}$. Therefore, the angle $3x = 120° + k \cdot 360°$ or $240° + k \cdot 360°$, k is an integer.
    To convert to radians, set up the proportion:

$$\frac{120°}{180°} = \frac{y}{\pi}$$

$$y = \frac{2}{3}\pi = \pi - \frac{\pi}{3} = 120°$$

$$\frac{240°}{180°} = \frac{z}{\pi}$$

$$z = \frac{4}{3}\pi = \pi + \frac{\pi}{3} = 240°.$$

Substitute these radian measurements for 120° and 240°.
In terms of radians:

$$3x = \pi - \frac{\pi}{3} + k \cdot 2\pi \text{ or } \pi + \frac{\pi}{3} + k \cdot 2\pi.$$

Divide by 3
Solution in radians:

$$x = \left\{ \frac{2\pi}{9}, \frac{4\pi}{9} \right\} + \frac{2k\pi}{3}$$

Convert these measures back to degrees.

$$\frac{\frac{2}{9}\pi}{\pi} = \frac{m}{180°} \qquad \frac{\frac{4}{9}\pi}{\pi} = \frac{n}{180°}$$

$$m = 40° = \frac{2}{9}\pi \text{ and } n = \frac{4}{9}\pi = 80°.$$

From before, $120° = \frac{2}{3}\pi$.

Solution: $x = \{40°, 80°\} + k \cdot 120°.$

● PROBLEM 35-22

Determine the non-negative values of x less than 2π for which

$$2 \cos^2 x + \sin x - 2 > 0.$$

<u>Solution:</u> We first transform the left member by means of the identity $\sin^2 x + \cos^2 x = 1$, thus

$$\cos^2 x = 1 - \sin^2 x$$

$$2\cos^2 x + \sin x - 2 > 0$$

$$2\left(1 - \sin^2 x\right) + \sin x - 2 > 0$$

$$2 - 2\sin^2 x + \sin x - 2 > 0,$$

a relation involving only one trigonometric function. Multiplying by -1 (or transposing), we have

$$2\sin^2 x - \sin x < 0,$$

or $(\sin x)(2\sin x - 1) < 0.$

Now for a product to be negative, then one factor must be negative and the other must be positive. There are two cases to be considered.

Case I		Case II
$\sin x > 0$	or	$\sin x < 0$
$2\sin x - 1 < 0$		$2\sin x - 1 > 0.$

For Case I, when $\sin x > 0$, the angle is in quadrant I or II and $x > 0$. Consider the second restriction, $2\sin x - 1 < 0$. Then $\sin x < \frac{1}{2}$. Thus, $x < \frac{\pi}{6}$. Combine these two restrictions: $0 < x < \frac{\pi}{6}$. Since $\frac{\pi}{6}$ is a reference angle in quadrant II, then $\frac{5}{6} < x < \pi$ since $0 < \sin x < \frac{1}{2}$.

For Case II, when $\sin x < 0$, the angle is in quadrant III or IV. Also, if $2\sin x - 1 > 0$, then $\sin x > \frac{1}{2}$. Thus $x > \frac{7}{6}\pi$ and $x < \frac{11}{6}\pi$, $\frac{7}{6}\pi < x < \frac{11}{6}\pi$.

● **PROBLEM** 35-23

Show that $\sin\left(\frac{1}{2}\pi + t\right) = \cos t$ for every number t.

<u>Solution:</u> Using the addition formula for the sine function, which states that $\sin(\alpha+\beta) = \sin\alpha\cos\beta + \cos\alpha\sin\beta$ and replacing α by $\frac{1}{2}\pi$ and β by t,

$$\sin\left(\frac{1}{2}\pi + t\right) = \sin\frac{1}{2}\pi\cos t + \cos\frac{1}{2}\pi\sin t \qquad (1)$$

Since $\sin\frac{1}{2}\pi = \sin 90° = 1$, and $\cos\frac{1}{2}\pi = \cos 90° = 0$, equation (1) becomes:

$$\sin(\tfrac{1}{2}\pi + t) = (1)\cos t + (0)\sin t$$
$$= \cos t + 0.$$

Thus,
$$\sin(\tfrac{1}{2}\pi + t) = \cos t.$$

● **PROBLEM** 35-24

Show that $\sec^2\theta - \tan^2\theta = 1$ is an identity.

<u>Solution:</u> Given a right triangle (see the accompanying figure) we have:
$$\tan\theta = \frac{\text{opposite side}}{\text{adjacent side}} = \frac{y}{x}$$

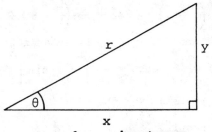

$$\sec\theta = \frac{1}{\cos\theta} = \frac{\text{hypotenuse}}{\text{adjacent side}} = \frac{r}{x}$$

Substitute these expressions into the identity:
$$\sec^2\theta - \tan^2\theta = \left(\frac{r}{x}\right)^2 - \left(\frac{y}{x}\right)^2$$
$$= \frac{r^2}{x^2} - \frac{y^2}{x^2}$$
$$= \frac{r^2 - y^2}{x^2}$$

By the Pythagorean Theorem, $r^2 = x^2 + y^2$; substitute x^2 for $r^2 - y^2$; since $x^2 + y^2 = r^2$,
$$x^2 = r^2 - y^2.$$

Thus, $\sec^2\theta - \tan^2\theta$
$$= \frac{x^2}{x^2}$$
$$= 1$$

● **PROBLEM** 35-25

Prove the following two identities:

    (1) $\cos\left(\dfrac{\pi}{2} - \theta\right) = \sin\theta$ (2) $\cos\theta = \sin\left(\dfrac{\pi}{2} - \theta\right)$.

<u>Solution:</u> To prove identity (1), we use the cosine difference formula,

which states: $\cos(u - v) = \cos u \cos v + \sin u \sin v$. Thus,

$$\cos\left(\frac{\pi}{2} - \theta\right) = \cos\frac{\pi}{2}\cos\theta + \sin\frac{\pi}{2}\sin\theta$$

Now, we must find the values for $\cos\frac{\pi}{2}$ and $\sin\frac{\pi}{2}$. Since $\frac{\pi}{2} = 90°$, and we know that $\cos 90° = 0$ and $\sin 90° = 1$, by substitution we have:

$$\cos\left(\frac{\pi}{2} - \theta\right) = 0\cdot\cos\theta + 1\cdot\sin\theta$$

$$= \sin\theta, \text{ the desired result.}$$

To prove identity (2) note that

$$\cos\left[\frac{\pi}{2} - \left(\frac{\pi}{2} - \theta\right)\right] = \cos\left(\frac{\pi}{2} - \frac{\pi}{2} + \theta\right) = \cos\theta .$$

Since $\cos\theta = \cos\left[\frac{\pi}{2} - \left(\frac{\pi}{2} - \theta\right)\right]$ we can use the difference formula, which states: $\cos(u-v) = \cos u \cos v + \sin u \sin v$. Thus, we have:

$$\cos\theta = \cos\left[\frac{\pi}{2} - \left(\frac{\pi}{2} - \theta\right)\right] = \cos\frac{\pi}{2}\cos\left(\frac{\pi}{2} - \theta\right) + \sin\frac{\pi}{2}\sin\left(\frac{\pi}{2} - \theta\right)$$

Now since $\cos\frac{\pi}{2} = 0$ and $\sin\frac{\pi}{2} = 1$, by substitution we obtain:

$$\cos\theta = 0\cdot\cos\left(\frac{\pi}{2} - \theta\right) + 1\cdot\sin\left(\frac{\pi}{2} - \theta\right)$$

$$= \sin\left(\frac{\pi}{2} - \theta\right)$$

which is the desired result.

● **PROBLEM 35-26**

If u and v are two numbers such that $u + v = \frac{1}{2}\pi$, show that $\sin^2 u + \sin^2 v = 1$

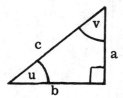

Solution: We know the following trigonometric identity, $\sin^2 u + \cos^2 u = 1$.

Now, if $u + v = \frac{\pi}{2}$, these angles are complementary, that is, the sum of these angles is $90° = \frac{\pi}{2}$. From the accompanying figure, it is seen that

661

$$\sin v = \frac{b}{c} = \cos u$$

Substitute this relation into the identity to obtain:

$$\sin^2 u + \cos^2 u = (\sin u)^2 + (\cos u)^2$$

$$= (\sin u)^2 + (\sin v)^2$$

$$= \sin^2 u + \sin^2 v = 1.$$

● PROBLEM 35-27

Show that $\tan^2 t + 1 = \sec^2 t$.

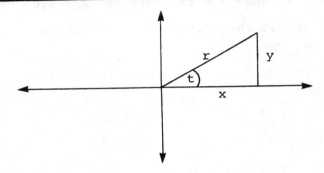

Solution: This equation is meaningless if the X-coordinate of $P(t)$ is 0, since, from the accompanying figure,

$$\tan t = \frac{\text{opposite side}}{\text{adjacent side}} = \frac{y}{x} = \text{undefined when } x = 0$$

and

$$\sec t = \frac{1}{\cos t} = \frac{1}{\dfrac{\text{adjacent side}}{\text{hypotenuse}}} = \frac{1}{\dfrac{x}{r}}$$

$$= \frac{r}{x} = \text{undefined when } x = 0$$

Furthermore, if the $\cos t$ is zero, then the $\sec t$ is undefined. We assume that it is not. In other words, $\cos t \neq 0$ for each number in which we are interested. Thus we may divide both sides of the elementary identity $\sin^2 t + \cos^2 t = 1$ by $\cos^2 t$ to obtain the equation:

$$\frac{\sin^2 t}{\cos^2 t} + 1 = \frac{1}{\cos^2 t}$$

Substitute $\tan t = \frac{\sin t}{\cos t}$ and $\sec t = \frac{1}{\cos t}$, to obtain:

$$\tan^2 t + 1 = \sec^2 t.$$

Prove that $\cos^4 B - \sin^4 B = \cos^2 B - \sin^2 B$ is an identity.

<u>Solution:</u> Note that $\cos^4 B = (\cos^2 B)^2$ and

$$\sin^4 B = (\sin^2 B)^2.$$

Thus, $\cos^4 B - \sin^4 B = (\cos^2 B)^2 - (\sin^2 B)^2$, which is the difference of two squares. Therefore, we may use the formula for the difference of two squares, $x^2 - y^2 = (x - y)(x + y)$, replacing x by $\cos^2 B$ and y by $\sin^2 B$, obtaining:

$$\cos^4 B - \sin^4 B = (\cos^2 B)^2 - (\sin^2 B)^2$$
$$= (\cos^2 B - \sin^2 B)(\cos^2 B + \sin^2 B).$$

Now, recall the trigonometric identity $\cos^2 B + \sin^2 B = 1$. Thus,

$$\cos^4 B - \sin^4 B = (\cos^2 B - \sin^2 B)(1)$$
$$= \cos^2 B - \sin^2 B.$$

Prove the identity $\cos^4 \beta - \sin^4 \beta = 1 - 2 \sin^2 \beta$.

<u>Solution:</u>

$$\cos^4 \beta - \sin^4 \beta = (\cos^2 \beta + \sin^2 \beta)(\cos^2 \beta - \sin^2 \beta)$$

For any angle θ,

$$\cos^2 \theta + \sin^2 \theta = 1 \text{ or } \cos^2 \theta = 1 - \sin^2 \theta.$$

Therefore,

$$\cos^4 \beta - \sin^4 \beta = (\cos^2 \beta + \sin^2 \beta)(\cos^2 \beta - \sin^2 \beta)$$
$$= (1)(\cos^2 \beta - \sin^2 \beta)$$
$$= \cos^2 \beta - \sin^2 \beta$$
$$= (1 - \sin^2 \beta) - \sin^2 \beta$$
$$= 1 - 2 \sin^2 \beta$$

Therefore, $\cos^4 \beta - \sin^4 \beta = 1 - 2 \sin^2 \beta.$

Prove the identity $1 + \sin 2x = (\sin x + \cos x)^2$.

Solution: To prove this identity, start with the right side of the equation.

$$(\sin x + \cos x)^2 = (\sin x + \cos x)(\sin x + \cos x)$$

$$= \sin^2 x + 2 \sin x \cos x + \cos^2 x$$

$$= \sin^2 x + \cos^2 x + 2 \sin x \cos x$$

$$= 1 + 2 \sin x \cos x, \text{ since}$$

$$\left(\sin^2 x + \cos^2 x\right) = 1.$$

But $\sin 2x = 2 \sin x \cos x$, and $(\sin x + \cos x)^2 = 1 + 2 \sin x \cos x$. Therefore $(\sin x + \cos x)^2 = 1 + \sin 2x$.

Prove the identity $\csc 2x = \dfrac{\csc x}{2 \cos x}$.

Solution: Starting with the right side of the identity,

$$\frac{\csc x}{2 \cos x} = \frac{1/\sin x}{2 \cos x}, \text{ since } \csc x = \frac{1}{\sin x}. \quad \text{Hence,}$$

$$\frac{\csc x}{2 \cos x} = \frac{1}{\sin x} \cdot \frac{1}{2 \cos x} = \frac{1}{2 \sin x \cos x}.$$

Using the double-angle formula, $\sin 2\alpha = 2 \sin \alpha \cos \alpha$,

$$\frac{\csc x}{2 \cos x} = \frac{1}{2 \sin x \cos x} = \frac{1}{\sin 2x}.$$

Again, since $\csc x = \dfrac{1}{\sin x}$, $\dfrac{\csc x}{2 \cos x} = \dfrac{1}{\sin 2x} = \csc 2x$. Since we have proved the left side equal to the right,

$$\frac{\csc x}{2 \cos x} = \csc 2x.$$

Prove the identity $\dfrac{\sin^2\theta + \cos^2\theta}{\cos^2\theta} = \sec^2\theta$.

Solution: The fraction $\dfrac{a+c}{b} = \dfrac{a}{b} + \dfrac{c}{b}$, by the definition of addition of fractions. Thus,

$$\frac{\sin^2\theta + \cos^2\theta}{\cos^2\theta} = \frac{\sin^2\theta}{\cos^2\theta} + \frac{\cos^2\theta}{\cos^2\theta}$$

$$= \left(\frac{\sin\theta}{\cos\theta}\right)^2 + 1 \ ,$$

since the fraction $\frac{a^2}{b^2} = \left(\frac{a}{b}\right)^2$ by definition, and $\frac{\cos^2\theta}{\cos^2\theta} = 1$.

$\frac{\sin\theta}{\cos\theta} = \tan\theta$ by definition. Thus we obtain:

$$\tan^2\theta + 1 \ ;$$

and using the trignometric identity $\tan^2\theta + 1 = \sec^2\theta$, we have:

$$\frac{\sin^2\theta + \cos^2\theta}{\cos^2\theta} = \sec^2\theta \ , \quad \text{our desired result.}$$

● **PROBLEM** 35-33

Prove the identity $\dfrac{1 - \sin^2\alpha}{\sin^2\alpha} = \cot^2\alpha.$

Solution: The left member is the more complicated of the two sides of the identity. Operating only on the left member, we obtain

$$\frac{1 - \sin^2\alpha}{\sin^2\alpha} = \frac{1}{\sin^2\alpha} - \frac{\sin^2\alpha}{\sin^2\alpha}$$

$$= \frac{1}{\sin^2\alpha} - 1.$$

Since $\csc\theta = \dfrac{1}{\sin\theta}$ or $\csc^2\theta = \dfrac{1}{\sin^2\theta}$ for any angle θ, then

$$\frac{1 - \sin^2\alpha}{\sin^2\alpha} = \csc^2\alpha - 1.$$

Also, since $\csc^2\theta - \cot^2\theta = 1$ or $\csc^2\theta - 1 = \cot^2\theta$ for any angle θ, then

$$\frac{1 - \sin^2\alpha}{\sin^2\alpha} = \cot^2\alpha.$$

● **PROBLEM** 35-34

Prove that
$$\frac{\cos^3 x - \cos x + \sin x}{\cos x} = \tan x - \sin^2 x$$

is an identity.

665

Solution: If we put each term of the numerator separately over the denominator, we have

$$\frac{\cos^3 x - \cos x + \sin x}{\cos x} = \frac{\cos^3 x}{\cos x} - \frac{\cos x}{\cos x} + \frac{\sin x}{\cos x}$$

But $\dfrac{\cos^3 x}{\cos^1 x} = \cos^{3-1} x = \cos^2 x$

$$\frac{\cos x}{\cos x} = 1$$

and $\dfrac{\sin x}{\cos x} = \tan x.$

Thus, replacing these values we obtain:

$$= \cos^2 x - 1 + \tan x.$$

Recall the identity $\sin^2 \theta + \cos^2 \theta = 1.$

Subtracting $\sin^2 \theta$ from both sides gives us

$$\cos^2 \theta = 1 - \sin^2 \theta,$$

and subtracting 1 from both sides we obtain

$$\cos^2 \theta - 1 = -\sin^2 \theta;$$

thus replacing $\cos^2 x - 1$ by $-\sin^2 x$ we have:

$$\frac{\cos^3 x - \cos x + \sin x}{\cos x} = -\sin^2 x + \tan x$$

$$= \tan x - \sin^2 x.$$

● **PROBLEM** 35-35

Show that $\tan t + \cot t = \csc t \sec t.$

Solution: Since $\tan t = \dfrac{\sin t}{\cos t}$ and $\cot t = \dfrac{\cos t}{\sin t}$, by substitution we have:

$$\tan t + \cot t = \frac{\sin t}{\cos t} + \frac{\cos t}{\sin t}$$

Since multiplying by $1 \equiv \dfrac{\sin t}{\sin t}$ and $1 \equiv \dfrac{\cos t}{\cos t}$

does not alter the value of either fraction, we perform this multiplication, and obtain:

$$\frac{\sin t}{\cos t} \left(\frac{\sin t}{\sin t} \right) + \frac{\cos t}{\sin t} \left(\frac{\cos t}{\cos t} \right)$$

$$= \frac{\sin^2 t}{\sin t \cos t} + \frac{\cos^2 t}{\sin t \cos t}$$

$$= \frac{\sin^2 t + \cos^2 t}{(\sin t)(\cos t)}$$

Now, using the trigonometric identity, $\sin^2 t + \cos^2 t = 1$, we obtain:

$$\frac{1}{(\sin t)(\cos t)} \; ;$$

and, since $\frac{1}{\sin t} = \csc t$ and $\frac{1}{\cos t} = \sec t$, we obtain the desired result:

$$\tan t + \cot t = \csc t \sec t.$$

● **PROBLEM** 35-36

Prove that

$$\frac{1}{\sec A - \tan A} = \sec A + \tan A$$

is an identity.

<u>Solution:</u> If we multiply numerator and denominator of the left member of the given equation by sec A + tan A, we have

$$\frac{1}{\sec A - \tan A} = \left(\frac{1}{\sec A - \tan A}\right)\left(\frac{\sec A + \tan A}{\sec A + \tan A}\right)$$

(Note that (sec A + tan A)/(sec A + tan A) = 1, and therefore does not alter the equation)

$$= \frac{\sec A + \tan A}{(\sec A - \tan A)(\sec A + \tan A)}$$

$$= \frac{\sec A + \tan A}{\sec^2 A - \tan^2 A}$$

Recall the trigonometric identity $1 + \tan^2 \theta = \sec^2 \theta$. Subtracting $\tan^2 \theta$ from both sides we obtain $1 = \sec^2 \theta - \tan^2 \theta$. Thus replacing $\sec^2 A - \tan^2 A$ by 1, we obtain:

$$\frac{1}{\sec A - \tan A} = \frac{\sec A + \tan A}{1} = \sec A + \tan A.$$

Prove the identity $\dfrac{1 - \cos \theta}{\sin \theta} = \dfrac{\sin \theta}{1 + \cos \theta}$.

Solution: One side of this identity is as complicated as the other so that it makes no difference which side is used. The illustration uses both sides.

(a)
$$\frac{1 - \cos \theta}{\sin \theta} = \frac{(1 - \cos \theta)}{\sin \theta} \frac{(1 + \cos \theta)}{(1 + \cos \theta)}$$

$$= \frac{1 - \cos \theta + \cos \theta - \cos^2 \theta}{\sin \theta (1 + \cos \theta)}$$

$$= \frac{1 - \cos^2 \theta}{\sin \theta (1 + \cos \theta)}$$

Since $\cos^2 \theta + \sin^2 \theta = 1$ or $\sin^2 \theta = 1 - \cos^2 \theta$, then

$$\frac{1 - \cos \theta}{\sin \theta} = \frac{1 - \cos^2 \theta}{\sin \theta (1 + \cos \theta)} = \frac{\sin^2 \theta}{\sin \theta (1 + \cos \theta)}$$

$$= \frac{\sin \theta}{1 + \cos \theta}$$

Note that this method starts with the left side of the identity to be proved. The following is another method which can be used to prove the identity.

(b)
$$\frac{\sin \theta}{1 + \cos \theta} = \frac{\sin \theta (1 - \cos \theta)}{(1 + \cos \theta)(1 - \cos \theta)}$$

$$= \frac{\sin \theta (1 - \cos \theta)}{1 + \cos \theta - \cos \theta - \cos^2 \theta}$$

$$= \frac{\sin \theta (1 - \cos \theta)}{1 - \cos^2 \theta}$$

Again, since $\sin^2 \theta = 1 - \cos^2 \theta$,

$$\frac{\sin \theta}{1 + \cos \theta} = \frac{\sin \theta (1 - \cos \theta)}{1 - \cos^2 \theta}$$

$$= \frac{\sin \theta (1 - \cos \theta)}{\sin^2 \theta}$$

$$= \frac{1 - \cos \theta}{\sin \theta}$$

Note that this second method starts with the right side of the identity to be proved.

Prove that

$$\frac{\cos A}{\csc A - 1} + \frac{\cos A}{\csc A + 1} = 2 \tan A$$

is an identity.

Solution: The left member is the more complicated; hence, we shall work with it and begin by performing the indicated addition. The lowest common denominator is $(\csc A - 1)(\csc A + 1) = \csc^2 A - 1$. Thus

$$\frac{\cos A}{\csc A - 1} + \frac{\cos A}{\csc A + 1} = \left(\frac{\csc A + 1}{\csc A + 1}\right) \frac{\cos A}{\csc A - 1}$$

$$+ \left(\frac{\csc A - 1}{\csc A - 1}\right) \frac{\cos A}{\csc A + 1}$$

$$= \frac{(\csc A + 1)\cos A}{\csc^2 A - 1} + \frac{(\csc A - 1)\cos A}{\csc^2 A - 1}$$

$$= \frac{\csc A \cos A + \cos A}{\csc^2 A - 1} + \frac{\csc A \cos A - \cos A}{\csc^2 A - 1}$$

$$= \frac{\csc A \cos A + \cos A + \csc A \cos A - \cos A}{\csc^2 A - 1}$$

$$= \frac{\csc A \cos A + \csc A \cos A}{\csc^2 A - 1}$$

Recall the trignometric identity $\csc^2 A - 1 = \cot^2 A$,

$$= \frac{2 \cos A \csc A}{\cot^2 A}$$

replace $\csc A$ by $\frac{1}{\sin A}$,

$$= \frac{(2 \cos A)/(\sin A)}{\cot^2 A}$$

replace $\frac{\cos A}{\sin A}$ by $\cot A$,

$$= \frac{2 \cot A}{\cot^2 A}$$

cancelling out $\cot A$,

$$= \frac{2}{\cot A}$$

Replace $\cot A$ by $\frac{1}{\tan A}$, $= \frac{2}{\frac{1}{\tan A}}$

Multiply numerator and denominator by $\tan A$,

$$= 2 \tan A$$

We have thus proved

$$\frac{\cos A}{\csc A - 1} + \frac{\cos A}{\csc A + 1} = 2 \tan A \quad \text{is an identity.}$$

Prove that the following equation is an identity:

$$\frac{1 - \sin x}{\cos x} = \frac{\cos x}{1 + \sin x}$$

<u>Solution:</u> This problem may be approached in a variety of ways. One method is based on the fact that two fractions are equal if their cross products are equal. That is,

$$\frac{1 - \sin x}{\cos x} = \frac{\cos x}{1 + \sin x} \text{ if } (1 - \sin x)(1 + \sin x)$$
$$= (\cos x)(\cos x).$$

$$(1 - \sin x)(1 + \sin x) = 1 - \sin^2 x$$

Now, recall the trigonometric identity $\cos^2 \theta = 1 - \sin^2 \theta$; thus $(1 - \sin x)(1 + \sin x) = \cos^2 x = (\cos x) \cdot$ $(\cos x)$, and since the cross products are equal, we have proven the original fractions equivalent.

Prove the identity: sec A csc A = tan A + cot A.

<u>Solution:</u> One approach to the proof of identities, when many functions are involved, is to express the given functions in terms of fewer functions. In this case, suppose we express each of the given trigonometric functions in terms of sine and cosine functions. We will work in parallel columns, with each side of the given equation:

Since sec A = 1/cos A and csc A = 1/sin A,

sec A csc A

$$= \frac{1}{\cos A} \cdot \frac{1}{\sin A}$$

$$= \frac{1}{\cos A \sin A}$$

Since tan A = sin A/cos A and cot A = 1/tan A = 1/sin A/cos A = cos A/sin A, tan A + cot A

$$= \frac{\sin A}{\cos A} + \frac{\cos A}{\sin A} .$$

To combine these fractions, we convert them into fractions with the least common denominator (LCD) cos A sin A; thus:

$$= \left(\frac{\sin A}{\sin A} \frac{\sin A}{\cos A}\right) +$$

$$\left(\frac{\cos A}{\cos A} \frac{\cos A}{\sin A}\right)$$

$$= \frac{\sin^2 A + \cos^2 A}{\cos A \sin A} \, .$$

Recall the trigonometric identity $\sin^2 A + \cos^2 A = 1$; replacing $\sin^2 A + \cos^2 A$ by 1 we obtain:

$$= \frac{1}{\cos A \sin A}$$

Now, since we have proved that both sides of the given equation are equal to the same expression, we are tempted to say that they are therefore equal to each other, and that we have therefore proved what we set out to prove.

We have indeed, except for one detail. We have not considered the values of A for which the given expressions and those which we substituted are meaningful.

This aspect of the proof of a trigonometric identity rarely leads to trouble, and may therefore usually be omitted. The careful student, however, will want to be prepared to investigate this question.

Thus we note that sec A and tan A are defined if and only if A is a real number of degrees not an odd multiple of 90; csc A and cot A are defined if and only if A is a real number of degrees not an even multiple of 90. Both sides of the equation are therefore defined if and only if A is a real number of degrees not an integer multiple of 90.

Each of the substitutions made in the parallel columns above is valid if A is such a number.

Therefore sec A csc A and tan A cot A are equal whenever both are defined, and the equation sec A csc A = tan A + cot A is an identity.

● **PROBLEM 35-41**

Show that tan (- v) = - tan v for every number v in the domain of the tangent function.

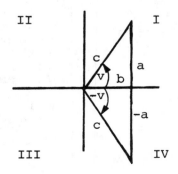

Solution: We have

$$\tan(-v) = \frac{\sin(-v)}{\cos(-v)}$$

Now, let us examine sin (- v) and cos (- v). We know that in drawing the negative angles we rotate clockwise on the coordinate axes.

Thus, the value of the negative angle is the same as that of the reference angle in quadrant I, but it may have a different sign. Therefore, sin v and sin (- v) have the value $\frac{a}{c}$, but for angle - v, the opposite side is - a, thus $\sin(-v) = \frac{-a}{c} = -\sin v$. Cos v and cos (- v) have the value $\frac{b}{c}$, and $\cos v = \frac{b}{c}$, $\cos(-v) = \frac{b}{c}$; thus, cos (- v) = cos v. Therefore, $\frac{\sin(-v)}{\cos(-v)}$ can be written as:

$$\frac{-\sin v}{\cos v} = -\tan v.$$

● **PROBLEM 35-42**

Prove that $\dfrac{\cos 2\theta}{\cos \theta} = \dfrac{1 - \tan^2 \theta}{\sec \theta}$.

Solution: Working on the left side of the equation, we note that cos 2θ can be rewritten as cos (θ + θ), to which we apply the formula for the cosine of the sum of two angles, cos (u + v) = cos u cos v - sin u sin v. Thus, cos 2θ = cos (θ + θ) = cos θ cos θ - sin θ sin θ = cos² θ - sin² θ.

Replacing cos 2θ by cos² θ - sin² θ we obtain,

$$\frac{\cos 2\theta}{\cos \theta} = \frac{\cos^2 \theta - \sin^2 \theta}{\cos \theta}$$

Divide numerator and denominator by cos² θ,

$$= \frac{\dfrac{\cos^2 \theta - \sin^2 \theta}{\cos^2 \theta}}{\dfrac{\cos \theta}{\cos^2 \theta}}$$

$$= \frac{\dfrac{\cos^2 \theta}{\cos^2 \theta} - \dfrac{\sin^2 \theta}{\cos^2 \theta}}{\dfrac{\cos \theta}{\cos^2 \theta}}$$

$$\frac{1 - \dfrac{\sin^2 \theta}{\cos^2 \theta}}{\dfrac{\cos \theta}{\cos^2 \theta}}$$

Recall that $\tan \theta = \sin \theta / \cos \theta$; thus

$$\tan^2 \theta = \left(\frac{\sin \theta}{\cos \theta}\right)^2 = \frac{\sin^2 \theta}{\cos^2 \theta} \ .$$

Substituting $\tan^2 \theta$ for $\sin^2 \theta / \cos^2 \theta$ we obtain,

$$= \frac{1 - \tan^2 \theta}{\dfrac{\cos \theta}{\cos^2 \theta}}$$

$$= \frac{1 - \tan^2 \theta}{\dfrac{1}{\cos \theta}} \ .$$

Since $1/\cos \theta = \sec \theta$, replace $1/\cos \theta$ by $\sec \theta$; thus

$$\frac{1 - \tan^2 \theta}{\sec \theta} \ .$$

Therefore, we have shown that

$$\frac{\cos 2\theta}{\cos \theta} = \frac{1 - \tan^2 \theta}{\sec \theta} \ .$$

● **PROBLEM 35-43**

Prove that $\dfrac{\cos \theta}{1 - \sin \theta} = \dfrac{1 + \sin \theta}{\cos \theta}$.

Solution: $\dfrac{\cos \theta}{1 - \sin \theta} = \dfrac{\cos \theta}{1 - \sin \theta} \cdot \dfrac{1 + \sin \theta}{1 + \sin \theta}$, since multiplication by 1 does not change the value of the fraction. Performing the multiplication we obtain:

$$\frac{\cos \theta (1 + \sin \theta)}{(1 - \sin \theta)(1 + \sin \theta)}$$

$$= \frac{\cos \theta (1 + \sin \theta)}{1 - \sin^2 \theta} \ .$$

Since we are dividing by $1 - \sin^2 \theta$, we must be sure $1 - \sin^2 \theta \neq 0$, that is $1 \neq \sin^2 \theta$. Thus $\sin \theta \neq 1, -1$.

We use the identity $1 - \sin^2 \theta = \cos^2 \theta$ to rewrite the fraction as follows: $\dfrac{\cos \theta (1 + \sin \theta)}{\cos^2 \theta} = \dfrac{\cos \theta (1 + \sin \theta)}{(\cos \theta)(\cos \theta)}$, since

$$\cos^2 \theta = (\cos \theta)(\cos \theta) \ .$$

Finally, $\dfrac{\cos \theta}{1 - \sin \theta} = \dfrac{1 + \sin \theta}{\cos \theta}$. Since we are dividing by $\cos \theta$ we must exclude all $\cos \theta = 0$. Hence $\cos \theta \neq 0$. Thus, we have proven that $\dfrac{\cos \theta}{1 - \sin \theta} = \dfrac{1 + \sin \theta}{\cos \theta}$, with the restrictions

$$\cos \theta \neq 0 , \sin \theta \neq 1, -1 \ .$$

673

Prove the identity $\sin^2\theta + \tan^2\theta = \sec^2\theta - \cos^2\theta$.

<u>Solution:</u> Since $\tan\theta = \dfrac{\sin\theta}{\cos\theta}$, and $\sec\theta = \dfrac{1}{\cos\theta}$, we can express this equation in terms of sine and cosine:

$$\sin^2\theta + \left(\frac{\sin\theta}{\cos\theta}\right)^2 = \left(\frac{1}{\cos\theta}\right)^2 - \cos^2\theta \ .$$

Since $\left(\dfrac{\sin\theta}{\cos\theta}\right)^2 = \dfrac{\sin^2\theta}{\cos^2\theta}$, and $\left(\dfrac{1}{\cos\theta}\right)^2 = \dfrac{1}{\cos^2\theta}$ we obtain

$$\sin^2\theta + \frac{\sin^2\theta}{\cos^2\theta} = \frac{1}{\cos^2\theta} - \cos^2\theta$$

We also know that $\cos^2\theta = 1 - \sin^2\theta$ **(a trigonometric identity).**
Substituting $1 - \sin^2\theta$ for $\cos^2\theta$:

$$\sin^2\theta + \frac{\sin^2\theta}{1-\sin^2\theta} = \frac{1}{1-\sin^2\theta} - \left(1 - \sin^2\theta\right)$$

In order to combine terms we multiply $\sin^2\theta$ and $\left(1-\sin^2\theta\right)$ by $\dfrac{1-\sin^2\theta}{1-\sin^2\theta}$ (which equals 1, and therefore does not change the value of the terms). Thus:

$$\frac{\sin^2\theta\left(1-\sin^2\theta\right)}{1-\sin^2\theta} + \frac{\sin^2\theta}{1-\sin^2\theta} = \frac{1}{1-\sin^2\theta} - \frac{\left(1-\sin^2\theta\right)\left(1-\sin^2\theta\right)}{1-\sin^2\theta}$$

Adding the fractions we obtain:

$$\frac{\sin^2\theta\left(1-\sin^2\theta\right) + \sin^2\theta}{1-\sin^2\theta} = \frac{1-\left(1-\sin^2\theta\right)\left(1-\sin^2\theta\right)}{1-\sin^2\theta}$$

Multiplying the terms in the numerator we obtain:

$$\frac{\sin^2\theta - \sin^4\theta + \sin^2\theta}{1-\sin^2\theta} = \frac{1-\left(1- 2\sin^2\theta + \sin^4\theta\right)}{1-\sin^2\theta}$$

$$\frac{2\sin^2\theta - \sin^4\theta}{1-\sin^2\theta} = \frac{1-\left(1-2\sin^2\theta + \sin^4\theta\right)}{1-\sin^2\theta}$$

$$\frac{2\sin^2\theta - \sin^4\theta}{1-\sin^2\theta} = \frac{1-1+2\sin^2\theta - \sin^4\theta}{1-\sin^2\theta}$$

$$\frac{2\sin^2\theta - \sin^4\theta}{1-\sin^2\theta} = \frac{2\sin^2\theta - \sin^4\theta}{1-\sin^2\theta}$$

Since $\sin^2\theta + \tan^2\theta = \dfrac{2\sin^2\theta - \sin^4\theta}{1-\sin^2\theta}$ and

$$\sec^2\theta - \cos^2\theta = \dfrac{2\sin^2\theta - \sin^4\theta}{1-\sin^2\theta}$$

$$\sin^2\theta + \tan^2\theta = \sec^2\theta - \cos^2\theta$$

because two expressions equal to the same expression are equal to each other.

● **PROBLEM 35-45**

Prove the identity $\dfrac{\sec x + 1}{\sec x - 1} = \cot^2 \dfrac{x}{2}$.

<u>Solution:</u> Starting with the left side of the identity,

$$\frac{\sec x + 1}{\sec x - 1} = \frac{\dfrac{1}{\cos x} + 1}{\dfrac{1}{\cos x} - 1} \qquad \text{since } \sec x = \frac{1}{\cos x}$$

Hence,

$$\frac{\sec x + 1}{\sec x - 1} = \frac{\dfrac{1}{\cos x} + \dfrac{\cos x}{\cos x}}{\dfrac{1}{\cos x} - \dfrac{\cos x}{\cos x}} \qquad \text{for } 1 = \frac{\cos x}{\cos x}$$

Combining fractions,

$$= \frac{\dfrac{1 + \cos x}{\cos x}}{\dfrac{1 - \cos x}{\cos x}}$$

Dividing by a fraction is equivalent to multiplying by its reciprocal, hence,

$$= \frac{1 + \cos x}{\cos x} \cdot \frac{\cos x}{1 - \cos x}$$

Cancelling $\cos x$ from numerator and denominator we obtain:

$$\frac{\sec x + 1}{\sec x - 1} = \frac{1 + \cos x}{1 - \cos x} = \left(\frac{+}{-}\sqrt{\frac{1 + \cos x}{1 - \cos x}}\right)^2$$

Looking at the right side of the identity:

from the formula for the tangent of a half angle,

$$\tan \tfrac{1}{2}\theta = \tan \frac{\theta}{2} = \frac{+}{-}\sqrt{\frac{1 - \cos \theta}{1 + \cos \theta}} , \quad \text{and}$$

$$\cot \tfrac{1}{2}x = \cot \frac{x}{2} = \frac{1}{\tan \dfrac{x}{2}} = \frac{1}{\dfrac{+}{-}\sqrt{\dfrac{1 - \cos x}{1 + \cos x}}}$$

$$= \frac{1}{\pm\sqrt{\dfrac{1 - \cos x}{1 + \cos x}}}$$

$$= (1)\left(\pm \frac{\sqrt{1 + \cos x}}{\sqrt{1 - \cos x}}\right) = \pm \frac{\sqrt{1 + \cos x}}{\sqrt{1 - \cos x}}$$

Hence, $\cot \dfrac{x}{2} = \pm \sqrt{\dfrac{1 + \cos x}{1 - \cos x}}$

Therefore, $\dfrac{\sec x + 1}{\sec x - 1} = \left(\cot \dfrac{x}{2}\right)^2 = \cot^2 \dfrac{x}{2}$.

● **PROBLEM 35-46**

Prove the identity

$$\tan\left(\frac{\pi}{4} + \frac{\theta}{2}\right) = \sec \theta + \tan \theta.$$

Solution: Factor out $\frac{1}{2}$.

$$\tan\left(\frac{\pi}{4} + \frac{\theta}{2}\right) = \tan \frac{1}{2}\left(\frac{\pi}{2} + \theta\right).$$

Then, apply the half-angle formula for the tangent.

$$\tan \frac{1}{2}\theta_1 = \pm\sqrt{\frac{1 - \cos \theta_1}{1 + \cos \theta_1}}.$$

Rationalize the denominator to obtain

$$\tan \frac{1}{2}\theta_1 = \pm\sqrt{\frac{1 - \cos \theta_1}{1 + \cos \theta_1}}\frac{\sqrt{1 - \cos \theta_1}}{\sqrt{1 - \cos_1}} = \frac{1 - \cos \theta_1}{\pm\sqrt{1 - \cos^2 \theta_1}}$$

$$= \frac{1 - \cos \theta_1}{\sqrt{\sin^2 \theta_1}}$$

$$\tan \frac{1}{2}\theta_1 = \frac{1 - \cos \theta_1}{\sin \theta_1}.$$

Replace θ_1 by $\frac{\pi}{2} + \theta$.

$$\tan \frac{1}{2}\left(\frac{\pi}{2} + \theta\right) = \frac{1 - \cos\left(\frac{\pi}{2} + \theta\right)}{\sin\left(\frac{\pi}{2} + \theta\right)}$$

Apply the formula for the sum of the sine of two angles

676

and the cosine of the sum of two angles.

$$\cos(\alpha + \beta) = \cos \alpha \cos \beta - \sin \alpha \sin \beta$$

$$\sin(\alpha + \beta) = \sin \alpha \cos \beta + \cos \alpha \sin \beta$$

$$\cos\left(\frac{\pi}{2} + \theta\right) = \cos \frac{\pi}{2} \cos \theta - \sin \frac{\pi}{2} \sin \theta$$

$$= 0(\cos \theta) - \sin \theta = -\sin \theta.$$

$$\sin\left(\frac{\pi}{2} + \theta\right) = \sin \frac{\pi}{2} \cos \theta + \cos \frac{\pi}{2} \sin \theta$$

$$= 1 \cos \theta + 0(\sin \theta)$$

$$= \cos \theta$$

Substitute these two results.

$$\tan \frac{1}{2}\theta_1 = \frac{1 - \cos \theta_1}{\sin_1} = \tan \frac{1}{2}\left(\frac{\pi}{2} + \theta\right) = \frac{1 - \cos\left(\frac{\pi}{2} + \theta\right)}{\sin\left(\frac{\pi}{2} + \theta\right)}$$

$$= \frac{1 - (-\sin \theta)}{\cos \theta} = \frac{1 + \sin \theta}{\cos \theta}$$

$$= \frac{1}{\cos \theta} + \frac{\sin \theta}{\cos \theta}$$

$$= \sec \theta + \tan \theta.$$

CHAPTER 36

COMPLEX NUMBERS

Basic Attacks and Strategies for Solving Problems in this Chapter. See pages 678 to 702 for step-by-step solutions to problems.

As mentioned in Chapter 1, *complex numbers* are numbers involving the unit imaginary number

$$i = \sqrt{-1}.$$

They are commonly written as sums (or differences) of real and imaginary numbers, such as $3 + 4i$, and this form is called the *Cartesian* or *rectangular* form. When adding (or subtracting) two complex numbers, the real parts are added (subtracted) independently of the imaginary parts.

Complex numbers are also associated with points on the Cartesian plane, where the real axis is identified with the x-axis and the imaginary axis with the y-axis. It is also possible to identify such a point with a vector from the origin. Sums and differences of complex numbers can then be computed by performing sums and differences of vectors.

A complex number is associated with another complex number, called its *conjugate*, which is its reflection across the x-axis. To obtain the conjugate of $a + bi$, the imaginary part is negated, producing $a - bi$. Note that a complex number multiplied by its conjugate results in a purely real number, since

$$(a + bi)(a - bi) = a^2 - b^2i^2 = a^2 - b^2(-1) = a^2 + b^2.$$

By associating a complex number with a point, one can use a distance and an angle as an alternate notation to locate the same point. This notation is sometimes referred to as the *trigonometric* notation. One determines the straight line distance (also called the *modulus* or *radius*) from the origin to the point, and also the angle that the ray from the origin to the point makes with the positive x-axis. Given that $a + bi$, to convert to trigonometric notation, one computes the distance

$$r = \sqrt{a^2 + b^2}$$

and the angle $\theta = \arctan {}^b/_a$. Given $[r, \theta]$, to convert to Cartesian notation, one computes $a = r \cos \theta$ and $b = r \sin \theta$. Thus,

$$a + bi = r \cos \theta + ir \sin \theta.$$

This can also be written in terms of a complex power of $e = 2.718281828....$ Using this notation,

$$r \cos \theta + ir \sin \theta = re^{i\theta}.$$

When multiplying two complex numbers in rectangular form, one follows the standard rules for multiplying two linear polynomials, remembering that $i^2 = -1$.

When multiplying two complex numbers in trigonometric form, one uses the following rule:

$$r_1(\cos \theta_1 + i \sin \theta_1) \cdot r_2(\cos \theta_2 + i \sin \theta_2)$$

$$= r_1 r_2(\cos [\theta_1 + \theta_2] + i \sin[\theta_1 + \theta_2])$$

(the moduli are multiplied and the angles are added).

When dividing two complex numbers in rectangular form, one usually multiplies the numerator and denominator by the conjugate of the denominator. This results in a new denominator which is purely real. The new numerator can then be multiplied out (as if it were two linear polynomials), simplified, and then divided term by term by the real denominator to produce the quotient.

When dividing two complex numbers in trigonometric form, one uses the following rule:

$$\frac{r_1(\cos\theta_1 + i\sin\theta_1)}{r_2(\cos\theta_2 + i\sin\theta_2)} = \frac{r_1}{r_2}(\cos[\theta_1 - \theta_2] + i\sin[\theta_1 - \theta_2]).$$

When raising a complex number in trigonometric form to a power, one uses the following rule (called *de Moivre's Theorem*):

$$[r(\cos \theta + i \sin \theta)]^n = r^n(\cos[n\theta] + i \sin [n\theta]).$$

If the power or root of a complex number in rectangular form is desired, it is often easier to translate the number to trigonometric form, obtain the answer and then translate it back to rectangular form. In some cases, the number raised to a power can be set equal to a complex variable, e.g., $x + iy$, and then techniques used in similar cases for solving purely real equations can be used (e.g., squaring both sides, setting real parts equal to each other and imaginary parts equal to each other, etc.).

These rules can be used to find all the roots of constants. One first converts a constant to the trigonometric form, e.g.,

$$1 = 1 \cdot (\cos 0° + i \sin 0°).$$

However, since $0°$ is coterminal with $360°$ and $720°$, 1 can also be written in terms of those two angles as well. To obtain, for example, all three cube roots of

1, de Moivre's theorem can be applied and we obtain

$$1^{\frac{1}{3}} = 1^{\frac{1}{3}}\left(\cos\frac{0}{3} + i\sin\frac{0}{3}\right) = 1.$$

However, we can also use the other expressions for 1 in terms of 360° and 720°. These produce the values of $\cos 120° + i \sin 120°$ and $\cos 240° + i \sin 240°$ as other possible cube roots of 1. Trying any other representations of 1 will only produce values of cube roots which are coterminal with one of those already obtained. There are three cube roots of one, and, in general, there are n n^{th} roots of k.

Step-by-Step Solutions to Problems in this Chapter, "Complex Numbers"

Find

(1) $(5+6i) - (9-4i)$

(2) $(4+8i) - (4+4i)$

graphically.

<u>Solution</u>:

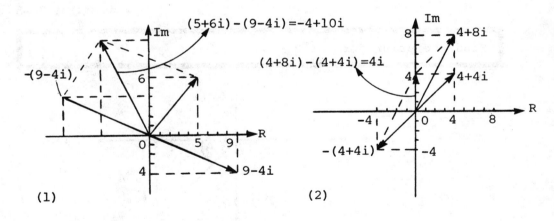

(1)   (2)

Express $4 + 4i$ in as many different forms as possible.

<u>Solution</u>: (1) $4 + 4i = \sqrt{4^2 + 4^2} \, (\cos\theta + i \sin\theta)$

where $\theta = \tan^{-1}\left(\dfrac{4}{4}\right) = 45^0$

The given complex number in its polar form is

$$4\sqrt{2} \, \cos(n \cdot 360^0 + 45^0) + i \, \sin(n \cdot 360^0 + 45^0)$$

(2)

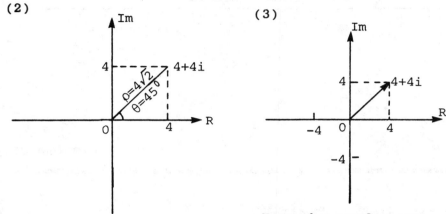

(3)

Here the complex number
is represented as a vector.

Here the complex number in its trigonometric form is
represented graphically.

● **PROBLEM 36-3**

Find the polar form of 3 - 4i.

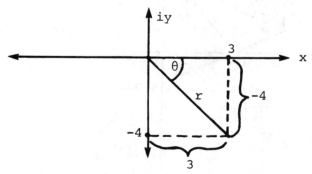

<u>Solution:</u> The given complex number is the cartesian co-
ordinate representation. We wish to transform this representation to the polar form, i.e. r(cos θ + i sin θ).
Consult the accompanying figure and notice that r can be
determined using the Pythagorean theorem, $r^2 = x^2 + y^2$,
and θ can be computed by the formula tan θ = y/x or
arctan y/x = θ, since we know that x = 3 and y = - 4.

Thus,
$$r = \sqrt{x^2 + y^2} = \sqrt{3^2 + (-4)^2} = \sqrt{9 + 16} = 5.$$
$\theta = \arctan - 4/3 = \arctan (-1.3333) = 306°52.25´$,
since $(3, -4)$ is a fourth quadrant point. Thus $3 - 4i = 5(\cos 306°52.25´ + i \sin 306°52.25´)$.

● **PROBLEM 36-4**

Express each of the following in trigonometric form.

(a) $- \sqrt{2} + \sqrt{2}i$ (b) $3 - 4i$ (c) $2 + i$

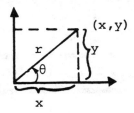

Solution: In the plane, a complex number is represented as $x + iy$.

Therefore the angle θ can be defined as $\arctan \frac{y}{x} = \theta$ or $\tan \theta = \frac{y}{x}$.

In part a the x coordinate is negative and the y coordinate is positive, therefore θ must lie in the second quadrant.

In part b the x is positive and the y is negative so θ lies in the fourth quadrant.

Finally in part c both x and y are positive implying that θ lies in the first quadrant.

The modulus or radius can be computed from the Pythagorean theorem $r^2 = x^2 + y^2$.

(a)   Tan $\theta = \frac{\sqrt{2}}{-\sqrt{2}} = -1$, and θ is in the second quadrant.

Since $\arctan 1 = 45°$, $\theta = 180° - 45° = 135°$.

$$r^2 = (-\sqrt{2})^2 + (\sqrt{2})^2 = 4 \quad \text{or} \quad r = \sqrt{4} = 2$$

Therefore, $- \sqrt{2} + \sqrt{2}i = 2 (\cos 135° + i \sin 135°)$.

(b)   Tan $\theta = \frac{-4}{3}$, and θ is in the fourth quadrant.

Since $\arctan 1.333 = 53° 10´$ (to the nearest 10´),

$$\theta = 360° - 53°10´ = 306°50´$$

680

$$r^2 = 3^2 + (- 4)^2 = 25 \quad \text{or} \quad r = 5.$$

Therefore, $3 - 4i = 5(\cos 306°50' + i \sin 306°50')$.

(c)   Tan $\theta = \frac{1}{2}$, and θ is in the first quadrant. Since arctan $\frac{1}{2} = 26°30'$ (to the nearest 10'),

$$\theta = 26°30'$$
$$r^2 = 2^2 + 1^2 = 5 \quad \text{or} \quad r = \sqrt{5}$$

Therefore, $2 + 1 = \sqrt{5}(\cos 26°30' + i \sin 26°30')$.

● PROBLEM 36-5

Find the value of $(4 - 4i) \cdot (\sqrt{3} - i)$ in polar form.

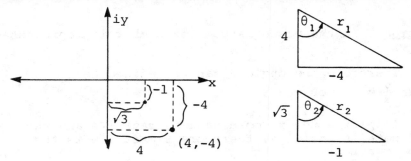

Solution: First change each cartesian representation to its polar representation. That is, we want to transform $x + iy$ to the form $r(\cos \theta + i \sin \theta)$. In the figure, notice that r can be determined using the Pythagorean theorem. And θ can be computed using the trigonometric functions.

$$r_1^2 = 4^2 + (- 4)^2 = 32$$

$$r_1 = \sqrt{32} = 4\sqrt{2}$$

$$r_2^2 = (\sqrt{3})^2 + (- 1)^2 = 3 + 1 = 4$$

$$r_2 = 2$$

Since all three sides of both triangles are known, any of the trigonometric functions can be used to determine θ. i.e. $\sin \theta = x/r$; $\cos \theta = y/r$; $\tan \theta = x/y$

Once θ_1 and θ_2 have been determined the multiplication can be performed according to the formula

$$r_1(\cos \theta_1 + i \sin \theta_1) \cdot r_2(\cos \theta_2 + i \sin \theta_2)$$

$$= r_1 r_2 \Big(\cos [\theta_1 + \theta_2] + i \sin [\theta_1 + \theta_2]\Big).$$

681

$$\tan \theta_1 = \frac{4}{-4} \rightarrow \theta_1 = \tan^{-1}(-1) = 315°, \text{ fourth quadrant.}$$

$$\tan \theta_2 = \frac{\sqrt{3}}{-1} \rightarrow \theta_2 = \tan^{-1}(-\sqrt{3}) = 330°, \text{ fourth quadrant.}$$

Changing $4 - 4i$ and $\sqrt{3} - i$ to polar form, we obtain $4 - 4i = 4\sqrt{2} (\cos 315° + i \sin 315°)$ and $\sqrt{3} - i = 2(\cos 330° + i \sin 330°)$. Thus, $(4 - 4i)(\sqrt{3} - i) = 4\sqrt{2} (\cos 315° + i \sin 315°) \cdot 2(\cos 330° + i \sin 330°) = 8\sqrt{2} (\cos 645° + i \sin 645°) = 8\sqrt{2}(\cos 285° + i \sin 285°)$.

● **PROBLEM 36-6**

Express each of the following in rectangular form, $a + bi$.   (a) $3(\cos 30° + i \sin 30°)$

(b) $10(\cos 180° + i \sin 180°)$

Solution: The complex numbers as given are in the trigonometric form

$$r(\cos \theta + i \sin \theta)$$

in part a, $\theta = 30°$,

(a) $3(\cos 30° + i \sin 30°) = \frac{3}{2}\sqrt{3} + \frac{3}{2} i$

in part b, $\theta = 180°$,

(b) $10(\cos 180° + i \sin 180°) = 10(-1 + i \cdot 0) = -10$

Check: $r^2 = x^2 + y^2$

part a: $(3)^2 = \left(\frac{3\sqrt{3}}{2}\right)^2 + \left(\frac{3}{2}\right)^2 = \frac{27}{4} + \frac{9}{4} = 9$

part b: $(10)^2 = (-10)^2$

● **PROBLEM 36-7**

Find $[2(\cos 30° + i \sin 30°)][8(\cos 60° + i \sin 60°)]$. Check by converting to rectangular form and multiplying.

Solution: The two complex numbers are written in the form $r(\cos \theta + i \sin \theta)$ and the rule for multiplying complex numbers in this form is:

$$r_1(\cos \theta_1 + i \sin \theta_1)r_2(\cos \theta_2 + i \sin \theta_2)$$

$$= r_1 r_2 (\cos \theta_1 \cos \theta_2 - \sin \theta_1 \sin \theta_2 + i \sin \theta_1 \cos \theta_2$$
$$+ i \cos \theta_1 \sin \theta_2)$$
$$= r_1 r_2 [\cos(\theta_1 + \theta_2) + i \sin(\theta_1 + \theta_2)]$$

$$[2(\cos 30° + i \sin 30°)][8(\cos 60° + i \sin 60°)]$$
$$= 16[\cos (30° + 60°) + i \sin (30° + 60°)]$$
$$= 16 (\cos 90° + i \sin 90°) = 16(0 + i) = 0 + 16i$$

Check $2(\cos 30° + i \sin 30°) = 2 (\tfrac{1}{2} \sqrt{3} + \tfrac{1}{2}i) = \sqrt{3} + i$

$8(\cos 60° + i \sin 60°) = 8(\tfrac{1}{2} + \tfrac{1}{2} \sqrt{3}i) = 4 + 4\sqrt{3}i$

$(\sqrt{3} + i)(4 + 4\sqrt{3}i) = 4\sqrt{3} - 4\sqrt{3} + i(4 + 12) = 0 + 16i.$

● **PROBLEM 36-8**

Compute $\left(\cos \dfrac{3\pi}{2} + i \sin \dfrac{3\pi}{2}\right)^6 .$

Solution: To raise the trigonometric representation of a complex number to a power, apply the rule:

$$w = r(\cos \theta + i \sin \theta)$$
$$w^n = \Big[r(\cos \theta + i \sin \theta)\Big]^n = r^n\big(\cos [n\theta] + i \sin [n\theta]\big);$$

here n = 6, r = 1. Thus,

$$\left(\cos \frac{3\pi}{2} + i \sin \frac{3\pi}{2}\right)^6 = \cos \frac{18\pi}{2} + i \sin \frac{18\pi}{2}$$
$$= \cos 9\pi + i \sin 9\pi. \text{ Recall the}$$

formula for determining coterminal numbers, $u = u \pm 2n\pi$; $u = 9\pi$, n = 4. Then, $9\pi = 9\pi - 2(4)\pi = 9\pi - 8\pi = \pi$. This means that on the unit circle both π and 9π begin at (1,0) and terminate at π. Thus, we have: cos π + i sin π, and since cos π = - 1 and sin π = o, = - 1 + i(0) = - 1. Note that cos $3\pi/2$ + i sin $3\pi/2$ is a number which can be raised to an even power to produce a negative product.

● **PROBLEM 36-9**

Calculate   (1) $(1 + i\sqrt{3})^{10}$

(2) $(1 - i)^{\frac{1}{9}}$

Solution: (1) Use De Moivre's theorem, which states that if n is any rational number, then

$$(r(\cos\theta + i \sin \theta))^n$$

$$= r^n(\cos n\theta + i \sin n\theta)$$

$$1 + i\sqrt{3} = \sqrt{1^2 + (\sqrt{3})^2} \ (\cos\theta + i \sin\theta)$$

where

$$\theta = \tan^{-1}\left(\frac{\sqrt{3}}{1}\right) = 60^0$$

Thus,

$$(1 + i\sqrt{3})^{10} = 2^{10}(\cos 10\cdot60^0 + i \sin 10\cdot60^0)$$

$$= 2^{10}(\cos 600^0 + i \sin 600^0)$$

$$= 1024 \left(\left(-\frac{1}{2}\right) + i \left(-\frac{\sqrt{3}}{2}\right)\right)$$

$$= -512 - i \ 512\sqrt{3}$$

(2) By De Moivre's theorem,

$$(1 - i)^{\frac{1}{9}} = (\sqrt{1^2 + (-1)^2})^{\frac{1}{9}} \ (\cos\frac{1}{9}\theta - i \sin\frac{1}{9}\theta)$$

where $\theta = 315^0$

$$(1 - i)^{\frac{1}{9}} = (\sqrt{2})^{\frac{1}{9}} \ (\cos 35^0 - i \sin 35^0)$$

$$= 2^{\frac{1}{18}}(0.819 - i \ 0.574)$$

$$= 0.851 - i \ 0.596$$

● **PROBLEM** 36-10

Find $\left[8\left(\cos\frac{\pi}{2} + i \sin\frac{\pi}{2}\right)\right] \div \left[2\left(\cos\frac{\pi}{6} + i \sin\frac{\pi}{6}\right)\right]$.

Solution: The complex numbers are written in the form r (cos θ + i sin θ).

Therefore, the division of these two numbers is performed by dividing the first modulus by the second, and subtracting the second angle from the first according to the formula:

$$\frac{r_1 (\cos\theta_1 + i \sin\theta_1)}{r_2 (\cos\theta_2 + i \sin\theta_2)}$$

$$= \frac{r_1}{r_2}\left[\cos(\theta_1 - \theta_2) + i \sin(\theta_1 - \theta_2)\right]$$

$$r_1 = 8$$

$$r_2 = 2$$

$$\frac{r_1}{r_2} = \frac{8}{2} = 4$$

$$\theta_1 = \pi/2$$

$$\theta_2 = \pi/6$$

$$\theta_1 - \theta_2 = \pi/2 - \pi/6 = \pi/3 = \text{ampltitude}$$

$$8 \left[\cos \frac{\pi}{2} + i \sin \frac{\pi}{2}\right] \div 2 \cos \left[\frac{\pi}{6} + i \sin \frac{\pi}{6}\right]$$

$$= 4 \left[\cos \frac{\pi}{3} + i \sin \frac{\pi}{3}\right]$$

Check $\quad 8 \left[\cos \frac{\pi}{2} + i \sin \frac{\pi}{2}\right] = 8(0 + i) = 8i.$

$$2 \left[\cos \frac{\pi}{6} + i \sin \frac{\pi}{6}\right] = 2 \left[\frac{\sqrt{3}}{2} + \frac{1}{2} i\right] = \sqrt{3} + i$$

$$\frac{8i}{\sqrt{3} + i} = \frac{8i}{\sqrt{3} + i} \cdot \frac{\sqrt{3} - i}{\sqrt{3} - i} = \frac{8 + 8\sqrt{3}i}{3 - i^2}$$

$$= \frac{8 + 8\sqrt{3}i}{4} = 2 + 2\sqrt{3}i$$

$$4 \left[\cos \frac{\pi}{3} + i \sin \frac{\pi}{3}\right] = 4 \left[\frac{1}{2} + \frac{\sqrt{3}}{2} i\right] = 2 + 2\sqrt{3}i$$

● **PROBLEM 36-11**

Show $\quad \dfrac{r_1(\cos \theta + i \sin \theta)}{r_2(\cos \phi + i \sin \phi)}$

$$= \frac{r_1}{r_2} [\cos (\theta - \phi) + i \sin (\theta - \phi)]$$

<u>Solution:</u> The trick is to multiply the original fraction by a fraction that is equivalent to one. The original value of the fraction will remain unchanged, but the complex number in the denominator now becomes a pure real. This is achieved by multiplying by a fraction whose numerator and denominator are the complex con-jugate of the complex number in the denominator of the original fraction.

number $\qquad\qquad$ conjugate

$r_2(\cos \phi + i \sin \phi) \qquad\qquad r_2(\cos \phi - i \sin \phi)$

$$\frac{r_1(\cos \theta + i \sin \theta)}{r_2(\cos \phi + i \sin \phi)}$$

$$= \frac{r_1(\cos \theta + i \sin \theta)}{r_2(\cos \phi + i \sin \phi)} \cdot \frac{r_2(\cos \phi - i \sin \phi)}{r_2(\cos \phi - i \sin \phi)}$$

$$= \frac{r_1 r_2 [(\cos \theta \cos \phi + \sin \theta \sin \phi) + i(\sin \theta \cos \phi - \cos \theta \sin \phi)]}{r_2^2(\cos^2 \phi - i^2 \sin^2 \phi)}$$

(Here we recognize the formulae for the cosine and sine of the difference of two angles. Also, we use the fact that $\cos^2 \phi - i^2 \sin \phi = \cos^2 \phi - (-1)\sin^2 \phi = \cos^2 \phi + \sin^2 \phi = 1$.)

$$= \frac{r_1 r_2 [\cos(\theta - \phi) + i \sin(\theta - \phi)]}{r_2^2 \cdot 1}$$

$$= \frac{r_1}{r_2} [\cos(\theta - \phi) + i \sin(\theta - \phi)]$$

● **PROBLEM 36-12**

Find the three cube roots of - 1.

Solution: The number - 1 is a pure real number. This can only occur if the amplitude of the angle in the $r(\cos \theta + i \sin \theta)$ representation of - 1 is a value for which $\sin \theta = 0$. This value occurs for either $\theta = \pi$ or π plus integral multiples of π. The real part of this number (equal to minus one) also occurs when $\theta = \pi$ or integral multiples of π.

$$- 1 = \cos \pi + i \sin \pi$$

$$= \cos(\pi + 2k\pi) + i \sin(\pi + 2k\pi)$$

$$(- 1)^{\frac{1}{3}} = \cos \left[\frac{\pi + 2k\pi}{3}\right] + i \sin \left[\frac{\pi + 2k\pi}{3}\right]$$

For k = 0, $\quad r_0 = \cos \frac{\pi}{3} + i \sin \frac{\pi}{3}$

For k = 1, $\quad r_1 = \cos \pi + i \sin \pi$

For k = 2, $\quad r_2 = \cos \frac{5\pi}{3} + i \sin \frac{5\pi}{3}$

Thus, the roots are

$$r_0 = \cos \frac{\pi}{3} + i \sin \frac{\pi}{3}$$

686

$$r_1 = \cos \pi + i \sin \pi$$

$$r_2 = \cos \frac{5\pi}{3} + i \sin \frac{5\pi}{3}$$

It is possible to check these roots by converting each to rectangular form and raising to the third power.

● **PROBLEM 36-13**

Find all the fifth roots of 2.

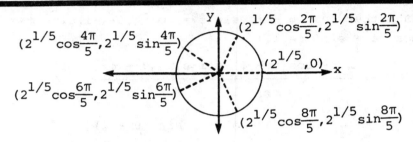

Solution: $2 = 2(\cos 0 + i \sin 0)$

Since 2 is a pure real number, its representation of the form $r(\cos \theta + i \sin \theta)$ has no complex part. Therefore, its amplitude must be equal to zero plus an integral multiple of π radians ($\theta = 0 + 2k\pi$). That is to say, the imaginary part must equal zero ($\sin \theta = 0$) and the real part must equal unity ($\cos \theta = 1$).

Therefore, $2 = 2 [\cos (0 + 2k\pi) + i \sin (0 + 2k\pi)]$

To raise a complex number to an integral power, n, the procedure is as follows:

$$(r[\cos(\theta + 2k\pi) + i \sin (\theta + 2k\pi)])^n$$

$$= r^n [\cos(n(\theta + 2k\pi)) + i \sin (n(\theta + 2k\pi))]$$

For finding n roots just raise the complex number to a rational power which corresponds, i.e. i/n:

$$(r[\cos(\theta + 2k\pi) + i \sin(\theta + 2k\pi)])^{1/n}$$

$$= r^{1/n} \left[\cos \left(\frac{\theta + 2k\pi}{n} \right) + i \sin \left(\frac{\theta + 2k\pi}{n} \right) \right]$$

Therefore, for the fifth roots of 2:

$$2^{\frac{1}{5}} = (2 [\cos(0 + 2k\pi) + i \sin(0 + 2k\pi)])^{\frac{1}{5}}$$

$$= 2^{\frac{1}{5}} \left(\cos \frac{2k\pi}{5} + i \sin \frac{2k\pi}{5} \right) \quad k = 0, 1, 2, \ldots$$

687

Now substitute values of k from 0 to 4, inclusive.

For k = 0, $\quad r_0 = 2^{\frac{1}{5}} (\cos 0 + i \sin 0)$

For k = 1, $\quad r_1 = 2^{\frac{1}{5}} \left(\cos \frac{2\pi}{5} + i \sin \frac{2\pi}{5} \right)$

For k = 2, $\quad r_2 = 2^{\frac{1}{5}} \left(\cos \frac{4\pi}{5} + i \sin \frac{4\pi}{5} \right)$

For k = 3, $\quad r_3 = 2^{\frac{1}{5}} \left(\cos \frac{6\pi}{5} + i \sin \frac{6\pi}{5} \right)$

For k = 4, $\quad r_4 = 2^{\frac{1}{5}} \left(\cos \frac{8\pi}{5} + i \sin \frac{8\pi}{5} \right)$

For k > 4, we obtain the same cycle of values. Hence, the fifth roots of 2 are those values designated as r_0, r_1, r_2, r_3, and r_4.

● **PROBLEM 36-14**

Find the 5th roots of - 1 + i.

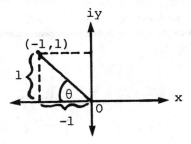

Solution: In the figure notice that we can determine r and θ so that we can transform the complex number from cartesian coordinates to a polar representation. Use the Pythagorean theorem to determine r.

$$r^2 = x^2 + y^2 = (-1)^2 + (1)^2 = 2$$

$$r = \sqrt{2}$$

θ can be computed by using the fact that tan θ = y/x or

$$\arctan \frac{y}{x} = \theta$$

$$\theta = \arctan \left(\frac{1}{-1} \right) = \arctan (-1)$$

i.e. $\dfrac{\sin \theta}{\cos \theta} = \dfrac{1}{-1}$; the angle whose sin is 1 and whose cos

688

is - 1 is 135° measured counterclockwise from 0°.

The 5th roots are calculated according to the rule

$$w^{1/n} = r^{1/n} \left[\cos \left(\frac{\theta + 2k\pi}{n} \right) + i \sin \left(\frac{\theta + 2k\pi}{n} \right) \right]$$

where $0 \leq k \leq n - 1$ inclusive

$$- 1 + i = \sqrt{2}(\cos 135° + i \sin 135°)$$

$$\sqrt[5]{-1 + i} = \left(2^{\frac{1}{2}} \right)^{1/5} \left[\cos \frac{135° + 2k\pi}{5} + i \sin \frac{135° + 2k\pi}{5} \right]$$

$$k = 0, 1, \ldots, 4$$

Thus,

$$w_0 = 2^{1/10}(\cos 27° + i \sin 27°)$$

$$w_1 = 2^{1/10}(\cos 99° + i \sin 99°)$$

$$w_2 = 2^{1/10}(\cos 171° + i \sin 171°)$$

$$w_3 = 2^{1/10}(\cos 243° + i \sin 243°)$$

$$w_4 = 2^{1/10}(\cos 315° + i \sin 315°)$$

● **PROBLEM** 36-15

Use De Moivre's theorem to find the value of $(-\sqrt{3} + i)^7$.

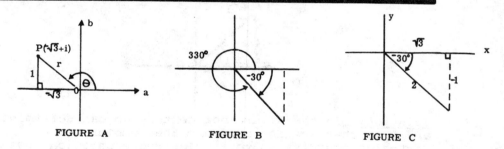

FIGURE A   FIGURE B   FIGURE C

<u>Solution:</u> Let a complex number $a + bi$ be expressed in polar form, $r(\cos \theta + i \sin \theta)$ where r is the radius vector and θ is the angle made between the x-axis and r. Then De Moivre's theorem states that if n is a positive integer,

$$[r(\cos \theta + i \sin \theta)]^n = r^n(\cos n\theta + i \sin n\theta) .$$

We must convert the given complex number to polar form. Plot the point P representing the complex number $(-\sqrt{3} + i)$ which is in the form $a + bi$, (see Figure A).

We need to find the length r of the radius vector of the complex number; $r^2 = 1^2 + (\sqrt{3})^2 = 1 + 3 = 4$. Since r is always positive, because we cannot have a negative length, $r = 2$. If $r = 2$, then the sine of the reference angle is $1/2$. Thus it is $30°$ and $\theta = 180° - 30° = 150°$. Substitute these two values into $r(\cos \theta + i \sin \theta)$.

689

Hence, $\sqrt{3} + i = 2(\cos 150° + i \sin 150°)$.

Applying De Moivre's theorem, we obtain

$$(-\sqrt{3} + i)^7 = [2(\cos 150° + i \sin 150°)]^7$$
$$= 2^7(\cos 7\cdot150° + i \sin 7\cdot150°)$$
$$= 128(\cos 1,050° + i \sin 1,050°)$$
$$= 128[(\cos 2\cdot360° + 330°) + i(\sin 2\cdot360° + 330°)]$$
$$= 128(\cos 330° + i \sin 330°)$$

The reference angle for $330°$ is $-30°$. (See Figure B). Then,

$$(-\sqrt{3} + i)^7 = 128[\cos(-30°) + i \sin(-30°)]$$

Note that $\cos(-30°) = +\dfrac{\sqrt{3}}{2}$ and $\sin(-30°) = -\dfrac{1}{2}$ since the cosine and sine functions are respectively positive and negative in quadrant IV, (see Figure C). Substitute these values, then:

$$(-\sqrt{3} + i)^7 = 128\left(\frac{\sqrt{3}}{2} - \frac{1}{2} i\right)$$
$$= 64\sqrt{3} - 64i .$$

● **PROBLEM** 36-16

Find the equations for $\sin 2\theta$ and $\cos 2\theta$ from the de Moivre equation with $n = 2$.

Solution: Let a complex number be expressed in polar form, $r(\cos \theta + i \sin \theta)$ where r is the radius vector and θ is the angle made between the x-axis and r. Then de Moivre's Theorem states that if n is any rational number, then

$$[r(\cos \theta + i \sin \theta)]^n = r^n(\cos n\theta + i \sin n\theta).$$

Furthermore we can see that:
$$(\cos \theta + i \sin \theta)^n = (\cos n\theta + i \sin n\theta) .$$

Also the complex exponential function is:
$$e^{i\theta} = \cos \theta + i \sin \theta .$$

If we substitute $n = 2$,
$$(\cos \theta + i \sin \theta)^2 = (\cos 2\theta + i \sin 2\theta) = \left(e^{i\theta}\right)^2$$
$$= e^{i\theta} \cdot e^{i\theta} = e^{i\theta+i\theta} = e^{2i\theta}$$

Expand the expression $(\cos \theta + i \sin \theta)^2$.
$$(\cos \theta + i \sin \theta)^2 = (\cos \theta + i \sin \theta)(\cos \theta + i \sin \theta)$$
$$= \cos^2\theta + 2i \sin \theta \cos \theta + i^2\sin^2\theta$$

Noting that $i^2 = -1$, we obtain:
$$(\cos \theta + i \sin \theta)^2 = \cos^2\theta - \sin^2\theta + 2i \sin \theta \cos \theta \qquad (1)$$

Furthermore by de Moivre's Theorem for the case $n = 2$, we have
$$(\cos \theta + i \sin \theta)^2 = \cos 2\theta + i \sin 2\theta \qquad (2)$$

Equate the right side of equations (1) and (2) to obtain:
$$\cos^2\theta - \sin^2\theta + 2i \sin \theta \cos \theta = \cos 2\theta + i \sin 2\theta$$

$$\cos^2\theta - \sin^2\theta + i2\sin\theta\cos\theta = \cos 2\theta + i\sin 2\theta$$

Equate the real and imaginary parts.

$$\cos^2\theta - \sin^2\theta = \cos 2\theta \qquad\qquad (3)$$

$$i2\sin\theta\cos\theta = i\sin 2\theta \qquad\qquad (4)$$

Note that, after dividing both sides by i, equation (4) becomes:

$$2\sin\theta\cos\theta = \sin 2\theta$$

Therefore, the expressions for $\sin 2\theta$ and $\cos 2\theta$ are:

$$\cos 2\theta = \cos^2\theta - \sin^2\theta$$

$$\sin 2\theta = 2\sin\theta\cos\theta .$$

OPERATIONS WITH COMPLEX NUMBERS

● **PROBLEM** 36-17

Express each of the following as the product of i and a real number.

(a) $2i^5$ (b) $\dfrac{-5}{i^7}$ (c) $\sqrt{-81}$

<u>Solution:</u> Recalling that $\sqrt{-1} = i$ $\left(\text{or} \ -1 = i^2\right)$:

(a) $2i^5 = 2 \cdot i^4 \cdot i = 2 \cdot 1 \cdot i = 2i$

(b) $\dfrac{-5}{i^7} = \dfrac{-5}{i^4 \cdot i^2} \cdot \dfrac{1}{i} = \dfrac{-5}{1 \cdot -1} \cdot \dfrac{1}{i} = \dfrac{-5}{1 \cdot -1} \cdot \dfrac{-i^2}{i} = -5i$

Note that $1 = -(-1) = -\left(i^2\right) = -i^2$. Hence $\dfrac{1}{i} = \dfrac{-i^2}{i}$

(c) $\sqrt{-81} = \sqrt{(-1)(81)} = \sqrt{-1} \cdot \sqrt{81} = 9i$

● **PROBLEM** 36-18

What is the conjugate of $3 - 2i$ and the conjugate of $5 + 7i$?

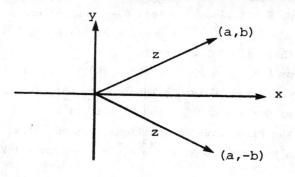

691

Any complex number may be interpreted as an ordered pair in the plane with the real component designated by the x value and the imaginary part designated by the y value. The conjugate of a complex number is that number which when multiplied by the original complex number yields a product which is purely real. Geometrically, the complex conjugate is a reflection of the complex number through the x-axis. The complex conjugate of 3 - 2i is 3 + 2i

i.e., $(3 - 2i)(3 + 2i) = 13$.

The conjugate of 5 + 7i is 5 - 7i

$$(5 + 7i)(5 - 7i) = 74 .$$

The conjugate of a pure real number a, which can be written a + 0i, is merely itself or a - 0i. Geometrically we see that the reflection of a real number is actually itself. The conjugate of a pure imaginary number bi is -bi. The conjugate of a complex number a + bi is a - bi.

● **PROBLEM 36-19**

Write each of the following in the form a + bi.

    a) (2 + 4i) + (3 + i)

    b) (2 + i) - (4 - 2i)

    c) (4 - i) - (6 - 2i)

    d) 3 - (4 + 2i)

Solution:

a) (2 + 4i) + (3 + i) = 2 + 4i + 3 + i

                    = (2 + 3) + (4i + i)

                    = 5 + 5i

b) (2 + i) - (4 - 2i) = 2 + i - 4 + 2i

                  = (2 - 4) + (i + 2i)

                  = -2 + 3i

c) (4 - i) - (6 - 2i) = 4 - i - 6 + 2i

                  = (4 - 6) + (-i + 2i)

                  = -2 + i

d)   3 - (4 + 2i) = 3 - 4 - 2i

692

$$= (3 - 4) - 2i$$

$$= -1 - 2i$$

● **PROBLEM** 36-20

Find the product $(2 + 3i)(- 2 - 5i)$.

Solution: Using the following method: product of first elements, + product of outer elements + product of inner elements + product of last elements:

$$(2 + 3i)(- 2 - 5i) = 2(- 2) + 2(- 5i) + 3i(- 2) + 3i(- 5i)$$

$$= - 4 - 10i - 6i - 15i^2$$

$$= - 4 - 16i - 15i^2$$

Recall $i^2 = - 1$, hence, $= - 4 - 16i - 15(- 1)$

$$= - 4 - 16i + 15$$

$$= 11 - 16i$$

The same result is obtained by using the distributive law.

$$(2 + 3i)(- 2 - 5i) = (2 + 3i)(- 2) - (2 + 3i)5i$$

$$= - 4 - 6i - 10i - 15i^2 = 11 - 16i.$$

In other words, if one multiples $2 + 3i$ and $- 2 - 5i$ as if they were polynomials and replaces i^2 by $- 1$, then the correct product is obtained.

● **PROBLEM** 36-21

Find the values of the following expressions:

    a. $(2 + 3i) + (6 - 2i)$

    b. $(2 - i)(1 + 3i)$

    c. $i - (2 + 3i)$

Solution: a) $(2 + 3i) + (6 - 2i) = 2 + 3i + 6 - 2i$

$$= 2 + 6 + 3i - 2i$$

$$= 8 + i$$

    b) $(2 - i)(1 + 3i) = 2(1) - i(1) + 2(3i) - i(3i)$

$$= 2 - i + 6i - 3i^2 \qquad (1)$$

693

Since $i^2 = -1$, equation (1) becomes:

$$(2 - 1)(1 + 3i) = 2 - i + 6i - 3(-1)$$
$$= 2 - i + 6i + 3$$
$$= 2 + 3 - i + 6i$$
$$= 5 + 5i$$

c) $i - (2 + 3i) = i - 2 - 3i$
$$= -2 + i - 3i$$
$$= -2 - 2i$$

● **PROBLEM 36-22**

Find the real and imaginary parts of

$$(2 + 3i) \div (3 + 4i)$$

<u>Solution:</u> In order to divide one complex number by another, the denominator must be converted to a real number. This can be done by multiplying the numerator and the denominator by the conjugate of the denominator. The complex numbers a + bi and a - bi are conjugates of each other. Therefore, the conjugate of the denominator 3 + 4i is 3 - 4i. Then,

$$\frac{2 + 3i}{3 + 4i} = \frac{(2 + 3i)(3 - 4i)}{(3 + 4i)(3 - 4i)}$$

$$= \frac{6 + 9i - 8i - 12i^2}{9 + \cancel{12i} - \cancel{12i} - 16i^2}$$

$$\frac{6 + i - 12(-1)}{9 - 16(-1)}, \quad \text{since } i^2 = -1$$

$$= \frac{6 + i + 12}{9 + 16}$$

$$= \frac{18 + i}{25}$$

$$= \frac{18}{25} + \frac{1}{25} i$$

Hence, the real part of the quotient is $\frac{18}{25}$ and the imaginary part is $\frac{1}{25} i$.

Simplify $\frac{3-5i}{2+3i}$.

Solution: To simplify $\frac{3-5i}{2+3i}$ means to write the fraction without an imaginary number in the denominator. To achieve this, we multiply the fraction by another fraction which is equivalent to unity, (so that the value of the original fraction is unchanged) which will transform the expression in the denominator to a real number. A fraction with this property must have the complex conjugate of the expression in the denominator of the original fraction as its numerator and denominator. The complex conjugate must be chosen because of its special property that when multiplied by the original complex number the result is real.

Note: a + bi; its complex conjugate is a - bi or they can be said to be conjugates of each other. To multiply notice that (a + bi)(a - bi) is the factored form of the difference of two squares. Thus we obtain

$$(a)^2 - (bi)^2; \quad i^2 = -1; \quad (a)^2 - (-1)(b)^2 \text{ or } a^2 + b^2.$$

$$\frac{3-5i}{2+3i} \cdot \frac{2-3i}{2-3i} = \frac{6-9i-10i+15i^2}{4-9i^2}$$

$$= \frac{6-19i-15}{4+9}$$

$$= \frac{-9-19i}{13} \text{ or } \frac{-9}{13} - \frac{19}{13}i$$

Since the resulting fraction has a rational number in the denominator, we have rationalized the denominator.

Simplify: (a) $4i - 7i^3$ (b) $\frac{2-3i}{5i}$

Solution: (a) Factor i in the expression to obtain,

$$i\left(4 - 7i^2\right)$$

$$i^2 = -1$$

and we obtain,

$$i(4 - 7[-1]) = i(4+7) = 11i$$

(b) Rationalize the denominator by multiplying

the original fraction by a fraction equivalent to unity
which will cause the imaginary expression in the denominator of the original fraction to be eliminated.

$\frac{i}{i}$ is suitable because $\frac{i}{i} = 1$ and $i^2 = -1$.

$$\frac{2 - 3i}{5i} = \frac{2 - 3i}{5i} \cdot \frac{i}{i} = \frac{2i - 3i^2}{5i^2} = \frac{2i + 3}{-5}$$

● **PROBLEM 36-25**

Expand $(2 + 3i)^3$.

Solution: The identity $(a + b)^3 = a^3 + 3a^2b + 3ab^2 + b^3$
is still valid in the case of complex numbers. Thus,
replacing a by 2 and b by 3i we obtain

$(2 + 3i)^3 = 2^3 + 3 \cdot 2^2(3i) + 3 \cdot 2(3i)^2 + (3i)^3$

$= 8 + 3 \cdot 4 \cdot 3 \cdot i + 3 \cdot 2(3^2i^2) + 3^3i^3$

$= 8 + 36i + 6(9)i^2 + 27i^3$

$= 8 + 36i + 54i^2 + 27i^3$.

Recalling that $i^2 = -1$, since $i = \sqrt{-1}$ and $i^2 = \sqrt{-1}\sqrt{-1} = -1$; and $i^3 = i^2(i) = (-1)i = -i$, we
obtain:

$= 8 + 36i + 54(-1) + 27(-i)$

$= 8 + 36i - 54 - 27i$

$= 9i - 46$.

● **PROBLEM 36-26**

Evaluate $x^2 - 2x + 6$ for $x = 3 + 2i$.

Solution: Substituting the given value, we get

$$x^2 - 2x + 6 = (3 + 2i)^2 - 2(3 + 2i) + 6$$

$$= (3 + 2i)(3 + 2i) - 6 - 4i + 6$$

Since $(a + b)(c + d) = ac + ad + bc + bd$

$$x^2 - 2x + 6 = (3)(3) + 6i + 6i + (2i)(2i) - 6 - 4i + 6$$

$$= 9 + 12i + (2i)^2 - 6 - 4i + 6$$

696

Since $(ab)^2 = a^2b^2$

$(2i)^2 = 2^2(i)^2$

By definition $i^2 = -1$, hence $(2i)^2 = 4(-1) = -4$ and $x^2 - 2x + 6$

$= 9 + 12i - 4 - 6 - 4i + 6$

Combine terms, $= 5 + 8i.$

Show that $\left[\dfrac{1}{\sqrt{2}} + \dfrac{1}{\sqrt{2}}i\right]^4 = -1.$

Solution: Factor out $\dfrac{1}{\sqrt{2}}$:

$$\left[\frac{1}{\sqrt{2}} + \frac{1}{\sqrt{2}}i\right]^4 = \left[\frac{1}{\sqrt{2}}(1 + i)\right]^4 = \left(\frac{1}{\sqrt{2}}\right)^4 (1 + i)^4 = \frac{1}{4}(1 + i)^4$$

Note: $\left(\dfrac{1}{\sqrt{2}}\right)^4 = \left(\dfrac{1}{\sqrt{2}}\right)\left(\dfrac{1}{\sqrt{2}}\right)\left(\dfrac{1}{\sqrt{2}}\right)\left(\dfrac{1}{\sqrt{2}}\right) = \dfrac{1}{2} \cdot \dfrac{1}{2} = \dfrac{1}{4}$

Now we apply the identity $(a + b)^4 =$

$(a + b)^2(a + b)^2 = (a + b)(a + b)(a + b)(a + b)$

$$= \left(a^2 + 2ab + b^2\right)\left(a^2 + 2ab + b^2\right).$$

Then,

$a^2 + 2ab + b^2$

$\underline{a^2 + 2ab + b^2}$

$a^4 + 2a^3b + a^2b^2$

$\qquad + 2a^3b + 4a^2b^2 + 2ab^3$

$\underline{\qquad\qquad + a^2b^2 + 2ab^3 + b^4}$

$(a+b)^4 = a^4 + 4a^3b + 6a^2b^2 + 4ab^3 + b^4.$

Thus, $a = 1$ $b = i.$

We obtain

$$(1 + i)^4 = 1^4 + 4i + 6i^2 + 4i^3 + i^4$$

substitute: $i^2 = -1$ and $i^3 = i^2 \cdot i = (-1)i = -i$ and $i^4 = i^2 \cdot i^2 = (-1)^2 = 1.$ Then,

$$(1 + i)^4 = 1 + 4i - 6 - 4i + 1 = -4.$$

Hence

$$\left(\frac{1}{\sqrt{2}} + \frac{1}{\sqrt{2}}i\right)^4 = \frac{1}{4}(-4) = -1.$$

● **PROBLEM 36-28**

Find $\sqrt{3 + 4i}$

Solution: Let $\sqrt{3 + 4i} = x + yi$ where x and y are real numbers. Square both sides of this equation:

$$(\sqrt{3 + 4i})^2 = (x + yi)^2$$

$$3 + 4i = (x + yi)(x + yi) \tag{1}$$

$$3 + 4i = x^2 + 2xyi + y^2 i^2 \tag{2}$$

Since $i^2 = -1$, equation (2) becomes:

$$3 + 4i = x^2 + 2xyi + y^2(-1)$$

$$3 + 4i = x^2 + 2xyi - y^2 \text{ or, by the commutative}$$

property of addition,

$$3 + 4i = x^2 - y^2 + 2xyi$$

Equate the real and imaginary parts of both members:

$$3 = x^2 - y^2 \tag{3}$$

$$4i = 2xyi \tag{4}$$

Dividing both sides of equation (4) by i:

$$\frac{4i}{i} = \frac{2xyi}{i}$$

$$4 = 2xy \tag{5}$$

Therefore, our equations are:

$$3 = x^2 - y^2 \tag{3}$$

$$4 = 2xy \tag{5}$$

Solving equation (5) for x:
dividing both sides by 2y,

$$\frac{4}{2y} = \frac{2xy}{2y}$$

$$\frac{2}{y} = x$$

Note that the above operation assumes that $y \neq 0$ since division by 0 is undefined. (Also, in our original expression $\sqrt{3 + 4i} = x + yi$, there is assumed to be an an imaginary part; namely, yi. If y were equal to 0,

698

then there would be no imaginary part since $yi = 0(i) = 0$.
Hence, y cannot equal 0.)

Substituting the expression for x in equation (3):

$$3 = \left(\frac{2}{y}\right)^2 - y^2$$

$$3 = \frac{4}{y^2} - y^2$$

Obtaining a common denominator of y^2 for the two terms on the right side of this equation:

$$3 = \frac{4}{y^2} - \frac{y^2(y^2)}{y^2}$$

$$3 = \frac{4}{y^2} - \frac{y^4}{y^2}$$

$$3 = \frac{4 - y^4}{y^2}$$

Multiplying both sides by y^2:

$$y^2(3) = y^2\left(\frac{4 - y^4}{y^2}\right)$$

$$3y^2 = 4 - y^4$$

Subtracting $\left(4 - y^4\right)$ from both sides:

$$3y^2 - \left(4 - y^4\right) = 4 - y^4 - \left(4 - y^4\right)$$
$$3y^2 - 4 + y^4 = 0$$

or $y^4 + 3y^2 - 4 = 0$

Factoring the left side of this equation as a product of two polynomials:

$$\left(y^2 + 4\right)\left(y^2 - 1\right) = 0$$

Whenever a product $ab = 0$ where a and b are any two numbers, either $a = 0$ or $b = 0$. Therefore,

either $y^2 + 4 = 0$ or $y^2 - 1 = 0$
$$y^2 = -4 \quad \text{or} \quad y^2 = 1$$
$$y = \pm\sqrt{1}$$
$$y = \pm 1$$

Note that there is no real solution to $y^2 = -4$ since there is no real number y whose square is -4.

Substituting $y = -1$ in equation (3):

$$3 = x^2 - (-1)^2$$
$$3 = x^2 - (1)$$
$$3 = x^2 - 1$$

699

Add 1 to both sides:
$$3 + 1 = x^2 - 1 + 1$$
$$4 = x^2$$

Take the square root of both sides:
$$\pm\sqrt{4} = x$$
$$\pm 2 = x.$$

Hence, the two solutions appear to be $(-2,-1)$ and $(2,-1)$.
Substituting $y = 1$ in equation (3):
$$3 = x^2 - (1)^2$$
$$3 = x^2 - 1$$

Add 1 to both sides:
$$3 + 1 = x^2 - 1 + 1$$
$$4 = x^2$$

Take the square root of both sides:
$$\pm\sqrt{4} = x \quad \text{or} \quad \pm 2 = x$$

Hence, the two additional solutions appear to be $(-2,1)$ and $(2,1)$. For the four solutions obtained:

when $(x,y) = (-2,-1)$, $\sqrt{3 + 4i} = -2 + (-1)i = -2-i$,
when $(x,y) = (2,-1)$, $\sqrt{3 + 4i} = 2 + (-1)i = 2 - i$,
when $(x,y) = (-2,1)$, $\sqrt{3 + 4i} = -2 + 1i = -2 + i$,
when $(x,y) = (2,1)$, $\sqrt{3 + 4i} = 2 + 1i = 2 + i$.

Checking the four solutions of $\sqrt{3 + 4i}$; namely, $-2-i$, $2-i$, $-2+i$, $2+i$, using equation 1:

for $-2-i$, $3 + 4i = (-2-i)(-2-i)$
$$3 + 4i = 4 + 2i + 2i + i^2$$
$$3 + 4i = 4 + 4i - 1. \quad \text{Note that } i^2 = -1.$$
$$3 + 4i = 3 + 4i \checkmark$$

Therefore, $-2-i$ is a solution to $\sqrt{3 + 4i}$.

For $2-i$, $3 + 4i = (2 - i)(2 - i)$
$$3 + 4i = 4 - 2i - 2i + i^2$$
$$3 + 4i = 4 - 4i - 1 \quad \text{since } i^2 = -1$$
$$3 + 4i \neq 3 - 4i$$

Therefore, $2 - i$ is not a solution to $\sqrt{3 + 4i}$.

For $-2+i$, $3 + 4i = (-2 + i)(-2 + i)$
$$3 + 4i = 4 - 2i - 2i + i^2$$
$$3 + 4i = 4 - 4i - 1 \quad \text{since } i^2 = -1$$
$$3 + 4i \neq 3 - 4i$$

Therefore, $-2 + i$ is not a solution to $\sqrt{3 + 4i}$.

For $2 + i$, $3 + 4i = (2 + i)(2 + i)$
$$3 + 4i = 4 + 2i + 2i + i^2$$
$$3 + 4i = 4 + 4i - 1 \quad \text{since } i^2 = -1$$

700

$$3 + 4i = 3 + 4i \checkmark$$

Therefore, $2 + i$ is a solution to $\sqrt{3 + 4i}$.

Hence, the only two solutions to $\sqrt{3 + 4i}$ are: $-2-i$ and $2+i$.

Show that $(a + bi) + (c + di) = (c + di) + (a + bi)$.

Solution: Use the assoiative, distributive, and commutative laws. Associate the corresponding components of the complex numbers, i.e., associate the real and imaginary parts respectively.

$$(a + bi) + (c + di) = (a + c) + (b + d)i$$

$$= a + c + bi + di$$

$$= a + bi + c + di$$

$$= (c + di) + (a + bi)$$

We would suspect that zero is still the additive identity, but zero is a real number. Recall that the real number 5 and the complex number $5 + 0i$ represent the same number. Then the additive identity should be $0 + 0i$. Let us see whether it is. Recall that adding the additive identity to a number does not change the number. Applying the definition of addition.

$$(a + bi) + (0 + 0i) = (a + 0) + (b + 0)i$$

$$= a + bi$$

This verifies that $0 + 0i$ is the additive identity.

Given $f(x) = x^3 + x + 1$, evaluate $f(1 + i)$.

Solution: $f(1 + i)$ indicates that $1 + i$ should be subsituted for x.

$$f(1 + i) = (1 + i)^3 + (1 + i) + 1$$

$$= (1 + i)(1 + i)(1 + i) + (1 + i) + 1$$

$$= \left(1 + 2i + i^2\right)(1 + i) + (1 + i) + 1$$

$$= 1 + 2i + i^2 + i + 2i^2 + i^3 + (1 + i) + 1$$

$$= 1 + 3i + 3i^2 + i^3 + (1 + i) + 1$$

$$= 1 + 3i + 3i^2 + 1^3 + 1 + i + 1$$

Note that $i^2 = -1, i^3 = i^2(i) = (-1)(i) = -i$. Then

$$f(1 + i) = 1 + 3i + 3(-1) + (-1)i + 1 + i + 1$$

$$= 3i.$$

CHAPTER 37

THE HYPERBOLIC AND INVERSE HYPERBOLIC FUNCTIONS

> **Basic Attacks and Strategies for Solving Problems in this Chapter. See pages 703 to 707 for step-by-step solutions to problems.**

The standard trigonometric functions are also sometimes called *circular trigonometric functions* since the identity

$$\sin^2 \alpha + \cos^2 \alpha = 1$$

can be rewritten as $y^2 + x^2 = 1$ (where $x = \cos \alpha$ and $y = \sin \alpha$), which is the equation of a circle.

There is another set of functions that are called *hyperbolic trigonometric functions*. These are defined in terms of $e = 2.718281828 \ldots$ and produce an identity that is the equation of a hyperbola. The basic two functions, *hyperbolic sine* (abbreviated sinh and pronounced *sinch*) and *hyperbolic cosine* (abbreviated cosh and pronounced *kosh*) are defined as follows:

$$\sinh x = \frac{e^x - e^{-x}}{2}$$

$$\cosh x = \frac{e^x + e^{-x}}{2}.$$

The other functions are defined similarly to the way the corresponding circular trigonometric functions are defined. In other words,

$$\tanh x = \frac{\sinh x}{\cosh x}, \quad \coth x = \frac{1}{\tanh x}. \quad \text{sech } x = \frac{1}{\cosh x}, \quad \text{and} \quad \text{csch } x = \frac{1}{\sinh x}.$$

The input value into hyperbolic trigonometric functions is usually termed an *argument* and is not called an *angle*, to avoid confusion with circular trigonometric functions. Using the definitions to $\sinh x$ and $\cosh x$, one obtains the identity

$$\cosh^2 x - \sinh^2 x = 1,$$

which is similar to the standard equation for a hyperbola centered at the origin. It also follows the $\sinh(-x) = -\sinh x$ and $\cosh(-x) = \cosh x$.

There also exist formulas for the hyperbolic sine and cosine of sums of arguments. For example,

$$\sinh(x + y) = \sinh x \cosh y + \cosh x \sinh y$$

and

$$\cosh(x + y) = \cosh x \cosh y + \sinh x \sinh y.$$

Formulas for double arguments and half arguments can be derived from the various identities similar to the way that equivalent formulas are derived in the case of the circular trigonometric functions.

As in the case of the circular trigonometric functions, one can denote the inverse hyperbolic trigonometric function either by the exponent of −1 or by using the prefix *arg*. Thus, given $y = \sinh x$, the inverse relationship is expressed as

$$x = \sinh^{-1} y \quad \text{or} \quad x = \text{argsinh} y.$$

Because the hyperbolic functions are defined in terms of e^x, the inverse hyperbolic functions can be defined in terms of $\ln x$, the inverse function of e^x. Thus, it can be shown that

$$\sinh^{-1} y = \ln(y + \sqrt{y^2 + 1}) \quad \text{and} \quad \cosh^{-1} y = \pm \ln(y + \sqrt{y^2 - 1}).$$

Step-by-Step Solutions to Problems in this Chapter, "The Hyperbolic and Inverse Hyperbolic Functions"

● PROBLEM 37-1

Graph   (1) $y = \sinh x$

(2) $y = \cosh x$

(3) $y = \tanh x$

<u>Solution</u>:

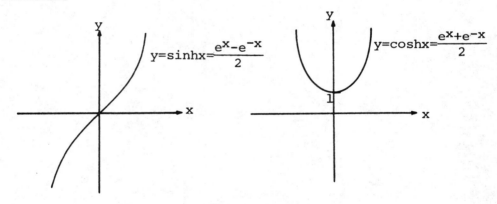

$$y = \sinh x = \frac{e^x - e^{-x}}{2}$$

$$y = \cosh x = \frac{e^x + e^{-x}}{2}$$

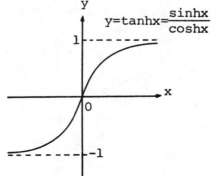

$$y = \tanh x = \frac{\sinh x}{\cosh x}$$

Calculate   (1)   cot h 4

                 (2)   csc h 4

                 (3)   sec h 4

                 (4) $\sin h^{-1} 4$

                 (5) $\cos h^{-1} 4$

Solution: (1) $\cot h = \dfrac{\cos h x}{\sin h x} = \dfrac{e^x + e^{-x}}{e^x - e^{-x}}$

$$\cot h\, 4 = \frac{e^4 + e^{-4}}{e^4 - e^{-4}} \doteq 1.0007$$

(2) $\csc h\, 4 = \dfrac{1}{\sin h\, 4} = \dfrac{2}{e^4 - e^{-4}}$

$$= 0.0367$$

(3) $\sec h\, x = \dfrac{1}{\cos h\, x} = \dfrac{2}{e^x + e^{-x}}$

$$\sec h\, 4 = \frac{2}{e^4 + e^{-4}} = 0.0366$$

(4) $\sin h^{-1} x = \ln(x + \sqrt{x^2 + 1})$

$\sin h^{-1} 4 = \ln(4 + \sqrt{4^2 + 1}) = 2.0947$

(5) $\cos h^{-1} x = \pm \ln(x + \sqrt{x^2 - 1})$

$\cos h^{-1} 4 = \pm \ln(4 + \sqrt{4^2 - 1})$

$$= \pm 2.0634$$

Show that   (1) $\cos h^2 x - \sin h^2 x = 1$

                (2) $\sin h\,(x+y) = \sin h\, x \,\cos h\, y$

                            $+ \cos h\, x \,\sin h\, y$

Solution: (1) $\cos h\, x = \dfrac{e^x + e^{-x}}{2}$

$$\sin h\, x = \frac{e^x - e^{-x}}{2}$$

$$\cos h^2 x - \sin h^2 x = \left(\frac{e^x + e^{-x}}{2}\right)^2 - \left(\frac{e^x - e^{-x}}{2}\right)^2$$

$$= \frac{(e^{2x} + 2e^x e^{-x} + e^{-2x}) - (e^{2x} - 2e^x e^{-x} + e^{-2x})}{4}$$

$$= \frac{4e^x e^{-x}}{4} = e^x e^{-x} = e^{x-x} = e^0 = 1$$

(2) $\sin h (x+y) = \dfrac{e^{x+y} - e^{-x-y}}{2}$

$$= \frac{e^x e^y - e^{-x} e^{-y}}{2}$$

$\sin h x \cos h y + \cos h x \sin h y$

$$= \left(\frac{e^x - e^{-x}}{2}\right)\left(\frac{e^y + e^{-y}}{2}\right) + \left(\frac{e^x + e^{-x}}{2}\right)\left(\frac{e^y - e^{-y}}{2}\right)$$

$$= \left[\frac{e^x e^y + e^x e^{-y} - e^{-x} e^y - e^{-x} e^{-y}}{4}\right]$$

$$+ \left[\frac{e^x e^y - e^x e^{-y} + e^{-x} e^y - e^{-x} e^{-y}}{4}\right]$$

$$= \frac{2e^x e^y - 2e^{-x} e^{-y}}{4} = \frac{e^x e^y - e^{-x} e^{-y}}{2}$$

$$= \frac{e^{(x+y)} - e^{-x-y}}{2} = \frac{e^{(x+y)} - e^{-(x+y)}}{2} = \sinh(x+y)$$

• PROBLEM 37-4

If $y = \sinh x = \frac{1}{2}(e^x - e^{-x})$, the inverse function is written $x = \sinh^{-1} y$. Similar notations are employed for the inverses of the remaining hyperbolic functions. Show that:

(a) $\sinh^{-1} y = \ln\left(y + \sqrt{1 + y^2}\right)$

(b) $\cosh^{-1} y = \pm\ln\left(y + \sqrt{y^2 - 1}\right)$

(c) $\tanh^{-1} y = \frac{1}{2} \ln \frac{1 + y}{1 - y}$.

<u>Solution:</u>   (a) To show that $\sinh^{-1} y = \ln\left(y + \sqrt{1+y^2}\right)$, we solve for x in the equation:

$$y = \frac{e^x - e^{-x}}{2}.$$

We have:

$$2y = e^x - e^{-x}$$

$$2y = e^x - \frac{1}{e^x}$$

$$2ye^x = e^{2x} - 1$$

$$e^{2x} - 2ye^x - 1 = 0$$

To solve for e^x, we use the quadratic formula, with $a = 1$, $b = -2y$ and $c = -1$, obtaining:

$$e^x = \frac{2y \pm \sqrt{4y^2 + 4}}{2} = y + \sqrt{1 + y^2}.$$

Therefore, $x = \ln\left(y + \sqrt{1 + y^2}\right) = \sinh^{-1} y$.

(b) If $y = \cosh x = \dfrac{e^x + e^{-x}}{2}$, solving for x gives $\cosh^{-1} y$. We have:

$$y = \frac{e^x + e^{-x}}{2}$$

$$2y = e^x + e^{-x}$$

$$2y = e^x + \frac{1}{e^x}$$

$$2ye^x = e^{2x} + 1$$

$$e^{2x} - 2ye^x + 1 = 0.$$

We use the quadratic formula to solve for e^x, letting $a = 1$, $b = -2y$ and $c = 1$, obtaining:

$$e^x = \frac{2y \pm \sqrt{4y^2 - 4}}{2} = y \pm \sqrt{y^2 - 1}.$$

Therefore,

$$x = \ln\left(y \pm \sqrt{y^2 - 1}\right) = \cosh^{-1} y.$$

(c) If $y = \tanh x = \dfrac{e^x - e^{-x}}{e^x + e^{-x}}$,

solving for x gives $\tanh^{-1} y$. We have

$$y = \frac{e^x - e^{-x}}{e^x + e^{-x}}$$

$$y\left(e^x + \frac{1}{e^x}\right) = e^x - \frac{1}{e^x}$$

$$ye^x + \frac{y}{e^x} = e^x - \frac{1}{e^x}$$

$$ye^{2x} + y = e^{2x} - 1$$

$$ye^{2x} - e^{2x} = -y - 1$$

$$e^{2x}(y - 1) = -y - 1$$

$$e^{2x} = \frac{-y - 1}{y - 1}$$

$$2x = \ln\left(\frac{-y - 1}{y - 1}\right).$$

Then

$$x = \frac{1}{2} \ln \frac{1 + y}{1 - y} = \tanh^{-1} y.$$

SECTION 3 — ANALYTIC GEOMETRY

CHAPTER 38

COORDINATES

Basic Attacks and Strategies for Solving Problems in this Chapter. See pages 708 to 727 for step-by-step solutions to problems.

Coordinates refer to the magnitudes used to locate a point in some *coordinate system* (where a plane or 3-dimensional space or some n-dimensional space). The coordinates along some axis should preserve the relative spacing of points in space, i.e., if one point is between two others in space, the magnitudes of the corresponding coordinates should reflect this fact.

The *projection of a point* onto one of the coordinate axes is merely the coordinate of the point corresponding to that axis. Thus, the projection of (3, 5) onto the x-axis is 3 and onto the y-axis is 5. The *projection of a line segment AB* onto one of the coordinate axes is the line segment obtained by projecting the endpoints A and B and identifying the line segment between the projected points.

Finding the midpoint between two points in the plane is the same as finding the midpoint on the line segment between these two points, and this is the same as finding the midpoints of the projections of this line segment onto each of the coordinate axes. Thus, given points P_1 and P_2 with coordinates (x_1, y_1) and (x_2, y_2), the midpoint is obtained by finding the midpoints along both axes, i.e., the midpoint

$$P_m \text{ is } \left(\frac{x_1 + x_2}{2}, \frac{y_1 + y_2}{2} \right).$$

Finding a point part of the distance along a line segment joining two points in the plane can be obtained by setting up a proportional equation (*cf.* Chapter 8). For example, if one wishes to find a point x_m between x_1 and x_2 such that the distance between x_m and x_1 is a and the distance between x_2 and x_m is b, one sets up the equation

$$\frac{x_m - x_1}{x_2 - x_m} = \frac{a}{b}$$

and solves for x_m. For a general line segment, one finds the appropriate points along the projections on both coodinate axes.

Finding the distance between two points is essentially the same as finding the hypotenuse of a right triangle. One first obtains the lengths of the projections of the line segment joining the two points onto each of the two coordinate axes, and then uses the Pythagorean theorem (*cf.* Chapter 28). In general, then, given points P_1, with coordinates (x_1, y_1), and P_2, with coordinates (x_2, y_2), the distance between P_1 and P_2 is

$$\sqrt{(x_2 - x_1)^2 + (y_2 - y_1)^2}.$$

The *slope* of a line is the change in the vertical height (y measure) that takes place given a specific horizontal displacement (x measure). Given two points on a line, P_1, with coordinates (x_1, y_1), and P_2, with coordinates (x_2, y_2), the slope of the line joining P_1 and P_2 is $\frac{y_2 - y_1}{x_2 - x_1}$. Given a linear equation, one can first pick two arbitrary x-values, and compute the corresponding y-values to obtain two points. Then one can compute the slope of the line.

The slope of a line can also be viewed as the tangent of the angle that the line makes when crossing the x-axis. Based on this fact, given the slopes of two lines, one can compute the angle made at the intersection of the two lines. One makes use of a trigonometric identity, namely that if m_1 and m_2 are the slopes of two lines, these values are equal to $\tan \alpha_1$ and $\tan \alpha_2$, where α_1 and α_2 are the respective angles of intersection of the two lines at the x-axis. The angle of intersection can be considered to be

$$\theta = \alpha_1 - \alpha_2.$$

We use

$$\tan \theta = \tan(\alpha_1 - \alpha_2) = \frac{\tan \alpha_1 - \tan \alpha_2}{1 + \tan \alpha_1 \tan \alpha_2} = \frac{m_1 - m_2}{1 + m_1 m_2}$$

to obtain the value of $\tan \theta$ from which we can determine the value of θ itself.

Three points in space are said to be *collinear* if they all lie on the same straight line. One can determine whether points are collinear in several different ways. If A, B, and C are points such that the projection of B on the x-axis is between the projections of A and C, then A, B, and C are collinear if

$$|AB| + |BC| = |AC|,$$

where $|AB|$ indicates the length of the line segment joining A and B together. An alternate way to determine collinearity is to use the fact that if three points are on the same line, the slopes of line segments joining any two of the three together must be equal.

The distance formula can be used to generate equations describing curves satisfying certain conditions. For example, suppose we want the curve such that any point on the curve is 2 units distant from point $(2, 3)$. If P is an arbitrary point on the curve and (x, y) are its coordinates, then we can use the distance formula to compute the distance from (x, y) to $(2, 3)$. We obtain the equation

$$2 = \sqrt{(x - 2)^2 + (y - 3)^2}.$$

This can be simplified by squaring both sides.

Given three points in space, P_1, with coordinates (x_1, y_1), $P_2(x_2, y_2)$ and $P_3(x_3, y_3)$, we can determine the angles of the triangle they form by first computing the slopes of each side and then using the tangent formula given above. The area of the triangle A is

$$\tfrac{1}{2} (x_1y_2 + x_2y_3 + x_3y_1 - x_1y_3 - x_2y_1 - x_3y_2).$$

● **PROBLEM 38-1**

Find the projection of AB on the coordinate axes where
the coordinates of the points A and B are (4,4) and
(9,10), respectively.

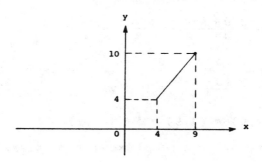

<u>Solution</u>: By definition, the projection of the line segment
joining points $P_1(x_1y_1)$ and $P_2(x_2,y_2)$ onto the x-axis or onto
any line parallel to the x-axis is (x_2-x_1), and the projection
of P_1P_2 onto the y-axis or any line parallel to the y-axis is
(y_2-y_1).

Therefore, the projection of AB onto the x-axis is
$(9-4) = 5$, and the projection of AB onto the y-axis is $(10-4)=6$,

Find the coordinates of the point $P(x,y)$ which divides
the following segments into the given ratios:

(1) $P_1(4,4)$, $P_2(8,8)$, $r_1 : r_2 = -1 : 2$

(2) $P_1(4,9)$, $P_2(9,10)$, $r_1 : r_2 = 0 : a$
where a is a real constant.

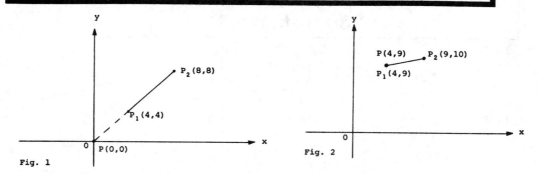

Fig. 1

Fig. 2

Solution: (1)   The point of division formula is

$$x = \frac{r_1x_2 + r_2x_1}{r_1 + r_2} \quad , \quad y = \frac{r_1y_2 + r_2y_1}{r_1 + r_2}$$

Hence, the point $P(x,y)$ has coordinates

$$x = \frac{(-1)(8) + 2\times4}{(-1) + 2} = 0 \quad \text{and}$$

$$y = \frac{(-1)(8) + 2\times4}{(-1) + 2} = 0$$

(2)   Using the point of division formula,

$$x = \frac{0\cdot9 + a\cdot4}{0 + a} = \frac{0 + a4}{a} = \frac{a4}{a} = 4 = x_1$$

$$y = \frac{0\cdot10 + a\cdot9}{0 + a} = 9 = y_2$$

So $P(x,y) = (4,9) = P_1(x_1,y_1)$.

Note that if $r_1 : r_2 = a : 0$ $P(x,y) = P_2(x_2,y_2)$.

● **PROBLEM** 38-3

Find the midpoint of the segment from $R(-3,5)$ to $S(2,-8)$.

Solution: The midpoint of a line segment from (x_1,y_1) to

(x_2, y_2) is given by

$$\left[\frac{x_1 + x_2}{2} , \frac{y_1 + y_2}{2} \right]$$

the abscissa being one half the sum of the abscissas of the endpoints and the ordinate one half the sum of the ordinates of the endpoints. Let the coordinates of the midpoint be $P(x_0, y_0)$. Then,

$$x_0 = \tfrac{1}{2}(-3 + 2) = -\tfrac{1}{2} \quad y_0 = \tfrac{1}{2}[5 + (-8)] = \tfrac{1}{2}(-3) = -\tfrac{3}{2}.$$

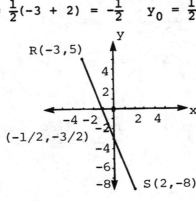

Thus, the midpoint is $P\left(-\tfrac{1}{2}, -\tfrac{3}{2}\right)$.

● **PROBLEM 38-4**

What are the coordinates of the midpoint of a line segment joining $P(-2,1)$ and $Q(6,4)$?

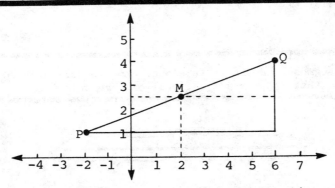

<u>Solution:</u> Let $M(\overline{x}, \overline{y})$ be the midpoint of a line segment joining $P(-2,1)$ and $Q(6,4)$. The x-coordinate of M is the average of the x-coordinates of P and Q. The y-coordinate of M is the average of the y-coordinates of P and Q:

$$M(\overline{x}, \overline{y}) = M\left[\frac{-2+6}{2} , \frac{1+4}{2} \right] = M\left(2, 2\tfrac{1}{2}\right).$$

Therefore, the coordinates of the midpoint M are $\left(2, 2\tfrac{1}{2}\right)$.
Plot the points P and Q, as illustrated in the figure.

Determine the coordinates of the midpoint of the line segment joining the points (3, -8) and (-7, 5).

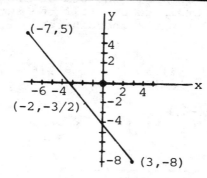

Solution: The coordinates of the desired midpoint are given by one half the sum of the abscissa and one half the sum of the ordinates. Thus,

$$x = 1/2 \ [3 + (-7)] = 1/2 \ (-4) = -2$$

$$y = 1/2 \ [(-8) + 5)] = 1/2 \ (-3) = -3/2$$

Hence the coordinates of the midpoint of the line segment joining these points is (-2, -3/2), as seen in our diagram.

A line segment AB is $7\frac{1}{2}$ in. long. Locate the point C between A and B so that AC is 3/2 in. shorter than twice CB.

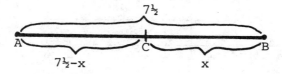

Solution: See the accompanying figure.

Let x = the length of CB in inches. Then $7\frac{1}{2}$ - x is the length of AC. We are told AC is 3/2 in. shorter than twice CB. Thus, AC = 2x - 3/2. Therefore

$$7\frac{1}{2} - x = 2x - \frac{3}{2}$$

$$\frac{15}{2} - x = 2x - \frac{3}{2}$$

Multiplying both members by 2,

$$15 - 2x = 4x - 3$$

$$-6x = -18$$
$$x = 3$$

Therefore CB = 3 and AC = $7\frac{1}{2}$ - 3 = $4\frac{1}{2}$. Hence, C is located $4\frac{1}{2}$ in. from
A and 3 in. from B.

● **PROBLEM** 38-7

Find the point Q that is 3/4 of the way from the point
P(- 4, - 1) to the point R(12, 11) along the segment PR.

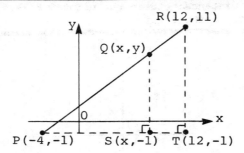

<u>Solution:</u> The figure illustrates the situation; we are
to find the numbers x and y, the coordinates of Q. To
find these two numbers, we might write the two equations

$\overline{PQ}$ = 3/4 $\overline{PR}$ and $\overline{QR}$ = ¼ $\overline{PR}$ in terms of x and y and solve.
Although this method will work, it is easier to use a
little geometry. If we introduce the auxiliary points
S(x, - 1) having the same x value as Q and same y value as

P, and T(12, - 1) having the same x value as R and the
same y value as P, (shown in the figure), we obtain the
similar triangles PSQ and PTR. We know these triangles
are similar because ∠ PSQ and ∠ PTR are right angles,
and ΔPSQ and ΔPTR have ∠ RPS in common. If 2 angles of
2 triangles are equal, their 3rd angles are equal. If
two triangles have the same angles they are similar and
their corresponding sides are proportional. Therefore,

$$\frac{\overline{PS}}{\overline{PT}} = \frac{\overline{PQ}}{\overline{PR}} \quad \text{and} \quad \frac{\overline{QS}}{\overline{RT}} = \frac{\overline{PQ}}{\overline{PR}} \tag{1}$$

From our figure, we see that $\overline{PS}$ = x - (- 4) = x + 4,
$\overline{PT}$ = 12 - (- 4) = 12 + 4 = 16, $\overline{QS}$ = y - (- 1) = y + 1,
and $\overline{RT}$ = 11 - (- 1) = 11 + 1 = 12, and it is a condition
of the problem that $\overline{PQ}/\overline{PR}$ = 3/4, since $\overline{PQ}$ = 3/4 $\overline{PR}$. Hence
replacing $\overline{PS}/\overline{PT}$ by $\frac{x + 4}{16}$ and $\overline{QS}/\overline{RT}$ by $\frac{y + 1}{12}$ in (1),
we obtain

$$\frac{x + 4}{16} = \frac{3}{4} \quad \text{and} \quad \frac{y + 1}{12} = \frac{3}{4} .$$

Cross multiplying,

$$4(x + 4) = 3(16) \quad \text{and} \quad 4(y + 1) = 12(3)$$

712

$$4x + 16 = 48 \qquad\qquad 4y + 4 = 36$$

$$4x = 32 \qquad\qquad\quad 4y = 32$$

$$x = 8 \qquad\qquad\qquad y = 8$$

Thus, the point Q that is 3/4 of the way from P(- 4, - 1) to R(12, 11) is (8,8).

● **PROBLEM 38-8**

What is the distance between the points P(-4,5) and Q(1,-7)? (Observe the accompanying figure).

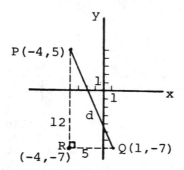

Solution: Observe the accompanying figure. P being 4 units to the left of the Y-axis and Q being 1 unit to the right, the horizontal distance between P and Q is 5 units. Similarly the vertical distance between P and Q is 12 units. The Pythagorean Theorem states that the sum of the squares of the legs of a right triangle equals the square of the hypotenuse. Thus, in right triangle PQR,

$$(\overline{PR})^2 + (\overline{RQ})^2 = (\overline{PQ})^2 .$$

$$(\overline{PQ})^2 = (12)^2 + (5)^2 = 144 + 25 = 169$$

Taking the square root of both sides, $\overline{PQ}$ = 13. Thus, the distance between (-4,5) and (1,-7), $\overline{PQ}$, is 13.

● **PROBLEM 38-9**

What is the distance between the points (2,3) and (7,11)?

Solution: Observe Figure A. The horizontal distance between (2,3) and (7,3) is 7 - 2 = 5. Thus $\overline{BC}$ = 5. Similarly, the vertical distance between (7,11) and (7,3) is 11 - 3 = 8, and $\overline{AC}$ = 8. The Pythagorean Theorem states that the sum of the squares of the legs of a right triangle equals the square of the hypotenuse. Thus, in right triangle

$$ABC, \quad (\overline{BC})^2 + (\overline{AC})^2 = (\overline{AB})^2 .$$

$$(\overline{AB})^2 = (5)^2 + (8)^2 = 25 + 64 = 89$$

713

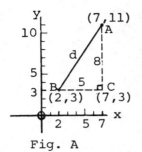

Fig. A

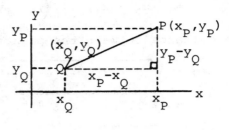

Fig. B

Taking the square root of both sides, $\overline{AB} = \sqrt{89}$. Thus, the distance between (2,3) and (7,11), $\overline{AB}$, is $\sqrt{89}$.

Generalizing, suppose we replace these points by P and Q with coordinates $\left(x_P, y_P\right)$ and $\left(x_Q, y_Q\right)$, respectively (see Figure B). Then what we have done in this problem would amount to using the following general formula for the distance between P and Q:

$$d(P,Q) = \sqrt{\left(x_P - x_Q\right)^2 + \left(y_P - y_Q\right)^2}$$

This formula continues to hold true in all possible positions of P and Q.

● **PROBLEM** 38-10

Find the distance from the origin to the point (x,y).

Solution: If P_1 is (0,0) and P_2 is (x,y), then to find the distance from the origin, which is point (0,0), and the point (x,y), apply the distance formula:

$$d = \sqrt{\left(x_2 - x_1\right)^2 + \left(y_2 - y_1\right)^2}$$

$$d = \sqrt{(x - 0)^2 + (y - 0)^2}$$

$$d = \sqrt{x^2 + y^2}.$$

● **PROBLEM** 38-11

Find the distance between the given pair of points, and find the slope of the line segment joining them.

(3, -5),(2, 4)

Solution: Let (3, -5) be P_1: $\left(x_1, y_1\right)$ and let (2, 4), be P_2: $\left(x_2, y_2\right)$. By the distance formula,

714

$$d = \sqrt{\left(x_2 - x_1\right)^2 + \left(y_2 - y_1\right)^2}, \text{ the}$$

the distance between the points (3, -5) and (2, 4) is:

$$d = \sqrt{(2 - 3)^2 + \left(4 - (-5)\right)^2}$$

$$= \sqrt{(-1)^2 + (4 + 5)^2}$$

$$= \sqrt{1 + (9)^2}$$

$$= \sqrt{1 + 81}$$

$$= \sqrt{82}$$

The slope of the line joining the points (3, -5) and (2, 4) is given by the formula:

$$\text{slope} = m = \frac{y_2 - y_1}{x_2 - x_1}$$

Again, let (3, -5) be P_1: $\left(x_1, y_1\right)$ and let (2, 4) be P_2: $\left(x_2, y_2\right)$. Then the slope is:

$$m = \frac{4 - (-5)}{2 - 3}$$

$$= \frac{4 + 5}{-1}$$

$$= \frac{9}{-1}$$

$$= -9$$

● **PROBLEM 38-12**

Given the three points P(4,3), Q(4,7), and R(7,3). Find the lengths of $\overline{PQ}$ and $\overline{PR}$.

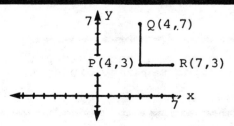

Solution: Points P and Q have the same x-coordinate and lie along a line parallel to the y-axis. Therefore the length of $\overline{PQ} = \left|y_P - y_Q\right|$. P and R have the same y-coordinate and lie along a line parallel to the x-axis. Hence the length of $\overline{PR} = |x_P - x_R|$.

$$\overline{PQ} = |3 - 7| = 4 \quad \text{and} \quad \overline{PR} = |4 - 7| = 3.$$

We could have used the distance formula

$$d = \sqrt{\left(x_1 - x_2\right)^2 + \left(y_1 - y_2\right)^2}$$

Then

$$\overline{PQ} = \sqrt{(4 - 4)^2 + (3 - 7)^2} = 4$$

$$\overline{PR} = \sqrt{(4 - 7)^2 + (3 - 3)^2} = 3$$

● **PROBLEM** 38-13

Use the distance formula to determine whether the points
A(0,-3), B(8,3), and C(11,7) are collinear.

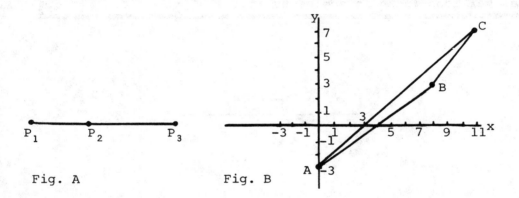

Fig. A

Fig. B

Solution: If three points P_1, P_2, P_3 are in such a position
that $\overline{P_1P_2} + \overline{P_2P_3} = \overline{P_1P_3}$ then the three points lie on a
straight line and we say that the points are collinear (Fig-
ure A).

Thus we find the distances between points A and B, A
and C, and B and C to determine whether the sum of any two
of these is equivalent to the third, making A, B, and C collinear. Using the formula for the distance between two
points $\left(x_1, y_1\right)$ and $\left(x_2, y_2\right)$,

$$d = \sqrt{\left(x_2 - x_1\right)^2 + \left(y_2 - y_1\right)^2}$$

The distance between (0,-3) and (8,3) is

$$d_1 = \sqrt{(8-0)^2 + [3-(-3)]^2} = \sqrt{8^2 + 6^2} = \sqrt{64 + 36} = \sqrt{100} = 10$$

The distance between (0,-3) and (11,7) is

$$d_2 = \sqrt{(11-0)^2 + [7-(-3)]^2} = \sqrt{(11)^2 + (10)^2} = \sqrt{121 + 100}$$

716

$$= \sqrt{221} \cong 14.74$$

The distance between (8,3) and (11,7) is

$$d_3 = \sqrt{(11-8)^2 + (7-3)^2} = \sqrt{(3)^2 + (4)^2} = \sqrt{9 + 16} = \sqrt{25} = 5$$

Since 5 + 10 = 15 and 15 > 14.74, the three points form a triangle as opposed to a straight line. Thus the points are not collinear.

Plotting the points on a graph, and attaching them we also observe that the points form a triangle, not a line. (Figure B).

● **PROBLEM 38-14**

Show that the triangle with (-3, 2), (1, 1), and (-4, -2) as vertices is an isosceles triangle.

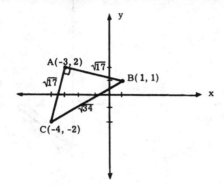

Solution: If we can show that two sides of the triangle are equal in length, then the triangle is isosceles. This can be done by applying the formula for the distance between two points, (x_1, y_1) and (x_2, y_2):

$$d = \sqrt{(x_1 - x_2)^2 + (y_1 - y_2)^2}$$

Let the given points be designated as A, B, and C respectively. Then

$$|AB| = \sqrt{(1 + 3)^2 + (1 - 2)^2} = \sqrt{17},$$

$$|AC| = \sqrt{(-4 + 3)^2 + (-2 - 2)^2} = \sqrt{17}.$$

Hence $|AB| = |AC|$, and the triangle is isosceles. Furthermore,

$$|BC| = \sqrt{(-4 - 1)^2 + (-2 - 1)^2} = \sqrt{34}.$$

Since $|BC|^2 = |AB|^2 + |AC|^2$ $\left(\sqrt{34}^2 = \sqrt{17}^2 + \sqrt{17}^2 \text{ or } 34 = 17 + 17\right)$, the Theorem of Pythagoras holds, and ABC is a right triangle, with the right angle at A. (See figure.)

Show that the points A (-2, 4), B(-3, -8), and C(2,2) are vertices of a right triangle.

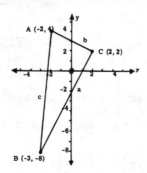

Solution: If triangle ABC is a right triangle, then $a^2 + b^2 = c^2$; that is, the sum of the square of the legs equals the square of the hypotenuse by the Pythagorean Theorem.

Thus we compute the distance from B to C which is side a,

the distance from C to A which is side b,

and the distance from A to B which is side c.

The formula for the distance between two points (x_1, y_1) and (x_2, y_2) is

$$\sqrt{(x_2 - x_1)^2 + (y_2 - y_1)^2}$$

Thus the distance from B to C, from (-3, -8) to (2,2), is

$$\sqrt{[2 - (-3)]^2 + [2 - (-8)]^2} = \sqrt{(2+3)^2 + (2+8)^2}$$

$$= \sqrt{5^2 + 10^2}$$

$$= \sqrt{25 + 100}$$

$$= \sqrt{125}$$

Hence side $a = \sqrt{125}$

The distance from C to A, from (2,2) to (-2,4), is

$$\sqrt{(-2 - 2)^2 + (4 - 2)^2} \qquad = \sqrt{(-4)^2 + 2^2}$$

$$= \sqrt{16 + 4}$$

$$= \sqrt{20}$$

Hence side $b = \sqrt{20}$

The distance from A to B, from (-2,4) to (-3, -8), is

$$\sqrt{[-3 - (-2)]^2 + (-8 - 4)^2} = \sqrt{(-3 + 2)^2 + (-12)^2}$$

$$= \sqrt{(-1)^2 + (-12)^2}$$

$$= \sqrt{1 + 144}$$

$$= \sqrt{145}$$

Hence side $c = \sqrt{145}$

Now, if triangle ABC is a right triangle, $a^2 + b^2 = c^2$.
Replacing,
$$a \text{ by } \sqrt{125}, \quad b \text{ by } \sqrt{20}, \quad \text{and } c \text{ by } \sqrt{145}$$
we obtain,

$$\left(\sqrt{125}\right)^2 + \left(\sqrt{20}\right)^2 = \left(\sqrt{145}\right)^2$$

Since, $\left(\sqrt{a}\right)^2 = \sqrt{a}\ \sqrt{a} = \sqrt{a} \cdot \sqrt{a} = \sqrt{a^2} = a$

$$\left(\sqrt{125}\right)^2 = 125$$

$$\left(\sqrt{20}\right)^2 = 20$$

and $\left(\sqrt{145}\right)^2 = 145$

Thus $a^2 + b^2 = c^2$ is equivalent to,
$$125 + 20 = 145$$
$$145 = 145$$

Therefore, triangle ABC is indeed a right triangle.

● **PROBLEM** 38-16

Find the equation for the set of points the sum of whose
distances from (4,0) and from (-4,0) is 10.

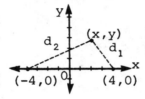

<u>Solution:</u> We find the desired equation by choosing an arbi-
trary point (x,y) and computing the sum of its distances from
(4,0) and (-4,0) (see accompanying figure). Applying the
distance formula for the distance between two points (a_1, b_1)

719

and (a_2,b_2), $d = \sqrt{(a_1-a_2)^2+(b_1-b_2)^2}$, we find that the distance from (x,y) to $(4,0)$ is

$$d_1 = \sqrt{(x-4)^2 + y^2}$$

and the distance from (x,y) to $(-4,0)$ is

$$d_2 = \sqrt{(x+4)^2 + y^2}.$$

We are given that the sum of the distances $d_1 + d_2 = 10$. Hence, the required equation for the set of points is

$$\sqrt{(x-4)^2 + y^2} + \sqrt{(x+4)^2 + y^2} = 10$$

$$\sqrt{(x-4)^2 + y^2} = 10 - \sqrt{(x+4)^2 + y^2}.$$

Squaring both sides,

$$\left(\sqrt{(x-4)^2 + y^2}\right)^2 = \left(10 - \sqrt{(x+4)^2 + y^2}\right)^2.$$

Since $(\sqrt{a})^2 = \sqrt{a}\,\sqrt{a} = \sqrt{a \cdot a} = \sqrt{a^2} = a$,

$$\left(\sqrt{(x-4)^2 + y^2}\right)^2 = (x-4)^2 + y^2.$$

Thus $\quad (x-4)^2 + y^2 = 100 - 20\sqrt{(x+4)^2 + y^2} + (x+4)^2 + y^2$

$$x^2 - 8x + 16 + y^2 = 100 - 20\sqrt{(x+4)^2 + y^2} + x^2 + 8x + 16 + y^2$$

Adding $-\left(100 + x^2 + 8x + 16 + y^2\right)$ to both members,

$$- 16x - 100 = - 20\sqrt{(x+4)^2 + y^2}$$

Dividing both sides by -4, $4x + 25 = 5\sqrt{(x+4)^2 + y^2}$

Squaring again,

$$(4x + 25)(4x + 25) = \left(5\sqrt{(x+4)^2 + y^2}\right)^2$$

$$16x^2 + 200x + 625 = 25\left[\sqrt{(x+4)^2 + y^2}\right]^2$$

$$16x^2 + 200x + 625 = 25\left[(x+4)^2 + y^2\right]$$

$$16x^2 + 200x + 625 = 25\left(x^2 + 8x + 16 + y^2\right)$$

$$16x^2 + 200x + 625 = 25x^2 + 200x + 400 + 25y^2$$

Adding $-(16x^2 + 200x + 400)$ to both members,

$$225 = 9x^2 + 25y^2.$$

Dividing both members by 225, we can write the last equation in the form

$$\frac{9x^2}{225} + \frac{25y^2}{225} = \frac{225}{225}$$

or
$$\frac{x^2}{25} + \frac{y^2}{9} = 1 \quad ,$$

which is the standard form of the equation of an ellipse. This is the desired equation.

● **PROBLEM** 38-17

Find the equation for the set of points the difference of whose distances from (5,0) and (-5,0) is 6 units.

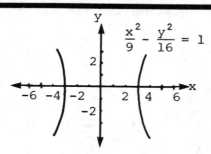

$$\frac{x^2}{9} - \frac{y^2}{16} = 1$$

<u>Solution:</u> We find the desired equation by choosing an arbitrary point (x,y) and computing the difference of its distance from (5,0) and (-5,0). Applying the distance formula for the distance between two points (a_1,b_1) and (a_2,b_2), $d = \sqrt{(a_1-a_2)^2 + (b_1-b_2)^2}$. From (x,y) to (5,0) is d_1, from (x,y) to (-5,0) is d_2.

$$d_1 = \sqrt{(x-5)^2 + y^2}; \quad d_2 = \sqrt{(x+5)^2 + y^2} \; ;$$

We are told that the difference of these distances, d_2-d_1, is 6. Hence, $\sqrt{(x+5)^2 + y^2} - \sqrt{(x-5)^2 + y^2} = 6.$

$$\sqrt{(x+5)^2 + y^2} = 6 + \sqrt{(x-5)^2 + y^2}$$

Squaring both sides:

$$\left(\sqrt{(x+5)^2 + y^2}\right)^2 = \left(6 + \sqrt{(x-5)^2 + y^2}\right)^2$$

$$\left(\sqrt{(x+5)^2 + y^2}\right)^2 = 36 + 12\sqrt{(x-5)^2 + y^2} + \left(\sqrt{(x-5)^2 + y^2}\right)^2$$

Since $\left(\sqrt{a}\right)^2 = \sqrt{a}\sqrt{a} = \sqrt{a \cdot a} = \sqrt{a^2} = a,$

$$\left(\sqrt{(x+5)^2 + y^2}\right)^2 = (x+5)^2 + y^2 \quad \text{and} \quad \left(\sqrt{(x-5)^2 + y^2}\right)^2 = (x-5)^2 + y^2 \; .$$

Thus we obtain,

$$(x+5)^2 + y^2 = 36 + 12\sqrt{(x-5)^2 + y^2} + (x-5)^2 + y^2$$

$$x^2 + 10x + 25 + y^2 = 36 + 12\sqrt{(x-5)^2 + y^2} + x^2 - 10x + 25 + y^2$$

Adding $-\left(x^2 + 25 + y^2\right)$ to both sides,

$$x^2 + 10x + 25 + y^2 - \left(x^2 + 25 + y^2\right) = 36 + 12\sqrt{(x-5)^2 + y^2} + x^2$$

721

$$- 10x + 25 + y^2 - \left(x^2 + 25 + y^2\right)$$

$$10x = 36 + 12\sqrt{(x-5)^2 + y^2} - 10x$$

Adding -36 + 10x to both sides,

$$10x - 36 + 10x = 36 + 12\sqrt{(x-5)^2 + y^2} - 10x - 36 + 10x$$

$$20x - 36 = 12\sqrt{(x-5)^2 + y^2}$$

Dividing both sides by 4,

$$5x - 9 = 3\sqrt{(x-5)^2 + y^2}$$

Squaring both sides,

$$(5x-9)^2 = \left(3\sqrt{(x-5)^2 + y^2}\right)^2$$

$$(5x-9)(5x-9) = 3^2\left(\sqrt{(x-5)^2 + y^2}\right)^2$$

$$25x^2 - 90x + 81 = 9[(x-5)^2 + y^2]$$

$$25x^2 - 90x + 81 = 9\left(x^2 - 10x + 25 + y^2\right)$$

$$25x^2 - 90x + 81 = 9x^2 - 90x + 225 + 9y^2$$

Adding $-\left(9x^2 - 90x + 9y^2\right)$ to both sides,

$$25x^2 - 90x + 81 - \left(9x^2 - 90x + 9y^2\right) = 9x^2 - 90x + 225 + 9y^2$$
$$- \left(9x^2 - 90x + 9y^2\right)$$

$$25x^2 - 9x^2 - 9y^2 + 81 = 225$$
$$16x^2 - 9y^2 + 81 = 225$$

Adding -81 to both sides,

$$16x^2 - 9y^2 = 144.$$

Dividing both sides by 144,

$$\frac{16x^2}{144} - \frac{9y^2}{144} = \frac{144}{144}$$

or

$$\frac{x^2}{9} - \frac{y^2}{16} = 1 ,$$

which is the standard form for the equation of a hyperbola.

From the form of the equation we can determine its graph. When the center is at the origin and its vertices are at (a,0) and (-a,0), the equation of the hyperbola is

$$\frac{x^2}{a^2} - \frac{y^2}{b^2} = 1 .$$

In this case, the vertices are (3,0) and (-3,0) since $a^2 = 9$ and $a = \pm 3$ (see figure).

However, when the vertices lie on the y-axis, the equation of the hyperbola is

$$\frac{y^2}{a^2} - \frac{x^2}{b^2} = 1.$$

The vertices would then be (0,a) and (0,-a).

722

Plot the points (1, - 2) and (5, 1) in the xy- plane.
What ordered pair corresponds to point C in the Figure?
If points A, B, and C of the Figure are three vertices
of a parallelogram, what are the coordinates of the
fourth vertex in the third quadrant?

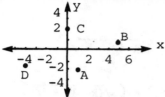

Solution: In the Figure, the point (1, -2) is 1 unit to
the right of the origin and 2 units below the x-axis.
Therefore it is located at point A. The point (5, 1) is
located at point B, 5 units to the right of the origin
and 1 unit above the x-axis.

The abscissa of point C is 0 and its ordinate is
2. Therefore, the ordered pair is (0, 2).

There are three possible locations for a fourth
vertex. In quadrant III the vertex is the intersection
of lines parallel to AB and BC, respectively. Since
point C is 5 units to the left of point B and 1 unit
above it, the required vertex D will be 5 units to the
left of point A and 1 unit above it. Its coordinates
are (- 4, - 1).

Simplify: $3a - 2\{3a - 2[1 - 4(a - 1)] + 5\}$.

Solution: When working with several sets of brackets and, or parentheses,
we work from the inside out. That is, we use the law of distribution
throughout the expression, starting from the innermost parentheses, and
working our way out. Hence in this case we have: $2\{3a - 2[1 - 4(a - 1)] + 5\}$
and we note that (a - 1) is the innermost parenthesis, so our first
step is to distribute the (-4). Thus, we obtain:

$$3a - 2[3a - 2(1 - 4a + 4) + 5].$$

We now find that (1 - 4a + 4) is in our innermost parentheses.
Combining terms we obtain:

$$(1 - 4a + 4) = (5 - 4a) ;$$

hence, $3a - 2[3a - 2(1 - 4a + 4) + 5] = 3a - 2[3a - 2(5 - 4a) + 5]$
and since (5 - 4a) is in the innermost parentheses we distribute the
(-2), obtaining:

723

$$3a - 2(3a - 10 + 8a + 5).$$

We are now left with the terms in our last set of parentheses, $(3a - 10 + 8a + 5).$ Combining like terms we obtain:

$$(3a - 10 + 8a + 5) = (11a - 5)$$

hence, $3a - 2(3a - 10 + 8a + 5) = 3a - 2(11a - 5)$.

Distributing the (-2), $= 3a - 22a + 10$

combining terms, $= -19a + 10$.

Hence $3a - 2\{3a - 2[1 - 4(a - 1)] + 5\} = -19a + 10$.

● **PROBLEM 38-20**

Find the slope of $f(x) = 3x + 4.$

Solution: Two points on the line determined by $f(x) = 3x + 4$ are $A(0,4)$ and $B(1,7).$

$$\frac{\text{difference of ordinates}}{\text{difference of abscissas}} = \frac{7 - 4}{1 - 0} = 3$$

Note that the ordinates are the y-coordinates and the abscissas are the x-coordinates. The slope determined by points A and B is 3. Hence, the slope of $f(x) = 3x + 4$ is 3. In general, the slope of a linear function of the form $f(x) = mx + b$ is m.

● **PROBLEM 38-21**

Show that the slope of the segment joining $(1,2)$ and $(2,6)$ is equal to the slope of the segment joining $(5,15)$ and $(10,35).$

Solution: The slope of the line segment, m, joining the points (x_1, y_1) and (x_2, y_2) is given by the formula

$$m = \frac{(y_2 - y_1)}{(x_2 - x_1)}$$

Therefore, the slope of the segment joining $(1,2)$ and $(2,6)$ is

$$\frac{6 - 2}{2 - 1} = \frac{4}{1} = 4.$$

The slope of the segment joining $(5,15)$ and $(10,35)$ is

$$\frac{35 - 15}{10 - 5} = \frac{20}{5} = 4.$$

Thus, the slopes of the two segments are equal.

724

Show that the slope of the segment joining (1,2) and (3,8) is equal to the slope of the segment joining (4,11) and (8,23)

Solution: The slope of the segment joining (1,2) and (3,8) is

$$\frac{8-2}{3-1} = \frac{6}{2} = 3 \ .$$

The slope of the segment joining (4,11) and (8,23) is

$$\frac{23-11}{8-4} = \frac{12}{4} = 3 \ .$$

Therefore, the slopes of the two segments are equal.

● **PROBLEM** 38-23

Given ℓ_1 and ℓ_2 as shown in Fig. 1,

find the angle θ.

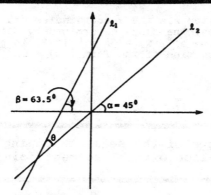

Solution: The slopes of ℓ_1 and ℓ_2 are $k_1 = \tan 63.5^0 \doteq 2$ and $k_2 = \tan 45^0 = 1$, respectively. The angle θ between any two lines whose slopes k_1 and k_2 are known is given by

$$\tan \theta = \frac{k_1 - k_2}{1 + k_1 k_2}$$

Therefore,

$$\tan \theta = \frac{k_1 - k_2}{1 + k_1 k_2} = \frac{2-1}{1 + (2 \times 1)} = \frac{1}{3}$$

$$\theta = \tan^{-1} \frac{1}{3} \doteq 18.4^0$$

● **PROBLEM** 38-24

Find the slope of a line ℓ_1 which makes an angle of 30^0 with a line ℓ_2, of slope 1.

Solution: Use the formula

$$\tan \theta = \frac{k_1 - k_2}{1 + k_1 k_2}$$

where θ is the angle between ℓ_1 and ℓ_2, and k_1 and k_2 are the slopes of ℓ_1 and ℓ_2, respectively.

Let the slope of ℓ_1 be k_1.

$$\tan \theta = \frac{k_1 - k_2}{1 + k_1 k_2}$$

$$\tan \theta (1 + k_1 k_2) = k_1 - k_2 \ ,$$

$$k_1 = \frac{\tan \theta + k_2}{1 - \tan \theta \, k_2}$$

$$= \frac{\frac{\sqrt{3}}{3} + 1}{1 - \left(\frac{\sqrt{3}}{3} \times 1\right)} = \frac{3 + \sqrt{3}}{3 - \sqrt{3}}$$

● **PROBLEM 38-25**

Find the interior angles of $\triangle ABC$, which has vertices $A(0,0)$, $B(4,-2)$, and $C(9,4)$. Find the area of $\triangle ABC$.

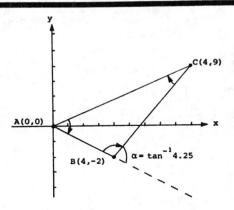

Solution: $\triangle ABC$ is shown in Fig. 1.

First, find the slope of each side of $\triangle ABC$ using the formula

$k = \frac{y_2 - y_1}{x_2 - x_1}$, where (x_1, y_1) and (x_2, y_2) are any two points on a side of $\triangle ABC$. Therefore,

$$k_{AB} = \frac{y_B - y_A}{x_B - x_A} = \frac{-2 - 0}{4 - 0} = -\frac{1}{2}$$

$$k_{AC} = \frac{4 - 0}{9 - 0} = \frac{4}{9} \ , \quad k_{BC} = \frac{4 - (-2)}{9 - 4} = \frac{6}{5}$$

726

Next, draw curved arrows to indicate the positive direction about each vertex as shown in Fig. 1. The head of the arrow is on the side whose slope is to be taken as k_1 in the formula $\tan \theta = \dfrac{k_1 - k_2}{1 + k_1 k_2}$, and the tail of the arrow is on the side whose slope is k_2. θ is the angle between these two sides.

The interior angles are then found as the following:

$$\tan A = \frac{k_1 - k_2}{1 + k_1 k_2} = \frac{k_{AB} - k_{AC}}{1 + k_{AB} k_{AC}} = \frac{-\frac{1}{2} - \frac{4}{9}}{1 + \left(-\frac{1}{2}\right)\frac{4}{9}}$$

$$= -\frac{17}{14} , \qquad \tan^{-1} -\frac{17}{14} = -50.5^0$$

$$A = 50.5^0$$

$$\tan B = \frac{\frac{6}{5} - \left(-\frac{1}{2}\right)}{1 + \left[\frac{6}{5} \times \left(-\frac{1}{2}\right)\right]} = 4.25$$

$$B = 180^0 - \tan^{-1} 4.25 = 180^0 - 76.76^0$$

$$= 103.24^0$$

Note that $\tan^{-1} 4.25 = 76.76^0$. Howvever, it's not B, as one can see from Fig. 1.

$$C = 180^0 - (A+B) = 180^0 - 50.5^0 - 103.24^0 = 26.26^0$$

Finally, the area of any triangle whose vertices are $P_1(x_1,y_1)$, $P_2(x_2,y_2)$, $P_3(x_3,y_3)$ is

$$A = \frac{1}{2} (x_1 y_2 + x_2 y_3 + x_3 y_1 - x_1 y_3 - x_2 y_1 - x_3 y_2)$$

Therefore, the area of $\triangle ABC$ is

$$A = \frac{1}{2} (0 \times (-2) + 4 \times 4 + 9 \times 0 - 0 \times 4 - 4 \times 0 - 9 \times (-2))$$

$$= 17$$

CHAPTER 39

STRAIGHT LINES AND FAMILIES OF STRAIGHT LINES

Basic Attacks and Strategies for Solving Problems in this Chapter. See pages 728 to 744 for step-by-step solutions to problems.

The general equation of a straight line involving two variables has the form

$$Ax + By + C = 0.$$

There are no powers of any variables. The general form can be transformed into a number of other forms, some of which are more useful for certain applications. Almost all forms of an equation of a line are derived from the formula for a slope of a line given in Chapter 38, i.e., given two points on a line, P_1, with coordinates (x_1, y_1), and P_2, with coordinates (x_2, y_2), the slope, m, of the line joining P_1 and P_2 is

$$\frac{y_2 - y_1}{x_2 - x_1}.$$

We also remember that a single straight line can only have one slope, no matter which points are chosen to compute it.

The *two point form* of the equation of a line is:

$$\frac{y_2 - y_1}{x_2 - x_1} = \frac{y - y_1}{x - x_1}.$$

The *point slope form* of the equation of a line is:

$$m = \frac{y - y_1}{x - x_1}.$$

The *slope intercept form* of the equation of a line is:

$$y = mx + b,$$

where b is the y-intercept.

The *two intercepts form* of the equation of a line is:

$$\frac{x}{a} + \frac{y}{b} = 1$$

where a is the x-intercept (and, as above, b is the y-intercept).

The *normal form* of the equation of a line is:

$$x \cos \theta + y \sin \theta - \rho = 0,$$

where θ is the angle the line makes with the x-axis (*cf.* Chapter 38) and ρ is the (positive) distance from the line to the origin. The general form can be transformed to the normal form by dividing by $\pm \sqrt{A^2 + B^2}$.

Two lines are parallel if they have the same slope. Unless two parallel lines are the same, they have no common point of intersection.

Two lines are perpendicular (intersect at right angles) if the slopes are negative reciprocals of each other, i.e.,

$$m_1 = -\frac{1}{m_2}.$$

This can be shown by using the formula for the tangent of the difference of two angles (*cf.* Chapter 38).

The *distance, d*, between a point, P_0, with coordinates (x_0, y_0) and a line

$$Ax + By + C = 0$$

is given by

$$\frac{|Ax_0 + By_0 + C|}{\sqrt{A^2 + B^2}}.$$

This gives a positive number as the distance (sometimes called the *undirected distance*). Removing the absolute value signs from the numerator will give the (signed) *directed distance*. Two points on the same side of the same line will have the same sign associated with their directed distances.

The coordinate axes (x-axis and y-axis) can be described as lines. The x-axis is the line $y = 0$ (having slope zero) and the y-axis is the line $x = 0$ (having an infinite slope).

Step-by-Step Solutions to Problems in this Chapter, "Straight Lines and Families of Straight Lines"

● PROBLEM 39-1

Write the equation of a line passing through points $P_1(5,10)$ and $P_2(0,4)$ in

    (1)   Point - slope form

    (2)   Slope - intercept form

    (3)   Two-point form

    (4)   Intercept form

    (5)   Form of general equation for straight lines.

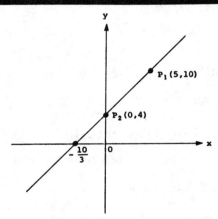

<u>Solution</u>: Since we are given two points on the line, the equation of the line in two-point form is the most easily obtained.

The two-point form is

$$y - y_1 = \frac{y_2 - y_1}{x_2 - x_1} \ (x - x_1). \quad \text{Hence}$$

728

$$y - 10 = \frac{4 - 10}{0 - 5} \ (x\text{-}5) = \frac{-6}{-5} \ (x\text{-}5) = \frac{6}{5} \ (x\text{-}5)$$

$(y - 10) = \frac{6}{5} \ (x\text{-}5)$ is the equation of the straight line in the point-slope form, where the slope of the line is $\frac{6}{5}$

To obtain the slope-intercept form of the line, $y = kx + b$,

$$y - 10 = \frac{6}{5} \ x - 6$$

$$y = \frac{6}{5} \ x - 6 + 10 = \frac{6}{5} \ x + 4$$

i.e., $\qquad y = \frac{6}{5} \ x + 4$

The intercept form of the equation of the line is

$\frac{x}{a} + \frac{y}{b} = 1$, where a and b are, respectively, the the x-intercept and the y-intercept of the line.

To find a, let y = 0.

$$0 = \frac{6}{5} \ x + 4 \ , \quad x = -\frac{10}{3} = a.$$

To find b, let x = 0.

$$y = 0 + 4 = 4 = b$$

Therefore, the equation of the line in intercept form is

$$-\frac{x}{\frac{10}{3}} + \frac{y}{4} = 1$$

The general equation of the straight line is

$$Ax + By + C = 0$$

From $\qquad y = \frac{6}{5} \ x + 4$, we obtain

$$5y = 6x + 20 \ , \quad 6x - 5y + 20 = 0.$$

The graph of the line is

● **PROBLEM 39-2**

Determine the constant A so that the lines 3x - 4y = 12 and Ax + 6y = -9 are parallel.

<u>Solution:</u> If two non-vertical lines are parallel, their slopes are

729

equal. Thus the lines $Ax + By + C = 0$ and $Ax + By + D = 0$ are parallel (since both have slope $= - A/B$). We are given two lines:

$$3x - 4y = 12 \qquad (1)$$

$$Ax + 6y = -9 \qquad (2)$$

We must make the coefficients of y the same for both equations in order to equate the coefficients of x. Multiply (1) by $-3/2$ to obtain

$$\frac{-3}{2}(3x - 4y) = -\frac{3}{2}(12)$$

$$-\frac{9}{2}x + 6y = -18 \qquad (3)$$

$$Ax + 6y = -9 \qquad (2)$$

Transpose the constant terms of (3) and (2) to the other side.

Adding 18 to both sides , $\qquad -\frac{9}{2}x + 6y + 18 = 0 \qquad (4)$

Adding 9 to both sides, $\qquad Ax + 6y + 9 = 0 \qquad (5)$

(4) and (5) will now be parallel if the co-efficients of the x-terms are the same. Thus the constant A is $-9/2$. Then equation (5) becomes $-9/2x + 6y + 9 = 0$. We can also express (5) in its given form, $Ax + 6y = -9$ or $- 9/2 x + 6y = -9$.

We also can write it in a form that has the same coefficient of x as (1), which clearly shows that they have equal slopes.

$$3x - 4y = 12 \qquad (1)$$

$$-\frac{9}{2}x + 6y = -9$$

Multiply the second equation by $- 2/3$ to obtain a coefficient of x equal to 3.

$$-\frac{2}{3}\left(-\frac{9}{2}x + 6y\right) = -\frac{2}{3}(-9)$$

$$3x - 4y = 6$$

Now equations (1), $3x - 4y = 12$, and the equation $3x - 4y = 6$ are parallel since the coefficients of x and y are identical.

● **PROBLEM 39-3**

Find the slope, the y-intercept, and the x-intercept of the equation $2x - 3y - 18 = 0$.

Solution: The equation $2x - 3y - 18 = 0$ can be written in the form of the general linear equation, $ax + by = c$.

$$2x - 3y - 18 = 0$$

$$2x - 3y = 18$$

To find the slope and y-intercept we derive them from the formula of the general linear equation $ax + by = c$. Dividing by b and solving for y we obtain:

$$\frac{a}{b}x + y = \frac{c}{b}$$

$$y = \frac{c}{b} - \frac{a}{b}x$$

where $-\frac{a}{b}$ = slope and $\frac{c}{b}$ = y-intercept.

To find the x-intercept, solve for x and let y = 0:

$$x = \frac{c}{b} - \frac{b}{a} y$$

$$x = \frac{c}{a}$$

In this form we have a = 2, b = -3, and c = 18. Thus,

$$\text{slope} = -\frac{a}{b} = -\frac{2}{-3} = \frac{2}{3}$$

$$\text{y-intercept} = \frac{c}{b} = \frac{18}{-3} = -6$$

$$\text{x-intercept} = \frac{c}{a} = \frac{18}{2} = 9$$

● PROBLEM 39-4

The equation $F = \frac{9}{5} C + 32$ relates the Fahrenheit and centigrade temperature scales. What do the numbers $\frac{9}{5}$ and 32 represent?

Solution: An equation in the form y = mx + b is a linear equation with slope m and y-intercept b. Thus, with F = 9/5 C + 32, 32 is the y-intercept and 9/5 is the slope. That is, the number 32 tells us that when the centigrade thermometer reads 0, the Fahnrenheit thermometer reads 32. The number 9/5 is the slope of the line we would obtain if we graphed our equation in an axis system in which centigrade temperatures are measured on the horizontal axis and Fahrenheit temperatures are measured on the vertical axis; that is, the number 9/5 is the number of units of Fahrenheit temperature rise per unit of centigrade temperature rise. If a body's temperature increases 1° C, then it increases $\frac{9}{5}$°F. If a body's temperature increases - 10° (decreases 10°) C, then it increases $\frac{9}{5}(-10)$° = -18°F.

● PROBLEM 39-5

The slope and one point of a line are given. Is the Y-intercept positive or negative?

(a) $m = \frac{22}{7}$, $(1, \pi)$ (b) $m = \sqrt{2}, (1, 1.414)$

Solution: a) The equation of a line is: y = mx + b, where m is the slope and b is the y-intercept. Given the slope m and one point of the line, the y-intercept b can be found. Thus it can be determined whether the y-intercept b is positive or negative. For the line with

731

slope $m = 22/7$ and which contains the point $(1,\pi)$:

$$y = mx + b$$
$$\pi = \frac{22}{7}(1) + b$$
$$\pi = \frac{22}{7} + b \qquad\qquad (1)$$

Since π is approximately $\frac{22}{7}$, equation (1) becomes:

$$\frac{22}{7} = \frac{22}{7} + b$$

Subtract $22/7$ from both sides to obtain:

$$\frac{22}{7} - \frac{22}{7} = \frac{\cancel{22}}{\cancel{7}} + b - \frac{\cancel{22}}{\cancel{7}}$$
$$0 = b$$

Hence, the y-intercept b is neither positive nor negative, since the y-intercept $b = 0$.

b) For the line with slope $m = \sqrt{2}$ and which contains the point $(1,1.414)$:

$$y = mx + b$$
$$1.414 = \sqrt{2}(1) + b$$
$$1.414 = \sqrt{2} + b \qquad\qquad (2)$$

Since $\sqrt{2}$ is approximately 1.414, equation (2) becomes:

$$1.414 = 1.414 + b$$

Subtract 1.414 from both sides to obtain:

$$1.414 - 1.414 = 1.\cancel{414} + b - 1.\cancel{414}$$
$$0 = b$$

Again, the y-intercept b is neither positive nor negative, since $b = 0$.

● **PROBLEM 39-6**

Find the equation for the line passing through (3,5) and (-1,2).

Solution A: We use the two-point form with $(x_1,y_1) = (3,5)$ and $(x_2,y_2) = (-1,2)$. Then

$$\frac{y - y_1}{x - x_1} = m = \frac{y_2 - y_1}{x_2 - x_1}.$$

$$\frac{y_2 - y_1}{x_2 - x_1} = \frac{2 - 5}{-1 - 3} \quad \text{thus} \quad \frac{y - 5}{x - 3} = \frac{-3}{-4}.$$

Cross multiply, $-4(y - 5) = -3(x - 3)$.

Distributing, $-4y + 20 = -3x + 9$

732

$$3x - 4y = -11.$$

Solution B: Does the same equation result if we let $(x_1, y_1) = (-1,2)$ and $(x_2, y_2) = (3,5)$?

$$\frac{y_2 - y_1}{x_2 - x_1} = \frac{5 - 2}{3 - (-1)} \quad \text{thus} \quad \frac{y - 2}{x + 1} = \frac{3}{4}.$$

Cross multiply, $4(y - 2) = 3(x + 1)$

$$3x - 4y = -11.$$

Hence, either replacement results in the same equation.

● **PROBLEM 39-7**

Find the equation of the line which passes through the points $(-3,1)$ and $(7,11)$.

Solution: The general equation for a line is $y = mx + b$, where m is the slope of the line and b is the y-intercept. Replacing $(-3,1)$ and $(7,11)$ for x and y in this equation, we obtain the equations $1 = m(-3) + b$, and $11 = m(7) + b$; or:

$$1 = -3m + b \tag{1}$$

and

$$11 = 7m + b \tag{2}$$

Thus, we solve equations (1) and (2) for m and b. Subtracting equation (2) from (1):

$$
\begin{array}{r}
1 = -3m + b \\
- \underline{(11 = 7m + b\ )} \\
-10 = -10m \\
m = 1
\end{array}
$$

Replacing m by 1 in equation (1) we solve for b:

$$1 = (-3)(1) + b$$
$$1 = -3 + b$$
$$b = 4$$

Hence the equation of the line passing through $(-3,1)$ and $(7,11)$ is $y = (1)x + 4$ or $y = x + 4$.

● **PROBLEM 39-8**

(a) Find the equation of the line passing through $(2,5)$ with slope 3.

(b) Suppose a line passes through the y-axis at $(0,b)$. How can we write the equation if the point-slope form is used?

Solution: (a) In the point-slope form, let $x_1 = 2$, $y_1 = 5$, and m = 3. The point-slope form of a line is:

$$y - y_1 = m(x - x_1)$$
$$y - 5 = 3(x - 2)$$
$$y - 5 = 3x - 6 \qquad \text{Distributive property}$$
$$y = 3x - 1 \qquad \text{Transposition}$$

(b)
$$y - b = m(x - 0)$$
$$y = mx + b.$$

● **PROBLEM** 39-9

Find the equation of the line passing through the points (2,5) and (6,2). Check the results graphically.

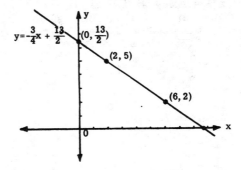

Solution: The general equation for a line is y = mx + b, where m is the slope of the line and b the y-intercept. Replacing (2,5) and (6,2) for x and y in this equation we obtain

$$5 = m(2) + b \qquad (1)$$
and
$$2 = m(6) + b \qquad (2)$$

Thus, we solve 5 = 2m + b and 2 = 6m + b for m and b:

subtracting equation (2) from (1):

$$5 = 2m + b$$
$$-\ (2 = 6m + b)$$
$$\overline{3 = -4m}$$
$$m = \frac{-3}{4}$$

Replacing m by $\frac{-3}{4}$ in equation (1) we solve for b:

$$5 = \left(\frac{-3}{4}\right)(2) + b$$
$$5 = \frac{-6}{4} + b$$
$$b = 5 + \frac{6}{4} = \frac{20}{4} + \frac{6}{4} = \frac{26}{4} = \frac{13}{2}$$

Hence, the equation of the line passing through (2,5) and (6,2) is

$$y = \frac{-3}{4} x + \frac{13}{2}.$$

This result is shown graphically in the accompanying figure.

734

If $f(x) = -2x - 5$, find the (a) slope, (b) x-intercept, and (c) y-intercept. (d) Graph the function.

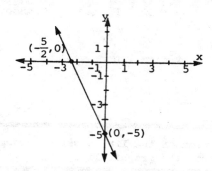

Solution: $f(x) = mx + b$ is called a linear function where m and b are constants. m is the slope of the line and b is the y-intercept of the line. In this case, $f(x) = -2x - 5$, m = -2 and b = -5. Therefore,

(a) slope: m = -2

The x-intercept is located on the x-axis where $f(x) = 0$. Then we solve for x.

$$f(x) = mx + b = 0$$
$$mx = -b$$
$$x = \frac{-b}{m} = \text{x-intercept}$$

Hence,

(b) x-intercept: $\frac{-b}{m} = \frac{-(-5)}{-2} = \frac{5}{-2} = \frac{-5}{2}$

(c) y-intercept: b = -5

(d) We can graph the function by locating the points where the graph crosses the y-axis, (0, -5), and the x-axis, $\left(-\frac{5}{2}, 0\right)$. Recall again that (0,b) is the y-intercept and that $\left(\frac{-b}{m}, 0\right)$ is the x-intercept

Discuss the graph of the function $y = -3x + 4$.

Solution: The graph is a straight line since it is of the form $y = mx + b$. The line intersects the y-axis at the point (0,4). That is, when x = 0 then y = 4. The y-intercept in this example corresponds to b = 4. The slope of the line is m = -3. This means that y decreases 3 units as x increases 1 unit, anywhere along the line.

Show that the graphs of $3x - y = 9$ and $6x - 2y + 9 = 0$ are parallel lines.

Solution: If the slopes of two lines are equal, the lines

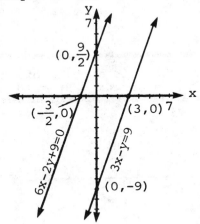

are parallel. Thus we must show that the two slopes are equal. In standard form the equation of a line is $y = mx + b$, where m is the slope.

Putting $3x - y = 9$ in standard form,

$$-y = 9 - 3x$$

$$y = -9 + 3x$$

$$y = 3x - 9.$$

Thus the slope of the first line is 3. Putting $6x - 2y + 9 = 0$ in standard form,

$$-2y + 9 = -6x$$

$$-2y = -6x - 9$$

$$y = 3x + \frac{9}{2}.$$

Thus the slope of this line is also 3. The slopes are equal. Hence, the lines are parallel.

To graph these equations pick values of x and substitute them into the equation to determine the corresponding values of y. Thus we obtain the following tables of values. Notice we need only <u>two</u> points to plot a line (2 points determine a line).

$$6x - 2y + 9 = 0 \qquad\qquad 3x - y = 9$$

$$y = 3x + \frac{9}{2} \qquad\qquad y = 3x - 9$$

x	0	$\frac{3}{2}$
y	$\frac{9}{2}$	0

x	0	3
y	-9	0

(See accompanying figure)

● PROBLEM 39-13

Determine whether there is a point of intersection of the graphs of 2x - 3y = 5 and 6x - 9y = 10.

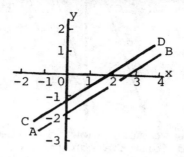

<u>Solution:</u> Geometric discussion. Rewriting the given linear equations in standard form, y = mx + b, the slope, m, can be read directly.

$$2x - 3y = 5 \qquad\qquad 6x - 9y = 10$$

$$-3y = 5 - 2x \qquad\qquad -9y = 10 - 6x$$

$$y = \frac{2}{3}x - \frac{5}{3} \qquad\qquad y = \frac{6}{9}x - \frac{10}{9}$$

$$m = \frac{2}{3} \qquad\qquad m = \frac{6}{9} = \frac{2}{3}$$

Recall that if the slope, m, of two lines are equal, the lines are parallel. This can be seen in the figure.

The graph of the first equation is the line AB through the point (1, -1) with slope $\frac{2}{3}$, and the graph of the second equation is the line CD through the point $\left(\frac{2}{3}, -\frac{2}{3}\right)$, with slope $\frac{2}{3}$. The lines are parallel, hence there is no point of intersection and the equations are inconsistent.

Algebraic discussion. If the members of the first equation are multiplied by 3, and if the members of the resulting equation are subtracted from the corresponding members of the second equation, we obtain

$$3(2x - 3y) = 3(5)$$

$$6x - 9y = 15$$

Then, $6x - 9y = 10$

$-(6x - 9y = 15)$

_____

$0 = -5$, which is impossible.

The steps that were taken were based on the assumption that the given equations had a solution. The fact that an impossible conclusion results proves that the assumption was false. In other words, the two equations have no common solution, and are therefore inconsistent.

● **PROBLEM 39-14**

Find the point of intersection of the graphs of $3x - y = 5$ and $9x - 3y = 15$.

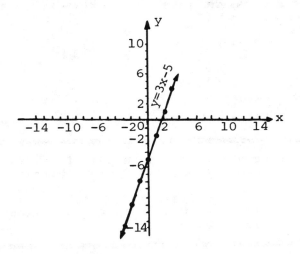

Solution: (1) $3x - y = 5$
(2) $9x - 3y = 15$

Divide the second equation by 3 to obtain:

(1) $3x - y = 5$

Thus any pair of values (x,y) which satisfies the first equation also satisfies the second equation. Hence the same straight line is the graph of both equations. It follows that there is no unique solution, but rather that every point on the common line is a solution. The two equations are dependent.

To solve the pair of dependent equations algebraically it is sufficient to assign an arbitrary value to x (or y), and then to solve for y (or x) in either equation.

$3x - y = 5$

$-y = 5 - 3x$ Add $-3x$ to both sides.

$y = 3x - 5$ Multiply by -1.

To graph this equation, choose values of x and obtain their corresponding values of y from $y = 3x - 5$. The following table is constructed.

738

x	-3	-2	-1	0	1	2	3
y	-14	-11	-8	-5	-2	1	4

We then plot the points found in the table and draw a smooth curve (which turns out to be a straight line) through them.

● **PROBLEM 39-15**

Find the equation of the line ℓ which is perpendicular to ℓ' : $x+y+4=0$, and passes through the point $P(0,0)$.

<u>Solution</u>: The slope of ℓ' is $k' = -\dfrac{A}{B} = -1$. Thus, the slope of ℓ must be $k = -\dfrac{1}{k'} = 1$. Using the point-slope form of the equation of a straight line $y - y_1 = k(x - x_1)$, one obtains the equation of ℓ as

$$y - 0 = 1 \cdot (x-0),$$

that is, $\qquad y = x.$

● **PROBLEM 39-16**

(1)   Find the equation of a line that is parallel to $Ax + By + C = 0$, where A, B, and C are arbitrary constants.

(2)   Prove that if two lines ℓ_1 and ℓ_2 are perpendicular to each other, then the slopes of the two lines have the following relationship: $k_1 k_2 = -1$.

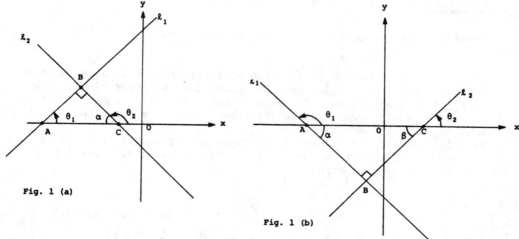

Fig. 1 (a)

Fig. 1 (b)

<u>Solution</u>:   (1)   Any line that is parallel to $Ax + By + C = 0$ must have the same slope as $Ax + By + C = 0$, which is $k = -A/B$.

739

Therefore, the equation of the lines which are parallel to $Ax + By + C = 0$ is $Ax + By + D = 0$, where D is an arbitrary constant.

(2) Let $k_1 = \tan\theta_1$ and $k_2 = \tan\theta_2$ (see Fig. 1).

From Fig. 1 (a), one has in $\triangle ABC$

$$\alpha = 180° - \theta_2 \quad \text{and} \quad \theta_1 + \alpha = 90°$$

That is,

$$\theta_1 + (180° - \theta_2) = 90°,$$

$$\theta_1 = \theta_2 - 90°$$

$$k_1 = \tan\theta_1 = \tan(\theta_2 - 90°)$$

$$= \tan\left[-(90° - \theta_2)\right]$$

$$= -\left[\tan(90° - \theta_2)\right] = -\left[\cot\theta_2\right]$$

$$= -\frac{1}{\tan\theta_2} = -\frac{1}{k_2}$$

Therefore,

$$k_1 k_2 = -1.$$

Similarly, from Fig. 1 (b), in $\triangle ABC$

$$\beta = \theta_2, \quad \alpha = 180° - \theta_1$$

$$\beta + \alpha = 90° = \theta_2 + 180° - \theta_1$$

$$\theta_1 = \theta_2 + 90°$$

$$k_1 = \tan\theta_1 = \tan(90° + \theta_2) = -\cot\theta_2$$

$$= -\frac{1}{\tan\theta_2} = -\frac{1}{k_2}.$$

● **PROBLEM 39-17**

(1) Find the undirected distance from $P(1,1)$ to the line $\ell : x + y - 10 = 0$.

(2) Find the undirected distance between $\ell_1 : x + 2y + 4 = 0$ and $\ell_2 : x + 2y - 9 = 0$.

Solution: (1) The directed distance d from a point $P(x_0, y_0)$ to a line $\ell: Ax + By + C = 0$ is given by

$$d = \frac{Ax_0 + By_0 + C}{\sqrt{A^2 + B^2}}$$

The undirected distance from P to ℓ is $|d|$.

Hence,

$$|d| = \left|\frac{1 \cdot 1 + 1 \cdot 1 - 10}{\sqrt{1^2 + 1^2}}\right| = 4\sqrt{2}$$

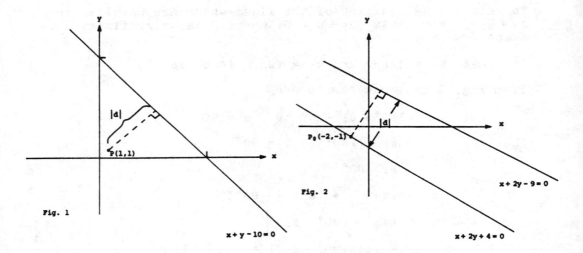

Fig. 1

Fig. 2

$x + y - 10 = 0$

$x + 2y - 9 = 0$

$x + 2y + 4 = 0$

(2) First find the undirected distance between either ℓ_1 or ℓ_2 and a point on ℓ_2 or ℓ_1. Point $P_0(-2,-1)$ is on the line ℓ_1 (since $P_0(-2,-1)$ satisfies $x + 2y + 4 = 0$, $(-2) + 2(-1) + 4 = 0$) The undirected distance from P_0 to ℓ_2 is

$$\left| \frac{1 \cdot (-2) + 2 \cdot (-1) - 9}{\sqrt{1^2 + 2^2}} \right| = \left| \frac{-13}{\sqrt{5}} \right| = \frac{13}{5} \sqrt{5}$$

which is the undirected distance between ℓ_1 and ℓ_2.

● **PROBLEM** 39-18

Find the equation of the line ℓ which is equidistant from $\ell_1 : x + y + 2 = 0$ and $\ell_2 : x + y - 2 = 0$.

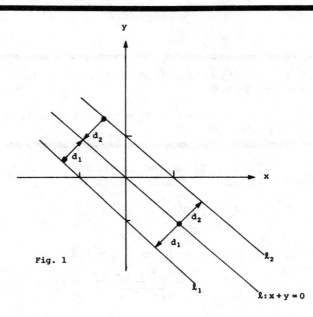

Fig. 1

741

Solution: As one sees from Fig. 1, ℓ_1 and ℓ_2 are parallel, ℓ must be between ℓ_1 and ℓ_2, and parallel to both ℓ_1 and ℓ_2. Furthermore, observe that the directed distances between ℓ_1 and ℓ, ℓ and ℓ_2 have the same magnitude but opposite sign, i.e., $d_1 = -d_2$.

Let a point $P(x_0, y_0)$ be a point on the line ℓ.

$$d_1 = \frac{1 \cdot x_0 + 1 \cdot y_0 + 2}{\sqrt{1^2 + 1^2}} = -d_2 = -\frac{1 \cdot x_0 + 1 \cdot y_0}{\sqrt{1^2 + 1^2}}$$

$$\frac{x_0 + y_0 + 2}{\sqrt{2}} = -\frac{x_0 + y_0 - 2}{\sqrt{2}}$$

$$(x_0 + y_0 + 2) = -(x_0 + y_0 - 2)$$

$$x_0 + y_0 = 0$$

Since $P(x_0, y_0)$ can be any point on ℓ, one can replace $P(x_0, y_0)$ by $P(x, y)$ to obtain the equation for ℓ, i.e., $x + y = 0$.

● **PROBLEM** 39-19

(1)   Discuss the graph of $Ax + By + C = 0$, where A, B, and C are arbitrary constants, except that both A and B may not be zero.

(2)   Reduce $Ax + By + C = 0$ to the normal form of the equation of a straight line.

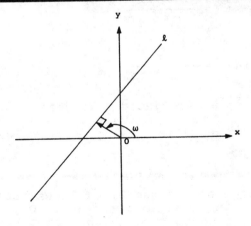

Solution: (1) $Ax + By + C = 0$ is the general equation of a straight line.

If $C = 0$, the line passes through the origin. If $B = 0$, the line is parallel to the y-axis, and if $A = 0$, the line is parallel to the x-axis.

The slope of the line is $-\frac{A}{B}$. The x-intercept and the y-intercept are $-\dfrac{C}{A}$ and $-\dfrac{C}{B}$, respectively.

(2) The normal form of the equation of a straight line not passing through the origin is

$$x \cos\omega + y \sin\omega - p = 0$$

where $p > 0$ is the length of the normal measured always away from the origin, and ω is the positive angle $(0 \leq \omega < 360^{0})$ measured from the positive end of the x-axis to the normal (see Fig. 1).

To reduce the equation $Ax + By + C = 0$ to its normal form, divide every term in the equation by $\pm\sqrt{A^2 + B^2}$, and decide the sign of the radical as the following:

(a) opposite to that of C, if $C \neq 0$

(b) to agree with that of B if $C = 0$ and $B \neq 0$.

(c) to agree with that of A, if $C = B = 0$.

So the normal form of the equation is

$$\frac{A}{\pm\sqrt{A^2 + B^2}}\, x + \frac{B}{\pm\sqrt{A^2 + B^2}} + \frac{C}{\pm\sqrt{A^2 + B^2}} = 0$$

where

$$p = -\left(\frac{C}{\pm\sqrt{A^2 + B^2}}\right),$$

$$\cos\omega = \frac{A}{\pm\sqrt{A^2 + B^2}},$$

and

$$\sin\omega = \frac{B}{\pm\sqrt{A^2 + B^2}}.$$

The sign of $\sqrt{A^2 + B^2}$ is decided as described.

● PROBLEM 39-20

Find the equation of the family of lines that satisfy the conditions (1) parallel to ℓ_0 : $x+2y+4 = 0$
(2) passing through the point $P_0(4,9)$.

Solution: (1) The slope of ℓ_0 is $-\dfrac{1}{2}$. Hence all lines with slope $-\dfrac{1}{2}$ will be in the family whose equation is required.

$$y = -\frac{1}{2}x + b \quad \text{is the equation of the family}$$
of lines that are parallel to ℓ_0, where b is the parameter.

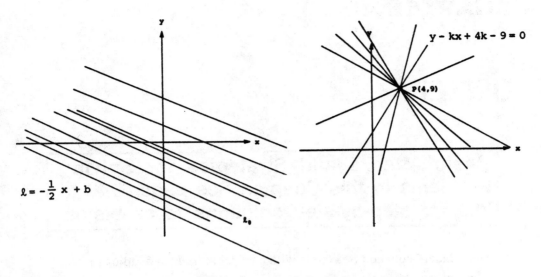

$$\ell = -\frac{1}{2} x + b$$

$$y - kx + 4k - 9 = 0$$

$$P(4,9)$$

(2) The equation of the family of lines passing through $P_0(4,9)$ is

$$y - 9 = k(x - 4)$$

with parameter k (the slope of the line). The equation can be written as

$$y - kx + 4k - 9 = 0$$

CHAPTER 40

CIRCLES

> **Basic Attacks and Strategies for Solving Problems in this Chapter. See pages 745 to 760 for step-by-step solutions to problems.**

The standard equation of a circle with a center at (h, k) and radius r is

$$(x - h)^2 + (y - k)^2 = r^2.$$

When expanded, the general form becomes

$$x^2 + y^2 + Dx + Ey + F = 0$$

(i.e., the coefficients of x^2 and y^2 must always be the same and there must be no xy term).

Given an arbitrary quadratic equation, if it is known that the curve describes a circle, the equation can be algebraically transformed (using the techniques of completing the square, *cf.* Chapter 13) into the standard equation from which the radius and the center can be determined.

There is at most one circle that will pass through any set of three points in the plane. As discussed in Chapter 12 in reference to a parabola, if three points are given (i.e., three sets of values for x and y), one can substitute these values into the general equation for a circle to obtain three linear equations in D, E, and F. This system can be solved using the techniques presented in Chapter 11 to obtain the values of D, E, and F. One could also use the standard form, but the resulting system of equations would include the squares of the parameters h, k, and r, which would make it more difficult to solve. If the three points are collinear, then no circle can pass through all of them except in the trivial sense that two of the points are the same.

Finding the intersection between a circle and another quadratic or even a linear equation is equivalent to finding the solution of a system of equations involving a quadratic. Techniques discussed in Chapter 13 can be used. The easiest technique is to substitute one equation into another.

Some problems involving finding the equation of a circle presuppose that the one solving the problem will draw some figures to visualize the information given and the information sought. When asked to find the equation of a line tangent to a circle, one should remember that a tangent line intersects a curve at only *one* point. Many problems involving equations of circles are not particularly difficult as much as they are algebraically involved.

● PROBLEM 40-1

Write equations of the following circles:

    (a) With center at (-1, 3) and radius 9.

    (b) With center at (2, -3) and radius 5.

<u>Solution:</u> The equation of the circle with center at (a,b) and radius r is

$$(x - a)^2 + (y - b)^2 = r^2.$$

(a) Thus, the equation of the circle with center at (-1, 3) and radius 9 is

$$[x - (-1)]^2 + (y - 3)^2 = 9^2$$

$$(x + 1)^2 + (y - 3)^2 = 81$$

(b) Similarly the equation of the circle with center at (2, -3) and radius 5 is

$$(x - 2)^2 + [y - (-3)]^2 = 5^2$$

$$(x - 2)^2 + (y + 3)^2 = 25$$

● PROBLEM 40-2

Find the center and radius of the circle

$$x^2 - 4x + y^2 + 8y - 5 = 0 \qquad (1)$$

<u>Solution:</u> We can find the radius and the coordinates of the center by completing the square in both x and y. To complete the square in either variable, take half the coefficient of the variable term (i.e., the x term or the y term) and then square this value. The resulting number is then added to both sides of the equation. Completing the square in x:

$$[\tfrac{1}{2}(-4)]^2 = [-2]^2 = 4$$

Then equation (1) becomes:

$$\left(x^2 - 4x + 4\right) + y^2 + 8y - 5 = 0 + 4,$$

or

$$(x - 2)^2 + y^2 + 8y - 5 = 4 \qquad\qquad (2)$$

Before completing the square in y, add 5 to both sides of equation (2):

$$(x - 2)^2 + y^2 + 8y - 5 + 5 = 4 + 5$$
$$(x - 2)^2 + y^2 + 8y = 9 \qquad\qquad (3)$$

Now, completing the square in y:

$$[\tfrac{1}{2}(8)]^2 = [4]^2 = 16$$

Then equation (3) becomes:

$$(x - 2)^2 + \left(y^2 + 8y + 16\right) = 9 + 16,$$

or

$$(x - 2)^2 + (y + 4)^2 = 25 \qquad\qquad (4)$$

Note that the equation of a circle is:

$$(x - h)^2 + (y - k)^2 = r^2 ,$$

where (h,k) is the center of the circle and r is the radius of the circle. Equation (4) is in the form of the equation of a circle. Hence, equation (4) represents a circle with center (2,-4) and radius = 5.

● **PROBLEM 40-3**

Discuss the graph of the equation $x^2 + y^2 = 25$.

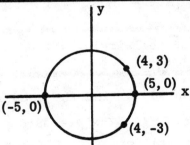

<u>Solution:</u> This is an equation of the form $x^2 + y^2 = r^2$, and therefore its graph is a circle with radius r = 5 and center at the origin (see figure). Note that the graph does not represent a function since, except for x = -5 or x = 5, each permissible value of x is associated with two values of y. For example, for x = 4, we have the ordered pairs (4,3) and (4,-3). The domain of this function is $\{x\mid -5 < x < 5\}$.   The range of this function is $\{y\mid -5 \lessgtr y \lessgtr 5\}$.

Graph the equation $2x^2 + 2y^2 - 13 = 0$.

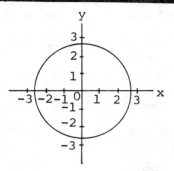

Solution: In order to verify that this is the equation of a circle, we put the equation in the standard form: add 13 to both sides of the given equation.

$$2x^2 + 2y^2 - 13 + 13 = 0 + 13$$

$$2x^2 + 2y^2 = 13$$

Divide both sides of this equation by 2.

$$\frac{2x^2 + 2y^2}{2} = \frac{13}{2}$$

or

$$x^2 + y^2 = \frac{13}{2}, \text{ which is the}$$

standard form for the equation of a circle with its center at the origin (0,0) and

$$\text{radius} = r = \sqrt{\frac{13}{2}} = \frac{\sqrt{13}}{\sqrt{2}} = \frac{\sqrt{13}}{\sqrt{2}} \frac{\sqrt{2}}{\sqrt{2}} = \frac{\sqrt{26}}{2},$$

Therefore, the radius of the circle is approximately $\frac{5.1}{2}$ or 2.55. The graph is represented in the Figure.

Find the equation of the circle that goes through the points (1,3), (-8,0) and (0,6).

Solution: We are asked to find the equation of a circle.

There are two forms: (1) The standard form $(x-h)^2 + (y-k)^2 = a^2$; (2) the general form $x^2 + y^2 + Dx + Ey + F = 0$. Both are equivalent; however, the standard form is more convenient when we are dealing with the coordinates of the center point (h,k) or the length of the radius a. As we can see, all three numbers h, k, and a, can be readily obtained given the standard form. The drawback of the standard form is its factored format which, in problems which do not ask for the center coordinates or the radius, is just extra work to calculate. In these cases, the multiplied out form, the general form, is more convenient.

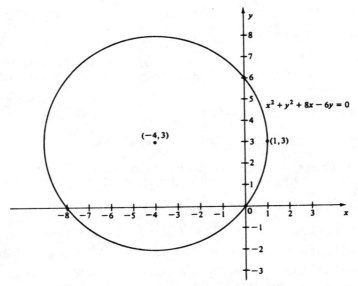

$x^2 + y^2 + 8x - 6y = 0$

Here we are asked to find the equation given three points. We are given three pairs of (x,y) and asked to find the three equation unknowns. If we were to use standard form, the unknowns would be h, k, and a. The unknowns would be squared. Moreover, to solve the equations we would have to multiply out the $(x-h)^2$ and $(y-h)^2$ terms. It would be long and difficult.

If we use the second method, though, the unknowns would be the unsquared terms D, E, and F. Furthermore, there are no multiplications (other than x^2 and y^2) necessary. Thus, even though we may wish to express the equation of the circle in standard form later on, it is easier to solve for the general form equation first.

Suppose the equation of the circle is $x^2 + y^2 + Dx + Ey + F = 0$. Because point (1,3) is on the circle it must satisfy the equation. Then,

(i) $1^2 + 3^2 + D(1) + E(3) + F = 0$

or $D + 3F + F = -10$.

Similarly since (-8,0) and (0,6) are also points on the circle, we have:

(ii) $(-8)^2 + 0^2 + D(-8) + E(0) + F = 0$ or

$-8D + F = -64$.

(iii) $0^2 + 6^2 + D(0) + E(6) + F = 0$ or $6E + F = -36$.

We thus have three equations and three unknowns.

(iv) $D + 3E + F = -10$

748

(v) $-8D + F = -64.$

(vi) $6E + F = -36.$

We make this system a system of two equations and two unknowns by eliminating D. Multiply (iv) by 8 and add it to (v). We have:

(vii) $24E + 9F = -144$

(viii) $6E + F = -36$ (see (vi))

Now eliminate E to form one equation and one unknown. Multiply (viii) by -4 and add the result to (vii) to obtain

(ix) $5F = 0$ or $F = 0.$

Substituting this result in (viii), we find E. Since $6E + F = -36$, then $6E + 0 = -36$ or $E = \frac{-36}{6} = -6$. Substituting the value for F in (v), we find $-8D + F = -64$. Thus, $-8D = -64$ or $D = \frac{-64}{-8} = 8$. The general form of the equation is

(x) $x^2 + y^2 + 8x - 6y = 0.$

Usually, it is standard practice to give the equation in standard form.

(xi) $(x^2 + 8x) + (y^2-6y) = 0$

(xii) $(x^2+8x+16) + (y^2-6y+9) - 16 - 9 = 0$

(xiii) $(x+4)^2 + (y-3)^2 = 25.$

● **PROBLEM 40-6**

Find the equation of the circle passing through the points (4,0), (-4,0), and (0,4).

<u>Solution</u>: The general equation of a circle is

$$x^2 + y^2 + 2\,Dx + 2\,Ey + F = 0$$

Thus, by substituting successively the coordinates of the given points and solving the system of equations

$$4^2 + 0^2 + 2D \cdot 4 + 2\,E \cdot 0 + F = 0$$

$$(-4)^2 + 0^2 + 2D \cdot (-4) + 2E \cdot 0 + F = 0$$

$$0^2 + 4^2 + 2D \cdot 0 + 2E \cdot 4 + F = 0,$$

one has $D = 0$, $E = 0$, and $F = -16$.

Therefore, the equation is

$$x^4 + y^2 - 16 = 0$$

● **PROBLEM** 40-7

Find the intersection between the circles

$x^2 + y^2 = 4$ and $x^2 + y^2 - 8y + 12 = 0$.

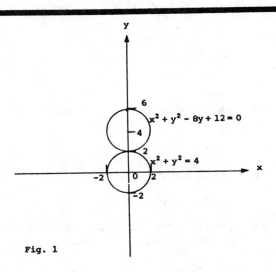

Fig. 1

<u>Solution</u>:   The graphs of the two circles are given in Fig. 1.

By solving the systems of equations

$$\begin{cases} x^2 + y^2 = 4 \\ x^2 + y^2 - 8y + 12 = 0, \end{cases}$$

one obtains $x = 0, \quad y = 2$.

Since there is only one solution, the two circles must be tangent.

● **PROBLEM** 40-8

Find the equation of the circle of radius $r = 9$, with center on $\ell : y = x$ and tangent to both coordinate axes.

<u>Solution</u>:   As seen in Fig. 1, there are two such circles, 0 and 0'.   Let the centers of 0 and 0' be (a,b) and (c,d), respectively.

750

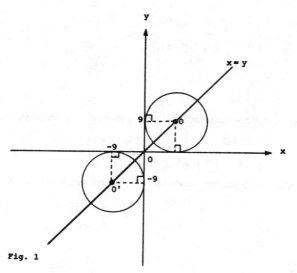

Fig. 1

Since points (a,b) and (c,d) are on ℓ: y = x, one has
a = b and c = d.

The equations will be

$$(x - a)^2 + (y - a)^2 = 9^2$$

and $(x - c)^2 + (y - c)^2 = 9^2$

Because the circles are tangent to both coordinate axes,
one obtains a = 9 = b and c = -9 = d. (See Fig. 1.) Therefore,
the equations are

$$(x - 9)^2 + (y - 9)^2 = 81 \quad \text{and}$$

$$(x + 9)^2 + (y + 9)^2 = 81$$

● **PROBLEM 40-9**

Find the equation of the circle that goes through the points
(1,2) and (3,4) and has radius a = 2.

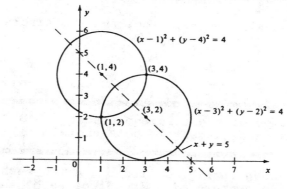

Solution: The two forms of the equation of a circle are the

standard and general forms. Because we are given the radius, we will use the standard form. Suppose the equation of the desired circle is $(x-h)^2 + (y-k)^2 = a^2$, where (h,k) is the center and a is the radius. We are given that the radius $a = 2$. Furthermore, because points $(1,2)$ and $(3,4)$ are points on the circle, they must satisfy the circle equation. This fact will allow us to find (h,k) and fully determine the circle. Hence,

(i) $(1-h)^2 + (2-k)^2 = 2^2$

(ii) $(3-h)^2 + (4-k)^2 = 2^2$.

Multiplying out, we obtain:

(iii) $1-2h + h^2 + 4 - 4k + k^2 = 4$

(iv) $9 - 6h + h^2 + 16 - 8k + k^2 = 4$.

Simplifying, we obtain

(v) $-2h + h^2 - 4k + k^2 = -1$

(vi) $-6h + h^2 - 8k + k^2 = -21$.

To deal with two unknowns with squared terms is too difficult. If we could find a relation between h and k, then we could substitute in equation (v) and obtain a quadratic equation in <u>one</u> variable. We can deal with this by using the quadratic equation.

To find a relation between h and k, note that if you subtract (vi) from (v), the squared terms subtract out. We thus obtain,

(vii) $+4h + 4k = 20$.

Thus, solving for h in terms of k, we have $h = \dfrac{20-4k}{4} = 5 - k$. Substituting this result in equation (v), we obtain:

(viii) $-2(5-k) + (5-k)^2 - 4k + k^2 = -1$.

Simplifying, we obtain

(ix) $k^2 - 6k + 8 = 0$.

Instead of using the quadratic formula, we see this can be factored. That is,

(x) $(k-4)(k-2) = 0$.

Thus, either $k = 2$ or $k = 4$. To solve for h, remember that $h = 5-k$. Therefore, if $k = 2$, $h = 5-2 = 3$ and if $k = 4$, $h = 5 - 4 = 1$. Thus, the two possible center points are $(h,k) = (3,2)$ or $(h,k) = (1,4)$. The two possible equations, therefore, are

(xi) $(x-1)^2 + (y-4)^2 = 4$

(xii) $(x-3)^2 + (y-2)^2 = 4.$

Note: It may seem odd that there are two possible answers.
but from a geometric viewpoint, it is logical. Given two
points and a radius length, there are two possible circles
that can be constructed, the centers of the circles lying on
opposite sides of the line determined by the two given
points.

● **PROBLEM** 40-10

Find the equation of the circle having radius 4, tangent to the line $\ell : x = -y$ at $(0,0)$.

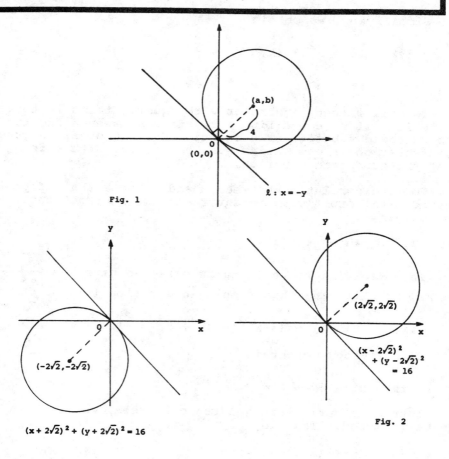

Fig. 1

Fig. 2

$(x+2\sqrt{2})^2 + (y+2\sqrt{2})^2 = 16$

Solution: The equation of the circle is

$$(x - a)^2 + (y - b)^2 = 4^2,$$ where (a,b) is the
center of the circle. From the undirected distance formula
for a point (a,b) and a line $x = -y$, one has

753

$$\left| \frac{(1 \cdot a) + (1 \cdot b) + 0}{\sqrt{1^2 + 1^2}} \right| = 4, \quad \left| \frac{a + b}{\sqrt{2}} \right| = 4$$

Since there are two unknowns, one more equation is needed. From Fig. 1, one obtains the slope of the radius through point (0,0) as

$$= \frac{1}{\text{slope of } \ell}$$

Thus

$$\frac{b - 0}{a - 0} = -\frac{1}{(-1)} , \quad \frac{b}{a} = 1$$

Solving

$$\begin{cases} \left| \dfrac{a + b}{\sqrt{2}} \right| = 4 \\[2mm] \dfrac{b}{a} = 1, \end{cases}$$

one obtains $a_1 = 2\sqrt{2}$, $b_1 = 2\sqrt{2}$, and $a_2 = -2\sqrt{2}$, $b_2 = -2\sqrt{2}$.

Therefore, the required equations are

$$(x - 2\sqrt{2})^2 + (y - 2\sqrt{2})^2 = 16$$

and

$$(x + 2\sqrt{2})^2 + (y + 2\sqrt{2})^2 = 16$$

as shown in Fig. 2.

● **PROBLEM** 40-11

Given that two circles $x^2 + y^2 + D_1 x + E_1 y + F_1 = 0$ and $x^2 + y^2 + D_2 x + E_2 y + F_2 = 0$ intersect at two points, show that the equation for the line determined by the points of intersection is $(D_1 - D_2)x + (E_1 - E_2)y + (F_1 - F_2) = 0$.

Solution: To show that the points of intersection lie on the equation $(D_1 - D_2)x + (E_1 - E_2) + (F_1 - F_2) = 0$, we show that any point that lies on both circles must lie on the line. Suppose points (x_1, y_1) and (x_2, y_2) lie on both circles. Then (x_1, y_1) must satisfy the equations for both circles. That is,

(i) $\quad x_1^2 + y_1^2 + D_1 x_1 + E_1 y_1 + F_1 = 0$

(ii) $\quad x_1^2 + y_1^2 + D_2 x_1 + E_2 y_1 + F_2 = 0.$

By subtraction property of equality, if equation (i)

and equation (ii) are true statements, then equation (i)
minus equation (ii) must be a true statement. Subtracting
(ii) from (i), we have:

(iii) $(D_1-D_2)x_1 + (E_1-E_2)y_1 + (F_1-F_2) = 0.$

Thus, point (x_1,y_1) satisfies the equation. $(D_1-D_2)x$
$+ (E_1-E_2)y + (F_1-F_2) = 0.$ Similarly, we can show (x_2,y_2)
satisfies the given equation. Thus, both points of intersection lie on the line with the given equation. Since two
distinct points determine a line, the line

$(D_1-D_2)x + (E_1-E_2)y + (F_1-F_2) = 0$

is the line determined by the points of intersection.

● PROBLEM 40-12

Find the equation of the line tangent to the circle
$x^2 + y^2 - 10x + 2y + 18 = 0$ and having a slope equal to 1.

Solution: Since we are given the slope of the tangent
line, to apply the slope-intercept form of a straight line

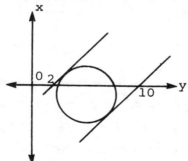

in finding the equation of the tangent line, we need to calculate the y-intercept.

The slope-intercept form is $y = mx + b$ where m is the
slope and b is the y-intercept. Hence, we can substitute
$m = 1$, the given, to find that all lines tangent with slope
equal to 1 will have the form $y = x + b$. Our task, now, is
to find b.

We can solve for b by substituting $y = x + b$ into the
equation for the circle and finding an equation for b by
considering the tangency characteristics. By substitution,
$x^2 + (x+b)^2 - 10x + 2(x+b) + 18 = 0$ and
$x^2 + (x^2+2xb+b^2) - 10x + 2x + 2b + 18 = 0$ which reduces to
$2x^2 + (2b-8)x + (b^2+2b+18) = 0.$ To derive an equation for
b, proceed to solve the quadratic equation for x. For the

755

general quadratic form $ax^2 + bx + c = 0$, $x = \dfrac{-b \pm \sqrt{b^2 - 4ac}}{2a}$.

There can be only one point of intersection between the circle and the line, since the line is a tangent. Hence, x can only have one value. Therefore, the discriminant, $b^2 - 4ac$ must equal zero. In our problem,

$$b^2 - 4ac = (2b-8)^2 - 4(2)(b^2+2b+18) = 0$$

$$4b^2 - 32b + 64 - 8b^2 - 16b - 144 = 0$$

$$-4b^2 - 48b - 80 = 0$$

$$-4(b^2 + 12b + 20) = 0$$

To find b such that $y = x + b$ is a tangent line, it is sufficient to solve,

$$b^2 + 12b + 20 = 0$$

$$(b + 10)(b + 2) = 0.$$

Hence, b = -10 or b = -2. Therefore, there are two possible lines tangent to the given circle with slope equal to one. Their equations are $y = x - 10$ and $y = x - 2$. The accompanying figure represents this circumstance.

● **PROBLEM 40-13**

Find the equation of the line drawn from the point (8,6) tangent to the circle $x^2 + y^2 + 2x + 2y - 24 = 0$.

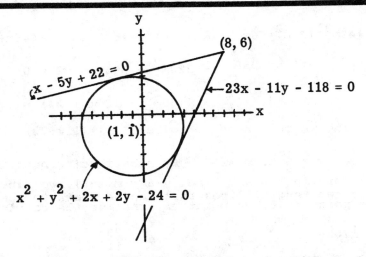

Solution: Given a point through which the line passes, we can fully describe the line only after we have found its slope. According to the point-slope form of a linear equation, the equations of the lines passing through (8,6) is given by $y - 6 = m(x - 8)$, where m is the slope. When m is the slope of the tangent, the equation, $y - 6 = m(x - 8)$, is the equation of the tangent. The equation can be rewritten as $y = mx - 8m + 6$.

To find the unique intersection point of the line and the circle, the tangency point, we can substitute $y = mx - 8m + 6$ into the equation for the circle and solve for x. Since the line is a tangent line, there can only be one value for x. This last fact will assist us in determining the slope of the line. Hence,

$$x^2 + (mx - 8m + 6)^2 + 2x + 2(mx - 8m + 6) - 24 = 0$$

$$x^2 + (m^2x^2 - 16m^2x + 12mx + 64m^2 - 96m + 36) + 2x + (2mx$$

$$- 16m + 12) - 24 = 0$$

which reduces to,

$$(m^2 + 1)x^2 - (16m^2 - 14m - 2)x + (64m^2 - 112m + 24) = 0.$$

By the quadratic formula,

$$x = \frac{-[-(16m^2-14m-2)] \pm \sqrt{[-(16m^2-14m-2)]^2 - 4(m^2+1)(64m^2-112m+24)}}{2(m^2+1)}$$

Since x is a coordinate of a tangency point, it can only take on one value. Hence, the discriminant must equal zero. By setting the discriminant equal to zero we can solve for m and, thereby, determine fully the equation of the tangent line.

$$[-(16m^2-14m-2)]^2 - 4(m^2+1)(64m^2-112m+24) = 0$$

$$(256m^4-448m^3+132m^2+56m+4) - (256m^4-448m^3+352m^2-448m+96) = 0$$

$$- 220m^2 + 504m - 92 = 0$$

$$- 4(55m^2 - 126m + 23) = 0.$$

By factoring we only need to solve $(5m-1)(11m-23) = 0$.

Hence, $m = \frac{1}{5}$ or $m = \frac{23}{11}$. By substituting back into our equation for the tangent line, $y-6 = m(x-8)$, we find that there are two tangent lines that can be drawn to the circle from the point (8,6), a point external to the circle.

$y - 6 = \frac{1}{5}(x-8)$ and $y - 6 = \frac{23}{11}(x-8)$

$5y - 30 = x - 8$ $11y - 66 = 23x - 184.$

Hence, $x - 5y + 22 = 0$ and $23x - 11y - 118 = 0$ are the equations of the tangent lines required in this problem. The figure shows the circle and its tangents.

Problem: Find the points of intersection (if any) of the circles C_1 and C_2 where

$$C_1: \quad x^2 + y^2 - 4x - 2y + 1 = 0$$

$$C_2: \quad x^2 + y^2 - 6x + 4y + 4 = 0.$$

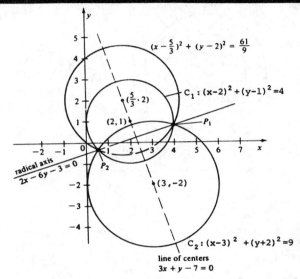

$$\left(x - \tfrac{5}{3}\right)^2 + (y - 2)^2 = \tfrac{61}{9}$$

$C_1: (x-2)^2 + (y-1)^2 = 4$

$\left(\tfrac{5}{3}, 2\right)$

$(2,1)$

P_1

radical axis
$2x - 6y - 3 = 0$

P_2

$(3, -2)$

$C_2: (x-3)^2 + (y+2)^2 = 9$

line of centers
$3x + y - 7 = 0$

Solution: The analytic method reduces this problem to a routine, although possibly messy, problem in algebra. We seek the points of intersection of the circle; thus, an obvious first step is to subtract the second equation from the first, giving:

$$2x - 6y - 3 = 0.$$

This is the equation of a straight line. The geometric significance of $2x - 6y - 3 = 0$ is that any points common to C_1 and C_2 must also lie on this line. It is a straight-forward process to find points of intersection of this line with either C_1 or C_2. We can solve for x in terms of y,

$$x = 3y + \frac{3}{2}$$

substitute back into the equation C_1, find the two real values of y, then use $x = 3y + \frac{3}{2}$ again to find the corresponding values of x. Using this or an equivalent procedure, we obtain as points of intersection

758

$$P_1\left(\frac{9}{4} + \frac{3\sqrt{135}}{20}, \frac{1}{4} + \frac{\sqrt{135}}{20}\right) \approx (3.99, 0.83) \quad \text{and}$$

$$P_2\left(\frac{9}{4} - \frac{3\sqrt{135}}{20}, \frac{1}{4} - \frac{\sqrt{135}}{20}\right) \approx (0.51, -0.33).$$

● **PROBLEM 40-15**

Write the equation of the family of all concentric circles whose common center is the point (-3,5). Draw three members of the family, specifying the value of the parameter in each case.

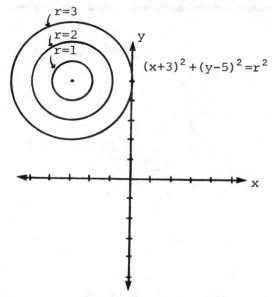

Solution: The standard form of the equation of all circles centered at (-3,5) is given by $(x+3)^2 + (y-5)^2 = r^2$ where r is the parameter equalling the radius length of the circle.

To draw three members of this family, let r take on three separate values and construct each circle. The accompanying graph contains the three circles.

When r = 1, the circle is given by $(x+3)^2 + (y-5)^2 = 1$. When r = 2, the circle is given by $(x+3)^2 + (y-5)^2 = 4$. When r = 3, the circle is given by $(x+3)^2 + (y-5)^2 = 9$.

● **PROBLEM 40-16**

Find the equation of the family of circles having (1) the common center at (0,0), and (2) radius 4 and center on the line x = 4.

759

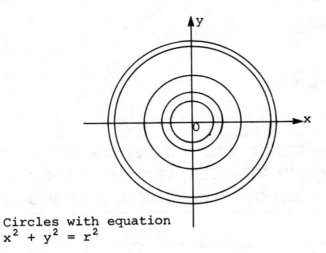

Circles with equation
$x^2 + y^2 = r^2$

<u>Solution</u>: (1) Use the standard form of the equation of the
circle

$$(x - a)^2 + (y - b)^2 = r^2 \quad \text{where } a = b = 0 \text{ for this}$$

problem. So, $x^2 + y^2 = r^2$ with parameter r is the required
equation.

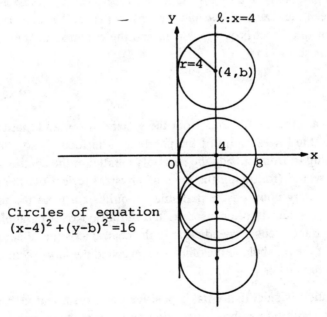

Circles of equation
$(x-4)^2 + (y-b)^2 = 16$

(2) Again, use the equation of the form $(x - a)^2 + (y - b)^2 = r^2$
where $r = 4$, and $a = 4$, since the center of the circle is on
the line $x = 4$.

Therefore, the required equation is

$$(x - 4)^2 + (y - b)^2 = 16, \text{ with b as the parameter.}$$

CHAPTER 41

PARABOLAS

Basic Attacks and Strategies for Solving Problems in this Chapter. See pages 761 to 774 for step-by-step solutions to problems.

The equation for a parabola (that opens upward, i.e., "concave upward" or "holding water") is usually written as

$$y = ax^2 + bx + c \quad \text{with} \quad a > 0.$$

This is the form of a quadratic equation as seen in Chapter 12. A common alternative equation (often called the *standard form*), however, is $(x - h)^2 = 4p (y - k)$, where (h, k) is the *vertex* of the parabola (the lowest point of an upward-opening parabola), and $|p|$ is the distance between the *focus* and the vertex and also between the *directrix* line and the vertex. (Thus, the focus is at $(h, k + p)$, the equation of the directrix is $y = k - p$, and the equation of the central axis is $x = h$.) It can be shown that

$$h = -\frac{b}{2a},$$

where a and b are the coefficients from the general form, and that this gives the x-coordinate of the lowest point of the parabola. The focus is so named because light rays coming in from outside the parabola parallel to the central axis (parallel to the y-axis) would (following the laws of physics) reflect off the parabola toward the focus. Any point on the parabola is equidistant from the focus and the directrix line (this fact is often used as the definition of a parabola). The line segment through the focus perpendicular to the central axis (parallel to the x-axis), from one arm of the parabola to the other arm, is called the *latus rectum*. The length of the *latus rectum* is $|4p|$.

If the equation is such that y has a positive coefficient and x^2 has a negative coefficient, the resulting parabola opens downward (i.e., "concave downward" or "spilling water"). In this case p is a negative number in the equations above. However, the coordinates of the focus and the equation of the directrix are the same.

Given an equation in which x has degree one and y is squared, the graph is also a parabola, either opening to the right (if the coefficients of x and y^2 are both positive or negative), or to the left (if one coefficient is negative). The standard equation then becomes

$$(y - k)^2 = 4p(x - h).$$

By referring to the definitions of the focus, vertex, directrix, and *latus rectum*, these can also be derived in the cases of parabolas whose axes are parallel to the x-axis.

To find the intersection of a parabola and a line, one uses the techniques discussed in Chapter 13. When the slope of a line plays a role in the problem, the equations presented in Chapters 38 and 39 may also be useful.

An equation may describe a parabola whose central axis is not parallel to either the x- or y-axis. Such an equation may also include an xy term and both an x^2 and y^2 term. Techniques for identifying such equations as parabolas usually involve the *rotation of axes* and are suggested in Chapter 44.

Step-by-Step Solutions to Problems in this Chapter, "Parabolas"

● **PROBLEM 41-1**

Draw the graphs of $f(x) = x^2$, $g(x) = 3x^2$, and also $h(x) = \frac{1}{2}x^2$ on one set of coordinate axes.

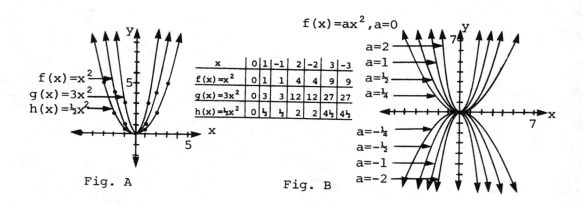

x	0	1	-1	2	-2	3	-3
$f(x)=x^2$	0	1	1	4	4	9	9
$g(x)=3x^2$	0	3	3	12	12	27	27
$h(x)=\frac{1}{2}x^2$	0	$\frac{1}{2}$	$\frac{1}{2}$	2	2	$4\frac{1}{2}$	$4\frac{1}{2}$

Fig. A   Fig. B

Solution: We construct a composite table showing the values of each function corresponding to selected values for x.

In the example, we graphed three instances of the function $f(x) = ax^2$, $a > 0$. For different values of a, how do the graphs compare? (Fig. A). Assigning a given value to a has very little effect upon the main characteristics of the graph. The coefficient a serves as a "stretching factor" relative to the y-axis. As a increases, the two branches of the curve approach the y-axis. The curve becomes "thinner". As a decreases, the curve becomes "flatter" and approaches the x-axis.

The graph of $f(x) = ax$, $a \neq 0$, is called a parabola. (Fig. B). The point (0,0) is the vertex, or turning point, of the curve; the y-axis is the axis of symmetry. The value of a determines the shape of the curve. For $a > 0$ the parabola opens upward and for $a < 0$ the parabola opens downward.

761

Show that the quadratic equation $y = 2x^2 - 20x + 25$ is the equation of a parabola.

Solution: If we can write $y = 2x^2 - 20x + 25$ in the standard form $y = a(x - h)^2 + k$, we will show that its graph is a parabola. The axis of symmetry will be the line $x = h$. The vertex will be at (h,k). Notice that the standard form has a term with a perfect square in it. We use a method known as completing the square to re-write

$$y = 2x^2 - 20x + 25$$

(a) Subtract the constant term, 25, from both members.

$$y - 25 = 2x^2 - 20x$$

(b) Factor 2 from the right member.

$$y - 25 = 2(x^2 - 10x)$$

(c) We need to add 25 within the parentheses since 25 is the square of one-half the coefficient of the term in x. If we do, then we must add 50 to the left member to maintain equality, since we have in fact added two times the quantity added in the parentheses $(2 \times 25 = 50)$ to the right member.

$$y + 25 = 2(x^2 - 10x + 25)$$

(d) The expression within the parentheses is now a perfect square.

$$y + 25 = 2(x - 5)^2$$

(e) Next we subtract 25 from each member.

$$y = 2(x - 5)^2 - 25$$

This is now in standard form $y = a(x - h)^2 + k$. Here $a = 2$, $h = 5$, and $k = -25$. Thus, the graph of $y = 2x^2 - 20x + 25$ is a parabola. Its axis of symmetry is the line $x = 5$; its vertex is at $(5,-25)$.

Change $y = ax^2 + bx + c$ into the form $(x - x_0)^2 = 4p(y - y_0)$, where x_0 and y_0 are real constants and a, b, c are real coefficients.

Solution: Dividing both sides of $y = ax^2 + bx + c$ by a,

$$\frac{y}{a} = x^2 + \frac{b}{a} x + \frac{c}{a}$$

$$\frac{y-c}{a} = x^2 + \frac{b}{a} x$$

Completing the square on the right side of the equation, one has

$$\frac{y-c}{a} = x^2 + \frac{b}{a}k x + \left(\frac{b}{2a}\right)^2 - \left(\frac{b}{2a}\right)^2$$

$$\frac{y-c}{a} + \frac{b^2}{4a^2} = \left(x + \frac{b}{2a}\right)^2$$

$$\frac{(y-c)}{a} + \frac{b^2}{4a^2} = \frac{4ay-4ac+b^2}{4a^2} = \left(x + \frac{b}{2a}\right)^2$$

$$\left(x + \frac{b}{2a}\right)^2 = \frac{4a}{4a^2}\left(y - c + \frac{b^2}{4a}\right)$$

Thus ,
$$\left(x + \frac{b}{2a}\right)^2 = \frac{1}{a}\left[y + \left(\frac{b^2}{4a} - c\right)\right]$$

is the required form , where

$$-\frac{b}{2a} = x_0, \quad \frac{4ac-b^2}{4-a} = y_0, \quad \text{and} \quad \frac{1}{4a} = p.$$

The vertex of the parabola is

$$V(x_0, y_0) \quad \text{or} \quad V\left(-\frac{b}{2a}, \frac{4ac-b^2}{4-a}\right), \text{the axis of the}$$

parabola is $x = -\frac{b}{2a}$, the focus is $F\left(-\frac{b}{2a}, \frac{1}{4a}\right)$, and the

directrix is $y = -\frac{1}{4a}$.

● **PROBLEM 41-4**

Discuss the graph of the parabola $(x-a)^2 = 4p(y-b)$, and
find its axis, focus, directrix, vertex and latus rectum.

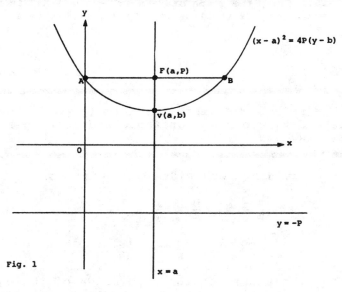

Fig. 1

Solution: The graph of $(x - a)^2 = 4p(y - b)$ is a parabola, as
shown in Fig. 1. It is concave up if $p > 0$, and concave down

763

if p < 0. Note that when p = 0, the graph is a straight line, x = a. Also, as |p| gets smaller, the parabola becomes narrower, whereas as |p| gets bigger, the parabola becomes wider.

The vertex of the parabola is V(a,b), its axis is x = a, the focus is F(a,p), the directrix is y = -p, and the latus rectum is the segment AB (as shown in Fig. 1), which is a part of the line y = p.

● **PROBLEM** 41-5

Construct the graph of the function defined by

$y = x^2 - 6x + 10$.

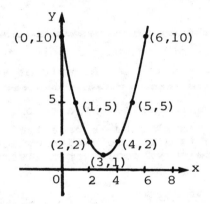

$Solution:$ We are given the function $y = x^2 - 6x + 10$. The most general form of the quadratic function is

$y = ax^2 + bx + c$ where a, b, and c are constants. If a is positive, the curve opens upward and it is U-shaped. If a is negative, the curve opens downward and it is inverted U-shaped.

Since a = 1 > 0 in the given equation, the graph is a parabola that opens upward. To determine the pairs of values (x,y) which satisfy this equation, we express the quadratic function in terms of the square of a linear function of x.

$$y = x^2 - 6x + 10 = x^2 - 6x + 9 + 1$$

$$= (x - 3)^2 + 1$$

y is least when x - 3 = 0 This is true because the square of any number, be it positive or negative, is a positive number. Therefore y would always be greater than or equal to one. Thus the minimum value of y is one when x - 3 = 0 or x = 3.

In order to plot the curve, we select values for x and calculate the corresponding y values. (See the table.)

764

x	$x^2 - 6x + 10 =$	y
0	$(0)^2 - 6(0) + 10$	10
1	$(1)^2 - 6(1) + 10$	5
2	$(2)^2 - 6(2) + 10$	2
3	$(3)^2 - 6(3) + 10$	1
4	$(4)^2 - 6(4) + 10$	2
5	$(5)^2 - 6(5) + 10$	5
6	$(6)^2 - 6(6) + 10$	10

The points and graphs determined by the table are shown in the accompanying figure.

● **PROBLEM 41-6**

Find the coordinates of the maximum point of the curve $y = -3x^2 - 12x + 5$, and locate the axis of symmetry.

<u>Solution:</u> The curve is defined by a second degree equation. The coefficient of the x^2 term is negative. Hence, the graph of this curve is a parabola opening downward. The maximum point of the curve occurs at the vertex and has the x-coordinate:

$$-\frac{\text{coefficient of x term}}{2\left(\text{coefficient of } x^2 \text{ term}\right)} = -\frac{b}{2a} = -\frac{-12}{2(-3)} = \frac{12}{-6} = -2.$$

For $x = -2$, $y = -3(-2)^2 - 12(-2) + 5 = 17$. Hence the coordinates of the vertex are $(-2,17)$. The curve is symmetric with respect to the vertical line through its vertex. The axis of symmetry of this curve is the vertical line through the point $(-2,17)$, i.e., the line $x = -2$.

● **PROBLEM 41-7**

Discuss the graph of the equation $y^2 = 12x$.

<u>Solution:</u> The equation written as $x = \frac{1}{12}y^2$ is a quadratic

765

equation with the coefficient of the y^2 term positive.
Therefore the graph is a parabola opening to the right.
Since $f(x) = -f(x)$ the parabola is symmetric with respect
to the x-axis. Point $(0,0)$ satisfies the equation and lies
on the axis of symmetry. Hence the vertex of the parabola
is at $(0,0)$, (see figure). The focus of the parabola lies
on the axis of symmetry, $y = 0$, at the point $(p,0)$ where
$4p$ = coefficient of x in the original equation: $4p = 12$,
$p = 3$. Therefore the focus is at $(3,0)$. The directrix is
the vertical line $x = -p = -3$. When $x = 3$, $y = 12x = 12(3)$
$= \pm 6$. Therefore the points $(3,6)$ and $(3,-6)$ are points on
the graph. The graph of this parabola is not the graph of
a function since for any given value of x there is more
than one corresponding value of y.

● **PROBLEM** 41-8

Find the equation of the tangent to the parabola
$y = x^2 - 6x + 9$, if the slope of the tangent equals 2.

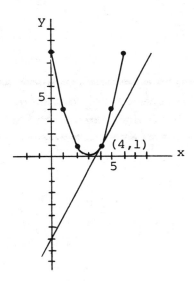

Solution: The equation of a straight line is $y = mx + k$
where m is the slope and k is the y-intercept. The
equation $y = 2x + k$ (1)
represents a family of parallel lines with slope 2, some
of which intersect the parabola in two points, others
which have no point of intersection with the parabola,
and just one which intersects the parabola in only one
point. The problem is to find the value of k so that the
graph of Equation 1 intersects the parabola in just one
point. If we solve the system

$y = 2x + k$ (1)

$y = x^2 - 6x + 9$ (2)

by substitution, we get for the first step

$2x + k = x^2 - 6x + 9$ or

766

$$x^2 - 8x + 9 - k = 0 \qquad (3)$$

This is a quadratic equation of the form $ax^2 + bx + c = 0$. The discriminant determines the nature of the roots when $ax^2 + bx + c = 0$. The condition that Equation 3 has but one solution is that the discriminant, $b^2 - 4ac$ equals 0. Therefore, if $a = 1$, $b = -8$, $c = 9 - k$, then $b^2 - 4ac = 64 - 4(9 - k) = 0$ or $k = -7$.

Substituting this value of k in equation 1, we have $y = 2x - 7$ which is the equation of the tangent to the given parabola when the slope of the tangent is equal to 2. The figure is the graph of the parabola and the tangent. The student may verify that the point of contact is (4, 1). This is shown by substituting (4, 1) into

$$y = 2x - 7 = x^2 - 6x + 9$$

$$1 = 2(4) - 7 = 4^2 - 6(4) + 9$$

$$1 = \qquad 1 = 1$$

● **PROBLEM** 41-9

Plot points of the curve corresponding to $y = \frac{1}{4}(x-2)^2$ for $x = 4, 3, 2, 1, 0$ and sketch the curve.

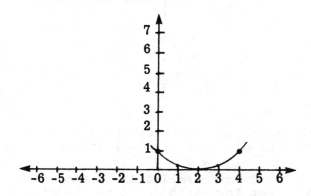

Solution: The equation of the curve is given to us as $y = \frac{1}{4}(x-2)^2$. In order to calculate the y coordinates corresponding to the given x coordinates, we substitute each x coordinate into the equation of the curve, and calculate y. This is done below:

$$x = 4 \quad , \quad y = \frac{1}{4}(4-2)^2 = 1$$

$$x = 3 \quad , \quad y = \frac{1}{4}(3-2)^2 = \frac{1}{4}$$

$$x = 2 \quad, \quad y = \tfrac{1}{4}(2-2)^2 = 0$$

$$x = 1 \quad, \quad y = \tfrac{1}{4}(1-2)^2 = \tfrac{1}{4}$$

$$x = 0 \quad, \quad y = \tfrac{1}{4}(0-2)^2 = 1.$$

We plot each point found above, connecting them by a smooth curve. (See figure). This gives us the parabola shown. Since the plotted points satisfy the given equation, the latter must be the equation of the parabola sketched.

● **PROBLEM** 41-10

Write the equation of the parabola whose focus has coordinates (0,2) and whose directrix has equation y = -2.(See figure)

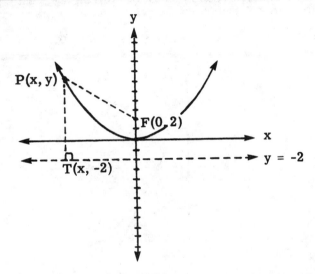

Solution: Since, by definition, each point lying on a parabola is equidistant from both the focus and directrix of the parabola, the origin must lie on the specific parabola described in the statement of the problem (see figure).

To find the equation of the parabola, choose a point P(x,y) lying on the parabola (see figure). By definition, then, the distance PT must equal the distance PF, where T lies on the directrix, directly below P. Since T also lies on y = -2, it has coordinates (x,-2). Using the distance formula, we find PF = PT

$$\sqrt{x^2 + (y-2)^2} = \sqrt{(y+2)^2}.$$

Squaring both sides of this equation, we obtain

$$x^2 + (y-2)^2 = (y+2)^2.$$

Expanding this, we get

$$x^2 + y^2 - 4y + 4 = y^2 + 4y + 4.$$

Subtracting y^2, $-4y$, and 4 from each side of this equation yields

$$x^2 = 8y.$$

Dividing both sides of this equation by 8 gives the equation of the parabola,

$$y = \frac{1}{8}x^2.$$

● **PROBLEM 41-11**

Find, both analytically and graphically, the points of intersection of the two curves whose equations are

$$2x + y - 4 = 0 \quad \text{and} \quad y^2 - 4x = 0.$$

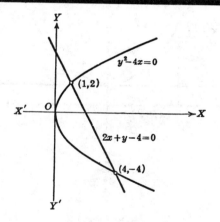

Solution: By using one equation to express y in terms of x, then substituting this into the second equation, we can form an equation in one variable, x. When we solve this equation, the values of x will be the abscissas of the points of intersection. The ordinates, y-values, can be found by substitution back into either original equation.

Since $2x + y - 4 = 0$, $y = 4 - 2x.$

Hence, by substitution $(4 - 2x)^2 - 4x = 0$

$$16 - 16x + 4x^2 - 4x = 0$$

which reduces to $x^2 - 5x + 4 = 0$.

Factoring, we obtain $(x - 4)(x - 1) = 0.$

769

This tells us x = 4 or x = 1.

Substituting these values into 2x + y - 4 = 0 gives us

y = 4 - 2(4)   and   y = 4 - 2(1)

y = - 4   y = 2

Hence, analytically we have found the points of intersection are (1, 2) and (4, - 4).

Graphically, the points of intersection can be
obtained by tracing the two given curves and reading the
coordinates of their intersection points. The graphs are
shown in the accompanying figure.

● **PROBLEM** 41-12

Consider the parabola $y^2 = 4px$. A tangent to the parabola
at point P_1 (x_1, y_1) is defined as the line that intersects
the parabola at point P_1 and nowhere else.

(a) Show that the slope of the tangent line is $\frac{2p}{y_1}$.

[Hint: Let the slope be m. Find the equation of the line
        passing through P_1 with slope m. What are the points
        of intersection of the tangent line and the parabola?
        For what values of m, would there be only one inter-
        section point?]

(b) Find the equation of the tangent line.

(c) Prove that the intercepts of the tangent line are
     $(-x_1, 0)$ and $(0, \frac{1}{2} y_1)$.

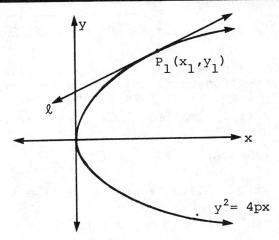

Solution: (a) We know three things about the tangent line ℓ.
First, ℓ is a line, therefore, its equation is of the form

770

$y = mx + b$. Second, P_1 is on ℓ. Therefore, (x_1, y_1) must satisfy the equation for ℓ. Therefore, $y_1 = mx_1 + b$. Third, ℓ is tangent to the parabola at point P_1. This implies two things: (1) P_1 is a point on the parabola. Thus, $y_1^2 = 4px_1$; and (2) P_1 is the only point common to both ℓ and the parabola.

To determine the line we must find m and b.

We first wish to find m. We know that ℓ and $y^2 = 4px$ intersect at only one point. Therefore, we solve for the intersection points of ℓ and $y^2 = 4px$ in terms of m. The value of m that permits only one point of intersection is the correct one.

Any point (X, Y) that is a point of intersection of the line and the parabola must satisfy the equations

   (i) $Y = mX + b$

   (ii) $Y^2 = 4pX$.

We have two equations in two unknowns X and Y. The first expresses Y in terms of X. Substituting this expression in (ii) does not change the validity of the second equation. In addition, it makes the second equation a single equation with a single unknown.

   (iii) $(mX + b)^2 = 4pX$.

Multiplying out and simplifying, we obtain:

   (iv) $m^2X^2 + (2mb - 4p)X + b^2 = 0$.

Then, by the quadratic formula, we can obtain a value for X in terms of m, p, and b.

   (v) $X = \dfrac{-(2mb-4p) \pm \sqrt{(2mb-4p)^2 - 4m^2b^2}}{2m^2}$

Because ℓ is a tangent, there should exist only one possible intersection point and, therefore, only one value of X. For equation (V) to have a single value, $\sqrt{(2mb - 4p)^2 - 4m^2b^2}$ must equal 0. In other words, for ℓ to be a tangent line,

   (vi) $(2mb-4p)^2 - 4m^2b^2 = 0$

   (vii) $4m^2b^2 - 16mbp + 16p^2 - 4m^2b^2 = 0$

   (viii) $16p^2 = 16mbp$ or $p = bm$.

Thus, we know $p = mb$. p is known; m and b are what we are solving for. To find values for m and b, we find another relationship between m and b.

Note that point $P_1(x_1, y_1)$ is a point on ℓ. Therefore, $y_1 = mx_1 + b$.

With this second set of simultaneous equations, $p = mb$ and $y_1 = mx_1 + b$, we can solve for m. Since $p = mb$, $b = \frac{p}{m}$. Substituting this in $y_1 = mx_1 + b$, we obtain $y_1 = mx_1 + \frac{p}{m}$. Multiplying by m, we obtain the expression $m^2 x_1 - my_1 + p = 0$. By the quadratic formula, we have $m = \dfrac{y_1 \pm \sqrt{y_1^2 - 4x_1 p}}{2x_1}$.

Note that (x_1, y_1) are points on the parabola; and therefore, $y_1^2 - 4x_1 p = 0$. The radical $\sqrt{y_1^2 - 4x_1 p} = 0$, and thus $m = \dfrac{y_1}{2x_1}$.

We are asked to show, though, that $m = \dfrac{2p}{y_1}$. Although not immediately obvious, this is the same as our answer:

$$m = \frac{y_1}{2x_1} = \frac{y_1}{2x_1} \cdot \frac{\frac{y_1}{4x_1}}{\frac{y_1}{4x_1}} = \frac{y_1^2/4x_1}{y_1/2} \left[\text{since } y_1^2 = 4px_1, \ p = \frac{y_1^2}{4x_1} \right]$$

$$m = \frac{p}{y_1/2} = \frac{2p}{y_1}$$

Thus the slope m of the tangent of the parabola $y^2 = 4px$ at point $P_1(x_1, y_1)$ is $\dfrac{2p}{y_1}$.

(b) The equation of the tangent line ℓ is $y = mx + b$. From part (a), we know $m = \dfrac{2p}{y_1}$. Furthermore, we know from part (a) that $p = mb$. Thus, $b = \dfrac{p}{m} = \dfrac{p}{(2p/y_1)} = \dfrac{y_1}{2}$, and the equation of ℓ becomes $Y = \left(\dfrac{2p}{y_1}\right)x + \left(\dfrac{y_1}{2}\right)$. Another acceptable form of the equation is obtained by letting $p = \dfrac{y_1^2}{4x_1}$: $Y = \left(\dfrac{y_1}{2x_1}\right)x + \dfrac{y_1}{2}$.

(c) Given the equation in part (b), we know that the y-intercept is given by b. Since $b = \dfrac{y_1}{2}$, the y-intercept is $(0, \dfrac{y_1}{2})$. An alternate method is to realize that the y-intercept is obtained by setting $x = 0$. Thus, $Y = \left[\left(\dfrac{y_1}{2x_1}\right) \cdot 0 + \left(\dfrac{y_1}{2}\right) = \dfrac{y_1}{2}\right]$.

To find the x-intercept, we set $y = 0$. Thus,

$$0 = \left(\dfrac{y_1}{2x_1}\right)X + \left(\dfrac{y_1}{2}\right), \quad \text{or} \quad \left(\dfrac{y_1}{2x_1}\right)X = -\dfrac{y_1}{2}.$$

Dividing by $\dfrac{y_1}{2}$ and multiplying by x_1, we obtain $X = -x_1$. Thus, $(-x_1, 0)$ is the x-intercept.

● **PROBLEM** 41-13

Discuss the rational integral equation
$$x^2 - 2xy + y^2 + 2x - 3 = 0,$$
and plot its graph.

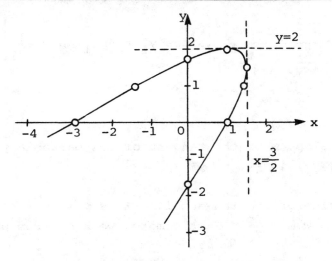

Solution: We solve this equation for y in terms of x.

$$x^2 - 2xy + y^2 + 2x - 3 = 0 \tag{1}$$

Subtract x^2 from both sides of equation (1):

$$\cancel{x^2} - 2xy + y^2 + 2x - 3 - \cancel{x^2} = 0 - x^2$$

$$-2xy + y^2 + 2x - 3 = -x^2 \tag{2}$$

Subtract $(2x-3)$ from both sides of equation (2):

$$-2xy + y^2 + 2x - 3 - (2x - 3) = -x^2 - (2x - 3)$$

$$-2xy + y^2 + \cancel{2x} - \cancel{3} - \cancel{2x} + \cancel{3} = -x^2 - 2x + 3$$

$$y^2 - 2xy = -x^2 - 2x + 3 \tag{3}$$

To complete the square on the left side of equation (3), take half the

773

coefficient of y and square it.

$$[\tfrac{1}{2}(-2x)]^2 = [-x]^2 = x^2 .$$

Add this value to both sides of equation (3):

$$y^2 - 2xy + x^2 = -x^2 - 2x + 3 + x^2$$
$$y^2 - 2xy + x^2 = -2x + 3$$
$$y^2 - 2xy + x^2 = 3 - 2x$$

or

$$(y - x)^2 = 3 - 2x .$$

Take the square root of each side:

$$\sqrt{(y-x)^2} = \pm \sqrt{3 - 2x}$$
$$y - x = \pm \sqrt{3 - 2x}$$

Add x to both sides:

$$y - x + x = x \pm \sqrt{3 - 2x}$$
$$y = x \pm \sqrt{3 - 2x} \qquad (4)$$

This explicit relation shows immediately that: (a) for every real value of x less than 3/2, there are two distinct real values of y; (b) for x = 3/2, there is only one y-value; namely, y = 3/2. (c) for x > 3/2, y is complex. Hence, we know that the graph of equation (1) will not extend to the right of the line x = 3/2. In addition, relation (4) enables us to compute the two values of y corresponding to each permissible value of x; thus, when x = 0, y = $\pm \sqrt{3}$, and when x = 1, we get y = 2 and y = 0. Substitute values for x and find the corresponding value of y. This is done in the following table:

x	$x \pm \sqrt{3-2x} = y$
-3	$-3 \pm \sqrt{3-2(-3)} = -3 \pm \sqrt{3+6} = -3 \pm \sqrt{9} = -3 \pm 3 = 0, -6$
$-\tfrac{3}{2}$	$-\tfrac{3}{2} \pm \sqrt{3-2\left(-\tfrac{3}{2}\right)} = -\tfrac{3}{2} \pm \sqrt{3+3} = -\tfrac{3}{2} \pm \sqrt{6} = -\tfrac{3}{2} \pm 2.45 = 0.95, -3.95$
0	$0 \pm \sqrt{3-2(0)} = 0 \pm \sqrt{3-0} = 0 \pm \sqrt{3} = \pm \sqrt{3} = \pm 1.73$
1	$1 \pm \sqrt{3-2(1)} = 1 \pm \sqrt{3-2} = 1 \pm \sqrt{1} = 1 \pm 1 = 2, 0$
$\tfrac{3}{2}$	$\tfrac{3}{2} \pm \sqrt{3-2\left(\tfrac{3}{2}\right)} = \tfrac{3}{2} \pm \sqrt{3-3} = \tfrac{3}{2} \pm \sqrt{0} = \tfrac{3}{2} \pm 0 = \tfrac{3}{2}$

These points may be plotted and joined by a smooth curve.

The graph obtained with the help of the above discussion is shown in the figure. By the methods of analytic geometry, it may be shown that this curve is a parabola, and additional characteristics of the curve may be determined.

CHAPTER 42

ELLIPSES

> **Basic Attacks and Strategies for Solving Problems in this Chapter. See pages 775 to 784 for step-by-step solutions to problems.**

An *ellipse* is frequently defined as the set of points (x, y) such that the sum of the distances from a point on the ellipse to two fixed points (called the *foci*) is a constant.

The general equation for an ellipse is

$$Ax^2 + Bx + Cy^2 + Dy + E = 0,$$

where $A \neq C$ (if $A = C$, this equation describes a circle). The standard equation for an ellipse is

$$\frac{(x-h)^2}{a^2} + \frac{(y-k)^2}{b^2} = 1.$$

We assume, for now, that a is larger than b. We then define the *eccentricity, e,* to be c/a, where

$$c = \sqrt{a^2 - b^2}.$$

Also, under this assumption, the longer axis of the ellipse is parallel to the x-axis. From the standard equation, the following can be determined: the center is (h, k), the foci are

$$(h \pm ae, k) = (h \pm c, k),$$

the vertices are

$$(h \pm a, k),$$

the length of the semimajor axis (i.e., half of the length of the longer axis) is a, and the length of the semiminor axis (i.e., half of the length of the shorter axis) is b.

Given some of the various quantities (e.g., the value of the eccentricity and the foci), it is possible, using the formulas given above, to deduce (h, k) and a and b, and then to derive the equation of the corresponding ellipse.

If the denominator under the y-term is larger than that under the x-term, it is called a, so that a is always the larger of a and b. In that case, the longer axis of the ellipse is parallel to the y-axis and the vertices and foci are on that axis (separated from the center by the same distance as in an ellipse with the other orientation).

The *latus rectum* is the line segment perpendicular to the major axis through a focus from edge to edge (as in the case of the parabola). In this case, there are two *latus recta*. The length is $\frac{2b^2}{a}$.

The area of an ellipse with a semimajor axis length of a and a semiminor axis length of b is $\pi a b$. This reduces to πr^2 for $y = b = r$ (in which case the ellipse becomes a circle).

● **PROBLEM 42-1**

Discuss the graph of $\frac{x^2}{25} + \frac{y^2}{9} = 1$.

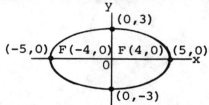

Solution: Since this is an equation of the form $\frac{x^2}{a^2} + \frac{y^2}{b^2} = 1$, with a = 5 and b = 3, it represents an ellipse. The simplest way to sketch the curve is to find its intercepts.

If we set x = 0, then

$$y = \sqrt{\left[1 - \frac{x^2}{25}\right]9} = \sqrt{\left[1 - \frac{0^2}{25}\right]9} = \pm 3$$

so that the y-intercepts are at (0,3) and (0,-3). Similarly, the x-intercepts are found for y = 0:

$$x = \sqrt{\left[1 - \frac{y^2}{9}\right]25}$$

$$= \sqrt{\left[1 - \frac{0^2}{9}\right]25}$$

$$= \pm 5$$

to be at (5,0) and (-5,0) (see figure). To locate the foci we note that

$$c^2 = a^2 - b^2 = 5^2 - 3^2$$

$$c^2 = 25 - 9 = 16$$

$$c = \pm 4.$$

The foci lie on the major axis of the ellipse. In this case it is the x-axis since a = 5 is greater than b = 3. Therefore, the foci are (+c,0), that is, at (-4,0) and (4,0). Therefore, the foci are at (-4,0) and (4,0). The sum of the distances from any point on the curve to the foci is 2a = 2(5) = 10.

● **PROBLEM 42-2**

In the equation of an ellipse,

$$4x^2 + 9y^2 - 16x + 18y - 11 = 0,$$

determine the standard form of the equation, and find the values of a, b, c, and e.

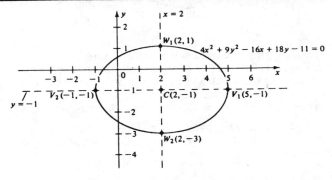

Solution: By completing the squares, we can arrive at the standard form of the equation, from which the values of the parameters can be determined. Thus,

$$4(x^2 - 4x + 4) + 9(y^2 + 2y + 1) = 36.$$

or $4(x - 2)^2 + 9(y + 1)^2 = 36.$ Dividing by 36,

$$\frac{(x - 2)^2}{9} + \frac{(y + 1)^2}{4} = 1.$$

Thus, the center of the ellipse is at (2, - 1). Comparing this equation with the general form,

$$\frac{x^2}{a^2} + \frac{y^2}{b^2} = 1, \quad \text{where } a > b, \text{ we see that } a = 3, b = 2.$$

$$c = \sqrt{a^2 - b^2} = \sqrt{5}.$$

Finally, $e = \frac{c}{a} = \frac{\sqrt{5}}{3} \approx 0.745.$

Find the equation of the ellipse which has vertices
V_1 (- 2, 6), V_2 (- 2, - 4), and foci F_1 (- 2, 4),
F_2 (- 2, - 2). (See figure.)

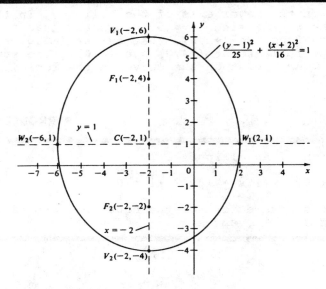

Solution: The major axis is on the line x = - 2, and
the center is at (- 2, 1). Hence, a, the length of the
semimajor axis, equals the difference between the y-
coordinates of V_1 (say) and the center, i.e., a = 5. From
the coordinates of the foci, c = 3. Since $h^2 = a^2 - c^2$,
b = $\sqrt{25 - 9}$ = 4, and the ends of the minor axis, on y = 1,
are at W_1 (2, 1) and W_2 (- 6, 1). The equation can now
be written, in the form

$$\frac{(y - k)^2}{a^2} + \frac{(x - h)^2}{b^2} = 1$$

$$\frac{(y - 1)^2}{25} + \frac{(x + 2)^2}{16} = 1.$$

The equation of an ellipse is $\frac{x^2}{a^2} + \frac{y^2}{b^2}$ = 1. Discuss what happens if a = b = r.

Solution: If a = b = r, the given equation becomes

$$\frac{x^2}{a^2} + \frac{y^2}{b^2} = \frac{x^2}{r^2} + \frac{y^2}{r^2} = 1. \qquad (1)$$

Multiplying the last branch of the equality in equation (1) by r^2 yields

$$x^2 + y^2 = r^2. \tag{2}$$

This is the equation of a circle with center at $C(0,0)$ and radius r. Hence, we see that a circle is a special case of an ellipse.

● PROBLEM 42-5

The latus rectum of an ellipse is the chord through either focus perpendicular to the major axis. Show that the length of the latus recta of ellipse

$$\frac{x^2}{a^2} + \frac{y^2}{b^2} = 1$$

is given by the formula $(2b^2)/a$.

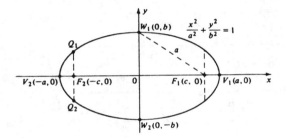

Solution: In the accompanying figure, ellipse

$$\frac{x^2}{a^2} + \frac{y^2}{b^2} = 1$$

has foci $(-c, 0)$ and $(c, 0)$ and the latus rectum $\overline{Q_1Q_2}$. We are asked to show $Q_1Q_2 = 2b^2/a$. To find the length of Q_1Q_2, we first find the coordinates of Q_1 and Q_2. To find $Q_1(x_1, y_1)$, note that both foci lie on the x-axis. Since the major axis is the line determined by the foci, the major axis is the x-axis. Further, since the latus recta are perpendicular to the major axis, this implies that the latus rectum Q_1Q_2 is a vertical line. Since every point on a vertical line has the same x-coordinate, then $Q_1(x_1,y_1)$ and $Q_2(x_2,y_2)$ must have the same x as $F_2(-c, 0)$. Thus, $x_1 = x_2 = -c$. To complete locating Q_1 and Q_2, we now find their y-coordinates. Note that Q_1 is a point on the ellipse and thus (x_1,y_1) must satisfy the equation

$$\frac{x^2}{a^2} + \frac{y^2}{b^2} = 1.$$

(i) $\qquad \dfrac{x_1{}^2}{a^2} + \dfrac{y_1{}^2}{b^2} = 1$

(ii) $\qquad \dfrac{(-c)^2}{a^2} + \dfrac{y_1{}^2}{b^2} = 1$

Solving for y_1, we obtain

$$\dfrac{y_1{}^2}{b^2} = 1 - \dfrac{c^2}{a^2} \quad \text{or,}$$

$$y_1 = b \sqrt{1 - \dfrac{c^2}{a^2}} = b\sqrt{\dfrac{a^2 - c^2}{a^2}} = \dfrac{b}{a}\sqrt{a^2 - c^2}.$$

In earlier problems, it was shown that a, b, and c must satisfy the equation $a^2 + b^2 = c^2$. Thus, $a^2 - c^2 = b^2$, and $\sqrt{a^2 - c^2} = b$. Therefore,

$$y_1 = \dfrac{b}{a}\sqrt{a^2 - c^2} = \dfrac{b^2}{a}$$

and the coordinates of Q_1 are $(-c, b^2/a)$.

By a similar method, we can show that the coordinates of Q_2 are $(-c, -b^2/a)$. Having the coordinates of Q_1 and Q_2, we can now solve for the length of the latus rectum Q_1Q_2.

$$Q_1Q_2 = \sqrt{(x_1 - x_2)^2 + (y_1 - y_2)^2}$$

$$= \sqrt{(-c - (-c))^2 + \left[\dfrac{b^2}{a} - \left(-\dfrac{b^2}{a}\right)\right]^2}$$

$$= \sqrt{0^2 + \left(\dfrac{2b^2}{a}\right)^2} = \sqrt{\left(\dfrac{2b^2}{a}\right)^2} = \left(\dfrac{2b^2}{a}\right)$$

Thus, the length of the latus rectum is $\dfrac{2b^2}{a}$

● **PROBLEM 42-6**

Consider a point $p_1(x_1, y_1)$ on the ellipse $b^2x^2 + a^2y^2 = a^2b^2$. A tangent to the ellipse at p_1 is a line through p_1 with no other point on the ellipse. Prove that if $y_1 \neq 0$, there is a tangent at p_1, its slope is m = $(-b^2x_1)/(a^2y_1)$ and its equation can be put in the form

$$\dfrac{x_1 x}{a^2} + \dfrac{y_1 y}{b^2} = 1.$$

Solution: This problem has two parts: (A) show that the tangent to the ellipse has slope m = $(-b^2x_1)/(a^2y_1)$, and (B) show the equation of the tangent is

$$\dfrac{x_1 x}{a^2} + \dfrac{y_1 y}{b^2} = 1.$$

(A) With the given information, we can determine the equation, $y = mx + z$, of the tangent line. (We use z instead of b for the y intercept to avoid confusion). To determine the equation, we must find the values of m and z. To solve for the two unknowns, we need a system of two equations in two unknowns.

The first equation derives from the given information that the tangent line ℓ contains point $p_1(x_1,y_1)$. Thus, (x_1, y_1) must satisfy the equation of ℓ or:

(i) $\qquad\qquad y_1 = mx_1 + z$

To find the second relationship, we use the fact that the tangent and the ellipse have only one common point.

Consider any line $y = mx + z$ intersecting the ellipse. We find the general formula for the intersection points in terms of m and z. Then we find the special relationship that must exist between m and z for there to be only one point of intersection. Combining this relation with the equation (i), we can then solve for our particular m and z.

Suppose (X, Y) are points of intersection of the line and the ellipse. Then X and Y must satisfy the equations:

(i) $Y = MX + Z$

(ii) $\dfrac{X^2}{a^2} + \dfrac{Y^2}{b^2} = 1.$

We have two simultaneous equations in two unknowns. Substituting the value Y in equation (i) into equation (ii), we obtain an equation totally in terms of X.

(iii) $\dfrac{X^2}{a^2} + \dfrac{(MX + Z)^2}{b^2} = 1$

Multiplying out and simplifying,

(iv) $\dfrac{X^2}{a^2} + \dfrac{M^2\,X^2}{b^2} + \dfrac{2MXZ}{b^2} + \dfrac{Z^2}{b^2} = 1$

(v) $X^2\left[\dfrac{1}{a^2} + \dfrac{M^2}{b^2}\right] + X\left[\dfrac{2MZ}{b^2}\right] + \left[\dfrac{Z^2}{b^2} - 1\right] = 0.$

Using the quadratic formula $x = \dfrac{-b \pm \sqrt{b^2 - 4ac}}{2a}$, we would normally obtain two values for X,

$x_1 = \dfrac{-b + \sqrt{b^2 - 4ac}}{2a}$ and $x_2 = \dfrac{-b - \sqrt{b^2 - 4ac}}{2a}.$

This would imply two different points of intersection (x_1, y_1) and (x_2, y_2). For there to be only one intersection point, and thus a tangent line, it must be true that $x_1 = x_2$ or $\dfrac{-b + \sqrt{b^2 - 4ac}}{2a} = \dfrac{-b - \sqrt{b^2 - 4ac}}{2a}.$

This is true only if $\sqrt{b^2 - 4ac} = 0$. Thus, we wish to find the relationship between M and Z that must always exist for $\sqrt{b^2 - 4ac} = 0$. In equation (v), $a = \left(\dfrac{1}{a^2} + \dfrac{M^2}{b^2}\right)$, $b = \dfrac{2MZ}{b^2}$, and $c = \dfrac{Z^2}{b^2} - 1$. Thus,

(vi) $0 = \sqrt{b^2 - 4ac}$

$$= \sqrt{\left(\dfrac{2MZ}{b^2}\right)^2 - 4\left(\dfrac{Z^2}{b^2} - 1\right)\left(\dfrac{1}{a^2} + \dfrac{M^2}{b^2}\right)}$$

(vii) $0^2 = \dfrac{4\,M^2Z^2}{b^4} - \dfrac{4\,Z^2}{a^2b^2} - \dfrac{4\,Z^2M^2}{b^4} + \dfrac{4}{a^2} + \dfrac{4\,M^2}{b^2}$

(viii) $\dfrac{Z^2}{a^2b^2} = \dfrac{1}{a^2} + \dfrac{M^2}{b^2}$

(ix) $Z^2 = b^2 + a^2M^2$.

Thus, if a line $y = Mx + z$ is to be tangent to the ellipse $\dfrac{x^2}{a^2} + \dfrac{y^2}{b^2} = 1$, then M and Z must satisfy the above equation. Since we wish the given ℓ to satisfy this equation, the slope m and intercept z must also satisfy equation (i). Thus, we have our two equations involving m and z.

(x) $y_1 = mx_1 + z$

(xi) $z^2 = b^2 + a^2m^2$

From equation (x), we have $z = y_1 - mx_1$. Substituting this result in (xi), we can solve for (xi) for **m.**

(xii) $(y_1 - mx_1)^2 = b^2 + a^2m^2$

(xiii) $y_1^2 - 2mx_1y_1 + m^2x_1^2 = b^2 + a^2m^2$

(xiv) $m^2(a^2 - x_1^2) + m(2x_1y_1) + (b^2 - y_1^2) = 0$.

Using the quadratic equation to solve for m, we obtain

(xv) $m = \dfrac{-(2x_1y_1) \pm \sqrt{4x_1^2y_1^2 - 4(b^2 - y_1^2)(a^2 - x_1^2)}}{2(a^2 - x_1^2)}$

(xvi) $m = \dfrac{-(2x_1y_1) \pm 2\sqrt{x_1^2y_1^2 - b^2a^2 + b^2x_1^2 + a^2y_1^2 - y_1^2x_1^2}}{2(a^2 - x_1^2)}$

(xvii) $m = \dfrac{-x_1y_1 \pm \sqrt{b^2x_1^2 + a^2y_1^2 - b^2a^2}}{a^2 - x_1^2}$

Note that x_1 and y_1 are points on the ellipse.

Therefore, $\dfrac{x_1{}^2}{a^2} + \dfrac{y_1{}^2}{b^2} = 1$ or $b^2x_1{}^2 + a^2y_1{}^2 - a^2b^2 = 0$.

Thus the expression under the radical is 0, and (xvii) becomes

(xviii) $m = \dfrac{-x_1y_1}{a^2-x_1{}^2}$

To get this into the desired form of the problem statement, note that since $\dfrac{x_1{}^2}{a^2} + \dfrac{y_1{}^2}{b^2} = 1$, $x_1{}^2 = a^2 - \dfrac{a^2}{b^2}y_1{}^2$

The denominator $a^2 - x_1{}^2$ thus becomes $\dfrac{a^2}{b^2}y_1{}^2$ and thus,

(xix) $m = \dfrac{-x_1y_1}{\dfrac{a^2}{b^2}y_1{}^2} = -\dfrac{b^2}{a^2}\dfrac{x_1}{y_1}$.

(B) The equation of the tangent is given by the formula $y = mx + z$ where $m = -\dfrac{b^2}{a^2}\dfrac{x_1}{y_1}$. To find z, remember from part (A) that $z^2 = b^2 + a^2m^2$. Thus,

$$z = \sqrt{b^2 + a^2m^2} = \sqrt{b^2 + a^2\left(+\dfrac{b^4}{a^4}\dfrac{x_1{}^2}{y_1{}^2}\right)}$$

$$= b\sqrt{1 + \dfrac{b^2}{a^2}\dfrac{x_1{}^2}{y_1{}^2}}$$

The equation of the tangent is

$$y = -\dfrac{b^2}{a^2}\dfrac{x_1}{y_1}x + b\sqrt{1 + \dfrac{b^2}{a^2}\dfrac{x_1{}^2}{y_1{}^2}}$$

To rework this into a more simplified form, note the

$$b\sqrt{1 + \dfrac{b^2}{a^2}\dfrac{x_1{}^2}{y_1{}^2}} \quad\text{equals}\quad \dfrac{b}{ay_1}\sqrt{a^2y_1{}^2 + b^2x_1{}^2}$$

$$= \dfrac{b}{ay_1}(ab)\sqrt{\dfrac{y_1{}^2}{b^2} + \dfrac{x_1{}^2}{a^2}} = \dfrac{b^2}{y_1}\sqrt{1} = \dfrac{b^2}{y_1} . \quad\text{Thus,}$$

(xx) $y = -\dfrac{b^2}{a^2}\dfrac{x_1}{y_1}x + \dfrac{b^2}{y_1}$

Multiplying both sides by $\dfrac{y_1}{b^2}$ and bringing the x term to the left we obtain:

(xxi) $\dfrac{x_1 x}{a^2} + \dfrac{y\, y_1}{b^2} = 1$.

Find the equation of the ellipse, given foci ($\pm$9,0) and eccentricity $\frac{3}{4}$.

Solution: Since the foci are on the x-axis, the major axis of the ellipse must also be on the x-axis.

$$c = 9$$

$$e = \frac{c}{a} = \frac{3}{4}, \quad a = \frac{c}{e} = \frac{9}{\frac{3}{4}} = 12.$$

Since $\quad a^2 - b^2 = c^2, \quad b = \sqrt{(12)^2-(9)^2} = \sqrt{63}$

Therefore, the equation required is

$$\frac{x^2}{144} + \frac{y^2}{63} = 1$$

Find the area of the ellipses

(a) $\frac{x^2}{9} + \frac{y^2}{25} = 1$ (b) $\frac{x^2}{144} + \frac{y^2}{256} = 1$ (c) $\frac{x^2}{64} + \frac{y^2}{49} = 1$

(d) $\frac{x^2}{81} + \frac{y^2}{16} = 1.$

Solution: The area of an ellipse whose equation is

$$\frac{x^2}{a^2} + \frac{y^2}{b^2} = 1 \quad \text{is} \quad \pi ab.$$

Applying this fact to the given equations, we find

(a) $\frac{x^2}{9} + \frac{y^2}{25} = 1, \quad a^2 = 9, \quad a = 3$

$$b^2 = 25, \quad b = 5$$

$$\text{Area} = \pi ab = 15\pi$$

(b) $\frac{x^2}{144} + \frac{y^2}{256} = 1, \quad a^2 = 144, \quad a = 12$

$$b^2 = 256, \quad b = 16$$

$$\text{Area} = \pi ab = 192\pi$$

(c) $\frac{x^2}{64} + \frac{y^2}{49} = 1, \quad a^2 = 64, \quad a = 8$

$$b^2 = 49, \quad b = 7$$

$$\text{Area} \quad = \pi ab = 56\pi$$

(d) $\dfrac{x^2}{81} + \dfrac{y^2}{16} = 1, \quad a^2 = 81, \quad a = 9$

$$b^2 = 16, \quad b = 4$$

$$\text{Area} \quad = \pi ab = 36\pi$$

CHAPTER 43

HYPERBOLAS

> **Basic Attacks and Strategies for Solving Problems in this Chapter. See pages 785 to 800 for step-by-step solutions to problems.**

A *hyperbola* is frequently defined as the set of points (x, y) such that the *difference* of the distances from a point on the hyperbola to two fixed points (called the foci) is a constant. (One can use this fact in conjunction with the formula for the distance between two points in the plane to derive a general equation for a hyperbola — cf. Chapter 38.)

The graph of the hyperbola has two sections, both bowl-shaped, opening away from each other, both totally contained within sections of the plane marked out by two straight lines intersecting at the center of the hyperbola. These two lines are called the *asymptotes*.

The general equation for a hyperbola is

$$Ax^2 + Bx + Cy^2 + Dy + E = 0,$$

where A and C have different signs.

One standard equation for a hyperbola is

$$\frac{(x-h)^2}{a^2} - \frac{(y-k)^2}{b^2} = 1.$$

(This results in a hyperbola whose central axis through the two foci is parallel to the x-axis.) We then define the *eccentricity*, e, to be c/a, where $c = \sqrt{a^2 + b^2}$. From the standard equation, the following can be determined: the center is (h, k), the foci are

$$(h \pm ae, k) = (h \pm c, k),$$

the vertices are

$$(h \pm a, k),$$

the length of the semi-transverse axis (i.e., the distance between the center and a vertex) is a, and the length of the "semi-conjugate" axis (perpendicular to the semi-transverse axis) is b. The two linear equations for the asymptotes can be multiplied together and written as

$$\frac{(x-h)^2}{a^2} - \frac{(y-k)^2}{b^2} = 0.$$

The directrices are $x = h \pm \frac{a}{e}$.

One can construct a rectangle between the two sections of a hyperbola with its center at the center of the hyperbola and bounded by the vertices and with corners on the asymptotes. This rectangle is sometimes called the *fundamental rectangle*. The dimensions of this rectangle are $2a$ and $2b$.

If the y-term has a positive sign and the x-term a negative, the denominator under the positive term is still associated with a and the standard form is written

$$\frac{(y-k)^2}{a^2} - \frac{(x-h)^2}{b^2} = 1.$$

In this case, the central axis is parallel to the y-axis. As in the case of the ellipse, the formulas for the other associated points, lines and concepts are appropriately modified.

It is also possible that the central axis is not parallel to either of the coordinate axes. If the central axis coincides with the diagonal lines $y = x$ or $y = -x$, the hyperbola has an equation of the form $xy = d$. If d is positive, the central axis coincides with $y = x$, and if d is negative, with $y = -x$.

Two hyperbolas are said to be *conjugate* if they share the same asymptotes. The conjugate

$$\frac{(x-h)^2}{a^2} - \frac{(y-k)^2}{b^2} = 1 \text{ is } -\frac{(x-h)^2}{a^2} + \frac{(y-k)^2}{b^2} = 1$$

and vice versa. These two hyperbolas share the same center and fundamental rectangle, their foci lie on the same circle (centered at the center of both hyperbolas), and the transverse axis of each coincides with the conjugate axis of the other.

● PROBLEM 43-1

Discuss the graph of $\frac{x^2}{9} - \frac{y^2}{9} = 1$.

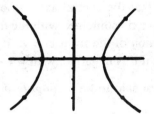

Solution: $\frac{x^2}{9} - \frac{y^2}{9} = 1$ is an equation of the form

$\frac{x^2}{a^2} - \frac{y^2}{b^2} = 1$ with a = 3 and b = 3. Therefore the graph is a hyperbola. The x-intercepts are found by setting y = 0:

$$\frac{x^2}{9} - \frac{0^2}{9} = 1$$

$$x^2 = 9$$

$$x = \pm 3.$$

Thus, the x-intercepts are at (-3,0) and (3,0). There are no y-intercepts since for x = 0 there are no real values of y satisfying the equation, i.e., no real value of y satisfies

$$\frac{0^2}{9} - \frac{y^2}{9} = 1$$

$$y^2 = -9, \quad y = \sqrt{-9}.$$

Solving the original equation for y:

$$y = \sqrt{\left(1 - \frac{x^2}{9}\right)(-9)} \quad \text{or} \quad y = \sqrt{x^2 - 9}$$

shows that there will be no permissible values of x in the interval $-3 < x < 3$. Such values of x do not yield real values for y. For $x = 5$ and $x = -5$ use the equation for y to obtain the ordered pairs (5,4), (5,-4), (-5,4), and (-5,-4) as indicated in the figure. The foci of the hyperbola are located at (±c,0), where

$$c^2 = a^2 + b^2$$

$$c^2 = 3^2 + 3^2 = 9 + 9 = 18$$

$$c = \pm\sqrt{18} = \pm 3\sqrt{2}.$$

Therefore, the foci are at $(-3\sqrt{2},0)$ and $(3\sqrt{2},0)$.

● **PROBLEM 43-2**

Draw the graph of xy = 6.

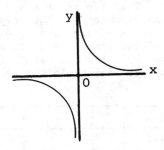

Solution: Since the product is positive the values of x and y must have the same sign, that is, when x is positive y must also be positive and when x is negative then y is also negative. Moreover, neither x nor y can be zero(or their product would be zero not 6), so that the graph never touches the coordinate axes. Solve for y and we obtain y = 6/x. Substituting values of x into this equation we construct the following chart:

x:	-6	-3	-2	-1	1	2	3	6
y:	-1	-2	-3	-6	6	3	2	1

The graph is obtained by plotting the above points and then joining them with a smooth curve, remembering that the curve can never cross a coordinate axis. The graph of the equation, xy = k, is a hyperbola for

all nonzero real values of k. If k is negative, then x and y must
have opposite signs, and the graph is in the second and fourth quadrants
as opposed to the first and third.

● PROBLEM 43-3

Graph the equation xy = -4.

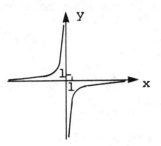

Solution: The graph of the equation, xy = c, is a hyperbola for all
non-zero real values of c. In this case, c = -4 and thus the particular equation is xy = -4. Since the product is negative, the values
of x and y must have different signs. If x is positive, then y
is negative and so part of the graph lies in quadrant IV. On the other
hand, if x is negative, then y is positive and the other part of
the hyperbola is located in quadrant II. If we solve for y, then
y = -4/x and x ≠ 0. Then the graph never touches the y-axis. Thus
the line x = 0 is an asymptote of the graph. On the other hand,
solving for x, we have x = -4/y and y ≠ 0. Thus the graph never
crosses the x-axis and the line y = 0 is an asymptote. We now pre-
pare a table of values by selecting values for x and finding the corresponding values for y. See table and graph.

x	$\frac{-4}{x}$ =	y
-3	$\frac{-4}{-3}$	$\frac{4}{3} = 1\frac{1}{3}$
-2	$\frac{-4}{-2}$	2
-1	$\frac{-4}{-1}$	4
1	$\frac{-4}{1}$	-4
2	$\frac{-4}{2}$	-2
3	$\frac{-4}{3}$	$\frac{-4}{3} = -1\frac{1}{3}$

● PROBLEM 43-4

Sketch the graph of the equation y = 2/x.

<u>Solution:</u> Substitute values for x and then find the cor-responding values for y. This is done in the following table.

x	$y = \frac{2}{x}$
-4	$-\frac{1}{2}$
-3	$-\frac{2}{3}$
-2	-1
-1	-2
$-\frac{1}{2}$	-4
$\frac{1}{2}$	4
1	2
2	1
3	$\frac{2}{3}$
4	$\frac{1}{2}$

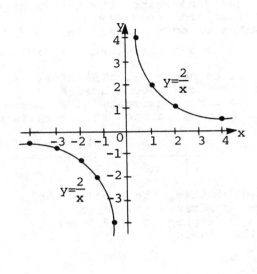

The graph is shown in the figure. This graph is an exam-ple of an equilateral hyperbola. Notice in the graph that, when x takes on larger and larger positive values, y gets closer and closer to 0. When x takes on larger and larger negative values, y also gets closer and closer to 0. Also, when x gets closer and closer to 0, y either takes on larger and larger positive values or larger and larger negative values. Note also that x cannot be 0, since $y = \frac{2}{x}$ $= \frac{2}{0}$ is not defined.

● PROBLEM 43-5

Discuss the graph of the function $y = \frac{12}{x^2}$.

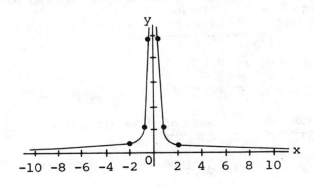

<u>Solution:</u> Intercepts: Since division by zero is not de-
fined, x cannot = 0, hence there is no y-intercept. Similarly, there is no x-intercept since y cannot = 0, because
no value of x allowed in the given equation yields a value
of y = 0. Symmetry: The curve is symmetric with respect
to the y-axis since the x-term appears squared in the given
function and hence f(x) = f(-x). Domain: There is no
limitation on x, except that x ≠ 0. Range: Since

$y = \dfrac{12}{x^2}$ is a positive number divided by a positive number,

y must be positive. Therefore, the curve exists only in
the first and second quadrants. Plotting: We note that,
in the first quadrant, as x increases y decreases. Several
points to illustrate this are listed in the following table.

x	0.5	1	2	3	...	10
y	48	12	3	1.3	...	0.12

After plotting these points, and tracing the curve in the
first quadrant, the second branch is drawn in quadrant II,
using the principle of symmetry. The curve is illustrated
in the figure.

● **PROBLEM** 43-6

Find the equation for the set of points the difference of whose
distances from (5,0) and (-5,0) is 6 units.

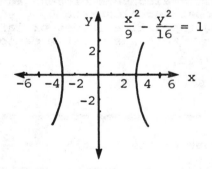

$$\frac{x^2}{9} - \frac{y^2}{16} = 1$$

<u>Solution:</u> We find the desired equation by choosing an arbitrary point
(x,y) and computing the difference of its distance from (5,0) and (-5,0).
Applying the distance formula for the distance between two points (a_1,b_1)
and (a_2,b_2), $d = \sqrt{(a_1-a_2)^2 + (b_1-b_2)^2}$. From (x,y) to (5,0) is d_1,
from (x,y) to (-5,0) is d_2.

$$d_1 = \sqrt{(x-5)^2 + y^2} \; ; \quad d_2 = \sqrt{(x+5)^2 + y^2} \; ;$$

We are told that the difference of these distances, d_2-d_1, is 6.
Hence, $\sqrt{(x+5)^2 + y^2} - \sqrt{(x-5)^2 + y^2} = 6$.

$$\sqrt{(x+5)^2 + y^2} = 6 + \sqrt{(x-5)^2 + y^2}$$

789

Squaring both sides:

$$\left(\sqrt{(x+5)^2 + y^2}\right)^2 = \left(6 + \sqrt{(x-5)^2 + y^2}\right)^2$$

$$\left(\sqrt{(x+5)^2 + y^2}\right)^2 = 36 + 12\sqrt{(x-5)^2 + y^2} + \left(\sqrt{(x-5)^2 + y^2}\right)^2$$

Since $\left(\sqrt{a}\right)^2 = \sqrt{a}\sqrt{a} = \sqrt{a \cdot a} = \sqrt{a^2} = a$,

$$\left(\sqrt{(x+5)^2 + y^2}\right)^2 = (x+5)^2 + y^2 \quad \text{and} \quad \left(\sqrt{(x-5)^2 + y^2}\right)^2 = (x-5)^2 + y^2 .$$

Thus we obtain,

$$(x+5)^2 + y^2 = 36 + 12\sqrt{(x-5)^2 + y^2} + (x-5)^2 + y^2$$

$$x^2 + 10x + 25 + y^2 = 36 + 12\sqrt{(x-5)^2 + y^2} + x^2 - 10x + 25 + y^2$$

Adding $-\left(x^2 + 25 + y^2\right)$ to both sides,

$$x^2 + 10x + 25 + y^2 - \left(x^2 + 25 + y^2\right) = 36 + 12\sqrt{(x-5)^2 + y^2} + x^2$$
$$- 10x + 25 + y^2 - \left(x^2 + 25 + y^2\right)$$

$$10x = 36 + 12\sqrt{(x-5)^2 + y^2} - 10x$$

Adding $-36 + 10x$ to both sides,

$$10x - 36 + 10x = 36 + 12\sqrt{(x-5)^2 + y^2} - 10x - 36 + 10x$$
$$20x - 36 = 12\sqrt{(x-5)^2 + y^2}$$

Dividing both sides by 4,

$$5x - 9 = 3\sqrt{(x-5)^2 + y^2}$$

Squaring both sides,

$$(5x-9)^2 = \left(3\sqrt{(x-5)^2 + y^2}\right)^2$$

$$(5x-9)(5x-9) = 3^2\left(\sqrt{(x-5)^2 + y^2}\right)^2$$

$$25x^2 - 90x + 81 = 9[(x-5)^2 + y^2]$$

$$25x^2 - 90x + 81 = 9\left(x^2 - 10x + 25 + y^2\right)$$

$$25x^2 - 90x + 81 = 9x^2 - 90x + 225 + 9y^2$$

Adding $-\left(9x^2 - 90x + 9y^2\right)$ to both sides,

$$25x^2 - 90x + 81 - \left(9x^2 - 90x + 9y^2\right) = 9x^2 - 90x + 225 + 9y^2$$
$$- \left(9x^2 - 90x + 9y^2\right)$$

$$25x^2 - 9x^2 - 9y^2 + 81 = 225$$
$$16x^2 - 9y^2 + 81 = 225$$

Adding -81 to both sides,

$$16x^2 - 9y^2 = 144 .$$

Dividing both sides by 144,

$$\frac{16x^2}{144} - \frac{9y^2}{144} = \frac{144}{144}$$

or

$$\frac{x^2}{9} - \frac{y^2}{16} = 1 \text{ ,}$$

which is the standard form for the equation of a hyperbola.

From the form of the equation we can determine its graph. When the center is at the origin and its vertices are at $(a,0)$ and $(-a,0)$, the equation of the hyperbola is

$$\frac{x^2}{a^2} - \frac{y^2}{b^2} = 1 \text{ .}$$

In this case, the vertices are $(3,0)$ and $(-3,0)$ since $a^2 = 9$ and $a = \pm 3$ (see figure).

However, when the vertices lie on the y-axis, the equation of the hyperbola is

$$\frac{y^2}{a^2} - \frac{x^2}{b^2} = 1.$$

The vertices would then be $(0,a)$ and $(0,-a)$.

● **PROBLEM 43-7**

Find the equation of the conjugate of the hyperbola $4x^2 - y^2 - 40 = 0$.

Solution: From $4x^2 - y^2 - 40 = 0$, one has

$$4x^2 - y^2 = 40, \qquad \frac{x^2}{10} - \frac{y^2}{40} = 1,$$

that is, $a = \sqrt{10}$, $b = 2\sqrt{10}$. Two hyperbolas having the transverse axis of each as the conjugate axis of the other are called conjugate hyperbolas. They have the same center, the same asymptotes, and their foci lie on a circle whose center is the common center of the hyperbolas.

The center of $\frac{x^2}{10} - \frac{y^2}{40} = 1$ is at $O(0,0)$, and the asymptotes are $y = \pm \frac{b}{a} x$, i.e., $y = \pm 2x$.

The conjugate of $4x^2 - y^2 - 40 = 0$ is easily obtained by switching the signs of $\frac{x^2}{a^2}$ and $\frac{y^2}{b^2}$. For example, we are given

$+ \frac{x^2}{10} - \frac{y^2}{40} = 1$. The conjugate is then $-\frac{x^2}{10} + \frac{y^2}{10} = 1$, i.e.,

$$\frac{y^2}{40} - \frac{x^2}{10} = 1.$$

This may be obtained directly from the conjugate of $4x^2 - y^2 - 40 = 0$, which is $-4x^2 + y^2 - 40 = 0$.

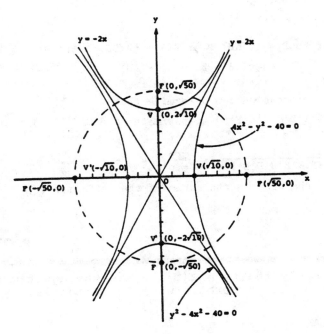

Prove that the ratio of the areas of any polygon circumscribed about a given circle is equal to the ratio of their perimeters.

Solution: Let ABCD and A'B'C'D'E'F' be any two polygons circumscribed about circle R. We wish to show

$$\frac{\text{Area ABCD}}{\text{Area A'B'C'D'E'F'}} = \frac{\text{Perimeter ABCD}}{\text{Perimeter A'B'C'D'E'F'}} .$$

First, we draw the segments connecting each vertex to the center of the inscribed circle, thus dividing the polygons into triangles. This will help us to determine an expression for area.

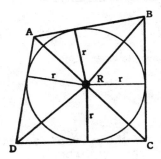 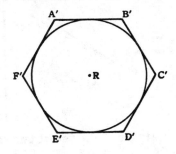

Area of ABCD =

area of $\triangle$ARB + area of $\triangle$BRC + area of $\triangle$DRC + area of $\triangle$ARD.

The area of a triangle is ½(base)(altitude). To find the altitude, note that each side is tangent to ⊙R. Therefore, the altitude to any side will be represented by a radius of ⊙R since a radius intersecting the tangent line at the point of tangency is perpendicular to the tangent.

Let r = the length of the radius of ⊙R. Then,

792

Area of ABCD = $\frac{1}{2}$r(AB) + $\frac{1}{2}$r(BC) + $\frac{1}{2}$r(CD) + $\frac{1}{2}$r(AD)

$\qquad$ = $\frac{1}{2}$r(AB + BC + CD + AD)

$\qquad$ ≈ $\frac{1}{2}$r(perimeter ABCD) = $\frac{1}{2}$rp.

By a similar process we can determine that

$\qquad$ Area of A'B'C'D'E'F' = $\frac{1}{2}$r(perimeter A'B'C'D'E'F') = $\frac{1}{2}$rp' .

Hence,

$$\frac{\text{Area of ABCD}}{\text{Area of A'B'C'D'E'F'}} = \frac{\frac{1}{2}rp}{\frac{1}{2}rp'} = \frac{p}{p'}$$

The ratio of the areas = the ratio of the perimeters.

● **PROBLEM** 43-9

Show informally that y = ±(b/a)x are the equations of the asymptotes of the hyperbola whose equation is

$$\frac{x^2}{a^2} - \frac{y^2}{b^2} = 1.$$

Solution: As we trace an hyperbola far to the right (or to the left), we notice that the hyperbola become arbitrarily

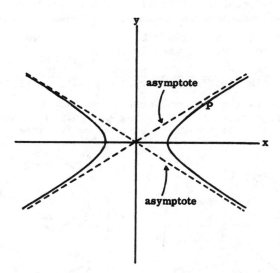

close to a set of intersecting straight lines (see figure). These lines are the asymptotes of the hyperbola.

Note from the figure that as P moves out along the hyperbola, it approaches the asymptotes. We may use this fact, along with the equation of the hyperbola, to find the equations of the asymptotes. First, we solve

$\frac{x^2}{a^2} - \frac{y^2}{b^2} = 1$ for y in terms of x. Adding $- \frac{x^2}{a^2}$

to both sides of this equation, we obtain

793

$$-\frac{y^2}{b^2} = -\frac{x^2}{a^2} + 1$$

Multiplying both sides by $-b^2$ yields:

$$y^2 = \frac{b^2}{a^2} x^2 - b^2 \tag{1}$$

Factoring b^2 from the right side of this equation, and taking the square root of both sides, we obtain

$$y = \pm b \sqrt{\frac{x^2}{a^2} - 1} \tag{2}$$

The $(-)$ sign appears in (2) because this is also a solution of equation (1). Now, as P moves out along the hyperbola, its x coordinate gets very large. Since a is a fixed constant; sooner or later $x > a$. This means that $x^2 \gg a^2$; or $\frac{x^2}{a^2} \gg 1$.

Hence, we may neglect 1 in comparison with $\frac{x^2}{a^2}$ in equation (2). We then find

$$y = \pm b \sqrt{\frac{x^2}{a^2}}$$

$$y = \pm \frac{b}{a} x$$

These are the equations of the asymptotes of the given hyperbola.

● **PROBLEM 43-10**

Consider the equation

$$x^2 - 4y^2 + 4x + 8y + 4 = 0.$$

Express this equation in standard form, and determine the center, the vertices, the foci, and the eccentricity of this hyperbola. Describe the fundamental rectangle and find the equations of the 2 asymptotes.

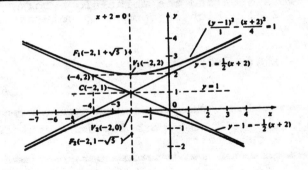

Solution: Rewrite the equation by completing the squares, i.e.,

$$(x^2 + 4x + 4) - 4(y^2 - 2y + 1) = -4$$

or $\quad (x + 2)^2 - 4(y - 1)^2 = -4$

or, dividing, rearranging terms,

$$\frac{(y - 1)^2}{1} - \frac{(x + 2)^2}{4} = 1.$$

The center, located at (h, k) in the equation

$$\frac{(y - k)^2}{a^2} - \frac{(x - h)^2}{b^2} = 1 \quad \text{is, therefore, at}$$

(- 2, 1). Furthermore, a = 1, b = 2. Thus,

$$c = \sqrt{1^2 + 2^2} = \sqrt{5}, \text{ and } e = \sqrt{5}.$$

The vertices are displaced ± a from the center while the foci are displaced ± c (along the transverse axis). Therefore, the vertices are (- 2, 1 ± 1) and the foci are (- 2, 1 ± $\sqrt{5}$).

By definition, the fundamental rectangle is the rectangle whose vertices are at (h ± b, k ± a). Hence, in this example, the coordinates of the vertices of the rectangle are (0, 2), (- 4, 2), (- 4, 0), and (0, 0). The equations of the two asymptotes are determined by finding the slopes of the lines passing through the center of the hyperbola and two of the vertices of its fundamental rectangle (see figure). Then,

$$m = \frac{\Delta y}{\Delta x} = \pm \tfrac{1}{2}$$

gives the two slopes and the point-slope form, choosing the point (- 2, 1) which is common to both asymptotes, gives

$$y - 1 = \pm \tfrac{1}{2}(x + 2).$$

● **PROBLEM 43-11**

Find the equation of the hyperbola with vertices V_1(8, 0), V_2(2, 0) and eccentricity e = 2.

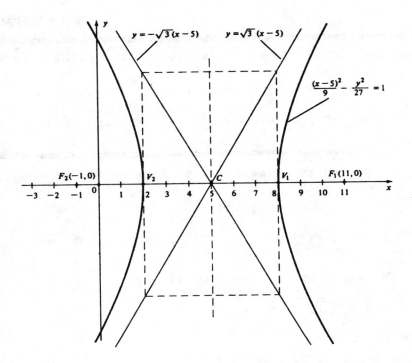

Solution: There are two basic forms of the equation of an hyperbola that is not rotated with respect to the coordinate axes:

$$\frac{(x - h)^2}{a^2} - \frac{(y - k)^2}{b^2} = 1 \quad \text{and} \quad \frac{(y - k)^2}{a^2} - \frac{(x - h)^2}{b^2} = 1.$$

Which form is appropriate depends upon whether the transverse axis is parallel to the x-axis or to the y-axis, respectively. To determine the equation of an hyperbola, then, it is necessary to first discover which equation applies, then to solve for the constants h, k, a, c. The information about the vertices implies that the transverse axis is the x- axis.

Thus, the first form of the equation for an hyperbola applies. In this case, the center, which is the average of the vertices is at (5, 0). The distance between the vertices is 2a = 6; therefore a = 3. In order to determine the value of b, we use the relation between eccentricity, e, c, and a: e = c/a. Thus, c = e · a = 2 · 3 = 6.

But b = $\sqrt{c^2 - a^2}$ = $\sqrt{6^2 - 3^2}$ = $\sqrt{27}$ ≈ 5.2 Substituting for h, k, a, b, we have

$$\frac{(x - 5)^2}{9} - \frac{y^2}{27} = 1,$$

or $3x^2 - y^2 - 30x + 48 = 0.$

Show that if the coordinates (x, y) of a point P satisfy

$$\frac{x^2}{9} - \frac{y^2}{16} = 1,$$

then $|F_1P - F_2P| = 6$, where $F_1(-5, 0)$ and $F_2(5, 0)$ are the foci.

Solution: We know that P(x, y) lies on the hyperbola

$$\frac{x^2}{9} - \frac{y^2}{16} = 1,$$

with foci at $F_1(-5, 0)$ and $F_2(5, 0)$. We must show that the distance $|F_1P - F_2P| = 6$.

We may write F_1P and F_2P as

$$F_1P = \sqrt{(x + 5)^2 + y^2} \qquad (1)$$

$$F_2P = \sqrt{(x - 5)^2 + y^2} \qquad (2)$$

where we have used the distance formula in (1) and (2). Hence,

$$|F_1P - F_2P| = \sqrt{(x + 5)^2 + y^2} - \sqrt{(x - 5)^2 + y^2} \qquad (3)$$

We cannot show that $|F_1P - F_2P| = 6$ unless we can relate x and y to some known value. We can do this because we know that the coordinates x and y satisfy

$$\frac{x^2}{9} - \frac{y^2}{16} = 1.$$

We must solve this equation for y^2 (or x^2), and substitute the result into equation (3). We start with

$$\frac{x^2}{9} - \frac{y^2}{16} = 1$$

Multiplying both sides by -16,

$$y^2 - \frac{16}{9} x^2 = -16$$

Adding $\frac{16}{9} x^2$ to both sides,

$$y^2 = \frac{16}{9} x^2 - 16 \qquad (4)$$

Substituting (4) in (3)

$$|F_1P - F_2P| = \sqrt{(x+5)^2 + \frac{16}{9} x^2 - 16} - \sqrt{(x-5)^2 + \frac{16}{9} x^2 - 16}$$

Expanding under the radicals in this last equation,

$$|F_1P - F_2P| = \sqrt{x^2 + 10x + 25 + \frac{16}{9}x^2 - 16} - \sqrt{x^2 - 10x + 25 + \frac{16}{9}x^2 - 16}$$

Simplifying,

$$|F_1P - F_2P| = \sqrt{\frac{25}{9}x^2 + 10x + 9} - \sqrt{\frac{25}{9}x^2 - 10x + 9}$$

Factoring $\frac{25}{9}$ from each term under each radical,

$$|F_1P - F_2P| = \sqrt{\frac{25}{9}\left(x^2 + \frac{90}{25}x + \frac{81}{25}\right)} - \sqrt{\frac{25}{9}\left(x^2 - \frac{90}{25}x + \frac{81}{25}\right)}$$

Now, $\left(x^2 + \frac{90}{25}x + \frac{81}{25}\right) = \left(x + \frac{9}{5}\right)^2$

$\left(x^2 - \frac{90}{25}x + \frac{81}{25}\right) = \left(x - \frac{9}{5}\right)^2$

Using the last 2 expressions in the equation for $|F_1P - F_2P|$ yields

$$|F_1P - F_2P| = \sqrt{\frac{25}{9}\left(x + \frac{9}{5}\right)^2} - \sqrt{\frac{25}{9}\left(x - \frac{9}{5}\right)^2}$$

Taking the indicated square roots, we obtain

$$|F_1P - F_2P| = \frac{5}{3}\left(x + \frac{9}{5}\right) - \frac{5}{3}\left(x - \frac{9}{5}\right)$$

$$|F_1P - F_2P| = \frac{90}{15} = 6$$

Notice that the term in x disappears. This is very convenient; if our final expression were to involve x, we would have to leave it in that form. There is no other equation given in the problem which relates x with known quantities.

● **PROBLEM 43-13**

By definition, if an hyperbola has foci $F_1(-c, 0)$ and $F_2(c, 0)$, and $P(x, y)$ is a point on the hyperbola, then $|PF_1 - PF_2| = k$, where k is a constant such that $k < F_1F_2 = 2c$.

Assuming that the above holds, and defining a constant such that $a = K/2$, and a constant b such that $b^2 = c^2 - a^2$, prove that the equation of the hyperbola is

$$\frac{x^2}{a^2} - \frac{y^2}{b^2} = 1.$$

<u>Solution</u>:   From the given facts,

$$|PF_1 - PF_2| = k = 2a. \tag{1}$$

By the distance formula,

$$PF_1 = \sqrt{(x + c)^2 + y^2} \tag{2}$$

$$PF_2 = \sqrt{(x - c)^2 + y^2} \tag{3}$$

Using (3) and (2) in (1), we obtain

$$\left|\sqrt{(x + c)^2 + y^2} - \sqrt{(x - c)^2 + y^2}\right| = 2a \tag{4}$$

Since any real number (positive or negative) times itself must be positive, we may write, using equation (4),

$$\left(\left|\sqrt{(x+c)^2+y^2} - \sqrt{(x-c)^2+y^2}\right|\right)^2 = \left(\sqrt{(x+c)^2+y^2} - \sqrt{(x-c)^2+y^2}\right)^2 = 4a^2.$$

Expanding the last branch of the last equation,

$$(x+c)^2+y^2+(x-c)^2+y^2-2\sqrt{((x+c)^2+y^2)((x-c)^2+y^2)} = 4a^2.$$

Expanding the last equation again, and regrouping terms

$$2x^2+ 2y^2+2c^2-2\sqrt{((x+c)^2+y^2)((x-c)^2+y^2)} = 4a^2.$$

Subtracting $- 2\sqrt{((x+c)^2+y^2)((x-c)^2+y^2)}$ and $4a^2$ from both sides of the last equation yields

$$2x^2+2y^2+2c^2-4a^2= 2\sqrt{((x+c)^2+y^2)((x-c)^2+y^2)}$$

Dividing both sides by 2

$$x^2 + y^2 + c^2 -2a^2 = \sqrt{((x + c)^2 + y^2)((x - c)^2 + y^2)}$$

Squaring both sides

$$(x^2 + y^2 + c^2 -2a^2)^2 = ((x + c)^2 + y^2)((x - c)^2 + y^2)$$

Expanding the right side

$$(x^2 + y^2 + c^2 -2a^2)^2 = (x^2 + y^2 +c^2 + 2cx)(x^2 + y^2 + c^2 - 2cx)$$

Expanding both sides using the distributive law,

$$(x^2+y^2+c^2)^2+4a^4-4a^2(x^2+y^2+c^2)=(x^2+y^2+c^2)^2-4c^2x^2$$

Cancelling like terms on both sides and expanding the remaining terms on the left side of the last equation, we find

$$4a^4 - 4a^2x^2 - 4a^2y^2 - 4a^2c^2 = - 4c^2x^2 \tag{5}$$

But, we know that

$$c^2 - a^2 = b^2$$

Adding a^2 to both sides of this equation,

$$c^2 = a^2 + b^2 \qquad\qquad (6)$$

Using equation (6) in equation (5)

$$4a^4 - 4a^2x^2 - 4a^2y^2 - 4a^4 - 4a^2b^2 = -4a^2x^2 - 4b^2x^2$$

Subtracting $-4a^2x^2$ and $-4b^2x^2$ from both sides of the last equation

$$-4a^2y^2 - 4a^2b^2 + 4b^2x^2 = 0$$

Dividing through by $4a^2b^2$ gives

$$-\frac{y^2}{b^2} - 1 + \frac{x^2}{a^2} = 0$$

Adding 1 to both sides of the last equation yields the desired result

$$\frac{x^2}{a^2} - \frac{y^2}{b^2} = 1.$$

CHAPTER 44

TRANSFORMATION OF COORDINATES

> **Basic Attacks and Strategies for Solving Problems in this Chapter. See pages 801 to 811 for step-by-step solutions to problems.**

There are two major methods for transforming coordinates: *translation,* by which the origin is moved and the equation of the figure is adjusted to correspond to the new origin, and *rotation,* by which the origin remains the same, but the coordinate axes are rotated.

Using translation, one can eliminate any linear (first degree) terms in the equation of a figure. The center (or vertex) of the figure becomes the new origin.

Using rotation, one can eliminate the cross-product term, xy. The coordinate axes become parallel (or perpendicular) to the central axis of the figure.

It is also possible to combine both techniques.

Translation is done by substituting x' for $x - a$ and y' for $y - b$ with appropriate values of a and b. The resulting equation should have only x' and y' in it, and the form of the equation should be the form with the center (or vertex, if a parabola) at the origin. For example, given

$$y = x^2 - 2x + 2 = (x - 1)^2 + 1$$

we can rewrite this as $y - 1 = (x - 1)^2$. Letting both a and b equal one, we substitute x' for $x - 1$ and y' for $y - 1$ to obtain $y' = x'^2$. The new parabola has its vertex at $(0', 0')$ in the new coordinate system which is $(1,1)$ in the original system. One can also determine a and b by substituting $x' + a$ for x and $y' + b$ for y in the original equation. After rearranging terms according to the new variables, one chooses a and b so that the coefficients of the linear terms x' and y' are zero.

Given the general form of a quadratic equation

$$Ax^2 + Bxy + Cy^2 + Dx + Ey + F = 0,$$

one can eliminate the xy term if one rotates the axes by an angle θ, where

$$\tan 2\theta = \frac{B}{A - C}.$$

The rotation is accomplished by means of the following substitution formulas:

$$x = x' \cos \theta - y' \sin \theta \quad \text{and} \quad y = x' \sin \theta + y' \cos \theta.$$

It can be shown that if A', B', C', D', E', and F' are the coefficients of a rotated equation, the quantity $B^2 - 4AC = B'^2 - 4A'C'$ and is said to be *invariant*. This expression is often called the *discriminant* of the quadratic. Given a general quadratic, it is possible to determine which curve it is by examining the discriminant.

- If $B^2 - 4AC < 0$ the curve is an ellipse.

- If $B^2 - 4AC = 0$ the curve is a parabola.

- If $B^2 - 4AC > 0$ the curve is a hyperbola.

Transform each of the following equations into another form, such that the new form of the equation contains no terms of the first degree.

(1) $x^2 + y^2 - 4x + 10y = 0$

(2) $xy + 4x - 2y + 10 = 0$

Solution: (1) By means of translation of the coordinate axes, one can find the required equations.

Method 1.

$$x^2 + y^2 - 4x + 10y$$

$= (x^2 - 4x) + (y^2 + 10y).$ Now, it is necessary to complete the square.

$$(x^2 - 4x + 2^2) + (y^2 + 10y + 5^2) - 4 - 25$$

$= (x - 2)^2 + (y + 5)^2 - 29 = 0$

Using the transformation

$$\begin{cases} x' = x - 2 \\ y' = y + 5, \end{cases} \quad \text{one obtains} \quad x'^2 + y'^2 = 29$$

as the required equation.

Method 2.

Let $x' = x - a$

$y' = y - b$,   and substitute

$$x = x' + a, \quad y = y' + b \quad \text{into the equation. Then}$$

$$(x' + a)^2 + (y' + b)^2 - 4(x' + a) + 10(y' + b)$$

$$= x'^2 + 2ax' + a^2 + y'^2 + 2by' + b^2 - 4x' - 4a + 10y' + 10b$$

$$= x'^2 + y'^2 + (2a - 4)x' + (2b + 10)y' + 10b - 4a + a^2 + b^2 = 0$$

Setting $2a - 4 = 0$

$$2b + 10 = 0$$

so that the above equation will contain no terms of first degree as required, one has a = 2, b = 5.

Therefore, by replacing x by x' + 2 and y by y' - 5, one can obtain the equation

$$x'^2 + y'^2 = 29$$

(2) Let $x = x' + a$

$$y = y' + b$$

$$xy + 4x - 2y + 10 = 0 \quad \text{becomes}$$

$$(x'+a) (y'+b) + 4(x'+a) - 2(y'+b) + 10$$

$$= x'y' + bx' + ay' + ab + 4x' + 4a - 2y' - 2b + 10$$

$$= x'y' + (b+4)x' + (a-2)y' + ab + 4a - 2b + 10$$

$$= 0$$

Setting $b + 4 = 0$ and $a - 2 = 0$ so that the above equation will have no terms of first degree, or b = -4 and a = 2, the transformation

$$x = x' + 2, \quad y = y' - 4$$

or

$$x' = x - 2, \quad y' = y + 4$$

reduces the given equation to x'y' + 18 = 0.

● **PROBLEM 44-2**

By a translation of the coordinate axes, transform the equation $x^2 - 4y^2 + 6x + 8y + 1 = 0$ into another equation to eliminate terms of the first degree. Plot the locus, and show both sets of axes.

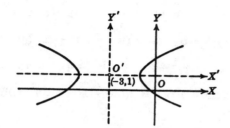

Solution: Since we are told that we must translate the coordinate axes in order to remove the first degree terms, let us review the rules for this procedure. Let us say that the translated origin is (h, k). If point P is represented before and after translation by (x, y) and (x',y'), respectively, then we can deduce that x = x' + h and y = y' + k and substitute these values to find the new equation.

In this problem we are not given the new coordinates of the translated origin. We must solve for these values given the restriction that the translated equation cannot contain a first order term. Hence, if we let (h, k) be the coordinates of the new origin, then, by substitution, we obtain the transformed equation.

$$(x' + h)^2 - 4(y' + k)^2 + 6(x' + h) + 8(y' + k) + 1 = 0$$

which, after expansion and collection of terms, assumes the form

(*) $x'^2 - 4y'^2 + (2h + 6)x' - (8k - 8)y' + h^2 - 4k^2$

$$+ 6h + 8k + 1 = 0$$

Therefore, in order that no first order term appears in the translated equation

$$2h + 6 = 0 \qquad and \qquad 8k - 8 = 0$$

Therefore,   h = - 3   and   k = 1.

Hence, the origin that will give us the required equation will be (- 3, 1)

We can substitute this into (*) to obtain

$x'^2 - 4y'^2 + (2(- 3) + 6)x - (8(1) - 8)y' + (- 3)^2$

$$- 4(1)^2 + 6(- 3) + 8(1) + 1 = 0$$

which reduces to $x'^2 - 4y'^2 - 4 = 0$.

This is the form of the equation of a hyperbola and the hyperbola, along with both sets of axes, is shown on the accompanying graph.

Transform the equation $X^3 - 3X^2 - Y^2 + 3X + 4Y - 5 = 0$ by translating the coordinate axes to a new origin at (1, 2). Plot the locus and show both sets of axes.

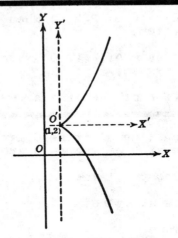

Solution: When translating the coordinate axes, each variable in the original equation becomes a function of the translation. If the axes are translated to a new origin at, say, (h, k) and any point P has coordinates of (X, Y) before the translation and (X', Y') after the translation, then we know $X = X' + h$ and $Y = Y' + k$.

These equations of transformation tell us that for every X and Y in the original equation we can substitute X' + h and Y' + k, respectively, to find the equivalent equation on the translated axes.

If we proceed with the substitution of X and Y into the given equation, we obtain

$$(X' + 1)^3 - 3(X' + 1)^2 - (Y' + 2)^2 + 3(X' + 1) +$$

$$4(Y' + 2) - 5 = 0$$

$$X'^3 + 3X'^2 + 3X' + 1 - (3X'^2 + 6X' + 3) - (Y'^2 + 4Y' + 4) + 3X' + 3 + 4Y' + 8 - 5 = 0$$

$$X'^3 - Y'^2 = 0$$

This is an easier equation to plot. Select several points in the (X',Y') system that satisfy the equation so that we can observe the shape of the locus.

X'	0	1	1	2	2
Y'	0	1	-1	$\sqrt{8}$	$-\sqrt{8}$

This locus is shown on the accompanying graph. Notice that it is symmetric around the X'-axis.

This is also the graph of the original equation relative to the X and Y axes.

Find the equations of transformation for a rotation of the coordinate axes through an angle $30°$, and use them to find

(1) the coordinate of P(2,4) when referred to the primed system;

(2) the coordinate of P'(0,0) when referred to the original system;

(3) the equation of the line $x + y + 4 = 0$.

Solution: The equation of transformation for the rotation of the coordinate axes of degree θ are

$$x = x'\cos\theta - y'\sin\theta \qquad (1)$$

$$y = x'\sin\theta + y'\cos\theta \qquad (2)$$

Here, $\theta = 30°$. Thus

$$x = \frac{\sqrt{3}}{2} x' - \frac{1}{2} y' \qquad (3)$$

$$y = \frac{1}{2} x' + \frac{\sqrt{3}}{2} y' \qquad (4)$$

(1) Given P(2,4), i.e., $x = 2$, $y = 4$

$$2 = \frac{\sqrt{3}}{2} x' - \frac{1}{2} y' \qquad (5)$$

$$4 = \frac{1}{2} x' + \frac{\sqrt{3}}{2} y \qquad (6)$$

Solving these two equations, one has

$$x' = 2 + \sqrt{3}, \quad y' = 2\sqrt{3} - 1$$

Therefore, in the prime system, point P' is P'($2+\sqrt{3}, 2\sqrt{3} - 1$).

(2) $x' = 0$, $y' = 0$. Substituting into equations (3) and (4),

$$x = \frac{\sqrt{3}}{2} \cdot 0 - \frac{1}{2} \cdot 0 = 0$$

$$y = \frac{1}{2} \cdot 0 + \frac{\sqrt{3}}{2} \cdot 0 = 0$$

Hence, in the original system, the point is P(0,0), which is the same point as in the primed system. Since the coordinates are unchanged, the origin is called an invariant point of the transformation.

(3) Substituting eqs. (3) and (4) into $x + y + 4 = 0$, one obtains

$$\left(\frac{\sqrt{3}}{2}\, x' - \frac{1}{2}\, y'\right) + \left(\frac{1}{2}\, x' + \frac{\sqrt{3}}{2}\, y'\right) + 4$$

$$= \left(\frac{\sqrt{3}+1}{2}\right) x' + \left(\frac{\sqrt{3}-1}{2}\right) y' + 4$$

$$= (\sqrt{3}+1)x' + (\sqrt{3}-1)y' + 8 = 0$$

So the required equation of the line in the primed system is

$$(\sqrt{3}+1)x' + (\sqrt{3}-1)y' + 8 = 0$$

● **PROBLEM 44-5**

What is the graph of the equation

$$4x^2 - 3xy + 4y^2 + 2x + 4y - 9 = 0 ?$$

Solution: The equation is given in the most general form of the second degree equation

$$Ax^2 + 2Bxy + Cy^2 + 2Dx + 2Ey + F = 0 \qquad (1)$$

when none of the coefficients A, B, C, D, E and F are zero. Both translation and rotation of the coordinate axes are needed to reduce the original equation to a more specified form. In general, one must first rotate the axes to eliminate the xy term, and obtain an equation in the form of

$$A'x'^2 + C'y'^2 + 2D'x' + 2E'y' + F' = 0$$

which is called the semi-reduced form of the second degree equation. Then, by translation, the semi-reduced form of the equation can be further simplified. In general,

(a) If $B^2 - AC < 0$, A' and C' have the same sign, and eq. (1) is a real ellipse.

(b) If $B^2 - AC = 0$, either $A' = 0$ or $C' = 0$, but not both; then eq.(1) represents a parabola or a pair of parallel lines.

(c) If $B^2 - AC > 0$, A' and C' differ in sign; then eq. (1) is an equation of a hyperbola or a pair of intersecting lines.

To eliminate the xy term, axes must be rotated with $\tan 2\theta = \frac{2B}{A-C}$ and $0 < \theta < 90^0$, $A \neq C$ (when $A = C$, $\theta = 45^0$).

For the given equation, $A = C$. Hence $\theta = 45^0$, and

$$\sin\theta = \frac{\sqrt{2}}{2}, \cos\theta = \frac{\sqrt{2}}{2} .$$

The equations of rotation are

$$x = x'\cos\theta - y'\sin\theta = \frac{\sqrt{2}}{2}(x' - y')$$ (2)

$$y = x'\sin\theta + y'\cos\theta = \frac{\sqrt{2}}{2}(x' + y')$$ (3)

$A' = A\cos^2\theta + 2B\sin\theta\cos\theta + C\sin^2\theta = 1$

$B' = \frac{1}{2}(C-A)\sin 2\theta + B\cos 2\theta = 0$

$C' = A\sin^2\theta - 2B\sin\theta\cos\theta + C\cos^2\theta = 7$

$D' = D\cos\theta + E\sin\theta = 3\sqrt{2}$

$E' = E\cos\theta - D\sin\theta = \sqrt{2}$

$F' = F = -9$

Or substituting eq. (2) and eq. (3) into the original given equation to obtain the same result,

$A'x'^2 + C'y'^2 + 2D'x' + 2E'y' + F'$

$= x'^2 + 7y'^2 + 6\sqrt{2}\,x' + 2\sqrt{2}\,y' - 9 = 0$ (4)

Next, eliminate the first degree terms of eq. (4).

$x'^2 + {}'7y'^2 \div 6\sqrt{2}x' + 2\sqrt{2}y' - 9$

$= (x'^2 + 6\sqrt{2}x' + (3\sqrt{2})^2) + 7\left(y'^2 + \frac{2\sqrt{2}}{7}\,y' + \left(\frac{\sqrt{2}}{7}\right)^2\right) - 9 - 18 - \frac{2}{7}$

$= (x' + 3\sqrt{2})^2 + 7\left(y' + \frac{\sqrt{2}}{7}\right)^2 - 27\frac{2}{7} = 0$

$$(x' + 3\sqrt{2})^2 + 7\left(y' + \frac{\sqrt{2}}{7}\right)^2 = \frac{191}{7}$$ (5)

Finally, using the translation equations

$$x'' = x' + 3\sqrt{3}$$

$$y'' = y' + \frac{\sqrt{2}}{7}\,,$$

one reduces eq.(5) to

$$x''^2 + 7y''^2 = \frac{191}{7}, \text{ or } \frac{x''^2}{\frac{191}{7}} + \frac{y''^2}{\frac{191}{49}} = 1$$

Therefore, eq.(1) represents an ellipse.

● **PROBLEM 44-6**

By a rotation of the coordinate axes, transform the equation

$$9x^2 - 24xy + 16y^2 - 40x - 30y = 0$$

into another equation lacking the cross product term. (Plot the locus and draw both sets of axes.)

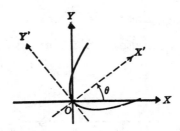

Solution: In a problem of this type where rotation of axes are required, we must apply the formulas

$$x = x' \cos \theta - y' \sin \theta \text{ and } y = x' \sin \theta + y' \cos \theta$$

where θ is the angle through which the axes are rotated and (x, y) and (x', y') are the coordinates of the point P before and after rotation, respectively. (This procedure is used principally to eliminate cross-terms in second degree equations.)

We are not given the required angle of rotation and must devise a procedure to transform the equation without this fact.

If we let θ = the angle of rotation, then by substitution into the original equation, we obtain

$$9(x' \cos \theta - y' \sin \theta)^2 - 24(x' \cos \theta - y' \sin \theta) \cdot$$

$$(x' \sin \theta + y' \cos \theta) + 16(x' \sin \theta + y' \cos \theta)^2$$

$$- 40(x' \cos \theta - y' \sin \theta) - 30(x' \sin \theta + y' \sin \theta) = 0.$$

After expansion and collection of terms, the equation assumes the form

$$*(9 \cos^2 \theta - 24 \cos \theta \sin \theta + 16 \sin^2 \theta)x'^2$$

$$+ (14 \sin \theta \cos \theta + 24 \sin^2 \theta - 24 \cos^2 \theta)x'y'$$

$$+ (9 \sin^2 \theta + 24 \sin \theta \cos \theta + 16 \cos^2 \theta)y'^2$$

$$- (40 \cos \theta + 30 \sin \theta)x' + (40 \sin \theta - 30 \cos \theta)y' = 0.$$

We are told that this equation is not supposed to have a x'y' term. Therefore,

$$14 \sin \theta \cos \theta + 24 \sin^2 \theta - 24 \cos^2 \theta = 0$$

Two standard trigonometric identities tell us $2 \sin \theta \cos \theta = \sin 2\theta$ and $\sin^2 \theta - \cos^2 \theta = - \cos 2\theta$. Hence, by substitution,

$$14 \sin \theta \cos \theta + 24(\sin^2 \theta - \cos^2 \theta) = 7 \sin 2\theta - 24 \cos 2\theta = 0.$$

and $\dfrac{\sin 2\theta}{\cos 2\theta} = \dfrac{24}{7}$ or $\tan 2\theta = \dfrac{24}{7}$.

All angles of rotation are between 0 and 90°. Hence,

$0 < 2\theta < 180°$, and tangent and cosine will agree in sign.

Again, by trigonometric identity,

$$\frac{1}{\cos^2 2\theta} = 1 + \tan^2 2\theta = 1 + \left(\frac{24}{7}\right)^2$$

$$\cos 2\theta = \sqrt{\frac{1}{\frac{49}{49} + \frac{576}{49}}} = \sqrt{\frac{1}{\frac{625}{49}}} = \frac{7}{25}$$

We can now make use of the half angle formulas of trigonometry to find the values of sin θ and cos θ which are needed to evaluate (*) and find the transformed equation.

$$\sin \theta = \sqrt{\frac{1 - \cos 2\theta}{2}} = \sqrt{\frac{1 - \frac{7}{25}}{2}} = \frac{3}{5}$$

$$\cos \theta = \sqrt{\frac{1 + \cos 2\theta}{2}} = \sqrt{\frac{1 + \frac{7}{25}}{2}} = \frac{4}{5}$$

If these values of sin θ and cos θ are substituted in (*), we have

$$\left[9\left(\frac{4}{5}\right)^2 - 24\left(\frac{3}{5}\right)\left(\frac{4}{5}\right) + 16\left(\frac{3}{5}\right)^2\right]x'^2$$

$$+ \left[14\left(\frac{4}{5}\right)\left(\frac{3}{5}\right) + 24\left(\frac{3}{5}\right)^2 - 24\left(\frac{4}{5}\right)^2\right]x'y'$$

$$+ \left[9\left(\frac{3}{5}\right)^2 + 24\left(\frac{3}{5}\right)\left(\frac{4}{5}\right) + 16\left(\frac{4}{5}\right)^2\right]y'^2$$

$$- \left[40\left(\frac{4}{5}\right) + 30\left(\frac{3}{5}\right)\right]x' + \left[40\left(\frac{3}{5}\right) - 30\left(\frac{4}{5}\right)\right]y' = 0$$

Hence, the required equation in reduced form is

$$25y'^2 - 50x' = 0 \qquad \text{or} \qquad y'^2 - 2x' = 0$$

This equation is in the standard form for a parabola. The parabola and axes are drawn on the accompanying coordinate grid.

● PROBLEM 44-7

Transform the equation $2x^2 + \sqrt{3}\, xy + y^2 = 4$ by rotating the coordinate axes through an angle of 30°. Plot the locus and show both sets of axes.

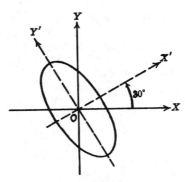

<u>Solution</u>: In this problem we will have to call on
several formulas derived with the help of certain key
trigonometric identities.

When the axes are rotated through an angle θ
about the origin as a fixed point and the coordinates
of any point P (x, y) are transformed into (x', y')
then the equations of transformation from the old to the '
new coordinates are given by

$$x = x' \cos \theta - y' \sin \theta$$

$$y = x' \sin \theta - y' \cos \theta.$$

The expression for x and y can be substituted into
the given equation to perform the transformation. In this
problem, θ = 30°. Hence, since

$$x = x' \cos 30° - y' \sin 30° = \frac{\sqrt{3}}{2} x' - \frac{1}{2} y'$$

and $y = x' \sin 30° + y' \cos 30° = \frac{1}{2} x' + \frac{\sqrt{3}}{2} y'$,

by substitution, we obtain:

$$2 \left(\frac{\sqrt{3}}{2} x' - \frac{1}{2} y' \right)^2 + \sqrt{3} \left(\frac{\sqrt{3}}{2} x' - \frac{1}{2} y' \right) \left(\frac{1}{2} x' + \frac{\sqrt{3}}{2} y' \right)$$

$$+ \left(\frac{1}{2} x' + \frac{\sqrt{3}}{2} y' \right)^2 = 4$$

Expanding this gives us

$$2 \left(\frac{3}{4} x'^2 - \frac{\sqrt{3}}{2} x'y' + \frac{1}{4} y'^2 \right) + \sqrt{3} \left(\frac{\sqrt{3}}{4} x'^2 + \frac{1}{2} x'y' - \frac{\sqrt{3}}{4} y'^2 \right)$$

$$+ \left(\frac{1}{4} x'^2 + \frac{\sqrt{3}}{2} x'y' + \frac{3}{4} y'^2 \right) = 4$$

which reduces to

$$\frac{10}{4} x'^2 + \frac{1}{2} y'^2 = 4.$$

Multiply by 2 to simplify further. Hence, we obtain

$$5x'^2 + y'^2 = 8.$$

This is the standard form for the equation of an ellipse. The ellipse and both sets of axes are shown on the accompanying graph.

CHAPTER 45

POLAR COORDINATES

> **Basic Attacks and Strategies for Solving Problems in this Chapter. See pages 812 to 831 for step-by-step solutions to problems.**

As mentioned in Chapter 36 (on complex numbers), one can locate a point in the plane either by rectangular coordinates or by means of an angle of orientation and a distance from the origin (trigonometric coordinates). We cannot only describe complex numbers in these two ways, but sets of points can often be described either by means of x, y coordinates (rectangular coordinates), or by means of an angle of orientation (θ) and a distance (ρ or r), which are called *polar coordinates*. For example, a circle centered at the origin with a radius of 1 can be described either as

$$x^2 + y^2 = 1 \quad \text{or as} \quad \rho = 1.$$

(Since for every θ, the distance from the origin must be a constant value of 1, θ does not appear in the polar equation.) The 45° line, $y = x$, can be written as $\theta = \pi/4$. (Since for any ρ, the orientation must remain constant on the 45° line, ρ does not appear in the polar equation.)

As in Chapter 29 (Definitions of Trigonometric Functions), angles are measured starting with the positive x-axis as zero and moving counterclockwise. Distances are measured along a straight line from a point to the origin. In an equation in which the distance ρ is a function of the angle θ, one should plot selected points (as with rectangular coordinates) and sketch a curve, in a circular path around the origin, starting at the positive x-axis.

One can convert from one coordinate system to another using the same rules mentioned in Chapter 36. By drawing a straight line from a point to the origin, one has a right triangle in which the hypotenuse

$$\rho = \sqrt{x^2 + y^2}$$

and the angle at the origin

$$\theta = \arctan \frac{y}{x}.$$

It also follows that

$$x = \rho \cos \theta \quad \text{and} \quad y = \rho \sin \theta.$$

One can discuss symmetry and other concepts for polar equations in a way similar to that for rectangular equations. For example, since ρ^2 has the same value for ρ or $-\rho$, an equation with ρ^2 and no linear occurrence of ρ will be symmetric about the origin (i.e., a polar symmetry). (This is true since if you "stand" at the origin and "walk" along the line $\theta = c$, an equation containing ρ^2 would be satisfied if you walked a distance of ρ or a distance [backwards] of $-\rho$.) Similarly, since $\cos \theta = \cos(-\theta)$, an equation with $\cos \theta$ and no other occurrence will be symmetric about the x-axis.

Step-by-Step Solutions to Problems in this Chapter, "Polar Coordinates"

● PROBLEM 45-1

Draw the graph of $\rho = 2 \cos \theta$.

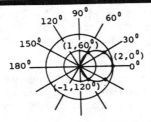

Solution: We assign values to θ and find the corresponding values of ρ, giving the following table:

θ	$\cos \theta$	$\rho = 2\cos \theta$
0^0	1	2
30^0	.87	1.74
60^0	.5	1
90^0	0	0
120^0	-.5	-1
150^0	-.87	-1.74
180^0	-1	-2

Values from 180^0 to 360^0 give the same points. (Check this.)

We then plot the points (ρ, θ) and draw a smooth curve through them. We get the graph of the figure. The equation which defines the path of P may involve only one of the variables (ρ, θ). In that case the variable which is not mentioned may have any and all values.

● PROBLEM 45-2

Transform the equation $x^2 + y^2 - x + 3y = 3$ to a polar equation.

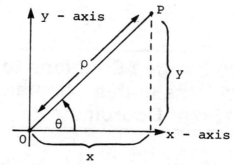

Solution: Ordinarily, when we wish to locate a point in a plane, we draw a pair of perpendicular axes and measure specified signed distances from the axes. The points are designated by pairs in terms of (x,y). These are called rectangular coordinates.

Another way is to designate a point in terms of polar coordinates. (ρ,θ) are the polar coordinates of a point P where ρ is the radius vector of P and θ is the angle that is made with the positive x-axis and the radius vector, OP. (See diagram.)

If P is designated by the coordinates (x,y) in rectangular coordinates and by (ρ,θ) in polar coordinates, then the following relationships hold:

$$\text{Cos } \theta = \frac{\text{adjacent side}}{\text{hypotenuse}} = \frac{x}{\rho} \quad \text{or} \quad x = \rho \text{ Cos } \theta$$

$$\text{Sin } \theta = \frac{\text{opposite side}}{\text{hypotenuse}} = \frac{y}{\rho} \quad \text{or} \quad y = \rho \text{ Sin } \theta$$

Now in this example, we replace x by ρ Cos θ and y by ρ Sin θ to obtain:

$$(\rho \text{ Cos } \theta)^2 + (\rho \text{ Sin } \theta)^2 - (\rho \text{ Cos } \theta) + 3(\rho \text{ Sin } \theta) = 3$$

$$\rho^2 \text{ Cos}^2 \theta + \rho^2 \text{ Sin}^2 \theta - \rho \text{ Cos } \theta + 3\rho \text{ Sin } \theta = 3$$

Factor out ρ^2 and $-\rho$.

$$\rho^2 (\text{Cos}^2 \theta + \text{Sin}^2 \theta) - \rho (\text{Cos } \theta - 3\text{Sin } \theta) = 3$$

Apply the identity $\text{Cos}^2 \theta + \text{Sin}^2 \theta = 1$. Then,

$$\rho^2 - \rho (\text{Cos } \theta - 3 \text{ Sin } \theta) = 3$$

● **PROBLEM 45-3**

Transform the equation $\rho = 2 \cos \theta$ to rectangular coordinates.

Solution:

$$\rho = \sqrt{x^2 + y^2}, \qquad \cos \theta = \frac{x}{\rho} = \frac{x}{\sqrt{x^2 + y^2}}$$

$$\rho = 2 \cos \theta$$

$$\sqrt{x^2 + y^2} = \frac{2x}{\sqrt{x^2 + y^2}}$$

$$x^2 + y^2 = 2x$$

● **PROBLEM 45-4**

Transform the equation   xy = 4 to polar coordinates.

<u>Solution:</u>

$$x = \rho \cos \theta$$

$$y = \rho \sin \theta$$

$$xy = 4$$

$$\rho \cos \theta \cdot \rho \sin \theta = 4$$

$$\rho^2 \cos \theta \sin \theta = 4$$

● **PROBLEM 45-5**

Transform the equation r = 4/(2 − 3 sin θ) to an equation in cartesian coordinates.

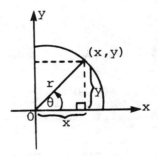

Solution: The given equation is in polar coordinates. This is another system of coordinates where a point (x, y) lies on a circle of radius r whose center is the origin. (see Figure.)

   We want to replace r and sin θ by rectangular co-ordinates. Observe the following needed substitutions which can be derived from the diagram.

$$\sin \theta = \frac{\text{opposite side}}{\text{hypotenuse}} = \frac{y}{r}$$

Pythagorean Identity $x^2 + y^2 = r^2$

Solving for r: $\sqrt{x^2 + y^2} = r$

Then we proceed as follows:

$$r = \frac{4}{2 - 3y/r} \qquad \text{replacing } \sin \theta \text{ by } y/r$$

$$= \frac{4r}{2r - 3y} \qquad \begin{array}{l}\text{simplifying the complex}\\ \text{fraction}\end{array}$$

$$2r - 3y = 4 \qquad \begin{array}{l}\text{multiplying each member by}\\ (2r - 3y)/r\end{array}$$

$$2\sqrt{x^2 + y^2} = 4 + 3y \qquad \begin{array}{l}\text{replacing } r \text{ by } \sqrt{x^2 + y^2} \text{ and}\\ \text{adding } 3y \text{ to each member}\end{array}$$

$$4x^2 + 4y^2 = 9y^2 + 24y + 16 \qquad \begin{array}{l}\text{equating the squares of each}\\ \text{member}\end{array}$$

$$4x^2 - 5y^2 - 24y = 16 \qquad \begin{array}{l}\text{adding } -9y^2 - 24y \text{ to each}\\ \text{member}\end{array}$$

● **PROBLEM 45-6**

Convert the equation $r = \tan \theta + \cot \theta$ to an equation in cartesian coordinates.

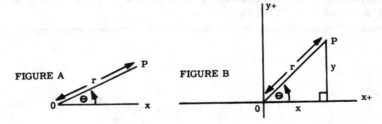

FIGURE A   FIGURE B

Solution: The given equation is expressed in polar co-ordinates (r, θ) where r is the radius vector, OP, and θ is the angle that r makes with the polar axis, OX. O is the fixed point called the pole. See figure A.

Since $\tan \theta \neq - \cot \theta$, then $r \neq 0$, and the graph of

$$r = \tan \theta + \cot \theta$$

does not pass through the pole. If r were equal to zero, then the curve would pass through (0,0). Therefore in the transformation of this equation to cartesian co-ordinates, we must remember that $(x, y) \neq (0,0)$. Now we must convert all expressions of r and θ into rectangular coordinates (x, y). If P is designated by the coordinates (x, y) in rectangular coordinates and by (r, θ) in polar coordinates, then the following relationships hold true: (see Figure B).

$$\tan \theta = \frac{\text{opposite side}}{\text{adjacent side}} = \frac{y}{x}$$

$$\cot \theta = \frac{\text{adjacent side}}{\text{opposite side}} = \frac{x}{y}$$

By the Pythagorean Identity $x^2 + y^2 = r^2$

Solve for r: $\qquad r = \sqrt{x^2 + y^2}$

Substitute these values for r, tan θ, and cot θ.

$$\sqrt{x^2 + y^2} = \frac{y}{x} + \frac{x}{y}$$

$$xy\sqrt{x^2 + y^2} = x^2 + y^2$$

Divide by $\sqrt{x^2 + y^2}$

$$xy = \frac{x^2 + y^2}{\sqrt{x^2 + y^2}}$$

Rationalize the denominator by multiplying the numerator and denominator by $\sqrt{x^2 + y^2}$

$$xy = \frac{x^2 + y^2}{\sqrt{x^2 + y^2}} \ \frac{\sqrt{x^2 + y^2}}{\sqrt{x^2 + y^2}}$$

$$xy = \frac{x^2 + y^2}{x^2 + y^2} \ \sqrt{x^2 + y^2}$$

$$xy = \sqrt{x^2 + y^2}$$

Squaring both sides, we obtain:

$$x^2y^2 = x^2 + y^2$$

where $x \neq 0$ and $y \neq 0$.

● **PROBLEM 45-7**

Discuss the graph of the equation
$$r = 4 \cos \theta - 2 \sin \theta .$$

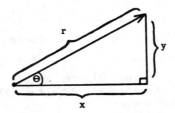

Solution: Instead of plotting points of the form (r,θ), we will write this equation in terms of cartesian coordinates x and y. Cartesian coordinates (x,y) and polar coordinates (r,θ) are related by the following equations: (noting the figure,)

$$\sin \theta = \frac{\text{opposite side}}{\text{hypotenuse}} = \frac{y}{r} \qquad (1)$$

$$\cos \theta = \frac{\text{adjacent side}}{\text{hypotenuse}} = \frac{x}{r} \qquad (2)$$

and by the Pythagorean Theorem,

$$r^2 = x^2 + y^2 . \qquad (3)$$

First, we multiply both sides of the given equation by r to obtain the equation

$$r^2 = 4r \cos \theta - 2r \sin \theta \qquad (4)$$

From equations (1), (2), and (3), the new equation (equation (4)) becomes:

$$x^2 + y^2 = 4\cancel{r}\left(\frac{x}{\cancel{r}}\right) - 2\cancel{r}\left(\frac{y}{\cancel{r}}\right)$$

$$x^2 + y^2 = 4x - 2y \qquad (5)$$

Subtract $(4x - 2y)$ from both sides of equation (5):

$$x^2 + y^2 -(4x - 2y) = 4x - 2y -(4x - 2y)$$

$$x^2 + y^2 - 4x + 2y = 0$$

$$x^2 - 4x + y^2 + 2y = 0 \qquad (6)$$

Now complete the square in both x and y. This is done by taking half the coefficient of the x term (or y term) and then squaring this value. The result is then added to both sides of equation (6). Completing the square in x:

$$[\tfrac{1}{2}(-4)]^2 = (-2)^2 = 4 .$$

Hence, equation (6) becomes:

$$\left(x^2 - 4x + 4\right) + y^2 + 2y = 0 + 4$$

or

$$\left(x^2 - 4x + 4\right) + y^2 + 2y = 4$$

or

$$\left(x - 2\right)^2 + y^2 + 2y = 4 \qquad (7)$$

Completing the square in y:

$$[\tfrac{1}{2}(2)]^2 = (1)^2 = 1.$$

Hence, equation (7) becomes:

$$\left(x - 2\right)^2 + \left(y^2 + 2y + 1\right) = 4 + 1$$

or

$$\left(x - 2\right)^2 + \left(y^2 + 2y + 1\right) = 5$$

or

$$(x - 2)^2 + (y + 1)^2 = 5 \qquad (8)$$

Also, note that the equation of a circle is:

$$(x - h)^2 + (y - k)^2 = r^2$$

where (h,k) are the coordinates of the center of the circle and r is the radius of the circle. Equation (8) is in the form for the equation of circle where:

$x - 2$ corresponds to $x - h$; i.e., $h = 2$,

$y + 1$ corresponds to $y - k$; that is, $y + 1 = y - k$

$$y + 1 - y = y - k - y$$

$$1 = -k$$
$$-1(1) = (-1)(-k)$$
$$-1 = k;$$

and r^2 corresponds to 5; that is; $r = \sqrt{5}$.

Therefore, the original equation given in polar coordinates (r,θ) represents a circle of center $(h,k) = (2,-1)$ and radius $= r = \sqrt{5}$.

● **PROBLEM** 45-8

Find the conditions which the equation of a polar figure must satisfy for the figure to be (a) symmetric about the polar axis, (b) symmetric about the 90°-axis, (c) symmetric about the pole.

Solution: We know that an object is symmetric about something if for every part of that object there is a corresponding part of the object on the opposite side of that something. In more concrete terms, we let P_1 be a point on the curve C; $\overline{P_1A}$ be the perpendicular from point P_1 to the line ℓ. Now we construct perpendicular $\overline{AP_2}$ congruent to $\overline{AP_1}$ in the opposite half plane. If for every P_1 on curve C, P_2 is also a point on C, then the curve C is said to be symmetric about the line ℓ.

Symmetry about a point has a similar definition. We let P_1 be a point on the curve, and O be the given point. We locate point P_4 such that O is the midpoint of $\overline{P_1P_4}$. If for every P_1 on the curve, P_4 is also on the curve, then the curve is said to be symmetric about the point O.

(a) Suppose P_1 is any point on the given curve; $\overline{P_1A}$ the altitude to the polar axis, and P_2 the point such that $P_1A = P_2A$ and $PA_2 \perp$ polar axis. Then the given curve is symmetric about the polar axis only if P_2 is also a point on the curve. This condition just stated is sufficient for symmetry.

However, we wish to find a condition that we could use given only the equation of the polar figure. Therefore, we express the condition in terms of r and θ.

Suppose P_1 has the coordinates (r_1, θ_1), and P_2 the coordinates (r_2, θ_2). To relate the coordinates of P_1 and P_2, note that $\triangle OAP_1 \cong \triangle OAP_2$ by the SAS Postulate $(\overline{OA} \cong \overline{OA}, \overline{P_1A} \cong \overline{P_2A}, \sphericalangle OAP_1$ and $\sphericalangle OAP_2$ are right angles). Thus, by corresponding parts, $\overline{OP_1} \cong \overline{OP_2}$ and $\sphericalangle P_1OA \cong \sphericalangle P_2OA$. Thus, $r_2 = OP_2 = OP_1 = r_1$.

In addition, remember that a polar angle is positive if directed counterclockwise from the base ($\theta = 0$) and negative if directed clockwise. Therefore, $\theta_2 = - m \sphericalangle P_2OA = - m \sphericalangle P_1OA = - \theta_1$. Thus, $P_2(r_1, - \theta_1)$. We now restate the condition.

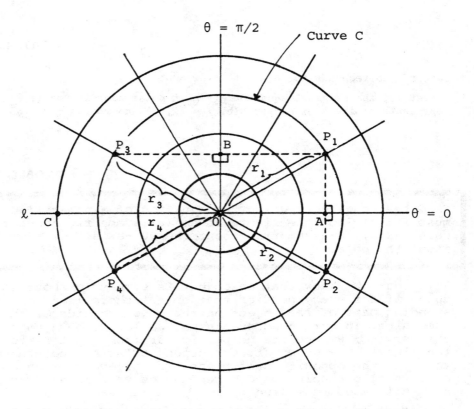

θ = π/2

Curve C

θ = 0

(c) In the figure, point P_1 is a point on the curve; $\overline{P_1O} \cong \overline{OP_4}$; and $\overrightarrow{P_1OP_4}$. Then the given curve is symmetric about the pole only if P_4 is also a point on the curve. Suppose the coordinates of P_1 and P_4 are (r_1, θ_1) and (r_4, θ_4).

Note that $\overline{P_1O} \cong \overline{OP_4}$. Therefore, $r_4 = OP_4 = OP_1 = r_1$. Furthermore, $\overrightarrow{P_1OP_4}$ and line $\theta = 0$ form opposite angles ∢ P_4OC and ∢ P_1OA. Because opposite angles are congruent, ∢ $P_1OA \cong$ ∢ P_4OC. Thus, $\theta_4 = m$ ∢ $P_4OC + m$ ∢ $COA =$
$= m$ ∢ $P_1OA + \Pi = \theta_1 + \pi$.

A graph is symmetric about the pole if, for every point (r, θ) on the graph, $(r, \theta_1 + \pi)$ is also on the graph.

A graph is symmetric about the polar axis, if every point (r, θ) on the graph, $(r, -\theta)$ is also on the graph.

(b) Suppose P_1 is any point on the given curve; $\overline{P_1B}$ the altitude to the 90°-axis; and P_3 the point such that $P_1B = P_3B$ and $\overline{P_3B} \perp$ 90°-axis. The given curve is symmetric about the 90°- axis only if P_3 is also a point on the curve.

Suppose P_1 has coordinates (r_1, θ_1) and $P_3(r_3, \theta_3)$. To relate the coordinates of the two points, note that $\triangle OP_1B \cong \triangle OP_3B$ by the SAS Postulate ($\overline{BP_3} \cong \overline{BP_1}$, $\overline{BO} \cong \overline{BO}$, and ∢ OBP_3 and ∢ OBP_1 are right angles). By corresponding parts,

819

$\overline{OP}_3 \cong \overline{OP}_1$ and $\sphericalangle$ BOP$_1 \cong \sphericalangle$ BOP$_3$. Thus, $r_3 = P_3O = P_1O = r_1$, and $\theta_3 = $
m $\sphericalangle$ P$_3$OA = m $\sphericalangle$ P$_3$OB + m $\sphericalangle$ BOA = m $\sphericalangle$ P$_3$OB + $\pi/2$ =
$(\pi/2 - $ m $\sphericalangle$ P$_1$OA$) + \pi/2 = \pi - $ m $\sphericalangle$ P$_1$OA = $\pi - \theta_1$. Thus, a
given curve is symmetric about the 90°-axis only if, for
every point (r, θ) on the graph, $(r, \pi - \theta)$ is also on the
graph.

● **PROBLEM** 45-9

Given the figure below, is it possible to add points to
the graph so that the final figure will have (a) symmetry
about the polar axis, (b) symmetry about the 90°-axis,
(c) symmetry about the pole? If so, add these points.

Solution: (a) For there to be symmetry about the polar
axis, point $(r, -\theta)$ must be on the completed graph whenever (r, θ) is a point on the graph. We are given that every
point in the figure of Figure 1 is a point of the final
graph. Consider point A in Figure 2, with coordinates
$(r', -\pi/4)$. For the final figure to be symmetric, point
$(r', -(-\pi/4)$ or $(r', \pi/4)$ must also be on the graph. Thus,
we add point A' $(r', \pi/4)$ to the graph. We continue until
every point in the lower figure has a corresponding point
in the upper half. The complete graph is symmetric about
the polar axis.

Note that the upper figure has exactly the same shape
as the lower figure except that it is upside down. This
characteristic can be exploited in graphing other polar

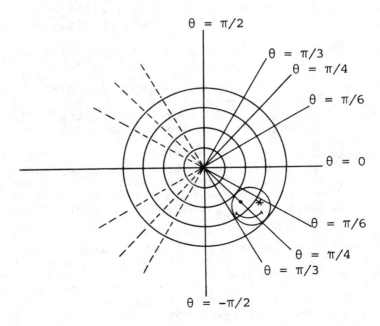

Fig. 1

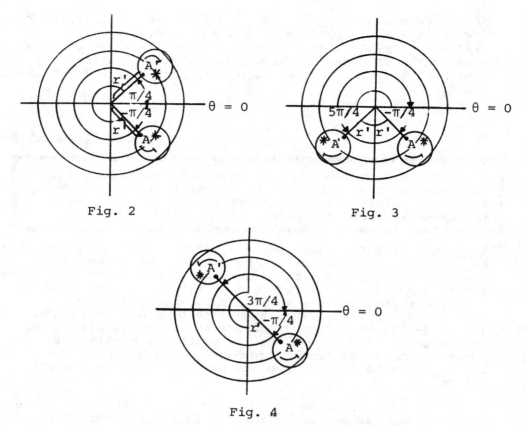

Fig. 2   Fig. 3

Fig. 4

figures. If we know the figure is symmetric about the polar
axis, we need only plot the half of the graph - from θ = 0
to π (instead of θ = 0 to 2π) and then sketching the other
half by "flipping upside down" the half we already have.

For there to be symmetry about the 90°-axis, the
point (r, π - θ) must be on the completed graph whenever
(r, θ)is a point on the graph. Proceeding as in part (a),
we add a point A'(r', π - θ) for every point A(r', θ) on
the original figure. The result (shown in Figure 3) is a
mirror image of the original figure (notice the position
of eyes) - the leftmost points are now the rightmost points.

This characteristic can be exploited in graphing other
polar figures. If we know that the figure is symmetric about
the 90°-axis, we need only plot the half of the graph -
from θ = - π/2 to π/2. We can then sketch the rest of the
figure by drawing the mirror image of the plotted half in
the left side plane.

For there to be symmetry about the pole, point
(r, π + θ) must be on the completed graph whenever (r, θ)
is on the graph. Proceeding as in part (a), we add a point
A'(r', π + θ) for every A(r', θ) on the original figure.
The result is a figure that can be obtained from the
original figure by (1) drawing the mirror image of the
original in the left plane, and (2) "flipping upside down"
the mirror image about the polar axis.

821

In exploiting the pole symmetry for graphing other polar figures, the two step flip may be rather hard to see. The better way is to graph any half of the graph (- 3π/4 ≤ θ < π/4, for example) and, for each point, locate the point directly opposite the points in the graph (the point such that the pole is the midpoint of the point and the graphed point).

With polar axis symmetry, we need only graph 0 ≤ θ ≤ π. With 90° axis symmetry, we need only graph - π/2 ≤ θ < π/2. If a figure has both 90°-axis and polar axis, note that if we plot 0 ≤ θ ≤ π/2, by the polar axis symmetry, we also know - π/2 ≤ θ ≤ 0. Thus, we know -π/2 ≤ θ ≤ π/2. Because of the 90°- axis symmetry, then we know π/2 ≤ θ ≤ 3π/2. Therefore, given 90°-axis and polar axis symmetry, we need only graph 0 ≤ θ ≤ π/2 to plot the whole graph.

Similarly, given all three types of symmetry, it can be shown that we need only plot 0 ≤ θ ≤ π/4.

● **PROBLEM 45-10**

Graph $r^2 = 4 \cos 2\theta$.

Solution: Before blindly plotting points, let us examine the equation for simplifying factors such as symmetries.

(1) Polar-Axis Symmetry. A figure is symmetric about the polar axis, if for every point (r, θ) on the graph (r, - θ) is also a point of the graph. Suppose (r₀, θ₀) is on the

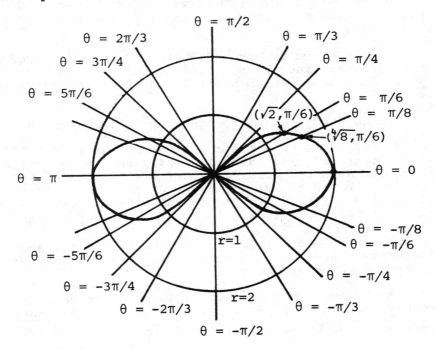

graph. Thus, $r_0^2 = 4 \cos 2\theta_0$. Since $\cos x = \cos (-x)$,
it must be true that $r_0^2 = 4 \cos (2\theta_0) = 4 \cos (-2\theta_0) = 4 \cos 2 (-\theta_0)$. Thus, if (r_0, θ_0) is a point on the graph,
then $(r_0, -\theta_0)$ is also on the graph. The figure $r^2 = 4 \cos 2\theta$ is symmetric about the polar axis.

(2) Symmetry about the 90°-axis. A figure is symmetric
about the 90-axis if, for every point (r, θ) on the graph,
$(r, \pi - \theta)$ is also a point on the graph. Suppose (r_0, θ_0)
is a point on the graph. Then we must show $r_0^2 = 4 \cos 2(\pi - \theta_0)$ is also a point on the graph. Working back-
wards, we use the identity $\cos 2\psi = 2 \cos^2 \psi - 1$ to show
that $r_0^2 = 4 \cos 2 (\pi - \theta_0)$ is true if and only if $r_0^2 = 4(2 \cos^2 (\pi - \theta_0) - 1)$. Since $\cos (\pi - \theta) = -\cos \theta$, $r_0^2 = 4(2 \cos^2 (\pi - \theta_0) - 1)$ is true if and only if $r_0^2 = 4(2 (-\cos \theta_0)^2 - 1) = 4 (2 \cos^2 \theta_0 - 1) = 4 \cos 2\theta_0$. But,
we know that, since (r_0, θ_0) is a point on the graph, it
must be true that $r_0^2 = 4 \cos 2\theta_0$. Thus, $r_0^2 = 4 \cos 2(\pi - \theta_0)$. Therefore, every other equation must also
be true. Thus, $(r, \pi - \theta)$ is a point on the graph if (r, θ)
is on the graph. The figure is thus symmetric about the
90°-axis.

(3) Symmetry with respect to the pole. A figure is
symmetric about the pole if, for every point (r, θ) on the
graph, $(r, \pi + \theta)$ is also a point on the graph. Working
backward as above, we obtain $r_0^2 = 4 \cos 2(\pi + \theta_0) = 4 (2 \cos^2 (\pi + \theta_0) - 1) = 4(2 [-\cos \theta_0]^2 - 1) = 4 (2 \cos^2 \theta_0 - 1) = 4 \cos 2\theta_0$. Since (r_0, θ_0) is given to
be on the graph, then $r_0^2 = 4 \cos 2\theta_0$ is a true statement;
and, thus, for every point (r_0, θ_0) on the graph, the point
$(r_0, \theta_0 + \pi)$ is also on the graph. The figure is symmetric
about the pole.

We now have three symmetries, and thus, we need only
consider values of θ between 0 and $\pi/4$. Two other important
results are:

(1) The curve is bounded. Note $\cos x \leq 1$. Therefore,
$4 \cos^2 2\theta \leq 4$. Since $r^2 = 4 \cos^2 2\theta$, it must be true that
$r \leq 2$. Since there is an upper limit to r, the curve is
bounded.

(2) To find the intercepts with the polar axis, either
$r = 0$, or $\theta = 0$, or $\theta = \pi$. If $r = 0$, then $0^2 = 4 \cos 2\theta$ or
$\cos 2\theta = 0$. Since only $\cos \pm \pi/2$ has a cosine of 0, $2\theta = \pm \pi/2$, or $\theta = \pm \pi/4$. Thus, $(0, \pi/4)$, and $(0, -\pi/4)$ are
intercepts with the polar axis, If $\theta = 0$, then $r^2 = 4 \cos 0 = 4$ or $r = \pm 2$ and $(2, 0)$ and $(-2, 0)$ are points on the graph.
Solving for r with $\theta = \pi$ yields the same points. Thus the
axis intercepts are $(\pm 2, 0)$, and $(0, \pm\pi/4)$.

Using the above information (the symmetries and the
intercepts), and with the points calculated below, we
obtain the accompanying graph.

For convenience of calculations, it is best to
express r in terms of θ. Thus, if $r^2 = 4 \cos 2\theta$, then
$r = 2 \sqrt{\cos 2\theta}$. For $\theta = \pi/8$, $r = 2 \sqrt{\cos \pi/4} = 2 \sqrt{\sqrt{2}/2} =$

$$\sqrt{4 \cdot \sqrt{2}/2} = \sqrt{2\sqrt{2}} = \sqrt{\sqrt{8}} = \sqrt[4]{8}.$$

Thus, ($\sqrt[4]{8}$, π/8) is a point on the graph for θ = π/6, r = 2 $\sqrt{\cos \pi/3}$ = 2$\sqrt{\frac{1}{2}}$ = $\sqrt{2}$.

Thus, ($\sqrt{2}$, π/6) is a point on the graph.

● **PROBLEM 45-11**

Graph r = sin 3θ.

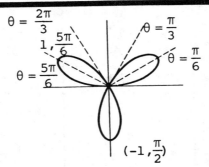

Solution: We graph the figure in five steps. First, we obtain the equation for r in term θ.

(1)   r = sin 3θ

Second, we find the boundary values of r and the symmetries. Note |sin 3θ| < 1. Thus, − 1 < r < 1. Because r = sin 3(− θ) = − sin 3θ ≠ sin 3θ, (r, − θ) being on the graph is not implied by (r, θ). Thus, there is no polar axis symmetry. Because r = sin 3(π − θ) = sin 3θ, (r, π − θ) is implied by (r, θ). Thus, there is symmetry about the 90°-axis. Because r = sin 3(π + θ) = − sin 3θ ≠ sin 3θ, (r, π + θ) is not implied by (r, θ), and there is no symmetry about the pole. Because there is symmetry about the 90°-axis, we need only consider values between − π/2 and π/2.

Third, we find the intercepts. The pole intercepts are found by setting r = 0: (0, 0), (0, ± π/3), and (0, ± 2π/3). The polar axis intercepts are found by setting θ = 0, and π: (0, 0) and (0, π). The 90°- axis intercepts are found by setting θ = ± π/2: (− 1, π/2) and (1, − π/2). [Note that these points are the same.]

Fourth, we find out the general behavior of the curve. Recall the shape of sin θ in rectangular coordinates.

θ	3θ	r = sin 3θ
Increasing from − π/2 to − π/3	Increasing from −3π/2 to −π	Decreasing from 1 to 0

824

$-\pi/3$ to $-\pi/6$	$-\pi$ to $-\pi/2$	0 to -1
$-\pi/6$ to 0	$-\pi/2$ to 0	-1 to 0
0 to $\pi/6$	0 to $\pi/2$	0 to 1
$\pi/6$ to $\pi/3$	$\pi/2$ to π	1 to 0
$\pi/3$ to $\pi/2$	π to $3\pi/2$	0 to -1

As a final step, we solve for particular points to make the sketch as precise as possible.

θ	3θ	$r = \sin 3\theta$
$\pi/18$	$\pi/6$	$\frac{1}{2} = .500$
$\pi/12$	$\pi/4$	$\sqrt{2}/2 = .707$
$\pi/9$	$\pi/3$	$\sqrt{3}/2 = .866$
$\pi/6$	$\pi/2$	$1 = 1.00$
$2\pi/9$	$2\pi/3$	$+ \sqrt{3}/2 = + .866$
$\pi/4$	$3\pi/4$	$+ \sqrt{2}/2 = .707$
$5\pi/18$	$5\pi/6$	$\frac{1}{2} = .500$
$\pi/3$	π	$0 = .000$

This curve is an example of a rose petal curve. The general equations of such a curve are

$$r = a \sin (n\theta) \quad \text{and} \quad r = a \cos (n\theta)$$

where n is a positive integer.

The number of leaves of the curve is equal to n if n is an odd integer. If n is even, the number of leaves is 2n. (See figure.)

● **PROBLEM 45-12**

Graph the curve $r = 2 + 2 \cos \theta$.

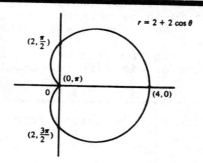

$r = 2 + 2 \cos \theta$

$(2, \frac{\pi}{2})$

$(0, \pi)$

$(4, 0)$

$(2, \frac{3\pi}{2})$

__Solution:__   In graphing any polar equation, we first re-write the equation in convenient form - which, in most cases, consists of expressing r in terms of θ. Here the equation is already given as r in terms of θ. There is no further simplification except for factoring out a 2:

(i)   r = 2(1 + cos θ).

The second step is to get a general picture about the figure - its symmetries, its bounding values, etc. A figure is bounded if there is a limit to the size of r. Note that |cos θ| is always ≤ 1. Thus, r ≤ 2 (1 + 1) or r ≤ 4. The figure is bounded.

There are three types of symmetry: (1) symmetry about the polar axis; (2) about the 90°-axis; and (3) about the pole. For a figure to be symmetrical about the polar axis, (r, - θ) must be a point on the figure each time (r, θ) is on the figure. Suppose (r, θ) is a point on the graph. Then, r = 2(1 + cos θ). Since cos θ = cos (- θ), it is also true that r = 2(1 + cos (- θ)). Thus, (r, - θ) is always a point on the graph whenever (r, θ) is a point on the graph. Thus, the figure is symmetric about the polar axis.

For symmetry about the 90°-axis, (r, π - θ) must be on the graph whenever (r, θ) is on the graph. Given r = 2(1 + cos θ), we wish to show r = 2(1 + cos (π - θ)). Re-writing cos (π - θ) as - cos θ, we have r = 2(1 - cos θ). Since the two expressions are not equivalent, the figure is not symmetric about the 90°-axis.

For symmetry about the pole, (r, π + θ) must be on the graph if (r, θ) is on the graph. Given r = 2(1 + cos θ), we would like to show r = 2(1 + cos (π + θ)). Since cos (π + θ) = cos π cos θ - sin π sin θ = - cos θ, this is equivalent to proving r = 2(1 - cos θ). Recall, cos π = - 1 and sin π = 0. Since the two expressions are not equivalent, there is no symmetry about the pole.

Because the figure is symmetric about the polar axis, we need only plot points from θ = 0 to θ = π.

The third step is to find the intercepts. A point is an intercept of the polar axis if the θ coordinate is 0 or π. We substitute these values in equation (i) and obtain (4, 0) and (0, π). For the 90°-axis intercepts, we set θ equal to π/2 and 3π/2. Thus, (2, π/2) and (2, 3π/2) are the 90°-axis intercepts. Finally, there are the pole intercepts for which r = 0. Substituting r = 0, we obtain (0, π).

We have the intercepts and the symmetries. The next step is to get some idea of the curve's behavior as θ changes. We do this by reviewing the basic pattern of the cos θ curve in rectangular coordinates.

θ	2 cos θ	r = 2 + 2 cos θ
increasing from 0 to π/2	decreases from 2 to 0	decreases from 4 to 2

increasing from $\pi/2$ to π	decreases from 0 to - 2	decreases from 2 to 0.

Finally, we plot some convenient points to make the graph more exact.

$\theta =$	$r = 2(1 + \cos \theta) =$
$\pi/6$	$2(1 + \sqrt{3}/2) \approx 3.73$
$\pi/4$	$2(1 + \sqrt{2}/2) \approx 3.414$
$\pi/3$	$2(1 + \frac{1}{2})\quad = 3$
$2\pi/3$	$2(1 - \frac{1}{2})\quad = 1$
$5\pi/6$	$2(1 - \sqrt{3}/2) \approx .268$

● **PROBLEM** 45-13

Determine the intercepts and symmetry of the polar curve $r = \theta$, for $\theta \geq 0$.

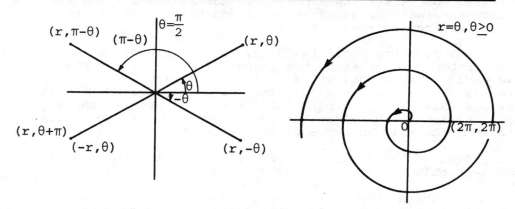

Solution: In polar coordinates points are given by specifying two numbers, r, the distance from the origin or pole, and θ, an angle of rotation. By convention, θ is measured from the horizontal axis, proceeding counter-clockwise. The horizontal axis is called the polar axis. See Figure 1.

Given an equation in terms of r and θ, we can graph the set of points whose values satisfy the equation.

An important case is that for which r can be written as an explicit function of θ, $r = f(\theta)$. In this case, we can find the intercepts of the curve with the polar axis by substituting integer multiples of π radians for θ, i.e. $\theta = n\pi$, n = 0, 1, 2, ... Similarly, we can find the intercept with the pole perpendicular to the

827

polar axis, the π/2 axis, by setting $\theta = (2n + 1) \, \pi/2$, (n = 0, 1, 2, ...). Negative values for θ need not be considered, because of the domain of the function in this problem. In other cases, we could consider intercepts for n = - 1, - 2, - 3, ...

In the present case, as θ increases, r increases without bound; the curve intercepts the polar axis infinitely often. Also, each value of n will determine a separate point on the π/2 axis. Thus, there are infinitely many intercepts for both axes.

On the other hand, let us examine the curve for various types of symmetry. The relevant types of symmetry are those with respect to the polar axis, with respect to the π/2 axis, and with respect to the pole.

The curve is not symmetric with respect to any of these:

A curve is only symmetric with respect to the polar axis if $f(\theta) = f(- \theta)$. In this case, $f(- \theta) = f(\theta)$ only for θ = π; hence this symmetry search fails.

Symmetry with respect to the π/2 axis is defined by $f(\pi - \theta) = f(\theta)$. This is not valid for the curve r = θ, e.g. $f(\pi - \pi/4) = 3\pi/4 \ne f(\pi/4) = \pi/4$.

Finally, symmetry with respect to the pole is defined by $f(\theta + \pi) = f(\theta)$, which is false for all values of this function, e.g., $f(\pi + \pi) = 2\pi \ne f(\pi) = \pi$. Hence, the graph shows no properties of symmetry whatsoever.

The curve, when sketched, is seen to be a spiral. If we had considered negative values for θ, we would have arrived at an oppositely oriented spiral. (See figure 2.)

● **PROBLEM 45-14**

Given the equation $x^2 - y^2 = 1$, translate to polar coordinates and draw a sketch of the curve using the polar equation.

Solution: To translate a Cartesian coordinate equation to polar coordinates, we use the following relations:

(i) $x = r \cos \theta;$ $y = r \sin \theta$

Substituting into the expression $x^2 - y^2 = 1$, we have

(ii) $(r \cos \theta)^2 - (r \sin \theta)^2 = 1.$

(ii) $r^2 (\cos^2 \theta - \sin^2 \theta) = 1$

Since $\cos 2\theta = \cos^2 - \sin^2 \theta$

(iii) $r^2 = \dfrac{1}{\cos 2\theta} = \sec 2\theta$

In graphing polar figures, we first rewrite the expression in convenient form: r in terms of θ, if possible.

(i) $r = \pm \sqrt{\sec 2\theta}$

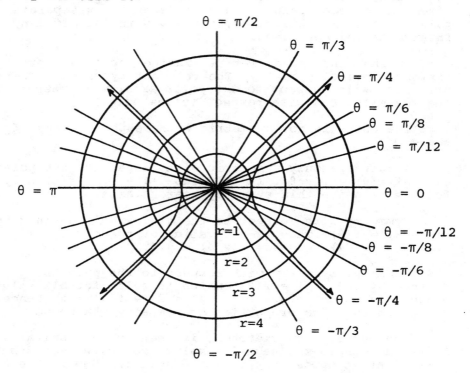

The second step is to find the upper bound on r and the symmetries. The square root function limits $\sqrt{\sec 2\theta}$ to values greater than 0. Even so, since $-\infty \le \sec 2\theta \le \infty$, $0 \le \sqrt{\sec 2\theta} \le \infty$. But, there is no maximum value for r.

Checking for symmetries, we find:

(1) Polar axis symmetry. $(r, -\theta)$ is a point on the graph only if $r = \pm \sqrt{\sec 2(-\theta)}$. Since $\sec \theta = \dfrac{1}{\cos \theta} = $
$\dfrac{1}{\cos (-\theta)} = \sec (-\theta)$, then $r = \pm \sqrt{\sec 2\theta}$. But this is always true if (r, θ) is a point on the graph. Since, $(r, -\theta)$ is implied by (r, θ), the graph is symmetrical about the polar axis.

(2) $90°$-axis symmetry. $(r, \pi - \theta)$ is a point on the graph only if

$$r = \pm \sqrt{\sec 2(\pi - \theta)} = \pm \sqrt{\dfrac{1}{\cos 2(\pi - \theta)}} = \pm \sqrt{\dfrac{1}{2\cos^2 (\pi - \theta) - 1}}$$

829

$$= \pm \sqrt{\frac{1}{2\,(-\cos\theta)^2 - 1}} = \pm \sqrt{\frac{1}{2\cos^2\theta - 1}} = \pm \sqrt{\frac{1}{\cos 2\theta}}$$

$$= \pm\sqrt{\sec 2\theta}.$$

However, $r = \pm \sqrt{\sec 2\theta}$ is always true if (r, θ) is on the graph. Thus, $(r, \pi - \theta)$ follows from (r, θ) and there is 90°-axis symmetry.

(3) Pole symmetry $(r, \pi + \theta)$ is a pole on the graph only if

$$r = \pm \sqrt{\sec 2(\pi + \theta)} = \pm \sqrt{\frac{1}{\cos 2(\pi + \theta)}}$$

$$= \pm \sqrt{\frac{1}{2\cos^2(\pi + \theta) - 1}} = \pm \sqrt{\frac{1}{2\,(-\cos\theta)^2 - 1}}$$

$$= \pm \sqrt{\frac{1}{2\cos^2\theta - 1}} = \pm \sqrt{\frac{1}{\cos 2\theta}} = \pm \sqrt{\sec 2\theta}.$$

Since $(r, \pi + \theta)$ is implied by (r, θ), the graph is symmetric about the pole.

Since the graph has all three symmetries, we need only consider θ between 0 and $\pi/4$.

For the third step, we find the intercepts. To find the polar axis intercepts, we set $\theta = 0$. (Since there is 90°-axis symmetry, we need not set $\theta = \pi$. Every axis intercept on the left corresponds to one on the right.) **We obtain** $(+1, 0)$ and $(1, \pi)$. To find the 90°-axis intercepts, we set $\theta = \pi/2$ (since there is polar axis symmetry, we need not set $\theta = -\pi/2$) and obtain $r = \pm \sqrt{-1}$, which has no real solution. There is no 90°-axis intercept. To find the pole intercepts, we set $r = 0$, and obtain $0 = \pm \sqrt{\sec 2\theta}$. Since secant never equals 0, there is no solution and the graph never intersects the pole.

For the fourth step, we find the general behavior of the graph.

θ	$\sec 2\theta$	$r = \pm \sqrt{\sec 2\theta}$
Increases from 0 to $\pi/4$	Increases from 1 to ∞	Increases from ± 1 to $\pm \infty$.

For the final step, we obtain a table of values.

θ	$\sec 2\theta$	$r = \pm \sqrt{\sec 2\theta}$
0	1	± 1
$\pi/12$	$\frac{2}{3}\sqrt{3}$	$\pm \sqrt{\frac{2}{3}\sqrt{3}} = \pm \sqrt{\sqrt{\frac{4}{3}}}$

		$= \pm \sqrt[4]{\dfrac{4}{3}} \cong \pm 1.07$
π/8	$\sqrt{2}$	$\pm \sqrt{\sqrt{2}} = \pm \sqrt[4]{2} \cong \pm 1.19$
π/6	2	$\pm \sqrt{2} \cong \pm 1.41$
π/4	∞	$\pm \infty$

CHAPTER 46

PARAMETRIC EQUATIONS

> **Basic Attacks and Strategies for Solving Problems in this Chapter. See pages 832 to 835 for step-by-step solutions to problems.**

As has been seen in Chapters 36 and 45, a point in space can be located by rectangular coordinates or by polar coordinates. In either system, *two* quantities must be given: x and y or ρ and θ. When used to describe a set of points such as a curve, it is possible that the quantities in either system are not related to each other in an equation, but are given, instead, in terms of another variable, called a *parameter*. In these cases, one has a set of two equations that together describe one curve and are called *parametric equations*.

Problems arising from natural phenomena frequently lend themselves to a description involving parametric equations. For example, it is relatively easy to describe the path of a projectile by two equations, height (y) as a function of elapsed time (t), and horizontal distance (x) as another function of elapsed time (t). Describing the path by a single equation is not immediately obvious.

In many cases, parametric equations can be transformed into a single equation with the parameter eliminated. For example, given $x = 3t$ and $y = 2t^2$, one can solve the x-equation (which is linear in t) for the parameter t (obtaining $t = x/3$) and then substitute it into the y-equation. This results (in this case) in

$$y = 2\left(\frac{x}{3}\right)^2 = \frac{2x^2}{9}.$$

Parametric equations involving trigonometric functions often make use of the trigonometric version of the Pythagorean theorem ($\sin^2 \theta + \cos^2 \theta = 1$) to produce a single equation.

Step-by-Step Solutions to Problems in this Chapter, "Parametric Equations"

Graph the parametric equations

(1) $\begin{cases} x = 2t \\ y = t + 4 \end{cases}$

(2) $\begin{cases} x = \cos\theta \\ y = \sin\theta \end{cases}$

(3) $\begin{cases} x = 4t \\ y = \dfrac{2}{t} \end{cases}$

<u>Solution</u>: (1) One of the methods for graphing these kinds of equations is to change them to regular equations without parameters. For instance, t in

$$\begin{cases} x = 2t & (1) \\ y = t + 4 & (2) \end{cases}$$

is eliminated as the following:

from eq.(1) $t = \dfrac{x}{2}$. Substituting in eq.(2),

$$y = \dfrac{x}{2} + 4, \quad x - 2y + 8 = 0 \text{ which is a straight}$$

line whose graph is easily constructed by choosing two points on the line, as shown in Fig. 1.

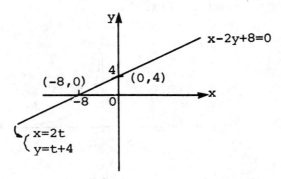

Note that another method for graphing would be using the table as shown below.

t	-2	-1	0	1	2
x	-4	-2	0	2	4
y	2	3	4	5	6

$$\begin{cases} x = 2t \\ y = t + 4 \end{cases}$$

(2) $\begin{cases} x = \cos\theta & \quad\quad\quad (3) \\ y = \sin\theta & \quad\quad\quad (4) \end{cases}$

Squaring eqs.(3) and (4) and adding them together, one obtains

$$\begin{cases} x^2 = \cos^2\theta \\ y^2 = \sin^2\theta \end{cases} , \quad x^2 + y^2 = \cos^2\theta + \sin^2\theta = 1$$

Hence, the graph is a circle of radius 1, with center O(0,0).

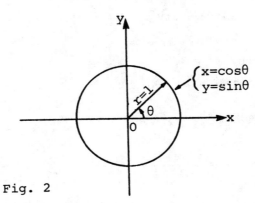

Fig. 2

(3) $\begin{cases} x = 4t & \quad\quad\quad (5) \\ y = \dfrac{2}{t} & \quad\quad\quad (6) \end{cases}$

From eq.(6), $t = \dfrac{2}{y}$. Substituting in eq.(5),

$$x = \dfrac{8}{y}, \quad xy - 8 = 0 .$$

833

The graph is shown in Fig. 3.

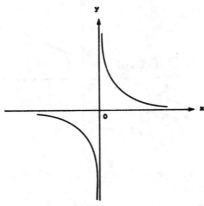

Fig. 3

Find the parametric equations for $x^3 + y^3 - 4xy = 0$.

   (Hint: Substitute $x = ty$.)

Solution: Substituting $x = ty$, where t is the parameter, one has

$$(ty)^3 + y^3 - 4(ty)y$$

$$= t^3y^3 + y^3 - 4ty^2$$

$$= (t^3+1)y^3 + 4ty^2 = 0$$

Dividing by $y^2 \neq 0$,

$$(t^3+1)y + 4t = 0$$

$$y = -\frac{4t}{t^3+1}.$$

Then $x = ty = -\frac{4t^2}{t^3+1}$

The parametric equations for the given equations are

$$\begin{cases} x = -\dfrac{4t^2}{t^3+1} \\ \\ y = -\dfrac{4t}{t^3+1} \end{cases}$$

Show that $x = 5 \cos \theta$ and $y = 3 \sin \theta$ satisfies

$$\frac{x^2}{25} + \frac{y^2}{9} = 1.$$

<u>Solution</u>: If $x = 5 \cos \theta$, $y = 3 \sin \theta$ satisfies the equation

$$\frac{x^2}{25} + \frac{y^2}{9} = 1,$$

then we will obtain an identity when we substitute these values of x and y into the equation. Doing this, we find

$$\frac{(5 \cos \theta)^2}{25} + \frac{(3 \sin \theta)^2}{9} = \frac{25 \cos^2 \theta}{25} + \frac{9 \sin^2 \theta}{9}$$

$$\cos^2 \theta + \sin^2 \theta = 1$$

$$1 = 1$$

Since this is an identity, the given values of x and y satisfy the given equation.

CHAPTER 47

SPACE-RELATED PROBLEMS

> **Basic Attacks and Strategies for Solving Problems in this Chapter. See pages 836 to 847 for step-by-step solutions to problems.**

Locating a point on a flat plane by means of two coordinate axes has been discussed in several earlier chapters, particularly Chapter 7 (Functions and Graphs) and Chapter 38 (Coordinates). Locating a point can be extended to three-dimensional space by introducing a third coordinate axis, usually called the z-axis. Every point in space is identified, therefore, by three coordinates (x, y, z). The standard way to depict the three three-dimensional axes (in two dimensions) is by imagining the xy-plane as being flat, with the positive x-axis coming toward the viewer (out from the page), the positive y-axis towards the right, and the positive z-axis pointing up.

Finding the distance between two points in (three-dimensional) space is possible by a double use of the Pythagorean theorem. This is usually written as one expression, so that the distance, d, between P_1 with coordinates (x_1, y_1, z_1) and P_2 with coordinates (x_2, y_2, z_2) is

$$\sqrt{(x_2 - x_1)^2 + (y_2 - y_1)^2 + (z_2 - z_1)^2}.$$

Solving equations in three variables follows the same rules as solving equations in one or two variables. If the values of certain variables are known, substitution into the equation can lead to an equation in fewer variables that is easier to solve.

The linear equation in three variables,

$Ax + By + Cz + D = 0,$

describes a *plane* in 3-dimensional space. Dividing the equation by $-D$ and renaming coefficients, we can rewrite the general equation as

$ax + by + cz = 1.$

It is also true that three points determine a plane. Given three points in space, one can substitute the values into the revised general equation of a plane and obtain three equations in three parameters, a, b, and c. This system can then be solved for values of a, b, and c, using the techniques of Chapter 11.

Two planes PL_1 (with equation

$$A_1x + B_1y + C_1z + D_1 = 0)$$

and PL_2 (with equation

$$A_2x + B_2y + C_2z + D_2 = 0)$$

are parallel if the coefficients of the variables are proportional, that is, if there is one constant K such that

$$A_1 = KA_2, \quad B_1 = KB_2, \quad \text{and} \quad C_1 = KC_2.$$

Given a plane

$$Ax + By + Cz + D = 0$$

(and choosing K to be 1), the plane

$$Ax + By + Cz + E = 0$$

is parallel to it for any E. One can determine E by substituting the coordinates of a point known to be in the plane for the variables x, y, and z.

The coordinate axes can be considered as subdividing space into eight sections called *octants*. The *first* octant is that in which all three axes have positive values. There is no standard numbering for the remaining seven octants. In space, the x-axis is the line determined by the double equation $y = z = 0$, the y-axis is the line $x = z = 0$, and the z-axis is the line $x = y = 0$. Intercepts of a plane with an axis can be determined by substituting 0 for the two appropriate variables (corresponding to the line that determines that axis), and then solving for the remaining variable.

Equations of solids in space often resemble equations of related figures in the plane. One standard way to determine what a solid looks like, given its equation, is to substitute an appropriate constant value for each variable, one at a time, and then analyze the resulting equations in two variables. Substituting a value for a variable (for example, $y = 2$) corresponds to taking a cross-section of the solid by imagining a plane slicing the solid through the appropriate axis at the given point (for example, perpendicular to the y-axis at the point 2).

The equation of a sphere with its center at (h, k, l) and a radius r is

$$(x - h)^2 + (y - k)^2 + (z - l)^2 = r^2.$$

Note that since a sphere sliced in any direction produces a cross-section of a circle, substituting an appropriate value for x or y or z into the equation of a sphere would produce the equation of a circle. The volume of a sphere,

$$V_S, \quad \text{is} \quad {}^4/_3\, \pi r^3.$$

When an equation in two variables (two dimensions) is expanded into space (three dimensions), it results in a figure called a *cylinder*. The most common cylinder is the right circular cylinder whose cross-section (perpendicular to its central axis) is a circle. (A can is a segment of a right circular cylinder.) In general, the cross-section of a cylinder can be any figure. The *directrix* of a cylinder is the curve formed by the intersection of the cylinder and a coordinate axis (i.e., it is the cross-sectional curve in one of the coordinate planes). To show that a solid (i.e., an equation in three variables) is a cylinder, one should take cross-sections (by letting an appropriate variable take on a constant value), and check to see if the curves are congruent. Note that one must take cross-sections perpendicular to all three coordinate axes to show that a solid is *not* a cylinder. Note also that it is possible for the axis of a cylinder to be a line not parallel to one of the three coordinate axes. In that case, the cross-sections might be congruent (e.g., all circles of a radius of one) but not located in the same place (i.e., all having different centers when projected onto a common coordinate plane).

● **PROBLEM** 47-1

Find the distance of the point (x, y, z) from the origin 0.

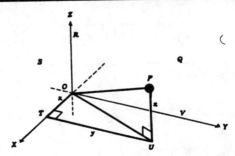

Solution: From the given diagram we see that point (x, y, z) is labeled P. Then OP is the distance of (x, y, z) from the origin 0. Thus, we wish to find OP. Referring to the figure, consider triangle OUP, in which the angle OUP is a right angle. From Pythagoras' theorem,

$$OP^2 = OU^2 + UP^2$$

Consider triangle OTU in which the angle OTU is a right angle. Using Pythagoras' theorem again,

$$OU^2 = OT^2 + TU^2$$

Substituting this value of OU^2 in the first equation

$$OP^2 = \left(OT^2 + TU^2\right) + UP^2$$

But OT = x, TU = y, UP = z, and so

$$OP^2 = x^2 + y^2 + z^2$$

The distance of the points (x, y, z) from the origin 0 is therefore

$$OP = \sqrt{x^2 + y^2 + z^2}.$$

● **PROBLEM 47-2**

Find the distance from (6, 3, -6) to (10, 0, 6).

Solution: The distance between any two points (x_1, y_1, z_1) and (x_2, y_2, z_2) in xyz-space is given by the formula:

$$d = \sqrt{(x_2 - x_1)^2 + (y_2 - y_1)^2 + (z_2 - z_1)^2}$$

Let (6, 3, -6) be P_1: (x_1, y_1, z_1) and let (10, 0, 6) be P_2: (x_2, y_2, z_2). Therefore, the distance between (6, 3, -6) and (10, 0, 6) is:

$$d = \sqrt{(10 - 6)^2 + (0 - 3)^2 + (6 - (-6))^2}$$

$$= \sqrt{(4)^2 + (-3)^2 + (6 + 6)^2}$$

$$= \sqrt{(4)^2 + (-3)^2 + (12)^2}$$

$$= \sqrt{16 + 9 + 144}$$

$$= \sqrt{25 + 144}$$

$$= \sqrt{169}$$

$$= 13$$

● **PROBLEM 47-3**

Find the solutions of the equation in xyz-space where the

    a. first and second components are zero,

    b. first and third components are zero, and

    c. second and third components are zero.

$$x^2 + 3y - z = 4$$

<u>Solution:</u>   a)   In this case, the first and second components are zero. In xyz-space, the point described is $(0, 0, z)$. Once the third component, z, is found, a solution will be found. Letting $x = 0$ and $y = 0$ in the given equation, since the first and second components are zero:

$$(0)^2 + 3(0) - z = 4$$

$$0 + 0 - z = 4$$

$$- z = 4$$

Multiplying both sides of this equation by -1:

$$-1(-z) = -1(4)$$

$$z = -4$$

Therefore, the solution in xyz-space is: $(0, 0, -4)$

b) In this case, the first and third components are zero. In xyz-space, the point described is $(0, y, 0)$. Once the second component, y, is found, a solution will be found. Letting $x = 0$ and $z = 0$ in the given equation, since the first and third components are zero:

$$(0)^2 + 3y - 0 = 4$$

$$0 + 3y - 0 = 4$$

$$3y = 4$$

Dividing both sides of this equation by 3:

$$\frac{3y}{3} = \frac{4}{3}$$

$$y = \frac{4}{3}$$

Therefore, the solution in xyz-space is: $\left(0, \frac{4}{3}, 0\right)$.

c) In this case, the second and third components are zero. In xyz-space, the point described is $(x, 0, 0)$. Once the first component, x, is found, a solution will be found. Letting $y = 0$ and $z = 0$ in the given equation, since the second and third components are zero:

$$x^2 + 3(0) - 0 = 4$$

$$x^2 + 0 - 0 = 4$$

$$x^2 = 4$$

Taking the square root of both sides:

$$\sqrt{x^2} = \sqrt{4}$$

$$x = \pm 2$$

Therefore, there are two solutions in xyz-space and they are (2, 0, 0) and (-2, 0, 0).

● **PROBLEM 47-4**

Show that the point $P_1(2, 2, 3)$ is equidistant from the points $P_2(1, 4, -2)$ and $P_3(3, 7, 5)$.

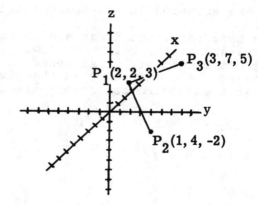

Solution: The distance, d, between any two points (x_1, y_1, z_1) and (x_2, y_2, z_2) is given by the formula

$$d = \sqrt{(x_2 - x_1)^2 + (y_2 - y_1)^2 + (z_2 - z_1)^2}$$

As a visual aid, let us plot the three points and draw the segments whose lengths we wish to show equal.

We are asked to show $P_1 P_2 = P_1 P_3$. By substituting into the formula given,

$$P_1 P_2 = \sqrt{(2 - 1)^2 + (2 - 4)^2 + (3 - (-2))^2}$$

$$= \sqrt{(1)^2 + (-2)^2 + (5)^2} = \sqrt{30}$$

$$P_1 P_3 = \sqrt{(2 - 3)^2 + (2 - 7)^2 + (3 - 5)^2}$$

$$= \sqrt{(-1)^2 + (-5)^2 + (-2)^2} = \sqrt{30}$$

Hence, $P_1 P_2 = P_1 P_3$.

In general, three points determine a plane. Find the equation of the plane determined by D(1, 2, 1), E(2, 0,3), and F(1, - 2, 0).

Solution: The equation of a plane in 3-space is of the form:

$$ax + by + cz = d$$

where a, b, and c cannot all equal zero.

We know that any three noncollinear points must determine a plane. Hence, by substituting the coordinates of D, E, and F into the general form of the plane, we can solve for a, b, and c which will fully determine the plane.

After substituting, we obtain:

(1) $a + 2b + c = d$

(2) $2a \qquad +3c = d$

(3) $a - 2b \qquad = d$

This is a linear system of equations of 4 unknowns in 3 equations which implies that there is no unique solution. This is not to say that we cannot find an equation of the plane. This means there is more than one equation of the plane. Suppose

$$2x + 3y + 4z = 1$$

is an equation of a plane. Multiplying the equation by 2 does not change the points that satisfy the equation. Thus,

$$4x + 6y + 8z = 2$$

is also an equation of the plane; also

$$6x + 9y + 12z = 3 \text{ and } 20x + 30y + 40z = 10.$$

More precisely, any equation of the plane $ax + by + cz + d = 0$ can be rewritten (if $d \neq 0$) as $a'x + b'y + c'z + 1 = 0$ where $a' = a/d$, $b' = b/d$, and $c' = c/d$. Written in this manner, it is more apparent that the plane is an equation in 3 unknowns - not 4.

To solve the system $ax + by + cz + d = 0$, we treat d as a constant. We then solve for a, b, and c, in terms of d. Thus, $a = a'd$, $b = b'd$, and $c = c'd$, where a', b', c' are constants. Substituting our results in the equation of the plane, we obtain:

$$a'dx + b'dy + c'dz + d = 0$$

Dividing through by unknown d, we obtain

a'x + b'y +c'z + 1 = 0.

We now solve the system for a, b, and c in terms of d. Adding (1) + (2) + (3), we obtain

(4)   4a + 4c = 3d

From (2), we have (5) $a = \dfrac{d - 3c}{2}$.

Substituting (5) in (4) and solving for c,

$$4 \left[\dfrac{d - 3c}{2}\right] + 4c = 3d$$

$$2d - 6c + 4c = 3d$$

$$- 2c = d$$

(6) $c = - \dfrac{d}{2}$.

By substitution of (6) in (5),

$$a = \dfrac{d - 3\left[- \dfrac{d}{2}\right]}{2} = \dfrac{d + \dfrac{3d}{2}}{2} = \dfrac{5d}{2} \cdot \dfrac{1}{2} = \dfrac{5d}{4}$$

Plugging this result into (3), we find

$$\dfrac{5d}{4} - 2b = d, \quad - 2b = - \dfrac{d}{4} \quad \text{or} \quad b = \dfrac{d}{8}$$

We can determine the equation of the plane by substituting the expressions for a, b, and c and then eliminating d, the as yet unspecified variable.

Hence, $\dfrac{5d}{4} x + \dfrac{d}{8} y - \dfrac{d}{2} z - d = 0$.

Divide by d ≠ 0

$$\dfrac{5}{4} x + \dfrac{1}{8} y - \dfrac{1}{2} z - 1 = 0$$

Multiply by 8, to simplify.

Therefore, the desired equation is
$$10x + y - 4z = 8.$$

● **PROBLEM 47-6**

Find the equation of the plane passing through the point (4, - 1, 1) and parallel to the plane 4x - 2y + 3z - 5 = 0.

Solution: The general form of the equation of a plane is

$$ax + by + cz + d = 0$$

Hence, to answer this question we must determine a, b, c, and d.

To find a, b, and c we must draw on an analogy from the two-dimensional case.

The equation of a line is given by

$$ex + fy + k = 0.$$

By the definition of slope, any line parallel to this line is of the form

$$ex + fy + \ell = 0.$$

Any two parallel lines have identical coefficients preceeding the variables in their equations.

While the slope concept is inapplicable to planes, the above rule of thumb can be extended to planes. Hence,

$ax + by + cz + d = 0$ is parallel to $ax + by + cz + e = 0$.

In this problem the plane parallel to $4x-2y+3z-5=0$ will have $a=4$, $b=-2$, and $c=3$.

We know $(4, -1, 1)$ is a point on the plane. This will allow us to find d, by substituting these coordinates into the following equation.

$$4x - 2y + 3z + d = 0.$$

It follows that, $4(4) - 2(-1) + 3(1) + d = 0$

$$16 + 2 + 3 + d = 0$$

Hence, $\qquad\qquad d = -21.$

Therefore, the equation of the required plane is

$$4x - 2y + 3z - 21 = 0.$$

● **PROBLEM 47-7**

Determine the intercepts of $4x + y + 2z = 8$ and sketch the portion of the graph in the octant of 3-space in which $x > 0$, $y > 0$, and $z > 0$.

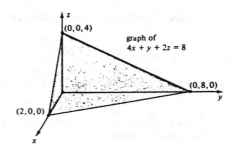

graph of
$4x + y + 2z = 8$

(0, 0, 4)

(0, 8, 0)

(2, 0, 0)

Solution: The given equation represents a plane in a 3-dimensional rectangular coordinate system.

The intercepts of the plane are given by the points $(x', 0, 0)$, $(0, y', 0)$, and $(0, 0, z')$, where the non-zero coordinate signifies the axis with which the plane intersects.

Thus, if $y = z = 0$ then $4x = 8$ and $x = 2$

if $x = z = 0$ then $y = 8$

if $y = x = 0$ then $2z = 8$ and $z = 4$.

The three intercepts are $(2, 0, 0)$, $(0, 8, 0)$, and $(0, 0, 4)$. The points have been plotted on the accompanying graph and the shaded region is the portion of the plane required by the question.

● **PROBLEM 47-8**

Find an equation of the sphere which has the segment joining $P_1(2, -2, 4)$ and $P_2(4, 8, -6)$ for a diameter.

Solution: The equation of a sphere with center at $C(h, k, \ell)$ and radius r is the graph of

$$(x - h)^2 + (y - k)^2 + (z - \ell)^2 = r^2.$$

(**Note** that this resembles the equation of a circle in two-space.)

The center of a sphere is located at the midpoint of a diameter and the length of the radius is the distance from the center of the sphere to the endpoint of the diameter.

Given the coordinates of the endpoints of the diameter, we can use the midpoint formula to obtain the coordinates of the center C.

Since, $P_1 = (2, -2, 4)$ and $P_2 = (4, 8, -6)$,

$$C = \left(\frac{4 + 2}{2}, \frac{8 - 2}{2}, \frac{-6 + 4}{2} \right), \text{ or } C = (3, 3, -1).$$

An application of the distance formula will allow us to obtain the radius length, CP_1. Hence

$$CP_1 = \sqrt{(3 - 2)^2 + (3 - (-2))^2 + ((-1) - 4)^2}$$

$$= \sqrt{(1)^2 + (5)^2 + (-5)^2} = \sqrt{51}.$$

By substituting $C = (3, 3, -1)$ and $CP_1 = \sqrt{51}$ back into the standard form of the equation of a sphere, we obtain the desired equation. Hence,

$$(x - 3)^2 + (y - 3)^2 + (z + 1)^2 = 51$$

is an equation of the sphere.

● PROBLEM 47-9

Find the center, radius, and volume of a sphere whose equation is

$$x^2 + y^2 + z^2 - 8x + 6y - 12z + 12 = 0.$$

Solution: The standard form of the equation of a sphere centered at (h, k, ℓ) with radius r is

$$(x - h)^2 + (y - k)^2 + (z - \ell)^2 = r^2.$$

Hence, by putting the given equation in this form we can read off the coordinates of the center and the length of the radius. We do this by completing the squares in x, y, and z.

Regrouping, we obtain

$$x^2 - 8x + y^2 + 6y + z^2 - 12z + 12 = 0$$

$$(x - 4)^2 + (y + 3)^2 + (z - 6)^2 = -12 + 61$$

$$= 49 = (7)^2$$

Hence, the center is $(4, -3, 6)$ and the radius is 7 units.

The volume, V, of a sphere is given by

$$V = \frac{4}{3} \pi r^3.$$

Consequently, by substitution, the volume of the sphere given by our value for the radius is

$$\frac{4}{3} \pi (7)^3 = (457.33) \pi.$$

Using $\pi = 3.14$, we obtain the volume 1436 cu. units.

Describe the geometric shape determined by the graph of
the equation $x^2 + y^2 = 1$ when plotted on a 3-dimensional
system of rectangular coordinates.

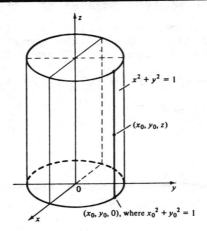

$x^2 + y^2 = 1$

(x_0, y_0, z)

$(x_0, y_0, 0)$, where $x_0{}^2 + y_0{}^2 = 1$

Solution: All points in a 3-dimensional system are
given by the triple (x, y, z). In this problem the determining equation contains only two variables. Consequently,
z can take on an infinite number of values corresponding to
the x and y values that satisfy the equation.

In the xy plane (z = 0), the equation traces out
a circle centered at the origin, (0, 0) with a radius of
1 unit.

For any value of z, say z = n, $-\infty < n < \infty$, the
given equation will always result in a circular graph
centered at (0, 0, n) with radius 1.

There are an infinite number of circles described
in this manner whose centers form a straight line, the
z-axis, and whose points determine a surface at a constant
1 unit from the z-axis. Hence, the geometric shape described is a right circular cylinder. (See the accompanying
figure for a graphical representation.)

Show that the equation $x^2 + y^2 + 2z^2 + 2xz - 2yz = 1$
represents a cylinder and find the equation of its
directrix.

Solution: A cylinder is a surface generated by a straight
line which moves so that it is always parallel to a given
fixed line and always passes through a given fixed curve
called the directrix. Since several curves would determine
the same cylinder, the directrix is taken to be the curve

formed by the intersection of the cylinder and a co-ordinate plane.

To prove that a given figure is a cylinder, we pass several parallel planes through the cylinder. If the curves formed by the intersection are congruent, then the figure is a cylinder. Furthermore, the curve formed by the intersection of the figure and the coordinate plane is the directrix.

To simplify calculations, we choose as our planes the planes parallel to the XY plane. Each such plane has the general form $z = k$. The intersections of the planes and the figure are the curves

(i) $x^2 + y^2 + 2k^2 + 2kx - 2ky = 1,$ $z = k.$

Rewriting, we obtain

(ii) $(x + k)^2 + (y - k)^2 = 1,$ $z = k.$

The equations of intersection (ii) are all circles of radius 1. Since parallel planes cut congruent curves with the figure, the figure is a cylinder.

To find the directrix, we find the intersection with the XY plane, $z = 0$.

(iii) $x^2 + y^2 = 1,$ $z = 0.$

● **PROBLEM 47-12**

A synchronous Earth satellite is one which is placed in a west-to-east orbit over the Equator at such an altitude that its period of revolution about Earth is 24 hours, the time for one rotation of Earth on its axis. Thus the orbital motion of the satellite is synchronized with Earth's rotation, and the satellite appears from Earth to remain stationary over a point on Earth's surface below. Such communication satellites as Syncom, Early Bird, Intelsat, and ATS are in synchronous orbits. Find the altitude for a synchronous Earth satellite.

Solution   The velocity can be found from the equation for circular orbital velocity. It can also be found by dividing the distance around the orbit by the time required; that is, $v = \dfrac{2\pi r}{t}$. Because the two velocities are equal,

$$\frac{2\pi r}{t} = \sqrt{\frac{GM}{r}}$$

$$\left(\frac{2\pi r}{t}\right)^2 r = GM$$

$$r^3 = \frac{GMt^2}{4\pi^2}$$

$$r = \sqrt[3]{\frac{GMt^2}{4\pi^2}}.$$

It is apparent that $t = 24$ hours. Substituting the other values yields

$$r = \sqrt[3]{\frac{1.24 \times 10^{12} \times 24^2}{4 \times 3.14^2}} = 10^4 \sqrt[3]{\frac{178.56}{9.86}}$$

$$= 10^4 = \sqrt[3]{18.1} = 26{,}260 \text{ mi.}$$

$$\text{Altitude} = 26{,}260 - 3{,}960 = 22{,}300 \text{ mi.}$$

$$v = \frac{2 \times 3.14 \times 26{,}260}{24} = 6{,}870 \text{ mi/hr.}$$

To understand orbits, we must know something of the nature and properties of the conic sections. They get their name, of course, from the fact that they may be formed by cutting or sectioning a complete right circular cone (of two nappes) with a plane. Any plane perpendicular to the axis of the cone cuts a section that is a circle. Incline the plane a bit, and the section formed is an ellipse. Tilt the plane still more until it is parallel to a ruling of the cone and the section is a parabola. Let the plane cut both nappes, and the section is a hyperbola, a curve with two branches. It is apparent that closed orbits are circles or ellipses. Open or escape orbits are parabolas or hyperbolas.

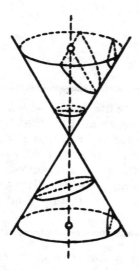

Another way of classifying the conic sections is by means of their eccentricity. If we represent the eccentricity by e, then a conic section is

A circle if $e = 0$,
An ellipse if $0 < e < 1$,
A parabola if $e = 1$,
A hyperbola if $e > 1$.

In actual practice, orbits that are exactly circular or parabolic do not exist because the eccentricity is never exactly equal to 0 or 1.

847

SECTION 4 — INTRODUCTION TO CALCULUS

CHAPTER 48

LIMITS

> **Basic Attacks and Strategies for Solving Problems in this Chapter. See pages 848 to 850 for step-by-step solutions to problems.**

A *limit* is a mathematical concept that assigns a value to an expression given a specific value for the variable it involves. The limit concept is expressed in notation as follows:

$$\lim_{x \to a} f(x) = b$$

indicates that, as the variable x gets closer and closer to the value of a, the limit value of the expression $f(x)$ is b.

The concept of a limit is most useful in situations where, for various reasons, an expression would not normally be considered to have a legitimate value, for example, when the denominator of a fraction is zero, or when both the numerator and denominator are zero. In some cases, one can envision the limit as the value the expression *should have* at a given point, but otherwise is undefined or illegal (by standard mathematical rules).

In the case of a standard polynomial, the value given by the limit and the value of the polynomial are the same.

In cases where evaluating a fractional expression at a point would lead to the undefined value of $^0/_0$, one first tries to factor the expression and divide out any common terms before evaluating. For example,

$$\lim_{x \to 0} \frac{x^2}{x} = \lim_{x \to 0} x = 0, \quad \text{but} \quad \frac{x^2}{x}$$

is not defined at $x = 0$.

In cases where evaluating a fractional expression at a point would lead to the numerator being zero and the denominator being some constant other than zero, the value of the limit is zero.

In cases where evaluating a fractional expression at a point would lead to the denominator being zero and the numerator being some constant other than zero, the value of the limit is infinity (∞) or negative infinity ($-\infty$).

It is also possible to compute a one-sided limit of an expression. In these cases we determine whether one approaches the limit point from the positive side (called a right-handed limit and denoted

$$\lim_{x \to a+} f(x))$$

or from the negative side (called the left-handed limit and denoted

$$\lim_{x \to a-} f(x)).$$

If a function has both a right-handed and a left-handed limit at a point and these have the same value, it is said to have a *limit* whose value is the same as both. For example, given

$$f(x) = 1/x, \lim_{x \to 0+} f(x) = \infty,$$

since for positive values of x, $f(x) = 1/x$ is always positive and the values grow as x gets closer to zero (from the positive side). On the other hand,

$$\lim_{x \to 0-} f(x) = -\infty,$$

since for negative values of x, $f(x) = 1/x$ is always negative and the values become smaller (i.e., they remain negative but their absolute value gets larger) as x gets closer to zero (from the negative side).

The greatest integer function (truncation function or "birthday" function), indicated by $[x]$, is the function that drops any fractional quantity over an integer. For example, $[\pi] = 3$ and $[2.999999999] = 2$. As an example,

$$\lim_{x \to 2+} [x] = 2 \quad \text{but} \quad \lim_{x \to 2-} [x] = 1.$$

Since the right- and left-handed limits at 2 are different, $[x]$ is not considered to have any limit at 2.

Step-by-Step Solutions to Problems in this Chapter, "Limits"

● **PROBLEM 48-1**

Find $\lim\limits_{x \to 2} f(x) = 2x + 1$

<u>Solution:</u>   As $x \to 2$, $f(x) \to 5$. Therefore,

$$\lim_{x \to 2} (2x + 1) = 5$$

● **PROBLEM 48-2**

Find $\lim\limits_{x \to 3} f(x) = \dfrac{x^2 - 9}{x + 1}$

<u>Solution:</u>

$$\lim_{x \to 3} \frac{x^2 - 9}{x + 1} = \frac{0}{4} = 0$$

● **PROBLEM 48-3**

Find $\lim\limits_{x \to 2} \left(\dfrac{x^2 - 5}{x + 3} \right)$

<u>Solution:</u>   For the numerator only

$$\lim_{x \to 2} (x^2 - 5) = 4 - 5 = -1$$

For the denominator only

$$\lim_{x \to 2} (x + 3) = 5$$

Therefore,

$$\lim_{x \to 2} \left(\frac{x^2 - 5}{x + 3} \right) = -\frac{1}{5}$$

Using the method of simple substitution, we find that $\lim_{x \to 3} f(x) = \infty$. There is no limit.

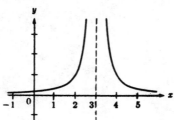

● **PROBLEM 48-4**

Find $\lim_{x \to 3} \dfrac{5x}{6 - 2x}$.

<u>Solution:</u> $\lim_{x \to 3} 5x = 15$

and

$$\lim_{x \to 3} [6 - 2x] = 0$$

Therefore,

$$\lim_{x \to 3} \frac{5x}{6 - 2x} = \frac{15}{0} = \infty .$$

The function has no limit.

● **PROBLEM 48-5**

Find:

(a) $\displaystyle \lim_{x \to 3+} \frac{x^2 + x + 2}{x^2 - 2x - 3}$

(b) $\displaystyle \lim_{x \to 3-} \frac{x^2 + x + 2}{x^2 - 2x - 3}$

(c) $\displaystyle \lim_{x \to 3} \frac{x^2 + x + 2}{x^2 - 2x - 3}$

849

Solution:

(a) $\lim\limits_{x \to 3^+} \dfrac{x^2 + x + 2}{x^2 - 2x - 3} = \lim\limits_{x \to 3^+} \dfrac{x^2 + x + 2}{(x - 3)(x + 1)}$

The limit of the numerator is 14. To find the limit of the denominator,

$$\lim\limits_{x \to 3^+} (x - 3)(x + 1) = \lim\limits_{x \to 3^+} (x - 3) \cdot \lim\limits_{x \to 3^+} (x + 1)$$

$$= 0 \cdot 4 = 0$$

Thus, the limit of the denominator is 0, as the denominator is approaching 0 through positive values. Consequently,

$$\lim\limits_{x \to 3^+} \dfrac{x^2 + x + 2}{x^2 - 2x - 3} = + \infty$$

(b) $\lim\limits_{x \to 3^-} \dfrac{x^2 + x + 2}{x^2 - 2x - 3} = \lim\limits_{x \to 3^-} \dfrac{x^2 + x + 2}{(x - 3)(x + 1)}$

As in part (a), the limit of the numerator is 14 here also. To find the limit of the denominator,

$$\lim\limits_{x \to 3^-} (x - 3)(x + 1) = \lim\limits_{x \to 3^-} (x - 3) \cdot \lim\limits_{x \to 3^-} (x + 1)$$

$$= 0 \cdot 4 = 0$$

In this case, the limit of the denominator is again zero, but since the denominator is approaching zero through negative values,

$$\lim\limits_{x \to 3^-} \dfrac{x^2 + x + 2}{x^2 - 2x - 3} = - \infty$$

(c) $\lim\limits_{x \to 3} \dfrac{x^2 + x + 2}{x^2 - 2x - 3} = |\infty|$

Since the right and left sides are not equal, the solution is not defined.

CHAPTER 49

CONTINUITY

> **Basic Attacks and Strategies for Solving Problems in this Chapter. See pages 851 to 853 for step-by-step solutions to problems.**

Continuity is a concept that enables us to determine whether there are any "holes" or "gaps" in the graph of an expression. For a function to be *continuous at a point c*, the following must hold:

- $f(c)$ must be defined,

- $\lim_{x \to c} f(x)$ must exist,

- $\lim_{x \to c} f(x) = f(c)$.

A function that is not continuous is one whose graph has a hole (i.e., is not defined at a point or has a value infinite in magnitude), or has a gap (i.e., is not defined over an interval), or has a jump (i.e., although defined at all points in an interval including a point, the function changes values abruptly at a certain point but is continuous afterward). Such a function is said to be *discontinuous*.

It could happen that $f(x)$ is not defined at c, for example, if

$$f(x) = \left| \frac{1}{x} \right|$$

and $c = 0$. It could happen that $\lim_{x \to c} f(x)$ does not exist, as occurs if the right-handed and left-handed limits do not agree. It could happen that a function's defined value at a point does not agree with the limit at that point, for example, if

$$f(x) = \frac{x^2 - 1}{x - 1} \quad \text{for} \quad x \neq 1 \quad \text{and} \quad f(1) = 3.$$

A function $f(x)$ is said to be *continuous over an open interval* (a, b) if it is continuous at each point in the interval. A function $f(x)$ is said to be *continuous over a closed interval* (a, b) if it is continuous over the open interval (a, b) and also if both

$$\lim_{x \to a+} f(x) = f(a) \quad \text{and} \quad \lim_{x \to b-} f(x) = f(b).$$

● **PROBLEM 49-1**

Investigate the continuity of the expression:

$$y = \frac{x^2 - 9}{x - 3} \quad \text{at} \quad x = 2.$$

Solution: For a function to be continuous at a point, in this case, 2, it must satisfy three conditions: (1) $f(2)$ is defined, (2) $\lim_{x \to 2} f(x)$ exists, (3) $\lim_{x \to 2} f(x) = f(2)$.

For $x = 2$, $y = f(x) = f(2) = \frac{(2)^2 - 9}{2 - 3} = 5$.

Also,

$$\lim_{x \to 2} \frac{x^2 - 9}{x - 3} = 5.$$

Therefore, the function is continuous at $x = 2$.

● **PROBLEM 49-2**

Determine the continuity of the following function and sketch its graph:

$$M(x) = \frac{|x|}{x}(x^2 - 1).$$

Solution: For $x > 0$, $|x| = x$ and for $x < 0$, $|x| = -x$. Therefore for $x > 0$

$$M(x) = \frac{x}{x}(x^2 - 1) = x^2 - 1 \qquad \text{- - - (1)}$$

for $x < 0$

$$M(x) = \frac{-x}{x}(x^2 - 1) = -(x^2 - 1) = 1 - x^2 - - - (2)$$

for x = 0

$$M(x) \Big|_{x=0} = \frac{x}{x}(x^2 - 1) \Big|_{x=0} = \frac{0}{0}, \text{ undefined} - - - (3)$$

To test the continuity of the function at x = 0, we might test whether

$$\lim_{x \to 0^+} M(x) = \lim_{x \to 0^-} M(x) = M(0).$$

It turns out that the former = - 1, the latter = + 1.

However, from (3), we know that M(0) is not even defined, therefore there is no need to calculate the limits, M(x) is not continuous at x = 0.

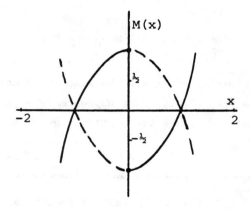

● **PROBLEM 49-3**

Determine whether the function $y = \frac{1}{x - 2}$ is continuous at (a) x = 0, (b) x = 1, and (c) x = 2.

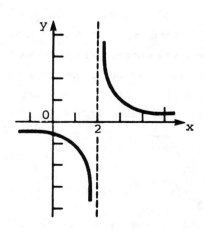

Solution: (a) $\lim\limits_{x \to 0} \dfrac{1}{x-2} = \lim\limits_{x \to 0} \dfrac{1}{0-2} = -\dfrac{1}{2}$

$$f(0) = -\dfrac{1}{2}.$$

Therefore,

$$\lim\limits_{x \to 0} \dfrac{1}{x-2} = f(0) = -\dfrac{1}{2},$$

and the function is continuous at $x = 0$, because the limit exists at $x = 0$ and $= f(0)$.

(b) $\lim\limits_{x \to 1} \dfrac{1}{x-2} = -1$

$$f(1) = -1.$$

The function is continuous at $x = 1$, because the limit exists at $x = 1$ and $= f(1)$.

(c) $\lim\limits_{x \to 2} \dfrac{1}{x-2} = \lim\limits_{x \to 2} \dfrac{1}{0} = \pm\infty$.

The limit does not exist at $x = 2$ because the function is not defined at this point. I.e., if we approach from the left the function approaches $-\infty$, from the right, it goes to $+\infty$.

● **PROBLEM 49-4**

If $h(x) = \sqrt{4 - x^2}$, prove that $h(x)$ is continuous in the closed interval $[-2, 2]$.

Solution: To prove continuity we employ the following definition: A function defined in the closed interval $[a, b]$ is said to be continuous in $[a, b]$ if and only if it is continuous in the open interval (a, b), as well as continuous from the right at a and continuous from the left at b. The function h is continuous in the open interval $(-2, 2)$. We must now show that the function is continuous from the right at -2 and from the left at 2. Therefore, we must show that $f(-2)$ is defined and $\lim\limits_{x \to -2^+} f(x)$ exists and that these are equal. Also, we must show that $f(2) = \lim\limits_{x \to 2^-} f(x)$. We have:

$$\lim\limits_{x \to -2^+} \sqrt{4 - x^2} = 0 = h(-2),$$

and

$$\lim\limits_{x \to 2^-} \sqrt{4 - x^2} = 0 = h(2).$$

Thus, h is continuous in the closed interval $[-2, 2]$.

CHAPTER 50

Δ-FUNCTION

Basic Attacks and Strategies for Solving Problems in this Chapter. See pages 854 to 856 for step-by-step solutions to problems.

The uppercase Greek letter delta, Δ, is used in conjunction with other variables to indicate *change*. Thus, Δx indicates the change in x, and Δy indicates the change in y. Δy is usually computed by evaluating $y = f(x)$ at two points separated by Δx, i.e.,

$$\Delta y = f(x + \Delta x) - f(x).$$

The slope of a line can be indicated as $\frac{\Delta y}{\Delta x}$. This can also be interpreted as the *average rate of change* of a section of a curve.

The expression that gives the slope of a curve can be computed by

$$\lim_{\Delta x \to 0} \frac{\Delta y}{\Delta x}.$$

This is called the *derivative* or *derived function* of the original function

$$y = f(x).$$

● PROBLEM 50-1

Find the slope of each of the following curves at the given point, using the Δ-method.

(a) $y = 3x^2 - 2x + 4$ at $(1,5)$

(b) $y = x^3 - 3x + 5$ at $(-2,3)$.

<u>Solution:</u> The slope of a given curve at a specified point is the derivative, in this case $\frac{\Delta y}{\Delta x}$, evaluated at that point.

(a) From the Δ-method we know that:

$$\frac{\Delta y}{\Delta x} = \frac{f(x + \Delta x) - f(x)}{\Delta x}.$$

For the curve $y = 3x^2 - 2x + 4$, we find:

$$\frac{\Delta y}{\Delta x} = \frac{3(x + \Delta x)^2 - 2(x + \Delta x) + 4 - (3x^2 - 2x + 4)}{\Delta x}$$

$$= \frac{3x^2 + 6x\Delta x + 3(\Delta x)^2 - 2x - 2\Delta x + 4 - 3x^2 + 2x - 4}{\Delta x}$$

$$= \frac{6x\Delta x + 3(\Delta x)^2 - 2\Delta x}{\Delta x}$$

$$= 6x + 3\Delta x - 2.$$

$$\lim_{\Delta x \to 0} \frac{\Delta y}{\Delta x} = \lim_{\Delta x \to 0} 6x + 3\Delta x - 2 = 6x - 2.$$

At $(1,5)$ $\frac{\Delta y}{\Delta x} = 4$ is the required slope.

(b) Again using the Δ-method, $\frac{\Delta y}{\Delta x}$ for the curve: $y = x^3 - 3x + 5$, can be found as follows:

$$\frac{\Delta y}{\Delta x} = \frac{f(x + \Delta x) - f(x)}{\Delta x} .$$

$$\frac{\Delta y}{\Delta x} = \frac{(x + \Delta x)^3 - 3(x + \Delta x) + 5 - (x^3 - 3x + 5)}{\Delta x}$$

$$= \frac{x^3 + 3x^2 \Delta x + 3x(\Delta x)^2 + (\Delta x)^3 - 3x - 3\Delta x + 5 - x^3 + 3x - 5}{\Delta x}$$

$$= \frac{3x^2 \Delta x + 3x(\Delta x)^2 + (\Delta x)^3 - 3\Delta x}{\Delta x}$$

$$= 3x^2 + 3x\Delta x + (\Delta x)^2 - 3 .$$

$$\lim_{\Delta x \to 0} \frac{\Delta y}{\Delta x} = \lim_{\Delta x \to 0} 3x^2 + 3x\Delta x + (\Delta x)^2 - 3 = 3x^2 - 3 .$$

At $(-2,3)$, $\frac{\Delta y}{\Delta x} = 9$ is the required slope.

● **PROBLEM 50-2**

Find the average rate of change, by the Δ process, for:
$$y = 1/x.$$

Solution:

$$y = f(x) = \frac{1}{x}$$

The average rate of change is defined to be $\frac{\Delta y}{\Delta x}$ with $\Delta y = f(x + \Delta x) - f(x)$.

Since

$$f(x) = \frac{1}{x} , \qquad f(x + \Delta x) = \frac{1}{x + \Delta x} ,$$

and

$$\Delta y = \frac{1}{x + \Delta x} - \frac{1}{x} = \frac{x - (x + \Delta x)}{x(x + \Delta x)}$$

$$= \frac{-\Delta x}{x(x + \Delta x)} .$$

Now,

$$\frac{\Delta y}{\Delta x} = \frac{-\ \Delta x}{x\ (x + \Delta x)\ \Delta x} = -\ \frac{1}{x\ (x + \Delta x)} .$$

Therefore, the average rate of change is

$$\frac{-\ 1}{x\ (x + \Delta x)} .$$

CHAPTER 51

THE DERIVATIVE

> **Basic Attacks and Strategies for Solving Problems in this Chapter. See pages 857 to 864 for step-by-step solutions to problems.**

The derivative of a function $y = f(x)$ with respect to the independent variable x is the function

$$y' = \lim_{\Delta x \to 0} \frac{f(x + \Delta x) - f(x)}{\Delta x}.$$

The alternate notations for y' are

$$\frac{dy}{dx} = f'(x) = D_x y.$$

The derivative gives a function that enables someone to find the slope of the original function $y = f(x)$ at a given point (x_1, y_1). The *slope* of a curve is defined as the slope of the line tangent to the curve at a given point and describes the instantaneous rate of change of the height of the curve with respect to the horizontal distance traversed.

If we know that the derivative of some function is $y' = 2x^2 + 1$, and that $(2, 3)$ is on the original curve, we can compute the slope of the original function at $x = 2$ by substituting into y' to obtain

$$2(2)^2 + 1 = 2 \cdot 4 + 1 = 9$$

and, using the point-slope form of the equation for a line, to obtain:

$$y - 3 = 9(x - 2).$$

The *normal line* to a curve at a point is the line perpendicular to the tangent line to the curve at that same point. As mentioned in Chapter 39, two lines are perpendicular if their slopes are negative reciprocals of each other.

When it is difficult to separate the variables in an equation and rewrite it in the standard form of $y = f(x)$, one must sometimes use the technique of *implicit differentiation*. One takes the derivative of each term separately with respect to the *independent* variable (e.g., x). Each time the derivative of the *dependent* variable (e.g., y) is taken, the derived term must include the derivative of that variable (e.g., y'). One then combines terms and solves for the derivative of the dependent variables (e.g., solves for y').

The derivative of the derivative is called the *second derivative* and is written

$$y'' = \frac{dy'}{dx} = \frac{d^2y}{dx^2} .$$

One can also compute higher order derivatives by repeatedly differentiating previous derivatives.

One of the standard examples for derivatives involves distance, velocity (also called speed), and acceleration. Such concepts are usually given as functions of a time variable, t. The derivative of distance, s, (i.e., rate of change of distance with respect to time) is the velocity, v, i.e.,

$$\frac{ds}{dt} = v.$$

The derivative of velocity (i.e., rate of change of velocity with respect to time) is the acceleration, a, i.e.,

$$\frac{dv}{dt} = a = \frac{d^2s}{dt^2} .$$

Certain observable facts can help solve problems involving distance, velocity, and acceleration. For example, when an object is thrown up, the velocity decreases until it stops rising and then it starts falling down. Thus, when the velocity is zero, the object is at its maximum distance above the ground. Therefore, to find the maximum height, one takes the derivative of the distance expression and sets it equal to zero, solving for the variable t. Then one substitutes that value for t into the original equation for distance.

Find the derivative of the function: $y = 2x^2 + 3x$, by the delta process.

<u>Solution:</u> By definition,

$$y'(x) = \lim_{\Delta x \to 0} \frac{f(x+\Delta x) - f(x)}{\Delta x}.$$

Since

$$f(x) = 2x^2 + 3x,$$

$$f(x+\Delta x) = 2(x+\Delta x)^2 + 3(x+\Delta x).$$

Substituting, we obtain:

$$y'(x) = f'(x) = \lim_{\Delta x \to 0} \frac{2(x+\Delta x)^2 + 3(x+\Delta x) - (2x^2 + 3x)}{\Delta x}.$$

Simplifying, we have:

$$f'(x) = \lim_{\Delta x \to 0} \frac{4x\Delta x + 2(\Delta x)^2 + 3\Delta x}{\Delta x}$$

$$= \lim_{\Delta x \to 0} 4x + 2\Delta x + 3.$$

Now, as $\Delta x \to 0$ the term: $2\Delta x$, drops out and we have

$$f'(x) = 4x + 3.$$

Using the delta method, find $f'(x)$ for the function:

$$f(x) = x^2 + \frac{1}{x}, \quad x \neq 0.$$

<u>Solution</u>: By definition,

$$f'(x) = \frac{dy}{dx} = \lim_{\Delta x \to 0} \frac{f(x+\Delta x)-f(x)}{\Delta x}$$

$$f(x+\Delta x) = (x+\Delta x)^2 + \frac{1}{x+\Delta x}$$

$$= x^2 + 2x\Delta x + (\Delta x)^2 + \frac{1}{x+\Delta x}.$$

$$f(x) = x^2 + \frac{1}{x}.$$

Subtracting the two expressions,

$$f(x+\Delta x) - f(x) = 2x\Delta x + (\Delta x)^2 + \frac{1}{x+\Delta x} - \frac{1}{x}$$

$$= 2x\Delta x + (\Delta x)^2 + \frac{x-(x+\Delta x)}{x(x+\Delta x)}$$

$$= \Delta x\left(2x + \Delta x - \frac{1}{x(x+\Delta x)}\right).$$

Dividing by Δx, we have:

$$\frac{f(x+\Delta x) - f(x)}{\Delta x} = 2x + \Delta x - \frac{1}{x(x+\Delta x)}.$$

Since

$$f'(x) = \lim_{\Delta x \to 0} \frac{f(x+\Delta x) - f(x)}{\Delta x},$$

we have,

$$\lim_{\Delta x \to 0} \left(2x + \Delta x - \frac{1}{x(x+\Delta x)}\right)$$

$$= 2x + 0 - \frac{1}{x(x+0)}$$

$$= 2x - \frac{1}{x^2}.$$

● **PROBLEM 51-3**

Find the derivative of $f(x) = 2/(3x+1)$, using the Δ-method.

<u>Solution</u>: By definition,

$$\frac{dy}{dx} = f'(x) = \lim_{\Delta x \to 0} \frac{f(x+\Delta x) - f(x)}{\Delta x}.$$

Since

$$f(x) = \frac{2}{3x + 1},$$

$$f(x+\Delta x) = \frac{2}{3(x+\Delta x) + 1}.$$

Substituting, we obtain:

$$f'(x) = \lim_{\Delta x \to 0} \frac{2/[3(x+\Delta x) + 1] - 2/(3x+1)}{\Delta x}$$

At this point we cannot substitute 0 for Δx because we would obtain a 0 in the denominator. Multiplying out and obtaining a common denominator we have:

$$\lim_{\Delta x \to 0} \frac{6x + 2 - 6x - 6\Delta x - 2}{(3x+3\Delta x+1)(3x+1)\Delta x}$$

$$= \frac{-6\Delta x}{(3x+3\Delta x+1)(3x+1)(\Delta x)}$$

$$= \lim_{\Delta x \to 0} \frac{-6}{(3x+3\Delta x+1)(3x+1)}$$

We can now substitute 0 for Δx, and we obtain:

$$- \frac{6}{(3x+1)^2} = f'(x).$$

● **PROBLEM** 51-4

Find the derivative of:

$$f(x) = \sqrt{2x - 1},$$

using the Δ-method.

Solution: By definition,

$$\frac{dy}{dx} = f'(x) = \lim_{\Delta x \to 0} \frac{f(x+\Delta x) - f(x)}{\Delta x}.$$

Since

$$f(x) = \sqrt{2x - 1},$$

$$f(x+\Delta x) = \sqrt{2(x+\Delta x) - 1}.$$

Substituting, we obtain:

$$f'(x) = \lim_{\Delta x \to 0} \frac{\sqrt{2(x+\Delta x) - 1} - \sqrt{2x - 1}}{\Delta x}.$$

Substituting 0 for Δx gives a 0 in the denominator. Therefore, we rearrange the numerator. Since multiplication by 1 does not change a term, we multiply

$$(\sqrt{2x + 2\Delta x - 1} - \sqrt{2x - 1})$$

by 1, writing 1 as:

859

$$\frac{\sqrt{2x + 2\Delta x - 1} + \sqrt{2x - 1}}{\sqrt{2x + 2\Delta x - 1} + \sqrt{2x - 1}}.$$

We have:

$$\frac{1}{\Delta x} \cdot \left(\sqrt{2x + 2\Delta x - 1} - \sqrt{2x - 1}\right)$$

$$= \frac{[\sqrt{2x+2\Delta x-1} - \sqrt{2x-1}][\sqrt{2x+2\Delta x-1} + \sqrt{2x-1}]}{\sqrt{2x+2\Delta x-1} + \sqrt{2x-1}} \cdot \frac{1}{\Delta x}.$$

● **PROBLEM 51-5**

Given $f(x) = x^{2/3}$, find $f'(x)$ using the Δ-method.

Solution: By definition,

$$\frac{dy}{dx} = f'(x) = \lim_{\Delta x \to 0} \frac{f(x+\Delta x) - f(x)}{\Delta x}$$

$$f(x) = x^{2/3}.$$

Therefore,

$$f(x+\Delta x) = (x+\Delta x)^{2/3}.$$

Substituting, we have:

$$\lim_{\Delta x \to 0} \frac{(x+\Delta x)^{2/3} - x^{2/3}}{\Delta x}.$$

We wish to rewrite this in a form in which Δx can approach the limit, 0. Direct substitution at this point would leave a 0 in the denominator. Since multiplication by 1 does not change the value, we multiply

$$\frac{(x+\Delta x)^{2/3} - x^{2/3}}{\Delta x}$$

by

$$\frac{(x+\Delta x)^{4/3} + (x+\Delta x)^{2/3}x^{2/3} + x^{4/3}}{(x+\Delta x)^{4/3} + (x+\Delta x)^{2/3}x^{2/3} + x^{4/3}},$$

which is equal to 1. Doing this we have,

$$f'(x) = \lim_{\Delta x \to 0}$$

$$\frac{\left[(x+\Delta x)^{2/3} - x^{2/3}\right]\left[(x+\Delta x)^{4/3} + (x+\Delta x)^{2/3}x^{2/3} + x^{4/3}\right]}{\Delta x\left[(x+\Delta x)^{4/3} + (x+\Delta x)^{2/3}x^{2/3} + x^{4/3}\right]}$$

$$= \lim_{\Delta x \to 0} \frac{(x+\Delta x)^2 - x^2}{\Delta x[x+\Delta x)^{4/3} + (x+\Delta x)^{2/3}x^{2/3} + x^{4/3}]}$$

$$= \lim_{\Delta x \to 0} \frac{x^2 + 2x(\Delta x) + (\Delta x)^2 - x^2}{\Delta x[(x+\Delta x)^{4/3} + (x+\Delta x)^{2/3}x^{2/3} + x^{4/3}]}$$

$$= \lim_{\Delta x \to 0} \frac{2x(\Delta x) + (\Delta x)^2}{\Delta x[(x+\Delta x)^{4/3} + (x+\Delta x)^{2/3}x^{2/3} + x^{4/3}]}$$

$$= \lim_{\Delta x \to 0} \frac{2x + \Delta x}{(x+\Delta x)^{4/3} + (x+\Delta x)^{2/3}x^{2/3} + x^{4/3}}.$$

Now, substituting 0 for Δx, we obtain:

$$\frac{2x}{x^{4/3} + x^{2/3}x^{2/3} + x^{4/3}}$$

$$= \frac{2x}{3x^{4/3}} = \frac{2x}{3x^{3/3}x^{1/3}}$$

$$= \frac{2}{3x^{1/3}}.$$

● **PROBLEM** 51-6

Find the equations of the tangent line and the normal to the curve: $y = x^2 - x + 3$, at the point (2,5).

Solution: Since the equation of a straight line passing through a given point can be expressed in the form: $y - y_1 = m(x - x_1)$, this is appropriate for finding the equations of the tangent and normal. Here $x_1 = 2$ and $y_1 = 5$. The slope, m, of the tangent line is found by taking the derivative, dy/dx, of the curve: $y = x^2 - x + 3$.

$$dy/dx = 2x - 1.$$

At (2,5), $dy/dx = 2(2) - 1 = 3$, therefore the slope, m, of the tangent line is 3. Substituting x_1, y_1 and m into the equation $y - y_1 = m(x - x_1)$ we obtain:

$$y - 5 = 3(x - 2),$$

as the equation of the tangent line, or

$$3x - y - 1 = 0.$$

Since the slope of the normal is given by: $m' = -1/m$, and since $m = 3$, the slope of the normal is $-1/3$. Substituting $x_1 = 2$, $y_1 = 5$ and the slope of the normal, $m' = -1/3$, into the equation: $y - y_1 = m'(x - x_1)$, we obtain:

$$y - 5 = -\frac{1}{3}(x - 2),$$

or,

$$x + 3y - 17 = 0.$$

This is the equation of the normal.

Find the equation of the line that is tangent to the parabola $y^2 = 2px$ at the point (x_1, y_1).

Solution: The equation of a straight line passing through a given point (x_1, y_1) can be expressed in the form $y - y_1 = m(x - x_1)$. To find the slope, m, of the line tangent to the curve: $y^2 = 2px$, we differentiate implicitly and obtain:

$$2yy' = 2p \quad \text{or} \quad y' = p/y;$$

now, $y' = m$, and therefore $m = p/y$. At (x_1, y_1) $\quad m = p/y_1$.

Substituting m into the equation: $y - y_1 = m(x - x_1)$, we obtain:

$$y - y_1 = \frac{p}{y_1}(x - x_1),$$

or,

$$y_1 y - y_1^2 = px - px_1.$$

Using the given equation, $y^2 = 2px$, we find that at (x_1, y_1) $y_1^2 = 2px_1$. Substituting $y_1^2 = 2px_1$ into the equation: $y_1 y - y_1^2 = px - px_1$, we obtain:

$$y_1 y - 2px_1 = px - px_1.$$

$$y_1 y = px + px_1.$$

$$y_1 y = p(x + x_1).$$

This is the required equation of the tangent line.

Find the equation of the tangent line to the ellipse: $4x^2 + 9y^2 = 40$, at the point $(1, 2)$.

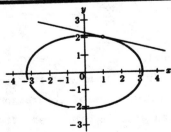

Solution: Since y is not given explicitly, the slope of the tangent to the ellipse at any point is best found by treating it as an implicit function. Differentiating, we have:

$$8x + 18yy' = 0,$$

from which
$$y' = -4x/9y.$$

Evaluating this derivative at the point (1,2), we have
$$y' = -4/18 = -2/9.$$

Thus the slope of the desired tangent line is $-2/9$.

The equation of a straight line at a given point can be expressed in the form $y - y_1 = m(x - x_1)$. Here $x_1 = 1$ and $y_1 = 2$, and the slope $m = -2/9$. Substituting, we obtain:
$$y - 2 = -2/9(x - 1).$$

$$9y - 18 = -2x + 2.$$

$$2x + 9y - 20 = 0, \text{ which}$$
is the equation of the tangent line.

The slope could also have been found by solving the equation of the curve for y, and then differentiating.

● PROBLEM 51-9

Find the equations of the tangent line and the normal line to the curve: $y = 3x^2$, at the point where $x = 3$.

Solution: The equation of a straight line passing through a given point (x_1, y_1) can be expressed in the form: $y - y_1 = m(x - x_1)$. Here $x_1 = 3$. Since $y = 3x^2$, $y_1 = 3(3^2) = 27$, therefore the given point is $(3, 27)$. The slope, m, of the line tangent to the curve: $y = 3x^2$, is found by taking $\frac{dy}{dx}$.

$$\frac{dy}{dx} = 6x.$$

At $(3, 27)$, $6x = 18$. Substituting $x_1 = 3$, $y_1 = 27$, and $m = 18$ into the equation: $y - y_1 = m(x - x_1)$, we obtain: $y - 27 = 18(x - 3)$. $18x - y - 27 = 0$. This is the equation of the tangent line.
The equation of the normal can also be expressed using $y - y_1 = m_1(x - x_1)$. But in this case the slope, m_1, is the negative reciprocal of the slope of the tangent line, or $m_1 = -1/m = -1/18$. Substituting $x_1 = 3$, $y_1 = 27$, and $m_1 = -1/18$, we find: $y - 27 = -1/18(x - 3)$, or, $x + 18y - 489 = 0$. This is the equation of the normal.

A ball is thrown vertically upward, and its distance from the ground is given by:

$$s = 104t - 16t^2$$

(a) Find the height to which the ball will rise, if s is expressed in feet when t is in seconds.
(b) Compute the acceleration.

<u>Solution:</u>   (a) The velocity is given by:

$$v = \frac{ds}{dt} = (104 - 32t)\,\text{ft/sec.,}$$

The ball rises until s is a maximum, or the velocity is zero, that is:

$$v = \frac{ds}{dt} = 104 - 32t = 0. \qquad\qquad (a)$$

Solving for t in the above expression gives the time to travel to the maximum height.

Thus, from equation (a),

$$t = \frac{104}{32} = 3\,\tfrac{1}{4}\ \text{sec.}$$

The height to which it **rises is the value**

of s for $t = 3\,\tfrac{1}{4}$ sec., or,

$$s = 104\left(3\,\tfrac{1}{4}\right) - 16\left(3\,\tfrac{1}{4}\right)^2$$

$$s = 169\ \text{ft.}$$

(b) The acceleration of the particle is the derivative of the expression for velocity:

$$a = \frac{dv}{dt} = -32\ \text{ft/sec}^2.$$

The acceleration is a negative quantity in this case. This implies that the velocity has been reduced with time, due to the gravitational force. This kind of negative acceleration is also called deceleration.

CHAPTER 52

DIFFERENTIATION OF ALGEBRAIC FUNCTIONS

> **Basic Attacks and Strategies for Solving Problems in this Chapter. See pages 865 to 869 for step-by-step solutions to problems.**

The definition of the derivative (involving the limit as Δx goes to zero, presented in Chapter 50) can be used to develop rules for differentiating various types of algebraic expressions. The following rules are the most common:

CONSTANT FUNCTION

The derivative of $y = c$ with respect to x, where c is a *constant*, is

$y' = 0.$

CONSTANT MULTIPLIER

The derivative of $y = cf(x)$ with respect to x, where c is a *constant*, is

$y' = cf'(x).$

SUM OF FUNCTIONS

The derivative of

$y = u(x) + v(x)$

with respect to x (where both u and v are functions of x) is

$y' = u'(x) + v'(x),$

where all derivatives are taken with respect to x.

POWER

The derivative of $y = x^n$ with respect to x is

$$\frac{dy}{dx} = y' = nx^{n-1}.$$

This rule applies even if n is a fraction or negative.

CHAIN RULE

The derivative of $y = f(u)$ with respect to x where $u = g(x)$ is

$$\frac{dy}{dx} = y' = \frac{dy}{du} \cdot \frac{du}{dx} = \frac{df}{du} \cdot \frac{dg}{dx}.$$

NOTE that the derivative of y is taken with respect to u while the derivative of u is taken with respect to x. The final expression is written all in terms of the desired independent variable. Often one needs first to make a substitution, for example, to rewrite an expression raised to a power as a new variable raised to a power. For example,

$$y = (x^2 + 3)^5$$

can be rewritten as $y = u^5$, where $u = x^2 + 3$. Its derivative is

$$y' = 5(x^2 + 3)^4 \cdot (2x) = 10x(x^2 + 3)^4.$$

PRODUCT

The derivative of $y = u \cdot v$ with respect to x, where both u and v are functions of x is

$$y' = uv' + vu',$$

where all derivatives are taken with respect to x.

QUOTIENT

The derivative of $y = {}^u/_v$ with respect to x, where both u and v are functions of x, is

$$y' = \frac{vu' - uv'}{v^2},$$

where all derivatives are taken with respect to x.

● PROBLEM 52-1

Find the derivative of: $y = x^{3b}$.

Solution: Applying the theorem for $d(u^n)$,

$$\frac{dy}{dx} = 3b \cdot x^{3b-1} .$$

● PROBLEM 52-2

Find the derivative of: $y = (2x^3 - 5x^2 + 4)^5$.

Solution: $D_x = \frac{d}{dx}$. This problem can be solved by simply applying the theorem for $d(u^n)$. However, to illustrate the use of the chain rule, make the following substitutions:

   $y = u^5$ where $u = 2x^3 - 5x^2 + 4$

Therefore, from the chain rule,

   $D_x y = D_u y \cdot D_x u = 5u^4(6x^2 - 10x)$

      $= 5(2x^3 - 5x^2 + 4)^4(6x^2 - 10x)$.

● PROBLEM 52-3

Find the derivative of:

$$v = \sqrt[4]{\frac{1}{y^7}} .$$

<u>Solution:</u>   Rewrite the expression to replace the radical symbol with an exponent, and apply the theorem for $d(u^n)$.

$$v = y^{-\frac{7}{4}}$$

$$\frac{dv}{dy} = -\frac{7}{4} \cdot y^{-\frac{7}{4}-1} = -\frac{7}{4} \cdot y^{-\frac{11}{4}}$$

$$= -\frac{7}{4} \sqrt[4]{\frac{1}{y^{11}}} \cdot$$

● PROBLEM 52-4

Find the derivative of:
$$f(\dot{x}) = \sqrt{x^2 + 1} \quad .$$

Solution:
$$f(x) = (x^2 + 1)^{\frac{1}{2}},$$

and

$$f'(x) = \frac{1}{2}(x^2 + 1)^{-\frac{1}{2}}(2x),$$

or

$$f'(x) = \frac{2x}{2\sqrt{x^2 + 1}} = \frac{x}{\sqrt{x^2 + 1}} \cdot$$

● PROBLEM 52-5

Find the sixth derivative of $y = x^6$.

Solution:

First derivative = $6x^{6-1} = 6x^5$

Second derivative = $5 \cdot 6x^{5-1} = 30x^4$

Third derivative = $4 \cdot 30x^{4-1} = 120x^3$

Fourth derivative = $3 \cdot 120x^{3-1} = 360x^2$

Fifth derivative = $2 \cdot 360x^{2-1} = 720x^1 = 720x$

Sixth derivative = $1 \cdot 720x^{1-1} = 720x^0 = 720$

The seventh derivative is seen to be zero, and therefore the function $y = x^6$ has seven derivatives.

Find the derivative of:

$$f(x) = (x^2 + 1)(1 - 3x).$$

Solution: Using the theorem for differentiating a product of terms, i.e.: $d(uv) = udv + vdu$

$$f'(x) = (x^2 + 1)(-3) + (1 - 3x)2x = -9x^2 + 2x - 3.$$

Find the derivative of:

$$f(x) = \frac{x^2 + 1}{1 - 3x}.$$

Solution: Using the theorem for differentiating a quotient of terms, i.e.:

$$\frac{d}{dx}\left(\frac{u}{v}\right) = \frac{v\frac{du}{dx} - u\frac{dv}{dx}}{v^2},$$

$$f'(x) = \frac{(1 - 3x)(2x) - (x^2 + 1)(-3)}{(1 - 3x)^2}$$

$$= \frac{-3x^2 + 2x + 3}{(1 - 3x)^2}.$$

Find the derivative of:

$$g(x) = \frac{x^3}{\sqrt[3]{3x^2 - 1}}.$$

Solution: First rewrite the expression to change the radical symbol to an exponent. This allows the application of the theorem for $d(u^n)$.

$$g(x) = x^3(3x^2 - 1)^{-1/3}.$$

Now, using the theorems for $d(uv)$ and $d(u^n)$ in succession,

$$g'(x) = 3x^2(3x^2 - 1)^{-1/3}$$

$$- \frac{1}{3}(3x^2 - 1)^{-4/3}(6x)(x^3)$$

$$= x^2 (3x^2 - 1)^{- 4/3} \left[3(3x^2 - 1) - 2x^2 \right]$$

$$= \frac{x^2 (7x^2 - 3)}{(3x^2 - 1)^{4/3}} \quad .$$

● PROBLEM 52-9

Find the derivative of:

$$f(x) = \frac{(x^2 + 1)\sqrt{x^2 - 1}}{3x + 2} \quad .$$

<u>Solution:</u> Using the theorems for d(uv) and $d\left(\frac{u}{v}\right)$,

$$f'(x) = \frac{(3x + 2)\left[(x^2 + 1)\frac{2x}{2\sqrt{x^2 - 1}} + \sqrt{x^2 - 1}\ 2x \right]}{(3x + 2)^2}$$

$$\frac{- (x^2 + 1)\ \sqrt{x^2 - 1}\ (3)}{(3x + 2)^2}$$

$$= \frac{(3x + 2)\left[x(x^2 + 1) + 2x(x^2 - 1) \right]}{(3x + 2)^2 \sqrt{x^2 - 1}}$$

$$\frac{- (x^2 + 1)(x^2 - 1)(3)}{(3x + 2)^2 \sqrt{x^2 - 1}}$$

$$= \frac{(3x + 2)(3x^3 - x) - 3x^4 + 3}{(3x + 2)^2 \ \sqrt{x^2 - 1}}$$

$$= \frac{6x^4 + 6x^3 - 3x^2 - 2x + 3}{(3x + 2)^2 \ \sqrt{x^2 - 1}} \quad .$$

A simpler solution results when logarithms are taken of both sides of the equation, as this reduces products and quotients to operations of additions and subtractions.

● PROBLEM 52-10

Find the derivative of:

$$f(x) = \left(\frac{2x + 1}{3x - 1} \right)^4 .$$

Solution: Applying first the theorem for $d(u^n)$ and then the theorem for $d\left(\dfrac{u}{v}\right)$,

$$f'(x) = 4\left(\frac{2x + 1}{3x - 1}\right)^3 \frac{(3x - 1)(2) - (2x + 1)(3)}{(3x - 1)^2}$$

$$= \frac{4(2x + 1)^3(-5)}{(3x - 1)^5}$$

$$= -\frac{20(2x + 1)^3}{(3x - 1)^5}.$$

CHAPTER 53

APPLICATIONS

> **Basic Attacks and Strategies for Solving Problems in this Chapter. See pages 870 to 877 for step-by-step solutions to problems.**

One of the major applications of derivatives (and the theory of slopes and rates of change related to them) involves solving maximum and minimum problems.

Problems involving *maxima* and *minima* points are solved by taking the derivative of a function and setting it equal to zero. A place where the derivative (i.e., slope) equals zero has a tangent parallel to the *x*-axis and *usually* (but not always) corresponds to a high point (maximum) or low point (minimum) on that curve.

The points where the derivative is zero (or fails to exist) are called *critical points* of the function.

If, in an interval (a, b), the slope of a function (i.e., $f'(x)$) is increasing as one moves from a to b, the graph of the function resembles a bowl opening upward, and the function is said to be *concave upward*. If, on the other hand, the slope of the function is decreasing, the graph of the function resembles a bowl opening downward, and the function is said to be *concave downward*. The point at which the slope stops decreasing and starts increasing (or stops increasing and starts decreasing, i.e., the point at which the concavity of the curve changes) is called a *point of inflection*.

The second derivative can also be used to determine certain qualities about the function. If $f''(x) > 0$ when x is a critical point (i.e., x such that $f'(x) = 0$), then x is a local minimum. If $f''(x) < 0$ when x is a critical point, then x is a local maximum. If $f''(x) = 0$, then x may be a point of inflection.

Word problems that ask for maximum or minimum quantities can be solved by first carefully translating the information given into an appropriate equation and then following the standard rules for finding the maximum or minimum of a function.

Sketching the graphs of curves can often be aided by determining maxima and minima points along with points of inflection and using that information to determine key points on a curve.

Step-by-Step Solutions to Problems in this Chapter, "Applications"

● PROBLEM 53-1

Find the maxima and minima of the function $f(x) = x^4$.

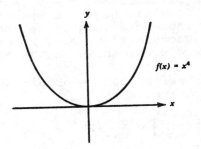

$f(x) = x^4$

Solution: To determine maxima and minima we find $f'(x)$, set it equal to 0, and solve for x to obtain the critical points. We find: $f'(x) = 4x^3 = 0$, therefore $x = 0$ is the critical value. We must now determine whether $x = 0$ is a maximum or minimum value. In this example the Second Derivative Test fails because $f''(x) = 12x^2$ and $f''(0) = 0$. We must, therefore, use the First Derivative Test. We examine $f'(x)$ when $x < 0$ and when $x > 0$. We find that for $x < 0$, $f'(x)$ is negative, and for $x > 0$, $f'(x)$ is positive. Therefore there is a minimum at $(0,0)$. (See figure).

● PROBLEM 53-2

Determine the critical points of $f(x) = 3x^4 - 4x^3$ and sketch the graph.

870

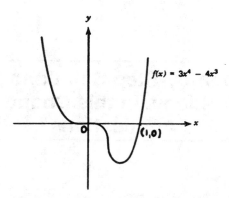

$f(x) = 3x^4 - 4x^3$

$(1,0)$

Solution: To determine the critical points we find
f'(x), set it equal to 0 and solve for x. These
are the abscissas of the critical points. Differentiating, we have $f'(x) = 12x^3 - 12x^2 = 12x^2(x-1)$.
Therefore, x = 0, 1 are the critical values. We now
examine f'(x) when x < 0, when 1 > x > 0, and when
x > 1 to determine whether there is a maximum, minimum or neither at each critical point. We find
that, when x < 0, f'(x) is negative. When 1 > x > 0,
f'(x) is also negative, and when x > 1, f'(x) is
positive. Because f'(x) changes sign from - to + at
x = 1, this is a minimum. At x = 0 there is no
change in sign, f'(x) is negative when x < 0 and
when 0 < x < 1, therefore this is neither a maximum
nor a minimum. Because f'(0) = 0, f has a horizontal tangent at x = 0, as shown in the figure.

Additional insight can be obtained by taking
the second derivative:

$$f''(x) = \frac{d}{dx}\left[f'(x)\right] = \frac{d}{dx}(12x^2 - 12x^2)$$

$$= 36x^2 - 24x$$

$$= + \text{ at } x = 1, \text{ a minimum}$$

$$\text{but } = 0 \text{ at } x = 0.$$

A maximum, on the other hand, would yield a negative
second derivative.

● **PROBLEM 53-3**

Where are the maxima and minima of $y = 3x^3 + 4x + 7$?

Solution: $y = 3x^3 + 4x + 7$, then $\frac{dy}{dx} = 9x^2 + 4 = 0$,
$x^2 = -\frac{4}{9}$,

$$x = \pm \sqrt{-\frac{4}{9}} = \pm \frac{2i}{3}.$$

871

This is an imaginary quantity. Since these are not real roots, in this example y has neither a maximum nor a minimum.

Find two positive numbers, the sum of which is 100, and the square of one number times twice the cube of the other number is to be a maximum.

Solution: We will let x = one number, then (100 - x) = the other. We have: $2(x^2)\left[(100 - x)^3\right] = f(x)$. Since both numbers are to be positive,

$$x \geq 0 \quad \text{and} \quad 100 - x \geq 0$$

or

$$100 \geq x \geq 0.$$

We want to maximize $f(x) = 2x^2(100 - x)^3$, where $0 \leq x \leq 100$.

We do this by finding f'(x), setting it equal to 0 and solving for x. We find:

$$f'(x) = 4x(100 - x)^3 - 6x^2(100 - x)^2$$

$$= 2x(100 - x)^2[2(100 - x) - 3x]$$

$$= 2x(100 - x)^2(200 - 5x) = 0.$$

Solving for x we have:

$$x = 0, \ x = 100, \ x = 40.$$

Therefore, the critical values are 0, 40, and 100. Since f(0) = 0, f(100) = 0, and f(40) > 0, the absolute maximum will be f(40) and the two desired numbers will be 40 and 60.

A piece of wire 20 inches long is to be cut and made into a rectangular frame. What dimensions should be chosen so that the area of the rectangle enclosed is maximal?

Solution: If we let two sides of the rectangle equal x, then the other two sides combined = 20 - 2x, therefore each side is 10 - x. The area of the rectangle = $A(x) = x(10-x) = 10x - x^2$. To maximize A we find $A'(x) = \frac{dA}{dx}$, set it equal to 0, and solve for x. We write:

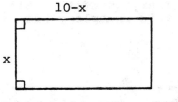

$$A'(x) = 10 - 2x = 2(5-x) = 0.$$

Therefore, x = 5. To show that this is a maximum value we choose a value less than 5, say 4, and a value greater than 5, say 6, and show that A'(x) changes sign from + to -.

We find: A'(4) = 2 and A'(6) =-2, therefore 5 is indeed a maximum value. Letting x = 5, we find the dimensions of the rectangle to be 5 and (10-5) or 5. Therefore, the rectangle is a square of area 25.

● **PROBLEM 53-6**

What rectangle of maximum area can be inscribed in a circle of radius r?

<u>Solution:</u> Let x = one side of the rectangle. The other side is obtained by the square root of the diagonal squared minus the square of the one side.

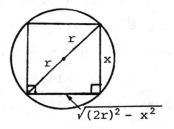

The diagonal is equal to 2r. The other side = $\sqrt{(2r)^2 - x^2} = \sqrt{4r^2 - x^2}$. (See diagram.)

$$A = x\sqrt{4r^2 - x^2} = \text{area of rectangle}$$

To obtain a maximum value we find $A' = \frac{dA}{dx}$ and set it equal to 0. Solving for x we obtain a critical value. We have:

$$\frac{dA}{dx} = x \cdot \frac{d}{dx}(\sqrt{4r^2 - x^2}) + \sqrt{4r^2 - x^2} \cdot 1 = 0$$

for a maximum or minimum.

$$x \cdot \frac{1}{2}(4r^2 - x^2)^{-1/2} \cdot (-2x) + \sqrt{4r^2 - x^2} = 0$$

or $-\dfrac{x^2}{(4r^2 - x^2)^{1/2}} + (4r^2 - x^2)^{1/2} = 0$

which reduces to

$$\dfrac{-x^2 + 4r^2 - x^2}{(4r^2 - x^2)^{1/2}} = 0$$

$$\dfrac{4r^2 - 2x^2}{(4r^2 - x^2)^{1/2}} = 0$$

Now, the denominator, $\sqrt{4r^2 - x^2}$ cannot be 0 because this is the value of a side of the rectangle. Therefore the numerator,

$$4r^2 - 2x^2 = 0$$

and

$$x^2 = 2r^2$$

or

$$x = r\sqrt{2} = \text{one side}$$

and

$$\left(4r^2 - x^2\right)^{1/2} = (4r^2 - r^2 \cdot 2)^{1/2} = (2r^2)^{1/2} = r\sqrt{2}$$

$$= \text{other side}$$

The figure is a square.

● PROBLEM 53-7

A closed tin can is to be made of a given quantity of material. For maximum volume, what are its dimensions?

Solution: The given quantity of material means that the total surface area (top, bottom, and side) is specified; call it S and let r = radius and h = height of the can. Then

$$S = 2\pi rh + 2\pi r^2.$$

Now the quantity to be maximized is the volume, and

$$V = \pi r^2 h.$$

We wish to eliminate the h in this equation, and obtain an expression using S, which is a constant, and r. We have: $S = 2\pi rh + 2\pi r^2$, therefore,

$$S_r = 2\pi r^2 h + 2\pi r^3.$$

Solving for $\pi r^2 h$, we substitute and obtain:

$$V = \frac{Sr}{2} - \pi r^3.$$

We now find $\frac{dV}{dr}$, equate it to 0 and solve for r to obtain the critical values. Doing this we have:

$$\frac{dV}{dr} = \frac{S}{2} - 3\pi r^2 = 0$$

$$r = \pm\sqrt{\frac{S}{6\pi}},$$

the critical values. But we reject the negative value. We use the Second Derivative Test to determine whether this value, $r = \sqrt{\frac{S}{6\pi}}$, is a maximum or minimum. We find:

$$\frac{d^2V}{dr^2} = -6\pi r,$$

which is negative for all (positive) values of r. Therefore, $r = \sqrt{\frac{S}{6\pi}}$ corresponds to a maximum. We now need to find the value of h. Using the equation, $S = 2\pi rh + 2\pi r^2$, we solve for h, obtaining:

$$h = \frac{S - 2\pi r^2}{2\pi r} = \frac{S - 2\pi \left(\frac{S}{6\pi}\right)}{2\pi\sqrt{\frac{S}{6\pi}}} = \frac{\frac{2}{3}S}{2\pi\sqrt{\frac{S}{6\pi}}}$$

$$= 2 \cdot \frac{S}{6\pi} \cdot \frac{1}{\sqrt{\frac{S}{6\pi}}} = 2r^2 \cdot \frac{1}{r} = 2r.$$

Hence the relative dimensions are h = 2r.

● **PROBLEM 53-8**

Sketch the graph of

$$f(x) = 6x^5 - 5x^3 = x^3(6x^2-5)$$

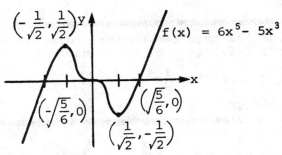

$$\left(-\frac{1}{\sqrt{2}}, \frac{1}{\sqrt{2}}\right)$$

$$f(x) = 6x^5 - 5x^3$$

$$\left(-\sqrt{\frac{5}{6}}, 0\right) \qquad \left(\sqrt{\frac{5}{6}}, 0\right)$$

$$\left(\frac{1}{\sqrt{2}}, -\frac{1}{\sqrt{2}}\right)$$

Solution: Note that since $f(x) = -f(-x)$ the function is an odd function.

If $\qquad f(x) = 0$ then $x = 0$, $x = \pm\sqrt{\frac{5}{6}}$

Since $\qquad f(x) = 6x^5 - 5x^3$,

$\qquad f'(x) = 30x^4 - 15x^2 = 15x^2(2x^2-1)$

$\qquad f''(x) = 120x^3 - 30x = 30x(4x^2-1)$

We consider $f'(x) = 15x^2(2x^2-1)$. If $f'(x) = 0$ then $x = 0$, $x = \pm\sqrt{\frac{1}{2}}$. There is no extreme at $x = 0$ since $f'(x)$ does not change sign there. The point $x = \sqrt{\frac{1}{2}}$ is a relative minimum and the point $x = -\sqrt{\frac{1}{2}}$ is a relative maximum.

Now we consider $f''(x) = 30x(4x^2-1)$. If $f''(x) = 0$ then $x = 0$, $x = \pm\frac{1}{2}$. The function f is concave up in the interval $x \geq \frac{1}{2}$ and down in the interval $0 \leq x \leq \frac{1}{2}$. Similarly, the function f is concave up in the interval $-\frac{1}{2} \leq x \leq 0$ and concave down for $x \leq -\frac{1}{2}$.

We make use of the fact that the function is an odd function. We first draw the graph for $x \geq 0$. Then, by folding the graph around the x-axis first and then around the y-axis, we obtain the graph for $x < 0$. By folding the graph around the x-axis, we cause all values of $f(x)$ to change sign. By folding the graph around the y-axis, we cause all values of $x \geq 0$ to become $x \leq 0$.

By the first fold $f(x)$ becomes $-f(x)$

By the second fold $-f(x)$ becomes $-f(-x)$

$f(x)$ becomes $-f(-x)$, indicative of the anti-symmetric property of the graph.

Determine the relative maxima, relative minima, and points of inflection of the function:

$$f(x) = \frac{1}{4}x^4 - \frac{3}{2}x^2.$$

Sketch the graph.

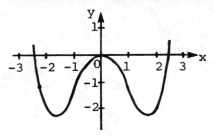

Solution: The derivatives are

$$f'(x) = x^3 - 3x \quad \text{and} \quad f''(x) = 3x^2 - 3.$$

The critical points are solutions of $x^3 - 3x = 0$. We obtain $x = 0, \sqrt{3}, -\sqrt{3}$. The Second Derivative Test tells us that

x = 0 is a relative maximum;

$x = \sqrt{3}, -\sqrt{3}$ are relative minima.

The possible points of inflection are solutions of $3x^2 - 3 = 0$; that is $x = +1, -1$. Since $f''(x)$ is negative for $-1 < x < 1$ and positive for $|x| > 1$, both $x = 1$ and $x = -1$ are points of inflection. We construct the table:

x	-2	$-\sqrt{3}$	-1	0	1	$\sqrt{3}$	2
f	-2	$-\frac{9}{4}$	$-\frac{5}{4}$	0	$-\frac{5}{4}$	$-\frac{9}{4}$	-2
f'	−	0	+	0	−	0	+
f"	+	+	0	−	0	+	+

The graph is symmetrical with respect to the y-axis.

CHAPTER 54

INTEGRAL

> **Basic Attacks and Strategies for Solving Problems in this Chapter. See pages 878 to 885 for step-by-step solutions to problems.**

An integral is an expression that indicates the inverse operation of finding a derivative. Thus, integration formulas are related to differentiation formulas. These formulas are also called *indefinite integrals* and always include the addition of the arbitrary constant in the resulting expression. Integrals are also related to areas under the graphs of a curve. Such integrals include *limits* on the integral sign that indicate the starting and stopping point for the area.

The following are some common integration formulas:

SUM

$$\int [u(x) + v(x)]\, dx = \int u(x)\, dx + \int v(x)\, dx$$

CONSTANT MULTIPLIER

$$\int cf(x)\, dx = c\int f(x)\, dx.$$

VARIABLE TO A CONSTANT POWER

$$\int u^n\, du = \frac{u^{n+1}}{n+1} + c,$$

for $n \neq -1$.

e RAISED TO A VARIABLE POWER

$$\int e^u\, du = e^u + c.$$

A CONSTANT RAISED TO A VARIABLE POWER

$$\int a^u\, du = \frac{a^u}{\ln a} + c,$$

for a an arbitrary constant.

RECIPROCAL OF A VARIABLE

$$\int \frac{du}{u} = \ln|u| + c.$$

PARTS

$$\int u \, dv = uv - \int v \, du.$$

Oftentimes, integration involves substituting a variable for a complicated expression. For example, given

$$\int \frac{x}{x^2 - 1} \, dx,$$

one should first substitute a variable for the more complicated part of the expressions, i.e., let

$$u = x^2 - 1.$$

Then

$$du = 2x \, dx \quad \text{or} \quad \frac{du}{2} = x \, dx.$$

We can substitute expressions in terms of u for the expressions in terms of x, thereby eliminating all occurrences of x in the integral and obtaining

$$\int \frac{1}{2} \cdot \frac{1}{u} \, du.$$

The answer is

$$\frac{1}{2} \ln|u| + c = \frac{1}{2} \ln|x^2 - 1| + c.$$

When evaluating definite integrals (with limits), after obtaining the expression, one substitutes the upper limit into the expression and subtracts the expression with the lower limit substituted into it. In other words, if

$$\int f(x) \, dx = F(x) + c,$$

then

$$\int_a^b f(x) \, dx = F(x) \Big]_a^b = F(b) - F(a).$$

When transforming variables, care must be taken to also properly transform the limits of integration.

Integrals can be used to find the areas between two curves. If $f_u(x)$ is the upper curve and $f_l(x)$ is the lower curve, then

$$\int_{x_1}^{x_2} [f_u(x) - f_l(x)]dx$$

gives the area between these two curves if x_1 and x_2 are the two points of intersection.

Areas can be used in rate of change problems. For example, since acceleration is the rate of change (i.e., derivative) of velocity, i.e.,

$$a = \frac{dv}{dt},$$

by integrating acceleration (with respect to time t), one can obtain the velocity.

● PROBLEM 54-1

Integrate the expression: $\int \sqrt{10^{3x}}\ dx$.

Solution: $\int \sqrt{10^{3x}}\ dx = \int 10^{3x/2}\ dx$. We wish to apply the formula $\int a^u du = \frac{a^u}{\ln a}$, where a is a positive constant > 0 and $\neq 1$. In this case we have $a = 10$, and $u = \frac{3}{2}x$. Then $du = \frac{3}{2}dx$; We write:

$$\int 10^{3x/2}\ dx = \frac{2}{3} \int 10^{3x/2}\ \frac{3}{2}dx.$$

Applying the formula for $\int a^u du$, we have:

$$\frac{2}{3} \cdot \frac{10^u}{\ln 10} + C = \frac{2 \cdot 10^{3x/2}}{3\ \ln 10} + C.$$

● PROBLEM 54-2

Integrate: $\int (x^2 + 4)^5\ 2x\ dx$.

Solution: We use the formula: $\int u^n du = \frac{u^{n+1}}{n+1} + C$, letting $u = x^2 + 4$, $du = 2x\ dx$, and $n = 5$. Then,

$$\int (x^2 + 4)^5\ 2x\ dx = \frac{(x^2 + 4)^6}{6} + C.$$

● PROBLEM 54-3

Integrate: $\int \frac{2x}{x + 1}dx$.

Solution: To integrate the given expression we manipulate the integrand to obtain the form $\int \frac{du}{u}$. This can be done as follows:

$$\int \frac{2x}{x+1} dx = 2 \int \frac{x}{x+1} dx$$

$$= 2 \int \left(\frac{x+1}{x+1} - \frac{1}{x+1} \right) dx$$

$$= 2 \int \left(1 - \frac{1}{x+1} \right) dx$$

$$= 2 \int dx - 2 \int \frac{dx}{x+1}.$$

Now, applying the formula $\int \frac{du}{u} = \ln u$, we obtain:

$$\int \frac{2x}{x+1} dx = 2x - 2 \ln(x+1) + C.$$

● PROBLEM 54-4

Integrate the expression: $\int x^2 e^{x^3} dx$.

Solution: To integrate this expression, we wish to consider the formula: $\int e^u du = e^u + C$. In this case we have $u = x^3$. Then $du = 3x^2 dx$. Applying the formula, we obtain:

$$\int x^2 e^{x^3} dx = \frac{1}{3} \int e^{x^3} (3x^2 dx) = \frac{1}{3} e^{x^3} + C.$$

● PROBLEM 54-5

Integrate: $\int (x^3 - 2x)^5 (3x^2 - 2) dx$.

Solution: Let $u = x^3 - 2x$, $du = (3x^2 - 2) dx$, and $n = 5$. Now we use the formula: $\int u^n du = \frac{u^{n+1}}{n+1} + C$. Therefore,

$$\int (x^3 - 2x)^5 (3x^2 - 2) dx = \frac{(x^3 - 2x)^6}{6} + C$$

● PROBLEM 54-6

$\frac{dy}{dx} = (2x^2 + 1)(x^2 - 3x + 2)$. Find $y = F(x)$.

Solution: First we multiply out the two factors. We obtain:

879

$$\frac{dy}{dx} = 2x^4 - 6x^3 + 5x^2 - 3x + 2,$$

which can be written as,

$$dy = (2x^4 - 6x^3 + 5x^2 - 3x + 2)dx.$$

We can now write

$$\int dy = \int (2x^4 - 6x^3 + 5x^2 - 3x + 2)dx$$

or,

$$y = \int (2x^4 - 6x^3 + 5x^2 - 3x + 2)dx$$

$$= 2\int x^4 \cdot dx - 6\int x^3 \cdot dx + 5\int x^2 \cdot dx - 3\int x \cdot dx$$

$$+ 2\int dx.$$

We can now integrate by applying the formula for $\int u^n du$, obtaining:

$$y = \frac{2x^5}{5} - \frac{3x^4}{2} + \frac{5x^3}{3} - \frac{3x^2}{2} + 2x.$$

● PROBLEM 54-7

Integrate the expression: $\int \sqrt{3x + 4}\ dx$.

Solution: $\int \sqrt{3x + 4}\ dx = \int (3x + 4)^{1/2}dx$.

In integrating this function, we use the formula:
$\int u^n du = \frac{u^{n+1}}{n+1} + C$, with $u = (3x + 4)$, $du = 3dx$, and

$n = \frac{1}{2}$. We obtain:

$$\int (3x + 4)^{1/2}dx = \frac{1}{3} \int (3x + 4)^{1/2}3dx$$

$$= \frac{1}{3}\left[\frac{(3x + 4)^{3/2}}{\frac{3}{2}}\right] + C$$

$$= \frac{2}{9}(3x + 4)^{3/2} + C.$$

● PROBLEM 54-8

Find the equation of the curve which has a horizontal tangent at the point (0,-1), and for which the rate of change, with respect to x, of the slope at any point is equal to $8e^{2x}$.

Solution: The slope of a curve at a point is the

derivative of the curve, $\frac{dy}{dx}$. The rate of change of the slope is the derivative of the slope, or the second derivative of the curve, $\frac{d^2y}{dx^2}$. Therefore, $\frac{d^2y}{dx^2} = 8e^{2x}$. To find the equation of the curve we:

    1) integrate the rate of change of the slope to obtain the slope, and

    2) integrate the slope to obtain the equation of the curve.

Integrating the rate of change we obtain, from:

$$\frac{d^2y}{dx^2} = 8e^{2x},$$

the expression:

$$\int \frac{d^2y}{dx^2} = \frac{dy}{dx} = \int 8e^{2x}dx.$$

$$\frac{dy}{dx} = \frac{1}{2} \int \left(8e^{2x}\right) 2dx.$$

$$\frac{dy}{dx} = 4e^{2x} + C_1 = m, \text{ the slope.}$$

Integrating the slope, m, to obtain the equation, we obtain:

$$\int \frac{dy}{dx} = y = \int 4e^{2x} + C_1 dx$$

$$y = \frac{1}{2} \int \left(4e^{2x} + C_1\right) 2dx$$

$$y = 2e^{2x} + C_1 x + C_2,$$

the equation of the curve.

To find the value of the constant C_1, we use the information given in the problem: that there is a horizontal tangent at the point $(0,-1)$.

A horizontal tangent occurs at a point where the slope is equal to zero. Therefore $\frac{dy}{dx} = 0$ or $4e^{2x} + C_1 = 0$, when $x = 0$ and $y = -1$. Substituting $x = 0$ into the equation: $4e^{2x} + C_1 = 0$, we find C_1 as follows:

$$\frac{dy}{dx} = 4e^{2x} + C_1 = 0$$

$$4e^{2(0)} + C_1 = 0$$

$$4 + C_1 = 0$$

● **PROBLEM 54-9**

Find the finite area above the x-axis and under the curve: $y = x^3 - 9x^2 + 23x - 15$.

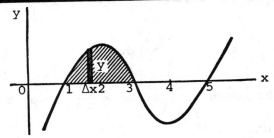

<u>Solution:</u> A sketch of the curve helps to determine the limits of the integral which give us the required area. From the diagram, the limits are the inter-sections of the curve and the x-axis, giving the limits: x = 1, x = 3, x = 5. But we are only con-cerned with the area above the x-axis and under the curve, therefore the limits we use are: x = 1 and x = 3. Now set up the integral. The area is equal to the integral of the upper function of x, $y = x^3 - 9x^2 + 23x - 15$, minus the lower function of x, y = 0 (the x-axis). Therefore we write:

$$A = \int_1^3 (x^3 - 9x^2 + 23x - 15) - (0)\,dx$$

$$= \int_1^3 (x^3 - 9x^2 + 23x - 15)\,dx$$

$$= \frac{x^4}{4} - 3x^3 + \frac{23x^2}{2} - 15x \Big]_1^3$$

$$= \left(\frac{81}{4} - 81 + \frac{207}{2} - 45\right) - \left(\frac{1}{4} - 3 + \frac{23}{2} - 15\right)$$

$$= 4.$$

● **PROBLEM 54-10**

Find the area bounded by the parabola: $y = x^2 - 3x$, and the line: y = x.

Solution: We draw a figure to show the area and the upper and lower bounds of the integral. Setting $y = x^2 - 3x = x$ to find the points of intersection, we have $x^2 = 4x$, $x = 0, 4$. From the diagram we see that the desired area is between $y = x$ and $y = x^2 - 3x$, with an upper bound of 4 and a lower bound of 0. Therefore,

$$A = \int_0^4 \left[x - (x^2 - 3x) \right] dx$$

$$= \int_0^4 [4x - x^2] dx$$

$$= 4 \int_0^4 x \, dx - \int_0^4 x^2 \, dx.$$

Applying the formula, $\int u^n \, du = \frac{u^{n+1}}{n+1}$, we obtain:

$$A = 4 \left[\frac{x^2}{2} \right]_0^4 - \left[\frac{x^3}{3} \right]_0^4$$

$$= 4 \left[\frac{16}{2} - 0 \right] - \left[\frac{4^3}{3} - 0 \right]$$

$$= 4(8) - \frac{64}{3}$$

$$= 32 - \frac{64}{3} = \frac{96}{3} - \frac{64}{3} = \frac{32}{3}.$$

Hence, the area bounded by the parabola $y = x^2 - 3x$ and the line $y = x$ is $\frac{32}{3}$.

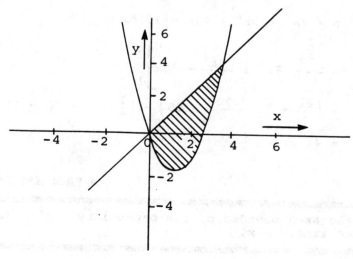

Find the area bounded by the parabola: $2y = x^2$, and the line: $y = x + 4$.

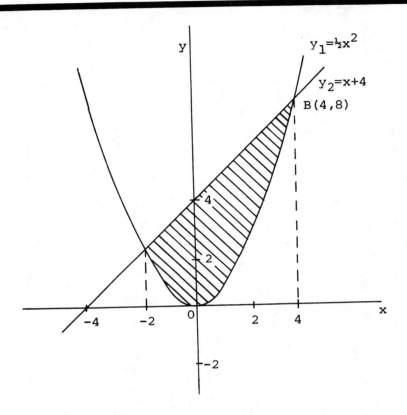

Solution: The limits of the integral which give the required area are the points of intersection of the two functions. To find the points of intersection, we set the two functions equal and solve for x.

$$\frac{x^2}{2} = x + 4. \qquad x^2 = 2x + 8. \qquad x^2 - 2x - 8 = 0.$$

$$(x - 4)(x + 2) = 0$$

$$x = 4 \quad x = -2.$$

The line: $y = x + 4$, is above the parabola: $y = \frac{x^2}{2}$.

Therefore, the area can be found by taking the integral of the upper function minus the lower function, or the line minus the parabola, from $x = -2$ to $x = 4$. Solving, we obtain:

$$A = \int_{-2}^{4} \left[(x + 4) - \frac{x^2}{2} \right] dx$$

$$= \left[\frac{x^2}{2} + 4x - \frac{x^3}{6}\right]_{-2}^{4}$$

$$= \frac{40}{3} + \frac{14}{3}$$

$$= 18.$$

SECTION 5 — ELEMENTARY STATISTICS

CHAPTER 55

ELEMENTARY STATISTICS

> **Basic Attacks and Strategies for Solving Problems in this Chapter. See pages 886 to 934 for step-by-step solutions to problems.**

A *frequency distribution* is a chart containing categories along with the number of items belonging to each category. A *histogram* is a bar graph in which the categories are listed on the horizontal axis and the heights of the bars correspond to the frequency of (i.e., number of items in) each category. A *frequency polygon* is a line graph connecting points (x_i, f_i), where x_i is the interval median of each category and f_i is the category frequency. A *cumulative frequency distribution* chart is a frequency distribution chart that contains an additional column listing the sums of the frequencies up to and including that row entry of the chart. An additional column is sometimes included that lists the *cumulative percentage*, that is, the percentage that the cumulative frequency is with respect to the total sample being studied. The cumulative percentage column can help determine *percentile levels*. A category (e.g., score, age, etc.) is said to be at the nth percentile if that category is greater than the lowest n percent of the total sample and less than the upper $100 - n$ percent. An *ogive* is the distributive curve of a cumulative frequency distribution. A cumulative percentage ogive is a line graph connecting points (x_i, f_i), where x_i is the interval median of each category and f_i is the *cumulative* frequency.

The *median* is the middle value in an ordered sample space, such that half of the items are below the median and half above it.

The *mean* is the "average," the value obtained by adding all individual values and dividing by the number of values. If x_i is one of the set of n values being analyzed, then the mean, $\bar{x}$, is computed as follows:

$$\bar{x} = \frac{\sum_{i=1}^{n} x_i}{n}.$$

The *mode* is the value that appears most frequently in the sample space.

The median, mean, and mode are also called *measures of central tendency*, since they each describe the center of the data in the sample space. In many sample spaces, all three values are somewhat close to each other. If there are a few widely scattered values, they may distort the value of the mean, but not the median or mode.

The *range* is the difference between the largest and smallest items in the sample data space.

The *variance* and *standard deviation* measure how widely dispersed the items in the sample space are from the mean. The variance, σ^2 or s^2, is computed as follows:

$$s^2 = \frac{\sum_{i=1}^{n}(x_i - \bar{x})}{n},$$

where $\bar{x}$ is the mean of the set of values x_i. The standard deviation is the square root of the variance.

The *coefficient of variation*, V, is $\frac{s}{\bar{x}}$ where s is the standard deviation and $\bar{x}$ is the mean.

A *Venn diagram* is a pictorial representation of the relationship between sets. It usually consists of several slightly overlapping circles drawn so that part of each circle does not overlap with any other, and part is overlapping with one or more other circles. This can be used to illustrate those items that are common to several sets, and those items that are distinctive to certain sets.

Statistics are related to probabilities. An overview of probability theory was presented in Chapter 23. Sometimes a statistic must be rephrased as a probability statement in order to solve a problem. For example, to say that "30% of students at the local college are from out of state" is the same as saying that "the probability that a student studying at the local college is from out of state is 0.30." There is a formula from probability theory, called Bayes' Theorem, that can sometimes be useful to compute an unknown probability from known values. If $B_1, B_2, ..., B_n$ are mutually exclusive events at least one of which must occur, then

$$P(B_i \mid A) = \frac{P(B_i) \cdot P(A \mid B_i)}{\sum_{j=1}^{n} P(B_j) \cdot P(A \mid B_j)}.$$

A *probability distribution of X* is a function that assigns probabilities to the values X may assume, such that the sum of all possible values is one, and each individual value is greater than (or equal to) zero.

The probability that a certain event Y will have a certain outcome (called a *success*) k times when Y is repeated n times (i.e., n "trials") is given by

$$P(X = k) = \binom{n}{k} p^k (1 - p)^{n-k},$$

where X is the combined event (i.e., n repetitions of every Y), and

$$p = P(Y)$$

is the probability of a success for any one trial. For example, $p(h) = \frac{1}{2}$ if we toss a coin and h indicates "heads." This is sometimes called a series of *Bernoulli trials*, or a *binomial distribution*.

The *expected value of event X, $E(X)$*, is given by

$$E(X) = \sum x_i \cdot P(X = x_i),$$

where x_i varies over all possible values of that event X can take on. The expected value can be considered as a version of the mean (defined above) when the probabilities vary according to the constituent events rather than always remaining $\frac{1}{n}$. Because of this, the expected value is also called the mean. The *variance*, *Var(X)*, (in this context) is computed as follows:

$$\sum [(x_i - E(X))^2 \cdot P(X = x_i)],$$

where x_i varies over all possible values of that event X can take on. This can also be written as

$$Var(X) = E(X^2) - [E(X)]^2,$$

where

$$E(X^2) = \sum x_i^2 \cdot P(X - x_i).$$

If X is a binomial distribution, the mean, μ, is $n \cdot p$ where (as above) n is the number of trials and p is the probability of success in a single trial. The variance, σ^2, is $np(1 - p)$ and the standard deviation, σ, (as before) is the square root of the variance.

The *normal distribution* is the familiar "bell-shaped" curve. The graph function (i.e., density function) of the normal distribution is

$$f(x) = \frac{1}{\sigma\sqrt{2\pi}} e^{-\frac{1}{2}\left(\frac{(x-\mu)}{\sigma}\right)^2}.$$

Often σ is chosen to be 1 and μ is 0. This curve is symmetric about μ, or the y-axis if $\mu = 0$.

The function $\Phi(x)$ is defined as

$$\Phi(x) = P(Z \le x),$$

where Z is the standard normal variable (i.e., the variable in the normal distribution function when $\sigma = 1$ and $\mu = 0$). In other words,

$$\Phi(x) = \int_{-\infty}^{x} \frac{1}{r\sqrt{2\pi}} e^{-\frac{1}{2}x^2} dx.$$

Sometimes, when a binomial distribution includes a large number of trials, it can be approximated by using a normal distribution. When this approximation takes place, often the range is increased by a half in either direction as a so-called "continuity correction."

The *moment generating function* (for a continuous variable x) is given by

$$M_x(t) = E(e^{tx}) = \int_{-\infty}^{\infty} e^{tx} f(x) dx,$$

where $f(x)$ is the distribution of (i.e., probability associated with) x. The mean is

$$E(x) = M'_x(t)\,|_{t=0}.$$

The variance can be computed by

$$Var(x) = E(x^2) - [E(x)]^2,$$

where

$$E(x^2) = M''_x(t)\,|_{t=0}.$$

The *Central Limit Theorem* asserts that, for large samples, the sampling distribution of the mean can be approximated closely with a normal distribution.

The interval (a, b) is said to be an *$n\%$ confidence interval* if we can assert with $n\%$ confidence that the interval (a, b) contains the parameter we are trying to estimate, that is, the probability that the parameter in (a, b) is $^n/_{100}$. Confidence intervals can be computed by reference to standard normal tables and the use of the standard normal quantity,

$$\frac{\overline{X} - E(\overline{X})}{\sqrt{Var(\overline{X})}},$$

where

$$E(\overline{X}) = \mu$$

(the true mean),

$$Var(\overline{X}) = \frac{\sigma^2}{n},$$

n is the size of the sample, and σ^2 is the true variance of the population.

The *Chi-squared distribution*, χ^2, can be used to determine confidence intervals for the standard deviation σ based on a sample standard deviation s. The parameter associated with the χ^2 distribution variable is called the *number of degrees of freedom*.

A *statistical hypothesis* is an assertion of conjecture about a parameter (or parameters) of a population. A *null hypothesis*, denoted by H_0, is a hypothesis set up primarily to see whether or not it can be rejected. The *alternative hypothesis* (i.e., alternative to the null hypothesis) is denoted by H_1 or H_A. We hypothesize about parameters, and not about statistics. The choice of whether to accept H_0 or H_A, however, must be based on (observable) statistics.

● PROBLEM 55-1

Twenty students are enrolled in the foreign language department, and their major fields are as follows: Spanish, Spanish, French, Italian, French, Spanish, German, German, Russian, Russian, French, German, German, German, Spanish, Russian, German, Italian, German, Spanish.

(a) Make a frequency distribution table.
(b) Make a frequency histogram.

Solution: The frequency distribution table is constructed by writing down the major field and next to it the number of students. A histogram follows.

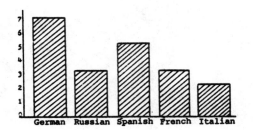

Major Field	Number of Students
German	7
Russian	3
Spanish	5
French	3
Italian	2
Total	20

In the histogram, the fields are listed and spaced evenly along the horizontal axis. Each specific field is represented by a rectangle, and all have the same width. The height of each, identified by a number on the vertical axis, corresponds to the frequency of that field.

● PROBLEM 55-2

The IQ scores for a sample of 24 students who are entering their first year of high school are:

115	119	119	134
121	128	128	152
97	108	98	130
108	110	111	122
106	142	143	140
141	151	125	126

(a) Make a cumulative percentage graph using classes of seven points starting with 96 - 102.
(b) What scores are below the 25th percentile?
(C) What scores are above the 75th?
(d) What is the median score?

Solution:

Interval	Interval Midpoint	Frequency	Cumulative Frequency	Cumulative Percentage
96-102	99	2	2	8.34
103-109	106	3	5	20.83
110-116	113	3	8	33.33
117-123	120	4	12	50.00
124-130	127	5	17	71.00
131-137	134	1	18	75.00
138-144	141	4	22	91.33
145-151	148	1	23	96.00
152-158	155	1	24	100.00

The frequency is the number of students in that interval. The cumulative frequency is the number of students in intervals up to and including that interval. The cumulative percentage is the percentage of students whose IQ's are at that level or below.

$$\text{Cumulative Percentage} = \frac{\text{Cumulative Frequency}}{24} \times 100 \text{ \%}$$

We will plot our graph using the interval midpoint as the x coordinate and the cumulative percentage as the y coordinate.

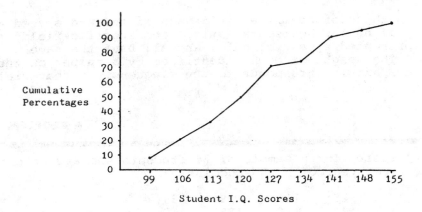

Cumulative
Percentages

Student I.Q. Scores

(b) The 25th percentile is defined to be a number that is exactly greater than the lowest 25 % of the scores. We want to know the score that is at least greater than (.25)24=6 other students. The 6 lowest scores are 97, 98, 106, 108, 108 and 110. We cannot use 111 as a 25th percentile since another student has that score so we use 110.5.

(c) The 75th percentile is the score which exceeds the lowest 75% of the population but is less than the top 25 % of scores. We want the score below 6 students and above 18. The 6 highest scores are 152, 151, 143, 142, 141, 140. The next highest is 134. As our 75 % percentile we can take any value between 134 and 140. We will take 137, the average of 134 and 140.

(d) The median is the value which half of the values of the population exced and half do not. There are 12 values ≤ 123 and 12 values ≥ 124. Therefore we take as our median 123.5, the average of these two values. The median is the 50th percentile.

● **PROBLEM 55-3**

The following data is a sample of the accounts receivable of a small merchandising firm.

37	42	44	47	46	50	48	52	90
54	56	55	53	58	59	60	62	92
60	61	62	63	67	64	64	68	
67	65	66	68	69	66	70	72	
73	75	74	72	71	76	81	80	
79	80	78	82	83	85	86	88	

Using a class interval of 5, i.e. 35 - 39,

(a) Make a frequency distribution table.
(b) Construct a histogram.
(c) Draw a frequency polygon.
(d) Make a cumulative frequency distribution.
(e) Construct a cumulative percentage ogive.

Class Interval	Class Boundaries	Tally	Interval Median	Frequency
35 - 39	34.5 - 39.5	/	37	1
40 - 44	39.5 - 44.5	//	42	2
45 - 49	44.5 - 49.5	///	47	3
50 - 54	49.5 - 54.5	////	52	4
55 - 59	54.5 - 59.5	////	57	4
60 - 64	59.5 - 64.5	//// ///	62	8
65 - 69	64.5 - 69.5	//// ///	67	8
70 - 74	69.5 - 74.5	//// /	72	6
75 - 79	74.5 - 79.5	////	77	4
80 - 84	79.5 - 84.5	////	82	5
85 - 89	84.5 - 89.5	///	87	3
90 - 94	89.5 - 94.5	//	92	2

We use fractional class boundaries. One reason for this is that we cannot break up the horizontal axis of the histogram into only integral values. We must do something with the fractional parts. The usual thing to do is to assign all values to the closest integer. Hence our above class boundaries. The appropriate histogram follows.

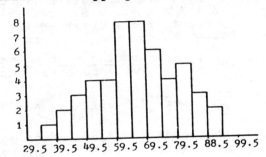

We now construct a frequency polygon as follows:

Plot points (x_i, f_i), where x_i is the interval median and f_i, the class frequency. Connect the points by successive line segments.

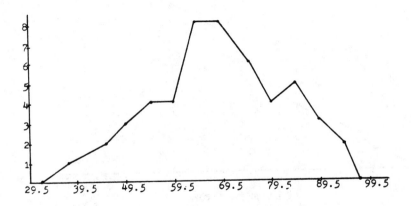

Accounts Receivable

Interval	Interval Median	Frequency (f_i)	Cumulative Frequency	Cumulative Percentage
35 - 39	37	1	1	2
40 - 44	42	2	3	6
45 - 49	47	3	6	12
50 - 54	52	4	10	20
55 - 59	57	4	14	28
60 - 64	62	8	22	44
65 - 69	67	8	30	60
70 - 74	72	6	36	72
75 - 79	77	4	40	80
80 - 84	82	5	45	90
85 - 89	87	3	48	96
90 - 94	92	2	50	100

The cumulative frequency is the number of values in all classes up to and including that class. It is obtained by addition. For example, the cumulative frequency for 65 - 69 is 1 + 2 + 3 + 4 + 4 + 8 + 8 = 30. The cumulative percentage is the percent of all observed values found in that class or below. We can use the formula -

$$\text{cumulative percentage} = \frac{\text{cumulative frequency}}{\text{total observations}} \times 100 \ \%.$$

For example, Cum. per. (65-69) $= \dfrac{30}{50} \times 100 \ \% = 60 \ \%.$

We construct the cumulative percentage ogive by plotting points (x_i, f_i) where x_i is the interval median and f_i is the cumulative frequency. Finally we connect the points with successive line segments.

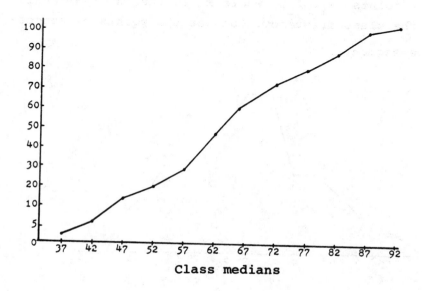

Class medians

Find the mean salary for four company employees who make $ 5/hr., $ 8/hr., $ 12/hr., and $ 15/hr.

Solution: The mean salary is the average.

$$\bar{X} = \frac{\Sigma \ x_i}{n} = \frac{\$ \ 5 + \$ \ 8 + \$ \ 12 + \$ \ 15}{4} = \frac{\$ \ 40}{4} = \$ \ 10/hr.$$

Find the mean length of five fish with lengths of 7.5 in., 7.75 in., 8.5 in., 8.5 in., 8.25 in.

Solution: The mean length is the average length.

$$\bar{X} = \frac{\Sigma \ x_i}{n} = \frac{7.5 + 7.75 + 8.5 + 8.5 + 8.25}{5} = \frac{40.5}{5} = 8.1 \quad in.$$

For this series of observations find the mean, median, and mode.

500, 600, 800, 800, 900, 900, 900, 900, 900, 1000, 1100

Solution: The mean is the value obtained by adding all the measurements and dividing by the number of measurements.

$$\bar{X} = \frac{\overset{\Sigma}{\underset{i}{}} \ x_i}{n}$$

$$\bar{X} = \frac{500+600+800+800+900+900+900+900+900+1000+1100}{11}$$

$$\bar{X} = \frac{9300}{11} = 845.45.$$

The median is the observation in the middle. We have 11, so here it is the sixth, 900.

The mode is the observation that appears most frequently. That is also 900, which has 5 appearances.

All three of these numbers are measures of central tendency. They describe the "middle" or "center" of the data.

Nine rats run through a maze. The time each rat took to traverse the maze is recorded and these times are listed below.

1 min., 2.5 min., 3 min., 1.5 min., 2 min., 1.25 min., 1 min., .9 min., 30 min.

Which of the three measures of central tendency would be the most appropriate in this case?

Solution: We will calculate the three measures of central tendency and then compare them to determine which would be the most appropriate in describing these data.

The mean, $\overline{X}$, is the sum of observations divided by the number of observations. In this case,

$$\overline{X} = \frac{1 + 2.5 + 3 + 1.5 + 2 + 1.25 + 1 + .9 + 30}{9}$$

$$= \frac{43.15}{9} = 4.79.$$

The median is the "middle number" in an array of the observations from the lowest to the highest.

0.9, 1.0, 1.0, 1.25, 1.5, 2.0, 2.5, 3.0, 30.0

The median is the fifth observation in this array or 1.5. There are four observations larger than 1.5 and four observations smaller than 1.5.

The mode is the most frequently occurring observation in the sample. In this data set the mode is 1.0.

mean, $\overline{X}$ = 4.79

median = 1.5

mode   = 1.0

The mean is not appropriate here. Only one rat took more than 4.79 minutes to run the maze and this rat took 30 minutes. We see that the mean has been distorted by this one large observation.

The median or mode seem to describe this data set better and would be more appropriate to use.

The staff of a small company sign a timesheet indicating the time they leave the office.

These times for a randomly chosen day are given below,

5:15	4:50	1:50
5:30	2:45	5:15
5:00	5:30	5:30
5:30	4:55	4:20
5:20	5:30	5:20
5:25	5:00	

How do the three measures of central tendency describe the data? How variable is the data? What is the variance and standard deviation?

Solution: The following table, lists the departure times in ascending order. It also aids in the computation of the moments. In order to compute averages, we must first convert each departure time to a score which can be meaningfully added and multiplied. We will then convert back to a time after the computations. Let 12:00 noon be zero and convert each departure time to the number of hours from 12 noon. Thus 5:00 PM would be converted to 5. 5:20 PM would be converted to $5 \frac{1}{3}$ hours or 5.33, etc.

The median is the middle observation, in this case the 9th observation. Thus the median is 5.25 = 5 1/4 or 5:15 PM.

The mode is the most frequent observation. For our sample, the mode is 5:30 PM which appears 5 times.

The variance and standard deviation describe the variation or dispersion in a sample. The formula for the sample variance is,

$$ s^2 = \frac{\Sigma(X_i - \overline{X})^2}{n} = \frac{\Sigma X_i^2 - n\overline{X}^2}{n} = \frac{\Sigma X_i^2}{n} - \overline{X}^2 $$

Converted Departure Time X_i	X_i^2
$1+\frac{50}{60}=1.83$	3.35
2.75	7.56
4.34	18.83
4.83	23.32
4.92	24.21
5 00	25.00
5.00	25.00
5.25	27.56
5.25	27.56
5.34	28.52
5.34	28.52
5.42	29.38
5.5	30.25
5.5	30.25
5.5	30.25
5.5	30.25
5.5	30.25

We can now compute the three measures of central tendency:

$$\overline{X} = \frac{\Sigma X_i}{n}$$

$$= \frac{\text{sum of observations}}{\text{number of observations}}$$

$$= \frac{82.77}{17} = 4.87$$

$$= 4\,\frac{87}{100} = 4\,\frac{52}{60}\ \text{PM}$$

$$= 4{:}52\ \text{PM}$$

$\Sigma X_i = 82.77$ $\Sigma X_i^2 = 420.06$. Thus,

$$s^2 = \frac{420.06}{17} - (4.87)^2$$

$$= .9925\ (\text{hours})^2 = 3573\ (\text{min.})^2.$$

The standard deviation is

$$s = \sqrt{s^2} = \sqrt{\frac{\Sigma X_i^2}{n} - \overline{X}^2} = \sqrt{.9925} = .9962\ \text{hours}$$

$$= 59.7\ \text{minutes}.$$

Note that the standard deviation is slightly preferable because it is expressed in meaningful units, minutes. The variance is expressed in $(\text{minutes})^2$.

Dividing $\Sigma(X - \overline{X})^2$ by n yields $\frac{\Sigma(X - \overline{X})^2}{n} = \frac{6}{8} = .75$.

Taking the square root, $s_x = \sqrt{\frac{\Sigma(X - \overline{X})^2}{n}} = \sqrt{0.75} = .866$.

Compute the standard deviations for the following sample of measurements, given that $\overline{X} = 8.0$ and

$$s = \sqrt{\dfrac{\sum\limits_{i=1}^{n} (X_i - \overline{X})^2}{n}} \quad \text{and } X = 5, 7, 8, 9, 11.$$

Solution: The standard deviation is a measure of dispersion about the sample mean $\overline{X}$.

$$s = \sqrt{\dfrac{\sum\limits_{i=1}^{n} (X_i - \overline{X})^2}{n}}$$

is one formula for the standard deviation, where the X_i are the actual observations and n is the number of observations in the sample.

In this case

$$s = \sqrt{\dfrac{(5-8)^2+(7-8)^2+(8-8)^2+(9-8)^2+(11-8)^2}{5}}$$

$$= \sqrt{\dfrac{3^2 + 1^2 + 0^2 + 1^2 + 3^2}{5}} = \sqrt{\dfrac{20}{5}} = \sqrt{4} = 2.$$

The radii of five different brands of softballs (in inches) are 2.03, 1.98, 2.24, 2.17, 2.08. Find the range, variance, standard deviation, mean deviation about the median, and coefficient of variation.

Solution: The range gives a measure of how dispersed our sample may be. It is defined as the difference between the smallest and largest observations. In this case the range equals 2.24 in. - 1.98 in. = 0.26 in.

To compute the variance, $s^2 = \dfrac{1}{n} \Sigma(X - \overline{X})^2$, we first need the mean, $\overline{X}$.

$$\overline{X} = \dfrac{\Sigma X}{n} = \dfrac{2.03 + 1.98 + 2.24 + 2.17 + 2.08}{5} = 2.10.$$

Variance $= \frac{1}{n} \Sigma (X - \overline{X})^2$. The computations involved are represented in tabular form.

X	X - X̄	$(X - \overline{X})^2$
2.03	2.03 - 2.10 = - .07	$(- .07)^2 = .0049$
1.98	1.98 - 2.10 = - .12	$(- .12)^2 = .0144$
2.24	2.24 - 2.10 = .14	$(+ .14)^2 = .0196$
2.17	2.17 - 2.10 = .07	$(.07)^2 = .0049$
2.08	2.08 - 2.10 = - .02	$(- .02)^2 = .0004$
		$\Sigma (X - \overline{X})^2 = .0442$

$$\frac{1}{n} \Sigma (X - \overline{X})^2 = \frac{.0442}{5} = .00884.$$

The standard deviation is the square root of the variance $s = \sqrt{.00884} = .094$.

Since we have 5 observations, the third from the lowest, 2.08, is the median. We will compute the mean deviation about the median with the aid of a table:

X	X - n	$|X - n|$
2.03	2.03 - 2.08 = - .05	$\|- .05\| = .05$
1.98	1.98 - 2.08 = - .10	$\|- .10\| = .10$
2.24	2.24 - 2.08 = .16	$\| .16\| = .16$
2.17	2.17 - 2.08 = .09	$\| .09\| = .09$
2.08	2.08 - 2.08 = 0	$\|0\| = 0$

Mean deviation about median $= \frac{\Sigma |X - n|}{n}$

$$= \frac{.05 + .10 + .16 + .09 + 0}{5} = \frac{.4}{5} = .08.$$

The Coefficient of Variation is defined as

$$V = \frac{s}{\overline{X}} .$$

Sometimes we want to compare sets of data which are measured differently. Suppose we have a sample of executives with a mean age of 51 and a standard deviation of 11.74 years. Suppose also we know their average IQ is 125 with a standard deviation of 20 points. How can be compare deviations? We use the coefficient of variation:

$$V_{age} = \frac{s}{\overline{X}} = \frac{11.74}{51} - .23; \quad V_{IQ} = \frac{s}{\overline{X}} = \frac{20}{125} = .16 .$$

We now know that there is more variation with respect to age.

In our example, $V = \frac{s}{\overline{X}} = \frac{0.094}{2.10} = .045.$

In an office, the employer notices his employees spend more time drinking coffee than working. He counts the number of coffee breaks each of his seven employees takes in the course of a day. The data are

1 , 1 , 2 , 2 , 3 , 5 , and 7 .

Find the mean, variance, standard deviation and the median number of coffee breaks a day.

Solution: To aid in these computations we present the following table,

X_i	$X_i - \bar{X}$	$(X_i - \bar{X})^2$
1	- 2	4
1	- 2	4
2	- 1	1
2	- 1	1
3	0	0
5	2	4
7	4	16

$$\bar{X} = \frac{\Sigma X_i}{n} = \frac{21}{7} = 3.$$

The median is the fourth observation, 2, the observation such that 3 other observations are higher and 3 are lower.

$$s^2 = \frac{\Sigma (X_i - \bar{X})^2}{n} = \frac{4 + 4 + 1 + 1 + 4 + 16}{7}$$

$$= \frac{30}{7} = 4.29.$$

Thus the variance is 4.29 and the standard deviation is,

$$s = \sqrt{s^2} = \sqrt{4.29} = 2.07.$$

The following measurements were taken by an antique dealer as he weighed to the nearest pound his prized collection of anvils. The weights were,

84, 92, 37, 50, 50, 84, 40, 98.

What was the mean weight of the anvils?

Solution: The average or mean weight of the anvils is

$$\overline{X} = \frac{\text{sum of observations}}{\text{number of observations}}$$

$$= \frac{84 + 92 + 37 + 50 + 50 + 84 + 40 + 98}{8}$$

$$= \frac{535}{8} = 66.88 \approxeq 67 \text{ pounds.}$$

An alternate way to compute the sample mean is to rearrange the terms in the numerator, grouping the numbers that are the same. Thus,

$$\overline{X} = \frac{(84 + 84) + (50 + 50) + 37 + 40 + 92 + 98}{8}.$$

We see that we can express the mean in terms of the frequency of observations. The frequency of an observation is the number of times a number appears in a sample.

$$\overline{X} = \frac{2(84) + 2(50) + 37 + 40 + 92 + 98}{8}.$$

The observations 84 and 50 appear in the sample twice and thus each observation has frequency 2.

In more general terms, the mean can be expressed as,

$$\overline{X} = \frac{\Sigma f_i X_i}{\text{number of observations}},$$

where X_i is the ith observation and f_i is the frequency of the ith observations.

The sum of the frequencies is equal to the total number of observations, $\Sigma f_i = n$.

Thus, $$\overline{X} = \frac{\Sigma f_i X_i}{\Sigma f_i}.$$

● **PROBLEM 55-13**

In a survey carried out in a school snack shop, the following results were obtained. Of 100 boys questioned, 78 liked sweets, 74 ice-cream, 53 cake, 57 liked both sweets and ice-cream. 46 liked both sweets and cake while only 31 boys liked all three. If all the boys interviewed liked at least one item, draw a Venn diagram to illustrate the results. How many boys liked both ice-cream and cake?

Solution: A Venn diagram is a pictorial representation of the relationship between sets. A set is a collection of objects. The number of objects in a particular set is the cardinality of a set.

To draw a Venn diagram we start with the following picture:

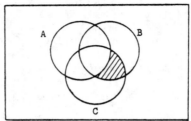

Each circle represents set A, B or C respectively. Let

A = set of boys who like ice-cream

B = set of boys who like cake

C = set of boys who like sweets.

The sections of overlap between circles represents the members of one set who are also members of another set. For example, the shaded region in the picture indicates the set of boys who are in sets B and C but not A. This is the set of boys who like both cake and sweets but not ice-cream. The inner section common to all three circles indicates the set of boys who belong to all three sets simultaneously.

We wish to find the number of boys who liked both ice-cream and cake. Let us label the sections of the diagram with the cardinality of these sections. The cardinality of the region common to all three sets is the number of boys who liked all three items or 31.

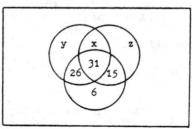

The number of boys who liked ice-cream and sweets was 57. Of these 57, 31 like all three leaving 26 boys in set A and set C but not set B. Similarly there are 15 boys in B and C, but not in A. 78 - 26 - 31 - 15 = 6 boys in C but not in A or B.

Let x = number of boys who are in A and B but not C

y = number of boys are in A but not B or C

z = number of boys who are in B but not A or C.

We know that the sum of all the labeled areas is 100 or

899

26 + 31 + 15 + 6 + x + y + z = 100

$$78 + x + y + z = 100.$$

Also, there are 74 boys total in set A or

x + y + 31 + 26 = 74   and 53 total in

set B or   x + z + 46 = 53.

Combining: x + y + z = 100 - 78 = 22

x + y = 74 - 57 = 17

x + z = 53 - 46 = 7 .

Substracting the second equation from the first gives z = 5 implying x = 2 and y = 15. Our answer is the number of boys in sets A and B = x + 31 = 33.

● **PROBLEM** 55-14

A survey was made of 100 customers in a department store. Sixty of the 100 indicated they visited the store because of a newspaper advertisement. The remainder had not seen the ad. A total of 40 customers made purchases; of these customers, 30 had seen the ad. What is the probability that a person who did not see the ad made a purchase? What is the probability that a person who saw the ad made a purchase?

Solution: In these two questions we have to deal with conditional probability, the probability that an event occurred given that another event occurred. In symbols, $P(A|B)$ means "the probability of A given B". This is defined as the probability of A and B, divided by the probability of B. Symbolically,

$$P(A|B) = \frac{P(A \cap B)}{P(B)} .$$

In the problem, we are told that only 40 customers made purchases. Of these 40, only 30 had seen the ad. Thus, 10 of 100 customers made purchases without seeing the ad. The probability of selecting such a customer at random is

$$\frac{10}{100} = \frac{1}{10} .$$

Let A represent the event of "a purchase", B the event of "having seen the ad", and $\bar{B}$ the event of "not having seen the ad."

Symbolically, $P(A \cap \bar{B}) = \frac{1}{10}$. We are told that 40 of the customers did not see the ad. Thus $P(\bar{B}) = \frac{40}{100} = \frac{4}{10}$.

Dividing, we obtain $\dfrac{1/10}{4/10} = \dfrac{1}{4}$, and, by definition of

conditional probability, $P(A|\bar{B}) = \dfrac{1}{4}$. Thus the probability

that a customer purchased given they did not see the ad is

$\dfrac{1}{4}$.

E₂ Purchases μ = 100 10 30 30 30 E, Ad Viewers

To solve the second problem, note that 30 purchasers
saw the ad. The probability that a randomly selected customer
saw the ad $\underline{\text{and}}$ made a purchase is $\dfrac{30}{100} = \dfrac{3}{10}$. Since 60
of the 100 customers saw the ad, the probability that a
randomly-picked customer saw the ad is $\dfrac{60}{100} = \dfrac{6}{10}$.
Dividing we obtain

$$P(A|B) = \frac{P(A \cap B)}{P(B)} = \frac{\dfrac{3}{10}}{\dfrac{6}{10}} = \frac{3}{6} = \frac{1}{2} .$$

● **PROBLEM 55-15**

In the St. Petersburg Community College, 30% of the men
and 20% of the women are studying mathematics. Further,
45% of the students are women. If a student selected at
random is studying mathematics, what is the probability
that the student is a woman?

$\underline{\text{Solution:}}$ This problem involves conditional probabilities.
The first two percentages given can be thought of as con-
ditional probabilitites; "30% of the men are studying
mathematics" means that the probability that a male student
selected at random is studying mathematics, is .3. Bayes'
formula allows us to use the probabilities we know to
compute the probability that a mathematics student is a
woman. Using the symbols M (the student is studying mathematics); W (the sudent is a woman); and N (the student is
not a woman), we write:

$$P(W|M) = \frac{P(W)\ P(M|W)}{P(W)\ P(M|W) + P(N)\ P(M|N)} , \text{ substituting}$$

$$= \frac{(.45)(0.2)}{(.45)(0.2) + (.55)(0.3)} = \frac{.09}{.09 + 0.165}$$

$$= \frac{.09}{.255} = \frac{6}{17} .$$

Thus, the probability that a randomly selected
math student is a woman equals $\dfrac{6}{17} = .353.$

901

Find the probability of drawing three consecutive face
cards on three consecutive draws (with replacement) from
a deck of cards.

Let: Event A: face card on first draw,
           Event B: face card on second draw, and
           Event C: face card on third draw.

Solution: This problem illustrates sampling with
replacement. After each draw, the card drawn is returned
to the deck. The deck, or "population" from which we draw
the second card is identical to the original deck. Thus
the drawing of a face card on the second draw is independent of the first draw. Similarly, the result of the third
drawing is independent of the first or second drawings.
This sampling without replacement implies that

$$P(ABC) = P(A) \cdot P(B) \cdot P(C).$$

But we know that the probability of drawing a face
card on any draw is

$$\frac{\text{number of face cards in deck}}{\text{number of cards in deck}} \quad \text{or} \quad \frac{12}{52}.$$

Therefore $P(ABC) = P(A)P(B)P(C)$

$$= \frac{12}{52} \cdot \frac{12}{52} \cdot \frac{12}{52} = \frac{27}{2197} = .012.$$

Let X be the random variable denoting the result of the single toss
of a fair coin. If the toss is heads, X = 1. If the toss results in
tails, X = 0.
       What is the probability distribution of X?

Solution: The probability distribution of X is a function which assigns
probabilities to the values X may assume.
       This function will have the following properties   if it defines a
proper probability distribution.
       Let $f(x) = Pr(X = x)$. Then $\Sigma \, Pr(X = x) = 1$ and $Pr(X = x) \geq 0$
for all x.
       We have assumed that X is a discrete random variable. That is, X
takes on discrete values.
       The variable X in this problem is discrete as it only takes on the
values 0 and 1.
       To find the probability distribution of X, we must find $Pr(X = 0)$
and $Pr(X = 1)$.
       Let $P_0 = Pr(X = 0)$ and $Pr(X = 1) = P_1$. If the coin is fair, the

events $X = 0$ and $X = 1$ are equally likely. Thus $P_0 = P_1 = p$. We must have $P_0 > 0$ and $P_1 > 0$.

In addition,
$$Pr(X = 0) + Pr(X = 1) = 1$$
or
$$P_0 + P_1 = p + p = 1$$
or
$$2p = 1$$

and
$$P_0 = P_1 = p = \tfrac{1}{2},$$
thus the probability distribution of X is $f(x)$: where
$$f(0) = Pr(X = 0) = \tfrac{1}{2} \text{ and}$$
$$f(1) = Pr(X = 1) = \tfrac{1}{2}$$
f (anything else) = $Pr(X = \text{anything else}) = 0$. We see that this is a proper probability distribution for our variable X.

$$\Sigma f(x) = 1 \quad \text{and}$$
$$f(x) \geq 0 .$$

● **PROBLEM 55-18**

If $f(x) = 1/4$, $x = 0,1,2,3$ is a probability mass function, find $F(t)$, the cumulative distribution function and sketch its graph.

<u>Solution:</u> $F(t) = \sum\limits_{x=0}^{t} f(x) = Pr(X \leq t)$. $F(t)$ changes for integer values of t. We have:

$$F(t) = 0 \qquad\qquad t < 0$$
$$F(t) = f(0) = 1/4, \quad 0 \leq t < 1$$
$$F(t) = f(0) + f(1) = 1/4 + 1/4 = 1/2 ,$$
$$\qquad\qquad\qquad\qquad 1 \leq t < 2$$
$$F(t) = f(0) + f(1) + f(2) \qquad 2 \leq t < 3$$
$$= \tfrac{1}{4} + \tfrac{1}{4} + \tfrac{1}{4} = 3/4 .$$
$$F(t) = \sum\limits_{x=0}^{t} f(x) = 1 \qquad\qquad 3 \leq t .$$

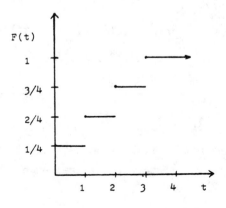

903

Given that the random variable X has density function

$$f(x) = \begin{cases} 2x & 0 < x < 1 \\ 0 & \text{otherwise} \end{cases}$$

Find $Pr(\frac{1}{2} < x < 3/4)$ and $Pr(-\frac{1}{2} < x < \frac{1}{2})$.

Solution: Since $f(x) = 2x$ is the density function of a continuous random variable, $Pr(\frac{1}{2} < x < 3/4)$ = area under $f(x)$ from $\frac{1}{2}$ to 3/4.

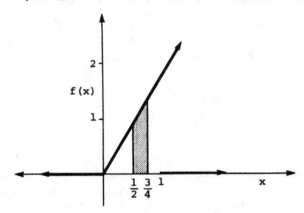

f(x) is indicated by the heavy line.

The area under $f(x)$ is the area of the triangle with vertices at (0,0), (1,0) and (1,2).
The area of this triangle is $A = \frac{1}{2} bh$ where b = base of the triangle and h is the altitude.

Thus, $A = \frac{1}{2}(1) \times 2 = \frac{2}{2} = 1$

proving that $f(x)$ is a proper probability density function.

To find the probability that $\frac{1}{2} < x < 3/4$ we find the area of the shaded region in the diagram. This shaded region is the difference in areas of the   right triangle with vertices (0,0), ($\frac{1}{2}$,0) and ($\frac{1}{2}$, f($\frac{1}{2}$)) and the area of the triangle with vertices (0,0), (3/4,0) and (3/4, f(3/4)).

This difference is $Pr(\frac{1}{2} < x < 3/4) = \frac{1}{2}(3/4)f(3/4) - \frac{1}{2}(\frac{1}{2})f(\frac{1}{2})$

$$= \frac{1}{2}[\frac{3}{4} \cdot \frac{6}{4} - \frac{1}{2} \cdot 1]$$

$$= \frac{1}{2}[\frac{9}{8} - \frac{1}{2}] = \frac{1}{2} \cdot \frac{5}{8} = \frac{5}{16} \quad .$$

The probability that $-\frac{1}{2} < x < \frac{1}{2}$ is
$Pr(-\frac{1}{2} < x < \frac{1}{2})$ = Area under $f(x)$ from $-\frac{1}{2}$ to $\frac{1}{2}$.
Because $f(x) = 0$ from $-\frac{1}{2}$ to 0, the area under $f(x)$ from $-\frac{1}{2}$ to 0 is 0. Thus

$\Pr(-\frac{1}{2} < x < \frac{1}{2}) = \Pr(0 < x < \frac{1}{2}) =$ area under $f(x)$ from 0 to $\frac{1}{2}$

$$= \frac{1}{2}(\frac{1}{2}) \ f(\frac{1}{2})$$

$$= \frac{1}{2}(\frac{1}{2}) \cdot 1 = \frac{1}{4} \ .$$

● **PROBLEM** 55-20

What is the probability of getting exactly 3 heads in 5 flips of a balanced coin?

Solution: We have here the situation often referred to as a Bernoulli trial. There are two possible outcomes, head or tail, each with a finite probability. Each flip is independent. This is the type of situation to which the binomial distribution,

$$P(X = k) = \binom{n}{k} \ p^k \ (1 - p)^{n-k},$$

applies. The a priori probability of tossing a head is $p = \frac{1}{2}$. the probability of a tail is $q = 1 - p = 1 - \frac{1}{2} = \frac{1}{2}$. Also $n = 5$ and $k = 3$ (number of heads required). we have

$$P(X = 3) = \binom{5}{3} \left(\frac{1}{2}\right)^3 \left(1 - \frac{1}{2}\right)^2 = \frac{5!}{3!2!} \left(\frac{1}{2}\right)^3 \left(\frac{1}{2}\right)^2$$

$$= \frac{5 \cdot \overset{2}{\cancel{4}} \cdot \cancel{3} \cdot \cancel{2} \cdot \cancel{1}}{\cancel{3} \cdot \cancel{2} \cdot \cancel{1} \cdot \cancel{2} \cdot \cancel{1}} \left(\frac{1}{2}\right)^5 = \frac{10}{2^5} = \frac{10}{32} = \frac{5}{16} \ .$$

● **PROBLEM** 55-21

Let X be a random variable whose value is determined by the flip of a fair coin. If the coin lands heads up X = 1, if tails then X = 0. Find the expected value of X.

Solution: The expected value of X, written E(X), is the theoretical average of X. If the coin were flipped many, many times and the random variable X was observed each time, the average of X would be considered the expected value.

The expected value of a discrete variable such as X is defined to be

$$E(X) = x_1 \ \Pr(X = x_1) + x_2 \ \Pr(X = x_2) \ \cdots \ +$$

$$x_n \ \Pr(X = x_n)$$

where $x_1, x_2, x_3, \cdots x_n$, are the values X may take on and $\Pr(X = x_j)$ is the probability that X actually equals the value x_j.

For our problem, the random variable X takes on only two values, 0 and 1. X assumes these values with

$$Pr(X = 1) = Pr(X = 0) = \frac{1}{2} .$$

Thus, according to our definition,

$$E(X) = 0 \cdot Pr(X = 0) + 1 \cdot Pr(X = 1)$$

$$= 0 \cdot \frac{1}{2} + 1 \cdot \frac{1}{2} = 0 + \frac{1}{2} = \frac{1}{2} .$$

● **PROBLEM 55-22**

Let Y = the Rockwell hardness of a particular alloy of steel. Assume that Y is a continuous random variable that can take on any value between 50 and 70 with equal probability. Find the expected Rockwell hardness.

Solution: The random variable Y has a density function that is sketched below.

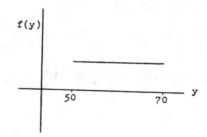

In order for Y to have a proper probability density function, the area under the density function must be 1. The area under the density function of Y is in the shape of a rectangle with length 20. Thus the height of the rectangle must be

$$f(y)(20) = 1 \qquad 50 < y < 70,$$

where the probability density function f(y) represents the width of this rectangle. Solving for f(y), we find the probability density function to be

$$f(y) = \frac{1}{20} \qquad 50 < y < 70 .$$

To find the expected value of a continuous random variable we use the technique of integration. Thus

$$E(Y) = \int_{50}^{70} yf(y) \, dy$$

$$E(Y) = \int_{50}^{70} y \frac{1}{20} \, dy \qquad = \frac{1}{20} \left(\frac{y^2}{2} \right) \Big|_{50}^{70}$$

$$= \frac{1}{20} \left[\frac{70^2 - 50^2}{2} \right] = \frac{(70 + 50)(70 - 50)}{40}$$

$$= \frac{70 + 50}{2} = \frac{120}{2} = 60 \; .$$

Thus, the expected Rockwell hardness of this alloy is 60.

● **PROBLEM 55-23**

Find the expected values of the random variables X and Y

if $Pr(X = 0) = \frac{1}{2}$ and $Pr(X = 1) = \frac{1}{2}$

and $Pr(Y = 1) = \frac{1}{4}$ and $Pr(Y = 2) = \frac{3}{4}$.

Compare the sum of E(X) + E(Y) with E(X + Y) if
$Pr(X = x, Y = y) = Pr(X = x)Pr(Y = y)$.

Solution: The expected value of X is

$$E(X) = 0 \cdot Pr(X = 0) + 1 \cdot Pr(X = 1) = (0)\left(\frac{1}{2}\right) + (1)\left(\frac{1}{2}\right) = \frac{1}{2} \; .$$

The expected value of Y is

$$E(Y) = 1 \cdot Pr(Y = 1) + 2 \cdot Pr(Y = 2) = \frac{1}{4} + \frac{6}{4} = \frac{7}{4} \; .$$

Thus, $E(X) + E(Y) = \frac{1}{2} + \frac{7}{4} = \frac{9}{4}$.

To find the expected value of the random variable
(X + Y) we need the joint distribution of X and Y. This
has been given to be

$$Pr(X = x, Y = y) = Pr(X = x) \, Pr(Y = y).$$

The distribution for X and Y is

$$Pr(X = 0, Y = 1) = \frac{1}{2} \cdot \frac{1}{4} = \frac{1}{8}$$

$$Pr(X = 0, Y = 2) = \frac{1}{2} \cdot \frac{3}{4} = \frac{3}{8}$$

$$Pr(X = 1, Y = 1) = \frac{1}{2} \cdot \frac{1}{4} = \frac{1}{8}$$

$$Pr(X = 1, Y = 2) = \frac{1}{2} \cdot \frac{3}{4} = \frac{3}{8}$$

$$E(X + Y) = \sum_x \sum_y (x + y) \, Pr(X = x, Y = y).$$

907

That is, the expected value of the random
able X + Y is the sum of the possible values that
r Y can assume times the probability that X + Y will
assume these values.

Thus, $E(X + Y) = (0 + 1) \Pr(X = 0, Y = 1)$ +

$(0 + 2) \Pr(X = 0, Y = 2)$ +

$(1 + 1) \Pr(X = 1, Y = 1)$ +

$(1 + 2) \Pr(X = 1, Y = 2)$;

$$E(X + Y) = 1 \cdot \frac{1}{8} + 2 \cdot \frac{3}{8} + 2 \cdot \frac{1}{8} + 3 \cdot \frac{3}{8}$$

$$= \frac{1}{8} + \frac{6}{8} + \frac{2}{8} + \frac{9}{8} = \frac{18}{8} = \frac{9}{4} \; .$$

Thus $E(X + Y) = E(X) + E(Y)$.

● **PROBLEM** 55-24

Given the probability distribution of the random
variable X in the table below, compute E(X) and Var (X).

x_i	$\Pr(X = x_i)$
0	$\frac{8}{27}$
1	$\frac{12}{27}$
2	$\frac{6}{27}$
3	$\frac{1}{27}$

Solution:

$$E(X) = \sum_i x_i \Pr(X = x_i) \quad \text{and} \quad \text{Var } X = E[(X - E(X))^2] \; .$$

Thus, $E(X) = (0) \Pr(X = 0) + (1) \Pr(X = 1)$

$+ (2) \Pr(X = 2) + (3) \Pr(X = 3)$

$$= (0) \frac{8}{27} + (1) \frac{12}{27} + (2) \frac{6}{27} + 3 \left(\frac{1}{27}\right)$$

908

$$= 0 + \frac{12}{27} + \frac{12}{27} + \frac{3}{27} = \frac{27}{27} = 1.$$

$$\text{Var } X = (0 - 1)^2 \text{ Pr}(X = 0) + (1 - 1)^2 \text{ Pr}(X = 1)$$

$$+ (2 - 1)^2 \text{ Pr}(X = 2) + (3 - 1)^2 \text{ Pr}(X = 3)$$

$$= (1^2) \frac{8}{27} + (0^2) \frac{12}{27} + (1^2) \frac{6}{27} + (2^2) \frac{1}{27}$$

$$= \frac{8}{27} + \frac{6}{27} + \frac{4}{27} = \frac{18}{27} = \frac{2}{3}.$$

● **PROBLEM** 55-25

Find the variance of the random variable X + b where X
has variance, Var X and b is a constant.

Solution:
$$\text{Var } (X + b) = E[(X + b)^2] - [E(X + b)]^2$$

$$= E[X^2 + 2bX + b^2] - [E(X) + b]^2$$

$$= E(X^2) + 2bE(X) + b^2 - [E(X)]^2 - 2E(X)b - b^2,$$

thus $\text{Var } (X + b) = E(X^2) - [E(X)]^2 = \text{Var } X.$

● **PROBLEM** 55-26

If Z is a standard normal variable, use the table of
standard normal probabilities to find:

(a) Pr(Z < 0)
(b) Pr(- 1 < Z < 1)
(c) Pr(Z > 2.54) .

Solution: The normal distribution is the familiar
"bell-shaped" curve. It is a continuous probability
distribution that it widely used to describe the
distribution of heights, weignts, and other characteristics.

The density function of the standard normal
distribution is

$$f(x) = \frac{1}{\sqrt{2\pi}} \exp \left(\frac{-x^2}{2} \right) \qquad - \infty < x < \infty .$$

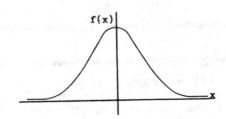

is a graph of this density. The probability of a standard normal variable being found in a particular interval can be found with the help of tables found in the backs of most statistics text books.

(a) To find the probability $Pr(Z < 0)$ we can take advantage of the fact that the normal distribution is symmetric about its mean of zero. Thus

$Pr(Z > 0) = Pr(Z < 0)$.   We know that

$Pr(Z > 0) + Pr(Z < 0) = 1$

because $Z > 0$ and $Z < 0$ are exhaustive events. Thus

$2Pr(Z < 0) = 1$ or $Pr(Z < 0) = \frac{1}{2}$.

(b) To find the $Pr(-1 < Z < 1)$ we use the tables of the standard normal distribution.

$Pr(-1 < Z < 1) = Pr(Z < 1) - Pr(Z < -1)$.

Reading across the row headed by 1 and down the column labeled .00 we see that $Pr(Z < 1.0) = .8413$.

$Pr(Z < -1) = Pr(Z > 1)$ by the symmetry of the normal distribution. We also know that

$Pr(Z > 1) = 1 - Pr(Z < 1)$.

Substituting we see,

$Pr(-1 < Z < 1) = Pr(Z < 1) - [1 - Pr(Z < 1)]$

$= 2Pr(Z < 1) - 1$ $= 2(.8413) - 1 = .6826$.

(c) $Pr(Z > 2.54) = 1 - Pr(Z < 2.54)$ and reading across the row labeled 2.5 and down the column labeled .04 we see that $Pr(Z < 2.54) = .9945$.

Substituting,

$Pr(Z > 2.54) = 1 - .9945 = .0055$.

● **PROBLEM 55-27**

Find $\Phi(-.45)$.

Solution: $\Phi(-.45) = Pr(Z \leq -.45)$,

where Z is distributed normally with mean 0 and variance 1.

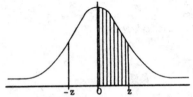

Let A(Z) = the area under the curve from 0 to Z.
From our table we find A(.45) = .17364 and by the symmetry
of the normal distribution,

A(- .45) = A(.45) = .17364.

We wish to find Φ(- .45), the shaded area below.

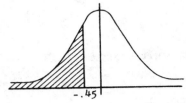

We know that Φ(0) = .5000 and from the diagram below
we know that Φ(0) - A(- .45) = Φ(- .45).

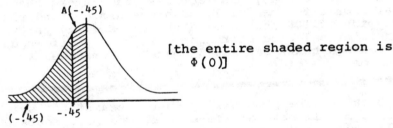

[the entire shaded region is
Φ(0)]

Substituting,

Φ(0) - A(- .45) = .5000 - .17364 = Φ(- .45)

and Φ(- .45) = .32636.

● **PROBLEM** 55-28

Given that the random variable x has density

$$f(x) = \frac{1}{\sqrt{18\pi}} \, e^{-(x^2 - 10x + 25)/18}, \quad -\infty < x < \infty.$$

Is this distribution normal? What is its maximum
value?

Solution: A normal distribution can be written in
the form

$$f(x) = \frac{1}{\sigma\sqrt{2\pi}} \, \exp\left[-\frac{(x - \mu)^2}{2\sigma^2}\right], \quad \text{when } -\infty < x < \infty.$$

Rewrite $(x^2 - 10x + 25)$ as $(x - 5)^2$ and 18 as $2 \cdot 9 = 2 \cdot 3^2$. Also $\sqrt{18\pi} = \sqrt{9 \cdot 2\pi} = 3\sqrt{2\pi}$. Thus $\mu = 5$ and $\sigma = 3$. Substitution gives

$$f(x) = \frac{1}{3\sqrt{2\pi}} \, e^{-(x - 5)^2/(2 \cdot 3^2)}.$$

The graph of a normal distribution reaches its maximum at $x = \mu$, so we substitute $x = 5$:

$$f(x) = \frac{1}{3\sqrt{2\pi}} \, e^{-(5 - 5)^2/2 \cdot 3^2} = \frac{1}{3\sqrt{2\pi}} \, e^0$$

$$= \frac{1}{3\sqrt{2\pi}} \, .$$

Thus the maximum value of $f(x)$ is $\dfrac{1}{3\sqrt{2\pi}}$.

● **PROBLEM** 55-29

A television company manufactures transistors that have an average life-span of 1,000 hours and a standard deviation of 100 hours. Find the probability that a transistor selected at random will have a life-span between 875 hours and 1,075 hours. Assume the distribution is normal.

Solution: The probability that a transistor selected at random will have a life-span between 875 hours and 1,075 hours can be expressed symbolically as $P(875 < X < 1075)$. Life-spans of transistors are normally distributed, but we must standardize X (the random variable which represents life-span) in order to use the standard normal table. We do this by subtracting its mean and dividing the resulting difference by its standard deviation. We are given that the mean (average life-span) is 1000 hours, and that the standard deviation is 100 hours.

Letting Z denote our standard normal random variable, $Z = \dfrac{X - \mu}{\sigma} = \dfrac{X - 1000}{100}$. We want to find the area under the standard normal curve between the Z-values for X = 875 and X = 1075, so we compute

$$Z(875) = \frac{875 - 1000}{100} = \frac{-125}{100} = -1.25, \quad \text{and}$$

$$Z(1075) = \frac{1075 - 1000}{100} = \frac{75}{100} = 0.75.$$

In terms of Z,

$$P(875 < X < 1075) = P(-1.25 < Z < .75).$$

912

Since some tables give areas under the standard normal curve only for positive Z-values, we put $P(-1.25 < Z < .75)$ in its equivalent form: $P(0 < Z < 0.75) + P(0 < Z < 1.25)$. The symmetry of the standard normal curve allows us to do this.

Reading the table we find $P(0 < Z < 0.75) = .2734$ and $P(0 < Z < 1.25) = .3944$. The total area between $Z = -1.25$ and $Z = .75$ is $.2734 + .3944 = .6678$, and this is the probability that a randomly selected transistor will function between 875 and 1075 hours.

● **PROBLEM** 55-30

A pair of dice *is* thrown 120 times. What is the approximate probability of throwing at least 15 sevens? Assume that the rolls are independent and remember that the probability of rolling a seven on a single roll is $\frac{6}{36} = \frac{1}{6}$.

Solution: The answer to this problem is a binomial probability. If X = number of sevens rolled, n = 120, then

$$Pr(X \geq 15) = \sum_{j=15}^{120} \binom{120}{j} \left(\frac{1}{6}\right)^j \left(\frac{5}{6}\right)^{120-j}.$$

This sum is quite difficult to calculate. There is an easier way. If n is large, $Pr_B(X \geq 15)$ can be approximated by $Pr_N(X \geq 14.5)$ where X is normally distributed with the same mean and variance as the binomial random variable. Remember that the mean of a binomially distributed random variable is np; n is the number of trials and p is the probability of "success" in a single trial. The variance of a binomially distributed random variable is $np(1 - p)$ and the standard deviation is $\sqrt{np(1 - p)}$.

Because of this fact $\dfrac{X - np}{\sqrt{np(1 - p)}}$ is normally distributed with mean 0 and variance 1.

$$np = (120)\left(\frac{1}{6}\right) = 20 \qquad \text{and}$$

$$\sqrt{np(1 - p)} = \sqrt{120\left(\frac{1}{6}\right)\left(\frac{5}{6}\right)} = \sqrt{\frac{50}{3}} = 4.08248.$$

Thus $Pr(X \geq 15) = Pr\left(\dfrac{X - 20}{4.08248} > \dfrac{14.5 - 20}{4.08248}\right)$

$= Pr(Z > -1.35) = 1 - Pr(Z < -1.35)$

$= 1 - \Phi(-1.35) = 1 - .0885$

$= .9115$.

913

$$Pr_B(X > 15) \underset{\sim}{\sim} Pr_N(X \geq 14.5).$$

The reason 15 has become 14.5 is that a discrete random variable is being approximated by a continuous random variable.

Consider the example below:

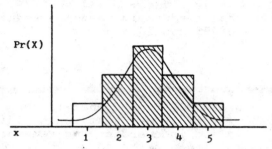

$Pr_B(2 \leq X \leq 5)$ = sum of the areas of the shaded rectangles. In approximating this area with a curve we must start at the edge of the first shaded rectangle and move to the edge of the last shaded rectangle. This implies

$$Pr_B(2 \leq X \leq 5) \underset{\sim}{\sim} Pr_N(1.5 \leq X \leq 5.5).$$

● **PROBLEM** 55-31

A multiple-choice test has 200 questions, each with 4 possible answers, of which only 1 is the correct answer. What is the probability that sheer guesswork yields from 25 to 30 correct answers for 80 of the 200 problems about which the student has no knowledge?

Solution: Let X be the number of correct answers in the 80 questions about which the student has no knowledge. If the student is guessing, the probability of selecting the correct answer is $\frac{1}{4}$. It may also be assumed that random guesswork will imply that each question is answered independently of any other question. With these assumptions, X is binomially distributed with parameters n = 80 and p = $\frac{1}{4}$. Hence $E(X) = np = 80 \cdot \frac{1}{4} = 20$ and

$$\sqrt{Var\ X} = \sqrt{np(1 - p)} = \sqrt{80 \cdot \frac{1}{4} \cdot \frac{3}{4}} = \sqrt{15} = 3.87.$$

We wish to find $Pr(25 \leq X \leq 30)$. This probability is found exactly to be

$$Pr(25 \leq X \leq 30) = \sum_{j=25}^{30} \binom{80}{j} \left(\frac{1}{4}\right)^j \left(\frac{3}{4}\right)^{80-j}.$$

This expression is quite tedious to calculate and we thus use the normal approximation to the binomial. Let Y be normally distributed with mean np = 20 and standard deviation $\sqrt{np(1 - p)} = 3.87$. Then

$$Pr(25 \leq X \leq 30) \cong Pr\left(25 - \frac{1}{2} \leq Y \leq 30 + \frac{1}{2}\right).$$

We add and subtract $\frac{1}{2}$ to improve the approximation of a discrete random variable by a continuous random variable.

To calculate,

Pr(24.5 < Y < 30.5) we standardize Y by subtracting the mean and then dividing by the standard deviation. Thus,

Pr(24.5 $\leq$ Y $\leq$ 30.5)

$$= Pr\left(\frac{24.5 - 20}{3.87} \leq \frac{Y - 20}{3.87} \leq \frac{30.5 - 20}{3.87}\right)$$

$$= Pr\left(\frac{4.5}{3.87} \leq Z \leq \frac{10.5}{3.87}\right)$$

where Z is normally distributed with mean 0 and standard deviation 1. From the table of the standard normal distribution,

$$Pr\left(\frac{4.5}{3.87} \leq Z \leq \frac{10.5}{3.87}\right) = Pr(1.163 \leq Z \leq 2.713)$$

$$= Pr(Z \leq 2.713) - Pr(Z \leq 1.163)$$

$$= .9966 - .8776 = .1190.$$

Thus, the approximate probability that the student answers between 25 and 30 questions by sheer guesswork is .1190.

● **PROBLEM** 55-32

Consider the exponential distribution $f(x) = \lambda e^{-\lambda x}$ for x > 0. Find the moment generating function and from it, the mean and variance of the exponential distribution.

Solution: By definition $M_x(t) = E(e^{tx})$

$$= \int_{-\infty}^{\infty} e^{tx} f(x) \, dx$$

$$= \int_{x=0}^{\infty} e^{tx} \lambda e^{-\lambda x} \, dx$$

$$= \int_0^\infty \lambda e^{(t - \lambda)x} \, dx = \lambda \int_0^\infty e^{(t - \lambda)x} \, dx$$

$$= \lambda \left[\frac{-1}{t - \lambda} \, e^{(t - \lambda)x} \right]_0^\infty$$

$$= \frac{\lambda}{\lambda - t} \left[e^{(t - \lambda)x} \right]_0^\infty$$

Consider $t < \lambda$. Then $\lambda - t > 0$ and $t - \lambda < 0$. Hence $e^{(t - \lambda)x} = e^{-kx}$ and $M_x(t) = \frac{\lambda}{\lambda - t} (0 - (-1)) = \frac{\lambda}{\lambda - t}$

for $t < \lambda$.

The mean is

$$E(x) = M_x'(t) \Big|_{t=0}$$

$$M_x'(t) = \frac{d}{dt} \left(\frac{\lambda}{\lambda - t} \right) = \lambda \left(\frac{d}{dt} \frac{1}{\lambda - t} \right)$$

$$= \lambda \left(\frac{-1}{(\lambda - t)^2} \right) \frac{d}{dt} (\lambda - t) = \frac{\lambda}{(\lambda - t)^2}$$

$$M_x'(0) = E(x) = \frac{\lambda}{(\lambda - 0)^2} = \frac{\lambda}{\lambda^2} = \frac{1}{\lambda} \, .$$

Also by the moment generating function's properties

$$E(x^2) = M_x''(t) \Big|_{t=0}$$

$$M_x''(t) = \frac{d}{dt} \frac{\lambda}{(\lambda - t)^2} = \lambda \frac{d}{dt} \frac{1}{(\lambda - t)^2} \, .$$

$$= \lambda \frac{-2}{(\lambda - t)^3} \frac{d}{dt} (\lambda - t) = \frac{2\lambda}{(\lambda - t)^3}$$

$$M_x''(0) = E(x^2) = \frac{2\lambda}{\lambda^3} = \frac{2}{\lambda^2} \, ,$$

Now $\text{Var}(X) = E(x^2) - (E(x))^2$

$$= \frac{2}{\lambda^2} - \left(\frac{1}{\lambda} \right)^2$$

$$= \frac{1}{\lambda^2} \, .$$

916

A population consists of the number of defective transistors in shipments received by an assembly plant. The number of defectives is 2 in the first, 4 in the second, 6 in the third, and 8 in the fourth.

(a) Find the mean $\bar{x}$ and the standard deviation s_x' of the given population.

(b) List all random samples, with replacement, of size 2 that can be formed from the population and find the distribution of the sample mean.

(c) Find the mean and the standard deviation of the sample mean.

Solution: The population is 2,4,6,8.

a) $\quad \bar{x} = \dfrac{\sum\limits_{i=1}^{n} x_i}{n} = \dfrac{\sum\limits_{i=1}^{4} x_i}{4} = \dfrac{2 + 4 + 6 + 8}{4} = \dfrac{20}{4} = 5$.

We will compute

$$s_x' = \sqrt{\dfrac{\sum\limits_{i=1}^{n}(x - \bar{x})^2}{n}} = \sqrt{\dfrac{(2-5)^2 + (4-5)^2 + (6-5)^2 + (8-5)^2}{4}}$$

$$= \sqrt{\dfrac{9 + 1 + 1 + 9}{4}} = \sqrt{\dfrac{20}{4}} = \sqrt{5} \ .$$

b) The following table should prove useful.

	Sample	Sample Mean
1.	2,2	2
2.	2,4	3
3.	2,6	4
4.	2,8	5
5.	4,2	3
6.	4,4	4
7.	4,6	5
8.	4,8	6
9.	6,2	4
10.	6,4	5
11.	6,6	6
12.	6,8	7
13.	8,2	5
14.	8,4	6
15.	8,6	7
16.	8,8	8

Collating the data we have

x = Sample Mean	N(x) = Number of times x occurs	F(x) = N(x)/n = N(x)/16
2	1	1/16
3	2	1/8
4	3	3/16
5	4	1/4
6	3	3/16
7	2	1/8
8	1	1/16

c) Sample Mean $= \dfrac{\sum\limits_{i=1}^{n} i^{th} \text{ Sample Mean}}{n}$

$= \dfrac{2+3+4+5+3+4+5+6+4+5+6+7+5+6+7+8}{16}$

$= \dfrac{80}{16} = 5$.

$s'_{\bar{x}} = \sqrt{\dfrac{\sum(sm - \overline{sm})^2}{n}}$

$= \sqrt{\dfrac{(2-5)^2 + 2(3-5)^2 + 3(4-5)^2 + 4(5-5)^2 + 3(6-5)^2 + 2(7-5)^2 + (8-5)^2}{16}}$

$= \sqrt{\dfrac{9+8+3+0+3+8+9}{16}} = \sqrt{\dfrac{40}{16}} = \sqrt{\dfrac{5}{2}}$.

● **PROBLEM 55-34**

Briefly discuss the Central Limit Theorem.

Solution: The theorem has to do with the means of large (greater than 30) samples. As the sample size increases, the distribution of the sample mean, $\bar{X}$, has a distribution which is approximately normal. This distribution has a mean equal to the population mean and a standard deviation equal to the population standard deviation divided by the square root of the sample size.

Since $\bar{X}$ is approximately normal, $\dfrac{\bar{X} - E(\bar{X})}{\sigma_{\bar{X}}} = \dfrac{\bar{X} - \mu}{\sigma/\sqrt{n}} = \dfrac{\sqrt{n}(\bar{X} - \mu)}{\sigma}$

will have a standard normal distribution.

● **PROBLEM 55-35**

Let X be $\chi^2(16)$. What is the probability that the random interval (X, 3.3X) contains the point x = 26.3? and what is the expected length of the interval?

918

Solution: We will begin by trying to transform the interval X < 26.3 < 3.3 X into an equivalent event with which we can more easily deal. It is clear that X must be less than 26.3. Examine now the right hand side of the inequality 26.3 < 3.3 X.   This is equivalent to X $>$ $\frac{26.3}{3.3}$ or X > 7.97.

Now we see Pr(X < 26.3 < 3.3 X) = Pr(7.97 < X < 26.3). Recall that X is χ^2(16), therefore:

Pr(7.97 < X < 26.3) = Pr(χ^2(16) < 26.3) - Pr(χ^2(16) < 7.97).

From the table of the Chi-square distribution, this equals .95 - .05 = .90.
The length of the interval is 3.3X - X = 2.3 X.

E (Length) = E(2.3 X) = 2.3 E(X)

by the linearity properties of expectation. Since X is χ^2(16), E(X) = 16 [E(χ^2(n))=n.].

E (Length) = 2.3(16) = 36.8.

● **PROBLEM 55-36**

Find a 95 per cent confidence interval for μ, the true mean of a normal population which has variance σ^2 = 100. Consider a sample of size 25 with a mean of 67.53.

Solution: We have a sample mean $\overline{X}$ = 67.53. We want to transform that into a standard normal quantity, for we know from the standard normal tables that

Pr(- 1.96 < Standard Normal Quantity < 1.96) = .95 .

$\dfrac{\overline{X} - E(\overline{X})}{\sqrt{Var\ (\overline{X})}}$ is a standard normal quantity.

Recall now that the expectation of a sample mean is μ, the true mean of a population. Also recall that the variance of a sample mean is $\dfrac{\sigma^2}{n}$ where σ^2 is the true variance of the population and n is the size of our sample. Applying this to our case,   E($\overline{X}$)= μ and $\sqrt{Var\,(\overline{X})} = \sqrt{\dfrac{\sigma^2}{n}} = \sqrt{\dfrac{100}{25}}$ = 2.

For our sample: $Pr\left[- 1.96 < \dfrac{\overline{X} - \mu}{2} < 1.96\right]$ = .95 .

Multiplying by 2: Pr(- 3.92 < $\overline{X}$ - μ < 3.92) = .95.
Transposing: Pr($\overline{X}$ - 3.92 < μ < $\overline{X}$ + 3.92) = .95 .

$\overline{X}$ - 3.92 < μ < $\overline{X}$ + 3.92 is our required confidence interval. If we insert our given sample mean, we come up with   67.53 - 3.92 < μ < 67.53 + 3.92

or $63.61 < \mu < 71.45.$

Barnard College is a private institution for women located in New York City. A random sample of 50 girls was taken. The sample mean of grade point averages was 3.0. At neighboring Columbia College a sample of 100 men had an average gpa of 2.5. Assume all sampling is normal and Barnard's standard deviation is .2, while Columbia's is .5. Place a 99% confidence interval on $\mu_{Barnard} - \mu_{Columbia}$.

Solution: The main idea behind all of these problems is the same. We want to find a standard normal pivotal quantity involving $\mu_B - \mu_C$ and use the fact, obtainable from the standard normal tables, that

$$Pr(-2.58 < \text{Standard Normal Quantity} < 2.58) = .99.$$

We want to find a quantity $\bar{B} - \bar{C}$, where $\bar{B}$ = Barnard's average and $\bar{C}$ = Columbia's average.

The sample mean from a normal population is normally distributed. $\bar{B} - \bar{C}$ is a difference in normal distributions and is thus also normal. Hence $\dfrac{(\bar{B} - \bar{C}) - E(\bar{B} - \bar{C})}{\text{S.D. } (\bar{B} - \bar{C})}$ is standard normal.

Recall that the expectation of a sample mean is μ, the expectation of the original distribution. Thus, $E(\bar{B}) = \mu_B$ and $E(\bar{C}) = \mu_C$.

$$E(\bar{B} - \bar{C}) = E(\bar{B} + (-\bar{C})) = E(\bar{B}) + E(-\bar{C}) = E(\bar{B}) - E(\bar{C})$$

by the linearity properties of expectation.

Hence $E(\bar{B} - \bar{C}) = E(\bar{B}) - E(\bar{C}) = \mu_B - \mu_C$.

Also, $\sigma = \sqrt{Var(\bar{B} - \bar{C})} = \sqrt{Var(\bar{B}) + Var(\bar{C})}$ since

$Var(ax + by) = a^2 Var(X) + b^2 Var(Y)$. Furthermore,

$$\sigma = \sqrt{\frac{\sigma_B^2}{n_B} + \frac{\sigma_C^2}{n_C}}$$ since the standard deviation

of a sample mean is $\dfrac{\sigma}{\sqrt{n}}$.

Now we see when substituting that

$$\Pr\left(-2.58 < \frac{(B - C) - (\mu_B - \mu_C)}{\sqrt{\frac{\sigma_B^2}{n_B} + \frac{\sigma_C^2}{n_C}}} < 2.58\right) = .99 .$$

We are given

$\overline{B} = 3.0$; $\overline{C} = 2.5$; $\sigma_B = .2$; $\sigma_C = .5$; $n_B = 50$; $n_C = 100$.

Inserting these values into the inequality, we obtain:

$$-2.58 < \frac{(3.0 - 2.5) - (\mu_B - \mu_C)}{\sqrt{\frac{(.2)^2}{50} + \frac{(.5)^2}{100}}} < 2.58.$$

Combining: $-2.58 < \dfrac{.5 - (\mu_B - \mu_C)}{\sqrt{.0033}} < 2.58.$

Multiplying by $\sqrt{.0033}$: $-.148 < .5 - (\mu_B - \mu_C) < .148.$

Subtracting .5: $-.648 < -(\mu_B - \mu_C) < -.352$.

Multiplying by -1: $.352 < \mu_B - \mu_C < .648$

Our required interval is

$.352 < \mu_B - \mu_C < .648$.

● **PROBLEM** 55-38

Consider a distribution $N(\mu, \sigma^2)$ where μ is known but σ^2 is not. Devise a method of producing a confidence interval for σ^2.

Solution: Let X_1, X_2, ..., X_n denote a random sample of a size n from $N(\mu, \sigma^2)$, where μ is known. The random variable

$$Y = \frac{\sum_1^n (X_i - \mu)^2}{\sigma^2}$$ is a chi-square with n degrees of

freedom. This is not to be confused with
$$\frac{\sum_1^n (X_i - \overline{X})^2}{\sigma^2}$$ which is $\chi^2(n - 1)$. We select a probability,

$1 - \alpha$, and for the constant n, determine values a and b, with a < b such that

$$Pr(a < Y < b) = 1 - \alpha .$$

Thus
$$P\left(a < \frac{\sum\limits_{1}^{n}(X_i - \mu)^2}{\sigma^2} < b\right) = 1-\alpha.$$

We will concern ourselves with the central inequality. Taking reciprocals, we obtain

$$\frac{1}{b} < \frac{\sigma^2}{\sum\limits_{1}^{n}(X_i - \mu)^2} < \frac{1}{a} .$$

Multiply through by $\sum\limits_{1}^{n}(X_i - \mu)^2$ and the interval is

$$\frac{\sum\limits_{1}^{n}(X_i - \mu)^2}{b} < \sigma^2 < \frac{\sum\limits_{1}^{n}(X_i - \mu)^2}{a} .$$

The interval $\left(\dfrac{\sum\limits_{1}^{n}(X_i - \mu)^2}{b} , \dfrac{\sum\limits_{1}^{n}(X_i - \mu)^2}{a}\right)$ is a

random interval having probability $1 - \alpha$ of including the unknown fixed point (parameter) σ^2. Once we perform the experiment and find that $X_1 = x_1$, $X_2 = x_2$, ..., $X_n = x_n$, then the particular interval we calculate is a $1 - \alpha$ confidence interval for σ^2.

You should observe that there are no unique numbers a and b, a < b, such that $Pr(a < Y < b) = 1 - \alpha$. A common convention, one which we will follow, is to find a and b such that $Pr(Y < a) = \frac{\alpha}{2}$ and $Pr(Y > b) = \frac{\alpha}{2}$. That way

$$Pr(a < Y < b) = 1 - \frac{\alpha}{2} - \frac{\alpha}{2} = 1 - \alpha.$$

● **PROBLEM 55-39**

In testing a hypothesis concerned with the value of a population mean, first the level of significance to be used in the test is specified and then the regions of acceptance and rejection for evaluating the obtained sample mean are determined. If the 1 percent level of signifi- cance is used, indicate the percentages of sample means in each of the areas of the normal curve, assuming that the population hypothesis is correct, and the test is two-tailed.

Solution: A level of significance of 1% signifies that when the population mean is correct as specified, the sample mean will fall in the critial areas of rejection only 1% of the time. Referring to the figure below, .005 or .5% of the sample means will fall in each area of rejection and 99% of the sample means will fall in the region of acceptance.

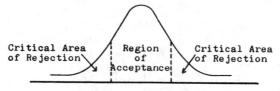

Critical Area
of Rejection

Region
of
Acceptance

Critical Area
of Rejection

● **PROBLEM 55-40**

A sample of size 49 yielded the values $\bar{x} = 87.3$ and $s^2 = 162$. Test the hypothesis that $\mu = 95$ versus the alternative that it is less. Let $\alpha = .01$.

Solution: The null and alternative hypotheses are given respectively by

$$H_0 : \mu = 95; \quad H_1 : \mu < 95 .$$

$\alpha = .01$ is the given level of significance.

Because the sample size is quite large (≥ 30), we can assume that the distribution of $\bar{X}$ is approximately normal. We are using the sample variance s^2 as an estimate of the true but unknown population variance and if the sample were not as large we would use a t-test.

The critical region consists of all z-scores that are less than $z_{.01} = -2.33$. The observed z-score is

$$z = \frac{\bar{X} - \mu}{\sqrt{s^2/n}} = \frac{87.3 - 95}{\sqrt{162/49}} = \frac{(-7.7)(7)}{\sqrt{162}} = -4.23 .$$

This observed score is in the critial region; thus we reject the null hypothesis and accept the alternative that $\mu < 95$.

● **PROBLEM 55-41**

Suppose it is required that the mean operating life of size "D" batteries be 22 hours. Suppose also that the operating life of the batteries is normally distributed. It is known that the standard deviation of the operating life of all such batteries produced is 3.0 hours. If a sample of 9 batteries has a mean operating life of 20 hours, can we conclude that the mean operating life of size "D" batteries is not 22 hours? Then suppose the standard deviation of the operating life of all such batteries is not known but that for the sample of 9 batteries the standard deviation is 3.0. What conclusion would we then reach?

Solution: Since the operating life of "D" batteries is normally distributed, and the standard deviation for all batteries is known, the sample mean will have a Z or normal distribution regardless of the sample size.
Hence for the first part of this problem, we will calculate

$$Z = \frac{\bar{X} - \mu}{\sigma_{\bar{X}}} \qquad \text{where} \qquad \sigma_{\bar{X}} = \frac{\sigma}{\sqrt{n}} .$$

The diagram for this problem is given below.

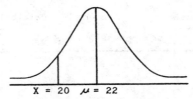

X = 20 μ = 22

The null hypothesis and alternate hypothesis are

$$H_0 : \mu = 22 \; ; \qquad H_1 : \mu \neq 22 .$$

The decision rule at $\alpha = .05$ is as follows: Reject H_0 if $Z > 1.96$ or $Z < -1.96$; accept H_0 if $-1.96 \leq Z \leq 1.96$. The value of 1.96 is chosen as the critial value because for this problem we have $\alpha = .05$ and a two-tailed test, and for the standard normal distribution, 2.5% of scores will have a Z-value greater than 1.96 and 2.5% of scores will have a Z-value less than -1.96.

For this set of data

$$\sigma_{\bar{X}} = \frac{3.0}{\sqrt{9}} = 1 \quad \text{and}$$

$$Z = \frac{20 - 22}{1} = -2.0 .$$

Therefore we will reject H_0 and conclude that the mean operating life of size "D" batteries is not 22 hours.

When the population standard deviation is not known the sample mean has a t-distribution with $n-1 = 8$ degrees of freedom. We calculate

$$t = \frac{\bar{X} - \mu}{S_{\bar{X}}}$$

and use the decision rule to reject H_0 if $t > 2.306$ or $t < -2.306$ (the critical t for 8 df's $\alpha = .05$, and a two-tailed test is 2.306) and to accept H_0 if $-2.306 \leq t \leq 2.306$. For this data

$$t = \frac{20 - 22}{1} = -2.0 .$$

So in this case we will accept H_0 that the mean operating life of size "D" batteries is 22 hours.

● **PROBLEM 55-42**

From appropriately selected samples, two sets of IQ scores are obtained. For group 1, $\bar{X} = 104$, $S = 10$, and $n = 16$; for group 2, $\bar{X} = 112$, $S = 8$, and $n = 14$. At the 5% significance level is there a significant difference between the 2 groups?

<u>Solution:</u>   For this problem, we have for our hypotheses

$H_0: \mu_1 - \mu_2 = 0$ $\qquad\qquad$ $H_1: \mu_1 - \mu_2 \neq 0.$

This problem can be depicted by the following diagram.

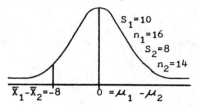

The statistic $\dfrac{(\overline{X}_1 - \overline{X}_2) - (\mu_1 - \mu_2)}{S_{\overline{X}_1 - \overline{X}_2}}$

has a t-distribution when $n_1 + n_2 \leq 30$.

We have $n_1 + n_2 - 2 = 28$ df's and so our decision rule for $\alpha = .05$ is: reject H_0 if $t > 2.048$ or $t < -2.048$; accept H_0 if $-2.048 \leq t \leq 2.048$. We must calculate

$$t = \frac{(\overline{X}_1 - \overline{X}_2) - (\mu_1 - \mu_2)}{S_{\overline{X}_1 - \overline{X}_2}} \quad \text{where}$$

$$S_{\overline{X}_1 - \overline{X}_2} = \sqrt{\frac{(n_1 - 1)S_1{}^2 + (n_2 - 1)S_2{}^2}{n_1 + n_2 - 2}} \; \sqrt{\frac{1}{n_1} + \frac{1}{n_2}} \; .$$

For the data of this problem,

$$S_{\overline{X}_1 - \overline{X}_2} = \sqrt{\frac{15(10)^2 + (13)(8)^2}{28}} \; \sqrt{\frac{1}{16} + \frac{1}{14}}$$

$$= \sqrt{\frac{15(100) + 13(64)}{28}} \; \sqrt{.0625 + .0714}$$

$$= \sqrt{\frac{1500 + 832}{28}} \; \sqrt{.1339}$$

$$= \sqrt{83.29} \; \sqrt{.1339} = 3.34 \qquad \text{and}$$

$$t = \frac{(105 - 112) - 0}{3.34} = \frac{-8}{3.34} = -2.40 \; .$$

Since $-2.40 < -2.048$, we reject H_0 and conclude that there is a significant difference between the scores of the 2 groups at the 5% level of significance.

● **PROBLEM** 55-43

A sports magazine reports that the people who watch Monday night football games on television are evenly divided between men and women. Out of a random sample of 400 people who regularly watch the Monday night game, 220 are men. Using a .10 level of significance, can be conclude that the report is false?

Solution: We have for this problem as our hypotheses:

H_0: $p = .50$, where p is the true population proportion.

H_1: $p \neq .50$.

The following diagram depicts the data of this problem, where $\bar{p}$ is the sample proportion of men who watch Monday night football.

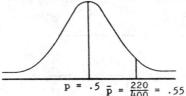

$$p = .5 \quad \bar{p} = \frac{220}{400} = .55$$

The statistic $(\bar{p} - p)/\sigma_p$ is approximately normally distributed with a mean of 0 and a standard deviation of 1. We calculate this value, which is called Z,

$$Z = \frac{\bar{p} - p}{\sigma_{\bar{p}}} \quad \text{where}$$

$$\sigma_{\bar{p}} = \sqrt{\frac{pq}{n}}$$

and compare the value of Z to a critical value. If Z lies beyond this critical value, we will reject H_0. For this problem, where we have $\alpha = 10\%$ and a two-tailed test, our critical value is 1.645, since for the normal distribution with mean of 0 and standard deviation of 1, 5% of scores will have a Z-value above 1.645 and 5% of scores will have a Z-value below -1.645. Therefore our decision rule is: reject H_0 if $|Z| > 1.645$, accept H_0 if $|Z| \leq 1.645$.

For the data of this problem,

$$\sigma_p = \sqrt{\frac{(.50)(.50)}{400}} \approx .025 \quad \text{and}$$

$$Z = \frac{.55 - .50}{.025} = \frac{.05}{.025} = 2.0 \ .$$

Since $2.0 > 1.645$, we reject H_0 and conclude that the report of the sports magazine is incorrect at a 10% level of significance.

● **PROBLEM 55-43**

A college has 500 women students and 1,000 men students. The introductory zoology course has 90 students, 50 of whom are women. It is suspected that more women tend to take zoology than men. In deciding to test this suspicion with the data of this class, what would the null and alternate hypotheses be? Is this a one-sample or a two-sample case?

Solution: Since it is suspected that women tend to take zoology more than men, we have a one-tailed test for this problem.

Of the 90 students in the zoology class, 50 or 5/9 are women.

We would like to know if this value of 5/9 is sufficiently greater than the proportion of women expected to take the course when men and women have equal preferences for the course. This proportion (the expected number) is 1/3 since 1/3 of the students in the college are women. Therefore our alternate hypothesis is $p > 1/3$. This leaves for our null hypothesis the possibility of $p \leq 1/3$.

This is a one-sample test; we are comparing a sample proportion for the number of women to a hypothesized population proportion. We are **not** comparing two sample proportions.

● **PROBLEM 55-44**

In 1964, 40% of shipments of U.S. cotton to Germany were sent to arbitration because of complaints about the quality being sub-standard, according to Time Magazine. Would it signify a real worsening of the situation if 20 out of the first 40 shipments in 1965 were likewise the cause of complaint, or might this difference be attributed to chance?

Solution: The sample proportion, $\bar{p}$, for this problem is $20/40 = .5$ and we wish to test whether this value is significantly greater than 40% or .4. We then let H_0, the null hypothesis, and H_1, the alternate hypothesis, be

$$H_0: \bar{p} = .40 \text{ , and}$$

$$H_1: \bar{p} > .40 \text{ .}$$

The statistic $(\bar{p} - p)/\sigma_p$ is approximately equal to the normal distribution with a mean of 0 and a standard deviation of 1.

We then test the statistic

$$Z = \frac{\bar{p} - p}{\sigma_{\bar{p}}} \quad \text{where}$$

$$\sigma_{\bar{p}} = \sqrt{\frac{pq}{n}} \text{ .}$$

Under H_0, $p = .40$, $1 - p = 2 = .60$ and

$$\sigma_{\bar{p}} = \sqrt{\frac{(.40)(.60)}{40}} = \sqrt{.006} = .077.$$

Therefore

$$Z = \frac{.50 - .40}{.077} = \frac{.10}{.077} = 1.30.$$

At $\alpha = .05$ and a one-tailed test, we would accept H_0 if the calculated value of Z is less than 1.96 Since 1.30 is less than 1.96, we accept H in this case and conclude that the difference between the 1965 value of 50% and the 1964 value of 40% can be attributed to chance.

An opinion survey in town A found that 73% of people considered architect's fees to be too high. A random sample of 30 people in town B were asked the same question 15 thought architect's fees to be too high. Is this proportion significantly different from that of town A?

Solution: Let us take as H_0, the null hypothesis, that the actual proportion, p, of people in town B who think that architect's fees are too high is .73. That is, we have

$H_0 : p = .73$, and $\qquad$ $H_1 : p \neq .73$.

But for this sample, $\bar{p} = 15/30 = .5$. We use the test statistic

$$Z = \frac{\bar{p} - p}{\sigma_{\bar{p}}} \qquad \text{where} \qquad \sigma_{\bar{p}} = \sqrt{\frac{pq}{n}} .$$

We do this because the statistic $(\bar{p} - p)/\sigma_{\bar{p}}$ is approximately normally distributed with a mean of 0 and a standard deviation of 1. Since $p = .73$, $q = 1 - p = .27$ and

$$\sigma_p = \sqrt{\frac{(.73)(.27)}{30}} = \sqrt{\frac{.197}{30}} = .081.$$

For $\alpha = .05$ and a two-tailed test, the critical value of Z is 1.96 and we will accept H_0 if the calculated value of Z lies between −1.96 and +1.96. Otherwise we will reject H_0.

For this set of data

$$Z = \frac{.50 - .73}{.081} = -\frac{.23}{.081} = -2.84$$

and we reject H_0.

For $\alpha = .01$, the critical value of Z is 2.57 and we will accept H_0 if the calculated value of Z lies between −2.57 and 2.57. Again we reject H_0. So for $\alpha = .05$ or $\alpha = .01$, we conclude that there is a significant difference between the proportion of people in town B and the proportion in town A who think that architects' fees are too high.

In one income group, 45% of a random sample of people express approval of a product. In another income group, 55% of a random sample of people express approval. The standard errors for these percentages are .04 and .03 respectively. Test at the 10% level of significance the hypothesis that the percentage of people in the second income group expressing approval of the product exceeds that for the first income group.

<u>Solution</u>: This problem may be depicted by the following diagram.

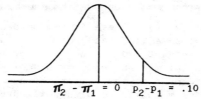

$$\pi_2 - \pi_1 = 0 \qquad P_2 - P_1 = .10$$

We would like to see whether $P_2 - P_1 = .10$ is far enough away from $P_2 - P_1 = 0$ to reject a null hypothesis that $\pi_2 - \pi_1 = 0$. Therefore our hypotheses are:

$$H_0 : \pi_2 - \pi_1 = 0 ,$$

where π_2 and π_1 are the population proportions for the second and first income group respectively.

$$H_1 : \pi_2 - \pi_1 > 0 .$$

The statistic

$$\frac{(P_2 - P_1) - (\pi_2 - \pi_1)}{\sigma_{P_2 - P_1}}$$

is approximately normally distributed with a mean of 0 and a standard deviation of 1.

We calculate this value, which is called Z:

$$Z = \frac{(P_2 - P_1) - (\pi_2 - \pi_1)}{\sigma_{P_2 - P_1}} \qquad \text{where} \qquad \sigma_{P_2 - P_1} = \sqrt{s^2_{P_1} + s^2_{P_2}}$$

and compare the value of Z to a critical value. If Z lies beyond this critical value, we will reject H_0. For this problem where we have $\alpha = 10\%$ and a one-tailed test, our critical value is 1.28, since for the normal distribution with mean of 0 and standard deviation of 1 10% of scores will have a Z-value above 1.28.

For $\alpha = .10$ and a one-tailed test our decision rule is: reject H_0 if $Z > 1.28$, accept H_0 if $Z \leq 1.28$.

We must calculate

$$Z = \frac{(P_2 - P_1) - (\pi_2 - \pi_1)}{\sigma_{P_2 - P_1}} \qquad \text{where} \qquad \sigma_{P_2 - P_1} = \sqrt{s^2_{P_1} + s^2_{P_2}} .$$

For the data of this problem

$$\sigma_{P_2 - P_1} = \sqrt{(.04)^2 + (.03)^2} = \sqrt{.0016 + .0009}$$

$$= \sqrt{.0025} = .05 \quad \text{and}$$

$$Z = \frac{(.55 - .45) - 0}{.05} = 2.0 .$$

Since $2.0 > 1.28$, we reject H_0 and conclude that the percentage of people in the second income group expressing approval of the product exceeds that for the first income group at the 10% level of significance.

Out of 57 men at a weekly college dance, 36 had been at the dance the week before, and of these, 23 had brought the same date on both occasions, and 13 had brought a different date or had come alone. Test whether the number of men who came both weeks with the same date is significantly different from the number who came both weeks but not with the same date. Use a 10% level of significance.

Solution: We test the difference between two proportions, P_1 and P_2, where $\bar{P}_1 = \frac{23}{57} = .40$ and $\bar{P}_2 = \frac{13}{57} = .23$. Our hypotheses are

$$H_0 : P_1 - P_2 = 0 \qquad\qquad H_1 : P_1 - P_2 \neq 0 .$$

This problem may be depicted by the following diagram.

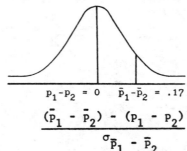

The statistic

$$\frac{(\bar{P}_1 - \bar{P}_2) - (P_1 - P_2)}{\sigma_{\bar{P}_1 - \bar{P}_2}}$$

is approximately normally distributed with a mean of 0 and a standard deviation of 1. We must calculate

$$Z = \frac{(\bar{P}_1 - \bar{P}_2) - 0}{\sigma_{\bar{P}_1 - \bar{P}_2}} \qquad \text{where}$$

$$\sigma_{P_1 - P_2} = \sqrt{\frac{p_1 q_1}{n_1} + \frac{p_2 q_2}{n_2}} .$$

For the data of this problem, using $\bar{P}_1$ and $\bar{P}_2$ as estimates of p_1 and p_2,

$$\sigma_{P_1 - P_2} = \sqrt{\frac{(.40)(.60)}{57} + \frac{(.27)(.73)}{57}}$$

$$= \sqrt{\frac{.24}{57} + \frac{.1971}{57}} = \sqrt{.0076684}$$

$$= .088$$

and

$$Z = \frac{(.40 - .23) - 0}{.088} = \frac{.17}{.088} = 1.93.$$

For a 10% level of significance and a two-tailed test, we have for our decision rule: reject H_0 if $Z > 1.65$ or $Z < -1.65$; accept H_0 if $-1.65 \leq Z \leq 1.65$. Since $1.93 > 1.65$, at the 10% level of significance we would reject H_0 and conclude that there is a significant difference between the number of men who had brought the same date on both occasions and the number who had brought a different date or came alone.

The Selective Service director of a certain state suspects that the proportion of men from urban areas who are physically unfit for military service is more than 5 percentage points greater than the proportion of physically unfit men from rural areas. He decides to treat the men called for a physical examination from urban areas during the next month as a random sample from a binomial population, and those from the rural areas as a random sample from a second binomial population. During the next month 3214 men were called from urban areas and 2011 from rural areas. There were 1078 physical rejects from the urban areas and 543 from rural areas.

Formulate the appropriate null and alternative hypotheses, and test the null hypothesis at the $\alpha = .05$ level.

Solution: Let π_1 = the true proportion of physically unfit from urban areas, and π_2 = the true proportion of physically unfit from rural areas.

The null and alternative hypotheses are

$$H_0 : \pi_1 - \pi_2 = .05 \qquad\qquad H_1 : \pi_1 - \pi_2 > .05 .$$

For this problem, the sample statistics, $\bar{p}_1$ and $\bar{p}_2$, are given by

$$\bar{p}_1 = \frac{X_1}{n_1} = \frac{1078}{3214} = .335 \quad \text{and}$$

$$\bar{p}_2 = \frac{X_2}{n_2} = \frac{543}{2011} = .270 ;$$

$$\bar{p}_1 - \bar{p}_2 = .335 - .270 = .065.$$

This problem may be depicted by the following diagram.

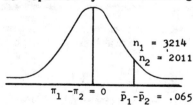

$$n_1 = 3214$$
$$n_2 = 2011$$

$$\pi_1 - \pi_2 = 0 \qquad \bar{p}_1 - \bar{p}_2 = .065$$

The statistic

$$\frac{(\bar{p}_1 - \bar{p}_2) - (\pi_1 - \pi_2)}{\sigma_{\bar{p}_1 - \bar{p}_2}}$$

is approximately a normal distribution with mean of 0 and standard deviation of 1.

For $\alpha = .05$ and a one-tailed test, we have for our decision rule: reject H_0 if $Z > 1.65$; accept H_0 if $Z \leq 1.65$.

We must calculate

$$Z = \frac{(\bar{p}_1 - \bar{p}_2) - (\pi_1 - \pi_2)}{\sigma_{\bar{p}_1 - \bar{p}_2}} \qquad \text{since the true proportions}$$

are unknown.

$\sigma_{\overline{P}_1 - \overline{P}_2}$ is estimated by: $\sqrt{\dfrac{\overline{P}_1(1-\overline{P}_1)}{n_1} + \dfrac{\overline{P}_1(1-\overline{P}_2)}{n_2}}$.

For this data of this problem we have

$$\sigma_{\overline{P}_1 - \overline{P}_2} = \sqrt{\frac{(.335)(.665)}{3214} + \frac{(.270)(.730)}{2011}}$$

and

$$= \sqrt{\frac{.228}{3214} + \frac{.197}{2011}} = .0130 \quad \text{and}$$

$$Z = \frac{(.335 - .270) - .05}{.0130} = \frac{.065 - .05}{.0130}$$

$$= \frac{.0150}{.0130} = 1.15 .$$

Since $1.15 < 1.65$, we accept H_0 and conclude that the proportion of men from urban areas who are physically unfit for military service is _not_ more than 5 percentage points greater than the proportion of physically unfit men from rural areas.

● **PROBLEM 55-49**

A new treatment plan for schizophrenia has been tried for 6 months with 54 randomly chosen patients. At the end of this time, 25 patients are recommended for release from the hospital; the usual proportion released in 6 months is $\frac{1}{3}$. Using the normal approximation to the binomial distribution, determine whether the new treatment plan has resulted in significantly more releases than the previous plan. Use a 0.05 level of significance.

Solution: The number of patients recommended for release from the hospital has a binomial distribution. We have p, the proportion of patients released from the hospital, equal to $\frac{1}{3}$, and q, the proportion of patients not released, equal to $1 - p$ or $\frac{2}{3}$. Since $np = (54)\left[\frac{1}{3}\right] = 18$ and $nq = (54)\left[\frac{2}{3}\right] = 36$, are both greater than or equal to 5, we may approximate this binomial distribution by the normal distribution in this case.

We use $\mu = np$ and $\sigma = \sqrt{npq}$ as the mean and standard deviation of this normal approximation.

Since $n = 54$, $p = \frac{1}{3}$ and $q = \frac{2}{3}$, we have $\mu = 54\left[\frac{1}{3}\right] = 18$

and $\sigma = \sqrt{54\left[\frac{1}{3}\right]\left[\frac{2}{3}\right]} = 3.46$.

The problem can now be depicted by the following diagram.

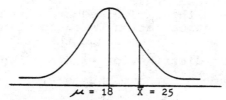

$\mu = 18 \qquad \bar{X} = 25$

We desire to see whether our sample mean of 25 exceeds 18 by enough for us to conclude that the actual population mean for the new treatment is greater than 18. Thus we choose as our alternate hypothesis

$H_1: \mu > 18.$

Hence, our null hypothesis is

$H_0: \mu = 18.$

The statistic $\dfrac{(x - \mu)}{\sigma}$ has a normal distribution with a mean of 0 and a standard deviation of 1.

We use a critical value of 1.645 for this problem, since we have $\alpha = .05$ and a one-tailed test, and for the normal distribution with mean of 0 and standard deviation of 1, 5% of scores will have a Z-value above 1.645. Therefore our decision rule is: reject H_0 if $Z > 1.645$; accept H_0 if $Z \le 1.645$.

We must now calculate

$Z = \dfrac{x - \mu}{\sigma}$.

Since the binomial distribution is a discrete distribution and the normal distribution is continuous, we will use 24.5 instead of 25 as the value for x in the above formula.

We calculate Z.

$Z = \dfrac{24.5 - 18}{3.46} = \dfrac{6.5}{3.46} = 1.88.$

Since 1.88 > 1.645, we reject H_0 and conclude that the population mean for the new treatment is greater than 18, and hence, it has resulted in significantly more releases than the previous plan.

● **PROBLEM** 55-50

A sample of size 10 produced a variance of 14. Is this sufficient to reject the null hypothesis that $\sigma^2 = 6$ when tested using a .05 level of significance? Using a .01 level of significance?

Solution: We use the χ^2 (chi-square) statistic to determine the value of a population variance when given a sample
variance.   We may do this because the test statistic
$\frac{(n-1)s^2}{\sigma^2}$ has a χ^2 distribution with n - 1 degrees of freedom.   Here we have as our hypotheses:

H_0: $\sigma^2 = 6$ H_1: $\sigma^2 \neq 6$.

     Since α = .05 and this is a two-tailed test, we will
reject H_0 if our calculated value for χ^2 is $> \chi^2_{.025}$ or
$< \chi^2_{.975}$. We will accept H_0 if $\chi^2_{.975} \leq$ calculated χ^2
$\leq \chi^2_{.025}$. The number of degrees of freedom for $\chi^2_{.025}$ and
$\chi^2_{.975}$ is n - 1, or in this case 9. Therefore, our decision
rule is: reject H_0 if calculated $\chi^2 > 19.023$ or $\chi^2 < 2.700$;
accept H_0 if $2.700 \leq \chi^2 \leq 19.023$.

     We now calculate χ^2 using the formula

$$\chi^2 = \frac{(n-1)s^2}{\sigma^2} \quad .$$

     Since n = 10, s^2 = 14, and σ^2 = 6 for this problem, we
have

$$\chi^2 = \frac{9(14)}{6} = 21.$$

     Since 21 > 19.023, we reject H_0 and conclude that
$\sigma^2 \neq 6$ at a 5% level of significance.

     For α = .01, we must compare our calculated χ^2 to
$\chi^2_{.005}$ and $\chi^2_{.995}$, again for 9 degrees of freedom. Our
decision rule now becomes: reject H_0 if calculated χ^2
> 23.589 or $\chi^2 < 1.735$; accept H_0 if $1.735 < \chi^2 < 23.589$.
Since our calculated χ^2 was 21, and $1.735 < 21 < 23.589$,
we would accept H_0 that σ^2 = 6 at a 1% level of significance.

INDEX

Numbers on this page refer to **PROBLEM NUMBERS**, not page numbers

Numbers on this page refer to **PROBLEM NUMBERS**, not page numbers